The Biology of the Laboratory Rabbit

Composite plate showing a few of the American breeds of rabbits in use today: (A) American Dutch, black–$aa\ du^d du^w$ (Mr. Frank Helberg, Branchbrook Rabbitry, Ludlow, Vermont); (B) New Zealand White–cc Mr. Lee K. Labelle, Acadia Farm, Ellsworth, Maine); (C) American Chinchilla–$c^{ch3}c^{bc3}$ (Mr. Richard Blowers, Hague, New York); (D) Californian–$aa\ c^Hc^H$ (Mr. Lee K. Labelle, Acadia Farm, Ellsworth, Maine); (E) Tan, black–a^ta^t (Mr. Lee K. Labelle, Acadia Farm, Ellsworth, Maine); (F) Silver Martin, black–$c^{ch3}c^{ch3}$ (Mrs. Joyce Krempa, Marble City Rabbitry, West Rutland, Vermont); (G) Harlequin–e^Je^J (Miss Lynn Lamoreux, Bar Harbor, Maine); (H) Champagne d'Argent–$aa,\ sisi$ (Mr. George S. Kopp, Pine Ridge Rabbitry, Clarendon Springs, Vermont); (I) Satin, siamese–sasa $c^{ch1}c^{ch1}$ (Mr. Richard Blowers, Hague, New York); (J) Rex, castor–r^1r^1 Cuozzo, Lyons & Noble Associates, Rutland, Vermont); (K) American Checkered Giant–$aa\ EnEn$ (Mrs. Cecile Bruneault, Valley View Rabbitry, Ira, Vermont); (L) Polish, white, blue-eyed–vv (Mr. Anthony Pisanelli, Rutland, Vermont).

THE BIOLOGY OF THE LABORATORY RABBIT

EDITED BY

Steven H. Weisbroth
Division of Laboratory Animal Resources
Health Sciences Center
State University of New York
Stony Brook, New York

Ronald E. Flatt
Department of Veterinary Pathology
College of Veterinary Medicine
Iowa State University
Ames, Iowa

Alan L. Kraus
Division of Laboratory Animal Medicine
School of Medicine and Dentistry
University of Rochester
Rochester, New York

ACADEMIC PRESS   New York and London   1974
A Subsidiary of Harcourt Brace Jovanovich, Publishers

ACADEMIC PRESS, INC.
111 Fifth Avenue, New York, New York 10003

United Kingdom Edition published by
ACADEMIC PRESS, INC. (LONDON) LTD.
24/28 Oval Road, London NW1

Library of Congress Cataloging in Publication Data

Weisbroth, Steven H
    The biology of the Laboratory rabbit.

    Includes bibliographies.
    1.   Rabbits.   2.   Rabbits–Diseases.   3.   Rabbits
as laboratory animals.   I.   Flatt, Ronald E., joint
author.   II.   Kraus, Alan L., joint author.
III.   Title.   [DNLM: 1.   Animals, Laboratory.
2.   Rabbits.   SF997.5.B2 W426b 1974]
QL737.L32W45   636'.93'22   73-18942
ISBN 0–12–742150–5

2476

Contents

List of Contributors

Numbers in parentheses indicate the pages on which the authors' contributions begin.

W. SHELDON BIVIN (73), Department of Animal Industries, Southern Illinois University, Carbondale, Illinois

BILL C. BULLOCK (155), Department of Comparative Medicine, Bowman Gray School of Medicine of Wake Forest University, Winston-Salem, North Carolina

THOMAS B. CLARKSON (155), Department of Comparative Medicine, Bowman Gray School of Medicine of Wake Forest University, Winston-Salem, North Carolina

CARL COHEN (167), The Center for Genetics, University of Illinois College of Medicine, Chicago, Illinois

LAURENCE M. CUMMINS (49), Department of Pathology, Abbott Laboratories, North Chicago, Illinois

RONALD E. FLATT (193,435), Department of Veterinary Pathology, College of Veterinary Medicine, Iowa State University, Ames, Iowa

HENRY L. FOSTER (179), The Charles River Breeding Laboratories, Wilmington, Massachusetts

RICHARD R. FOX (1,337), Rabbit Genetics Laboratory, The Jackson Laboratory, Bar Harbor, Maine

FRITZ P. GLUCKSTEIN (453), National Library of Medicine, Bethesda, Maryland

KARL W. HAGEN* (23), Endoparasite Vector Pioneering Research Laboratory, Veterinary Sciences Research Division, U.S. Department of Agriculture, Agricultural Research Service, Pullman, Washington

DANIEL D. HARRINGTON (403), Department of Veterinary Science, University of Kentucky, Lexington, Kentucky

HOWARD A. HARTMAN (91), Biological Research Department, Section of Toxicology, Sandoz-Wander, Inc., East Hanover, New Jersey

CHARLES E. HUNT (403), Department of Comparative Medicine, University of Alabama Medical Center, Birmingham, Alabama

CARLOS KOZMA (49), Division of Drug Safety Evaluation, Abbott Laboratories, Chicago, Illinois

ALAN L. KRAUS (287,435), Division of Laboratory Animal Medicine, School of Medicine and Dentistry, University of Rochester, Rochester, New York

NOEL D. M. LEHNER (155), Arteriosclerosis Research Center, Bowman Gray School of Medicine of Wake Forest University, Winston-Salem, North Carolina

J. RUSSELL LINDSEY (377), Department of Comparative Medicine, University of Alabama Medical Center, Birmingham, Alabama

WILLIAM MACKLIN (49), Department of Toxicology and Experimental Pathology, Wellcome Research Laboratories, Research Triangle Park, North Carolina

C. J. MARE (237), Department of Veterinary Microbiology and Preventive Medicine, College of Veterinary Medicine, Iowa State University, Ames, Iowa

RUSSELL MAUER (49), Agricultural Research Division, Abbott Laboratories, North Chicago, Illinois

STEVEN P. PAKES (263), Comparative Medicine and Animal Resources Center, University of Texas, Health Science Center at Dallas, Dallas, Texas

EDWARD H. TIMMONS (73), Vivarium, Southern Illinois University, Carbondale, Illinois

ROBERT G. TISSOT (167), The Center for Genetics, University of Illinois College of Medicine, Chicago, Illinois

*Deceased.

STEVEN H. WEISBROTH (331,435), Division of Laboratory Animal Resources, Health Sciences Center, State University of New York, Stony Brook, New York

RICHARD B. WESCOTT (317), Department of Veterinary Pathology, College of Veterinary Medicine, Washington State University, Pullman, Washington

Preface

The American College of Laboratory Animal Medicine (ACLAM) was founded in 1957 to encourage education, training, and research in laboratory animal medicine. The organizational goals include professional certification and continuing education as main areas of concern. This book represents part of a program developed by ACLAM to further the educational goals of the college. The idea arose during the fall of 1969 as an outgrowth of recognition by members of the College that need existed for a serious effort that would review and summarize literature pertaining to the use of the rabbit as an experimental substrate in the scientific process. Many of the chapters were delivered in part during the course of two symposia. The chapters dealing with disease were presented as a group at the annual meeting of the American Veterinary Medical Association in Detroit, Michigan in July of 1971. Those chapters related to normative biology and research utilization were presented as part of the annual meeting of the Federation of American Societies of Experimental Biology in Atlantic City, New Jersey in April of 1972.

Disease has been especially emphasized for several reasons. These include the fact that naturally occurring disease continues to exert intercurrent effects that limit the value of the rabbit as a research tool. Recognition of these effects and reduction of their impact are major tasks of those engaged in the practice of laboratory animal medicine. Prominent also among reasons for emphasis on disease is increasing awareness of the value of certain rabbit diseases as models of considerable interest in comparative medicine. Students of rabbit disease particularly will recognize that the literature dealing with this subject has been scattered throughout the entire range of scientific proceedings and scholarly and trade journals with no recent comprehensive effort to gather it into the format of an organized review. That is the purpose of this work.

The editorial intention has been to attempt to make this book useful to the widest possible audience. We have assembled a distinguished group of contributors, each chosen on the basis of eminence in the topical area of contribution. Their charge has been to present the subject matter at a level of quality sufficient to establish the book as an authoritative reference work not only for those in laboratory animal medicine but also for those in the general scientific community. At the same time our goal has been to provide students of veterinary medicine and that community concerned with the applied sciences of laboratory animal care and rabbit husbandry with a text of interest to them as well.

The attempt to capture the respect and interest of an audience with such varied backgrounds has not been a simple task. We therefore welcome comments with regard to errors of fact or interpretation and significant omissions. Suggestions related to topical areas not included in this edition or areas that should be presented with changes in emphasis in future editions will also be appreciated. To the extent that the book is used by the breadth of the intended audience, the editors will gauge their success.

STEVEN H. WEISBROTH
RONALD E. FLATT
ALAN L. KRAUS

The Biology of the Laboratory Rabbit

CHAPTER 1

Taxonomy and Genetics*

Richard R. Fox

I. TAXONOMY AND GEOGRAPHICAL DISTRIBUTION OF RABBIT POPULATIONS

*This work was supported in part by NIH Research Grants HD-01496 from the National Institute of Child Health and Human Development, RR-00251 from the Division of Research Resources, and by an allocation from the Sagamore Foundation.

I wish to thank Drs. Paul B. Sawin, Sheldon Dray, and Rose G. Mage, Mr. Roscoe F. Cuozzo, and Mrs. Dorcas D. Crary for their invaluable help in the preparation of this manuscript.

The European rabbit (*Oryctolagus cuniculus*) occurs on the European continent in three forms: wild, feral, and domestic. In North America, however, only the domestic and the feral forms exist. The wild or ancestral type probably evolved in the Iberian Peninsula and spread to other regions of the Mediterranean. The familiar domestic form is typified by a great variety of breeds and strains which are used for meat, fancy and laboratory animal production. The feral rabbit is a reversion from the domestic to the wild type and examples may be found on the Farallon Islands off the coast of San Francisco, the San Juan Islands in the Juan de Fuca Straits in Washington, the Channel Islands off Santa Barbara, California, and Isla del Flores near Montevideo, Uruguay. The only population of wild *Oryctolagus* in the Americas has spread from the island of Tierra del Fuego, Chile, northward into the mainland several hundred miles north of Santiago.

A. Taxonomy

Laboratory rabbits, descendents of the European wild rabbit, *Oryctolagus cuniculus*, along with other rabbits, hares, and pikas, were originally classified as members of the order Rodentia, or rodents. However, instead of the rodents' four incisor or chisel teeth, rabbits have six. The additional pair is reduced in size and placed directly behind the large pair in the upper jaw. These little teeth are rounded and lack a cutting edge. They are only moderately useful and cannot be seen without opening the mouth and looking in back of the large upper incisor teeth. However, they constitute the scientific basis for placing these animals in a separate order, the Lagomorpha (51). The zoological position of the lagomorphs (51, 220) can best be seen in outline form (Table I).

B. Geographical Distribution

The order Lagomorpha is comprised of two major families, Ochotonidae (pika) and Leporidae (rabbits and hares), with many genera and species native to all parts of the world (5, 6, 10, 98, 102, 117, 168, 174, 220). The pikas, sometimes called rock rabbits or mouse hares, are small, tailless members with short, broad, rounded ears, chunky bodies, and short legs. The front pair of legs is but little shorter than the rear pair. This is in contrast to the usually larger rabbits and hares with their long ears and relatively long hind limbs. Major genera of the Leporidae include *Lepus* (hares), *Oryctolagus* (true rabbits), and *Sylvilagus* (cottontail rabbits). The Idaho pygmy rabbit, *Brachylagus*, is the smallest rabbit, not only in the Americas but in the world. The Mexican pygmy rabbit, *Romerolagus*, is slightly larger and actually more like a pika than a rabbit. Examples of some of the current major species in each genus, including their natural distribution, may be seen in Table II.

Members of the order Lagomorpha have found their way by natural means to most parts of the world except Madagascar, New Zealand, and Australia. Also rabbits were distributed to various portions of the world by early sailing vessels whose masters wished to have a readily available source of meat at various points on their voyages. In addi-

TABLE I

TAXONOMIC OUTLINE OF THE GENUS *ORYCTOLAGUS*[a]

Phylum Cordata: Animals with notochord and gills
Subphylum Craniata (vertebrata):   Chordates with organized head region
Class Mammalia: Warm-blooded craniates with hair coat. Young nourished from mammary glands
Subclass Theria: Viviparous mammals
Infraclass Eutheria: Placental mammals
Cohort Glires: (Lagomorpha + rodentia)
Order Lagomorpha: Gnawing placentals with chisellike incisors of which there is in the upper jaw a small second pair directly behind the main pair. Coitus-induced ovulation seems to be general in this order
Family Ochotonidae (Pika)
Genus *Ochontona*
Family Leporidae (Rabbits and hares)
Subfamily Paleolaginae
Genus *Pronolagus*
Genus *Pentalagus*
Genus *Romerolagus*
Subfamily Leporinae
Genus *Lepus*
Subgenus *Poelagus*
Genus *Sylvilagus*
Genus *Oryctolagus*
Genus *Nesolagus*
Genus *Brachylagus*
Genus *Caprolagus*

[a]References (51, 220).

tion, the European rabbit, *Oryctolagus cuniculus*, was transported to Australia and New Zealand, and the European hare, *Lepus europaeus*, was established at several points in North America and is now abundant in the northwestern portion of America.

II. ORIGIN AND DOMESTICATION OF THE RABBIT

A. Historical Considerations

Nachtsheim, in his book, "Vom Wildtier zum Haustier" (151), has given the most comprehensive review of the early history of the domestication of the rabbit. Our knowledge of the natural spread of the wild rabbit in prehistoric times is somewhat limited as very little has been preserved from early periods. Rabbit bones are small, light, and fragile and are often overlooked in archeological diggings. Also predators often leave little for history. From records from the early tertiary strata it appears that at this time the Leporidae were lacking in Europe but were present in America and Asia. Hares and rabbits, or their predecessors, probably migrated from Asia to Europe during the early tertiary. From the end of the Pleiocene to the beginning

TABLE II
Geographical Distribution of Members of the Order Lagomorpha[a]

Family	Subfamily	Genus	Species	Common name	Distribution
				Pikas	
Ochotonidae		*Ochontona*	*princeps*	Rocky Mt.	Slides and rock piles in Rocky Mountains of British Columbia to Arizona
			alpina	Gray headed	High elevations of California, Oregon, Idaho, Nevada, and Utah
			collaris	Collared	Northwestern Canada and Alaska
			pusilla	Himalayan	Himalaya. Largest species of the Pika
			hypoborea		Altai Mountains of Siberia
			daurica		Manchuria
			roylei		Northern India
				Rabbits and hares	
Leporidae	Paleolaginae	*Pronolagus*	*crassi-caudatus*	Rock hare	Central and Southern Africa
			randensis		Central and Southern Africa
		Romerolagus	*diazzi*	Volcano rabbit	High tablelands in the vicinity of Mexico City
		Pentalagus			Asia
	Leporinae	*Lepus*		Hares	
			arcticus	Arctic	American Arctic
			othus	Alaskan	Alaskan tundra
			timidus	Blue, mountain, varying or alpine	Alpine regions from 5000–10,000 ft in Northern Europe, Scandinavia, Scotland, and Ireland
			groenlandicus	Greenland	Greenland
			tschukschorum		Northern Siberia
			europaeus	European	Common hare of Europe and introduced in North America in nineteenth century
			catrolagus	Hispid	Himalayan foothills
			townsendii	White-tailed jack or prairie hare	Central Saskatchewan to northern New Mexico and from western Wisconsin to central Washington and the Sierra Nevadas of California
			callotis	White sided jack	Southern Arizona to Oaxaca, Mexico
			insularis		Espiritu Santo Island in the Gulf of California
			californicus	Black-tailed jack	State of Washington east to Nebraska and south into Mexico
			gaillardi	Gaillard's jack	These two species are also called black-tailed jackrabbits but are found in small areas of Mexico, Arizona, and New Mexico
			alleni	Antelope jack	
			americanus	Snowshoe or varying hare	Northeastern U.S. and Canada
			washingtonii	Snowshoe or varying hare	Western highlands to central New Mexico and California
			bairdii	Snowshoe or varying hare	Southeast in mountains as far as Virginia
			mexicanus		Mexico
			flavigularis		Mexico
			capensis		South Africa
			atlanticus		South Africa
			saxatilis	Mountain hare	South Africa
			whytei		South Africa
			salai		South Africa
			nigricollis	Black naped	South Africa
		Poelagus	*marjorita*	Grass hare	Uganda to the Anglo Egyptian Sudan (resembles *Oryctolagus* and is often considered a subgenus of *Lepus*)

Table II (*continued*)

Family	Subfamily	Genus	Species	Common name	Distribution
		Oryctolagus	*cuniculus*	European or true rabbit	Central and southern Europe and Northern Africa (introduced throughout the world, in particular Australia and New Zealand. They never became established in North America)
		Nesolagus		Sumatra rabbit	Related to *Oryctolagus*
		Sylvilagus		Cottontails	
			floridanus	Eastern	Atlantic coast to southeast Wyoming, extreme southern Arizona, and Central America to Costa Rica
			nuttallii	Mountain or Nuttall's	Western North Dakota to central New Mexico, west to the Cascades and the Sierra Nevada
			transitionalis	New England	New England
			audubonii	Desert	Southwestern U.S. and northwestern Mexico
			palustris	Marsh	Atlantic and Gulf coastal regions from southeast of Virginia to southern Florida and southern Alabama
			aquaticus	Swamp	Western Gulf coast, Alabama to southern Texas to northern Alabama, southern Illinois, and southern Oklahoma
			bachmani	Brush	The Pacific Slope from the western foothills of the Sierra Nevada to the coast of Oregon and Southern California
			mansuetus	Brush	Closely related to *S. bachmani*, found on San José Island
			graysoni		Tres Maries Islands
			braziliensis	Forest	Central America
			cunicularis	Mexican	Mexico
			insonus	Omitteme	Forested park of the Sierra Madre del Sur between 7000 and 10,000 ft in Mexican state of Guerrero
		Brachylagus	*idahoensis*	Pygmy rabbit	Central Nevada to southern Idaho and in southeastern Oregon and northeastern California. (This rabbit is sometimes classed with the cottontails but is a distinct genus)
		Caprolagus		Bristly rabbit	Asia

[a]References (5, 6, 10, 98, 102, 117, 168, 174).

of the Ice Age, they apparently became widely distributed in Europe. Remains have been found in France, Belgium, and Germany. During the period of glaciation the rabbit was pushed southward to southwestern Europe and the western portions of North Africa.

The Phoenicians were the discoverers of the rabbit in historical times. In their journeys to the coast of Africa and the Iberian Peninsula in 1100 B.C. they observed numerous creatures similar to their cliff terriers, the description of which resembles our rabbit very closely. Because of the numbers of these terrier-like creatures they named the coast or island as the land of this creature and called it "i-shephan-im." Later, this name for the Iberian Peninsula was renamed in the Latin form by the Romans "Hispania." The Hebrew word "Saphan or Shaphan," for cliff terrier

(*Hyrax syriacus*), was later incorrectly translated by Luther in his Bible translation to the word rabbit. Early accounts of the rabbit in the Greek and Roman literature, including the writings of Xenophon and Aristotle, speak primarily of the hare, up through the fourth century B.C. In the second century B.C. the Greek historian, Polybios, observed rabbits among other wild animals coming from Corsica and called them the most graceful cuniculi. Cuniculi are what the Romans called subterranean passages and mines such as they dug during sieges of city states. There is some question as to whether the rabbits were named after the mines or the mines after the rabbit holes. In the first century B.C., the Roman, Varro, called Spain the homeland of the rabbit and recommended keeping the rabbits in leporaria, or walled rabbit gardens. Varro described these leporaria as being surrounded by high plastered walls designed to keep out predators and having trees and bushes for shade and for protection from predatory birds. Some consider this the start of domestication but the rabbits were still very wild. However, they did breed in these enclosed areas in contrast to the hare and were caught easily and killed for meat. Pliny also designated Spain as the original home of the rabbit and stated that in the Balearic Islands during the first century A.D. fetuses taken from pregnant does and newborn young, when uneviscerated, were considered delicacies. Since rabbit meat at this time was highly regarded, animals were distributed to various islands during the first few centuries A.D. Strabo reported that a single pair was placed on one of the Balearic Islands, probably Mallorca, at the time of Christ's birth. By Pliny's time, the rabbit was observed on both Mallorca and Minorca but had not yet spread to the small island group of Pityusen. In many areas, however, the rabbit became a menace by multiplying rapidly and destroying the vegetation. Finally, the inhabitants of the Balearic Islands asked Emperor Augustus for aid in controlling these field-destroying gnawers. Muzzled ferrets were used to drive the rabbits out of their burrows so they could be hunted. Under Emperor Hadrian a rabbit was represented on gold and silver coins of the empire as a symbol of Spain, indicating how much the rabbit was considered characteristic of the Iberian Peninsula.

Rabbits were placed on islands by masters of sailing vessels during the Middle Ages to provide a source of food along their various seafaring routes. In these favorable environments, and lacking their usual predators, the rabbits greatly increased in numbers and often did more harm than good, and the experiences of the inhabitants of the Balearic Isles in antiquity were repeated over and over again. A prime example used by Darwin was the small island of Porto Santo where rabbits were introduced in 1418 and in a short period of time took over the island. Also they evolved into a new species which would no longer interbreed with other rabbits. In 1859, a single pair of *Oryctolagus* was released in Victoria, Australia, by an English colonist. By 1890 the rabbit population was estimated at 20,000,000 and had become a frightful plague. All imaginable means, both public and private, were used to eliminate this species. In one decade the government of New South Wales spent about 4 million dollars trying unsuccessfully to eradicate the rabbit from the portion of Australia. The rabbit succeeded best following man-made distribution of the various species in those climates resembling their original home. Thus, in Australia and New Zealand, *Oryctolagus cuniculus* populated very quickly and became a serious pest. Fortunately for North America, climatic conditions and native predatory animals have been more than a match for the introduced rabbit, *Oryctolagus*, which has nowhere established itself in the wild state (20).

B. Domestication

The actual process of domestication may have started when the Romans learned that the rabbit, in contrast to the hare, propagated itself very easily within the leporaria, or rabbit gardens. Thus, the original simple holding pen used for the hares now became a breeding pen for the propagation of the rabbits. These rabbit gardens had many uses, depending on their sizes. From the sixteenth century, we learned that Queen Elizabeth of England had a leporarium and King Henry IV of France had a large rabbit enclosure which he used as a hunting area. The hunting of rabbits for sport was also taken up by women of this time, in part because this was an easy and safe hunting sport. These rabbit gardens persisted in Europe for many centuries after the fall of the Roman Empire, with frequent escapes aiding in the spread of the species in the wild.

True domestication, with breeding in captivity, probably started in the monasteries during the sixteenth century. Under domestication, the coat has varied greatly in color and to a minor extent in length and texture. By 1700, 7 different mutant types were known, namely nonagouti, brown, albino, dilute, yellow, silver, and Dutch spotting (217). It seems surprising that with this knowledge the formulation of Mendel's laws had to wait until his experiments with peas. Also, Nachtsheim (151) refers to changes in brain weights, spinal cord, various sense organs, the eyes, and the ability to hear, associated with domestication. The ear length has varied considerably from approximately 7 cm long in wild rabbits to varied lenghts in the domesticated races, ranging up to the grotesque lop-eared rabbit having an ear length of approximately 25 cm. There has been a diminution in the number of taste buds in the tongue. All in all, there are many changes which have occurred during domestication since the critical senses essential for survival are not needed under the confinement that we have with the cage system employed today. These differences are also reflected in muscles, heart size, capacity of the stomach, and bone weight; body weight has varied considerably from an approximate 2 kg for wild rabbits to the hermelin dwarf of approximately 1 kg on the one hand and to the German giant of 8 kg or more on the other. Sirks (221) quoted from a letter written by Leeuwenhoek in 1683 showing that he was well aware of the dominance of the wild-type coat over albinism, nonagouti, etc. However, little use was made of this until after the rediscovery of Mendel's laws. In the period 1700–1850, two new mutations for coat color and the factor causing angora hair became known (140). After 1850, and especially during the early twentieth century, after Mendel's principles of recombination became known, the differentiation of new races or strains of rabbits varying with coat color, body size, and hair morphology increased rapidly. Currently there are well over 50 well-established breeds of domestic rabbits, but there is still a potential for many additional combinations of genes to be utilized, depending on the needs and desires of the rabbit fancier or the investigator using rabbits in research.

C. Breed Formation

The European rabbit, *Oryctolagus cuniculus*, has, however, been the only species that has yielded to domestication. All of the laboratory rabbits and all of the rabbits bred by the fanciers for coat color or hair characteristics are descended from it. The early breeders took advantage of the large variation in body size, coat colors, and coat characteristics to establish, by selection, a variety of new breeds valuable for both meat and fur and some especially for pets. A variety of coat colors and body sizes including some breeds of rabbits in common use today are illustrated in the frontispiece.

Serious attempts to analyze the genetics of the domestic rabbit did not start until after the rediscovery of Mendelism and will be discussed later. It is possible to describe many of the differences in coat color and hair characteristics represented by the various breeds by present knowledge of genetics. Table III lists and illustrates this by giving the currently recognized American breeds of rabbits as found in the 1966–1970 Standards of Perfection of the American Rabbit Breeders Association and the currently known genotype of these breeds. Details on the gene symbols involved are in Table V.

An extensive listing of the British, Continental, and American breeds by Robinson in 1958 (189) shows that many of the genotypes in the American breeds are duplicates of those found in the other countries. Moreover, many additional combinations are seen elsewhere. To establish breeds of rabbits, one starts basically with a wild type rabbit (i.e., "normal" white bellied agouti) and by substitutions of mutant alleles at specific loci obtains a variety of color types and patterns which can be selected for by the animal breeder. Substitution at the A locus of a^t for normal agouti A gives a black and tan rabbit with a white belly, black nonagouti on the dorsal surface, whitish eye circles, and tan on the foot pads, under tail, and edge of white belly. Replacement by nonagouti a results in a solid black animal. At the B locus substitution by b replaces the black eumelanin with brown. Agouti brown rabbits are cinnamon colored, whereas nonagouti brown are solid brown colored as in the case of the Havana rabbit.

Mutations at the C locus reduce pigmentation. The c^{ch3}, dark chinchilla, will reduce only yellow pigment to white. The c^{ch2} will, in addition, reduce black to a sepia-brown and the c^{ch1}, or light chinchilla, will further reduce pigment to pale brown. Next in the series the Himalayan gene, c^H, restricts all pigment to the extremities, as in the

case of the Californian or Himalayan rabbits. An interesting phenomenon associated with the Himalayan allele is its temperature sensitivity. If a portion of the hair is shaved and the rabbit placed in an environment of 25°C or more, the new hair growth will resemble the old. If, however, the temperature is, say, 10°C, a two step process, first enzyme production, then oxidation, will result in the new hair being pigmented on the tips. Pigment in the eye is also reduced in all mutant alleles at this locus. The c, or albino allele, eliminates all pigment.

The dilute locus, when homozygous dd, results in a pigment change from either black to a blue-gray, brown to lilac, or yellow to cream, depending on the other genes present. At the E locus dominant black, E^D, tends to reduce or eliminate the agouti band of phaeomelanin and darken the belly. An $E^D E^D$ is indistinguishable from a nonagouti black (aa), $E^D e$ is very similar to $E^D E^D$ in its coloration, and $E^D E$ rabbits are "agouti-black." In contrast ee animals are a fawn color. Another allele, e^J, results in Japanese brindling as is seen in the Harlequin rabbit. E^S has also been reported as a weaker edition of E^D (189).

The Vienna white allele, v, modifies the wild-type phenotype by removing all pigment from the hair and from the anterior surface of the iris (195), resulting in a blue-eyed white rabbit. It was thought at one time that the presence of this gene in homozygous condition was necessary for epileptic seizures but in our colony just as many Vv rabbits develop epilepsy-like seizures as do vv rabbits.

In addition to these six sets of alleles for color, many other factors modify the color, including the width of the agouti band (23), Dutch (180) and English spotting (15), and silvering (133). Further variations in phenotype may be seen by modifying the hair morphology. Loci include rex (33, 123), angora (12), wuzzy (198), satin (32), wirehair (198), waved (175), furless (25), and naked (108). Innumerable combinations are possible, depending on the fancy of the animal breeder (189, 217).

III. GENETICS OF THE RABBIT

Prior to 1900, only minimal work had been done on the genetics of the rabbit. The dominance of wild type over nonagouti and other genes was evident as early as 1683 (221); however, this was believed to be associated with paternal inheritance since the crosses were made only with tame white females and wild colored males. A few color genes and hair morphology genes were known (151), but it was not until after the rediscovery, about 1900, of Mendel's paper that he had presented before the Natural History Society of Brünn in 1865 that the science of genetics was actually applied to the rabbit. Initially, certain genes were recombined into the progenitors of some of the current breeds, selection being based on particular

TABLE III
GENOTYPES OF AMERICAN BREEDS OF RABBITS[a]

Breed	Ideal mature weight in lb ♂/♀	Genotype[b]	Breed	Ideal mature weight in lb ♂/♀	Genotype[b]
American White	9/10	cc	New Zealand		
American Blue	9/10	$aa\ dd$	Red	10/11	ee
American Sable	8/9	$aa\ c^{chl}c^{chl}$	White	10/11	cc
American Standard chinchilla	6.5/7	$c^{ch3}c^{ch3}$	Black	10/11	aa
American chinchilla	10/11	$c^{ch3}c^{ch3}$	Palomino		
American Giant chinchilla	13.5/14.5	$c^{ch3}c^{ch3}$	Golden	9/10	
English Angoras			Lynx	9/10	$bb\ dd$
White	6/7	$ll\ cc$	Polish		
Black	6/7	$ll\ aa$	White	2.5/2.5	cc
Blue	6/7	$ll\ aa\ dd$	Black	2.5/2.5	aa
Fawn	6/7	$ll\ ee$	Chocolate	2.5/2.5	$aa\ bb$
French Angoras			Rex		
White	8/8	$ll\ cc$	White	8/9	$rr\ cc$
Black	8/8	$ll\ aa$	Black	8/9	$rr\ aa$
Blue	8/8	$ll\ aa\ dd$	Blue	8/9	$rr\ aa\ dd$
Fawn	8/8	$ll\ ee$	Castor	8/9	rr
Belgian Hare	8/8	ww	Chinchilla	8/9	$rr\ c^{ch3}$
Beverens			Opal	8/9	$rr\ dd$
White	9/10	vv	Lynx	8/9	$rr\ bb\ dd$
Blue	9.5/10.5	$aa\ dd$	Sable	8/9	$rr\ aa\ c^{chl}c^{chl}$
Black	9.5/10.5	aa	Seal	8/9	$rr\ c^{ch2}c^{ch2}$
Californian	9/9.5	$aa\ c^{H}c^{H}$	Red	8/9	$rr\ ee$
Champagne d'Argent	10/10.5	$aa\ sisi$	Lilac	8/9	$rr\ aa\ bb\ dd$
Creme d'Argent	9/10	$ee\ sisi$	Havana	8/9	$rr\ aa\ bb$
American Checkered Giant	11/12	$aa\ EnEn$	Californian	8/9	$rr\ aa\ c^{H}c^{H}$
American Dutch			Satin		
Black	4.5/4.5	$du^{d}du^{w}\ aa$	Black	9/9.5	$sasa\ aa$
Blue	4.5/4.5	$du^{d}du^{w}\ aa\ dd$	Blue	9/9.5	$sasa\ aa\ dd$
Chocolate	4.5/4.5	$du^{d}du^{w}\ aa\ bb$	Havana	9/9.5	$sasa\ aa\ bb$
Tortoise	4.5/4.5	$du^{d}du^{w}\ aa\ ee$	Red	9/9.5	$sasa\ ee$
Steel-gray	4.5/4.5	$du^{d}du^{w}E^{D}e$	Chinchilla	9/9.5	$sasa\ c^{ch3}$
English Spots			Copper	9/9.5	$sasa\ bb$
Black	7/7	$Enen\ aa$	Siamese	9/9.5	$sasa\ c^{chl}c^{chl}$
Blue	7/7	$Enen\ aa\ dd$	Satin		
Chocolate	7/7	$Enen\ aa\ bb$	White	9/9.5	$sasa\ cc$
Gray	7/7	$Enen$	Californian	9/9.5	$sasa\ aa\ c^{H}c^{H}$
Tortoise	7/7	$Enen\ aa\ ee$	Siamese Sable	6/6	$aa\ c^{chl}c^{chl}$
Lilac	7/7	$Enen\ aa\ bb\ dd$	Silvers		
Flemish Giants			Gray	6/6	$sisi\ aa$
Steel gray	14/15	$E^{D}E$	Fawn	6/6	$sisi\ ee$
Light gray	14/15		Brown	6/6	$sisi\ ww$
Sandy	14/15	ww	Silver Fox		
Black	14/15	aa	Blue	9.5/10.5	$sisi\ aa\ dd$
Blue	14/15	$aa\ dd$	Black	9.5/10.5	$sisi\ aa$
White	14/15	cc	Silver Martin		
Fawn	14/15	ee	Black	7.5/8.5	$a^{t}a^{t}\ c^{ch3}$
Florida White	5/5	cc	Blue	7.5/8.5	$a^{t}a^{t}\ c^{ch3}\ dd$
Havana			Chocolate	7.5/8.5	$a^{t}a^{t}\ c^{ch3}\ bb$
Brown	6/6	$aa\ bb$	Sable	7.5/8.5	$a^{t}a^{t}\ c^{chl}c^{chl}$
Blue	6/6	$aa\ dd$	Tans		
Harlequin	8	$e^{J}e^{J}$	Black	4.5/5	$a^{t}a^{t}$
Himalayan	3.5	$aa\ c^{H}c^{H}$	Blue	4.5/5	$a^{t}a^{t}\ dd$
Lops (French) varied + white	10/11	$Enen$	Chocolate	4.5/5	$a^{t}a^{t}\ bb$
Lops (English) varied + white	10/11	$Enen$	Lilac	4.5/5	$a^{t}a^{t}\ bb\ dd$
Lilac	6.5/7	$aa\ bb\ dd$			

[a] References (23, 158, 189, 217, 223).
[b] Only the mutated genes are indicated in list. Genotypes indicated are the most common ones.

colors and on hair type, depending on the fancy of the animal breeder. Study of the inheritance of coat color has been of interest to many people because of the obvious variations that may be seen. To list all the references here would be impossible, but the literature on the inheritance of coat color in the rabbit can best be seen in the review articles of Castle (23, 27), Sawin (198), Robinson (189), and Searle (217).

A. Size Inheritance

At about the turn of the century, studies involving body size and body weight (size being different from weight) were attempted. MacDowell (124, 125) made a series of linear measurements on a large and a small strain. He then crossed these strains and obtained F_1, F_2, and backcrosses to the larger parent. He observed that bone growth was less subject to environmental variations than soft tissue weights and presented *prima facie* evidence for a polygenic basis of size inheritance. Castle (14), in an appendix to MacDowell's 1914 paper, showed that size inheritance was determined by general size genes. Wright (234), following a more sophisticated statistical analysis of the data, contended that, although Castle was correct that general factors affecting size were present, other lesser influences were also at work. He did not differentiate whether these were genetic or environmental. Later, Castle (17), using bone measurements, body weight, and ear length, tackled this problem again with animals of Polish, Himalayan, and Flemish races of as "pure" stock as possible. He made standard matings for genetic analysis and was able to support his earlier findings and point out that genetic influence was primarily general in nature and not correlated with albinism, dilute, yellow, or the angora loci. Also, sex appeared to have little effect on these measurements. From his data he was able to show that there were at least 10 to 12 pairs of chromosomes in the rabbit. This, we will see later, is an underestimate. Wright (237), using a more refined technique, reanalyzed both MacDowell's 1914 data and the 1922 data of Castle. Wright was in general agreement with the earlier reports and concluded that genetic factors were prominent in the general size influences, but were also present to a lesser extent in the regional and specific size factors. These data, when reanalyzed by Tanner and Burt (227) using factor analysis, resulted in conclusions similar to Wright's.

Concurrently, Pease (172) and Punnett and Bailey (179), employing crosses between large and small stocks of rabbits, reported two studies using body weight and ear length as the major criteria of size. Pease used the turning point in the growth curves as an estimate of the age of puberty and as his criterion for mature body weight. He reported that the age at maturity was influenced by genetic factors but not affected by sex, litter size, or season, either season of birth or season of maturity. Castle (22) criticized Pease's report based in part on the variability in weight of Pease's stocks and in part on the fact that with 14-day intervals between weighings it would be difficult to accurately determine the point of maturity. Castle provided data to prove his point.

B. Embryological Variants

Robb (184, 185) and Castle (24) investigated the nature of the growth curves of body and organ weights in a large strain and in a small strain of rabbits. Robb pointed out that, while differences in size and in organ weights between breeds of rabbits did occur, the relationship of organ size to total body mass was a constant. However, Castle in 1932 refuted this conclusion and maintained that "growth rates are not identical in the large and small races at any time, either subsequent to birth or prior to it." This was particularly evident in the papers of Castle and Gregory (29, 30) and Gregory and Castle (95) in which they showed that the size genes of the rabbit act by altering rate of development. Eggs of a genetically large-bodied race of rabbits underwent cleavage more rapidly than the eggs of a small-bodied race, so that a larger embryonic disc was formed and larger bodied young were born. These continued to grow at a more rapid rate until maturity. The parental influence on size was exerted through sperm and egg alike, but there was some maternal influence which was greater than the paternal influence in affecting the ultimate body size. Castle believed that this might act through the cytoplasm; however, another very realistic possibility is the factor of how much the doe is actually lactating and hence giving the young a greater (or lesser) initial start in life. Castle also summarized in his 1932 paper (24): "The important general conclusion to which all of our studies on size inheritance in rabbits point is that differences in adult body size are determined primarily by different growth potentials inherent in the gametes (eggs and sperm) of each race. The effects of these differences in growth potential are manifested first in differences in rate of segmentation of the fertilized egg, then in differences in the size of the blastocyst and of the embryonic area which develops upon it, later in differences in size of the young at birth and in (percentage) growth rate subsequent to birth, and finally in a more prompt and complete arrest of growth at puberty." A succinct review of size inheritance has been given by Castle (28).

C. Growth and Morphology

Sawin, who started his scientific career studying the hereditary variations of the chinchilla rabbit involving an allele for full color, the three chinchilla alleles, the Himalayan allele, and the albino allele of the C-locus (194, 195), made his major contribution to the study and understanding of the process of growth. He used the skeleton of the rabbit as a grid on which to record changes in growth associated with specific gene or genome differences. Originating with a paper on homeotic variations in the axial skeleton (196), a series of about 40 papers have been published by Sawin and associates on the morphogenetic studies of the rabbit. They deal with qualitative and quantitative variations of skeleton and soft tissues in several strains of rabbits differing in size and conformation and with respect to the effects of three specific dwarf genes, including two chondrodystrophies. For complete bibliography see Sawin and Gow (207). Tendencies for changes in rib and presacral vertebrae number (197), in numbers of crural insertions of the diaphragm (211), and in patterns of incidence of primary and secondary centers of ossification known to be associated more often with some of the strains than with others have demonstrated the genetic control of the localized gradients of vertebral growth (54, 201). However, none of the variations studied have been attributed to single, specific genes. Studies of strain differences in gradient growth pattern and in F_1 and backcross generations of reciprocal crosses have revealed the nature of growth interaction in different genomes (203), and the mechanism by which such epigenetic variations arise. They show that the gradients can be analyzed by standard genetic procedures and also demonstrate the importance of the ontogenetic approach in direction and in time. The importance of these findings has more recently been confirmed by studies of the *Da* chondrodystrophy gene (207, 213). In the most recent studies, in which attention is focused on specific types of mating (208), and on additional epigenetic variants as reference points, the interaction of additional genetic growth influences becomes manifest (210, 212) and the relation of gene, genome, gradient, and specific functions are more clearly portrayed. A further analysis of the sex and strain correlations between strains III and X, reported in early communications in the morphogenetic series by Sawin and his co-worker Latimer, has revealed gradient pattern and differences in correlation associated with the functions of locomotion and posture, suggesting a possible newer approach to the study of growth (205). Zarrow, Ross, and Denenberg, in collaboration with Sawin, summarize the effects of hereditary factors on maternal behavior in the rabbit and its endocrine basis (191, 239, 240).

D. Pathology

Workers at The Rockefeller Institute (now the Rockefeller University) have also contributed greatly to the knowledge of the rabbit, in part from the genetic standpoint but predominantly by studying the pathology of those mutations and diseases, including neoplasia, of the rabbit bearing on problems of human constitutional disease. Major contributions to the genetics of the rabbit from The Rocke-

feller Institute include the works of Greene, Pearce, Brown, and their co-workers. Their contributions include hereditary variations of the skull (86, 92), the modifying influence of breed of rabbit on such conditions as toxemia of pregnancy (89), rabbit pox (88), and uterine adenocarcinoma (91), and the description of specific mutations of the rabbit: pituitary dwarfism (90, 93), brachydactylia (87, 94), achondroplasia (170), and osteopetrosis (171).

Nachtsheim's contributions to the genetics of the rabbit have been extremely important also, particularly in relation to the pathology of many of the mutant genes. His genetic contributions include the inheritance of shaking palsy or tremor (142), hypoplasia pelvis (143), supernumerary incisors (144), audiogenic seizures (145), rex^2 and rex^3 (33), absent incisors (147), lethal muscle contracture (148), marbled eye (149), and hydrops fetalis (150). Nachtsheim also reported a dwarf gene in 1937 (146) which, based on a series of test matings and gross morphology, appears to be the same as Greene's *Dw* gene and the dwarf reported by Kröning (112). The Pelger anomaly, while not originally reported by Nachtsheim, has been extensively studied by him (152) and his co-workers (100). Nachtsheim's extensive contributions to the pathology of inherited diseases of the rabbit may be seen in two reviews: one in 1937 (145) and a more recent monograph in 1958 (154). Two other excellent review papers covering the influence of genes in the pathology of the rabbit are Sawin (198) and Robinson (189).

E. Immunology

Another area of study in the rabbit which has been of importance in early genetic analysis has been in the field of immunology, where it was observed by Hulot and Raymond in 1901 that rabbits could produce hemolysins and agglutinins against red blood cells of other rabbits (46). Genetic analysis was initiated by Cameron and Snyder (11), Castle and Keeler (31), and Keeler and Castle (106, 107). On the basis of this work rabbits were classified into four blood types. A number of different blood group systems have been studied since then and these are best reviewed in the papers of Cohen (43, 44, 46) and in Chapter 7 of this volume. In 1958 Cohen clarified some of the confusion arising from the varied nomenclature used by different authors and listed the symbols that had been reported for the same blood groups (45). Recently Cohen and Tissot (48) reported two new isoantibodies in the rabbit. Added to these early studies on blood groups are more recent studies starting with Oudin (161, 162) and involving a series of reports, the genetics of which have come in good part from the laboratories of Mage, Dray, Oudin, Dubiski, Stormont, and their co-workers, ranging from γ-globulins (49, 50, 65, 66, 84, 99, 111, 127, 132, 163–165) to heme-binding proteins (96), low density lipoproteins (1–3), red cell esterases (97, 215), α_2-macroglobulin (8, 109), and α_1-aryl esterases (4). Some of these loci are still not clearly defined as to whether they constitute two closely linked loci or a single composite locus, e.g., the low density lipoproteins (3).

F. Inbreeding

The need for uniform research stocks was recognized early. It was Castle, in fact, (17) who showed in 1922 the importance of obtaining races of rabbits as uniformly true breeding as possible to study the inheritance of differences in size. Inbreeding may be defined as the process of mating together individuals which are more closely related than the average of the particular population. Since the early 1930's, rabbits have been inbred to varying degrees. The process of inbreeding rabbits is really no different from that of inbreeding mice or any other species. Rabbits may be inbred by a number of different breeding systems; the simplest is the brother-sister inbreeding (Fig. 1A) where the change in the theoretical level of inbreeding can be predicted from a table by Wright (236). In this system it is possible to intersperse father-daughter or mother-son matings (Fig. 1B) with no change in the theoretical rate of inbreeding provided that no one animal is used more than two generations and that these generations are consecutive. Animals may be inbred by half sib mating (Fig. 1C) although the rate of change is somewhat slower. The process of line breeding (Fig. 1D) is commonly practiced in a commercial rabbit breeding program, particularly where a male is of some specific type, either coat color or conformation, that is of interest to the breeder. Also, in many laboratories, especially where the maintenance of specific mutant genes in a particular stock is essential, it is sometimes necessary to deviate from a regular pattern of inbreeding. While this does not constitute any real problem, it does slow down the process of inbreeding and makes it more difficult to estimate the theoretical level of inbreeding achieved.

Level of inbreeding is usually estimated by the coefficient of inbreeding, a value that ranges from zero to 1. This coefficient defines that proportion of loci for which an original or base population was heterozygous but which, through the various means of inbreeding, theoretically has become homozygous. In a random breeding or panmictic base population, one might make the estimate that 50% of the loci were homozygous by chance and the other 50% were heterozygous. Hence, the process of inbreeding from this base population would then tell what proportion of the remaining 50% of the loci became homozygous through the process of inbreeding.

There are three major ways of determining the theoretical level of inbreeding. First, if the breeding program is a regular pattern, such as either full sib mating, half sib mating, or double first cousin mating, etc., the theoretical coefficient of inbreeding has been calculated by Wright (235, 236) and the numerical values can be seen in Table IV.

TABLE IV

COEFFICIENTS OF INBREEDING FOLLOWING A VARIETY OF REGULAR MATING SYSTEMS

Generation	Self-fertilization[a]	Full sib[b]	Half sib[c]	Double first cousins
0	0.000	0.000	0.000	0.000
1	0.500	0.250	0.124	0.124
2	0.750	0.375	0.218	0.188
3	0.875	0.500	0.304	0.250
4	0.938	0.594	0.380	0.312
5	0.968	0.672	0.448	0.368
6	0.984	0.734	0.510	0.418
7	0.992	0.786	0.562	0.464
8	0.996	0.826	0.610	0.508
9	0.998	0.860	0.654	0.548
10	0.999	0.886	0.692	0.584
11		0.908	0.726	0.618
12		0.926	0.756	0.648
13		0.940	0.782	0.676
14		0.952	0.806	0.702
15		0.960	0.830	0.726
16		0.968	0.846	0.748
17		0.974	0.864	0.768
18		0.980	0.878	0.788
19		0.984	0.892	0.804
20		0.986	0.904	0.820
∞	1.000	1.000	1.000	1.000

[a] Repeated backcrosses to an inbred line will result in the same rate of increase in homozygosity, i.e., same coefficient of inbreeding.

[b] Parent offspring matings may be interspersed with no change in theoretical expectations as long as no individual used more than two generations.

[c] Mating of one male in each line with an indefinite number of half sisters which are also half sisters to each other. If females are full sibs the coefficient of inbreeding will go up slightly more rapidly (235).

Gen.

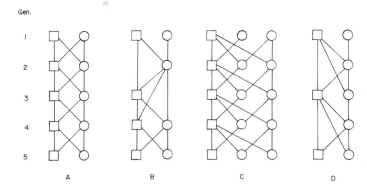

Fig. 1. Schematic diagrams of regular systems of breeding where squares denote males and circles denote females: A, Full sib brother by sister mating; B, full sib with parent offspring interspersed; C, half sib mating scheme where females are full sibs; and D represents a type of line breeding where selected males are used more than one generation. This may be varied greatly depending on the situation.

In a half sib mating program, where the females are half sibs, the theoretical level of inbreeding may be taken from the table. If, however, the females are full sibs this will increase the rate of homozygosity slightly. Conversely, if a single first cousin mating system is used, the coefficient of inbreeding will be slightly less than half the value given for double first cousins. If, however, the pattern is irregular, then one can use either Wright's method of path coefficients (238), or the method of Cruden based on the relationship of the parents (58). Both of these processes involve considerable computation, when a large number of generations is involved. However, Li and Roderick (122) have computerized the Cruden method. Their program has the advantage that overlapping generations, which often occur in an irregular system of breeding, can be taken into account.

G. Research Stocks

With the development of a wide variety of rabbit breeds (Table III), marked differences occurred in body size and in specific genes for coat color, etc., based on the needs and desires of animal breeders and research workers. Since the early 1930's rabbits have been inbred to varying degrees as have mice. Initially, there was no formal listing of the stocks of rabbits that were available. Instead, individual investigators had their own stocks with their own names associated with them. In 1963, Jay (105) started a list of the stocks that were then under the direction of C. K. Chai, C. Cohen, M. Lurie, and P. B. Sawin. However, in this early listing little was done to actually characterize the different strains with respect to anything but the barest morphological variations. More recently a series of papers reporting data on some of the physiological variants in strains maintained at The Jackson Laboratory has been published (37, 38, 65, 76, 77, 81, 83, 115, 116, 203, 204). These reports cover such things as antibody production, serum allotypes, response to teratogens, blood pressures, biochemical variations, and skin grafting.

H. Mutations

During the process of inbreeding, segregation of many recessive genes has been observed. Since many of the mutant genes are covered in depth in other sections of this book, depending on their morphologic grouping, I have listed here alphabetically all of the known genes of the rabbit, giving their gene symbols, names, and appropriate references (Table V). The first reference insofar as possible gives the original description of the gene and the data showing its Mendelian inheritance, and the second, one of the more current articles on the particular mutation, enabling one to search the literature for additional references on the subject. Since there has been some discrepancy in gene symbols for particular mutations, I have attempted to verify all the original references to ensure that, barring duplication, the symbolism is not only in keeping with the recommendations of Dunn, Grüneberg, and Snell (67) on mouse genetic nomenclature, but also is that gene designation given by the original authors.

In addition to those Mendelian characters mentioned in Table V, a series of other hereditary deviants has been reported which, for one reason or another, is not yet classed as due to single genes. In some cases the stocks were lost before sufficient matings could be made to prove the precise mode of inheritance. Sometimes the traits were polygenic. These hereditary deviants include: anophthalmia, Nachtsheim (145); avitaminosis, L. Pearce (unpublished); brachygnathia, Nachtsheim (143); conjoined twins, Chai and Crary (40); cretinoid, L. Pearce (unpublished); droopy ear, Chai and Clark (39); lop ear, Castle and Reed (34); nasal horns and cranioschisis, Nachtsheim (154); oxycephaly, Greene (86); scoliosis, Sawin and Crary (200); and thoracogastroschisis, Crary (52) and Sawin (198).

I. Linkage

As the process of inbreeding progressed and new breeds were formed, based on hereditary differences in coat color and morphology, the question of linkage of some of the specific genes involved became important. Since the first discovered case of linkage in the rabbit was reported by Castle (19) to exist between the *b* locus and the *c* locus, a series of papers have been published covering six linkage groups and 17 known loci. Some variation in crossover frequencies has been observed associated with sex (21, 26). Two articles, one by Castle and Sawin (35) and the other by Robinson (188), review all but the most recent addition by Bauer and Bennett (7). Spendlove and Robinson (222) recently showed independent assortment of the *sa* gene with the *c*, the *e*, and the *En* loci and with the sex chromosomes. A map of the currently known linkages is given in Fig. 2.

In linkage groups I, IV, and V, the linear order of some of the genes is uncertain (Fig. 2). We know, for example, that in linkage group IV the *a* locus and the *w* locus are linked by approximately 29.9 crossover units and that the *a* locus and the *Dw* locus are linked by approximately 14.7 units, but we do not know the linkage relationship of *w* to *Dw*. It has been suggested that the *Dw* gene is located between the *a* locus and the *w* locus but there has been no concrete evidence to prove this point, as pointed out by Robinson (188). The same holds true in linkage group V with reference to the location of the *f* locus and in linkage group I with the recent addition of linkage between the *bu* locus and the *c* locus.

TABLE V
RECOGNIZED GENE MUTATIONS IN THE RABBIT (*Oryctolagus cuniculus*)

Locus	Alleles	Description	Investigator (year)	Ref.
A-antigen	*An*	Absence α-agglutinin	Stuart *et al.* (1936)	(225)
	an	Presence α-agglutinin	Sawin *et al.* (1944)	(209)
Absent incisors	I^2	Absence of secondary incisors	Nachtsheim (1938)	(147)
	i^2	Presence of secondary incisors	Nachtsheim (1958)	(154)
Achondroplasia	*Ac*	Normal	Brown and Pearce (1945)	(9)
	ac	Disproportionate dwarf	Shepard and Bass (1971)	(218)
			Bargman *et al.* (1972)	(6a)
Acrobat	*Ak*	Normal	Letard (1935)	(120)
	ak	Walks on forelegs only	Nachtsheim (1958)	(154)
Adrenal hyperplasia	*Ah*	Normal	Fox and Crary (1972)	(74)
	ah	Hyperplasia of the adrenal		
Agouti	*A*	Gray with white belly	Castle (1930)	(23)
	a^t	Black and tan	Cleffmann (1953)	(42)
	a	Black (nonagouti)		
Angora	*L*	Hair length normal	Castle (1903)	(12)
	l	Hair long	Crary and Sawin (1953)	(56)
Ataxia	*Ax*	Normal	Sawin *et al.* (1942)	(199)
	ax	Loss of coordination	Robinson (1970)	(186)
Alkaline lipolytic activity	Ala^F	Fast migration of soluable esterase from adipose tissue in starch gel	Cortner and Schnatz (1970)	(50a)
	Ala^f	Slow migration of soluble esterase in starch gel		
Paralytic tremor	*Pt*	Normal	Osetowska (1967)	(159)
	pt	Resembles Parkinson's trembling. A new sex linked ataxia		
Atropinesterase	*As*	Presence of enzyme which destroys atropine	Sawin and Glick (1943)	(206)
	as	Nonenzyme	Lévy (1946)	(121)
Audiogenic seizures	*Ep*	Normal	Nachtsheim (1937)	(145)
	ep	Epileptic-like seizures	Nellhaus (1965)	(156)
Blood groups				
Hg	Hg^A	Antigen A		
	Hg^D	Antigen D (+JK)	Cameron and Snyder (1933)	(11)
			Castle and Keeler (1933)	(31)
	Hg^F	Antigen F (+ K)	Keeler and Castle (1933)	(106)
Hb	Hb^B	Antigen B		
	Hb^M	Antigen M		
Hc	Hc^C	Antigen C	Cohen (1955)	(44)
	Hc^L	Antigen L	Cohen (1958)	(45)
He	*He*	Presence of E	Cohen *et al.* (1964)	(47)
	he	Lack of E	Cohen and Tissot (1965)	(48)
Hh	*Hh*	Presence of H		
	hh	Lack of H		
Brachydactyly	*Br*	Normal	Greene (1935)	(87)
	br	Absence of nails, digits, limbs	Inman (1941)	(103)
Brown	*B*	Black	Punnett (1912)	(178)
	b	Brown (chocolate)	Castle (1940)	(27)
Buphthalmia	*Bu*	Normal	Vogt (1919)	(230)
	bu	Buphthalmus or hydrophthalmus	Sheppard *et al.* (1971)	(219)
Cataract	*Cat-1*	Normal	Nachtsheim and Gürich (1939)	(155)
	cat-1	Cataract of lens	Nachtsheim (1958)	(154)
	Cat-2	Cataract of lens	Ehling (1957)	(68)
	cat-2	Normal	Nachtsheim (1958)	(154)
Color	*C*	Fully colored	Castle (1905)	(13)
	c^{ch3}	Dark chinchilla, yellow absent, blue eyes	Castle (1921)	(16)
	c^{ch2}	Medium chinchilla, yellow absent, black diluted, eyes brown, pigment thermolabile	Sawin (1932)	(195)
	c^{ch1}	Light chinchilla, further dilution of black, eyes brown, pupil pinkish	Castle (1926)	(21)

Table V (*continued*)

Locus	Alleles	Description	Investigator (year)	Ref.
	c^H	Himalayan albinism, pigment thermolabile and restricted to extremities	Castle (1905) Voloss-Mialhe (1950)	(13) (231)
	c	Albino, total lack of pigment, eyes pink	Castle (1905)	(13)
Chondrodystrophy	Cd	Normal	Fox and Crary (1971)	(73)
	cd	Disproportionate dwarf		
Congenital luxation	Lu	Normal	DaRosa (1945)	(60)
	lu	Luxation of the hip		
Cyclopia	Cy	Normal	Menschow (1934)	(136)
	cy	Cyclopian monster	Nachtsheim (1958)	(154)
Dachs	Da	Viable chondrodystrophic dwarf	Crary and Sawin (1952)	(55)
	da	Normal	Sawin and Hamlet (1970)	(210)
Dilution	D	Black and yellow intense	Castle *et al.* (1909)	(36)
	d	Dilution of black to blue and red to yellow, acts in restricting pigment to medulla of hair	Castle (1940)	(27)
Dominant white spotting	En	English marking-black herringbone on white background		(15)
	en	Self-colored	Robinson (1955)	(187)
Dwarf	Dw	Proportionate (pituitary) dwarf	Greene *et al.* (1934)	(93)
	dw	Normal	Latimer and Sawin (1963)	(118)
Dwarf	Nan	Normal	Nachtsheim (1937)	(146)
	nan	Proportionate dwarf-nanosomia (may be same as Dw on different genome)	Suchalla (1943) Kröning (1939) Schnecke (1941)	(226) (112) (216)
Dwarf	Zw	Normal	Degenhardt (1960)	(64)
	zw	Proportionate dwarf (zwergwuchs) (may be same as Dw on different genome)		
Extension	E^D	Extension of black pigment to belly tends to obscure agouti and produce steel coat color	Punnett (1912)	(178)
	E^S	Steel—weaker edition of E^D	Robinson (1958)	(189)
	E	Normal gray		
	e^J	Japanese—mosaic distribution of black and yellow	Castle (1924)	(18)
	e	Coat yellow, white belly	Castle (1940)	(27)
Furless	F	Normal	Castle (1933)	(25)
	f	Furless	Nachtsheim (1958)	(154)
Hemolytic anemia	Ha	Normal	Fox *et al.* (1970)	(78)
	ha	Hemolytic anemia	Fox *et al.* (1971)	(80)
Hydrocephalus	Hy	Normal	Nachtsheim (1939)	(148)
	hy	High cranial vault, soft and easily depressed, excess of fluid in cranial ventricles	DaRosa (1946)	(61)
Hydrops fetalis	Hd	Erythroblastosis fetalis	Nachtsheim (1947)	(150)
	hd	Normal	Helmbold (1956)	(101)
Hypogonadia	Hg	Normal	Sawin and Crary (1962)	(202)
	hg	Absence of germ cells	Fox and Crary (1971)	(72)
Hypoplasia pelvis (originally called spastic paralysis)	Hyp	Normal	Nachtsheim (1936)	(143)
	hyp	Hypoplasia of ischium	Nachtsheim (1958)	(154)
Lethal muscle contracture	Mc	Normal	Nachtsheim (1939)	(148)
	mc	Contracture and atrophy of muscles	Sawin (1955)	(198)
Lymphosarcoma	Ls	Normal	Fox *et al.* (1970)	(78)
	ls	Lymphosarcoma	Fox *et al.* (1970)	(79)
Mandibular prognathism	Mp	Normal	Fox and Crary (1971)	(71)
	mp	Overgrown incisors due to malocclusion		
Naked	N	Normal	Kislovsky (1928)	(108)
	n	Naked	David (1932)	(63)
Osteopetrosis	Os	Normal	Pearce and Brown (1948)	(171)
	os	Abnormal bone and tooth development	Pearce (1950)	(169)

Table V (*continued*)

Locus	Alleles	Description	Investigator (year)	Ref.
Pelger	*Pg*	Chondrodystrophic dwarf, primary effect on nuclei of leukocytes	Undritz (1939)	(229)
	pg	Normal	Harm (1955)	(100)
Pelt loss	*Ps-1*	Normal	Nachtsheim (1937)	(146)
	ps-1	Absence of wool hairs, less pronounced than furless	Nachtsheim (1958)	(154)
	Ps-2	Normal	Nachtsheim (1954)	(153)
	ps-2	Absence of underwool hair, coat thicker than in *ps-1*	Nachtsheim (1958)	(154)
Recessive white, spotting	*Du*	Self-colored	Punnett and Pease (1925)	(180)
	du^d	Dark Dutch, minimal amounts of white spotting, beginning on nose, forehead and extremities	Castle (1940)	(27)
	du^w	Extensive white spotting	Castle (1940)	(27)
Red cell esterases	*Es-1^A*, *Es-1^B*	⎫	Grunder *et al.* (1965)	(97)
	Es-2^A, *Es-2^B*	⎬ Appear closely linked	Schiff and Stormont (1970)	(215)
	Es-3^A, *Es-3^B*	⎭	Schiff (1970)	(214)
Red eye	*Re*	Normal	Magnussen (1952)	(128)
	re	Red eye color	Magnussen (1954)	(129)
Renal agenesis	*Na*	Normal	DaRosa (1943)	(59)
	na	Absence of one kidney		
Renal cysts	*Rc*	Normal	Fox *et al.* (1971)	(75)
	rc	Cortical renal cysts		
Rex^1	*R^1*	Hair length normal	Lienhart (1927)	(123)
	r^1	Hair and vibrissae short and curled		
Rex^2	*R^2*	Hair length normal	Castle and Nachtsheim (1933)	(33)
	r^2	Hair and vibrissae short and curled		
Rex^3	*R^3*	Hair length normal	Castle and Nachtsheim (1933)	(33)
	r^3	Hair and vibrissae short and curled	Castle (1940)	(27)
Satin	*Sa*	Normal	Castle and Law (1936)	(32)
	sa	Absence of medulla of hair	Spendlove and Robinson (1970)	(222)
Serum allotypes				
Aa	*Aa^1*, *Aa^2*, *Aa^3*	IgA, IgM, IgG (heavy chain Fd-fragment)	Oudin (1960)	(164)
			Dray *et al.* (1963)	(65)
	Aa^{11}, *Aa^{12}*	IgG (heavy chain, hinge region)	Mandy and Todd (1968)	(131)
			Mandy and Todd (1969)	(132)
	Aa^8, *Aa^{10}*	IgG (heavy chain, Fc-fragment)	Hamers and Hamers-Casterman (1967)	(99)
	Aa^{14}, *Aa^{15}*	IgG (heavy chain, Fc-fragment)	Dubiski (1969)	(66)
	Aa^{31}	Ig? (heavy chain, Fab-fragment)	Knight *et al.* (1971)	(111)
Ab	*Ab^4*, *Ab^5*	IgG, IgA, IgM (κ-light chains)	Oudin (1960)	(164)
	Ab^6, *Ab^9*		Dray *et al.* (1963)	(65)
			Oudin (1966)	(165)
Ac	*Ac^7*, *Ac^{21}*	IgG, IgA, IgM (λ-light chains)	Mage *et al.* (1968)	(127)
			Gilman-Sachs *et al.* (1969)	(84)
Af	*Af^{71}*, *Af^{72}*, *Af^{73}*	IgA	Conway *et al.* (1969)	(49)
	Af^{74}, *Af^{75}*	IgA	Conway *et al.* (1969)	(50)
Hph	*Hph^1*, *Hph^2*	Haptoglobin	Chiao and Dray (1969)	(41)
Hbp	*Hbp^1*, *Hbp^2*, *Hbp^3*	Heme-binding proteins	Grunder (1966)	(96)
Lpj	*Lpj^1*, *Lpj^2*	High density lipoprotein	Dray (1971)	personal communication
Lpq	*Lpq^1*, *Lpq^2*	Low density lipoprotein	Albers and Dray (1968)	(1)
			Albers and Dray (1969)	(3)
Lpq	*Lpq^3*, *Lpq^4*	Low density lipoprotein (may be linked with *Lpq^1* and *Lpq^2*)	Albers and Dray (1969)	(2)
			Albers and Dray (1969)	(3)
Ess	*Ess^1*, *Ess^2*	α$_1$-Aryl esterase	Albers *et al.* (1969)	(4)
Mtz	*Mtz^1*, *Mtz^2*	α$_2$-Macroglobulin	Knight and Dray (1968)	(109)
			Knight and Dray (1968)	(110)
	Mtz^3, *Mtz^4*	α$_2$-Macroglobulin (may be linked with *Mtz^1*, *Mtz^2*)	Berne *et al.* (1970)	(8)

Table V (*continued*)

Locus	Alleles	Description	Investigator (year)	Ref.
Shaking palsy	*Tr*	Normal	Nachtsheim (1934)	(142)
(tremor)	*tr*	Continuous trembling and convulsions	Nachtsheim (1958)	(154)
Silvering	*Si*	Normal	Marchlewski (1924)	(133)
	si	Silver—probably polygenic	Quevedo and Chase (1957)	(181)
Spina bifida	*Sb*	Normal	Crary *et al.* (1966)	(53)
	sb	Spina bifida occulta totalis		
Supernumerary	*Isup*	Normal	Nachtsheim (1936)	(144)
incisors	*isup*	Extra incisors	Rohloff (1945)	(190)
Syringomyelia	*Sy*	Normal	Ostertag (1930)	(160)
	sy	Asymmetrical spastic paralysis	Nachtsheim (1958)	(154)
Vienna white	*V*	Self-colored	Pap (1921)	(167)
	v	White, blue eyed	Castle (1940)	(27)
Waved	*Wa*	Normal	Pickard (1941)	(175)
	wa	Hair waved, manifested only in rex rabbits		
Wide band	*W*	Normal agouti band	Sawin (1932)	(195)
	w	Subterminal agouti band double in width	Wilson and Dudley (1946)	(233)
Wirehair	*Wh*	Absence of wool hair	Sawin (1955)	(198)
	wh	Normal		
Wuzzy	*Wu*	Normal	Sawin (1955)	(198)
	wu	Hair sticky and matted	Crary and Sawin (1959)	(57)
Yellow fat	*Y*	White fat	Pease (1928)	(173)
			Castle (1940)	(27)
	y	Yellow fat	Wilson and Dudley (1946)	(233)

Note: A mutation for distal foreleg curvature was published by Pearce (169a) posthumously. This condition is a "bowing" deformity of the radius and ulna resulting in a seallike or "flipper" position of the forepaws. I propose the symbol *fc* for this autosomal recessive gene for "foreleg curvature" reported by Dr. Pearce.

J. Chromosomes: Numbers and Morphology

Following Painter's (166) original work in 1926 showing that the diploid chromosome number in the rabbit was 44, a series of confirmatory reports was made (for list, see Makino, 130). Then in 1956, Melander (135), using lung cells from 19-day-old rabbit embryos, showed clearly the position of the centromeres, enabling idiogram analysis. The chromosomes were, therefore, easily paired but individual identification was difficult. Sarkar *et al.* (193), using cultured cells from the rabbit cornea, separated the chromosomes into two major groups based on relative length and arm ratios. Teplitz and Ohno (228) suggested that the X and Y chromosomes were both small acrocentric ones. McMichael *et al.* (126) found that three breeds of domestic rabbits all had a common karyotype but differed with Teplitz and Ohno regarding the morphology of the X and Y chromosomes. Myers *et al.* (141), using bone marrow cells, separated the rabbit chromosomes into seven groups based on total length and arm to leg ratios. Kuhlmann (113), using blastocysts, also observed seven groups differing, however, from Myers *et al.* At the same time, Nichols *et al.* (157), using a variety of tissues, gave an excellent karyotype, however, with only four morphologically different groups based on length and arm ratios. Ray and Williams (182), using cells from leukocyte cultures, revealed seven distinct

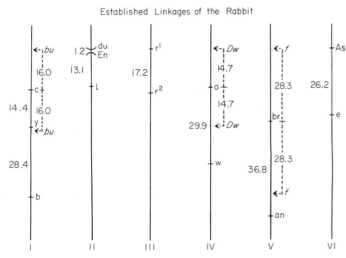

Fig. 2. Established linkages of the rabbit showing number of crossover units between loci. Dotted lines also denote linkage but in linkage group I, for example, the dotted lines show that the linkage between the *bu* locus and the *c* locus is 16 units, however, it is not known on which side of *c* that *bu* is located. Known linkage groups include: I—c, bu, y, b; II—du, En, l; III—r¹, r²; IV—a, Dw, w; V—br, f, an; and VI—As, e.

groups differing slightly, however, from previous reports. Dave *et al.* (62) placed the chromosomes into groups based on arm ratios alone but their groups differed from those of Nichols *et al.* Dave *et al.* (62) also reported the X chromosome to be the fifth largest. Pruniéras *et al.* (177) suggested that the X chromosome varied morphologically from one situation to the next. A recent report by Issa *et al.* (104) failed to confirm Dave's or Pruniéras's statement regarding the X chromosome and showed 11 morphologically different groups. As can be seen from these reports, there is still lack of agreement as to the rabbit karyotype. It is anticipated that with either the quinacrine mustard fluorescent technique (138) or a new procedure using Giemsa (E. M. Eicher, personal communication) the chromosomes can be identified individually as they have been now for the mouse (138).

Computer analysis of rabbit chromosomes has recently been accomplished. The first paper describes the procedures used in measuring the chromosomes (224) and the second, the program (82). Figure 3 gives the rabbit karyotype prepared in the manner of Issa *et al.* (104).

K. Techniques for Karyotyping

Following the early techniques of Melander (135), using a squash procedure with acetic orcein stain on embryonic lung tissue, a series of reports has been published with descriptions employing a variety of tissues and techniques. Rothfels and Siminovitch (192), using monkey kidney cells from tissue culture, reported an air-drying technique utilizing acetic orcein or Giemsa stain that has been used by McMichael *et al.* (126) on the rabbit with cultured tumor cells. Mark (134) also used tumor cells. However, he prepared his chromosomes with a squash technique and stained them with acetic orcein. Bone marrow, which can be obtained easily from the live rabbit with negligible trauma (114), has been prepared by air-drying and Giemsa staining (141), by squashing followed by acetic orcein staining (157), and by a blaze dry technique where a variety of stains were employed (119). Sarkar *et al.* (193) used cells removed from the corneal epithelium, cultured them for about one week, and then fixed and stained them by the air-drying method of Rothfels and Siminovitch (192). The culture of leukocytes obtained from the peripheral blood has been of considerable interest due to the ease and simplicity of obtaining blood from adult rabbits, and even from newborns. Moorhead *et al.* (139) reported an air-drying technique using 10 ml of human blood which, while usable for adult rabbits, precluded the use of young or newborn material. Nichols *et al.*, in a later communication, also used this technique (157) modified for the rabbit. Williams and Ray in the same year (232) immunized their rabbits by injecting 5 ml of human AB serum and subsequently obtained

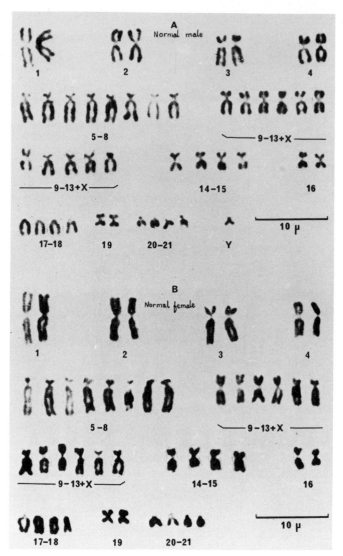

Fig. 3. Karyotype of the rabbit. In the manner of Issa *et al.* (104). Reprinted with permission of Dr. M. Issa.

approximately 10 ml of blood. After culturing this blood they used the air-drying procedure of Rothfels and Siminovitch (192) to prepare their chromosomes. Ray and Williams later published another paper on the rabbit idiogram using this technique (182). Two microculture techniques using whole rabbit blood have been reported, both employing an air-dry procedure (176, 183). One of the most recent techniques has been reported by Issa *et al.* (104) in which they used cultured cells from either rabbit blastocysts or adult tissues, air-dried their preparations, and subsequently stained them with lactic acetic orcein coupled with autoradiography.

Meiotic preparations include an ovarian squash technique using Giemsa stain (228) and a series of air-drying techniques of testicular material using Giemsa stain (70), lactic aceto-orcein stain (137), or a variety of stains (69).

IV. INBRED STRAINS

The production of inbred strains (defined as having an inbreeding coefficient of greater than or equal to that resulting from 20 generations of brother by sister inbreeding) has been successful in mice, rats, and guinea pigs and more recently in rabbits (38). Some have claimed that the larger species cannot be inbred, as various attempts have been made in many species. However, it has been stated that for every inbred line of mice now in existence there were probably many more populations started (85). Hence, with a low degree of success the economics of inbreeding a larger species becomes a limiting factor. The initiation of the inbreeding that produced the inbred strains of rabbits reported by Chai (38) goes back to 1952 when Dr. Paul B. Sawin produced the F_1, F_2, and F_3 generations from stock composed of 62.5% Dutch stock obtained from The Rockefeller Institute (carrying the *ac*, *cr*, and *ep* genes), 6.25% race X, 7.8125% race III, 4.6875% race III$_c$, and 18.75% miscellaneous New Zealand White and New Zealand Red stock. Races X, III, and III$_c$ have been used extensively by Sawin in his studies of growth and maternal behavior. These F_3 animals were then transferred to Dr. Carl Cohen who continued the inbreeding for 3 to 4 generations at which time Dr. C. K. Chai took over the colony and continued the inbreeding to establish the ACCR-B (B-line) and the ACCR-Y (Y-line) inbred strains (38).

Problems encountered with the inbreeding of rabbits were similar to those observed with mice and other mammalian species. These problems, the result of inbreeding depression, include decreased litter size and overall productivity, increased susceptibility to disease and infection, and greater mortality. The decrease in litter size in rabbits has been shown to be in good part due to increased embryonic mortality, as the number of ova shed in the inbred lines is similar to noninbred lines (38).

Inbred lines are valuable for medical research in that they provide genetic uniformity in the biological material. This means that often fewer animals are needed to achieve a specified level of precision. Moreover, the results are very repeatable when the animals are maintained under a controlled environment. However, inbreds are more subject to environmental influences and in certain experiments the use of a hybrid between two inbred strains is preferred. The F_1 hybrid has the advantage over the inbred of increased vigor coupled with the same degree of genetic uniformity and repeatability of an inbred and, therefore, makes an excellent experimental animal for certain experiments. However, it will not breed true as genetic segregation will occur in F_2 generations. Sources of inbred, incipient inbred, and mutant-bearing rabbits are recorded in the list of Genetic Stock Centers (*Genetics*, September 1965) and supplemented quarterly by the ILAR newsletter published by the Institute of Laboratory Animal Resources, National Academy of Sciences, National Research Council, 2101 Constitution Avenue, Washington, D.C. 20418. This list covers current references on animal models for biomedical research and is an excellent source.

REFERENCES

1. Albers, J. J., and Dray, S. (1968). Identification and genetic control of two rabbit low-density lipoprotein allotypes. *Biochem. Genet.* **2**, 25–35.
2. Albers, J. J., and Dray, S. (1969). Identification and genetic control of two new low-density lipoprotein allotypes: Phenogroups at the *Lpq* locus. *J. Immunol* **103**, 155–162.
3. Albers, J. J., and Dray, S. (1969). Allelic exclusion and phenogroup expression in individual molecules of rabbit low-density lipoprotein allotypes. *J. Immunol.* **103**, 163–169.
4. Albers, L. V., Dray, S., and Knight, K. L. (1969). Allotypes and isozymes of rabbit α_1-aryl esterase. Allelic products with different enzymatic activities for the same substrates. *Biochemistry* **8**, 4416–4424.
5. Asdell, S. A. (1964). "Patterns of Mammalian Reproduction," 2nd ed. Cornell Univ. Press, Ithaca, New York.
6. "Audubon Nature Encyclopedia" sponsored by the National Audubon Society. (1965). Vol. 5, pp. 840–843, Hares; Vol. 6, pp. 1099–1100, Lagamorphs; Vol. 9, pp. 1623–1630, Rabbits. Curtis Publ., New York.
6a. Bargman, G. J., Mackler, B., and Shepard, T. H. (1972). Studies of oxidative energy deficiency. 1. Achondroplasia in the rabbit. *Arch. Biochem. Biophys.* **150**, 137–146.
7. Bauer, E. J., Jr., and Bennett, J. (1964). Linkage of *bu* and *c* in the domestic rabbit. *Genetics* **50**, 234 (abstr.).
8. Berne, B. H., Dray, S., and Knight, K. L. (1970). Identification and genetic control of the Mt-3 and Mt-4 allotypes of rabbit serum α_2-macroglobulin. *J. Immunol.* **105**, 856–864.
9. Brown, W. H., and Pearce, L. (1945). Hereditary achondroplasia in the rabbit. I. Physical appearance and general features. *J. Exp. Med.* **82**, 241–260.
10. Cahalane, V. H. (1961). "Mammals of North America," Chapter 18, pp. 577–625. Macmillan, New York.
11. Cameron, R. D., and Snyder, L. H. (1933). The inheritance of isohemagglutinogens in rabbits. *Ohio J. Sci.* **33**, 50–54.
12. Castle, W. E. (1903). The heredity of "angora" coat in mammals. *Science* **18**, 760–761.
13. Castle, W. E. (1905). Heredity of coat characters in guinea pigs and rabbits. *Carnegie Inst. Wash. Publ.* **23**, 1–78.
14. Castle, W. E. (1914). Nature of size factors as indicated by a study of correlation. *Carnegie Inst. Wash. Publ.* **196**, 51–55.
15. Castle, W. E. (1919). Studies of heredity in rabbits, rats, and mice. *Carnegie Inst. Wash. Publ.* **288**, 4–28.
16. Castle, W. E. (1921). Genetics of the chinchilla rabbit. *Science* **53**, 387–388.
17. Castle, W. E. (1922). Genetic studies of rabbits and rats. *Carnegie Inst. Wash. Publ.* **320**, 1–57.
18. Castle, W. E. (1924). Genetics of the Japanese rabbit. *J. Genet.* **14**, 225–229.
19. Castle, W. E. (1924). On the occurrence in rabbits of linkage in inheritance between albinism and brown pigmentation. *Proc. Nat. Acad. Sci. U.S.* **10**, 486–488.
20. Castle, W. E. (1925). Heredity in rabbits and guinea pigs. *Bibliogr. Genet.* **1**, 418–458.
21. Castle, W. E. (1926). Studies of color inheritance and of linkage in rabbits. *Carnegie Inst. Wash. Publ.* **337**, 1–47.

22. Castle, W. E. (1929). A further study of size inheritance in rabbits with special reference to the existence of genes for size characters. *J. Exp. Zool.* **53**, 421–454.

23. Castle, W. E. (1930). "The Genetics of Domestic Rabbits." Harvard Univ. Press, Cambridge, Massachusetts.

24. Castle, W. E. (1932). Growth rates and racial size in rabbits and birds. *Science* **76**, 259–260.

25. Castle, W. E. (1933). The furless rabbit. *J. Hered.* **24**, 81–86.

26. Castle, W. E. (1936). Further data on linkage in rabbits. *Proc. Nat. Acad. Sci. U.S.* **22**, 222–225.

27. Castle, W. E. (1940). "Mammalian Genetics." Harvard Univ. Press, Cambridge, Massachusetts.

28. Castle, W. E. (1941). Size inheritance. *Amer. Natur.* **75**, 488–498.

29. Castle, W. E., and Gregory, P. W. (1929). The embryological basis of size inheritance in the rabbit. *J. Morphol. Physiol.* **48**, 81–104.

30. Castle, W. E., and Gregory, P. W. (1931). The effects of breed on growth of the embryo in fowls and rabbits. *Science* **73**, 680–681.

31. Castle, W. E., and Keeler, C. E. (1933). Blood group inheritance in the rabbit. *Proc. Nat. Acad. Sci. U.S.* **19**, 92–98.

32. Castle, W. E., and Law, L. W. (1936). Satin, a new hair mutation of the rabbit. *J. Hered.* **27**, 235–240.

33. Castle, W. E., and Nachtsheim, H. (1933). Linkage interrelations of three genes for rex (short) coat in the rabbit. *Proc. Nat. Acad. Sci. U.S.* **19**, 1006–1011.

34. Castle, W. E., and Reed, S. C. (1936). Studies of inheritance in lop-eared rabbits. *Genetics* **21**, 297–309.

35. Castle, W. E., and Sawin, P. B. (1941). Genetic linkage in the rabbit. *Proc. Nat. Acad. Sci. U.S.* **27**, 519–523.

36. Castle, W. E., Walter, H. E., Mullenix, R. C., and Cobb, S. (1909). Studies of inheritance in rabbits. *Carnegie Inst. Wash. Publ.* **114**, 70 pp.

37. Chai, C. K. (1968). The effect of inbreeding in rabbits. *Transplantation* **6**, 689–693.

38. Chai, C. K. (1969). Effects of inbreeding in rabbits. Inbred lines, discrete characters, breeding performance, and mortality. *J. Hered.* **60**, 64–70.

39. Chai, C. K., and Clark, E. M. (1967). Droopy-ear, a genetic character in rabbits. *J. Hered.* **58**, 149–152.

40. Chai, C. K., and Crary, D. D. (1971). Conjoined twinning in rabbits. *Teratology* **4**, 433–444.

41. Chiao, J. W., and Dray, S. (1969). Identification and genetic control of rabbit haptoglobin allotypes. *Biochem. Genet.* **3**, 1–13.

42. Cleffmann, G. (1953). Untersuchungen über die Fellzeichnung des Wildkaninchens. Ein Beitrag zur Wirkungsweise des Agutifaktors. *Z. Indukt. Abstamm.-Vererbungsl.* **85**, 137–162.

43. Cohen, C. (1955). Blood group factors in the rabbit. *J. Immunol.* **74**, 432–438.

44. Cohen, C. (1955). Blood group factors in the rabbit. II. The inheritance of six factors. *Genetics* **40**, 770–780.

45. Cohen, C. (1958). On blood groups and confusion in the rabbit. *Transplant. Bull.* **5**, 21–23.

46. Cohen, C. (1962). Blood groups in rabbits. *Ann. N.Y. Acad. Sci.* **97**, 26–36.

47. Cohen, C., DePalma, R. G., Colberg, J. E., Tissot, R. G., and Hubay, C. A. (1964). The relationship between blood groups and histocompatibility in the rabbit. *Ann. N.Y. Acad. Sci.* **120**, 356–361.

48. Cohen, C., and Tissot, R. G. (1965). Blood groups in the rabbit. Two additional isoantibodies and the red cell antigens they identify. *J. Immunol.* **95**, 148–155.

49. Conway, T. P., Dray, S., and Lichter, E. A. (1969). Identification and genetic control of three rabbit γA immunoglobulin allotypes. *J. Immunol.* **102**, 544–554.

50. Conway, T. P., Dray, S., and Lichter, E. A. (1969). Identification and genetic control of f4 and f5 γA immunoglobulin allotypes. *J. Immunol.* **103**, 662–667.

50a. Cortner, J. A., and Schnatz, J. D. (1970). Alkaline lipolytic activity of rabbit adipose tissue: genotypes and their inheritance. *Biochem. Genet.* **4**, 529–537.

51. Craigie, E. H. (1948). "Bensley's Practical Anatomy of the Rabbit," 8th ed. Univ. of Toronto Press, Toronto.

52. Crary, D. D. (1951). A thoraco-gastroschisis in the rabbit. *Anat. Rec.* **109**, 368 (abstr.).

53. Crary, D. D., Fox, R. R., and Sawin, P. B. (1966). Spina bifida in the rabbit. *J. Hered.* **57**, 236–243.

54. Crary, D. D., and Sawin, P. B. (1949). Morphogenetic studies of the rabbit. VI. Genetic factors influencing the ossification pattern of the limbs. *Genetics* **34**, 508–523.

55. Crary, D. D., and Sawin, P. B. (1952). A second recessive chondroplasia in the domestic rabbit. *J. Hered.* **43**, 254–259.

56. Crary, D. D., and Sawin, P. B. (1953). Some factors influencing the growth potential of the skin in the domestic rabbit. *J. Exp. Zool.* **124**, 31–62.

57. Crary, D. D., and Sawin, P. B. (1959). Inheritance and hair morphology of the "wuzzy" mutation in the rabbit. *J. Hered.* **50**, 31–34.

58. Cruden, D. (1949). The computation of inbreeding coefficients for closed populations. *J. Hered.* **40**, 248–251.

59. DaRosa, F. M. (1943). Agenesia de um rim, uma nova mutação no coelho. *Rev. Med. Vet. (Lisboa)* **38**, 349–363.

60. DaRosa, F. M. (1945). Uma nova mutação, luxação congénita da anca no coelho. *Rev. Med. Vet. (Lisboa)* **40**, 1–23.

61. DaRosa, F. M. (1946). Hidrocefalia, uma nova mutação no coelho. *Rev. Med. Vet. (Lisboa)* **41**, 1–55.

62. Dave, M. J., Takagi, N., Oishi, H., and Kikuchi, Y. (1965). Chromosome studies on the hare and the rabbit. *Proc. Jap. Acad.* **44**, 244–248.

63. David, L. T. (1932). External expression and comparative dermal histology of hereditary hairlessness in mammals. *Z. Zellforsch. Mikrosk. Anat.* **14**, 616–719.

64. Degenhardt, K. H. (1960). Die genetische und morphologische Analyse spezieller Entwicklungsstörungen in einem Stamm ingezüchteter Hermelin-Kaninchen. *Akad. Wiss. Lit., Mainz, Abh. Math.-Naturwiss. Kl.* No. 12. pp. 919–988.

65. Dray, S., Young, G. O., and Gerald, L. (1963). Immunochemical identification and genetics of rabbit γ-globulin allotypes. *J. Immunol.* **91**, 403–415.

66. Dubiski, S. (1969). Immunochemistry and genetics of a "new" allotypic specificity Ae^{14} of rabbit γG immunoglobulins: Recombination in somatic cells. *J. Immunol.* **103**, 120–128.

67. Dunn, L. C., Grüneberg, H., and Snell, G. D. (1940). Report of the Committee on Mouse Genetic Nomenclature. *J. Hered.* **31**, 505–506.

68. Ehling, U. (1957) Untersuchungen zur kausalen Genese erblicher Katarakte beim Kaninchen. *Z. Konstitutionslehre* **34**, 77–104.

69. Eicher, E. M. (1966). An air-drying procedure for mammalian male meiotic chromosomes, following softening in gluconic acid and cell separation by an ethanol-acetic mixture. *Stain Technol.* **41**, 317–321.

70. Evans, C. P., Breckon, G., and Ford, C. E. (1964). An air-drying method for meiotic preparations for mammalian testes. *Cytogenetics* **3**, 289–294.

71. Fox, R. R., and Crary, D. D. (1971). Mandibular prognathism in the rabbit: Genetic studies. *J. Hered.* **62**, 23–27.

72. Fox, R. R., and Crary, D. D. (1971). Hypogonadia in the rabbit: Genetic studies and morphology. *J. Hered.* **62**, 163–169.

73. Fox, R. R., and Crary, D. D. (1971). A new recessive chondrodystrophy in the rabbit. *Teratology* **4**, 245–246. (abstr.).

74. Fox, R. R., and Crary, D. D. (1972). A lethal recessive gene for adrenal hyperplasia in the rabbit. *Teratology* **5**, 255 (abstr.).

75. Fox, R. R., Krinsky, W. L., and Crary, D. D. (1971). Hereditary cortical renal cysts in the rabbit. *J. Hered.* **62**, 105–109.

76. Fox, R. R., Laird, C. W., Blau, E. M., Schultz, H. S., and Mitchell, B. P. (1970). Biochemical parameters of clinical significance in rabbits. I. Strain variations. *J. Hered.* **61**, 261–265.

77. Fox, R. R., Meier, H., and Crary, D. D. (1971). Genetic predisposition to tumors in the rabbit. *Naturwissenshaften* **9**, 457.

78. Fox, R. R., Meier, H., Crary, D. D., Myers, D. D., Norberg, R. F., and Laird, C. W. (1970). Hereditary lymphosarcoma and anemia in rabbits. *Teratology* **3**, 200 (abstr.).

79. Fox, R. R., Meier, H., Crary, D. D., Myers, D. D., Norberg, R. F., and Laird, C. W. (1970). Lymphosarcoma in the rabbit. Genetics and pathology. *J. Nat. Cancer Inst.* **45**, 719–729.

80. Fox, R. R., Meier, H., Crary, D. D., Norberg, R. F., and Myers, D. D. (1971). Hemolytic anemia associated with thymoma in the rabbit. Genetic studies and pathological findings. *Oncology* **25**, 372–382.

81. Fox, R. R., Schlager, G., and Laird, C. W. (1969). Blood pressure in thirteen strains of rabbits. *J. Hered.* **60**, 312–314.

82. Geldermann, H., Stranzinger, G., and Paufler, S. (1970). Automatisierung der chromosomenanalyse mit dem Elektronenreckner. *Humangenetik* **9**, 325–360.

83. Gill, T. J., III. (1965). Studies on synthetic polypeptide antigens. XIV. Variations in antibody production among rabbits of different inbred strains. *J. Immunol.* **95**, 542–545.

84. Gilman-Sachs, A., Mage, R. G., Young, G. O., Alexander, C., and Dray, S. (1969). Identification and genetic control of two rabbit immunoglobulin allotypes at a second light chain locus, the *c* locus. *J. Immunol.* **103**, 1159–1167.

85. Green, E. L., and Doolittle, D. P. (1963). Systems of mating used in mammalian genetics. *In* "Methodology in Mammalian Genetics" (W. J. Burdette, ed.), pp. 3–41. Holden-Day, San Francisco, California.

86. Greene, H. S. N. (1933). Oxycephaly and allied conditions in man and in the rabbit. *J. Exp. Med.* **57**, 967–976.

87. Greene, H. S. N. (1935). Hereditary brachydactylia and associated abnormalities in the rabbit. *Science* **81**, 405–407.

88. Greene, H. S. N. (1935). Rabbit pox. IV. Susceptibility as a function of constitutional factors. *J. Exp. Med.* **62**, 305–329.

89. Greene, H. S. N. (1938). Toxemia of pregnancy in the rabbit. II. Etiological considerations with especial reference to hereditary factors. *J. Exp. Med.* **67**, 369–388.

90. Greene, H. S. N. (1940). A dwarf mutation in the rabbit. *J. Exp. Med.* **71**, 839–856.

91. Greene, H. S. N. (1941). Uterine adenomata in the rabbit. III. Susceptibility as a function of constitutional factors. *J. Exp. Med.* **73**, 273–292.

92. Greene, H. S. N., and Brown, W. H. (1932). Hereditary variations in the skull of the rabbit. *Science* **76**, 421–422.

93. Greene, H. S. N., Hu, C. K., and Brown, W. H. (1934). A lethal dwarf mutation in the rabbit with stigmata of endocrine abnormality. *Science* **79**, 487–488.

94. Greene, H. S. N., and Saxton, J. A., Jr. (1939). Hereditary brachydactylia and allied abnormalities in the rabbit. *J. Exp. Med.* **69**, 301–314.

95. Gregory, P. W., and Castle, W. E. (1931). Further studies on the embryological basis of size inheritance in the rabbit. *J. Exp. Zool.* **59**, 199–211.

96. Grunder, A. A. (1966). Inheritance of a heme-binding protein in rabbits. *Genetics* **54**, 1085–1093.

97. Grunder, A. A., Sartore, G., and Stormont, C. (1965). Genetic variation in red cell esterases of rabbits. *Genetics* **52**, 1345–1353.

98. Hall, E. R., and Kelson, K. R. (1959), "The Mammals of North America," Vol. I, pp. 246–288. Ronald Press, New York.

99. Hamers, R., and Hamers-Casterman, C. (1967). Evidence for the presence of the Fc allotypic marker As8 and the Fd allotypic marker Asl in the same molecules of IgG. *Cold Spring Harbor Symp. Quant. Biol.* **32**, 129–132.

100. Harm, H. (1955). Zur Klassifizierung und Vererbung der Neutrophilen-Kernform bei Mensch und Kaninchen. *Blut* **1**, 3–25.

101. Helmbold, W. (1956). Über Blutkörperchenantigene des Kaninchens in Zusammenhang mit dem Hydropsproblem. I. Die Isolierung von 10 spezifischen Antikörpern aus Isoimmunseren und die Bestimmung des korrespondierenden Antigene. *Blut* **2**, 9–31.

102. "Illustrated Encyclopedia of Animal Life." (1961). Mammals. Vol. 3, pp. 235–247. Greystone Press, New York.

103. Inman, O. R. (1941). Embryology of hereditary brachydactyly in the rabbit *Anat. Rec.* **79**, 483–505.

104. Issa, M., Atherton, G. W., and Black, C. E. (1968). The chromosomes of the domestic rabbit, *Oryctolagus cuniculus. Cytogenetics* **7**, 361–375.

105. Jay, G. E., Jr. (1963). Genetic strains and stocks. *In* "Methodology in Mammalian Genetics" (W. J. Burdette, ed.), pp. 83–123. Holden-Day, San Francisco, California.

106. Keeler, C. E., and Castle, W. E. (1933). A further study of blood groups of the rabbit. *Proc. Nat. Acad. Sci. U.S.* **19**, 403–411.

107. Keeler, C. E., and Castle, W. E. (1934). Blood group inheritance in rabbits. *J. Hered.* **25**, 433–439.

108. Kislovsky, D. A. (1928). Naked—a recessive mutation in the rabbit. *J. Hered.* **19**, 438–439.

109. Knight, K. L., and Dray, S. (1968). Identification and genetic control of two rabbit α_2-macroglobulin allotypes. *Biochemistry* **7**, 1165–1171.

110. Knight, K. L., and Dray, S. (1968). Contribution of allelic genes to the formation of individual α_2-macroglobulin molecules. *Biochemistry* **7**, 3830–3835.

111. Knight, K. L., Gilman-Sachs, A., Fields, R., and Dray, S. (1971). Allotypic determinants on the Fab fragment of rabbit *Aa* locus negative IgG-immunoglobin. *J. Immunol.* **106**, 761–767.

112. Kröning, F. (1939). Ein neuer Fall von erblichem Zwergwuchs beim Kaninchen. *Biol. Zentralbl.* **59**, 148–160.

113. Kuhlmann, W. (1964). Die Strahlenempfindlichkeit bestimmter Oozyten-Meiosephasen, und die chromosomalen Verhältnisse beim Kaninchen *Oryctolagus cuniculus. Akad. Wiss. Lit., Mainz, Abh. Math. Naturwiss.* Kl. No. 6, pp. 267–318.

114. Laird, C. W., and Fox, R. R. (1964). A method for obtaining rabbit bone marrow for chromosome studies. *Proc. Soc. Exp. Biol. Med.* **115**, 751–752.

115. Laird, C. W., Fox, R. R., Mitchell, B. P., Blau, E. M., and Schultz, H. S. (1970). Effect of strain and age on some hematological parameters in the rabbit. *Amer. J. Physiol.* **218**, 1613–1617.

116. Laird, C. W., Fox, R. R., Schultz, H. S., Mitchell, B. P., and Blau, E. M. (1970). Strain variations in rabbits: Biochemical indicators of thyroid function. *Life Sci.* **9**, 203–214.

117. "Larousse Encyclopedia of Animal Life." (1967). Section on Mammals, pp. 518–521. McGraw-Hill, New York.

118. Latimer, H. B., and Sawin, P. B. (1963). Morphogenetic studies of the rabbit. XXXV. Comparison of the weights and linear measurements in normal and heterozygous dwarf rabbits of race X. *Anat. Rec.* **146**, 85–92.

119. Lee, M. R. (1969). A widely applicable technique for direct processing of bone marrow for chromosomes of vertebrates. *Stain Technol.* **44**, 155–158.

120. Letard, E. (1935). Une mutation nouvelle chez le lapin. *Bull. Acad. Vet. Fr.* **8**, 608–610.

121. Lévy, J. (1946). Transmission héréditaire de la tropanolestérase. *C. R. Soc. Biol.* **140**, 823–825.

122. Li, F. H. F., and Roderick, T. H. (1970). Computer calculation of Wright's inbreeding coefficient by Cruden's method. *J. Hered.* **61**, 37–38.

123. Lienhart, R. (1927). Apropos d'une récente mutation chez le lapin domestique, le lapin Castorrex. *C. R. Soc. Biol.* **97**, 386–388.

124. MacDowell, E. C. (1914). Size inheritance in rabbits. *Carnegie Inst. Wash. Publ.* **196**, 7–49.

125. MacDowell, E. C. (1914). Multiple factors in Mendelian inheritance. *J. Exp. Zool.* **16**, 177–194.

126. McMichael, H., Wagner, J. E., Nowell, P. C., and Hungerford, D. A. (1963). Chromosome studies of virus-induced rabbit papillomas and derived primary carcinomas. *J. Nat. Cancer Inst.* **31**, 1197–1215.

127. Mage, R. G., Young, G. O., and Reisfeld, R. A. (1968). The association of the c7 allotype of rabbits with some light polypeptide chains which lack *b* locus allotypy. *J. Immunol.* **101**, 617–620.

128. Magnussen, K. (1952). Beitrag zur Genetic und Histologie eines isolierten Augen-Albinismus beim Kaninchen. *Z. Morphol. Anthropol.* **44**, 127–135.

129. Magnussen, K. (1954). Beitrag zur Genetik und Histologie eines isolierten Augen-Albinismus beim Kaninchen. II. *Z. Morphol. Anthropol.* **46**, 24–29.

130. Makino, S. (1951). "An Atlas of the Chromosome Numbers in Animals." Iowa State Coll. Press, Ames.

131. Mandy, W. J., and Todd, C. W. (1968). Allotypy of rabbit immunoglobulin: An agglutinating specificity. *Vox Sang.* **14**, 264–270.

132. Mandy, W. J., and Todd, C. W. (1969). Characterization of allotype All in rabbits: A specificity detected by agglutination. *Immunochemistry* **6**, 811–823.

133. Marchlewski, T. (1942). [A case of polimery in coat color of rabbits.] *Bull. Int. Acad. Polon. Sci. Lett., Cl. Sci. Math. Natur., Ser.* pp. 697–714.

134. Mark, J. (1965). Chromosome analyses of Rous tumors in the rabbit. *Hereditas* **53**, 165–170.

135. Melander, Y. (1956). The chromosome complement of the rabbit. *Hereditas* **42**, 432–435.

136. Menschow, G. B. (1934). Fattori letali nel coniglio cincilla. *Riv. Coniglicolt.* **6**, 8–9.

137. Meredith, R. (1969). A simple method for preparing meiotic chromosomes from mammalian testis. *Chromosoma* **26**, 254–258.

138. Miller, O. J., Miller, D. A., Kouri, R. E., Allderdice, P. W., Dev, V. G., Grewal, M. S., and Hutton, J. J. (1971). Identification of the mouse karyotype by quinacrine fluorescence, and tentative assignment of seven linkage groups (Mus musculus/autosomes/X and Y chromosomes). *Proc. Nat. Acad. Sci. U.S.* **68**, 1530–1533.

139. Moorhead, P. S., Nowell, P. C., Mellman, W. J., Battips, D. M., and Hungerford, D. A. (1960). Chromosome preparations of leukocytes cultured from human peripheral blood. *Exp. Cell Res.* **20**, 613–616.

140. Müntzing, A. (1959). Darwin's views on variation under domestication in the light of present-day knowledge. *Proc. Amer. Phil. Soc.* **103**, 190–220.

141. Myers, L. B., O'Leary, J. L., and Fox, R. R. (1964). Classification of chromosomes in normal and ataxic rabbits. *Neurology* **14**, 1058–1065.

142. Nachtsheim, H. (1934). Schüttellähmung—ein Beispiel für ein einfach mendelndes rezessives Nervenleiden beim Kaninchen. *Erbarzt* **1**, 36–38.

143. Nachtsheim, H. (1936). Die Genetik einiger Erbleiden des Kaninchens, verglichen mit ähnlichen Krankheiten des Menschen. *Deut. Tieraerztl. Wochenschr.* **44**, 742–746.

144. Nachtsheim, H. (1936). Erbliche Zahnanomalien beim Kaninchen. *Zuechtungskunde* **11**, 273–287.

145. Nachtsheim, H. (1937). Erbpathologie des Kaninchens. *Erbarzt* **4**, 25–30 and 50–55.

146. Nachtsheim, H. (1937). Erbpathologische Untersuchungen an Kaninchen. *Z. Indukt. Abstamm.-Vererbungsl.* **73**, 463–466.

147. Nachtsheim, H. (1938). Erbpathologie der Haustiere. I. Organe des äusseren Keimblattes. *Fortschr. Erbpathol.* **2**, 58–104.

148. Nachtsheim, H. (1939). Erbleiden des Nervensystems bei Säugetieren. *In* "Handbuch der Erbbiologie des Menschen" (K. H. Bauer, E. Hanhart, and J. Lange, eds.), pp. 1–58. Springer-Verlag, Berlin and New York.

149. Nachtsheim, H. (1943). Ergebnisse und Probleme der vergleichenden und experimentellen Erbpathologie. *Jena Z. Med. Naturwiss.* **76**, 81–108.

150. Nachtsheim, H. (1947). Ein erbliche fetale Erythroblastose beim Tier und ihre Beziehungen zu den Gruppenfaktoren des Blutes. *Klin. Wochenschr.* **1947**, 590–592.

151. Nachtsheim, H. (1949). "Vom Wildtier zum Haustier," 2nd ed. Parey, Berlin.

152. Nachtsheim, H. (1950). The Pelger anomaly in man and rabbit. A Mendelian character of the nuclei of the leucocytes. *J. Hered.* **41**, 131–137.

153. Nachtsheim, H. (1954). Die Mutabilität menschliche Gene. *Proc. Int. Congr. Genet., 9th. 1953* pp. 139–154.

154. Nachtsheim, H. (1958). Erbpathologie der Nagetiere. *In* "Pathologie der Laboratoriumstiere" (P. Cohrs, R. Jaffe, and H. Meesen, eds.), pp. 310–452. Springer-Verlag, Berlin and New York.

155. Nachtsheim, H., and Gürich, H. (1939). Erbleiden des Kaninchenanges. I. Erbliche Nahtbändchentrübung der Linse mit nochfalgendem Kernstar. *Z. Konstitutionslehre* **23**, 463–483.

156. Nellhaus, G. (1965). Experimental epilepsy in rabbits: Failure to detect a difference in phenylalanine metabolism in convulsant rabbits. *Proc. Soc. Exp. Biol. Med.* **120**, 259–260.

157. Nichols, W. W., Levan, A., Hansen-Melander, E., and Melander, Y. (1965). The idiogram of the rabbit. *Hereditas* **53**, 63–76.

158. Official Guide Book. (1968). American Rabbit Breeders Ass., Inc., Pittsburgh, Pennsylvania.

159. Osetowska, E. (1967). Nouvelle maladie héréditaire du lapin de laboratoire. *Acta Neuropathol.* **8**, 331–344.

160. Ostertag, B. (1930). Die Syringomyelie als erbbiologisches Problem. *Verh. Deut. Pathol. Ges.* **25**, 166–174.

161. Oudin, J. (1956). Réaction de précipitation spécifique entre des sérums d'animaux de même espèce. *C. R. Acad. Sci.* **242**, 2489–2490.

162. Oudin, J. (1956). L'allotypie de certain antigènes protéidiques du sérum. *C. R. Acad. Sci.* **242**, 2606–2608.

163. Oudin, J. (1960). Allotypy of rabbit serum proteins. I. Immunochemical analysis leading to the individualization of seven main allotypes. *J. Exp. Med.* **112**, 107–124.

164. Oudin, J. (1960). Allotypy of rabbit serum proteins. II. Relationships between various allotypes: Their common antigenic specificity, their distribution in a sample population; genetic implications. *J. Exp. Med.* **112**, 125–142.

165. Oudin, J. (1966). Genetic regulation of immunoglobulin synthesis. *J. Cell. Physiol.* **67**, Suppl. 1, 77–108.

166. Painter, T. S. (1926). Studies in mammalian spermatogenesis. VI. The chromosomes of the rabbit. *J. Morphol. Physiol.* **43**, 1–43.

167. Pap, E. (1921). Ueber Vererbung von Farbe und Zeichnung bei dem Kaninchen. *Z. Indukt. Abstamm.-Vererbungsl.* **26**, 185–270.

168. Park, E. (1960). Rabbits, hares, and the pika. *In* "Wild Animals of North America," Chapter 21, pp. 294–309. National Geographic Society. Washington, D.C.

169. Pearce, L. (1950). Hereditary osteopetrosis of the rabbit. III. Pathologic observations; skeletal abnormalities. *J. Exp. Med.* **92**, 591–600.

169a. Pearce, L. (1960). Hereditary distal foreleg curvature in the rabbit. I. Manifestations and course of the bowing deformity: Genetic studies. II. Genetic and pathological effects. *J. Exp. Med.* **111**, 801–830.

170. Pearce, L., and Brown, W. H. (1945). Hereditary achondroplasia in the rabbit. III. Genetic aspects, general considerations. *J. Exp. Med.* **82**, 281–295.

171. Pearce, L., and Brown, W. H. (1948). Hereditary osteopetrosis of the rabbit. I. General features and course of the disease; genetic aspects. *J. Exp. Med.* **88**, 579–596.

172. Pease, M. S. (1928). Experiments on the inheritance of weight in rabbits. *J. Genet.* **20**, 261–309.

173. Pease, M. (1928). Yellow fat in rabbits, a linked character? *Z. Indukt. Abstamm.- Vererbungsl.* **2**, Suppl., 1153–1156.

174. Petter, F. (1959). Eléments d'une révision des lièvres africains du sous-genre *Lepus. Mammalia* **23**, 41–67.

175. Pickard, J. N. (1941). Waved—a new coat type in rabbits. *J. Genet.* **42**, 215–222.

176. Plummer, B. H., and Fox, R. R. (1966). A whole-blood microculture technique for cytological examination of neonatal rabbits. *Proc. Soc. Exp. Biol. Med.* **122**, 868–870.

177. Pruniéras, M., Jacquemont, C., and Mathivon, M. F. (1965). Etudes sur les relations virus-chromosomes. V. Le caryotype du lapin domestique. *Ann. Inst. Pasteur, Paris* **109**, 465–471.

178. Punnett, R. C. (1912). Inheritance of coat colour in rabbits. *J. Genet.* **2**, 221–238.

179. Punnett, R. C., and Bailey, P. G. (1918). Genetic studies in rabbits, I. On the inheritance of weight. *J. Genet.* **8**, 1–25.

180. Punnett, R. C., and Pease, M. S. (1925). On the pattern of the Dutch rabbit. *J. Genet.* **15**, 375–412.

181. Quevedo, W. C., Jr., and Chase, H. B. (1957). Histological observations on the silvering process in the Champagne d'Argent rabbit. *Anat. Rec.* **129**, 87–95.

182. Ray, M., and Williams, T. W. (1966). Karyotype of rabbit chromosomes from leucocyte cultures. *Can. J. Genet. Cytol.* **8**, 393–397.

183. Razavi, L. (1965). An inexpensive and simple method for preparing chromosome spreads. *Proc. Soc. Exp. Biol. Med.* **118**, 717–719.

184. Robb, R. C. (1929). On the nature of hereditary size limitation. I. Body growth in giant and pigmy rabbits. *Brit. J. Exp. Biol.* **6**, 293–310.

185. Robb, R. C. (1929). On the nature of hereditary size limitation. II. The growth of the parts in relation to the whole. *Brit. J. Exp. Biol.* **6**, 311–324.

186. Robinson, N. (1970). Enzyme changes in the hereditary ataxic rabbit. *Acta Neuropathol.* **14**, 326–337.

187. Robinson, R. (1955). Viability of dominant white in the rabbit. *J. Hered.* **46**, 266.

188. Robinson, R. (1956). A review of independent and linked segregation in the rabbit. *J. Genet.* **54**, 358–359.

189. Robinson, R. (1958). Genetic studies of the rabbit. *Bibliogr. Genet.* **17**, 229–558.

190. Rohloff, R. (1945). Entwicklungsgeschichtliche Untersuchungen über erbliche Anomalien der Incisiven bei Oryctolagus cuniculus, L.; zugleich Mitteilung von Beobachtungen an den Rudimentärzahnchen. Inauguraldissertation, University of Berlin (unpublished).

191. Ross, S., Sawin, P. B., Zarrow, M. X., and Denenberg, V. H. (1963). Maternal behavior in the rabbit. *In* "Maternal Behavior in Mammals" (H. L. Rheingold, ed.), Chapter 3, pp. 94–121. Wiley, New York.

192. Rothfels, K. H., and Siminovitch, L. (1958). An air-drying technique for flattening chromosomes in mammalian cells grown *in vitro. Stain Technol.* **33**, 73–77.

193. Sarkar, P., Basu, P. K., and Miller, I. (1962). Karyologic studies on cells from rabbit cornea and other tissues grown *in vitro. Invest. Ophthalmol.* **1**, 33–40.

194. Sawin, P. B. (1932). Hereditary variation of the chinchilla rabbit in coat and eye color. *J. Hered.* **23**, 39–46.

195. Sawin, P. B. (1932). Albino allelomorphs of the rabbit with special reference to blue-eyed chinchilla and its variations. *Carnegie Inst. Wash. Publ.* **427**, 15–50.

196. Sawin, P. B. (1937). Preliminary studies of hereditary variations in the axial skeleton of the rabbit. *Anat. Rec.* **69**, 407–408.

197. Sawin, P. B. (1946). Morphogenetic studies of the rabbit. III. Skeletal variations resulting from the interaction of gene determined growth forces. *Anat. Rec.* **96**, 183–200.

198. Sawin, P. B. (1955). Recent genetics of the domestic rabbit. *Advan. Genet.* **7**, 183–226.

199. Sawin, P. B., Anders, M. V., and Johnson, R. B. (1942). "Ataxia," a hereditary nervous disorder of the rabbit. *Proc. Nat. Acad. Sci. U.S.* **28**, 123–127.

200. Sawin, P. B., and Crary, D. D. (1955). Congenital scoliosis in the rabbit. *Anat. Rec.* **121**, 2 (abstr.).

201. Sawin, P. B., and Crary, D. D. (1956). Morphogenetic studies of the rabbit. XVI. Quantitative racial differences in ossification pattern of the vertebrae of embryos as an approach to basic principles of mammalian growth. *Amer. J. Phys. Anthropol.* **14**, 625–648.

202. Sawin, P. B., and Crary, D. D. (1962). Inherited hypogonadia in the rabbit. *Anat. Rec.* **142**, 325 (abstr.).

203. Sawin, P. B., and Crary, D. D. (1964). Genetics of skeletal deformities in the domestic rabbit (*Oryctolagus cuniculus*). *Clin. Orthop. Relat. Res.* **33**, 71–90.

204. Sawin, P. B., Crary, D. D., Fox, R. R., and Wuest, H. M. (1965). Thalidomide malformations and genetic background in the rabbit. *Experientia* **21**, 672–677.

205. Sawin, P. B., Fox, R. R., and Latimer, H. B. (1970). Morphogenetic studies of the rabbit. XLI. Gradients of correlation in the architecture of morphology. *Amer. J. Anat.* **128**, 137–146.

206. Sawin, P. B., and Glick, D. (1943). Atropinesterase, a genetically determined enzyme in the rabbit. *Proc. Nat. Acad. Sci. U.S.* **29**, 55–59.

207. Sawin, P. B., and Gow, M. (1967). Morphogenetic studies of the rabbit XXXVI. Effect of gene and genome interaction on homeotic variation. *Anat. Rec.* **157**, 425–435.

208. Sawin, P. B., Gow, M., and Muehlke, M. (1967). Morphogenetic studies of the rabbit. XXXVII. Genome, gradient growth pattern and malformation. *Amer. J. Anat.* **121**, 197–216.

209. Sawin, P. B., Griffin, M. A., and Stuart, C. A. (1944). Genetic linkage of blood types in the rabbit. *Proc. Nat. Acad. Sci. U.S.* **30**, 217–221.

210. Sawin, P. B., and Hamlet, M. (1970). Morphogenetic studies of the rabbit. XL. Growth gradient interaction and function in morphology. *J. Morphol.* **130**, 397–420.

211. Sawin, P. B., and Hull, I. B. (1946). Morphogenetic studies of the rabbit. II. Evidence of regionally specific hereditary factors influencing the extent of the lumbar region. *J. Morphol.* **78**, 1–26.

212. Sawin, P. B., and Muehlke, M. (1969). Morphogenetic studies of the rabbit. XXXVIII. Epigenetic probing of the genome. *Amer. J. Anat.* **125**, 233–246.

213. Sawin, P. B., and Trask, M. (1965). Morphogenetic studies of the rabbit. XXXV. Pleiotropic effects of the dachs gene and the gradient growth pattern. *J. Morphol.* **117**, 87–114.

214. Schiff, R. (1970). The biochemical genetics of rabbit erythrocyte esterases: Histochemical classification. *J. Histochem. Cytochem.* **18**, 709–721.

215. Schiff, R., and Stormont, C. (1970). The biochemical genetics of rabbit erythrocyte esterases: Two new esterase loci. *Biochem. Genet.* **4**, 11–23.

216. Schnecke, C. (1941). Zwergwuchs beim Kaninchen und seine Vererbung. *Z. Konstitutionsehre* **25**, 427–457.

217. Searle, A. G. (1968). "Comparative Genetics of Coat Colour in Mammals." Academic Press, New York.

218. Shepard, T. H., and Bass, G. L. (1971). Organ-culture studies of achondroplastic rabbit cartilage: Evidence for a metabolic defect in glucose utilization. *J. Embryol. Exp. Morphol.* **25**, 347–363.

219. Sheppard, L. B., Shanklin, W. M., Harris, T. H., and Fox, R. R. (1971). A histologic study of regenerating epithelium of normal and buphthalmic rabbit cornea. *Ophthalmol. Res.* **2**, 116–125.

220. Simpson, G. G. (1945). The principles of classification and a classification of mammals. *Bull. Amer. Mus. Natur. Hist.* **85**, 1–350.

221. Sirks, M. J. (1959). Leeuwenhoek on dominance in rabbits. *Genetica* **30**, 292.

222. Spendlove, W. H., and Robinson, R. (1970). A linkage test with *satin* in the rabbit. *Genetica* **41**, 635–637.

223. Standard of Perfection. (1966–1970). American Rabbit Breeders Ass., Inc., Pittsburgh, Pennsylvania.

224. Stranzinger, G., Geldermann, H., and Paufler, S. (1970). Eine halbautomatische chromosomenmessmethode mit dem Koordinatenmesstisch "Pencel Follower." *Humangenetik* **10**, 44–50.

225. Stuart, C. A., Sawin, P. B., Wheeler, K. M., and Battey, S. (1936). Group-specific agglutinins in rabbit serums for human cells. I. Normal group-specific agglutinins. *J. Immunol.* **31**, 25–29.

226. Suchalla, H. (1943). Variabilität und Erblickeit von Schädelmerkmalen bei Zwerg und Riesenrassen; dargestellt an Hermelin und Widder-Kaninchen. *Z. Morphol. Anthropol.* **40**, 274–333.

227. Tanner, J. M., and Burt, A. W. A. (1954). Physique in the inframammalia: A factor analysis of body measurements of dairy cows. *J. Genet.* **52**, 36–51.

228. Teplitz, R., and Ohno, S. (1963). Postnatal induction of ovogenesis in the rabbit. *Exp. Cell Res.* **31**, 183–189.

229. Undritz, E. (1939). Das Pelger-Huëtsche Blutbild beim Tier und seine Bedeutung für die Entwichlungsgeschichte des Blutes. *Schweiz. Med. Wochenschr.* **1939**, 1177.

230. Vogt, A. (1919). Vererbter Hydrophthalmus beim Kaninchen. *Klin. Monatsbl. Augenheilk.* **63**, 233.

231. Voloss-Mialhe, C. (1950). Rôle du sympathique dans la pigmentation du lapin Himalaya. *C. R. Soc. Biol.* **144**, 19–20.

232. Williams, T. W., and Ray, M. (1965). A method for culturing leukocytes of rats and rabbits. *Cytogenetics* **4**, 365–368.

233. Wilson, W. K., and Dudley, F. J. (1946). Fat colour and fur colour in different varieties of rabbit. *J. Genet.* **47**, 290–294.

234. Wright, S. (1918). On the nature of size factors. *Genetics* **3**, 367–374.

235. Wright, S. (1921). Systems of mating. II. The effects of inbreeding on the genetic composition of a population. *Genetics* **6**, 124–143.

236. Wright, S. (1921). Systems of mating. V. General considerations. *Genetics* **6**, 167–178.

237. Wright, S. (1932). General, group and special size factors. *Genetics* **17**, 603–619.

238. Wright, S. (1934). The method of path coefficients. *Ann. Math. Stat.* **5**, 161–215.

239. Zarrow, M. X., Gandelmun, R., and Denenberg, V. H. (1971). Prolactin: Is it an essential hormone for maternal behavior in the mammal? *Horm. Behav.* **2**, 343–354.

240. Zarrow, M. X., Sawin, P. B., Ross, S., and Denenberg, V. H. (1962). Maternal behavior and its endocrine basis in the rabbit. *In* "Roots of Behavior" (E. L. Bliss, ed.), pp. 187–197. Harper, New York.

CHAPTER 2

Colony Husbandry

Karl W. Hagen

I. INTRODUCTION

The European rabbit (*Oryctolagus cuniculus*) is found on the European continent in three forms; the wild form, a feral form, and a domestic form. In North America, however, we are acquainted only with the domestic and feral. The wild or ancestral form probably evolved in the Iberian peninsula and spread to other regions of the Mediterranean. The domestic form, the one with which we are most familiar, is typified by a great variety of breeds and strains which are used for meat and laboratory animal production. The feral form may be described as a reversion from domestic to wild. Examples of this form may be found on the Farallon Island off the coast of San Francisco, on the San Juan

Islands in the Juan de Fuca straits in Washington, the Channel Islands off Santa Barbara, California, and on Isla del Flores near Montevideo, Uruguay. The only population of wild European rabbits in the Americas is in and spreading from the island of Tierra del Fuego, Chile. They have spread north into the mainland several hundred miles north of Santiago.

The European rabbit is a member of the order Lagomorpha which consists of two families: the Ochotonidae (pikas) and the Leporidae (rabbits and hares). *Oryctolagus cuniculus* is the only species of the entire order which has been domesticated and as such the species has been introduced into every country of the world as a source of meat or fur or for use as a laboratory animal.

The actual domestication of the European rabbit probably took place in the monasteries during the middle ages (114). By the middle of the seventeenth century the rabbit was completely domesticated and rabbit raising was active in England and continental Europe. *Oryctolagus cuniculus* is one of the more successful mammals of the world; it is both prolific and adaptable and appears to be equally at home on all five continents.

Most of the fancy breeds have been developed within the last 100 years and only since the 1900's has domestic rabbit raising been recognized in the United States. The first commercial colonies were started in southern California, but it was not until meat rationing during World War II that the infant industry blossomed. Today approximately 250,000 people are engaged in some phase of the rabbit business and animals are being produced in almost every state of the nation. Meat processors, serving many major cities, market over 50,000,000 lb annually. Over 500,000 rabbits per year are being used in medical research in addition to by-products being utilized for pharmaceutical and vaccine production. Over the years the breeds have been improved to the compact, blocky animal compared with the long, rangy, low meat yield type. Production has been increased from less than 65 lb of meat per year per doe to a minimum of 120 lb per doe and 200 lb per doe is not unlikely in the future. Feed conversion has been reduced from 5 to 6 lb/lb of meat to 3.5–4, and disease diagnosis and control has been effective in many areas.

Biomedical research has become complex and laboratory animals used in research have become more well-defined. Producers of rats and mice have long recognized the need for improvement and have invested in activities resulting in animals far superior to any heretofore available. Rabbit producers have not always availed themselves of the newer knowledge in this field and their animals consequently are usually not raised under optimum conditions. A start now is being made by commercial producers to house animals in controlled environments and to maintain a good state of health. Buildings should be constructed to control extremes of temperatures, cages should be of all wire or metal cons-

truction, and water and feeding systems should be of the type which cannot become easily contaminated. Professional assistance should be sought in determining and maintaining the health status of the colony. Some of the specific considerations which give meaning to these objectives are discussed in this section.

II. SEXUAL BEHAVIOR AND BREEDING

A. Sexual Behavior in Males

1. Courtship

In the wild, courtship chasing between adults is an early form of sexual activity. Caged rabbits, of course, are denied this opportunity and male courtship antics are restricted to tail flagging and enurination.

Tail flagging is a common form of behavior. The buck elevates his hindquarters, walks stiff-legged, and lays his tail flat across his back. In this way he displays the underside of the tail, which in colored breeds will be white. Various flagging movements of the tail may be performed as the buck circles about in the cage. This behavior is almost entirely limited to the male. Besides supplying a visual stimulus, it may also provide an olfactory stimulus from the inguinal glands.

Bucks will emit a jet of urine (enurination) at a partner in a display of sexual aggressiveness. The buck may turn his hindquarters toward the doe and eject a small amount of urine which is then followed by some form of circling about the cage. Enurination is most frequently noted in bucks caged in proximity to each other. A buck will circle wildly or throw his hindquarters around in such a manner that a jet of urine can be directed at a buck in an adjacent cage. Enurination may also occur as the result of surprise or being startled, as some animal caretakers may have discovered.

2. Copulation

When a receptive doe is placed in a buck's hatch, she will quickly raise her hindquarters to allow copulation. The Buck mounts the doe and grasps the female's body with a foreleg on each flank. Intromission is usually accomplished after 8–12 rapid copulatory movements and ejaculation follows on the first intromission. Following ejaculation, the buck falls off backward or sideways and may emit a characteristic cry. The cry is regarded as an indication of pain and may, however, arise from either the buck or doe. A vigorous buck may attempt to copulate again. Slow-motion photography reveals that at the time of ejaculation, both hind feet are off the ground and being unable to compensate for this imbalance, the buck falls off backward or to the side. A gelatinous substance in the ejaculate may form a vaginal plug which is lost shortly after copulation, but no cervical seal is formed (9).

The fluid portion of the ejaculate usually ranges from 0.5 to 1.5 ml while the sperm density/milliliter ranges from 0.5×10^8 to 3.5×10^8 (42). Fructose, 40 to 400 mg/100 ml, occasionally glucose, and 215 to 370 mg/100 ml glyceryl-phosphorylcholine are found in the fluid portion (42). Several enzymes have been identified, and having a higher level of catalase than in other animals, rabbit sperm is resistant to hydrogen proxide (65).

3. Sex Drive

Copulation has been described as an intricate phenomenon mediated by sensory receptors and any impairment of the senses may be expected to affect sexual performance. Factors such as age, frequency of mating, nutritional state, and heredity have been found to affect sex drive. In experimental studies on sex drive, Hafez (32) found that a sequence of exploration, smelling, jumping, chin rubbing, mounting, gripping with teeth, pelvic oscillations, exploratory movement of erect penis, intromission, and orgasm with ejaculation was fairly constant in all individuals. But the rate and persistency of the pattern varied considerably from male to male and in the same male at different times. From this study seven degrees of sex drive were recognized which ranged from aggressive with immediate mounting and ejaculation to offensive reaction with general smelling of the skin, biting, and no ejaculation.

Semen samples collected daily for up to 13 months showed considerable variation between individuals and in the same individual at different times. Maximum sperm counts occurred on the average every 3.16 days (20). Sperm count was related to motility and total volume but not to amount of gel in the semen. Doggett and Ett (21) evaluated sex drive on the basis of reaction time to a dummy female, numbers of ejaculations before exhaustion, and interest in a dummy female, but found no significant correlations.

Hafez found that sperm motility was not related to a degree of sex drive but that aggressive mating was associated with a larger volume of ejaculate, less sperm concentration, and a higher percentage of live sperm (32). Studies by Oloufa *et al.* (77) indicated that high environmental temperatures (91°F) had an adverse effect on all semen characteristics and that thyroxine improved the fertility of animals kept intermittently at high temperatures (77).

B. Sexual Behavior in Females

1. Estrus

Wild rabbits show a definite period of anestrus and a seasonal variation in reproductive activity but domestic do not show a regular estrus cycle. They do, however, vary in their receptivity and a certain rhythm does exist (33). On occasion a doe will refuse to mate but will readily accept another buck. She may accept a buck once but refuse him a second time.

Hamilton (40) reported cyclic changes in certain cells from vaginal smears and thin sections of cervical and vaginal tissues. These cycles recurred at 4 to 6-day intervals and were accompanied by changes in blood estrogen levels. Meyers and Poole found that domestic does periodically became sexually attracted to males in intervals of 4–6 days, a shorter intervals than the 7-day cycle they reported in the wild (*Oryctolagus*) rabbit (67, 68). These changes in reproductive activity should perhaps be more correctly referred to as estrus behavior rather than a clear-cut estrus cycle.

Blount states that a sexually receptive doe has a congested vulva characterized by a purplish color (9). Full sexual receptivity may then be indicated by a congested, moist, purple vulva, restlessness, trying to join other rabbits in adjacent cages, and chin rubbing. The vaginal smear technique is not satisfactory.

2. Ovulation and Conception

Rabbits do not ovulate spontaneously; ovulation occurs 10–13 hours after coitus or after the injection of a luteinizing hormone. Ovulation may be induced experimentally by electrical or manual stimulation or as the result of an orgasm induced by contact with other females. The rabbit resembles the cat, ferret, some other mustelids, and the shrew in its ovulatory pattern. Some does fail to ovulate following coitus. This may be the result of a deficiency of luteinizing hormone from the pituitary gland (27).

Ovulation is induced by the intravenous injection of 20–25 IU of human chorionic gonadotropin. The percentage of ovulation is correlated with body weight (95). Stormshak and Casida induced ovulation in pseudopregnant rabbits with 0.01 mg of a purified luteinizing hormone (33).

Experimental studies indicate changes in reproductive traits are more dependent on changes in temperature (5, 77) than changes in light conditions (5, 52). Data collected in southern California indicate that the conception rate exhibited a close inverse relationship with the maximum temperature (90). A major drop in conception rate from May to June accompanies a rise in maximum temperature (Table I). The low figures for conception, total litters and young, and total young born alive occurred after 3 months of the highest temperature. After the temperature fell below 75°F these traits immediately improved suggesting restoration of fertility once the maximum temperature drops below a critical level (Fig. 1.).

3. Pregnancy

a. LENGTH OF GESTATION. The gestation period is 31–32 days (41, 97), but is believed to vary with season (5).

TABLE I
MONTHLY CONCEPTION RATE[a]

Month of service	No. services	Average maximum temperature °F	Conception rate (%)			Total
			1 to 4	5 to 8	9 to 12	
January	186	63	83	72	87	80
February	194	66	70	88	77	77
March	182	71	71	85	71	75
April	190	75	78	81	56	75
May	172	77	77	87	74	80
June	209	85	70	64	69	68
July	216	90	68	64	60	65
August	212	90	63	65	61	64
September	277	87	54	51	43	52
October	269	80	63	52	61	60
November	230	73	63	67	58	63
December	226	70	73	69	71	72
All months	2563		68	69	65	68

[a] Modified from Sittmann et al. (90).

Table II shows the frequency distribution of 2886 New Zealand White litters in respect to size of litter and length of gestation period (98). It will be noted that 85.5% of the litters were born on days 31 and 32. No normal litters were born under 29 days. Larger litters are carried a shorter time than smaller-sized litters.

Pregnancy can be determined by palpating the developing fetuses *in utero*. Holding the ears and loose skin over the shoulders in the left hand, the right hand is placed between the hind legs and in front of the pelvis. At 12–14 days of pregnancy fetuses can be felt as they slip between the fingers during palpation of the uteri. With experience one can determine pregnancy at 10–12 days after breeding. One can also identify retained fetuses, abscesses or cysts,

TABLE II
LENGTH OF THE GESTATION PERIOD AND NUMBERS OF YOUNG IN LITTERS OF NEW ZEALAND WHITE RABBITS[a, b]

No. in litter	Days of gestation and nos. of litters						
	29	30	31	32	33	34	35
1				6	3	6	2
2				13	19	7	
3			8	34	28	9	
4			15	47	24	7	
5		1	27	78	22	3	
6		2	63	84	24	6	
7		7	104	145	22	2	
8		8	163	149	24		
9	1	13	216	171	26	3	
10	2	27	234	156	18		
11	3	33	227	119	10	1	
12	1	10	132	73	13		
13	1	14	73	38			
14		4	41	19	1		
15		4	16	10	1		
16		1	5	1	1		
17		2	2				
18			1				
	8	126	1327	1143	238	44	2
%	0.3	4.4	45.9	39.6	8.3	1.5	>0.1

[a] Templeton (98).
[b] Total no. litters = 2886. Total no. young = 25,683. Ave. young/litter = 8.8.

inflammed or congested uteri, and retained afterbirths by palpation. Pregnancy can also be determined by "test mating" with a buck, but some does will accept service when pregnant and others may refuse when not pregnant. Increased thickness of the mammary glands is also an indication of pregnancy. Radiographic confirmation is possible after the eleventh day of pregnancy (33).

b. PARTURITION. Parturition usually occurs during the early morning hours. Sawin found 68% of all litters were born between 5 AM and 1 PM and only 8% were born between 9 PM and 5 AM (86). Both anterior and breech presentations are normal. Kindling requires less than 30 minutes. Sometimes young are born several hours or even days apart. The maximum time between appearance of the young and survival is less than 3 days. Fetuses that are retained beyond 35 days die and if not expelled prevent future pregnancy (1). Labor may be induced in such does with oxytocin.

The onset of parturition depends on a weakening or withdrawal of the progesterone block of the myometrium and a sudden release of oxytocin. The length of the gestation period may depend on the life span of the corpora lutea. The cause for the sudden release of oxytocin is unknown but the withdrawal of the progesterone block in the central nervous system could be involved.

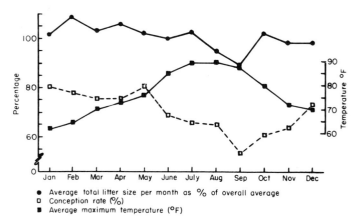

● Average total litter size per month as % of overall average
□ Conception rate (%)
■ Average maximum temperature (°F)

Fig. 1. Effect of season on reproductive capacity. Modified from Sittman et al. (90).

c. Litter Size and Sex Ratio. Litter size may vary with the stock, order of the litter, season of the year, and age of the female. Templeton (98) reported the data given in Table II which shows the average litter size to be 8.8 young. In collecting this data no consideration was given to season, age of doe, or order of litter. The colony, however, was closed. Rollins *et al.* analyzed the breeding records of 1472 litters of randomly bred New Zealand White rabbits and found the average litter was 7.1 live young (81). Their data suggested no consistent seasonal pattern of variation but did suggest an age-of-doe effect. Litter sequence analysis indicated numbers born alive were greatest in second and third litters, fewer in fourth and fifth litters, and smallest in first litters.

Seasonal variation in litter size was investigated by Sittmann *et al.* (90), and monthly average litter size is shown in Table III. The total litter size and numbers born alive in the litter decreased from a high in the month of February to a low in September. The average litter size for all months was 7.39. Their data indicated total litter size exhibited an inverse relationship to the monthly maximum temperature (Fig. 1).

Hafez (33) states that in general more males are born than females. From 226 litters of wild rabbits from 21 days of gestation to full term, Brambell (10) recorded 506 males and 534 females, a ratio of 48.6:51.4. In another study 191 males were found to 203 females, a ratio almost identical to the previous report.

d. Fetal Length and Weights. The length of the fetus (crown–rump length) can be shown as a linear function of gestational age for at least part of the intrauterine development. A second relationship exists regarding fetal weight. Combining this information one may arrive at a generalized estimate of the embryonic age. Points on Figs. 2 and 3 are average measurements obtained from 430 young

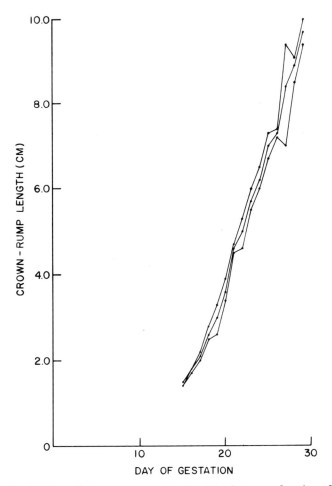

Fig. 2. Plot of the crown–rump length of the fetus as a function of time of gestation (38). High and low measurements are plotted above and below the average. Greatest variation is noted on twenty-seventh day.

produced by 50 timed litters of New Zealand White rabbits (38). Spencer and Coulombe pointed out that Haase's rule, in certain forms, applied to part of the intrauterine growth of the rat, chick, and rabbit (99). Haase's equation is $L = dT + f$, where L is a length, T is time of gestation, and d and f are constants. For the rabbit between 9.5 and 15 days of gestation the equation is: R (crown–rump length) = $0.70D - 5.53$ when Minot's (69) data for the rabbit *in utero* is used. Spencer and Coulombe (94) in plotting the crown–rump length of the rabbit fetus as drawn by Minot used the illustration size rather than the actual size of the fetus. The illustration was 4 times natural size. When this factor is considered, the formula applies and the crown–rump length of a 15-day fetus (1.3 cm) compares favorably with that figure (1.4 cm) shown in Fig. 2.

e. Postpartum Breeding. Rabbits are usually bred 6–8 weeks after kindling when the young are weaned, but does will mate immediately following parturition. If young are removed following delivery, sexual receptivity con-

TABLE III
AVERAGE LITTER SIZE BY MONTH OF BIRTH[a]

Month	No. litters	Total litter size (average)
January	246	7.52
February	191	8.07
March	235	7.65
April	199	7.75
May	209	7.46
June	206	7.30
July	196	7.51
August	212	6.97
September	170	6.49
October	183	7.45
November	214	7.22
December	186	7.17
All months	2447	7.39

[a] Modified from Sittmann *et al.* (90).

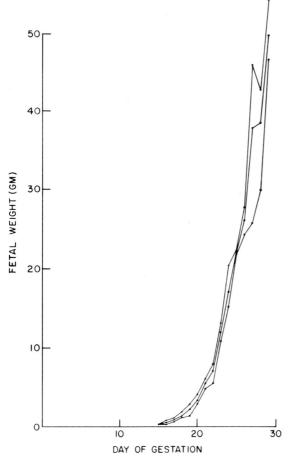

Fig. 3. Plot of the fetal weight as a function of the time of gestation (38). High and low weights are plotted above and below the average.

injection of a luteinizing hormone. Pseudopregnancy lasts 16–17 days during which time the doe is unable to conceive (99). After 18–22 days the doe may pull hair and attempt to make a nest. When this false pregnancy has terminated, the doe will resume normal reproductive behavior. The development of the mammary glands and changes in the reproductive tract are most pronounced in the first 10 days of pseudopregnancy. At 15 days the organs are involuting and at 18 days the corpora lutea are disintegrating and the uterus approaches normal size of a nonpregnant doe. Asdell and Salisbury (6) reported the stimulus for mammary development during pseudopregnancy to be in the ovary and not the uterus. Corpora lutea secrete progesterone causing the uterus and mammary glands to grow.

4. Lactation

a. MAMMARY GLANDS.   The mammary glands are distributed within the fatty tissue along the ventrolateral aspects of the thoracic and inguinal regions (Fig. 4). The glands, usually 8 in number, develop rapidly during the last week of pregnancy. Milk letdown is usually delayed until after kindling, but can be stimulated by prolactin injections. The average daily milk yield is 160–200 gm during the first litter then increases to 170–220 in subsequent litters (33). Maximum milk production takes place 2 weeks after kindling, remains high for another week,

tinues for at least 36 days. Postpartum receptivity varies with the intensity of lactation. Hammond and Marshall (41) found that 100% of rabbits will mate after parturition, 71% on the fourth day, 42% on the eighth day, and 11% on the twelfth day. Following that time all does refused service until the young were 50–60 days old and feeding themselves. Sexual receptivity is lost then as long as the mammary glands are active.

The postpartum pregnancy rate appears to be inversely related to litter size since the lactating rabbit seldom conceives when mated before the eighth day after parturition unless only 1 or 2 young remain (41). Nursing reduces the size of the ovaries and follicles and a reduced pregnancy rate results from the regression of corpora lutea and a lowering of pituitary follicle stimulating hormone. The stress of pregnancy is less than that of lactation, therefore, large numbers of young can be obtained by fostering young and rebreeding the does immediately after parturition.

f. PSEUDOPREGNANCY.   False pregnancy may be caused by an infertile mating, sexual excitement caused by one doe mounting or being mounted by another, and by an

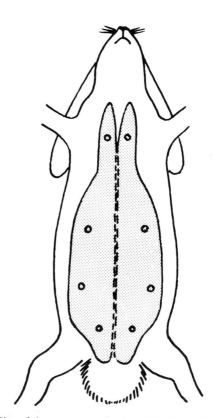

Fig. 4. Size of the mammary glands and location of the nipples.

and declines during the fourth week. A doe may lactate for an additional 2–4 weeks. The amount of milk produced and the duration of lactation will depend on diet, numbers of suckling young, and genetic factors. High milk yield is a factor to be considered when selecting animals for breeding stock.

Rabbit milk is richer than cow's milk and contains 12.3% protein, 13.1% fat, 1.9% lactose, and 2.3% minerals (93). The weight of the litter at 21 days is often used as an indicator of milk production. At 8 weeks, however, the weight of the litter is no longer related to milk yield as the young are consuming approximately 90% of their intake in the form of plant proteins.

b. TIME AND FREQUENCY. Seton (89) had suggested that rabbits and hares nurse their young only once a day but this observation had not been experimentally verified. Zarrow et al. (112) allowed Dutch Belted rabbits free access to their young, access once a day, and access twice a day. Their observations indicated that the doe nursed the young only once a day and body weight curves from each group overlapped to a great extent. The amount of time the doe spent with her young varied from 2.7 to 4.5 minutes and ended when the doe moved away from the young. Nursing generally took place during the early morning and not in the evening.

The observations that rabbits nurse only once a day is contrary to the earlier procedure of feeding hand-reared infants every 2–4 hours (see Chapter 8). With an adequate diet, a once-a-day feeding schedule should provide normal growth rates. The protein content of cow's milk can be increased by adding calcium caseinate (1:20) during the first week. The amount should be gradually increased until the young can eat fresh green feeds 3 weeks later.

5. Retrieving

Retrieving, while common among most mammals, does not occur among rabbits. Ross et al. (82) found that when rabbits were given opportunities to retrieve their young this behavior was not observed. This can perhaps be explained when one considers the daily nursing pattern of the species. The young are kept hidden from predators in a nest from which it is difficult to escape. The doe visits the nest once a day for a brief period of time while she nurses the litter. Under these circumstances it is not necessary for the female to retrieve her young.

6. Nest Box Behavior

Several days prior to parturition, the doe will collect the hay, straw, excelsior, or similar material provided her and carries it to the nest box. With this material a nest is constructed in what in nature would be her burrow. Hair from her bdoy is plucked out and interwoven with the nesting material. The center of the nest is hollowed out to hold the newborn young. As the young are born the doe eats the placenta and membranes and severs the umibical cord much in the same manner as the rat. Abnormalities associated with nest box behavior are failure to build a nest, birth of the young outside of the nest, scattering of the young, and cannibalism.

The nest is built only under conditions of pregnancy, pseudopregnancy, or hormonal manipulations. Denenberg et al. (19) found the quality of the nest improved from the first through third litter and leveled off at the fourth. The mechanism is not known but could be a function of a learning process or systematic endocrine changes. A unique feature of the nest is the maternal hair used as a lining material. The hair becomes relatively loose near the time of parturition and the degree of loosening has been measured by Sawin et al. (87). Both hair and straw contribute to the survival of the young, but the maternal hair is the more important element (113).

The quality of the nest was also found to improve between the first and fourth litters. Related to nest quality was the percentage of live-born young suckled on the first day, the time of nest building, a lack of scattering, and cannibalism. There are breed and strain differences in the time of nesting and quality of the nest. These differences are explained by relationship between pituitary and ovarian hormones at the termination of pregnancy (83).

Hafez et al. (34) reported maternal food intake had no effect on quality of the nest, but viability of the young decreased in both overweight and underweight does. Cannibalism was not affected by maternal food intake.

7. Artificial Insemination (AI)

a. SEMEN COLLECTION. Artificial vaginas (AV) have been described by Hafez (32) and Bredderman et al. (11). The AV used by Hafez (32) was made from 5 inches × $1\frac{1}{4}$ inches radiator hose. The inner lining was $1\frac{1}{2}$ inches surgical drainage tubing allowing internal pressure and temperature to be regulated. A 5-ml test tube was used for collection and was fitted to the AV with a bored rubber stopper. The required internal pressure and temperature may vary among bucks and can be adjusted through a water filling hold in the outside rubber casing. The temperature inside the AV is usually about 40°C but should not raise above 45°C or the semen may become contaminated with urine. The AV is generally used in conjunction with a teaser female or a rabbit skin worn over the arm. When the rabbit skin is placed on the arm, the AV is held in such a manner to allow easy access for the buck.

Collections can be made twice a week but daily collections have been made for as long as a year without a decline in sex drive or fertility. Daily collections yield 2 to 4 times as many sperm as weekly collections. However, semen production and sex drive will decrease if the bucks are used excessively (33). New Zealand White bucks produce more sperm than the Dutch Belted (2), but there are individual differences in frequency and output. Hafez found semen volume, sperm concentration, and mobility were higher when a white fur was used with the AV than when a gray or spotted fur was used (32).

b. STORAGE. Undiluted semen should be used within 12 hours after collection, however, semen diluted in glucose-yolk-citrate will remain viable for 4 days if kept at 10°C (33). Physiological saline or Ringer's solution may be used as dilutents and the ejaculate may be diluted as much as 10 times and still be used successfully. Sperm number rather than the dilution of seminal plasma is more important (109). Fox and Burdick (25), Sawada and Chang (85), and O'Shea and Wales (78) describe techniques for diluting and storing semen.

c. INSEMINATION. The insemination dose should contain 20 to 50 × 10⁶ spermatozoa in 0.3 to 0.7 ml of sperm suspension (33). One million sperm cells are required for optimal fertility. Wales *et al.* (109) obtained litters using semen diluted 1:10,000 with a calcium-free Krebs Ringer phosphate solution.

An insemination pipette can be made from 15 to 20-cm length of rubber, soft plastic or glass tubing with an outside diameter of 6–7 mm. The pipette is introduced into the vagina while the doe is held on her back and the inseminate is deposited in the cervical area of the vagina. The technique is described in detail by Hafez (33). Semen may also be deposited in the uterus or oviduct during laparotomy (75). Intraperitoneal insemination may be performed 12 to 16 hours before ovulation and the semen is deposited in the peritoneal cavity with a 20-gauge needle (72).

When artificial insemination (AI) is used instead of natural mating, the doe must be stimulated so that ovulation will occur. This can be done by mating with a vasectomized buck but the method is not completely reliable. The stimulus for ovulation which may be given 5 hours before to several hours after insemination is an injection of luteinizing hormone (20–25 IU). The marginal ear vein is convenient for giving this injection. Several does may be inseminated the same day to obtain numerous young of the same age.

d. CONCEPTION RATE. The conception rate and litter size resulting from artificial insemination is not different from that obtained following natural mating (73). Beatty (8) found artificial insemination appeared more effective when semen from several males were used in the inseminate even if the total number of spermatozoa per inseminate was constant. Conception rates were similar following uterine or vaginal insemination if the females were receptive. In pseudopregnant does the conception rate is lower in vaginal

rather than uterine insemination. Repeated injections of luteinizing hormone may result in antibody formation and a drop in conception. This can be avoided by using an alternative source of luteinizing hormone or by mating with a vasectomized buck.

III. FEEDS AND FEEDING BEHAVIOR

A. Basic Requirements

Thirty years ago the standard diet for domestic rabbits was composed of whole grains, a plant protein supplement, and alfalfa hay or some other legume roughage. Today most rabbits are fed a pelleted feed made up of grains, hay, and certain supplements in what is believed to be a balanced diet. There is apparently no one way of feeding rabbits, a situation caused in part by the lack of knowledge regarding specific requirements for maintenance, growth, and reproduction, as well as the demand for rapid development and conditioning for market. This section serves as a practical guide to rabbit nutrition. Details of actual requirements can be found in Chapter 16.

Data on nutrient requirements reveal many generalities and a lack of precise figures. Rations for dry does, herd bucks, and developing young should provide the following (16):

Crude protein	12–15%
Fat	2–3.5%
Fiber	20–27%
Nitrogen-free extract	43–47%
Ash or mineral	5–6.5%

Rations for pregnant does and lactating does with litters should contain more protein and provide:

Crude protein	16–20%
Fat	3–5.5%
Fiber	15–20%
Nitrogen-free extract	44–50%
Ash or mineral (TDN)	4.5–6.5%

1. Total Digestible Nutrients (TDN)

Specific values for TDN are not available, however, extensive feeding trials indicate that there are certain levels adequate for maintenance, growth, reproduction, and lactation. The total feed requirement for maintenance of does or bucks was found to be 4.0, 3.8, and 3.7 of body weight for 5-, 10-, and 15 lb rabbits, respectively. Lactating does with litters required 4.8, 4.6, and 5.6% of body weight for similar live weights, while 6.7% total daily feed was adequate for normal growth of does and bucks averaging 6.5 lb live weight (12).

2. Net Energy

Specific energy requirements have not been established but it would appear that 500–600 Calories/lb of feed are adequate for maintenance and reproduction. Lee (58) calculated the theoretical maintenance requirement to be 438 Calories per day for a rabbit weighing 5.0 kg. Changes in energy levels of rabbit rations are not directly reflected in changes in the rate of gain since the rabbit can increase feed consumption to satisfy energy and nutrient requirements when fed a low energy feed (7).

3. Protein

The requirement of 12–15% crude protein for dry or nonlactating does and developing young and 16–20% for lactating does with litters is reasonably well established (16). These requirements are based on feed consumption of New Zealand White rabbits and can be adjusted for other breeds of different size by the equation of Lee (58). These estimates are above optimum but insure adequate levels. Templeton (101) states there is no danger in feeding high levels of protein if the ration is adequate in all other ingredients. A low protein diet supplemented with urea will not support growth (76).

4. Fat

Most commercial rabbit pellets indicate a 2–3% fat content on the feed tag, however, the amount required is open to question. Templeton and Kellogg (102) recommended 2–5.5%. Thacker (103) utilized rations containing 5–25% fat in the form of vegetable oil and reported animals on 10–25% fat produced greater weight gains than those on 5%. These increased gains were associated with higher caloric intake and increased feed consumption. Palatability and feed utilization increased as the fat in the diet increased, but the digestibility of dry matter, ether extract material, or crude protein did not. Rabbits could benefit from higher levels of fat than are now being fed (12).

5. Fibers

No recommendations are made by the National Research Council (74) on the optimum level, but on the basis of extensive feeding trials, Templeton and Kellogg (102) recommend 20–27% fiber for dry does, herd bucks, and developing young and 14–20% for lactating does with litters. Fiber deficiency may be a factor in "fur eating" since supplementary greens or hay often eliminates this problem (101). Fifteen percent fiber is probably the minimal level that should be fed to rabbits.

6. Minerals

Requirements have been estimated for only a few minerals. Approximately 1.0% calcium and 0.5% phos-

phorus are considered adequate for normal growth and reproduction (12). The magnesium requirement has been estimated at 30–40 mg % of diet (56) while 0.3 mg of manganese per rabbit per day was sufficient for normal bone development (91). Cobalt is required only in trace amounts as the estimated daily requirement is less than 0.1 μg per animal (106). Sodium is supplied as sodium chloride and Templeton (97) has estimated that 0.1–0.17 lb/100 lb of concentrates provides an adequate amount.

7. Vitamins

Rabbits require certain of the known vitamins. The actual requirements have been estimated for only a few. Vitamin A deficiency is prevented by 500 to 1000 IU of vitamin A per day (64). Controversy exists regarding the importance of vitamin D. While symptoms of rickets have been produced in rabbits fed a deficient diet, Jarl (50) demonstrated interference with calcification was temporary and he concluded growing rabbits did not require a supplemental supply of vitamin D. Eppstein and Margulis (24) estimated a requirement of 0.32 mg DL-α-tocopherol/kg body weight daily to cure or prevent muscular dystrophy. Hogan and Hamilton (44) suggested the addition of vitamin K to simplified diets to insure normal reproduction. Dietary requirements for the water-soluble vitamins are not available. Normal growth on simplified diets containing dried yeast or water extract of dried liver as the sole source of all water-soluble vitamins was obtained by Hogan and Hamilton (44). There is evidence that rabbits synthesize certain of the B vitamins. Rabbits do not require a dietary source of vitamin C(43,45).

8. Water

The necessity of a plentiful supply of fresh, clean water must be emphasized. A doe and her litter may consume a gallon of water in 24 hours, depending upon the temperature. Resting animals may drink 4–6 oz daily but consumption may increase in hot, dry weather. Water deficiency retards growth, development, and lactation. Feeding moist feeds as a substitute for giving rabbits fresh water is an unsound practice. An automatic pipeline watering system is the most satisfactory method of providing a constant supply of fresh water. In addition it saves both time and labor. In a feeding trial conducted in cold weather the water consumption of rabbits weighing 5–7 kg, being fed a concentrated pellet, was 570 ml per rabbit per day (73). A water consumption study in southern California indicated a commercial rabbitry with an automatic water system was using one gallon of water for each 3 lb of pelleted ration consumed (61).

B. Protein Supplements

Soybean, peanut, sesame, cottonseed, and linseed meals are rich in protein and are desirable for balancing rabbit rations. These meals are used in mashes and pelleted feeds but cannot be mixed with whole grains because of settling out. Soybean seeds contain approximately 36% protein and 18% fat. After extraction the meal may have as much as 45% protein and 1 to 5% fat. Soybean meal supplement was found to be 6.8% more efficient than peanut meal (51). Linseed meal is also efficientt but more is required to bring the ration to the same protein level as soybean meal and the costs therefore exceed other rations. Sesame meal is the least efficient of these protein supplements. Caution should be used when using cottonseed meal as untreated meal contains gossypol, a substance toxic to rabbits. Degossypolized cottonseed meal should be used and is a suitable replacement for soybean meal at levels up to 7% of the diet (16). Cottonseed meal containing 0.04% free gossypol proved toxic when fed as 20% of the diet and mortality occurred when the calculated intake per animal was not more than 100 mg free gossypol (46). Gossypol is absorbed slowly, acts as a cumulative toxin, and the effects are increased due to the habit of coprophagy. Soybean meal appears to be the most satisfactory protein supplement for rabbits. A typical ration may contain 18% soybean meal and 4% linseed meal. The use of several protein supplements helps to insure the availability of lysine, arginine, and methionine, amino acids found essential for the growth of young rabbits (28).

C. Hay and Roughage

The legume hays such as alfalfa, clover, lespedeza, cowpea, vetch, kudzu, and peanut are palatable and make good feed for rabbits. The carbonaceious hays, timothy and prairie, and hays made from Johnson grass, Sudan grass, or Dallas grass are less palatable but still valuable when legume hays are not readily available. The grass hays ordinarily contain only about half as much protein as the legume hays and, if they are used, the protein supplements must be increased. Hays furnish bulk and fiber in addition to nutrients. Alfalfa hay usually comprises 40–50% of the pelleted ration.

D. Grains and Milled Feeds

Oats, wheat, barley, grain sorghums, buckwheat, and rye can be used as whole grains or as milled feeds. Corn can be used in a milled or cracked form. These grains are all similar in their nutritive value and can be substituted on a pound-for-pound basis. When rabbits are given a free choice of several varieties, the first choice will be oats followed by wheat, grain sorghums, and barley depending on freshness. Milled wheat products such as bran, middlings, shorts, and red dog flour may be included in mash mixtures or pellets. A standard ration may contain ground milo or barley, 18%; red wheat bran, 15%; and ground oats, 4%.

E. Feeding Methods

Two methods of feeding are in general practice in commercial rabbitries. In one, a measured amount of feed is placed in the appropriate container each day. The other utilizes a hopper or self-feeder which holds several days feed and the rabbits can eat it *ad libitum*. Crocks or troughs may also be used for full feeding but will have to be filled more often. The hopper feeding system saves time and labor and if the hopper is properly constructed prevents waste and contamination. Full feeding generally requires less feed than hand fed rabbits to produce a pound of live weight, however, hand feeding provides the operator with an opportunity to check feed consumed by each litter daily.

1. Pregnant and Lactating Does

It is important to recognize the different feed requirements for pregnant or lactating does as compared with the requirements for dry does and herd bucks. In feeding a pregnant doe it is advisable that she receive good quality hay and all of the concentrates she can eat. These ingredients may be supplied in the form of a grain pellet plus hay, or more commonly a complete pelleted ration. Following parturition the doe may be fed in the same manner as during pregnancy until the young are weaned at approximately 8 weeks. The amount of feed required will increase as the young leave the nest box and start consuming solid food. In general a doe and litter of 8 will consume approximately 100 lb of feed from the time of diagnosis of pregnancy by palpation (12–14 days gestation) to weaning at 8 weeks.

2. Dry Does and Herd Bucks

Dry does, herd bucks, and developing junior does require less concentrates to sustain growth and it is important to prevent the animal from becoming too fat. Mature does and herd bucks can be maintained on a hay diet. A complete alfalfa pellet plus 1% salt, coarse alfalfa crumbles, or turkey grind alfalfa crumbles in a satisfactory feed. Herd bucks that are in service can be maintained on high quality hay but it is advisable to supplement this diet with 2–3 oz of complete pellet daily. If the herd is being fed only a complete pellet, then the amount of feed must be restricted. With the medium weight breeds (9–12 lb at maturity), 4–6 oz of complete pellet should be sufficient. For developing junior does, the feed intake must be regulated so that they will grow and be in good condition when ready for breeding. A restricted intake of 4–6 oz of complete pellet daily should allow the junior doe to develop normally.

After mating, junior or mature does can be maintained on hay pellets alone, or if only a complete pellet is fed their intake should be restricted to 4–6 oz daily. When they are found to be pregnant by palpation, then they may be full fed. If a doe fails to conceive, rebreed and feed hay only or restricted amounts of complete pellet until pregnant.

3. Creep Feed

Creep feeding has been used for both purebred and commercial herds of lambs, calves, and weaning pigs. The economics of using creep feeding in a commercial rabbitry was reported first by Guthrie (29). He found creep feeding resulted in greater profit based on a 5 litter-per-year production schedule and allowed for an accelerated breeding program (9 liter per year) with a profit increase of 22%.

A commercially available creep feed provides 22% protein, 5% fat, and 13.4% fiber and is fed *ad libitum*. Feed is provided in conventional cage feeders, but a sheet metal hood wired to the cage floor prohibits the doe access to the ration. Composition of creep feed (26) is given below.

Alfalfa meal	37.5%
Soybean meal	10.0%
Oats	10.0%
Red wheat bran	15.0%
Distillers corn solubles	13.5.%
Skimmed milk	11.5%
Vegetable oil	1.25%
Salt	0.5%
Vitamin premix	7.5 lb
Anise oil powder	4.5 lb
Trace minerals	0.75% (2lb)
Vitamin A	1 lb

Young rabbits given creep feed in addition to their normal diet were found to obtain maximum growth and they gained significantly more weight during the 14- to 56-day period than did litter without creep feed.

F. Weaning and Weaned Weights

Under most commercial programs the young are weaned at 8 weeks and at that time should weigh approximately 4 lb each. Smaller litters (under 5 young) can be weaned at an earlier age and the doe rebred. Under accelerated breeding programs where does are bred less than 35 days following parturition, it is advisable to wean the young at 6 or 7 weeks of age. A study of randomly bred New Zealand White rabbits indicated there were significant season and age–of-doe effects on individual weaning weights (15). Individual weaning weights of a doe's first litter (3.72 lb) were lower than litters 2 through 5 (3.95 lb) and litters 6 through 10 (3.90 lb). With respect to numbers weaned there were no significant age-of-doe or seasonal effect and total litter weights were affected only by season. March and April produced maximum weaned weights (3.95 lb), while the months of July through October produced the lowest (3.57 lb) weaned weights for a doe's first litter. September and

October produced lowest (3.81 lb) weights for later litters. These findings are based on production records collected in southern California where seasonal changes are not as marked as in other sections of the United States.

G. Coprophagy

The normal physiological procedure of coprophagy by the rabbit has been well documented. It was first reported by Morot (70, 71) and later confirmed by Madsen (63), Taylor (96), Eden (23), and well reviewed by Thompson and Worden (105). Reingestion, a term perhaps more appropriate for this behavior, has been reported in wild *Oryctolagus* (92), in the hare, *Lepus europaeus* (22, 111), and in the cottontial, *Sylvilagus nuttallii* (53). It seems likely reingestion is a normal aspect of lagomorph biology.

The coprophagous pellets differ from noningested fecal pellets in physical consistency, size, and color. They are covered with mucus. Most of the pellet is composed of secretions from the cecum. These soft pellets are taken directly from the anus and swallowed intact. Coprophagy is practiced at night by domestic rabbits and by wild rabbits during the day while they are in their burrows. Young rabbits, less than 3 weeks old, are coprophagous since soft pellets can be found in the cardiac portion of the stomach (110).

Soft pellets contain twice the protein and half the fiber of the hard fecal pellet (48). Thacker and Brandt (104) found coprophagy improved the utilization of nitrogen probably through a reingestion of bacterial protein synthesized in the cecum. Kulwich *et al.* showed that coprophagy in conjunction with fermentation in the large intestine supplied an abundance of certain B vitamins and played an important role in the incorporation of sulfur in the soft tissues (54, 55).

H. Colony Medication in Feed and Water

It is common practice for commercial rabbit production to include antibiotics or other supplements in the feed or water for young growing rabbits. When small groups of specific pathogen-free rabbits are used, no benefits from antibiotics are noted; however, when large numbers of animals and pathogens are involved the improvement in livability is evident.

Sulfonamides have been widely used in the control of hepatic coccidiosis. Jankiewicz (49) reported *Eimeria stiedae* effectively controlled with the daily ingestion of 0.625 gm succinylsulfathiazole (sulfasuxadine). Sulfamethazine was effective when used at 0.05 to 1.0% in the feed (47). Peterson (79) found that 0.02% sulfamerazine in the drinking water was also effective. Sulfaquinoxaline in the water at levels of 0.05 and 0.02% controlled liver coccidiosis (18, 79), but 0.03% sulfaquinoxaline in the feed controlled the infection only if medication was initiated not more than 4 days following infection (62). This level produced no significant increase in weight gains or improvement in feed conversion (35).

Intestinal coccidiosis was effectively treated with 50 mg sulfaquanadine/100 gm of diet fed daily for 2 weeks (31). Both sulfamonomethoxine and sulfadimethoxine (75 mg/kg body weight) given orally for 7 days were found to control intestinal and hepatic coccidiosis (107). Sulfadimethoxine was particularly effective and produced negative oocyst counts for *E. stiedae* after 7 days medication.

Leeuwenberg and Whitmore (60) fed 10 and 20 gm/ton of chlortetracycline-B$_{12}$ feed supplement of a large commercial colony. They obtained a significant decrease in young mortality and an improvement in feed conversion. Lawrence and McGinnis (57) found no beneficial effect on growth when graded levels up to 50 gm/ton oxytetracycline (Terramycin) was added to the feed. Chas. Pfizer and Co. reported trials using 25 gm/ton of oxytetracycline, chlortetracycline (Aureomycin), diamine penicillin, and bacitracin (80). They found no significant differences in rates of gain or feed conversion and mortality was very low in all lots. Templeton (100) tested 10 and 20 gm/ton chlortetracycline against a control ration and found no increase in rate of gain on either level but enteritis mortality was reduced by 75%. Feed conversion was 3.93 lb of rabbit weaned in the control lot and 3.72 in the 10-gm lot, but when litters were analyzed in which there was no mortality, no differences in feed conversion was found. Subsequent attempts to lower mortality enteritis from using chlortetracycline were unsuccessful and indicated inconsistent results using this antibody. Huang *et al.* showed that neither oxytetracycline nor chlortetracycline produced growth improvements and were also ineffective when used with sulfathaladine, methionine, or in purified rations containing casein (98). They concluded growth was an indication of protein level and not the effects of antibiotic supplements. Continuous and temporary supplimentation with chlortetracycline (100 gm/ton) and oxytetracycline (50 gm/ton) had no significant effect on growth except in the doe's first litter (14). Antibiotic supplementation also lowered the incidence and mortality due to enteritis but had no effect on the young once they had contracted enteritis.

Zinc bacitracin (50 gm/ton) was added to control rations and found to reduce enteritis mortality at a significant level but had no effect on feed conversion or growth of the young as indicated by the weaned weights (13).

One sulfonamide, sulfaquinoxoline, was found to reduce respiratory mortality resulting from pasteurellosis (35) but it had no effect on enteritis losses. Sulfamethazine (100 gm/ton) alone or combined with chlortetracycline (100 gm/ton) and procaine penicillin (50 gm/ton) reduced the number of *Pasteurella* isolations from nasal turbinates of both mature and young animals, but enteritis mortality was almost 4 times the control (39).

Furazolidone (50 gm/ton) increased weaned weights by 7% when the mortality factor was removed but enteritis mortality was unaffected (36). In a subsequent trial, the same level furazolidone reduced the incidence of enteritis and all but eliminated mortality due to pneumonia (38). Nitrofurazone and furazolidone used separately and in combination did not prevent the development of liver lesions due to *E. stiedae* but the combination did have a detrimental effect on the life cycle as many infertile forms of the parasite were found in the bile tract (37).

The Food and Drug Administration permits three medications in rabbit pellets. These are shown in the tabulation below.

Medication	Concentration	Action
Furazolidone	0.0055% (50 gm/ton)	For the prevention of *Pasteurella*-type pneumonia and enteritis when fed continuously
Furazolidone plus sulfaquinoxaline	0.0055% 0.0125%	For the prevention of *Pasteurella*-type pneumonia, enteritis, and liver-type coccidiosis when fed continuously
Oxytetracycline	10 gm/ton	As an aid in stimulating growth and improving feed efficiency
Sulfaquinoxaline	0.025%	As an aid in preventing coccidiosis (concentrations of 0.1% used for 2 weeks when fed continuously)

Soluble medications and vitamins can be added to the water system. The most convenient methods involves the use of an automatic water proportioner added to the water line (Fig. 6). There are several types available but the two that are commonly used are the Vineland Crown proportioner and the Gland-O-Lac. The Crown proportioner accurately mixes 1 oz of water soluble material with each gallon of water at a rate up to 6 gal per minute. As water is consumed, slowly or quickly, the proportioner meters the soluble additions at a consistent level. A 50 gm/ton feed level can be replaced by adding 1 gm active ingredients to 15 gal of water.

IV. COMMERCIAL HERD MANAGEMENT

A. The Rabbitry and Equipment

1. Buildings

The type of building required will depend largely on climatic conditions. In planning such a facility emphasis should be placed on comfort of the rabbits and convenience for the caretaker. The building should have a simple, clean design, protect the animals from winds, rain, direct sunlight and ect., provide light and fresh air. In mild climates, hutches may be placed under a lath superstructure providing protection from the direct hot sun. In hot climates, cooling measures must be provided in addition to shade. This can be accomplished by the use of overhead sprinklers combined with adequate drainage to reduce the humidity. Where strong winds and stormy weather prevail, a building should be protected more completely on the weather side and curtains or panels can be used to close up the structure during inclement weather. In extremely cold conditions, even more protection will be needed.

The size of the building will depend on the production desired and can be of almost any dimension which will accomodate rows of hutches with at least $3\frac{1}{2}$ ft of aisle space. Depending on arrangement of hutches and aisles, a building 18 ft × 84 ft or one 38 ft × 64 ft will house 100 production females. Hutches should be arranged so that they face one another since time is saved by working two rows at one time. Aisles should be run in multiples of two (2, 4, or 6) instead of 3 or 5. This way the caretaker ends up at the starting end of the building.

Cement floors are desirable for several reasons. They are easier to clean and permit easier movement of feed carts and other pieces of rolling equipment. Cement floors are usually flushed with water to clean, but this may add to the problem of reducing humidity in the building. Cost and availability of water may also be a limiting factor. If floors are not concrete, drain pipes should be placed under the rows of hutches to carry off the excess moisture that collects from the watering system, the animals, and from flushing the cement aisles. When drain pipes are under hutches, the manure beds stay relatively dry, odors are reduced, and manure removal can be handled more efficiently.

A new concept in rabbit housing is being developed (30), and is referred to as "controlled ventilation." This type of building calls for tight construction, insulation, thermostatically controlled air flow, and efficient, low cost cooling. The cement is confined to perimeter foundations, 36-inch wide aisles, and work and loading areas. The building is steel-trussed and built on 10-ft centers. By using aluminum siding on both exterior and interior walls, a roof and ceiling, and total insulation, the heat from the summer sun is reflected and the interior heat is retained in winter.

Since moisture must be kept to a minimum in a closed building, drain pipes are placed under each water system line so that each manure bed has drainage. This allows moisture from water lines and rabbit urine to drain from the beds and escape from the building.

Fans have improved rabbit house environment by providing positive ventilation when needed and can be regulated to change the amount of air flow to an automatic operation. By installing three 36-inch exhaust fans with 2-speed motors and back draft shutters, the air can be changed in the building once every 12 minutes at minimum efficiency to once every 90 seconds at maximum efficiency. Where high summer temperatures interfere, an evaporative cooling system can be added to the fan system. By combining exhaust fans at one end of the building with a series of wet fibrous pads in a continuous section located in one side wall at the opposite end of the building, air is drawn in through the wet pads. This cooled air passes through the building absorbing heat and is then exhausted from the unit by the fans. The pad system should be placed on the side the summer winds prevail and since more water is required than is evaporated, a water circulation system is necessary. Each exhaust fan should have its own thermostat, and by setting each one a few degrees apart, the fans turn on and off by stages depending on the temperature. Water flow to the wet pads is similarly controlled.

A "fan-jet system" (Acme Engineering and Manufacturing Corp., Muskogee, Oklahoma) automatically supplies fresh air as needed and distributes the air throughout the building. It helps to maintain a more uniform temperature and humidity and provides circulation of air within the unit without drafts. Air inlets are mounted in the end walls and the fan-jet pressurizing fan is mounted inside. A plastic tube is attached to the discharge end of the pressurizing fan and runs the length of the building suspended from the ceiling. The plastic tube has outlet holes to produce air circulation. A constant running fan-jet unit recirculates the air in the building maintaining a uniform temperature and humidity. During summer months fresh air is pulled through the wet pad area and cooled. The air inlet is closed and the fan-jet recirculates the cooled air. During winter months, the pad area is covered with sheet plastic and fresh air is introduced from the outside through the air inlet shutters and circulated by the fan-jet tubes. This fresh air is mixed with the warmer building air before it reaches the rabbits and prevents sudden changes in temperature. When the desired temperature and humidity conditions are reached the exhaust fans and air inlets shutters automatically close preventing the loss of needed heat. The fan-jet continues to run maintaining constant recirculation and air movement in the building.

This system automatically provides a controlled environmental climate for all seasons of the year and a more profitable operation should result. Higher animal production, better quality, improved feed conversion, and lower mortality rates have been attributed to this improved concept of housing.

2. Hutches

The construction of the hutch will depend to a degree on the type of building to be utilized. The type of hutch and hutch material will be different for rabbits raised outdoors as compared to rabbits raised indoors. Climate and weather conditions will dictate how much protection will be required. Single decked or tier decked hutches will influence

Fig. 5. All-metal, wire floor commercial type hutch feeder and automatic water system. (Courtesy Don Guthrie, Small Stock Industries.)

construction also. Single decked hutches are easier and cheaper to build and they are also easier to keep clean. Multiple decked hutches require more substantial construction and therefore, cost more and are harder to keep clean, but they require less space and less walking for the caretaker.

The fundamental requirement of the rabbit hutch (Fig. 5) is that it should be constructed of all metal materials. This means the sides, front, back, and top should be of some sort of wire. With the invention of welded wire, there is no need to use wood in cage construction. Wood has many undesireable characteristics when used for hutch construction. It warps, rabbits like to chew on it, and being porous it is difficult to keep clean. For the sides, back, front, and top, a

welded wire fabric either 1 × 1 inch or 1 × 2 inch is generally used. Using clips and pliers these wire panels can be assembled so the cage is strong enough to support the rabbits in it. Under no circumstances should poultry netting be used in the construction of a hutch. It rusts and rabbits can chew through it.

The floor of the hutch also should be made of wire. A $\frac{1}{2}$ × 1 inch, 14 or 16 guage galvanized wire is preferable. Care must be taken to see that the smooth side of the welded wire is put upward in the cage. It may be advisable to use a supporting metal rod to prevent the floor from *sagging* in the middle. All breeds of rabbits in all climates can be successfully raised on wire floors. No part of this floor should be solid. A loose board should not be placed in the cage in the mistaken belief that the rabbit will be more comfortable. Wire floors are easier to keep clean and dry, a feature that makes solid floors completely unsatisfactory.

The size of the rabbit hutch was formerly based on the belief that one square foot of floor space was required for each pound of live weight. A 10-lb doe would then require 10 ft^2 of floor space. Most commercial hutches now provide $7\frac{1}{2}$ ft^2 in a hutch 36 inches wide and 30 inches deep. The height of the hutch should be at least 16 inches and need not be more than 24 inches. Since most commercial rabbitries use feeders mounted on the outside of the hutch, automatic water systems and nest boxes which are removed when the young are about 3 weeks old, all of the inside hutch area can be used by the doe and her litter.

Two types of door openings are used. One has the conventional front opening door and the other has the top opening door. The front opening door requires a latch to keep it closed and a hanger to keep it open. This type of cage should be suspended at a height above the aisle floor convenient for the caretaker to reach all corners. The top opening hutch has some advantages. The full front panel of the cage is available for the feeder and when the nest box is placed along the front panel it can be inspected without opening the cage door. Since this type is suspended lower, 24–30 inches above the floor, one can check the feeders more readily.

If cages are single decked, the tops should be of wire. If they are tiered, then only the top cage should have a wire top. The other top (or tops) must be solid to prevent urine and fecal material from falling through. These solid tops can serve two purposes in that they function as a dropping pan as well as the top for the hutch below. Again galvanized metal or plastic should be used and they should slope to the back.

Cages should be suspended from ceiling trusses and if each cage is suspended independently one person can lift a cage off the rods and remove the cage from the building for cleaning or repair. Water lines should hang from separate rods to facilitate easy removal of the cage as required.

3. Feed and Water

Using crocks, cans, or other open containers for feeders should be considered a temporary measure because pellets provided in such containers become contaminated by the young and feed is wasted. Metal feeders are commercially available and are convenient and efficient for both young and adult rabbits. Metal feeders are attached to the front panel of the cage and can be filled outside without opening the hutch door. Floor space inside the hutch is saved and since they are easily removed the feeders can be conveniently cleaned. By placing these feeders not over 4 inches above the floor, young rabbits are able to reach the feed. Capacity of feeders vary from 2 to 7 lb some are also

made with covers, and provision for attaching a hutch card identification can be provided.

Storing and handling bagged feed creates a problem in any commercial rabbitry. Bulk feed tanks help to solve these problems and can be set up outside or be built into the building under cover. Gravity loading chutes or spiral augers facilitates filling the feed cart with pellets. A screen at the bottom of each chute is helpful in removing dust or feed fines before the feed goes into the cart. This saves time normally spent cleaning feeders.

Automatic or pipeline watering systems are widely used in commercial rabbitries. They eliminate the labor and time involved in washing, cleaning, and filling watewcontainers. Fresh, clean potable water is supplied at all times. In cold climates, however, the system must be protected against freezing. This can be done by heat cables wrapped around or running through the water pipes. Conventional systems consist of a pressure-reducing tank with a float valve, a $\frac{1}{2}$ inch supply pipe (metal or plastic), a dewdrop valve for each hutch, and drinking valves. The pressure-reducing tank is installed 1 ft or more above the hutch so that sufficient pressure is maintained in the supply lines to have proper tension on the dewdrop valves. If there is too much pressure the rabbits cannot trip the valve, but too little tension will cause them to leak. The supply line should be installed 9 inches from the hutch floor for medium to large breeds and 7 inches for the smaller breeds. The pipe should be hung outside and at the back of the hutch and short nipples are used to bring the dewdrop valve into the cage. When hutches are back to back, one water line may be used for both rows of hutches. The water line may also be used to help support the cages; however, it is advantageous to suspend cages and water lines independently. This way each cage can be removed without attachment to the water line. Correct positioning of the dewdrop valve is important; they should point to the hutch at a 45° angle. If the valve points straight out (parallel to the floor) it is likely to leak. If the valve points straight down animals have difficulty drinking from the valve. Small operations can use a semiautomatic system. Water lines can be serviced by large rustproof containers which are filled by hand once or twice a day. Automatic watering systems should be checked frequently, especially when a rabbit is placed into a cage that has been unoccupied for several days. When valves are not used even for a few days minerals in the water may cause them to stick. Liquid proportioners may be used to medicate through the drinking water (Fig. 6).

4. Nest Boxes

Nest boxes should supply seclusion for the doe at kindling and comfort and protection for the young. Nest boxes vary in size but in general measure 18–22 inches long, 11–12 inches wide, 12 inches high at the back, and 9 inches high in

Fig. 6. Water proportioner for chlorinating water or adding medicants to drinking water. (Courtesy Don Guthrie, Small Stock Industries.)

front. In warmer climates, an open box is satisfactory but in colder areas a partly covered nest box is recommended. Nest boxes are usually made of wood with metal trimmed edges to prevent chewing. Metal nest boxes are used but in some climates they retain the cold temperature or collect water condensation. All types of nest boxes should supply adequate drainage and ventilation. To assist the doe in preparing the nest, some sort of bedding material must be supplied besides her own fur. Most kinds of straw will do. but rice straw may be preferred in some areas. In the Pacific Southwest a toxic plant (woody-pod Milkweed) grows among the wheat and oats. When the straw is harvested, the toxic plant may be cut and baled at the same time. The dried leaves are palatable and the doe will eat the pieces in the straw bedding. Excelsior, shredded paper, or wood shavings are suitable types of bedding.

5. Electric Cart

An electric cart is an essential piece of equipment for any large, efficiently operated commercial rabbitry. In any livestock production there are operations that must be done regularly, quickly, and efficiently. The electric cart is battery driven, can be used all day, and then recharged at night. A feed tank can be placed on the cart and used to carry large

amounts of feed from the bulk tanks to cages throughout the rabbitry. By replacing the feed tank with partitioned wire baskets, the cart can be used to wean animals, carry them to weighing scales, and then to holding pens. The cart can be used to carry does to buck cages for breeding or to put in and take out nest boxes. The cart saves many steps and much time each day. A blade can be attached to the front of the cart and used to remove manure from the beds under the cages. This versatile piece of equipment may be the key to top efficiency and increased profits.

B. Managing the Herd

1. Handling

Small rabbits may be lifted and carried by grasping the loin region gently but firmly. Place the heel of the hand toward the tail of the rabbit. This method prevents bruising the animal. To lift and carry a medium weight rabbit, grasp the fold of skin over the shoulders and support the rabbit by placing the other hand under its rump. Lift and carry heavier breeds in the same manner. If the rabbits struggle, hold it firmly under your arm. Never lift rabbits by the ears or legs.

2. Mating

The proper age of bucks and does for the first mating depends on breed and individual development. Smaller breeds develop more rapidly and on the average can be bred when 4–5 months old. The medium breeds are mated at 5–6 months and the large breeds at 8–10 months. In commercial production it is general practice to hold the bucks a month longer than the does for the first mating. Some difficulties are experienced if does are held too long before being bred.

One buck is kept for 8–12 does. A mature, vigorous buck can be used for 2–3 successive days. Normally a buck is used for breeding 2–3 times a week. Buck breeding records should be kept to establish his breeding performance.

The doe should be taken to the buck's hutch. There may be difficulties if the buck is taken to the doe. The doe may object to having another animal in her hutch and injure the buck or resist copulation. Some bucks are slow to mate in a strange hutch. Mating should occur almost immediately on placing the doe in the buck's cage. After the buck mounts the doe and falls over on his side the mating has been completed and the doe should be returned to her hutch.

Some does are reluctant to breed and they may be restrained for mating. To restrain such a doe, hold the ears and fold of skin over the shoulders in one hand and place the other hand under the body and between the hind legs. By placing a finger on each side of the vulva and pushing forward, the doe's tail is thrown over its back. The doe's hindquarters are elevated to the normal height for service. Not all of the does restrained for service will conceive but the technique will help to increase the number of kindlings. It has been suggested that forced mating of does refusing to mate naturally is as effective as natural mating or artificial insemination. This is contrary to the findings of Casady *et al.* (17). Their findings indicated an average of 25% pregnancies from forced matings and a definite seasonal pattern. Pregnancy ranged from 16% in October to 35% in April as compared with an average of 71% for natural matings. Large commercial rabbitries do not force-mate due to the increased time and labor involved. When does are reluctant to breed, the females are left in the buck's cage for 4–6 hours. During this time an increasing percent of animals will accept service and the labor involved is far less than that required for force-mating.

3. Pregnancy Determinations

Pregnancy can be determined quickly and accurately by palpation 12–14 days after mating. Test mating, i.e., placing the doe in the buck's hutch, is not a reliable method since some does will accept service when pregnant and others may refuse when they are not pregnant. The doe may be palpated in her own hutch or, if more convenient, she may be placed on a table covered with burlap or carpeting to prevent slipping. The ears and fold of skin over the shoulders are held in one hand. The other hand is placed between the hind legs but ahead of the pelvis. At 12–14 days following breeding the fetuses have developed to a palpable size and thus can be felt between the fingers when the hand is moved forward and backward with slight pressure. Caution must be used not to exert too much pressure.

There is less danger of injuring the fetuses when palpating at 12–14 days than at a later date. When one improves his technique, pregnancy can be accurately determined as early as the seventh or eighth day. Diagnosing pregnancy at this early age makes it possible to ship bred does a considerable distance before they are due to kindle. Females found not to be pregnant can be rebred without delay.

4. Kindling

A nest box should be placed in the hutch 27–28 days after the doe is mated. This allows the doe time to prepare the nest for her young. The nest should be examined on the day of birth and the young counted and recorded on the hutch card. If the environmental temperature is high and the doe has pulled an excessive amount of fur, the cover should be adjusted to keep the young comfortable. This excess fur can be saved and used in nest boxes where does have failed to pluck sufficient fur. In extremely low temperatures nest boxes should be insulated. A false bottom placed over a layer of wood shavings will insulate the bottom of a nest box in cold weather.

Some females will not pull fur and may kindle the young on the wire floor. These young can often be saved by arranging the bedding material to make a nest and putting the young on it. Kindling on the wire may be forgiven with the doe's first litter, but if this happens with successive litters, the doe should be removed from the production herd.

5. Care of the Young

On the day of kindling the litter should be inspected and any deformed, undersized, or dead should be removed. Litters vary in size but for commercial purposes 8–10 young are left with the doe. Surplus rabbits can be transferred from large litters to foster mothers with small litters to balance litter size. Surplus young should be transferred to foster mothers within a few days of birth; however, transfer is possible as late as 2 weeks. Litters born as far apart as 3 days may be mixed but it is preferable to mix litters of the same age. When mixing litters of different ages, younger rabbits should be added to an older litter. The survival of fostered young to weaning age is about 60%.

Nest boxes should be inspected daily to make sure the young are being adequately fed and cared for by the doe. Occasionally a doe fails to produce milk and the young will starve within 2–3 days unless the condition is noted and the young foster-nursed. When the young start leaving the nest box, usually at 2–3 weeks, there is no value in leaving the nest box in the hutch any longer.

C. Identification

Each breeding rabbit should be marked for identification. Tattooing the ears is a satisfactory method; it is permanent and does not disfigure the ear. Ear tags and clips are not totally satisfactory because they can tear out and disfigure the ear.

A tattoo device constructed like a pair of pliers is most convenient. Letters or numbers can be interchanged in the holder and various sizes are available to fit the breed of rabbit. After the tattoo marks are made on the inside of the ear, India ink is rubbed into the marks. Very young rabbits can be marked with an aqueous solution of acriflavine, picric acid, or Gentian Violet. This dye-marking process may require repeated applications. When animals are weaned, it is a convenient time to mark the does and bucks to be retained for replacement stock.

D. Records and Record Keeping

A practical record keeping system is needed to keep track of breeding, kindlings, weaning operations, and cost. The information is used in culling unproductive animals, selecting replacement stock, and identifying productive crosses and strains. A simple but comprehensive record system shows the result of each mating and indicates the productive capacity of both buck and doe and follows the litter growth through weaning to evaluate the doe and herd performance.

Because the rabbitry produces a single basic product and has none of the design on style changes found in other industries, basic records can be quite simple. In order to know what the animals are doing, one should note (1) number of does bred, (2) number of conceptions, (3) number of kindlings, (4) number of young left with does, (5) number of young weaned, and (6) weight of weaned young.

All of the essential information may be accumulated on daily cards, monthly cards, and a yearly report.

1. Hutch Card

The first consideration must be given to the hutch card, which shows the origin, age, and performance of every animal. The essential features of a hutch card are shown in (Fig. 7). The card for doe 301 indicates she was born on 12/12/69; her dam was 604 and sire 394. On 6/1/70 she was bred to buck 418, she was palpated on 6/15 and had her first litter on 7/2. Eleven young were born alive and none born dead. The remarks column show 3 were transferred to doe 421, leaving doe 301 with 8 in the nest box. At 8 weeks, 8 young were weaned and the total litter weighed 30.2 lb. Doe 301 was then rebred to buck 418 on 8/24/70, approximately a week before the young were weaned. Palpation following the third breeding to buck 418 indicated the doe was not pregnant; the doe was rebred immediately.

2. Buck Breeding Record

The buck breeding record is closely related to the hutch card and contains the same details (Fig. 8). On a single card the buck's breeding and production record is available. From this information the decision to retain or replace a

BUCK BREEDING RECORD

Buck No. 418

Breed NZW   Sire 107
Date born 11-5-69   Dam 918

| Doe | Location | Date Bred | Result of breeding | | | Weaned | |
| | | | Kindled | | Passed | | |
			Alive	Dead	Date	Number	Weight
301	A	6/1/70	11	0		8	30.2
307	B	6/3	9	0		7	28.3
415	A	6/8	5	1		6	23.7
505	A	6/12	8	0		8	32.1
411	C	6/13	—	—	6/27	0	—

Fig. 8. Buck breeding record card.

production animal can be made. Bucks with frequent or prolonged periods of low conception records should be replaced in the breeding herd.

3. Performance Record

Records covering the daily operations for the month can be tabulated and totaled on a performance record (Fig. 9). In general, the basic information on a given doe is recorded and from these figures top production animals can be identified as well as those animals at the lower end of the scale. The keeping of such records should lead to valid decisions regarding feed, strain characteristics, management practices, and other factors affecting the cost of each unit produced. Accurate records serve as a guide for action and indicate the financial position and progress of the rabbitry; good records are vital..

A computerized record keeping and breeding system has been developed by Pacific Sciences, Inc. of Bellevue, Washington. This system provides all the standard management services necessary to operate a colony and provides for special sorting and statistical correlations for genetic and experimental data. Some of the services include herd evaluations, financial records, weight-growth histories, litter histories, work task lists, and pedigrees. Records are kept on a single daily record sheet and computer output listings are provided weekly.

E. Meat Production

1. Inspection

The Federal Meat Inspection Act does not include rabbits in their requirements. They do, however, have

RABBIT HUTCH CARD
WESTERN FARMERS ASSOCIATION

Hutch No. A - 10   Ear No. 301
Breed NZW   Date Born 12/12/69   No. in Litter 9
Name   Sex F   Remarks
Sire 394   Dam 604

| Date Bred | Buck | Date Tested | Date Kindled | No. of Young | | At 56 Days | | | REMARKS |
				Kindled	Left	No. Left	Weight		
6/1/70	418	6/15	7/2	11-0	8	8	30.2		3 trans. to 421
8/24	418	9/9	9/24	9-0	8	8	31.0		1 trans. to 300
11/16	418	11/30	passed	—	—		—		
11/30	421	12/14	12/30	9-1	8	8	32.0		1 destroyed
3/21/71	421	3/7	3/24	11-0	8	7	28.0		3 trans to 415

Fig. 7. Hutch card for keeping doe's record.

INDIVIDUAL RECORD - RABBIT BREEDING COLONY

RABBIT #	SEX	BIRTH DATE	PARENTS		COAT COLOR	REMARKS
			FEMALE	MALE		
709	F	3/12/71	645	671	NZ W	

DATE MATED	MATED WITH	LITTER BORN	NUMBER BORN	SEX		NUMBER WEANED	REMARKS
				F	M		
9/28/71	436	10/29	9	5	4	8	♀ DOA 11/7
11/31/71	517	12/31	7	3	4	7	
2/5/72	680						

Fig. 9. Summary record performance card. (Courtesy of Sheldon Scher.)

regulations governing the grading and inspection of rabbits and their edible products and specifications for grading. The regulations are to be found under Title 7—Agriculture, Chapter 1—Consumer and Marketing Service, USDA, Part 54—Grading and Inspection of Domestic Rabbits and Edible Products thereof; and United States Specification for Classes, Standards, and Grades with respect thereto. To date only one rabbit processor has a Federally inspected operation.

State and some local authorities have influence on the slaughter of rabbits and inspection is based on the Food, Drug and Cosmetic Act. This act provides that the premises shall be inspected at intervals for sanitation purposes. Future slaughterhouse inspection will probably be conducted under State Wholesome Meats Acts. It is expected then that slaughter inspection will be done on a continuous basis with inspection being done both before and during slaughter.

The minimum Federal standards for sanitation, facilities, and operating procedures require separate rooms for certain operations (Sec. 54.210–54.232):

1. Receiving and feeding live animals
2. Killing and skinning operations
3. Eviscerating, chilling, and packing operations (Fig. 10)

Fig. 10. Evisceration and inspection operation. (Courtesy Pel-Freez Rabbit Meat Inc., Rogers, Ark.)

4. Inedible products departments
5. Refuse room
6. Storage and supply room
7. Service and machinery room

Building construction requires moisture-proof floors, smooth walls impervious to moisture to a height of 6 ft, and moisture-resistant ceilings. All drains and gutters must be trapped and vented and constructed so that sewage cannot back up and flood the floors. Hot water must be available for sanitation purposes. Ample light must be provided and all rooms are to be adequately ventilated.

2. Slaughter and Postmortem Inspection

A preferred method of slaughtering a rabbit is by dislocating the neck. Hold the animal by its hind legs with the left hand. Place thumb of the righthand on the neck just back of the ears and the four fingers extended uner the chin. Press down with the thumb and raise the animal's chin by a quick movement and dislocate the neck. Another method is to stun the animal with a sharp blow at the skull. In both instances the head is immediately severed and the carcass is allowed to bleed out.

Each carcass is opened to expose the viscera for inspection and evisceration is not completed until after inspection. Each carcass in which any lesion or condition which might render it unfit for human consumption is found must be retained for a final examination. Each carcass or part thereof found unfit must be condemned. Diseases or conditions which require condemnation (Sec. 54.130) are tularemia, anthrax, pyemia, septicemia, leukemia, enteritis, peritonitis, sarcomatosis, metritis, necrobacillosis, tuberculosis, pseudotuberculosis, advanced snuffles, generalized tumors, suppurating abscesses, icterus, advanced mange, or carcasses contaminated by volatile oils, paints, poisons, or gases. Rabbits from pathological laboratories shall be condemned.

3. Classes and Quality

Domestic rabbits are classed as fryers or young rabbits and roasters or mature rabbits. A fryer is a young carcass weighing not less than $1\frac{1}{2}$ lb and rarely more than $3\frac{1}{2}$ lb processed from a rabbit usually less than 12 weeks of age. A roaster is a mature carcass of any weight but usually over 4 lb processed from a rabbit usually 8 months of age or older.

Three grades of quality exist. To be of A quality the carcass (Sec. 54.275):

a. is short, thick, well-rounded, and full-fleshed.
b. has a broad back, broad hips, and broad deep-fleshed shoulders and firm muscle texture.
c. has a fair quantity of interior fat in the crotch and over the inner walls of the carcass, and a moderate amount of fat round the kidneys.
d. is free from evidence of incomplete bleeding such as more than an occasional slight coagulation in a vein. Is free from any evidence of reddening of the flesh due to fluid in the connective tissues.
e. is free from all foreign material (hair, dirt, bone particles, etc.).
f. is free from broken bones, flesh bruises, defects, and deformities.

To be of B quality the carcass (Sec. 54.276):

a. is short, thick, fairly well-rounded, and fairly full-fleshed.
b. has a fairly broad back and hips, fairly broad and deep-fleshed shoulders, and fairly firm muscle texture.

c. has at least a small amount of interior fat in the crotch, over the inner walls of the carcass, and around the kidneys.
d. is free from evidence of incomplete bleedings as in A, d.
e. is free from all foreign material as in A, e.
f. is free from broken bones, flesh bruises, defects, and deformation.

A carcass that does not meet the requirements of A or B quality may be of C quality and such carcass (Sec. 54.277):

a. may be long, rangy, and fairly well-fleshed.
b. may have a thin narrow back and hips and soft flabby muscle texture.
c. may show very little evidence of exterior fat.
d. may show very slight evidence of reddening of the flesh due to blood in the connective tissues.
e. is free from all foreign material.
f. may have moderate bruises, defects, and deformities; may have more than one broken bone, may have a portion of the carcass removed due to serious bruises. Discoloration due to bruising in the flesh shall be free from clots.

F. Labor and Economics

1. Labor

Rabbit management studies conducted in California (108) during 1962 and 1963 revealed that the hours of labor per doe per year ranged from 2.3 to 27.7 with an average of 6.4 hours. These rabbitries studied varied in size from 30 to 800 working does with an average of 250. The labor source reported was a husband and wife team. Based on the 6.4-hour average, a 40-hour weekly work schedule would allow care for 350 working does. Leeuwenburg (59) believes that one able-bodied man can take care of 750 hutches but this would require more than the modern concept of a 40-hour work week. This would be possible only if the rabbitry was well constructed and well managed, and would include all the labor-saving devices. From this example then a 200-doe rabbitry would require about 2 hours labor per day. A man working full time with proper equipment and good breeding stock should be able to care for 400 to 500 working does.

2. Investment

The costs for housing and equipment vary considerably depending on location. The 1962–1963 management studies in California (108) indicated the average retail price for an all-wire hutch to be $5.07 per doe. Accessory equipment such as nest boxes, feeders, and medication tank added $4.11 per doe, or a total cost of $9.18 per doe. The building and related electrical and watering equipment cost $8.39 per doe. The total cost, less land, was then $17.57 per doe. Breeding stock costs averaged $5.67 per doe. The total investment including stock, building, and equipment but less land will require about $23.25 per unit or per doe. Using a ratio of 128 units or hutches required for 100 working does, the total investment for a 100 doe rabbitry would be about $3,000.00 plus land costs.

A more recent study conducted in the Ozark Plateau Region (88) indicated an investment range from $2.50 to a high cost of $61.54 per doe. Neither these high or low figures should be considered representative because the large investment results from conversion to an expensive building which the low cost rabbitry was built entirely of scrap materials. Average costs for a 100-doe rabbitry or larger amounted to $28.37 per doe including land costs.

3. Returns

The normal profit on live rabbits is between 5 and 10 ¢ per pound. A 4 lb rabbit will realize a profit of 20–40 ¢. In estimating normal weekly production, figure one-half fryer per week per doe; therefore, 100 does should produce at least 50 fryers per week on a year round basis. The average financial returns over feed costs will be about $7.00 to $9.00 per doe per year, or 60–75¢ per doe per month. Efficient operations utilizing good equipment and good breeding stock should be able to increase the number of fryers to 60–70/100 does per week and increase the returns to $1.00 to $1.25 per month per doe.

The average good doe should raise at least 30–35 fryers per year. At 4 lb each, this means the doe will produce 120–140 lb of meat per year. There are does that will raise 35–40 fryers per year. With accelerated breeding programs, 50 fryers per year becomes possible. A doe must produce a minimum of 120 lb of meat per year to justify costs. A good rabbit will average 120–160 lb per year as a minimum.

Net profits depend to a great extent on the efficiency or productivity of the working doe. The conclusion that rabbitry size automatically determines production efficiency because maximum production will not be maintained when size level exceeds the operator's ability was not supported by the Ozark Plateau study (88). The rabbitry reporting the highest net profit per doe was a 100-doe operation. The production per doe was 133 lb of meat per year, which is about average but nowhere near the production of 156 lb reported by one of the larger operators who had a smaller net income per doe. The high profit operator achieved a good return because of lower unit costs ($15.85) while the higher producer had a cost per doe of $20.33. Average net profit reported was $8.97 per doe per year.

V. LABORATORY MANAGEMENT

A. Equipment

Most laboratory rabbits are not produced under indoor laboratory conditions at present. This usually means that attempts to control the environment and sanitization of the colony are minimal, and that the means of providing housing, water supply, refuse removal, and general housekeeping are extremely varied. However, with due attention to the principles of good husbandry and management, vigorous and healthy rabbits can be produced. It is recommended that insofar as possible, rabbits be housed so as to control extremes of environment, and that all cages and feed equipment be of metal construction to facilitate sanitation; water systems be of dewdrop type (protected from freezing electrically when necessary); and that floors and wall be impervious to liquid, moisture, and be vermin-proof.

The following standards are based on present-day knowledge and experience in the care and maintenance of rabbits and are the results of the Committee on Standards, Institute of Laboratory Animal Resources. These recommendations are to be applied to colonies maintained for experimental purposes as well as for production colonies for laboratory use.

In general, the facilities, equipment, and husbandry procedures should be designed and operated to afford maximum environmental control, optimal comfort and welfare for animals, and minimal opportunity for the transmission of diseases and parasites from one animal to another and from group to group.

The physical facilities shall be so designed and constructed that clean and soiled material and equipment will be separated and should afford minimal variation in temperature, humidity, and ventilation, thus providing the animals with optimal conditions for their comfort and welfare.

Operating procedures should be performed according to the recommendations of these standards and shall consist of those practices that will include the best information and experience in nutrition, genetics and animal breeding, care, maintenance, colony management, and disease control, so that the best possible environmental control is accomplished, and optimal conditions for the comfort and welfare of the animals are provided.

1. Cages

Cages shall be fabricated of a smooth, corrosion-resistant material. It shall be impervious to liquids and moisture, easily sanitized and/or sterilized. Materials that are considered acceptable include galvanized metal, stainless steel, or other stainless metal alloys, aluminum (hard alloys), and magnesium. Cages should be movable so as to facilitate transportation to a cage washing area. The painting of cages and racks is not recommended.

Rabbit cages may have bottoms of expanded metal or of a galvanized wire mesh with a recommended size of 1 inch by $\frac{1}{2}$ inch, or $\frac{5}{8}$ inches square. Wire mesh floors should be smooth and free of sharp projections. The galvanizing process should take place after welding.

The minimum cage area for a nursing female and her

litter should be 1080 square inches (2½ ft by 3 ft) including the space for a nest box. The following table is presented as a guide to the allowable floor space for holding rabbits:

Weight	Square inches per animal	Maximum population/cage
3 to 5 lb	180	6
Over 5 lb	360	3

Males over 5 lb, and females over 6 lb should be caged separately.

Where it is impossible to transport cages to a central washroom cages may be cleaned in place by means of a portable, high-pressure washing apparatus which uses a hot detergent-disinfectant solution. Alternatively, steam jennies or propane fire torches can be used.

Racks (stands) shall be fabricated of a smooth, corrosion-resistant material. They should be impervious to liquids and moisture, easily cleaned, sanitized, and sterilized. If racks are fabricated of pipe or tubular material, all openings shall be closed without crevices. Racks may be of fixed or of portable design, arranged to facilitate cleaning activities. Wood is not acceptable.

Nest boxes may be of wood, with edges and corners metal lined, and with bottom inside dimensions not less than 9 inches × 17 inches.

2. Feed, Bedding, and Water

Feed and bedding are generally used in animal facilities without prior treatment, and hence are potential sources of contamination for diseases, parasites, and abnormal hormonal stimulation. It is, therefore, urged that producers and users be aware of these problems and exercise care in their purchase and storage.

Purchasers should request the vendor's assurance, and receive guarantees, that the pelleted feed that he uses is (a) free of additives containing drugs, hormones, antibiotics, and pesticides; and (b) free from animal and vermin contaminations.

Feed should not be accepted by the production colony management unless it is delivered in clean and sealed containers made of new material.

Feed received should not be accepted unless it is marked with the milling date. It should be fed as soon as possible after receipt.

Feed shall be stored in a clean, dry, cool, vermin-free area in covered containers with tightly fitting lids, or in their original sealed containers.

Feed shall be supplied, preferably, on a daily basis. Under no circumstances shall any cage be provided with more than one week's supply of these prepared foods.

Feed that is present in food hoppers or cages when cages are scheduled for washing should be destroyed.

Feed supplements, in the form of hay, may be used, but it should be stored in buildings or metal containers which are vermin-free, bird-proof, and to which dogs, cats, and other animals do not have access.

Hay should be placed in a rack for immediate consumption.

Food hoppers shall be constructed of any durable material other than wood, shall be resistant to the gnawing of rabbits, and corrosion resistant when exposed to acids, alkalis, detergents, moisture, liquids, and excreta. They shall be easily sanitized and/or sterilized. They shall be designed to permit easy access, yet not permit any entry of rabbits. They should be mounted 4 inches from the floor of the cage. Food hoppers showing rust spots or other signs of deterioration shall not be used.

The food hopper must be designed so that food will be available to the rabbits *only* from the lower third of the hopper.

Bedding must be used in nest boxes. It may be of straw, hay, excelsior, wood shavings, or other similar nonedible material.

Bedding should not be purchased from a source whose storage facilities are not adequately protected from vermin and animal contamination. Evidence of such protection should be determined wherever possible by visits to the supplier's premises.

Bedding should be stored in dry, vermin-proof, frequently sanitized containers in storage areas separate from rabbit rooms, to which dogs, cats, and other animals have no access.

Automatic watering systems which do not allow backflow are recommended. Water bottles may be used. They should be formed of clear glass or sterilizable plastic. Bottles should be mounted or suspended in a manner that will prohibit contact by the caged animal. Sipper tubes should be fabricated of metal, preferably stainless steel. Open water containers should not be used.

3. Environmental Controls

Outside windows are not required. It is suggested that a viewing port of transparent glass be installed for inspection purposes.

Lights shall be of a type and in a location that simplifies cleaning. Lighting fixtures, switches, and convenience outlets shall be designed and constructed so they will not afford shelter for rodents and vermin.

Air conditioning—temperature, humidity, ventilation, and air filtration control—is recommended for all indoor rabbit quarters. Recirculation of air is not recommended. Animal room temperatures should be maintained between 60°F and 70°F. Relative humidity levels should be maintained between 40 and 60%. All rabbit rooms should be equipped with humidity and temperature alarm systems.

Graphic recorders are useful for tabulating systems performance.

Duct work, preferably, should be located above the ceiling or flush mounted. However, if installed below the ceiling, the upper surface of the duct work shall be sealed to the ceiling or suspended at least 6 inches below the ceiling to facilitate cleaning. The diffusers and exhaust openings must be so located and controlled as to avoid drafts. Exhaust openings must be located close to floor level and at a distance from the intake openings. Openings in ventilation grillwork should be of a size that will not permit the entrance of rodents. The supply of air should be such as to provide a minimum of 10 air changes per hour. The air pressure within clean spaces should be greater than in public and refuse areas.

Supplemental exhaust fans, if used, should be permanently mounted in external window or wall openings and screened. Their frames shall be sealed to the building structure. They should be located at a distance from the air intake.

B. Colony Operations

The freedom of the initial stock from disease is required for the successful operation of a rabbit colony. Thereafter, procedures in the operation of a colony are of utmost importance, for they are the means by which outside contaminants are denied access to the colony. The latest design in physical facilities and equipment is no better than the procedures used in operating them. It is, therefore, strongly recommended that producers and users alike constantly review (and revise) their operating procedures, so that optimal conditions of care and management are provided for the animals.

REFERENCES

1. Adams, C. E., Aitken, F. C., and Worden, A. N. (1967). The rabbit. *In* "The UFAW Handbook on the Care and Management of Laboratory Animals" (UFAW, ed.), pp. 396–448. Livingstone, Edinburgh.
2. Amann, R. P. (1966). Effect of ejaculation frequency and breed on semen characteristics and sperm output of rabbits. *J. Reprod. Fert.* **11**, 291–293.
3. Anonymous. (1967). "Standards for the Breeding, Care, and Management of Laboratory Rabbits." A Report of the Committee on Standards. Institute of Laboratory Animal Resources, National Research Council, Washington, D.C.
4. Anonymous. (1971). "Feed Additive Compendium; pp. 242, 322, and 395. Miller Publ. Co., Minneapolis, Minnesota.
5. Asdell, S. A. (1946). "Patterns of Mammalian Reproduction." (Comstock), Ithaca, New York.
6. Asdell, S. A., and Salisbury, G. W. (1933). The cause of mammary development during pseudopregnancy in the rabbit. *Amer. J. Physiol.* **103**, 595–599.
7. Barboreck, J. (1953). Effect of the addition of fiber on the energy metabolism of the rabbit. Thesis, University of Zurich.
8. Beatty, R. A. (1960). Fertility of mixed semen from different rabbits. *J. Reprod. Fert.* **1**, 52–60.
9. Blount, W. P. (1945). "Rabbits' Ailments." Fur and Feather, Idle, Bradford, England.
10. Brambell, F. W. R. (1942). Intra-uterine mortality of the wild rabbit, *Oryctolagus cuniculus* (L). *Proc. Roy. Soc., Ser. B* **130**, 462–479.
11. Bredderman, R. H., Foote, R. H., and Yassen, A. M. (1964). An improved artificial vagina for collecting rabbit semen. *J. Reprod. Fert.* **7**, 401–403.
12. Casady, R. B., and Gildow, E. M. (1959). Rabbit nutrition. *Proc. Anim. Care Panel* **9**, 9–30.
13. Casady, R. B., Hagen, K. W., Bertrand, J. E., and Thomas, H. G. (1964). Effect of zinc bacitracin of the incidence of enteritis in young rabbits. *Clin. Med.* **71**, 871–875.
14. Casady, R. B., Hagen, K. W., and Sittmann, K. (1964). Effects of high level antibiotic supplementation in the ration on growth and enteritis in young domestic rabbits. *J. Anim. Sci.* **23**, 477–480.
15. Casady, R. B., Rollins, W. C., and Sittmann, D. B. (1962). Effect of season and age of dam on weaning weights, number weaned and total litter weight of hutch-raised domestic rabbits. *Small Stock Mag.* **46**, No. 7, 23.
16. Casady, R. B., Sawin, P. B., and Van Dam, J. (1966). Commercial rabbit raising. *U.S., Dep. Agr. Agr. Hand.* **309**, 1–69.
17. Casady, R. B., Suitor, A. E., and Mize, K. E. (1961). Seasonal breeding performance and the use of forced matings in hutch-raised domestic rabbits. *Small Stock Mag.* **45**, 5.
18. Chapman, M. P. (1948). The use of sulfaquinoxaline in the control of liver coccidiosis in domestic rabbits. *Vet. Med.* **43**, 375–379.
19. Denenberg, V. H., Zarrow, M. X., and Ross, S. (1969). The behavior of rabbits. *In* "The Behavior of Domestic Animals" (E. S. E. Hafez, ed.), pp. 417–437. Williams & Wilkins, Baltimore, Maryland.
20. Doggett, V. C. (1956). Periodicity in the fecundity of male rabbits. *Amer. J. Physiol.* **187**, 445–450.
21. Doggett, V. C., and Ett, J. C. (1958). Libido and the relationship to periodicity in the fecundity of male rabbits, *Fed. Proc.*, **7**, *Fed. Amer. Soc. Exp. Biol.* **17**, 36.
22. Drane, R. (1895). *Trans. Cardiff Natur. Soc.* **27**, 101.
23. Eden, A. (1940). Coprophagy in the rabbit. *Nature (London)* **145**, 36–37.
24. Eppstein, J. H., and Margulis, S. (1941). Minimum requirements of rabbits for DL, alpha-tocopherol. *J. Nutr.* **22**, 415–424.
25. Fox, R. R., and Burdick, J. F. (1963). Preservation of rabbit spermatozoa: Ethylene glycol vs glycerol for frozen semen. *Proc. Soc. Exp. Biol. Med.* **113**, 853–856.
26. Fox, R. R., and Guthrie, D. (1968). The value of creep feed for laboratory rabbits. *Lab. Anim. Care* **18**, 34–38.
27. Fox, R. R., and Krinsky, W. L. (1968). Ovulation in the rabbit related to dosage of human chorionic gonodotrophin and pregnant mare's serum. *Proc. Soc. Exp. Biol. Med.* **127**, 1222–1227.
28. Gaman, E., and Fisher, H. (1970). The essentiality of arginine, lysine, and methionine for the growing rabbit. *Nutr. Rep. Int.* **1**, 57–64.
29. Guthrie, D. (1966). Creep feeding for increased production. *Small Stock Mag.* **50**, No. 5, 19.
30. Guthrie, D. (1969). New concepts in rabbit housing. *Small Stock* **53**, No. 5, 7.
31. Habermann, R. T., Williams, F. P., and Thorp, W. T. S. (1954). Identification of some internal parasites of laboratory animals. *U.S. Pub. Health Serv., Publ.* **343**, 13–16.
32. Hafez, E. S. E. (1960). Sex drive in rabbits. *Southwest. Vet.* **14**, 46–49.
33. Hafez, E. S. E. (1970). Rabbits. *In* "Reproduction and Breeding Techniques for Laboratory Animals" (E. S. E. Hafez, ed.), pp. 273–298. Lea & Febiger, Philadelphia, Pennsylvania.
34. Hafez, E. S. E., Lindsay, D. R., and Moustafa, L. A. (1966). Some maternal factors affecting nest building in the domestic rabbit. *Z. Tierpsychol.* **6**, 691–700.

35. Hagen, K. W. (1958). The effects of continuous sulfaquinoxaline feeding on rabbit mortality. *Amer. J. Vet. Res.* **19**, 494–496.

36. Hagen, K. W. (1958). The effects of low levels of furazolidone on rabbit mortality. *Proc. Nat. Symp. Nitrofuraus Agr.,* 2nd, 19   pp. 118–121.

37. Hagen, K. W. (1961). Hepatic coccidiosis in domestic rabbits treated with 2 nitrofuran compounds and sulfaquinoxaline. *J. Amer. Vet. Med. Ass.* **138**, 99–100.

38. Hagen, K. W. (1964). Unpublished data.

39. Hagen, K. W. (1967). Effect of antibiotic-sulfonamide therapy on certain microorganisms in the nasal turbinates of domestic rabbits. *Lab. Anim. Care* **17**, 77–80.

40. Hamilton, C. E. (1951). Evidences of cyclic reproductive phenomena in the rabbit. *Anat. Rec.* **110**, 557–571.

41. Hammond, J., and Marshall, F. H. A. (1925). "Reproduction in the Rabbit." Oliver & Boyd, Edinburgh.

42. Hamner, C. E. (1970). The semen. *In* "Reproduction and Breeding Techniques for Laboratory Animals" (E. S. E. Hafez, ed.), pp. 56–73. Lea & Febiger, Philadelphia, Pennsylvania.

43. Harris, L. J. (1956). Vitamin C economy of rabbits. *Brit. J. Nutr.* **10**, 373–382.

44. Hogan, A. G., and Hamilton, J. W. (1942). Adequacy of simplified diets for guinea pigs and rabbits. *J. Nutr.* **23**, 533–543.

45. Hogan, A. G., and Ritchie, W. S. (1934). Nutritional requirements of rabbits and guinea pigs. *Mo., Agr. Exp. Sta., Res. Bull.* **219**.

46. Holley, K. T. (1955). Cottonseed meal in swine and rabbit rations. *Ga, Agr. Exp. Sta., Mimeo. Ser.* [N.S.] **12**, 1–26.

47. Horton-Smith, C. (1947). The treatment of hepatic coccidiosis in rabbits. *Vet. J.* **103**, 207–213.

48. Huang, T. C., Ulrich, H. E., and McCay, C. A. (1954). Antibiotics, growth, food utilization, and the use of chromic oxide in studies with rabbits. *J. Nutr.* **54**, 621–630.

49. Jankiewicz, H. A. (1945). Liver coccidiosis prevented by sulfasuxadine. *J. Parasitol.* **31**, Suppl., 15.

50. Jarl, F. (1948). [Experiments with vitamin D to rabbits.] Undersokningar ovar D-vitaminets betydelse i kanineraas foderstater. *Husdjursforsokonst Medd.* **29**, 1–40.

51. Kellogg, C. E., Templeton, G. S., and Suitor, A. E. (1949). Feed required to produce 6-pound rabbits after weaning and conditions affecting their carcass grades and cuts. *U. S., Dep. Agr., Circ.* **819**, 1–20.

52. Kihlstrom, J. E. (1956). The effect of artificial light upon the production of male sperm antagglutin in rabbits. *Int. Congr. Anim. Reprod. Pap.* **3**, 17.

53. Krull, W. (1943). Corpophagy in the wild rabbit, *Sylvilagus nuttallii grangeri* (Allen). *Vet. Med.* **38**, 72.

54. Kulwich, R., Pearson, P. B., and Lankenau, A. H. (1954). Effect of coprophagy on S^{35} uptake by rabbits after ingestion of labeled sodium sulfate. *Arch. Biochem.* **50**, 180–187.

55. Kulwich, R., Struglia, L., and Pearson, P. B. (1953). The effect of coprophagy on the excretion of B vitamins by the rabbit. *J. Nutr.* **49**, 639–645.

56. Kunkel, H. O., and Pearson, P. B. (1948). Magnesium in the nutrition of the rabbit. *J. Nutr.* **36**, 657–666.

57. Laurence, J. M., and McGinnis, J. (1952). The effect of terramycin on the growth of rabbits. *Arch. Biochem.* **37**, 164–166.

58. Lee, R. C. (1939). Size and basal metabolism of the adult rabbit. *J. Nutr.* **18**, 489–500.

59. Leeuwenburg, P. (1970). Commercially yours. *Small Stock Mag.* **54**, 9.

60. Leeuwenburg, P., and Whitmore, R. (1951). Results of the use of aureomycin-B_{12} feed supplement. *Small Stock Mag.* **35**, 29.

61. Ludwig, L. (1964). Lou-Jay Rabbitry, Vista, California (personal communication).

62. Lund, E. E. (1954). The effect of sulfaquinoxaline on the course of *Eimeria stiedae* infections in the domestic rabbit. *Exp. Parasitol.* **3**, 497–503.

63. Madsen, H. (1939). Does the rabbit chew the cud? *Nature (London)* **143**, 981–982.

64. Mann, I., Pirie, A., Tansley, K., and Wood, C. (1946). Some effects of vitamin A deficiency on the eye of the rabbit. *Amer. J. Ophthalmol.* **29**, 801–815.

65. Mann, T. (1964). "The Biochemistry of Semen and of the Male Reproductive Tract." Methuen, London.

66. Menzies, W., and Moss, A. (1960). The mating of domestic rabbits. *J. Anim. Tech. Ass.* **11**, 7–9.

67. Meyers, K., and Poole, W. E. (1958). Sexual behavior cycles in the wild rabbit *Oryctolagus cuniculus* (L.) *CSIRO Wildl. Res.* **3**, 144–145.

68. Meyers, K., and Poole, W. E. (1962). Oestrus behavior cycles in the rabbit, *Oryctolagus cuniculus* (L.) *Nature (London)* **195**, 358–359.

69. Minot, C. S. (1907). The problem of age, growth, and death. III. The rate of growth. *Pop. Sci. Mon.* **71**, 193–216.

70. Morot, C. H. (1882). Memoire relatif aux pelotes stomacales des leporides. *Rec. Med. Vet.* **59**, 635–646.

71. Morot, C. H. (1911). [Stomach pellets of rabbits. On their origin, their nature, and their function.] *Bull. Soc. Hist. Natur. Autun.* **24**, 44–49.

72. Mroueh, A., and Mastroianni, L. (1966). Insemination via the intraperitoneal route in rabbits. *Fert. Steril.* **17**, 76–82.

73. Napier, R. A. N. (1963). Rabbits. *In* "Animals for Research" (W. Lane-Petter, ed.), pp. 323–364. Academic Press, New York.

74. National Research Council, Committee on Animal Nutrition (1966). Nutrient requirements of rabbits. *Nat. Acad. Sci.—Nat. Res. Counc., Publ.* **1194**.

75. Noyes, R. W., Adams, C., and Walton, A. (1959). The passage of spermatozoa through the genital tract of female rabbits after ovariectomy and oestrogen treatment. *J. Endocrinol.* **18**, 165–174.

76. Olcese, O., and Pearson, P. B. (1948). Value of urea in the diet of rabbits. *Proc. Soc. Exp. Biol. Med.* **69**, 377–379.

77. Oloufa, M. M., Bogart, R., and McKenzie, F. F. (1951). Effect of environmental temperatures and the thyroid gland on fertility in the male rabbit. *Fert. Steril.* **2**, 223–228.

78. O'Shea, T., and Wales, R. G. (1969). Further studies on the deep freezing of rabbit spermatozoa in reconstituted skim milk powder. *Aust. J. Biol. Sci.* **21**, 831–833.

79. Peterson, E. H. (1950). The prophylaxis and therapy of hepatic coccidiosis in the rabbit by the administration of sulfonamides. *Vet. Med.* **45**, 170–172.

80. Pfizer and Co., Inc. (1952). "Antibiotics in the Nutrition of the Rabbit," Mimeo. rep.

81. Rollins, W. C., Casady, R. B., Sittman, K., and Sittman, D. B. (1963). Genetic variance component analysis of litter size and weaning weight of New Zealand White rabbits. *J. Anim. Sci.* **22**, 654–657.

82. Ross, S., Denenberg, V. H., Frommer, G. P., and Sawin, P. B. (1959). Genetic, physiological, and behavioral background of reproduction in the rabbit. V. Nonretrieving of neonates. *J. Mammal.* **40**, 91–96.

83. Ross, S., Sawin, P. B., Zarrow, M. X., and Denenberg, V. H. (1963). Maternal behavior in the rabbit. *In* "Maternal Behavior in Mammals" (H. L. Rheingold, ed.), pp. 94–121. Wiley, New York.

84. Rubin, H. B., and Azrin, N. H. (1937). Temporal patterns of sexual behavior in rabbits as determined by an automatic recording technique. *J. Exp. Anim. Behav.* **10**, 219–231.

85. Sawada, Y., and Chang, M. C. (1964). Motility and fertilizing capacity of rabbit spermatozoa after freezing in a medium containing dimethyl sufoxide. *Fert. Steril.* **15**, 222–229.

86. Sawin, P. B. (1950). The rabbit. *In* "The Care and Breeding of Laboratory Animals" (E. J. Farris, ed.), pp. 153–181. Wiley, New York.

87. Sawin, P. B., Denenberg, V. H., Ross, S., Hafter, E., and Zarrow,

M. X. (1960). Maternal behavior in the rabbit. Hair loosening during gestation. *Amer. J. Physiol.* **198**, 1099–1102.

88. Scott, T. H. (1967). A study of profit maximization in selected commercial rabbitries. *Small Stock Mag.* **51** (Feb.), 5, 31; (Mar.), 5, 26; (April), 5, 22 and 23; (May-June), 5, 26; (July), 5, 23.

89. Seton, E. T. (1909). "Lives of Game Animals." Doubleday, Garden City, New York.

90. Sittmann, D. B., Rollins, W. C., Sittmann, K., and Casady, R. B. (1964). Seasonal variation in reproductive traits of New Zealand White rabbits. *J. Reprod. Fert.* **8**, 29–37.

91. Smith, S. E., and Ellis, G. H. (1947). Studies on the manganese requirements of rabbits. *J. Nutr.* **34**, 33–42.

92. Southern, H. N. (1940). Coprophagy in the wild rabbit. *Nature (London)* **145**, 262.

93. Spector, W. S. (1956). "Handbook of Biological Data." Saunders, Philadelphia, Pennsylvania.

94. Spencer, R. B., and Coulombe, M. J. (1965). Fetal weight—gestational age relationship in several species. *Growth* **29**, 165–171.

95. Staples, R. E., and Holtkamp, D. E. (1966). Influence of body weight upon corpus luteum formation and maintenance of pregnancy in the rabbit. *J. Reprod. Fert.* **12**, 221–224.

96. Taylor, E. L. (1939). Does the rabbit chew the cud? *Nature (London)* **143**, 982–983.

97. Templeton, G. S. (1939). Feed requirements of rabbits. *World's Poultry Congr. Exposition* [*Proc.*] *7th, 1939* pp. 468–471.

98. Templeton, G. S. (1939). Length of gestation period in domestic rabbits. *Small Stock Mag.* **23**, 3.

99. Templeton, G. S. (1940). Pseudopregnancy in domestic rabbits. *U.S., Fish Wildl. Serv., Circ.* **4**, 1–13.

100. Templeton, G. S. (1953). Rabbit mucoid enteritis. *Small Stock Mag.* **37**, 2–3.

101. Templeton, G. S. (1955). "Domestic Rabbit Production." Interstate Printers and Publishers, Danville, Illinois.

102. Templeton, G. S., and Kellogg, C. E. (1950). Rabbit production farmers bull. *U.S., Dep. Agr., Tech. Bull.* **1730**, 1–58.

103. Thacker, E. J. (1956). The dietary fat level in the nutrition of the rabbit. *J. Nutr.* **58**, 243–249.

104. Thacker, E. J., and Brandt, C. S. (1955). Coprophagy in the rabbit. *J. Nutr.* **55**, 375–386.

105. Thompson, H. V., and Worden, A. N. (1956). "The Rabbit." Collins, London.

106. Thompson, J. F., and Ellis, G. H. (1947). Is cobalt a dietary essential for the rabbit? *J. Nutr.* **34**, 121–128.

107. Tsunoda, K., Imai, S., Tsutsumi, Y., and Inoue, S. (1968). Clinical effectiveness of sulfamonomethoxine and sulfadimethoxine in spontaneous coccidial infections in rabbits. *Jap. J. Vet. Sci.* **30**, 109–117.

108. Van Dam, J. (1963). "Annual Summary 1963 Rabbit Management Study." University of California Agricultural Extension Service.

109. Wales, R. G., Martin, L., and O'Shea, T. (1965). Effect of dilution rate and the numbers of spermatozoa inseminated on the fertility of rabbits ovulated with chorionic gonodotrophin. *J. Reprod. Fert.* **10**, 69–78.

110. Watson, J. S. (1954). Breeding season of the wild rabbit in New Zealand. *Nature (London)* **174**, 608.

111. Watson, J. S., and Taylor, R. H. (1955). Reingestion in the hare *Lepus europaeus* Pal. *Science* **121**, 314.

112. Zarrow, M. X., Denenberg, V. H., and Anderson, C. O. (1965). Rabbit: Frequency of suckling in the pup. *Science* **150**, 1835–1836.

113. Zarrow, M. X., Farooq, A., Denenberg, V. H., Sawin, P. B., and Ross, S. (1963). Maternal behavior in the rabbit: Endocrine control of maternal nest building. *J. Reprod. Fert.* **6**, 375–383.

114. Zeuner, F. E. (1963). "A History of Domestic Animals." Hutchinson, London.

C H A P T E R   3

Anatomy, Physiology, and Biochemistry of the Rabbit

Carlos Kozma, William Macklin, Laurence M. Cummins, and Russell Mauer

I. INTRODUCTION

The purpose of this chapter is to describe the anatomy, physiology, and clinical biochemistry of the rabbit. Normal data on clinical chemical, hematological, and physiological properties for different stocks or breeds of rabbits are presented in tabular form. The intent of these tables is to present reference material and there is little elaboration on their content. The physiology section is devoted mainly to the description of the cardiovascular, respiratory, and reproductive systems. Clinical biochemistry is involved with the morphology of blood and discussions on normal biochemical constituents. Brief mention is made about age, sex, and breed variations where such differences occur.

II. ANATOMY—NEW ZEALAND WHITE RABBIT

A. External Features

The body is well haired with both underfur and guard hairs being present. Naked areas of skin are located on the tip of the nose, a small portion of the scrotum of the male, and the inguinal spaces in both sexes.

1. Head

Prominent external features would include the large pinnae or external ears with their readily visualized vasculature. The cleft upper lip and the undivided lower lip together with the rather anterior lateral commisures form the relatively small external opening of the mouth. Sensory hairs or vibrissae are prominent. The external nares are ovoid in shape and are confluent with the cleft in the upper lip. The eyes are directed more nearly laterally than those of most mammals.

2. Neck

It is otherwise unremarkable except for the dewlap which is more prominent in some breeds than others and is not infrequently affected with a moist dermatitis due to continual hydration of the area.

3. Trunk

The trunk is divided into the thorax, abdomen, and back or dorsum. Ventral to the base of the short tail is the anus with deep hairless depressions, the inguinal spaces on either side. Ventral to the anus in the female is the vulva enclosed by skin folds and containing the clitoris in its ventral wall. The penis surrounded by the prepuce is present in the analogous area in the male. The scrotal sacs, containing the testes, lie on either side. In the female, the four or five pairs

of nipples are present on the ventral surface of the abdomen and thorax. Development of nipples does not occur in male rabbits.

4. Extremities

The well-developed posterior extremity is divided into the thigh, leg, and foot and has four fully developed digits. The anterior extremity is comprised of the upper arm, forearm, and hand and possesses five digits.

B. Skeleton

1. Axial Skeleton

a. SKULL. The skull is comprised of distinct bones and a few cartilages characterized by immovable articulations (sutures) with the exception of the mandible. The sutures become gradually less obvious with age but can still be visualized in old animals.

The skull can be divided into the cranial portion which encloses the brain and contains the auditory capsules and the facial portion which includes the structures related in the main to the jaws and palate.

The posterior (nuchal) surface of the skull is formed by the composite occipital bone which is formed by developmental fusion of the supraoccipital, paired lateral exoccipitals, and the ventral basioccipital. It is pierced by the foramen magnum which at each side bears the occipital condyles for articulation with the altas. The nuchal surface of the skull is separated from the dorsal surface by a sharp ridge having a median projection, the external occipital protuberance. The dorsal ligament of the neck and occipital muscles attach to the latter.

The dorsal surface of the skull is formed by the dorsal surface of the occipital (supraoccipital), the small interparietal, which separates the supraoccipital from the pair of parietal bones, the paired frontal bones, and the nasal bones which roof the nasal cavities. The frontal bones have anterior and posterior dorsolaterally situated supraorbital processes and also extend ventrally to form part of the orbital wells.

The inverted cone-shaped mastoid portion of the periotic or petramastoid bone with its pitted surface is visible at the posteriolateral aspect of the cranium. Enclosed in it and its petrous portion is the inner ear. The petrous portion is concealed by the mastoid portion and the prominent, round tympanic bulla in front of the latter. The bulla is formed by the tympanic bone and contains the large tympanic cavity, encloses the 3 ossicles of the middle ear, and continues dorsally into a short wide bony tube with a large opening which in the natural state would be continued into the aperture of the external ear. The whole tube is the external acoustic meatus.

Anterodorsal to the preceding bones is the squamosal bone which forms a large portion of the lateral wall of the cranium. It has a prominent projection, the zygomatic process, which forms the caudal end of the zygomatic arch. The mandibular fossa, in which the head of the mandible fits, is formed by the hollowed out ventral side of the zygomatic process. Anterior to the squamosal is the alisphenoid and anterior to it is the orbitosphenoid which is pierced by the optic foramen. The cephalic wall of the orbit is formed ventrally by the maxilla and dorsally by the lacrimal bone which extends beyond the orbital rim. The maxilla extends forward from the orbit. Its lateral zygomatic process forms the anterior aspect of the zygomatic arch and is fused with the anterior end of the zygomatic or jugal bone which forms the main portion of the arch. The alveolar process or ventral portion of the maxilla contains the roots of the cheek teeth. Anterior to the maxilla is the premaxilla in which are inserted the roots of the incisors. It joins medially with its opposite member and bounds the anterior ventral and lateral part of the nasal cavity. Its frontal process extends dorsocaudally along the lateral aspect of the nasal bone.

The ventral aspect of the cranium is formed by the medially situated basioccipital, basisphenoid, and presphenoid bones proceeding cephalad from the foramen magnum. The basisphenoid is perforated by the round foramen cavernosum which leads into the interior of the bone. The hypophyseal fossa is located on its dorsal surface (the floor of the cranial vault).

The ventral aspect of the skull is continued by the paired palatine bone whose caudal ends are notched and articulate with the two laminae of the pterygoid process of the alisphenoid bone. Part of the thin dorsal extension of the palatine bone can be seen in the wall of the orbit. The palatine continues forward medial to the alveolar process of the maxilla with which it articulates and spreads medially to join its opposite member and thus forms the caudal portion of the hard palate. The hard palate is continued anteriorly by the ventral portions of the maxilla and premaxilla. These are pierced by the long and narrow paired incisive foramina which are continuous across the median line in their caudal third.

In the bisected skull the structure of the nasal cavity can be observed. The cavity is divided into right and left fossae by the median, vertical cartilaginous nasal septum (cartilaginous portion of the mesethmoid). This is continuous caudally with the perpendicular plate of the ethmoid bone which consists also of the cribiform plate exposed to the anterior cranial fossa and the paired lateral ethmoidal labyrinths. Posteriorly the ventral portion of the cartilaginous septum is supported by the vomer bone. The scroll-like turbinate bones are situated on the lateral walls of the nasal cavity. The maxilloturbinalor concha inferior are rigid masses of bone which occupy the anterior portion of

the lateral wall of the nasal fossa. The other paired turbinates are represented by the nasoturbinals and the ethmoturbinals.

The mandible is composed of paired members united by a fibrous or fibrocartilaginous symphysis. The horizontal portion which bears the teeth comprises the body and the posterior vertical portion extending from the angle comprises the ramus.

For convenience the hyoid apparatus is mentioned with the skull even though it is derived from embryonic visceral arches. It is comprised of a median hyoid bone and the paired greater and lesser cornua which articulate medially with the former and are connected laterally by two muscles to the jugular process of the occipital bone. The apparatus is embedded in the base of the tongue for which it functions as a support and lies between the angles of the mandible.

b. VERTEBRAL COLUMN. The usual vertebral formula is $C_7T_{12}L_7S_4C_{16}$ but 13 thoracic vertebrae are present in some animals. The first two cervical vertebrae are modified as in other species, the spinous processes are enlarged in the thoracic region, the sacral vertebrae are fused to form the sacrum, and the last nine coccygeal vertebrae are devoid of arches and thus solid.

c. RIBS. Normally there are 12 pairs of ribs with each consisting of a dorsal bony portion and a shorter, ventral cartilaginous portion. The head of the rib articulates with the corresponding thoracic vertebra and with the posterior part of the body of the vertebra cephalad to it. The costal cartilages of the first seven ribs (true ribs) articulate with the sternum, while those of the last five ribs (false ribs) do not. The costal cartilage of the eighth rib is attached to that of the seventh and that of the ninth to the eighth. The cartilages of the last three ribs (floating ribs) lie free as they are not attached to any others.

d. STERNUM. The sternum is composed of six distinct sternebrae. The first is the manubrium sterni and the sixth is the xiphoid process with a thin but broad plate of cartilage at its caudal end.

2. Appendicular Skeleton

a. PECTORAL GIRDLE AND LIMB. The pectoral girdle consists of the scapula and the small clavicle. Its only direct attachment to the axial skeleton is through the sternoclavicular ligament with the major attachments being provided by muscles. The humerus, radius, and ulna are as in other species. The carpus consists of two rows of bones with four in proximal row and five in the distal. The bones of the distal row articulate with the five metacarpal bones. There are five digits each comprised of three phalanges with the exception of the first digit (pollex) which has two. Sesamoid bones are located on the volar surface of the foot, occurring in transverse pairs at the metacarpophalangeal articulations and in linear pairs at the articulations of the second and third phalanges. The pisiform carpal bone (1 of the 9) is also a sesamoid bone.

b. PELVIC GIRDLE AND LIMB. The pelvic girdle is made up of the paired coxal or innominate bones formed by the fusion of the ilium, ischium, and pubis and is firmly attached to the vertebral column at the iliosacral articulation. The coxal bones are ventrally united via the pelvic symphysis involving both the pubis and ischium. A small accessory bone, the os acetabuli, is present in the rabbit thus excluding the pubis from the acetabulum which is formed by the former bone plus the ilium and ischium.

The femur articulates distally with the tibia alone. The fibula is represented as a thin, bladelike bone along the lateral side of the tibia. The fibula is fused with the tibia distally for somewhat more than half their length. It attaches proximally to the lateral condyle of the tibia and its only free portion is joined to the tibia by the interosseous ligament.

The tarsal bones are six in number arranged in three rows with the proximal row containing two large bones; the middle row, one bone; and the distal row, three bones.

Metatarsals two through five are well developed while the first metatarsal is reduced to an inconspicuous splinter. The four digits of the hind limb each possess three phalanges.

The sesamoid bones of the pelvic limb are represented by the patella, the three sesamoid bones situated in muscle at the posterior of the knee joint, and the small ones on the plantar surface of the foot and situated at the metatarsophalangeal joints and at the joints between the second and third phalanges.

C. Abdominal Viscera

1. Liver and Gallbladder

The liver with its anterior surface applied to the diaphragm presents four lobes. A deep medium cleft divides it into right and left lobes which are subdivided into anterior and posterior lobules. Further subdivision of the right posterior lobule is seen in some animals. The quadrate lobe is a subdivision of the light lobe and lies medial to the gallbladder. The caudate lobe is small and circular with a thick extension or stalk.

The gallbladder is situated in a deep depression on the caudal surface of the right anterior lobule. The hepatic ducts unite to form the common bile duct which receives the cystic duct from the gallbladder and enters the dorsal surface of the duodenum immediately posterior to the pylorus.

2. Pancreas

The pancreas is diffuse and lies in the mesoduodenum, which joins the limbs of the duodenal loop, and is often associated with a good deal of fat. The single pancreatic duct opens into the posterior portion of the ascending limb of the duodenum and thus is well separated from the opening of the bile duct.

3. Spleen

The relatively flat, elongate spleen is connected by the mesogastrium to the left side of the dorsal surface of the greater curvature of the stomach.

4. Gastrointestinal Tract

The simple stomach of the rabbit which lacks specialized regions is thin-walled and large. The small intestine consists of the very long duodenum and the mesenterical small intestine or jejunum and ileum with the terminal portion of the latter expanded as the rounded sacculus rotundus. The ileum possesses a thinner wall and thus is more transparent than the more vascular jejunum and duodenum. The very large, thin-walled, coiled cecum terminates in the thick-walled light-colored vermiform process or cecumappendix. The cecum is characterized by a spirally arranged constriction related to internal folding of the mucosa (spiral valve). The first part of the colon is replaced by or structured like the cecum and constitutes the ampulla caecalis coli. The colon is characterized by sacculations, or haustra, and the presence of taeniae coli. The colon is continued through the pelvis as the rectum terminating in the anus. The paired rectal or anal gland is an elongated organ enveloping the rectum a short distance anterior to the anus and producing a secretion delivered to the latter. As in other species Peyer's patches are found in the small intestine. The minute honeycombed external appearance of the sacculus rotundus is due to the presence of a large number of lymph follicles (ileocecal tonsil) in this segment and the vermiform process is likewise characterized by abundant lymphatic tissue which imparts the light color to its wall. The mesenteric lymph glands are aggregated forming a compact mass covering the left side of the superior mesenteric artery. Small nodes are also present near the posterior end of the mesoduodenum, associated with the portal vein near the lesser curvature of the stomach, and near the junction of the splenic and superior mesenteric veins.

5. Adrenals

The paired adrenals are located anterior and medial to the kidneys and appear flattened against the dorsal body wall. The left adrenal is caudolateral to the origin of the superior mesenteric artery and the right adrenal is in quite close proximity to the postcava. In this same general area the small, curved, gray-brown superior mesenteric ganglion can often be located lying against the left side of the superior mesenteric artery in the mesentery. Slightly anteroventral is the coeliac ganglion lying between the coeliac and superior mesenteric arteries.

D. Urogential System

1. Urinary System

The kidneys are smooth and bean-shaped with the right kidney being situated more anterior than the left. On cut surface the single renal papilla can be seen. The ureter is easily followed from the renal hilus to the bladder. The urethra in the male extends from the bladder to the tip of the penis while in the female the short urethra opens into the vagina.

2. Male Genital Organs

The testes present no unusual features. The somewhat dilated terminal portion (ampulla) of the ductus deferens enters the ventral wall of the seminal vesicle dorsal to the bladder. The medially situated seminal vesicle opens into the prostatic segment of the urethra immediately in front of an oval elevation, the colliculus seminalis. The vesicular gland and prostrate gland lie in the dorsal wall of the more posterior part of the seminal vesicle which they largely cover and compress. The pair of vesicular gland ducts enter the urethra at either side of the seminal colliculus and the 4–6 minute prostatic ducts enter the urethra at either side and just posterior to them. The variable in number paraprostatic glands are minute fingerlike projections of the urethral lining imbedded in the outer part of its wall at either side of the base of the seminal vesicle. The small paired bulbourethral glands form a bilobed swelling in the dorsal wall of the urethra in which they are imbedded, immediately posterior to the prostate.

The penis does not present a glans penis as such but rather its free extremity is known as the pars libera. Immediately dorsolateral to the body of the penis and subcutaneously are located at each side a white inguinal gland and a closely associated brown inguinal gland which empty their secretions into the hairless inguinal spaces.

3. Female Genital Organs

The ovaries of the rabbit are elongated ellipse-shaped bodies. The ostium of the uterine tube with its fibriae, one of which is attached to the anterior end of the ovary, tends to enclose the margin of the ovary so that ovulated ova are almost certain to enter the funnel. The right and left uteri are separate for their entire length. The uteri unite to form the vagina with the cervix marking the junction. When the urethra enters the vagina, a common tube, the urogenital sinus or vestibulum results. The external margins of the urogenital sinus form the vulva. The bulbourethral gland, which is similar to that of the male, is situated in the dorsal wall of the vestibulum. The clitoris lies along the ventral surface of the urogenital sinus and its glans clitoris projects into the urogenital aperture.

E. Thoracic Viscera

The thymus which usually retains considerable size in the adult rabbit lies partly ventral to the heart and extends forward to the thoracic inlet. The paired pleural cavities containing the lungs are separated by the mediastinum with the space of the latter largely occupied by the heart. The posterior (dorsal) mediastinum is occupied by the esophagus, thoracic aorta, bronchi, the pulmonary vessels, and lymph nodes. The heart of the rabbit differs from some other mammalian hearts in that the right atrioventricular valve is composed of only two rather than three cusps making the term tricuspid valve inappropriate. The right lung is divided into apical, cardiac, and diaphragmatic lobes with the latter consisting of a large lateral and a smaller intermediate lobe. The diaphragmatic lobe of the left lung is not subdivided and the superior lobe is very imperfectly represented. Consequently, the left lung is only about $\frac{2}{3}$ the size of the right. The diaphragm is pierced from dorsal to ventral by the hiatus aorticus, the hiatus esophagus, and the foramen venae cavae.

F. Oral Cavity, Pharynx, and Larynx

The dental formula of the rabbit is

$$i\frac{2}{1}, c\frac{0}{0}, pm\frac{3}{2}, m\frac{3}{3}.$$

The second upper incisor is smaller than and set in behind the first. The tiny openings of the nasopalatine ducts are located about a millimeter behind the secondary incisors. The large tongue presents paired, pinhead size vallate papilla in pits on the posterior surface, ridged foliate papillae anterolateral to the latter, and fungiform papillae on the softer anterior dorsal surface of the tongue. Microscopic taste buds are located on the various papillae. The soft palate, which is long in the rabbit, divides the pharynx into oral and nasal portions with the latter receiving the opening of the eustachian tube. The paired tonsils are located on the anterior margin of the tonsillar fossa or sinus which is a deep depression in the lateral wall of the pharynx near the base of the tongue. The epiglottis guarding the entrance to the glottis projects upward into the pharyngeal cavity. The cartilages of the larynx consist of the thyroid, cricoid, paired arytenoid, corniculate, and epiglottic cartilages. The vocal folds are rudimentary in the rabbit.

G. Head and Neck

The diffuse white or brownish parotid gland is situated on the lateral surface immediately behind the angle of the mandible. It expands ventrally beneath the mandible and dorsally to cover the lateral aspect of the base of the external ear. Its duct runs anteriorly across the lateral surface of the masseter muscle in association with the branches of the facial nerve, eventually opening into the oral cavity opposite the last upper molar. A lymph gland is embedded in the posteriodorsal aspect of the parotid. The submaxillary or submandibular salivary gland is somewhat compact and rounded or oval and is situated at the medial side of the extreme ventral portion of the mandible. The superficial mandibular gland is a flattened oval cutaneous gland closely applied to the ventrolateral surface of the mandible covered by the anterior part of the platysma. Small lymph glands of irregular occurrence are usually found ventral to the mandible. The sublingual gland lies ventral to the tongue with its dorsal surface crossed by the duct of the submandibular gland.

The broad buccinator muscle encloses the cheek arising from the alveolar borders of the upper and lower jaw and inserting forward in the lip. By dividing this muscle the superior buccal gland, a long, narrow band of loosely connected lobules is exposed dorsally and the inferior buccal gland ventrally. The facial lymph gland overlies the dorsal edge of the buccinator muscle and is covered laterally by another muscle, the zygomaticus minor.

The aperture of the nasolacrimal duct is easily observed on the anteromedial aspect of the lower eyelid. Other structures of the orbit besides the eyeball include the large Harderian gland which consists of a larger pale gray-red lobe and a smaller, almost white lobe. The gland lies in the anterior portion of the orbit with its duct opening on the internal surface of the third eyelid. The smaller and darker loblated lacrimal gland lies close to the skull in the temporal angle of the orbit. Several fine ducts from the gland communicate with the caudal part of the inner surface of the upper eyelid. The infraorbital gland is related to the lacrimal and lies in the ventral part of the orbit. A small white or yellow gland, the zygomatic, lies in the anteroventral angle of the orbit immediately medial to the zygomatic arch and ventral to the anterior end of the infraorbital gland.

By removing the ventral musculature of the neck from the sternum to the mandible the trachea is exposed. The heavy, annular cricoid cartilage is just cephalad to the first tracheal ring and just caudal to the saddle-shaped thyroid cartilage of the larynx. All tracheal rings are incomplete dorsally. The lobes of the thyroid gland are situated on the lateral aspect of the trachea, just behind the cricoid cartilage and are connected ventrally by a thin isthmus. The elongated deep cervical lymph node is located dorsolateral to the thyroid cartilage close to the internal jugular vein. The major vessels of the neck are the common carotid artery which courses along the side of the trachea and the internal jugular vein, lateral to the common carotid artery. The vagus nerve lies between these two vessels and is the largest of the four nerves accompanying the carotid artery. The other three nerves are the ramus decendens of the hypoglossal nerve on the ventral surface of the artery, the cervical portion of the sympathetic trunk on the dorsal surface medial to the vagus, and the ramus cardiacus of the vagus or depressor nerve on the dorsal surface of the artery on the medial side of the sympathetic trunk. The esophagus accompanies the trachea dorsally.

H. Brain and Spinal Cord

The dura matter is adherent to the inner surface of the cranium and extends down between the cerebral hemispheres as the Falx cerebri and between the cerebellum and cerebrum as the tentorium cerebelli. The large, nearly smooth cerebral hemispheres are separated by the longitudinal cerebral fissure (sagittal fissure) at the bottom of which is the corpus callosum connecting the cortex of the two hemispheres. The olfactory bulbs, at the cephalic end of each cerebral hemisphere, receive the endings of the sensory olfactory or first cranial nerve. Proceeding caudally on the dorsal aspect of the brain, the pineal body can be located. It originates just posterior to the corpus callosum and extends in fingerlike fashion in a posterodorsal direction to nearly the posterior end of the cerebral hemispheres. The corpora quadrigemina are situated between and beneath the posterior ends of the cerebral hemispheres. The anterior pair are the superior colliculi and the posterior pair the inferior colliculi. The convoluted cerebellum conceals the posterior part of the corpora quadrigemina, and consists of the median unpaired vermis, a cerebellar hemisphere at each side and the stalked paraflocculus arising ventrolaterally beneath each hemisphere. The paraflocculus is frequently damaged in removal as it is almost completely enclosed by the dorsal portion of the petrosal bone. Underlying the cerebellum and roofing the fourth ventricle are the thin anterior medullary velum and posterior medullary velum. The medulla oblongata, the most posterior portion of the brain, lies ventral to the cerebellum and constricts as it continues posteriorly into the spinal cord.

In ventral view the following structures are noted proceeding from anterior to posterior: the olfactory bulbs are replaced caudally by a white band of fibers, the olfactory tracts, which terminate in the posteroventral part of the cerebral hemispheres or pyriform lobes. The optic chiasma forms a median crosslike elevation, the posterior portion of which enters the optic tracts and the anterior portion into the bases of the optic nerves. The pituitary (hypophysis) which is usually left behind when removing the brain, lies behind the optic chiasma. What usually remains is the slit-like cavity of the infundibulum which connects the pituitary with the brain. The tuber cinecum is represented by the gray tissue surrounding the aperture just mentioned. It and the mamillary body, which projects behind it are part of the hypothalamus. The ventral part of the midbrain is occupied by a pair of thick ridges, the cerebral penduncles which are separated by the faint median interpeduncular fossa, just behind the mamillary body. The third cranial or oculo-motor nerve emerges from the ventral surface of the peduncle. The cut end of the proximal portion of the fourth cranial or trochlear nerve can be observed on the lateral surface of the peduncle. The pons is represented as a broad band extending transversely across the ventral aspect of the brain posterior to the peduncles, and continuing upward into the cerebellar peduncles. A median depression, the sulcus basilaris, occupied by the basilar artery divides the surface into two parts. The medulla oblongata is that portion of the brain caudal to the pons. The anterior median fissure of the spinal cord continues anteriorly and ends at the posterior margin of the pons in the faint depression known as the foramen caecum. The pyramids of the medulla are represented by narrow bands on each side of the midline extending back from the posterior margin of the pons and continuous with the anterior funiculi of the spinal cord. The trapezoid body is a smaller, superficial transverse band located in the angle formed by the lateral margin of the pyramid with the posterior margin of the pons. The larger sensory and smaller motor roots of the trigeminal nerve can be seen at the lateral border of the pons. The slender abducent nerve arises by several very delicate rootlets along the lateral margin of the anterior end of the pyramid. The seventh (facial) and eight (acoustic) cranial nerves arise from the lateral margin of the trapezoid body. The glossopharyngeal, vagus, and spinal accessory (cranial nerves 9, 10, 11) arise by several roots arranged in a linear series along the lateral margin of the medulla. The twelfth or hypoglossal nerve arises by several roots from the ventral surface of the caudal medulla at the lateral margin of the pyramid.

By cutting through the cerebellar peduncles and severing the anterior and posterior medullary vela the cerebellum is removed exposing the fourth ventricle. The posterior medullary velum supports a chorioid plexus. The fourth ventricle connects with the third ventricle by the aqueduct of Sylvius through the midbrain. The dorsal portion of the third ventricle can be seen by spreading the posterior tips of the cerebral hemispheres and pulling away the pineal body and its tissue attachment. This thin membrane forming the root of the ventricle is the site of another chorioid plexus. The lateral margins of the aperture revealed are largely formed by small spindle-shaped masses, habenulae, one on each side which are united caudally by the habenular commissure. The intermediate mass broadly connecting the thalami can be seen from the dorsal surface as crossing and largely filling the cavity. The thalami are thick masses of nervous tissue which form the lateral walls of the ventricle. The dorsal portion of the thalamus forms the somewhat oval projection, the lateral thalamic tubercle, lateral to the habenula and the anterior tubercle of the thalamus, a faint small elevation in the angle between the lateral tubercle and the anterior portion of the aperture of the third ventricle. The lateral geniculate bodies are the most lateral parts of the dorsal surface at each side of the lateral tubercle. The medial geniculate bodies are less prominent and lie posteromedial to the lateral bodies.

The midline structures are exposed by division of the cerebrum in half through the sagittal fissure and the lateral ventricle is exposed by removing the dorsal wall of one cerebral hemisphere. The corpus callosum is shown in section. It ends anteriorly in a somewhat club-shaped expansion and bends downward posteriorly forming the splenium. The latter is attached to the body of the fornix. The fornix is composed of a pair of greatly curved bands fused for a short distance in the midline to form the body of the fornix. The fornix begins in the hippocampus and ends in the mamillary body. The septum pellucidum is seen between the body of the fornix and the anterior portion of the corpus callosum enclosing the lateral ventricle medially. The hippocampus presents as a somewhat oblique, convex ridge on the posteromedial floor of the lateral ventricle. A similar ridge, having a smaller ventricular exposure, the corpus striatum is anterolateral in position on the ventricular floor. Between the two the pigmented vascular tissue of the chorioid plexus of the lateral ventricle may be seen.

The spinal cord is enclosed in the meninges as is the brain, but in the former the dura mater is not firmly attached to the surrounding bone, resulting in the epidural space. The spinal cord presents slight enlargements in both the cervical and lumbar areas representing the areas from which the nerves to the limbs arise. The ventral rami of cervical nerves four to eight and thoracic nerve one contribute to the brachial plexus and the ventral rami of lumbar nerves four to seven and sacral nerves one to three combine to form the lumbosacral plexus. The number of spinal nerves by regions are C_8, T_{12}, L_7, S_4, C_6. The spinal cord contracts to a slender filament at about the middle of the sacrum, the filum terminale, which continues caudal to the base of the tail. The posterior spinal nerves originate cephalad to their

TABLE I
ORGAN WEIGHTS OF YOUNG ADULT RABBITS[a]

Breed	Sex	Sample size	Statistic	Body weight (in kg)	Brain	Liver	Spleen	Adrenal	Kidneys	Ovaries	Testes without epididymus	Thyroid	Thymus	Heart
New Zealand White	M	23	Mean	2.775	0.364	2.870	0.042	0.0098	0.521		0.109	0.0055	0.145	0.203
			S.D.[b]	0.198	0.035	0.417	0.0040	0.0032	0.059		0.029	0.0019	0.044	0.021
			S.E.[c]	0.0413	0.007	0.087	0.0009	0.0007	0.012		0.006	0.0004	0.009	0.004
	F	21	Mean	2.541	0.374	3.275	0.037	0.0095	0.510	0.0072		0.0063	0.156	0.200
			S.D.	0.235	0.045	0.593	0.0070	0.0023	0.055	0.0021		0.0021	0.053	0.0200
			S.E.	0.051	0.010	0.129	0.0016	0.0005	0.012	0.0005		0.0005	0.011	0.0044
Dutch Belted	M	20	Mean	1.500	0.560	3.390	0.0320	0.0104	0.730		0.115		0.150	0.270
			S.D.	0.310	0.089	0.950	0.0106	0.0029	0.074		0.064		0.050	0.030
			S.E.	0.069	0.020	0.210	0.0024	0.0007	0.017		0.014		0.011	0.007
	F	13	Mean	1.480	0.560	3.200	0.0420	0.0118	0.720	0.0099			0.130	0.280
			S.D.	0.199	0.056	0.505	0.0140	0.0037	0.102	0.0050			0.052	0.025
			S.E.	0.055	0.016	0.140	0.0038	0.0010	0.028	0.0016			0.015	0.007

[a] Expressed in grams per 100 gm body weight.
[b] S.D. equals standard deviation.
[c] S.E. equals standard error.

exit from the vertebral column, thus forming the cauda equina which accompanies the filum terminale caudally.

I. Organ Weights

Table I presents data on clinically normal adult New Zealand White rabbits. Organs listed were freed of fat and adhering connective tissue before weighing.

III. PHYSIOLOGY

A. Introduction

In evolutionary development the rabbit has evolved as one of the major sources of meat for many beasts of prey by adaptations in physiology which allow it both to escape capture and to reproduce prolifically.

B. Vision

The field of vision is that spatial area at which the complete visual image of the eye is formed. However, in all domestic animals, the vision fields overlap centrally in a certain area. There is only a small nasal area in each field of the rabbits' eyes which overlaps. In this area, the images perceived must superimpose perfectly in the cerebral cortex or double vision will occur.

Fields of vision in the rabbit (101):

Divergence between visual axes	150°–170°
Panoramic field	360°–or less
Binocular field	10°–35° (9° at rear)

The wide panoramic field of vision created by the lateral positioning of the eyes as well as the small binocular field to the rear when the animal raises its head affords the rabbit a very wide range of vision. The prominent eyes are another feature of the rabbit contributing to this wide panoramic field of vision. These adaptations give the rabbit an advantage in escaping predators.

C. Respiration

It is difficult to measure normal values for the compartments of the lung because many of the various subdivisions require cooperative and voluntary acts for their measurement. However, the functional residual capacity and tidal volume can be measured accurately and are about 11.3 ml and 15.8 ml, respectively, varying with the size of the rabbit (30, 48) (Table I). The lung compliance in the anesthetized rabbit is 6.0 ml/cm H_2O and the chest wall compliance is 9.4 ml/cm H_2O.

The respiratory rate varies from 32 to 60 breaths/minute. For a 2.4-kg rabbit it is 39 breaths/minute. The total work done as calculated by the area between the volume axis and the inspiratory portion of the pressure-volume curve is 1502 (798–2500) gm·cm/minute. Therefore, work/body weight in this case is 0.62 (0.39–1.09) gm·cm/minute/gm and work per unit ventilation is 2370 (1960–3280) gm·cm/min/liter/minute (30). Resting volume of ventilation is given as the product of the tidal volume and respiratory

TABLE II
VARIOUS RESPIRATORY MEASUREMENTS IN THE RABBIT

Description	Value	Ref.
Lung compliance (ml/cm H_2O)		30
Absolute	3.5–10.8	30
Per gm of lung	0.44–1.04	30
Per ml of lung	0.19–0.41	30
Chest wall compliance (ml/cm H_2O)		
Absolute	8.2–10.6	30
Per gm of lung	0.94–1.20	30
Per ml of lung	0.40–0.60	30
Lung resistance (cm H_2O/liter/sec)		30
Absolute	15.3–42	30
Per gm of lung	159–445	30
Per ml of lung	400–732	30
Work of breathing (gm·cm/min)	798–2500	30
Compliance ratio (chest/lung)	0.95–2.43	30
Time constant	0.087–0.193	30
Respiratory rate (breaths/min)	51 (32–60)	48, 112
Tidal volume (ml)	21 (19.3–24.6)	48
Minute volume (liter/min)	1.07 (0.37–1.14)	48
Oxygen consumption (cc O_2/gm fresh tissue)	640–850	59

frequency in breaths per minute. In the rabbit this is about 1.0 liter per minute.

For a 1.5-kg rabbit, the oxygen capacity is 42.5 μl/ml at pH 7.1, P_{50} is 49 mm Hg, log P_{50} pH is 0.75, and carbonic anhydrase (C.A.E.U./μl rbc) is 13.8 (101). Normal respiratory values appear in Table II.

D. Circulation

Various circulatory values and references are reported in Table III. Fox, Schlager, and Laird (40) have reported

TABLE III
VARIOUS CIRCULATORY VALUES IN THE RABBIT

Description	Value	Ref.
Whole blood volume (ml/kg body weight)	55.6–57.3	7
Plasma volume (ml/kg body weight)	38.8 (27.8–51.4)	7
Erythrocyte volume (ml/kg body weight)	16.8–17.5	
Erythrocyte dimensions (dry film) μm	7.5 (6.5–7.5)	7
Blood pressures (mm Hg)		
Adult		
Systolic	110 (90–130)	40, 80, 89
Diastolic	80 (60–91)	40, 89
Newborn		
Systolic	35	52
Diastolic	1	52
Whole blood pH	7.35 (7.21–7.57)	58, 116
Heart rate (beats/min)	306–333	80
Newborn	220	52

the blood pressure and pulse rates in both male and female rabbits of thirteen inbred strains (Table IV). Even though highly significant strain differences were found in blood pressure, pulse rate, and body weight, there were no significant correlation patterns either between or within strains. The authors reported that at 30 days of age the ACEP rabbits had been stimulated with 90 dB of sound resulting in an audiogenic seizure. Thus, such a hyperactive strain of rabbit has a high mean blood pressure. Similar values for these parameters were found in the ACCR (Y) and (B) strains which have common ancestry with the ACEP strain. However, lower values were reported for the AC strain which also has common ancestry.

It is reported that in the electrocardiogram of the rabbit, the P wave is low or even negative in lead I yet always positive in leads II–III (71). In the apex lead it measures 0.1–0.15 mV and 0.03–0.04 seconds. Q and R are detectable in the apex lead and lead I, whereas leads II and III detect R and S. In newborn rabbits, the QRS axis shows right deviation until the tenth day of life. T_I and T_{II} are almost always upright and T_{III} may be inverted in some animals. The RS–T deviation never exceeds 1 mm. The following values have been given for the various intervals of the ECG of the rabbit (71):

P–R	0.05–0.10 seconds
QRS	0.015–0.04 seconds
AT	0.12 seconds (at heart rate of 150–365)

Also, between Q–Q and Q–T, the following relation was reported:

$$0.20 - 0.20$$
$$0.30 - 0.22$$
$$0.40 - 0.23$$
$$0.60 - 0.24$$
$$1.00 - 0.25$$

The Q–T value was given at 2.8 R–R (71). Amplitude of the normal rabbit ECG is 0.01 mV.

E. Blood Coagulation

In the literature are numerous reports on coagulation studies in rabbits showing discrepancies between the authors. These differences are due in part to the use of different methods. It is known that there is not a significant species-specificity among the mammals on clotting times, clot retraction, Hageman factor, and thromboplastin reaction. Comparison of the relative prothrombin levels shows that the rabbits have 89% assuming a value of dog 100% and man 89%. These values correspond to values found and reported by other investigators (72, 73) (Tables V, VI, VII).

The prothrombin-proconvertin concentration in rabbit reported by Didisheim *et al.*, were 250 "units" in two minutes activation time (33). The same authors also

TABLE IV

MEANS AND STANDARD ERRORS OF BLOOD PRESSURE AND PULSE RATE FOR INCIPIENT INBRED STRAINS OF RABBITS[a]

Rabbit sex and strain	Mean arterial pressure (mm Hg)	Systolic blood pressure (mm Hg)	Diastolic blood pressure (mm Hg)	Pulse pressure (mm Hg)	Pulse rate (BPM)	Body weight (gm)
Males						
ACCR (B)	99 ± 2	126 ± 3	86 ± 2	40 ± 2	277 ± 11	1834 ± 54
ACEP (ep/ep)	95 ± 5	116 ± 7	84 ± 4	44 ± 4	247 ± 17	2091 ± 57
A	94 ± 4	118 ± 5	82 ± 4	37 ± 2	251 ± 9	2353 ± 58
WH	93 ± 3	113 ± 4	84 ± 2	29 ± 3	302 ± 10	2404 ± 83
ACCR (Y)	93 ± 3	117 ± 4	81 ± 3	35 ± 2	271 ± 12	2472 ± 64
AX	92 ± 2	112 ± 3	82 ± 2	30 ± 2	250 ± 15	3364 ± 53
AX$_{bubu}$	92 ± 2	113 ± 3	82 ± 1	31 ± 2	248 ± 14	3523 ± 92
III$_{mo}$	91 ± 2	109 ± 3	82 ± 2	27 ± 2	256 ± 11	2958 ± 82
III$_c$	90 ± 2	109 ± 3	80 ± 2	29 ± 3	287 ± 16	3684 ± 57
C	86 ± 2	108 ± 3	75 ± 2	33 ± 2	229 ± 8	2571 ± 120
X	84 ± 4	102 ± 5	75 ± 3	26 ± 2	260 ± 13	2099 ± 76
AC	81 ± 4	101 ± 5	70 ± 3	31 ± 2	242 ± 14	2371 ± 108
OS	73 ± 4	90 ± 4	65 ± 4	25 ± 1	234 ± 12	2610 ± 64
Females						
ACCR (B)	103 ± 4	130 ± 5	90 ± 3	41 ± 3	246 ± 10	2098 ± 48
ACEP (ep/ep)	104 ± 4	129 ± 5	91 ± 4	38 ± 3	240 ± 9	2098 ± 111
A	96 ± 3	123 ± 4	82 ± 2	40 ± 3	263 ± 6	2950 ± 103
WH	91 ± 3	112 ± 3	81 ± 2	31 ± 2	274 ± 7	2626 ± 126
ACCR (Y)	92 ± 2	113 ± 2	81 ± 2	32 ± 2	256 ± 11	2434 ± 89
AX	90 ± 3	108 ± 3	81 ± 3	27 ± 2	211 ± 10	3608 ± 92
AX$_{bubu}$	86 ± 3	108 ± 3	74 ± 3	34 ± 2	242 ± 15	3638 ± 157
III$_{mo}$	85 ± 3	103 ± 4	77 ± 3	26 ± 2	209 ± 17	2873 ± 108
III$_c$	91 ± 3	110 ± 3	81 ± 3	29 ± 2	277 ± 12	3733 ± 211
C	87 ± 3	107 ± 4	77 ± 2	31 ± 2	257 ± 10	3104 ± 117
X	81 ± 3	98 ± 4	73 ± 2	25 ± 2	274 ± 5	2172 ± 55
AC	80 ± 4	98 ± 5	71 ± 4	45 ± 2	251 ± 11	2413 ± 122
OS	75 ± 4	91 ± 5	67 ± 3	24 ± 2	241 ± 10	2825 ± 93

[a]From ref. 38.

reported that relatively "normal" amounts of thromboplastic activity were generated in the rabbit, the average percent of human activity from two determinations being 58%.

F. Digestion and Metabolism

The average absolute lengths of the small intestine, cecum, and colon have been given as 3.56, 0.61, and 1.65 m, respectively. In the rabbit the intestine is 10 times longer than the body length itself (101).

The range of rectal temperatures is 38.6° to 40.1°C, the average being 39.5°C (101). The capillary density in skeletal muscle is 2.1 capillaries/fiber (25) and the mitochondrial density is 2.96 mitochondria/gm liver (26).

The total catecholamines in adrenal glands of rabbit are reported as 0.48 mg/gm of whole adrenal gland with only 2% norepinephrine (101). The ratio of epinephrine to norepinephrine produced by the adrenal medulla varies with age. At birth, the adrenal medulla contains about

70%, whereas the adult tissue contains 2% norepinephrine. Differential release of these two hormones has been observed in the rabbit (101).

G. Reproduction

The age at which rabbits reach puberty is somewhat dependent on breed and strain, varying anywhere from four to 12 months. The smaller, such as "Polish" rabbits, may be bred at four months, whereas the heavier "Flemish" rabbits reach puberty between 9 and 12 months.

1. Spermatogenesis

The male rabbit tends to mature slower than the female, with motile spermatozoa appearing in the ejaculate at four months. Adult levels of sperm production are reached at about seven or eight months (6). Spermatozoa pass through the epididymis in four to seven days and mature after reach-

TABLE V

COAGULATION STUDIES IN RABBIT EMPLOYING STANDARD
TEST SYSTEMS

Clotting time, 37°C	
Glass (min)	5
Silicone (min)	15
Clot retraction	4+
Clot lysis	0
Platelet count ($\times 10^3/mm^3$)	750
Prothrombin (%)	95
Proconvertin (%)	315
Proaccelerin (%)	2000
AHF (%)	300
PTC (%)	100
Hagemen (%)	100
Fibrinogen (%)	143

TABLE VI

SPECIFICITY OF TISSUE THROMBOPLASTINS[a]

Source of thromboplastin	Prothrombin times of rabbit plasma (sec)
Brain	
Human	13.8
Cow	25.2
Sheep	27.4
Dog	58
Cat	19.6
Rabbit	8.2
Raccoon	20.4
Opossum	28.4
Chicken	24.0
Duck	26.4
Beef lung	6.8
Soluplastin (horse)	7.0
0.85% NaCl	63

[a]From ref. 33.

TABLE VII

SPECIFICITY OF THROMBIN–FIBRINOGEN
REACTION[a]

Source of thrombin	Rabbit fibrinogen (clotting time in seconds)
Human	16.0
Dog	20.0
Cat	14.6
Rabbit	13.2
Raccoon	18.5
Opossum	300
Chicken	15.0

[a]From ref. 33.

ing the middle corpus region primarily in the lower half of the corpus epididymis where they survive longer than in the female tract (51). Prior to fertilization, the semen must age (capacitation) for at least four to six hours in the female reproductive tract (8, 24).

2. Ovulation

The female is an induced ovulator which means she remains in estrus until copulation which triggers the induction of ovulation. Sterile copulation results in a pseudopregnancy lasting approximately 16 days and fertile copulation results in pregnancy and a gestation of 30 to 35 days depending on breed.

The female has been used extensively for reproductive studies because of some of its unique reproductive physiology. Ovulation occurs $9\frac{3}{4}$ to $13\frac{1}{2}$ (approximately 10) hours postcopulation (pc) or luteinizing hormone (LH) injection (56). Very precise timing can be obtained for studies of egg maturation, fertilization, cleavage, and implantation.

3. Embryo Viability

The ova and embryo can be manipulated extensively *in vitro* without apparent loss of viability making embryo culture and transfer useful tools for reproductive studies in this species. During its passage down the fallopian tube the embryo is covered with a mucin layer varying in thickness from 65 to 129 μm (45).

4. Uterine Anatomy

The rabbit has a duplex uterus, two completely separate uterine horns, and two cervixes. In studies involving embryo transfer each animal may serve as its own control, i.e., treated embryos can be transferred to one uterus or fallopian tube and compared to control embryos transferred to the opposite tract. With two separate cervixes the embryos do not migrate from one horn to the other as in livestock, dogs, and cats.

5. Placentation

The rabbit has the hemochorial type of placentation which allows the closest contact between maternal and fetal circulation, similar to humans. Experimental studies on placental transfer of drugs, metabolites, steroids, etc. would be more applicable to humans if studied in the rabbit than in the pig, judged on the basis of placentation similarities.

6. Reproductive Physiology

There are extensive studies and reviews of the physiological parameters of reproduction in the rabbit, many of which are summarized in Table VIII.

Copulation stimulates the release of LH immediately,

TABLE VIII
REPRODUCTIVE DATA FOR THE RABBIT

Description	Value	Ref.
Preovulatory LH peak (hr pc)	1–2	93
Systemic LH levels (ng/ml serum)		93
Before mating	22	
30 min pc	154	
60 min pc	1,110	
90 min pc	885	
120 min pc	1,343	
300 min pc	245	
1 day pc	40	
5 day pc	38	
Ovulation time (hr pc)	$9\frac{3}{4}$–$13\frac{1}{2}$	106
Fertilization time (hr after ovulation)[a]	1–2	26
Embryo cleavage rate (hr pc)		
First cleavage spindle	24	21
2-cell	21–25	20
4-cell	25–32	22
8-cell	32–40	23
16-cell	40–47	47
32-cell	48	86, 103
Blastocoele formation	75–96	74
Site of fertilization	Ampulla	25
Passage rate down fallopian tube		
Ovary to ampullary-isthmic junction[b] (min)	8.4	57
Embryo passage into uterus (hr pc)	72–75	11
Implantation (days pc)	7–$7\frac{1}{2}$	75
Gestation period (days)	30–35 (32)	23
Sperm capacitation time (hr)	6	8, 24
Ejaculate volume (ml)	0.63–2.3	9, 49
Sperm conc. (million/ml)	263	9
Sperm transit time to site of fertilization (hr)	3	13
Fertilizable life of ovum (hr)	6	53
Fertilizable life of sperm (hr)	30	54
Ovarian vein progesterone (μg/ml)		
15 days pc	2.4	79
2 days pc	0.4	79

[a] Sperm penetration of the ovum.
[b] Transport in the cumulus clot.

with peak serum levels being reached 1–2 hours pc (93). The peak serum levels as measured by radioimmunoassay were 1110 and 1343 ng/ml serum, at one and two hours pc, respectively, compared to a basal reading of 22 ng/ml serum (93). At approximately 10 hours pc ($9\frac{3}{4}$—$13\frac{1}{2}$), ovulation occurs. The ova are shed in a cumulus clot which is moved quickly into the ampulla via the fimbria. By 10 minutes after ovulation the ova have reached the ampullary–isthmic junction (57), where they remain for 48 hours before entering the isthmic portion of the tube (44). Sperm penetration occurs within 1–2 hours of ovulation and the first cleavage is completed 21 to 25 hours pc. Cleavage continues in the fallopian tube to the late morula stage, and from 72 to 75 hours pc the embryo enters the uterus. Blastocoele formation occurs 72–96 hours pc, and from

day four to day seven the blastocyst expands to a diameter of 4.5 to 6 mm (32). Implantation occurs 7 to $7\frac{1}{2}$ days pc.

The fluids in the reproductive tract (Table VIII) are dynamic in their response to the changing hormonal environment. The secretion rate decreases during pregnancy, the decrease evident by day two pc and plateaued at 50% of the estrus rate by day three (78). Progesterone has been shown to produce a similar decrease in secretion rate (77). Estrogen produces increases in secretion rate in the castrate animal to rates similar to the estrous doe (77). Ovum denudation and degeneration of the corona cells appear to be conditioned by components of the tubal fluid (117). Greenwald (43) has shown that the secretion of mucin by the rabbit oviduct can be modified by exogenous hormones. Estrogen reduced deposition of mucin by the rabbit oviduct lumen; progesterone increased it.

Ovarian vein progesterone levels increase to a high of 2.36 μg/ml in midpregnancy and then gradually decline to a low of 0.41 μg/ml two days before parturition (79). Mammary development is noticeable over the last week of gestation and the fur also begins to loosen in the flank and mammary area (92). The doe pulls the fur for use in nest building. After the young are two days old they apparently nurse only once a day (118).

The young are born relatively immature. Average litter size ranges from less than four to 8–10 depending on the breed (51). The young are eating dry food by three weeks and are weaned at eight weeks of age.

The following list gives litter sizes for various breeds of rabbits (51):

Polish	4
Angora, Beveren, Chinchilla Havana, Harlequin	4–5
American Chinchillas, Beaver, Belgian, Dutch, Flemish Lilac, Rex. Sable, Squirrel, Vienna White	6–7
Chinchilla, Flemish Giant, French Lop, New Zealand (Red, White)	8–10

Normal reproductive rabbit data appears in Tables VIII and IX.

IV. BIOCHEMISTRY

A. Clinical Chemistries

1. Blood

The values for certain biochemical constituents in the blood of different strains of rabbits are summarized in Table X. Both sex and strain differences occur in several of the clinical values. Serum cholesterol and urea nitrogen levels ($p < 0.01$)* in Table X appear to be greater in female than in male New Zealand White rabbits. Although the number of Polish rabbits tested was small, urea nitrogen

TABLE IX

CONCENTRATIONS OF VARIOUS COMPONENTS IN
REPRODUCTIVE TRACT FLUID

Component	Value	Ref.
Tubal fluid		
Na (mg/ml)	2.9–3.3	55, 61, 78
Cl (mg/ml)	3.3–4.1	55, 61, 77
Ca (mg/ml)	0.32	61, 78
K (mg/ml)	0.22–0.24	55, 61, 78
Bicarbonate (mg/ml)	1.76–3.1	55, 105
Polysaccharine (mg/ml)	0.37	55
Total protein (mg/ml)	2.43–2.73	55, 61
Drug matter (mg/ml)	8.28	55
Phosphate (μg/ml)	5.98	55
Mg (μg/ml)	3.43	55
Zn (μg/ml)	6.48	55
Lactic acid (μg/ml)	31.35–500	55, 61, 78
pH	7.8–8.2	55, 105
Secretion volume (ml/day)	0.4–3.18	7, 46, 50, 55, 77
Amino acids (mg%)		
Lysine	Present	55
Aspartic acid	Present	55
Serine	1.1	46, 55
Glycine	16.7	46, 55
Glutamic acid	3.6	46, 55
Alanine	3.2	46, 55
Valine	Present	55
Leucine	Present	55
Glucose (mg%)	25.7	55
Uterine fluid		
Na (mEq/liter)	110	76
Cl (mEq/liter)	55	76
K (mEq/liter)	10.7	76
Amino acids (mg%)		
Alanine	6.4	46
Glutanic acid	11.9	46
Glycine	26.8	46
Serine	10.4	46
Threonine	4.0	46
5-Nucleotidase	Present	70
ATPase	Present	70
Purine nucleosidase	Present	70
Bicarbonate (ml CO_2/100 ml)	99–120	76, 105
Glucose (mg/100 ml)	12–30	76
Lactic acid (mg/100 ml)	4–5	76

values were significantly greater ($p < 0.01$) in the Polish strain when compared to values obtained from New Zealand White or Dutch Belted rabbits.

Strain differences also appear to exist in glucose, total protein, sodium, and chloride levels (Tables X and XI). When analyzed under similar conditions, total protein levels in Dutch Belted rabbits are significantly greater than the protein levels of New Zealand White rabbits ($p < 0.01$) and less than those of Polish rabbits. New Zealand White rabbits have the highest blood glucose levels, followed, in decreasing order, by the Dutch Belted and Polish rabbits. Serum sodium ($p < 0.01$) and chloride ($p < 0.05$) are slightly greater in Dutch Belted rabbits when compared to the New Zealand White rabbits (Table XI).

Differences in biochemical parameters among eleven strains of rabbits bred at the Jackson Laboratory (Bar Harbor, Maine) have been reported by Fox *et al.* in 1970 (39). In addition to the strain differences already discussed, significant differences were found in inorganic phosphorus concentrations and in the activities of serum alkaline phosphatase (Table XII).

The protein fractions found in the blood of normal rabbits are shown in Table XIII. The albumin fraction represents about 55 to 65% of the total protein concentration and appears to be greater in the female than male New Zealand White rabbit (68). Kozma *et al.* reported that the female New Zealand rabbit has a lower α-globulin fraction than the male (68). Allen and Watson (4), however, reported no sex differences in serum protein patterns. Wostmann (111) compared the serum proteins of conventionally bred rabbits to germfree rabbits. Differences appear to occur only in the γ-globulin fraction. Germfree rabbits show only a trace (0–2% of the total protein) of γ-globulin when compared to conventional rabbits (10–12%). Allen and Watson (4) reported specific changes in the serum of rabbits with coccidiosis or kidney lesions.

There are significant differences in published clinical chemistry values for the same strain of rabbit. For example, the serum cholesterol and glucose levels in New Zealand White rabbits published by Burns and deLannoy (15) are only about half the values reported by Fox *et al.* (39).

Factors which may have been related to these discrepancies are physiological differences among the breeding stocks of a particular strain, the different methods used for chemical analysis, and seasonal or circadian changes in blood constituents. Differences in the methods of bleeding may also contribute to the variations in reported values. Bito and Eakins (10) have reported changes in the chemical composition of blood plasma of rabbits anesthetized with various agents. The effects of ether, however, were not tested. Circadian changes in urea nitrogen levels were observed in rabbits by Fox and Laird (38) but there appeared to be little variation in the levels of total protein, glucose, inorganic phosphorus, or alkaline phosphatase activity.

An interesting phenomena occurs when determining blood glucose levels in rabbits. Even after 96 hours of fasting, no significant change from the zero hour sample is seen, indicating that short periods of fasting have no effect on glucose levels. This evidently occurs because there is a bolus of food that is being continually digested in the intestine throughout this fasted period (31).

2. Liver and Kidney Function Tests

Sulfobromophthalein (BSP) and indocyanine green (ICG) disappearance from the plasma are useful for deter-

TABLE X
VALUES FOR CERTAIN BIOCHEMICAL CONSTITUENTS IN THE BLOOD OF NORMAL RABBITS

Constituent	Rabbit stock	Sex	No. of rabbits	Sample	Mean ± SEM	Range	Ref.
Cholesterol (mg/100 ml)	New Zealand White	MF	89	Plasma	45 ± 18 (total)	—	12
					22 ± 13 (free)	—	
					23 ± 12 (ester)	—	
	New Zealand White	MF	142	Serum	26.7 ± 1.3	5.7–71.0	15
	New Zealand White	M	29	Serum	42 ± 3	20–83	16
		F	26		76 ± 3		
	Dutch Belted	M	25	Serum	55 ± 3	35–82	16
		F	18		60 ± 2	39–70	
	Polish	MF	3	Serum	64 ± 14	50–92	16
Creatinine (mg/100 ml)	New Zealand White	MF	165	Serum	1.59 ± 0.34	0.80–2.57	15
	New Zealand White	M	31	Serum	1.4 ± 0.04	1.0–1.9	16
		F	26		1.2 ± 0.02	1.0–1.4	
	Dutch Belted	M	26	Serum	1.22 ± 0.03	0.8–1.7	16
		F	19		1.25 ± 0.05		
	Polish	MF	5	Serum	1.78 ± 0.28	1.3–2.9	16
Urea nitrogen (mg/100 ml)	New Zealand White	MF	147	Serum	19.23 ± 0.41	9.17–31.73	15
	New Zealand White	M	30	Serum	17.0 ± 0.8	10–28	16
		F	25		20.0 ± 0.7	12–28	
	Dutch Belted	M	25	Serum	22.1 ± 1.4	13–40	16
			19		23.3 ± 1.2	16–38	
	Polish	MF	3	Serum	38.0 ± 6.0	26–45	16
Glucose (mg/100 ml)	New Zealand White	MF	98	Serum	73.39 ± 0.97	50–93.18	15
	New Zealand White	MF	6	Plasma	98.7 ± 8.6	—	10
	New Zealand White	M	30	Serum	144 ± 2	127–156	16
		F	25		135 ± 3	112–160	
	Dutch Belted	M	24	Serum	129 ± 3	69–159	16
		F	18		126 ± 3	102–149	
	Polish	MF	4	Serum	104 ± 6	87–115	16
Total protein (gm/100 ml)	New Zealand White	M	29	Serum	6.3 ± 0.1	5.3–7.9	16
		F	25				
	Dutch Belted	M	26	Serum	7.2 ± 0.2	5.7–9.7	16
		F	19		7.1 ± 0.1	6.4–9.0	
	Polish	M					
		F	4	Serum	8.5 ± 0.5	7.7–9.8	16
Total lipids (mg/100 ml)	New Zealand White	MF	4	Serum	328 ± 8	—	41
	New Zealand White	—	14	Plasma	390 ± 80	—	107
		—	89	Plasma	243 ± 89	—	12
Total phospholipid (mg/100 ml)	New Zealand White	—	12	Plasma	113 ± 29	—	107
		—	89	Plasma	78 ± 33	—	12

mining liver function. Plasma clearance of BSP appears to be primarily determined by the biliary excretion rate. The rate of disappearance of BSP from plasma in a New Zealand White rabbit given an intravenous dose of 60 mg/kg is 1.8 mg/minute/kg (66). This rate of BSP clearance is similar to that of the rat but faster than the dog. The rabbit clears ICG faster than both the rat and dog (67). Half of an intravenous dose of 32 mg/kg of ICG is cleared by the rabbit in seven minutes.

Glomerular filtration and tubular secretion rates of various endogenous and exogenous substances are useful indicators of kidney function. The volume of plasma cleared of various substances by the rabbit kidney are listed in Table XIV (5).

3. Cerebrospinal Fluid

Small amounts of clear cerebrospinal fluid can be withdrawn from the cesterna magna by puncture at the atlanto-occipital site. Values for chemical constituents of cerebral spinal fluid of rabbits are listed in Table XV (16, 60).

B. Urine Content

The amount of urine excreted daily by an animal depends on many factors including food and water consumption, activity, and environmental temperature. A 24-hour urine volume of an adult rabbit with access to food and water varies within a wide range (20–350 ml/kg), with an average

TABLE XI
BLOOD ELECTROLYTES IN NORMAL RABBITS

Electrolyte	Rabbit stock	Sex	No. of rabbits	Blood fraction	Mean ± SEM	Range	Ref.
Sodium (mEq/liter)	New Zealand White	MF	148	Serum	125.41 ± 0.79	100–145	15
	New Zealand White	M	31	Serum	141.0 ± 0.8	133–153	16
		F	25		139.0 ± 0.7	131–145	
	New Zealand White	MF	24	Plasma	131.47 ± 7.12	—	10
	Dutch Belted	M	26	Serum	147.3 ± 0.9	138–156	16
		F	18		147.7 ± 0.8	144–158	
	Polish	MF	3	Serum	133.0 ± 10.0	114–146	16
Potassium (mEq/liter)	New Zealand White	MF	143	—	5.06 ± 0.93	3.00–7.20	15
	New Zealand White	M	31	Serum	5.1 ± 0.1	3.6–6.9	16
		F	26	Serum	5.3 ± 0.1	4.2–6.6	
	New Zealand White	MF	Av. of 4 groups	Plasma	4.00 ± 0.2	—	10
	Dutch Belted	M	26	Serum	5.9 ± 0.1	4.4–7.4	
		F	19				
	Polish	MF	3	Serum	5.4 ± 0.2	5.0–5.7	15
	New Zealand White	MF	150	—	10.0 ± 0.2	—	15
	New Zealand White	M		Serum	14.2 ± 0.6	11.8–16.5	16
		F		Serum	15.9 ± 0.4	13.1–18.9	
	Dutch Belted	M	26	Serum	14.5 ± 0.3	9.5–17.3	
		F	19		15.0 ± 0.3	13.5–16.7	
	Polish	MF	3	Serum	15.6 ± 1.4	13.5–18.3	
	New Zealand White	MF	99	—	9.65 ± 0.69	—	15
	New Zealand White	M	31	Serum	10.6 ± 0.8	97–114	16
		F	25		10.2 ± 1.0	93–110	
Chloride (mEq/liter)	Dutch Belted	M	26	Serum	108.7 ± 0.8	102–120	
		F	19		109.1 ± 1.3	95–118	
	Polish	MF	3	Serum	103.0 ± 7.1	89–112	
Phosphorus (mg/100 ml)	New Zealand White	MF	148	—	5.47 ± 0.10	—	15
Phosphate (mg/100 ml)	New Zealand White	M	31	Serum	6.2 ± 0.2	4.4–7.8	16
		F	26		6.2 ± 0.3	3.8–8.6	
	Dutch Belted	M	26	Serum	5.9 ± 0.3	3.1–815	
		F	19		6.1 ± 0.4	3.2–9.2	
	Polish	MF	3	Serum	4.7 ± 0.3	4.0–5.1	
Magnesium (mg/100 ml)	—	—	7	Whole blood	5.4 ± 0.74	—	37
			7	Serum	3.2 ± 0.59	—	
	New Zealand White	MF	24	Plasma	1.27 ± 0.07	—	10

excretion rate of about 130 ml/kg. The specific gravity of urine ranges from 1.003 to 1.036, with an average of 1.015. The maximum urine concentrating ability by the rabbit is about 1.9 osmoles/liter (62). Rabbit urine is normally alkaline with an average pH of about 8.2.

Normal rabbit urine has a yellow tint and is turbid in appearance. Ketones and occult blood are normally not detected by chemical analysis; however, occasional traces of protein or glucose may be present. Microscopic examination of urine sediment is characterized by an absence or small amount of epithelial cells and bacteria. Crystals are usually found in large amounts in the urine. The types of crystals have been identified as ammonium magnesium phosphate (triple phosphate), calcium carbonate monohydrate, and anhydrous calcium carbonate (37a). Casts are not seen in normal rabbit urine sediment. However,

a rare amount of white or red blood cells may occasionally be observed.

The values for some urinary excretion products for adult New Zealand White rabbits are listed in Table XVI.

C. Feces

Rabbits normally excrete both hard and soft feces; about 80% of the total excreted being of the hard type (114). Soft feces are produced by the initial digestion of food. The soft feces are normally consumed by the rabbit and excreted as the hard type (36, 102). By consuming the soft feces, the rabbit apparently makes use of nutritive products resulting from metabolism of intestinal flora. Germfree rabbits, however, do not consume their soft type feces.

TABLE XII
ACTIVITIES OF SOME ENZYMES IN THE SERUM OF NORMAL RABBITS

Enzyme (unit)	Rabbit stock	Sex	No. of rabbits	Enzyme activity		Ref.
				Mean ± SEM	Range	
Acid phosphatase (IU)	New Zealand White	M	18	110.3 ± 14.3	94.8–134.4	99
Alkaline phosphatase (King-Armstrong)	New Zealand White	M	31	10.6 ± 0.6	4.9–15.6	16
		F	25	9.9 ± 0.5	6.5–15.3	
	Dutch Belted	M	20	8.2 ± 1.4	2.1–16.0	16
		F	15	12.1 ± 1.0	6.6–19.7	
	Polish	MF	5	5.1 ± 0.2	4.3–5.6	16
Acid protease (*M* tyrosine/ml of serum/hour)	New Zealand White	M	18	1.29 ± 0.15	—	99
Lactic dehydrogenase (sigma/ml)	Polish	M	6	243 ± 50	—	41
Glutamic oxaloacetic transaminase (Karmen)	New Zealand White	M	26	42 ± 6	14–113	
		F	27	44 ± 5	16–108	16
	Dutch Belted	M	22	35 ± 3	16–75	
		F	19	40 ± 5	18–108	16
	Polish	MF	3	76 ± 15	48–101	16
Glutamic pyruvic transaminase (Wroblewski LaDue)	New Zealand White	M	27	35 ± 3	17–67	16
		F	25	33 ± 2		
	Dutch Belted	M	21	22 ± 2	12–37	16
		F	18	28 ± 1	16–37	
	Polish	MF	5	32 ± 4	25–46	16

TABLE XIII
PROTEIN FRACTIONS IN THE BLOOD OF NORMAL RABBITS

Rabbit strain	Sex	No. of rabbits	Sample	Total protein	% of total protein				Ref.
					Albumin	Globulin	β-Globulin	Globulin	
	M	8	Plasma	7.2 ± 0.11	57.08	15.00	14.72	13.19	63
Several	MF	18	Serum		64.00	13.01	12.77	9.90	4
Germfree		5	Serum	5.85 ± 0.15	56.07	20.34	22.22	2.05	111
Conventional		6	Serum	6.99 ± 0.27	52.50	17.31	18.31	11.73	
		(3–4 mo. old)							
New Zealand White	M		Serum	5.20 ± 0.79	56.94	8.90	9.94	11.97	68
	F	S	Serum	5.7	65.78	6.90	10.95	9.25	

The composition of soft and hard feces in conventionally bred and germfree Dutch rabbits was studied by Yoshida *et al.* (115) and is shown in Table XVII.

V. HEMATOLOGY

A. Morphology of Peripheral Blood Cells

1. Erythrocytes

The reported means of erythrocyte diameter fall generally in the 6.5 to 7.5 μm range, with the values for adult rabbits near the lower end of the scale (3, 42, 94, 96). The thickness is reported as a consistent 2.4 μm (96). Erythro-cytes of the newborn exceed 9 μm in diameter and show adult values at 20 to 120 days of age (69). Polychromasia and 2 to 4% reticulocytes are consistent findings related to the relatively short life span and consequent rapid turnover. Numerous crenated forms characterize smears of rabbit blood (96). Normal erythrocyte values appear in Table XVIII.

2. Thrombocytes

Thrombocytes appear singly and in groups in stained blood films. They are oblong or oval to round bodies of 1 to 3 μm in diameter and stain (Wright's) an intense violet centrally (collection of azurophilic granules) with a pale blue to colorless periphery (42, 95, 96).

TABLE XIV
RENAL CLEARANCE OF VARIOUS SUBSTANCES IN RABBITS

Substance	Clearance rate
Inulin	7.0 (5.0–8.4) ml/min/kg
Creatinine	3.2 (2.2–4.2) ml/min/kg
p-Aminohippurate	60.1 (21.5–199.0) ml/min
Diodine	36.2 (30.9–45.7) ml/min
Urea	14.3 (3–28) ml/min/m^2 of Body surface area

TABLE XV
VALUES FOR CONSTITUENTS OF RABBIT CEREBROSPINAL FLUID (CSF)

Constituent	Concentration in CSF
Glucose	75 mg/100 ml
Urea nitrogen	20 mg/100 ml
Creatinine	017 mg/100 ml
Cholesterol	33 mg/100 ml
Total protein	59 mg/100 ml
Alkaline phosphatase	5.0 units (King-Armstrong)
Carbon dioxide	41.2–48.5 ml%
Sodium	149 mEq/liter
Potassium	3.0 mEq/liter
Chloride	127 mEq/liter
Calcium	5.4 mg/100 ml
Magnesium	2.2 mEq/liter
Phosphate	2.3 mg/100 ml
Lactic acid	1.4–4.0 mg/100 ml
Nonprotein N	5.6–16.8 mg/100 ml

TABLE XVI
VALUES FOR CERTAIN URINARY EXCRETION PRODUCTS OF MALE NEW ZEALAND WHITE RABBITS

Product	No. of animals	Mean ± SEM (range)
Sodium (mEq/kg/day)	19	1.41 ± 0.13 (0.31–2.69)
Potassium (mEq/kg/day)	19	8.67 ± 0.69 (4.46–15.69)
Chloride (mEq/kg/day)	19	3.5 ± 0.4 (0.8–8.2)
Creatinine (mg/kg/day)	19	44.3 ± 3.7 (22.4–85.0)
Calcium (mg/kg/day)	19	96.0 ± 11.2 (12.1–193.2)
Phosphate (mg/kg/day)	16	14.0 ± 2.3 (5.0–32.8)

3. Leukocytes

a. NEUTROPHILS (PSEUDOEOSINOPHIL). While some reports suggest that the neutrophil is the predominant white cell in peripheral blood (42, 84), the majority regard the lymphocyte as the predominant cell (18, 35, 87, 98). The brilliant red staining of the granules with eosin caused Ehrlich to call these cells pseudoeosinophils. Other names that have been used are amphophil and heterophil. Regardless of the names the cell is given it is analogous to the neutrophil of other species. The diameter is 7 to 11 μm

and a polymorphous nucleus is present which stains a combination of light purple and light blue. The cytoplasm contains numerous smaller granules and a variable number of large, superimposed, granules.

The presence of "drumsticks" in a small percentage of neutrophils, primarily of the female, occurs in rabbits as in other mammals. The Pelger or Pelger-Huet anomaly of leukocytes was initially reported in rabbits in the 1930's and, according to Schermer, all cases reported can be traced back to one animal (96). A 1950 review discusses this anomaly which manifests as a failure in normal maturation of leukocytes, primarily neutrophils (81).

b. EOSINOPHILS. The eosinophil can be distinguished from the pseudoeosinophil by its larger size and characteristic granules. Eosinophils are 10–15 μm in diameter, the granules are three to four times the size of the granules of the pseudoeosinophil, fill the cytoplasm (sometimes obscuring the nucleus) and stain intensely acidophilic (42, 95, 96).

c. BASOPHIL. Among common laboratory animals the rabbit alone may show relatively large numbers of basophils in the peripheral blood. While some animals may reveal only a few or none, others may have counts representing up to 30% of a differential (95, 96). The basophil is similar in size to the neutrophil. The nucleus stains a light purple and is often obscured by the purple and black metachromatic cytoplasmic granules.

d. LYMPHOCYTE. The most common cell type found in the peripheral blood of rabbits is the lymphocyte which presents no differences in staining characteristics from those of other species. Azurophilic granules are sometimes present in the cytoplasm. Both large and small lymphocytes are observed: The small lymphocytes are similar in size to the erythrocytes, while the larger ones are as large as neutrophils (96). Normal leukocyte and platelet values are listed in Table XIX.

e. MONOCYTES. The monocyte is the largest cell found in normal rabbit blood. Unlike some other species, granules are not normally present in the cytoplasm but do occur under toxic conditions (110). Monocytes can usually be readily distinguished from large lymphocytes by their amoeboid nuclear patterns, diffuse, light-staining nuclei containing a few vacuoles, and the absence of the nonstaining perinuclear cytoplasm seen in lymphocytes (42, 95).

B. Erythrocyte Parameters

1. Erythrocyte Life Span

The erythrocyte life span as determined by techniques employing ^{55}Fe, ^{59}Fe, or ^{15}N averages around 50 days, with

TABLE XVII
COMPOSITION OF FECES AND CECAL CONTENTS OF DUTCH RABBITS[a]

Component	Cecal contents		Soft feces		Hard feces		Diet Composition
	Conventional	Germfree	Conventional	Germfree	Conventional	Germfree	
Moisture (fresh)	74.3	85.2	44.7	74.3	17.5	29.9	24.9%
Crude ash (dry)	8.6	6.4	7.7	6.7	6.2	5.8	5.4
Crude fat (dry)	1.0	0.9	1.3	0.7	1.4	0.9	4.9
Crude protein (dry)	42.3	26.5	39.7	21.8	20.3	13.3	26.1
Crude fiber (dry)	24.4	27.2	26.4	33.6	47.4	44.0	16.1
Nitrogen-free extract (dry)	23.8	39.1	24.9	37.2	24.8	36.0	47.5
Nonprotein N (dry)	2.3	3.0	1.3	2.4	0.3	1.0	—
Calcium (dry)	0.8	0.5	1.0	0.7	1.2	0.9	0.7
Phosphorus (dry)	1.4	0.6	1.4	0.9	1.1	1.0	0.5

[a]Averaged percentage in fresh or dry fecal matter of 2 or 3 rabbits wearing collars to prevent coprophagy; from the data of Yoshida et al. (115).

TABLE XVIII
ERYTHROCYTE AND RELATED PARAMETERS IN THE RABBIT[a]

RBC × 10^6	Hb (%)	PVC (%)	Reticulocytes (%)	RBC diameter (μm)	Ref.	Notes
4.5–7.0				5–8 (6.7)	94	
5.98 ± 0.78					18	
5.4 ± 0.55	11.18 ± 1.06		2.64 ± 1.22		87	
		36–48				
(5.61)	12.1	(41)		(6.9)	42	
3.0–8.0						
(5.2 ± 0.63)	8.4–15.5				84	
4–6.4			1–7			
(5.25)	(12.4)		(5)		96	
4.2–7.1	10.8–16.0	35–50				
(5.31)	(13.1)	(40.0)			16	c
5.36	13.35				91	
5.67 ± 0.77	12.0 ± 1.38				35	
5–6	12–12.5	36–38	2–4		95	
6.27	13.1	39.7		6.7–6.9	109	
4.5–7.0	8–15	33–50	2–3			
(5.7)	(11.9)	(41.5)	(2.2)	6.5–7.5	3	
6.33–7.21	13.95–15.56	40.28–44.99			38	c
4.33–6.82	9.37–13.94	31.59–50.0		6.4–9.3	69	d
	11.91–16.3	36.67–43.53			69	
5.30 ± .14	14.21 ± .35	41.79 ± .91	34.0		16	b,e,f
5.45 ± .13	13.98 ± .25	41.18 ± .71	33.9			b,e,g

[a]Values are given in percentage and/or absolute numbers of cells per cubic millimeter. All ± values are S.D. except as noted.
[b]± is S.E.
[c]New Zealand Whites (NZW).
[d]Inbred NZW.
[e]Dutch Belted.
[f]Females.
[g]Males.

TABLE XIX

Leukocyte and Platelet Counts in the Rabbit[a]

WBC × 10³	Neutrophils (pseudo-eosino-phils)	Lymphocytes	Monocytes	Eosinophils	Basophils	Platelets × 10³	Ref.	Notes
4.0–13.0	30–50	30–50	2–16	0.5–5.0	2–8		94	
(7.9)	(43.4)	(41.8)	(9.0)					
7.07 ± 1.88	37.8 ± 13.18	53.26 ± 14.0					87	
5.8–15.4								
(10.7)	39.1 ± 10.8	56.4 ± 13.3	1.5 ± 1.4	1.1 ± 0.8	3.6 ± 2.1		18	
5.2–12.0						126–1000		
(8.0)	8–50	20–90	1–4	1–3	0.5–30	(222)	96	
(3.1–11.3)	2–53	46–98	0–3	0–5	0–4	113.8–657.6		
6.2	(20)	(77)	(1)	(1)	(1)	(335.6)	16	c
10.05 ± 1.73	3.193 ± 1.682	6.861 ± 1.723					35	
7.0–18.7								
(11.1)	34.4	56.7	4.7	3.3	1.2		98	
8.86	42.0	40.5	8.1	1.24	8.06	481	91	
3.2–23.5	1500–15,390	630–9900	72–5405	0–1760	0–5359			
(9.56)	(45.4)	(31.8)	(10.5)	(2.2)	(9.9)		84	
5.0–21.5	32–59	20–68	1.5–16	1.0–4.0	2.4–9.0	170–798		
(8.6)	(45.0)	(38.4)	(9.5)	(1.6)	(6.3)	(552.5)	(42)	
6.0–13.0	36–52	30–52	4–12	0.5–3.5	2–7			
(9.0)	(46)	(39)	(8)	(2)	(5)	533	3	
1.38–13.78							69	d
8.03–13.28	6.38–19.88	73.00–90.88	0.73–3.25	0.13–1.63	1.13–3.63		38	c
7.01 ± .48	23.42 ± 2.51	71.53 ± 3.01	2.37 ± .43	0.89 ± .20	1.74 ± .58			b,e,f
7.14 ± .74	36.29 ± 2.94	58.47 ± 2.79	2.47 ± .40	2.00 ± .57	1.35 ± .43		16	b,e,g

[a]Values are given in percentage and/or absolute numbers of cells per cubic millimeter. All ± values are S.D. except as noted.
[b]± is S.E.
[c]New Zealand White (NZW).
[d]Inbred NZW.
[e]Dutch Belted.
[f]Females.
[g]Males.

values reported as low as 45 and high as 68 days (14, 17, 82). This relatively short life span is associated, as it is in other small species of laboratory animals, with a consistently significant number of reticulocytes. The polychromasia and anisocytosis observed in rabbit erythrocytes have a similar basis (95). A summary of normal hematological values on the New Zealand rabbit from our laboratory is presented in Table XX.

2. Erythrocyte Resistance

The erythrocyte resistance to hemolysis, or osmotic fragility, in hypotonic saline is expressed in minimum and maximum values. The minimum resistance is that concentration of sodium chloride at which the onset of hemolysis is observed, while the maximum is that concentration resulting in hemolysis of all cells. Most reported minimum values range from 0.50 to 0.58%, maximum range from 0.25 to 0.52% (3, 85, 94). Coldman *et al.* (28) put the range at 0.46 to 0.33% between minimum and maximum values, which more closely agrees with the results obtained con-

sistently in our laboratory (16) than with those listed in a standard reference (3).

3. Erythrocyte Sedimentation Rate

Little settling occurs in rabbit blood with resultant low reported values for millimeters sedimentation after one hour. Reported values fall between 1 and 4 ml (3, 96, 109).

C. Morphology and Differentiation of Bone Marrow Cells

Methods are available for the quantitative determination of marrow cells (34, 64, 114). However, the most widely used procedure is the differential count of aspirated marrow smears. To the degree the smear is representative of the entire marrow, meaningful information can be secured relative to the myeloid to erythroid ratio, the presence of abnormal cells, and assessment of the normality of maturation. The problems in morphological identification are concerned primarily with the earlier states of the erythroid cells. However, there are less problems in rabbits

TABLE XX

NORMAL HEMATOLOGICAL VALUES FOR ADULT NEW ZEALAND WHITE RABBITS[a]

Constituent	Sex[b]	No. of rabbits	Mean ± S.D.	Range
Hemoglobin (gm%)	MF	160	11.6 ± 1.1	9.4–13.8
Hematocrit (% PVC)	MF	182	36.3 ± 3.2	29.8–42.7
RBC × 10⁶/cmm	MF	169	4.84 ± 0.49	3.86–5.82
WBC × 10³/cmm	MF	179	7.25 ± 2.31	2.63–11.87
Neutrophils (%)	MF	158	33.5 ± 10.8	11.9–55.10
Lymphocytes (%)	MF	160	62.6 ± 11.4	39.8–85.4
Monocytes (%)	MF	112	1.6 ± 1.9	0–5.4
Eosinophils (%)	MF	96	0.74 ± 0.93	0–2.6
Basophils (%)	MF	158	2.46 ± 1.99	0–6.4
Myeloblasts (%)	MF	24	0.08 ± 0.28	0–0.6
Normocytes (%)	MF	24	0.33 ± 0.56	0–1.5

[a]From ref. 1.
[b]M equals male; F equals female.

than in rats. When more specific answers are required, additional histochemical techniques could be employed (88, 113). Published color plates can be of benefit in morphological identification, also (96).

Published myeloid to erythroid ratios or normal rabbits are generally around 1:1 (34, 96). The results of Sabin *et al.* (91) reveal a significant departure in one-week-old rabbits with a M:E ratio of 0.19:1 which was back to 1.09:1 by four weeks of age. Table XXI is considered a representative normal of the differential marrow cellularity of the normal rabbit.

D. Sources of Variation in Hematological Values

1. Age

It is widely recognized that total leukocyte numbers in the newborn are quite low and gradually increase to maximum values by the end of the first half year (27, 69, 83, 95, 96). An ensuing gradual decrease in the lymphocytes occurs, concomitant with the involution of the thymus (96). The newborn also has a smaller number of erythrocytes, adult numbers being reached by three to four months (42, 69, 96).

A drop in erythrocyte numbers during the first week of life was reported by one group (91) and between 10 and 20 days by a more recent group of investigators (69). Another report cites a rise in numbers during the first week of life (96).

Earlier reports disagree concerning the effect of age on hemoglobin levels which likely reflects the range of minimum ages samples (42). A recent paper (69) records significant differences. Adult levels were present at birth, followed by a significant drop by 20 days, then a gradual

TABLE XXI

DIFFERENTIAL CELL DISTRIBUTION IN THE MARROW OF NORMAL RABBITS[a]

Cell type	Minimum (%)	Maximun (%)	Average (%)
Erythroid series			
Rubriblasts	0.2	0.8	0.2
Prorubricytes	0.2	2.0	0.6
Basophilic rubricytes	0.4	10.8	5.5
Polychromatophilic rubricytes	10.9	26.6	18.9
Metarubricytes	6.6	24.3	16.7
Total erythrocytic			41.9
Granulocytic series			
Myeloblasts	0.2	1.6	0.7
Progranulocytes	0.1	1.6	0.6
Myelocytes	1.1	1.6	0.6
Metamyelocytes	2.8	10.0	7.4
Band pseudoeosinophils	10.8	33.6	23.2
Segmenter pseudoeosinophils	2.0	9.0	5.3
Basophils	0.1	2.4	0.7
Eosinophils	0.2	2.4	1.4
Total granulocytic[b]			42.4
Other cells in the marrow			
Megakaryocytes	0.1	0.3	0.1
Lymphocytes	4.1	21.3	12.6
Monocytes	0.4	3.6	1.6
Plasma cells	0.1	1.2	0.2
RE nuclei	0.2	1.7	1.0
Hemocytoblasts	0.1	0.8	0.2

[a]From ref. 34.
[b]Myeloid to erythroid ratio equals 1.01:1.0.

rise to adult levels. A similar pattern follows for the hematocrit (42, 69) and is similar to that reported for the mouse and human (90, 108). Newborn levels are not reached in the adult (69).

The values for mean corpuscular hemoglobin and mean corpuscular volume are highest in the newborn, gradually reaching normal adult standards.

2. Sex

There are less clear-cut differences related to sex. Little variation in erythrocyte numbers between sexes was reported by Wintrobe *et al.* (109), though there was a suggestion of slightly higher hemoglobin levels in females. While some workers report higher erythrocyte and hemoglobin values for males, others have found no striking differences (16, 96). A recent report describes significantly higher hemoglobin and hematocrit values in adult males (69). However, the sex differences were not observed in all strains and were not significant from 0 to 120 days of age. The same workers reported no effect due to sex in total leukocyte numbers, but noted an age-sex interaction.

3. Breed and Strain

Breed differences in hematological parameters have been suggested but are generally accepted to be of little consequence (16, 19). A 1970 paper provides results for hemoglobin and hematocrit values in eleven inbred strains and one hybrid strain of adult New Zealand White rabbits in which significant effects due to strain, sex, and strain-sex interaction were observed (69).

4. Seasonal and Diurnal

Seasonal changes in erythrocyte, hemoglobin, and leukocyte values have been suggested by Gardner (42) who reported higher values present in the spring. However, laboratory bred and raised rabbits would not likely manifest such differences. Schermer (96) indicates a lack of agreement relative to rhythmic diurnal variations in leucocytes. Fox and Laird (38) reveal not only diurnal variation in total leukocytes, but also in all components of the differential count as well as hemoglobin. The hemoglobin effect was observed only in males. The highest levels were present at eight in the morning and the lowest in late afternoon and evening. The same situation prevailed for the erythrocyte numbers. A similar tendency was observed in the hematocrit, though not at significant levels. The significant diurnal effect on total leukocytes produced a high point at eight in the morning and a low between four and eight P.M. An inverse relationship was observed for lymphocytes and segmented neutrophils: The highest lymphocyte and lowest neutrophil values were recorded at four in the morning and the reverse at eight in the evening. Monocytes and basophils were the lowest in the early morning and highest about 12 hours later while eosinophils were highest in the early afternoon and lowest about four in the morning. The eosinophil effect was suggested by the authors to be correlated with the circulating adrenal glucocorticoid levels and the nocturnal nature of rabbits. Should this be correct, it is of interest that the lymphocytes, which are similarly responsive to adrenal glucocorticoids, reached their highest relative and total numbers when eosinophil numbers were the lowest and vice versa. Statistics suggested less likelihood of the lymphocyte values being related to chance than the values recorded for the eosinophils.

5. Nutritional Aspects

While digestive leukocytosis occurs in rabbits, it should be of no consequence when the common practice of supplying one day's food at a time to permit more or less continuous feeding is followed. Schermer reports that rabbits show inappreciable fluctuations in their hematological parameters as a result of nutrition or environment (96).

6. Trauma

Cold stress, even for short periods, was reported to result in increases in platelet and erythrocyte counts, packed cell volume, clotting time, total plasma proteins, and β-globulins. Decreases were observed in serum albumin and prothrombin time (101).

A recent report describes a method for repeated blood sampling of rabbits (65). Aside from the technique, investigators should review the literature relative to the effects of repeated sampling on hematologic parameters. The discussion in Schermer (96) and the references he cites reveal the great variations such procedures may produce as well as additional sources of variation.

E. Blood Volume

Blood volume of the domestic rabbit as measured by the T-1824 (Evans Blue) dye method is reported by various authors at $5.7 \pm .48$, 7.0, and 6.98 ± 0.91 ml/100 gm body weight (2, 7, 29). These differences are possibly related to the age of the animals as Table IV illustrates. Measured red cell volume, plasma volume, and blood volume all decrease significantly from birth to four months of age (young adult), with the greatest decreases noted in the plasma and blood volume.

REFERENCES

1. Abbott Laboratories. (1970). Unpublished data.
2. Aikawa, J. K. (1950). Fluid volumes and electrolyte concentrations in normal rabbits. *Amer. J. Physiol.* **162**, 695–702.
3. Albritton, E. C. (1952). "Standard Values in Blood." Saunders, Philadelphia, Pennsylvania.
4. Allen, R. C., and Watson, D. F. (1958). Paper electrophoretic analysis of rabbit serum as an aid in the selection of experimental rabbits. *Amer. J. Vet. Res.* **19**, 1001–1003.
5. Altman, P. L., and Dittmer, D. S. (1968). "Metabolism." Fed. Amer. Soc. Exp. Biol. Bethesda, Maryland.
6. Amann, R. P., and Lambiase, J. T., Jr. (1967). The male rabbit. I. Changes in semen characteristics and sperm output between puberty and one year of age. *J. Reprod. Fert.* **14**, 329–332.
7. Armin, J., Grant, R. T., Pels, H., and Reeve, E. B. (1952). The plasma, cell and blood volumes of albino rabbits as estimated by the dye (T 1824) and ^{32}P marked cell methods. *J. Physiol (London)* **116**, 59–73.
8. Austin, C. R. (1951). Observations on the penetration of the sperm into the mammalian egg. *Aust. J. Sci. Res., Ser. B* **4**, 581–569.
9. Beatty, R. A. (1960). Fertility of mixed semen from different rabbits. *J. Reprod. Fert.* **1**, 52–60.
10. Bito, L. Z., and Eakins, K. E. (1969). The effect of general anesthesia on the chemical composition of blood plasma of normal rabbits. *J. Pharmacol. Exp. Ther.* **169**, 277–286.
11. Boving, B. G. (1956). Rabbit blastocyst distribution. *Amer. J. Anat.* **98**, 403.
12. Boyd, E. M. (1942). Species variation in normal plasma lipids estimated by oxidative micromethods. *J. Biol. Chem.* **143**, 131–132.

13. Braden, A. W. H. (1953). Distribution of sperms in the genital tract of the female rabbit after coitus. *Aust. J. Biol. Sci.* **6**, 693–705.

14. Brown, I. W., Jr., and Eadie, G. S. (1953). An analytical study of *in vivo* survival of limited populations of animal red blood cells tagged with radioiron. *J. Gen. Physiol.* **36**, 327–343.

15. Burns, K. F., and deLannoy, C. W., Jr. (1966). Compendium of normal blood values of laboratory animals, with indication of variations. 1. Random-sexed populations of small animals. *Toxicol. Appl. Pharmacol.* **8**, 429–437.

16. Burroughs Welcome. (1971). Unpublished data.

17. Burwell, E. L., Brickley, B. A., and Finch, C. A. (1953). Erythrocyte life span in small animals. Comparison of two methods employing radioiron. *Amer. J. Physiol.* **172**, 718–724.

18. Bushnell, L. D., and Bangs, E. F. (1926). A study of the variation in number of blood cells of normal rabbits. *J. Infec. Dis.* **39**, 291–301.

19. Casey, A. E., Rosahn, P. D., Hu, C. K., and Pearce, L. (1934). Hereditary variations in the blood cytology of normal rabbits. *Science* **79**, 189–190.

20. Chang, M. C. (1948). Probability of normal development after transplantation of fertilized rabbit ova stored at different temperatures. *Proc. Soc. Exp. Biol. Med.* **68**, 680–683.

21. Chang, M. C. (1948). Transplantation of fertilized rabbit ova: The effect on viability of age, *in vitro* storage period, and storage temperature. *Nature (London)* **161**, 978–979.

22. Chang, M. C. (1950). Development and fate of transferred rabbit ova or balstocyst in relation to the ovulation time of recipients. *J. Exp. Zool.* **114**, 197.

23. Chang, M. C. (1951). Fertility and sterility as revealed in the study of fertilization and development of rabbit eggs. *Fert. Steril.* **2**, 205–222.

24. Chang, M. C. (1951). Fertilizing capacity of spermatozoa deposited into the Fallopian tube. *Nature (London)* **168**, 697–698.

25. Chang, M. C. (1955). Mammalian fertilization and the possibilities of its control. *Proc. Int. Conf. Planned Parenthood, 5th, 1955,* pp. 168–175.

26. Chang, M. C., and Adams, C. E. (1962). Fate of rabbit ova fertilized by hare spermatozoa. *Ist. Sper. Ital. Fecond. Artif.* **25**, 1–6.

27. Cheng, S. C. (1930). Leucocyte counts in rabbits; observations on influence of various physiological factors and pathological conditions. *Amer. J. Hyg.* **11**, 449–533.

28. Coldman, M. D., Gent, M., and Good, W. (1969). The osmotic fragility of mammalian erythrocytes in hypotonic solutions of sodium chloride. *Comp. Biochem. Physiol.* **31**, 605–609.

29. Courtice, F. C. (1943). The blood volume of normal animals. *J. Physiol. (London)* **102**, 290.

30. Crosfill, M. L., and Widdicombe, J. G. (1961). Physical characteristics of the chest and lungs and the work of breathing in different mammalian species. *J. Physiol. (London)* **158**, 1–14.

31. Cummins, L. M. Abbot Laboratories. (1972). Unpublished data.

32. Daniel, J. C., Jr. (1964). Early growth of rabbit trophoblast. *Amer. Natur.* **98**, 85–98.

33. Didisheim, P., Hattori, K., and Lewis, J. H. (1959). Hematologic and coagulation studies in various animal species. *J. Lab. Clin. Med.* **53**, 866–875.

34. Dikovinova, N. V. (1957). The absolute number of cells in bone marrow and myelograms of normal rabbits. *Byull. Eksp. Biol. Med.* **44**, 1129.

35. Dougherty, T. F., and White, A. (1944). Influence of hormones on lymphoid tissue structure and function. The role of the pituitary adrenotrophic hormone in the regulation of the lymphocytes and other cellular elements of the blood. *J. Endocrinol.* **35**, 1–14.

36. Eden, A. (1940). Coprophagy in the rabbit: Origin of "night" feces. *Nature (London)* **145**, 628–632.

37. Eveleth, D. F. (1937). Comparison of the distribution of magnesium in blood cells and plasma of animals. *J. Biol. Chem.* **119**, 289–292.

37a. Flatt, R. E., and Carpenter, A. B. (1971). Identification of crystalline material in urine of rabbits. *Amer. J. Vet. Res.* **32**, 655–658.

38. Fox, R. R., and Laird, C. W. (1970). Diurnal variations in rabbits: Hematological parameters. *Amer. J. Physiol.* **218**, 1609–1612.

39. Fox, R. R., Laird, C. W., Balw, E. M., Schultz, H. S., and Mitchell, B. P. (1970). Biochemical parameters of clinical significance in rabbits. 1. Strain variations. *J. Hered.* **61**, 261–265.

40. Fox, R. R., Schlager, G., and Laird, C. W. (1969). Blood pressure in thirteen strains of rabbits. *J. Hered.* **60**, 312–314.

41. Garbus, J., Highman, B., and Altland, P. D. (1967). Alterations in serum enzymes and isoenzymes in various species induced by epinephrine. *Comp. Biochem. Physiol.* **22**, 507–516.

42. Gardner, M. V. (1947). The blood picture of normal laboratory animals. A review of the literature 1936–1946. *J. Franklin Inst.* **243**, 498–502.

43. Greenwald, G. S. (1958). Endocrine regulation of the secretion of mucin in the tubal epithelium of the rabbit. *Anat. Rec.* **130**, 477–496.

44. Greenwald, G. S. (1961). A study of the transport of ova through the rabbit oviduct. *Fert. Steril.* **12**, 80–95.

45. Greenwald, G. S. (1962). The role of the mucin layer in development of the rabbit blastocyst. *Anat. Rec.* **142**, 407–415.

46. Grégoire, A. T., Gongsakdi, D., and Rakoff, A. E. (1961). The free amino acid content of the female rabbit genital tract. *Fert. Steril.* **12**, 322–327.

47. Gregory, P. W. (1930). The early embryology of the rabbit. *Contrib. Embryol. Carnegie Inst.* **21**, 141.

48. Guyton, A. C. (1947). Measurement of the respiratory volumes of laboratory animals. *Amer. J. Physiol.* **150**, 70–77.

49. Hafez, E. S. E. (1960). Sex drive in rabbits. *Southwest. Vet.* **14**, 46–49.

50. Hafez, E. S. E. (1963). The uterotubal junction and the luminal fluid of the uterine tube in the rabbit. *Anat. Rec.* **145**, 7–12.

51. Hafez, E. S. E. (1970). Rabbits. *In* "Reproduction and Breeding Techniques for Laboratory Animals" (E. S. H. Hafez, ed.), p. 274. Lea & Febiger, Philadelphia, Pennsylvania.

52. Hamilton, W. F., Woodbury, R. A., and Woods, E. B. (1937). The relation between systemic and pulmonary blood pressures in the fetus. *Amer. J. Physiol.* **119**, 206–212.

53. Hammond, J. (1934). The fertilization of rabbit ova in relation to time. A method of controlling the litter size, the duration of pregnancy and the weight of the young at birth. *J. Exp. Biol.* **11**, 140.

54. Hammond, J., and Asdell, S. A. (1926). The vitality of the spermatozoa in the male and female reproductive tract. *J. Exp. Biol.* **4**, 155–185.

55. Hamner, C. E., and Williams, W. L. (1965). Composition of rabbit oviduct secretions. *Fert. Steril.* **16**, 170–176.

56. Harper, M. J. K. (1961). The time of ovulation in the rabbit following the injection of luteinizing hormone. *J. Endocrinol.* **22**, 147–152.

57. Harper, M. J. K. (1965). Transport of eggs in cumulus through the ampulla of the rabbit oviduct in relation to day of pseudopregnancy. *Endocrinology* **77**, 114–123.

58. Hawkins, J. A. (1924). The acid-base equilibrium of the blood of normal guinea pigs, rabbits and rats. *J. Biol. Chem.* **61**, 147–155.

59. Heilbrunn, L. V. (1952). "An Outline of General Physiology." Saunders, Philadelphia, Pennsylvania.

60. Himwich, W., and Himwich, H. E. (1970). Cerebral circulation, blood-brain barrier and cerebrospinal fluid. *In* "Dukes' Physiology of Domestic Animals" (M. J. Swenson, ed.), 8th ed., pp. 258–289. Cornell Univ. Press, Ithaca, New York.

61. Holmdahl, T. H., and Mastroianni, L., Jr. (1965). Continuous collection of rabbit oviduct secretions at low temperature. *Fert. Steril.* **16**, 587–595.

62. Houpt, T. R. (1970). Water, electrolytes and acid-base balance. *In* "Dukes' Physiology of Domestic Animals" (M. J. Swenson, ed.), 8th ed., 745–766. Cornell Univ. Press, Ithaca, New York.

63. Hudgins, P. C., Cummings, M. M., and Patnode, R. A. (1956). Electrophoretic distribution of serum proteins in rabbit, guinea pig and rat following BCG administration. *Proc. Soc. Exp. Biol. Med.* **92**, 75–77.

64. Hulse, E. V. (1964). Quantitative cell counts of the bone marrow and blood and their secular variations in the normal adult rat. *Acta Haematol.* **31**, 50–63.

65. Jacobs, P., and Adriaenssens, L. (1970). A simple method for repeated blood sampling in small animals. *J. Lab. Clin. Med.* **75**, 1013–1016.

66. Klaassen, C. D., and Plaa, G. L. (1967). Species variation in metabolism, storage, and excretion of sulfobromophthalein. *Amer. J. Physiol.* **213**, 1322–1326.

67. Klaassen, C. D., and Plaa, G. L. (1969). Plasma disappearance and biliary excretion and indocyanine green in rats, rabbits and dogs. *Toxicol. Appl. Pharmacol.* **15**, 374–384.

68. Kozma, C. K., Pelas, A., and Salvador, R. A. (1967). Electrophoretic determination of serum proteins of laboratory animals. *J. Amer. Vet. Med. Ass.* **151**, 865–869.

69. Laird, C. W., Fox, R. R., Mitchell, B. P., Blau, E. M., and Schultz, H. S. (1970). Effect of strain and age on some hematological parameters in the rabbit. *Amer. J. Physiol.* **218**, 1613–1617.

70. Leone, E., Libonati, M., and Lutwak-Mann, C. (1963). Enzymes in the uterine and cervical fluid and in certain related tissues and body fluids of the rabbit. *J. Endocrinol.* **25**, 551–552.

71. Lepschkin, E., and Wilson, F. N. (1951). "Modern Electrocardiography." Williams & Wilkins, Baltimore, Maryland.

72. Lewis, J. H., and Didisheim, P. (1957). Differential diagnosis and treatment in hemorrhagic disease. *AMA Arch. Intern. Med.* **100**, 157–168.

73. Lewis, J. H., Ferguson, J. H., Fresh, J. W., and Zucker, M. B. (1957). Primary hemorrhagic diseases. *J. Lab. Clin. Med.* **49**, 211–232.

74. Lewis, W. H., and Gregory, P. W. (1929). Cinematographs of living developing rabbit-eggs. *Science* **69**, 226–229.

75. Lutwak-Mann, C. (1962). Glucose, lactic acid and bicarbonate in rabbit blastocyst fluid. *Nature (London)* **193**, 653–654.

76. Lutwak-Mann, C. (1962). Some properties of uterine and cervical fluid in the rabbit. *Biochim. Biophys. Acta* **58**, 637–639.

77. Mastroianni, L., Jr., Beer, F., Shah, U., and Clewe, T. H. (1961). Endocrine regulation of oviduct secretions in the rabbit. *Endocrinology* **68**, 92–100.

78. Mastroianni, L. Jr., and Wallach, R. C. (1961). Effect of ovulation and early gestation on oviduct secretions in the rabbit. *Amer. J. Physiol.* **200**, 815–818.

79. Mikhail, G., Noall, M. W., and Allen, W. M. (1961). Progesterone levels in the rabbit ovarian vein blood throughout pregnancy. *Endocrinology* **69**, 504–509.

80. Mott, J. C. (1965). Haemorrhage as a test of the function of the cardiovascular system in rabbits of different ages. *J. Physiol. (London)* **181**, 728–752.

81. Nachtsheim, H. (1950). Pelger-anomaly in man and rabbit: A mendelian character of the nuclei of the leucocytes. *J. Hered.* **41**, 131–137.

82. Neuberger, A., and Niven, J. S. F. (1951). Haemoglobin formation in rabbits. *J. Physiol. (London)* **112**, 292–310.

83. Pearce, L. (1948). Hereditary osteopetrosis of the rabbit. II. X-Ray, hematologic and chemical observations. *J. Exp. Med.* **88**, 597–620.

84. Pearce, L., and Casey, A. E. (1930). Studies in the blood cytology of the rabbit. I. Blood counts in normal rabbits. *J. Exp. Med.* **51**, 83–97.

85. Perk, K., Frei, Y. F., and Herz, A. (1964). Osmotic fragility of red blood cells of young and mature domestic and laboratory animals. *Amer. J. Vet. Res.* **25**, 1241–1248.

86. Pincus, G. (1936). "The Eggs of Mammals." Macmillan, New York.

87. Pintor, P. P., and Grassini, V. (1957). Individual and seasonal spontaneous variations of haematological values in normal male rabbits; statistical survey. *Acta Haematol.* **17**, 122–128.

88. Ramsell, T. G., and Yoffey, J. M. (1961). The bone marrow of the adult male rat. *Acta Anat.* **47**, 55–65.

89. Rodbard, S. (1940). Direct arterial pressure in the unanesthetized rabbit. *Amer. J. Physiol.* **129**, 448 (abstr.).

90. Russell, E. S., and Bernstein, S. E. (1966). Blood and blood formation. *In* "Biology of the Laboratory Mouse" (E. L. Green, ed.), 2nd ed. McGraw-Hill, New York.

91. Sabin, F. R., Miller, F. R., Smithburn, K. C., Thomas, R. M., and Hummel, L. E. (1936). Changes in the bone marrow and blood cells of developing rabbits. *J. Exp. Med.* **64**, 97–120.

92. Sawin, P. B., Denenberg, V. H., Ross, S., Hafter, E., and Zarrow, M. X. (1960). Maternal behavior in the rabbit: Hair loosening during gestation. *Amer. J. Physiol.* **198**, 1099–1102.

93. Scaramuzzi, R. J., Blake, C. A., Papkoff, H., Hilliard, J., and Sawyer, C. H. (1972). Radioimmunoassay of rabbit luteinizing hormone; serum levels during various reproductive states. *Endocrinology* **90**, 1285–1291.

94. Scarborough, R. A. (1931). The blood picture of normal laboratory animals. *Yale. J. Biol. Med.* **3**, 64–80.

95. Schalm, O. W. (1965). "Veterinary Hematology," 2nd ed. Lea & Febiger, Philadelphia, Pennsylvania.

96. Schermer, S. (1967). "The Blood Morphology of Laboratory Animals," 3rd ed., Davis, Philadelphia, Pennsylvania.

97. Schmidt-Nielsen, K., and Pennycuik, P. (1961). Capillary density in mammals in relation to body size and oxygen consumption. *Amer. J. Physiol.* **200**, 746–750.

98. Scott, J. M., and Simon, C. E. (1924). Experimental measles, 1. Thermic and leucocytic response of the rabbit to inoculation with the virus of measles, and their value as a criteria of infection. *Amer. J. Hyg.* **4**, 559–604.

99. Shtacher, G. (1969). Selective renal involvement in the early development of hypercalcemia and hypophosphatemia in VX-2 carcinoma bearing rabbits. Studies on serum and tissues alkaline phosphatase and renal handling of phosphorus. *Cancer Res.* **29**, 1512–1518.

100. Smith, R. E. (1956). Quantitative relations between liver mitochondria metabolism and total body weight in mammals. *Ann. N. Y. Acad. Sci.* **62**, 405–421.

101. Sutherland, G. B., Trapani, I. L., and Campbell, D. H. (1958). Cold adapted animals. II. Changes in the circulating plasma proteins and formed elements of rabbit blood under various degrees of cold stress. *J. Appl. Physiol.* **12**, 367–372.

102. Swenson, M. J. ed. (1970). "Duke's Physiology of Domestic Animals," 8th ed. Cornell Univ. Press, Ithaca, New York.

103. Thacker, E. J., and Brandt, C. S. (1955). Coprophagy in the rabbit. *J. Nutr.* **55**, 375–385.

104. Venge, O. (1950). Studies of the maternal influence on the birth weight in rabbits. *Acta. Zool. (Stockholm)* **31**, 1–148.

105. Vishwakarma, P. (1962). The pH and bicarbonate-ion content of the oviduct and uterine fluids. *Fert. Steril.* **13**, 481–485.

106. Walton, A., and Hammond, J. (1928). Observations on ovulation in the rabbit. *Brit. J. Exp. Biol.* **6**, 190.

107. Westerman, M. P., Wiggans, R. G., and Mao, R. (1970). Anemia and hypercholesterolemia in cholesterol-fed rabbits. *J. Lab. Clin. Med.* **75**, 893–902.

108. Wintrobe, M. M. (1961). "Clinical Hematology," 5th ed. Lea & Febiger, Philadelphia, Pennsylvania.

109. Wintrobe, M. M., Shumacker, H. B., Jr., and Schmidt, W. J. (1936). Values for number, size, and hemoglobin content of erythrocytes in normal dogs, rabbits, and rats. *Amer. J. Physiol.* **114**, 502–507.

110. Witts, L. J., and Webb, R. A. (1927). Monocytes of rabbit in B. monocytogenes infection; study of their staining reactions and histogenesis. *J. Pathol. Bacteriol.* **30**, 687–712.

111. Wostmann, B. S. (1961). Recent studies on the serum proteins of germ-free animals. *Ann. N. Y. Acad. Sci.* **94**, 272–293.

112. Wright, C. I. (1934). The respiratory effects of morphine, codeine and related substances. I. The effect of codeine, isocodeine, allo-pseudocodeine and pseudocodeine on the respiration of the rabbit. *J. Pharmacol. Exp. Ther.* **51**, 327–342.

113. Yam, L. T., Li, C. Y., and Crosby, W. H. (1971). Cytochemical identification of monocytes and granulocytes. *Amer. J. Clin. Pathol.* **55**, 283–290.

114. Yoffey, J. M., and Parnell, J. (1944). The lymphocyte content of rabbit bone marrow. *J. Anat.* **78**, 109–112.

115. Yoshida, T., Pleasants, J. R., Reddy, B. S., and Wostmann, B. S. (1968). Efficiency of digestion in germ-free and conventional rabbits. *Brit. J. Nutr.* **22**, 723–727.

116. Young, I. M. (1952). CO_2 tension across the placental barrier and acid-base relationship between fetus and mother in the rabbit. *Amer. J. Physiol.* **170**, 434–441.

117. Zamboni, L., Hongsanand, C., and Mastroianni, L., Jr. (1965). Influence of tubal secretions on rabbit tubal ova. *Fert. Steril.* **16**, 177–184.

118. Zarrow, M. X., Denenberg, V. H., and Anderson, C. O. (1965). Rabbit: Frequency of suckling in the pup. *Science* **150**, 1835–1836.

CHAPTER 4

Basic Biomethodology

W. Sheldon Bivin and Edward H. Timmons

I. INTRODUCTION

Basic biomethodologic techniques are those that permit one to monitor and record experimental data. All parts of a biological experiment must be dependable if the results are going to be significant. For the rabbit to be a useful research tool, it must be a specifically defined animal with known physiological parameters. This is accomplished by using an animal that is genetically defined (Chapter 1), housed in an acceptable environment, and maintained by means of rigidly husbandry controlled standards (Chapter 2).

This chapter describes biological methods for the investigator who is about to start his first rabbit experiment and progresses to more specialized research techniques which may be useful to the experienced laboratory worker.

II. HANDLING AND RESTRAINT

The importance of proper animal handling and restraint requires careful consideration. Improper techniques can result in human and animal injuries and nonvalid experimental results. In many cases, a frightened, mishandled or environmentally stressed rabbit will not present the same values for parameters that are being measured as will a properly handled animal.

A. Physical Methods

Rabbits should always be handled firmly but gently (81). They are prone to kick their hind legs and can inflict the handler with painful scratches if proper precautionary measures are not taken (41). A rabbit should be picked up by obtaining a firm grip on the loose skin over the scruff of the neck with one hand, and use the other hand to support the animal's hindquarters and control back leg movement (Fig. 1). If head control is desirable, the ears can be laid posteriorly and held with the same hand that is used for holding the neck scruff (Fig. 2). It is not advisable to pick up rabbits by the loin area because this often causes violent thrashing and subsequent back injuries.

B. Mechanical Methods

There are many rabbit-restraining devices that are in common usage. The two most common devices are the rabbit-restraining board (Fig. 3) and variants of a restraint box (Fig. 4). The rabbit board is used primarily for restraint during cardiac punctures or other manipulations involving procedures on the ventral surface of the animal. Care must be taken to prevent exciting the rabbit when it is being

Fig. 1. Illustrates proper restraint utilizing the scruff of the neck and support of the hindquarters.

placed on the board. The rear legs should be secured firmly enough to prevent excessive animal movement, but not so tight as to cause discomfort or back injuries. The head can be completely immobilized by wrapping the board with a cloth bandage (74). Restraint boxes are used primarily for manipulations about the ear or head. Many variations in box designs are available, but all types serve for rabbit body control while the ears are exposed.

Special animal holders have been described which will restrain rabbits and allow for eye and ear manipulations (8), intraperitoneal injections (3), and for the administration of intravenous fluids over prolonged periods (50).

III. SAMPLING TECHNIQUES

A. Blood Sampling

Blood sampling is commonly done by using the large, readily accessible arteries and veins of the rabbit's ears. The animal is placed in a restraining box, petroleum jelly is lightly applied medial to the site of the venepuncture. A 23- or 25-gauge hypodermic needle is inserted into the

Fig. 2. Illustrates proper head restraint. Grasp the ears in conjunction with the scruff of the neck.

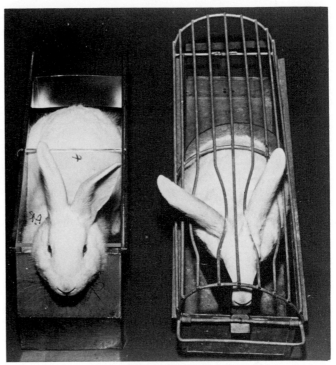

Fig. 4. Two variations of a rabbit-restraining box.

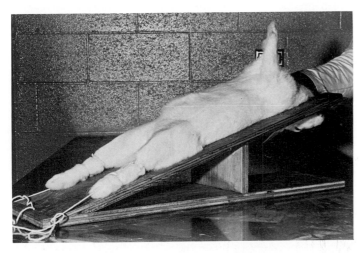

Fig. 3. A rabbit-restraining board.

auricular vein and removed. Blood for hematological examination can then be collected by capillary tubes or placed directly onto microscope slides. After the sample is collected, blood flow from the puncture wound is stopped by applying pressure directly to the wound. If white blood cell counts are not being done, the ear can be dampened with xylol which causes an inflammatory response resulting in blood vessel engorgement and dilatation. The ear should be wiped with alcohol and water after the completion of the bleeding.

The auricular artery may be used when larger samples are required. The needle is inserted into the artery with the tip directed toward the base of the ear. Blood can be collected into a syringe or by the use of vacuum devices (27).

Hoppe *et al.* (34) have described a simple technique for bleeding rabbit ear veins which allows for repeated sampling of large quantities (30 to 50 ml) of blood. This method utilizes a modified 500-ml boiling flask with a vacuum reservoir. The marginal ear vein is punctured and the ear is placed in the neck of the flask and blood collection is facilitated by the vacuum. This technique can be used on rabbits of various ages with only minor modifications to the collecting equipment.

B. Cardiac Puncture

The cardiac puncture is sometimes used for exsanguination. The rabbit should be anesthetized and placed on a rabbit board. If blood is to be collected under aseptic conditions, proper animal preparation, equipment sterility, and operative techniques must be employed. The cardiac impulse is palpated for specific location and then blood is collected with an 18-gauge $1\frac{1}{2}$-inch needle (39). One approach that can be used is to insert the needle at approximately a 30° angle immediately posterior to the xyphoid cartilage. It is generally advantageous for the inexperienced person to examine the thoracic cavity of a carcass and become familiar with heart size, location, and specific body reference points before attempting this procedure.

C. Urine Collection

Urine collection over long periods of time is done most satisfactorily utilizing a metabolism cage. This is a cage that has a collection device incorporated into the drop pan. An investigator must consider the effects that the pan metal and other debris will have on urine parameters that are being measured.

Urine can also be collected by catheterization. A male rabbit should be used if possible when this is the preferred collection method. The male should be supported in a sitting position and a lubricated, sterile 9 French flexible catheter should be passed into the urethra for about 2 cm. The initial insertion should be in a downward direction followed by a manual depression of the catheter and penis and then the catheter tip is eased into the bladder. Firm abdominal pressure with a cupped hand will help empty the bladder after the catheter is inserted (40).

Manual expression of urine can sometimes be done satisfactorily, but this is more successful on an anesthetized rabbit. If distended, the bladder can be palpated and pressure applied in a firm steady manner. Excessive pressure may cause bladder rupture, especially if there is urinary tract blockage.

IV. METHODS OF COMPOUND ADMINISTRATION

A. Oral

The simplest method of administering compounds orally to a rabbit is by incorporating the material into the feed or water. If it is important that precise amounts of substance be given this method may be unsatisfactory because of consumption variables, spillage, and wastage. For a rabbit to volumtarily consume materials, the substance must not have a disagreeable odor or taste, and it must be in a form that can be consumed.

Small volumes of liquid materials can be administered by placing the tip of a syringe in the corner of the rabbit's mouth and slowing introducing the material. If the material is tasteless or injected too rapidly, there is the possibility of causing inhalation pneumonia.

The most accurate means of insuring oral compound administration is by placing the material directly into the stomach. To accomplish this, the animal should be restrained and a speculum used to control mouth movements.

This can be as simple as a tongue depressor stick with a hole drilled in the center. Place the depressor stick on edge immediately behind the rabbit's incisor teeth with the hole lined up with the esphageal opening. Another method, using a plexiglass speculum, has been described that does not cause any mouth irritation and can be safely used without a restraining box (16). An 8 French infant feeding tube is passed through the speculum toward the oral pharynx until the gag reflex is elicited. At this time the catheter is introduced into the esophagus and passed on to the stomach. The investigator must then determine that the catheter is in the stomach and not in the bronchial tree. This can be done by confirming that there is no air passage in the tube corresponding to respiratory movements. A syringe is then attached to the free end of the tube and the liquid compound is administered. It is advisable to flush the tube with water to assure complete material delivery and then withdraw the tube while the empty syringe is still attached in order to trap the remaining water in the catheter and prevent possible inhalation.

A small animal balling gun can be made that will permit the administration of materials in capsules or pill form. This device consists of a polypropylene tube and a soft aluminum wire insert to act as a plunger. A loop should be formed in the wire to prevent overinsertion. This complete unit is placed into the mouth and when the gag reflex occurs, the rod is simultaneously advanced, pushing the capsule from the tube into the oral pharynx. The balling gun is removed and the rabbit's mouth is held closed until swallowing occurs (61).

B. Intravenous

Although the cephalic and recurrent tarsal veins are used, the ear veins, as discussed under Section III,A, are the veins of choice for intravenous injections. The ear should be cleaned and prepared for an aseptic injection. Digital pressure should then be applied at the base of the ear to distend the vein. Details of the actual needle insertion vary greatly among investigators. The following description is a guide for the individual who is inexperienced in administering intravenous injections.

Do not attempt to insert the needle directly into the vein from above. Pressure that is applied in order for the needle to penetrate the skin will flatten the vein making entry difficult, if not impossible. The needle should penetrate the skin just beside and parallel to the blood vessel. The needle point is then in subcutaneous tissue beside a still distended vein. The needle is then inserted into the vein with the beveled edge up and is threaded into the vein to assure that the injected material goes intravenously and not subcutaneously (35). Precautions must be taken to prevent the injection of air bubbles into rabbit veins, or a subsequent fatality may occur due to air emboli.

C. Intramuscular

Intramuscular injections should be made into the bodies of large muscles. Care must be taken to avoid hitting large blood vessels, nerves, and bone. After the needle is inserted, the syringe plunger should be withdrawn lightly to confirm that no blood can be aspirated. If blood does appear, the position of the needle tip should be changed slightly. Two good locations for intramuscular injection are the gluteal muscles and the large thigh muscles. The thigh muscles should be held firmly with one hand while the other hand is used to make the injection.

D. Intraperitoneal

Intraperitoneal injections should preferably be made when the rabbit's stomach and bladder are empty. If an investigator suspects an enlarged liver, kidney, or spleen, special care must be taken to prevent damaging these organs on injection. Accidental injection into the intestinal tract rarely occurs if precautions are taken to prevent injection into the cecum. To make the injection, the rabbit should be held with the hindquarters elevated and the needle inserted just lateral to the midline and just posterior to the area of the umbilicus. A one-inch needle is used and inserted toward the spine.

E. Intradermal

Intradermal injections are at times required for diagnostic and test procedures. The dermal injection sites are commonly located in the loin and flank areas in a position that is difficult for the rabbit to reach with his hind feet. The selected area should be clipped and remaining hair removed with a depilatory cream. The area is then prepared with an appropriate antiseptic sterilized and injections are made using a 25-gauge needle and tuberculin syringe. The skin is stretched and care is taken so that the needle is inserted only into the dermis. A small bleb is formed on injection. When the needle is removed, the investigator should look for seepage of the test material. If this occurs, the shaft of the needle should be inserted further into the skin during injections.

F. Other Routes

Research protocol often dictates that compounds be administered by routes other than the ones discussed. In specialized situations an investigator may want to consider topical applications, rectal injections, conjunctival sac instillation, and compound insertion into other body orifices.

G. Special Techniques in the Neonate

Complicating factors in administering compounds to a newborn rabbit include the small size of the animal and delicacy of the skin, membranes, and blood vessels. Intravenous administration can be done using a $\frac{5}{8}$ inch 30-gauge needle inserted into the external jugular vein. The maximum suggested inoculum is 0.001 ml/gm body weight. Intraperitoneal injections can be done by using a $\frac{5}{8}$ inch 27-gauge needle inserted into the lower left quadrant of the abdomen and 3 mm from the midline. The injected volume should not exceed 0.01 ml/gm body weight.

Subcutaneous injections can be given under the loose skin over the dorsal aspect of the neck. A 27-gauge needle is satisfactory and the suggested maximum volume is 0.01 ml/mg body weight. Gastric intubation can be done in a day-old rabbit by using an 18-gauge oral feeding needle. Care should be exercised to prevent traumatic stomach injuries (24).

V. ANESTHESIOLOGY

A. Principles of Anesthesia

Anesthesia is required for most types of experimental or therapeutic surgical procedures. However, rabbits are probably one of the most difficult and unpredictable species of laboratory animals to anesthetize. Three major factors are involved: (a) The respiratory center of the rabbit is very sensitive to the paralyzing action of anesthetics. (b) The range between anesthetic and lethal doses is extremely narrow. (c) Inter-rabbit variability to the depressant action of conventional anesthetics is so great that the doses for surgical anesthesia virtually have to be individualized for different animals (12, 23, 46, 60).

Because there is variability between individual rabbits within a breed, several criteria should be followed for in the selection of a particular animal for surgery. Long-eared rabbits are usually preferred because of the accessibility of the large marginal ear veins for injection. Although the New Zealand White breed is probably the most widely used, mixed breeds are reported to give the most satisfactory results (21). A major factor for consideration is the health of the rabbit. It is important that clinical signs of diseases (Chapters 11–20, this volume) be recognized. Clinical signs indicating respiratory tract disease, dehydration, and other metabolic disorders result in a poor anesthetic risk.

After a healthy rabbit has been selected for a particular surgical procedure, a decision must be made as to the plane of anesthesia required. The stages of anesthesia are well described in the literature (88). In order to evaluate the plane of anesthesia, the following clinical signs should be continuously monitored: (a) rate and depth of respiration;

(b) color of the oral mucosa; (c) corneal reflex (although this may be difficult at times as the rabbit usually closes his eyes during anesthesia); (d) paw or pedal reflex (determined by pinching the web of the paw to create a pain stimulus); and (e) posterior stretch reflex. These five check points can be successfully used to determine the depth of anesthesia in any individual rabbit (23). It is imperative that any anesthetic agent used be administered to produce a desired plane of anesthesia and that one never attempts to administer a precalculated dose without very close animal observation.

B. Preanesthetic Agents and Procedures

It is the purpose of a preanesthetic to produce calmness, analgesia to some degree, ataxia, lack of righting reflex, and a cataleptoid state. With an animal in this state, there is less apprehension to the induction of general anesthesia, and in some cases these preanesthetics tend to prolong anesthesia. In addition, their use may reduce the dosage required of the general anesthetic by as much as 50%. A number of preanesthetic agents have been tested and a summary of their use follows.

1. Chlorpromazine Hydrochloride

Chlorpromazine hydrochloride has been utilized as an effective preanesthetic agent in rabbits. It is administered in intramuscular doses ranging from 25 to 100 mg/kg of body weight, although the most widely used dosage is 25 mg/kg of body weight. This agent is usually administered approximately thirty minutes prior to the use of the selected general anesthetic and, in the case of pentobarbital, it will triple or quadruple the usual duration of surgical anesthesia. It is important to remember that the dosage of pentobarbital is reduced about 50% when using chlorpromazine as a preanesthetic. Chlorpromazine has the added advantage of stimulating respiration, thus partially overcoming the respiratory depression caused by barbiturates (17).

2. Diazepam and Propiopromazine

Diazepam and propiopromazine are classified as psychosedative drugs (60). They may be used singly or in combination to produce a safe level of preanesthesia. Diazepam is usually administered in a dosage of 5–10 mg/kg body weight and propiopromazine in a dosage of 5 mg/kg body weight. The usual route of administration is by intramuscular injection into the dorsal aspect of the thigh. There seems to be less irritation and pain in movement associated with this location rather than using the caudal aspect of the thigh (32).

3. Paraldehyde

Paraldehyde is another agent which has been used as a preanesthetic agent and, at times, as a general anesthetic. Its recommended dosage when used as a preanesthetic is 1 ml/kg of body weight to be given by stomach tube (38). It can be injected into the muscle; however, it is essential that a deep intramuscular injection be used to prevent serious irritation and skin sloughing (55). When combined with other preanesthetic agents such as diazepam or propiopromazine, its dosage should be reduced to 0.3 ml/kg body weight (60).

4. Fentanyl and Droperidol

Fentanyl and Droperidol* have been used in combination to produce sedation. A dosage of 0.18 ml/kg body weight or less will produce tranquilization and may be useful in conjunction with barbiturate or inhalant anesthetics (78).

5. Phencyclidine and Ketamine HCl

Phencyclidine† and CI-634‡ have been shown to produce catalepsy in the rabbit. These two preanesthetics are incapable of producing sufficient analgesia for major surgical operations; however, they do enhance analgesia and anesthesia in rabbits given other hypnotic anesthetic agents. Although they have similar actions, CI-634 appears to be a shorter acting drug than phencyclidine, and it produces a greater central depression and relaxation of the skeletal musculature. The recommended dosage for CI-634 varies from 5 to 50 mg/kg body weight, with a usual dosage of 20 mg/kg. It is administered intramuscularly or intravenously with apparently the same effect, that is, it will produce immobilization within two to five minutes. For best results it is recommended that this drug be combined with chloral hydrate (250 mg/kg) to produce anesthesia. Efforts to combine it with thiamylal sodium or paraldehyde have not been promising (12).

C. Intravenous Anesthesia

For agents which are intended to reach the central nervous system and produce narcosis or anesthesia, the intravenous route is obviously more direct than the one through the respiratory tract. However, care must be exercised in using this route because there is no rapid means

---

*Innovar-Vet, McNeil Laboratories, Fort Washington, Pennsylvania.
†Sernylan, Parke, Davis and Company, Detroit, Michigan.
‡Chemistry Department, Parke, Davis and Company, Ann Arbor, Michigan.

of detoxifying or excreting the anesthetic agent once it has been given.

1. Pentobarbital Sodium

Pentobarbital sodium is perhaps the most widely used barbiturate (58). The dosage is generally 20–40 mg/kg body weight and should be administered in the following manner: (a) Gently place the rabbit in a restraint box. (b) Slowly inject one-half to three-quarters of the calculated dose of a 2% pentobarbital solution intravenously over a two-minute period. (c) Keep the needle in the vein as you remove the rabbit from the restraint box and lay it on its side. (d) Slowly continue the injection until the desired plane of anesthesia is reached. (e) Try to avoid giving a second dose of pentobarbital to prolong anesthesia. The normal duration of anesthesia is 30–45 minutes. However, if the animal begins to awaken, it is best not to give more barbiturate but rather to supplement with an inhalation anesthetic (60).

Even though this method has been used successfully, many investigators have been troubled with excessive death losses when using this agent (14, 17, 21, 23, 53). Because of the extremely variable dose range, the abundant fat deposits characteristic of rabbits, and the narrow margin between an adequate surgical dose and lethal suppression of the respiratory center, it appears that pentobarbital is a high risk surgical anesthetic for the rabbit (23). However, if a barbiturate is desired, a drug similar to pentobarbital in action is available which appears to have a less depressing effect (73).

2. Phenobarbital

Phenobarbital is one of the long-acting barbiturates and is capable of anesthetizing a rabbit for 24 hours or longer. It is usually injected intraperitoneally at a dosage of 100 mg/kg body weight. Sometimes a combination of phenobarbital and pentobarbital gives more satisfactory results than either alone. It is important to remember that all of the long-acting barbiturates are slowly excreted, unchanged by the kidney glomeruli, and therefore should not be used on rabbits with renal damage (60).

3. Urethane

Urethane is another widely used intravenous anesthetic. This agent has a marked advantage over pentobarbital because it has a wider margin of safety and will provide stable anesthesia for a longer period of time, i.e., at least five to six hours (21). Urethane, in ordinary anesthetic doses of 1.5 gm/kg body weight, has been shown to have an adverse affect on blood and blood vessels in rabbits. So, even though urethane will produce a relatively safe extended period of

surgical anesthesia, one should be aware of the acute pathological changes which may accompany the use of this drug (11, 76).

4. Chloral Hydrate

Chloral hydrate is a drug which has been used alone and in conjunction with several other agents such as paraldehyde, barbiturates and the phencyclidine derivatives. At the present time it would appear that the most acceptable combination is the use of CI-634 (20 mg/kg body weight) and chloral hydrate (250 mg/kg body weight) given intravenously. The outstanding feature of this combination is the consistency in duration of anesthesia for different rabbits. Most rabbits recover within four hours and there appear to be no side effects (12).

5. Magnesium Sulfate

Magnesium sulfate can be administered intravenously, intramuscularly or subcutaneously to produce anesthesia in a dose of 5 ml/kg body weight of a 20% solution. The duration is usually three-quarters to one and one-half hours, but can vary widely. If an overdosage is apparent, intravenous calcium gluconate or a combination of neostigmine and pentylenetetrazol will effectively counteract the anesthetic effect (55).

6. "Equi-Thesin"

Equi-Thesin* is a combination of chloral hydrate, magnesium sulfate, pentobarbital sodium propylene glycol, and alcohol which has been shown to give satisfactory surgical anesthesia. It is usually administered following the use of a preanesthetic such as propiopromazine hydrochloride and/or paraldehyde. "Equi-Thesin" is given intravenously at a dosage of 1–3 ml/kg body weight, but it should always be given to effect. Dosage will depend to a great extent on how well the preanesthetic agents have worked (32).

D. Inhalation Anesthesia

The wide distribution of inhalation anesthetics throughout the entire body exerts a controlling influence on the rate of uptake or elimination of the anesthetic by the brain. Since most gaseous and volatile anesthetic agents behave as inert gases, the anesthetist should have a thorough understanding of the processes involved in the exchange of inert gases in the body (88). Because controlled intravenous surgical anesthesia in the rabbit is difficult to attain, inhalation anesthetics are regaining favor. This is especially true

---

*Equi-Thesin, Jensen-Salsbery Laboratories, Kansas City, Missouri.

since the development of nonexplosive agents such as Methoxyflurane* and Halothane.†

1. Methods

Before discussing these and other agents, an understanding of the different methods of inhalation anesthesia is needed. There are four basic methods: (1) the open method, (2) the semi-open method, (3) the closed method with carbon dioxide absorption, and (4) the semi-closed method.

a. THE OPEN METHOD. This is by far the simplest and is used to volatilize agents such as chloroform and ether. Gauze or cotton is held over the nostrils and a few drops of the agent are dropped onto absorbent material. The essential feature of this method is that there is a free flow of air between the mask and the face at all times, and the room must be well ventilated.

b. THE SEMI-OPEN METHOD. This method differs somewhat from the open method in that all of the inspired air is forced through a mask on which the vaporization agent occurs. Today, the open and semi-open methods are commonly used, but with them it is difficult to maintain a stable plane of anesthesia and the newer anesthetic agents are very expensive to use this way (88).

c. THE CLOSED METHOD. This method used with carbon dioxide absorption requires expensive equipment and a thorough knowledge of its use. However, with the use of this specialized equipment, surgical anesthesia can become a very satisfying technique. The basic principle involved is that exhaled anesthetic gases remain unchanged. Therefore, if carbon dioxide is removed and sufficient oxygen is added to fulfill metabolic requirements, the same anesthetic gasses can be rebreathed to maintain anesthesia (88). Although there are numerous machines available commercially, some investigators have attempted to make their own equipment (85). If a large number of surgical procedures are planned, the added expense of acquiring this equipment can be easily outweighed by the added ease, convenience, and safety associated with closed circuit anesthesia. In addition, the amount of anesthetic used is greatly reduced with this equipment and can therefore result in a significant cost saving factor over a period of time.

d. THE SEMI-CLOSED METHOD. This procedure allows gases and vapors to flow from the anesthetic apparatus into a reservoir bag from which the animal inhales. Most of the exhaled gases pass out through an expiratory valve into the atmosphere and are lost. Of course this means additional anesthetic agents must be utilized to produce the same result as with a closed method. However, the cost of the equipment is less, and this may be a desirable method under some circumstances.

2. Choice of Chemical Agent

a. ETHER. Ether is the most widely used volatile anesthesic agent, since it is inexpensive and simple to administer, and since recovery from its effects is rapid. The administration of ether should be carried out with special caution to avoid overdosage. Rabbits invariably hold their breath as long as possible at the first sniff of ether and then take a deep breath, which at times can result in depression of heart action. If the heart stops, efforts at resuscitation are seldom successful. To avoid this problem, the anesthetic cone should be lifted periodically until the rabbit begins to breathe regularly (60).

Ether is an excellent supplement to other anesthetic agents such as pentobarbital, thiamylal, and some of the narcotics. The investigator should be aware of the explosive properties of ether, which means it should not be used around electrical equipment, unless it is specially grounded, or near open flames. An added precaution is that ether may cause pulmonary edema in cases of prolonged anesthesia (26).

b. CHLOROFORM. This chemical agent has a distinct advantage over ether in that it is noninflammable. However, it does have a narrower margin of safety and can be toxic to the liver and kidney (60).

Because of the recent interest in the use of nonexplosive anesthetics, the use of methoxyflurane and halothane will be discussed in some detail.

c. METHOXYFLURANE. Usage of this anesthetic, via the open drop method, has been advocated in two recent publications, (48, 56). In both cases, premedication was provided with acepromazine (1 mg/kg body weight) given intramuscularly about $1\frac{1}{2}$ hours before inhalation anesthesia. This technique was utilized on both immature and mature rabbits and provided safe, reliable surgical anesthesia. In older rabbits, surgical anesthesia is best maintained by mixing oxygen, nitrous oxide, and methoxyflurane. A high concentration of methoxyflurane in the induction mask during induction carries with it the added risk of respiratory depression. Careful attention is required, and if depression occurs, the mask must be removed immediately.

Although some investigators have stated that the use of closed circuit anesthesia is too difficult and offers few advantages (60), others insist that it is relatively easy to perform and offers several advantages (13, 33, 85). The inherent safety features of methoxyflurane include its low degree of volatility which allows the rabbit to alter the depth

---

* Metofane, Pitman-Moore Co., Indianapolis, Indiana.

† Fluothane, Imperial Chemical Industries, Ltd., Milbank, London, S.W.1.

of anesthesia by a reflex change of rate and depth of breathing. It also produces a profound level of muscle relaxation and analgesia followed by a rather short recovery period. For a closed system to be effective, intubation of the animal is suggested. This has been a difficult technique for some, as the rabbit has a long curved mouth with a narrow opening which makes visualization of the eipiglottis and larynx difficult. However, a mouth speculum is available which facilitates intubation and this technique can be performed routinely (33).

The rabbit is confined in a holding box with the head exposed, and the mouth is opened using the speculum. The tongue is grasped with a smooth clamp and pulled forward and to one side of the mouth. An endotracheal tube (17 French) is then passed into the pharynx. When the tube reaches the epiglottis, resistance is encountered. Care should be taken in advancing the tube into the trachea. Check for a flow of air to be sure that it has not entered the esophagus and, when this has been determined, fasten the endotracheal tube to the lower jaw with a piece of gauze. Once the tube has been positioned, it is best to wait a few minutes to allow the rabbit to return to normal respiration. At this time, the anesthesia machine can be connected to the endotracheal tube. One should closely observe the rabbit during the induction. Most rabbits tend to resist inhalation anesthetics and may even hold their breath for extended periods of time at the first breath of the anesthetic agent. When the rabbit does breathe, he may take in excessive quantities of the anesthetic, so care must be taken to regulate the amount of anesthetic being administered. When the desired plane of anesthesia has been reached, the respiratory reflex of the rabbit will usually regulate the amount of anesthesia within safe limits (33). This type of surgical anesthesia has been successfully used for such procedures as hysterotomy, ovariohysterectomy, splenectomy, thoracic duct isolations, and other major surgical operations. The recovery phase is generally smooth, without excitement, and is complete within 30–60 minutes.

d. HALOTHANE. This is another nonexplosive inhalation anesthetic. This product must be used with a closed system because it has a much narrower margin of safety than methoxyflurane. However, the technique has been used effectively (13, 85). Again, an endotracheal tube is needed and, if there is trouble in passing it, the animal may be sedated with 1–2 ml of a 2% solution of an ultra-short acting barbiturate such as thiamylal sodium. Extreme care should be taken during the induction period to avoid an overdosage as was mentioned with methoxyflurane. Although flow meters are required to control the amount of halothane administered in dogs, cats, and larger animals, Hoge *et al.* (33) made no attempt to use them for rabbits. He maintained that an adequate mixture of oxygen and anesthetic agent could be regulated by watching the re-

breathing bag. With good technique and adequate equipment, halothane can provide an excellent plane of surgical anesthesia followed by a relatively short recovery period.

Halothane has also been adapted for use within germfree isolators. A 2–4% mixture of halothane and oxygen can be passed into the wall of an isolator, fitted on each side with a high efficiency gas line filter. A funnle-type attachment is used inside the barrier to administer the anesthetic to a particular animal. The animal should be held near the exhaust outlet so that escaping anesthetic gases do not involve the other animals within the barrier. This is actually an open-circuit method which requires increased amounts of the anesthetic agent and thus can be expensive. However, with the many problems associated with germfree work, halothane may be well worth the additional cost (13).

e. CYCLOPROPANE. This anesthetic can be mixed with oxygen and used within a closed system just as was described for methoxyflurane or halothane. It is not recommended that this anesthetic be used in an open system as it is explosive and can be dangerous. This anesthetic, if properly administered, will produce a safe level of surgical anesthesia (33).

E. Other Anesthesia and Associated Agents

1. Hypnosis

Hypnosis is a somnolent state in which the rabbit can be placed for ease in handling during simple laboratory procedures such as intravenous or intramuscular injections, subcutaneous injections, some cannulations, and radiography (68). Some investigators have been dissatisfied with this technique, stating that the duration of hypnosis is very short and extraneous stimuli such as loud noises may arouse the animal from its hypnotic trance (49). In more recent years, experiments have been conducted which suggest that when the technique of hypnosis is properly employed, the rabbit exhibits excellent long-term analgesia for most minor surgical procedures, and there is almost no problem with disruption of the hypnotic trance (29). The basic technique of hypnosis is performed as follows: The rabbit should be grasped firmly but gently by the skin over the withers with the left hand, supporting the hind quarters with the right hand, and placed on a table. The technique of Gruber and Amato (29) requires that the animal's legs be tied to a restraining board with the animal in a dorsal recumbency. However, others have demonstrated similar results and the limbs were not restrained. Once the animal is placed on its back, the head should be grasped by an assistant and gentle traction applied. At this point a gentle monotone voice may be utilized for soothing effects as well as a gentle stroking of the abdominal region with the palm of the hand. After a minute or two, the animal will begin to relax and its breathing will become less rapid and deeper. The pupils will show some contraction, while the eyelids remain open. If the breathing starts to accelerate or there is movement of the limbs, the research procedure should be stopped and the hypnotic technique reapplied. In difficult cases, the tying of all four limbs and applying gentle traction on the head may produce desired results (29). Usually more time is required to induce hypnosis the first time, with each succeeding procedure on a particular animal requiring less time. If the hypnotic trance is too short for the desired procedure, the administration of tranquilizers, chlorpromazine and meprobamate, will greatly enhance the hypnotic state. Pentobarbital can also be used to enhance the duration of the trance, but it apparently will not potentiate the depth of the trance (49).

2. Spinal Anesthesia

Spinal anesthesia can be used very effectively when cutaneous desensitization is a critical portion of the experiment (7, 54). Although most local anesthetics will work, the use of xylocaine or lidocaine hydrochloride (1%) mixed with epinephrine 1:100,000 and an equal volume of a 50% solution of dextrose in water will produce a safe and acceptable level of spinal anesthesia. The technique is as follows: (a) The loin area is clipped and depilated with a commercial depilatory agent 24 hours prior to anesthesia. (b) The rabbit is then placed in a squeeze box and compressed to arch the vertebral column. This technique separates the dorsal vertebral spines and tends to displace the cord towards the ventral surface of the lumbar area. (c) Locate the twelfth rib by palpation and count caudally to locate the cranial aspect of the second lumbar spine. (d) Then use a 21-gauge, 1½-inch spinal needle, bevel directed caudally, and introduce it perpendicularly through the supraspinous ligament. The needle is inserted ventrally and cranially until it passes through the dura. (e) Aspirate a small amount of spinal fluid to make sure the needle has penetrated the dura; however, extreme care must be taken not to injure the cord itself. If in doubt it is much better to inject the fluid extradurally rather than risk injury to the cord. Extradural anesthesia will still produce a satisfactory level, but the duration will be shorter. The usual dosage of such a hyperbaric mixture is 1 ml (15).

3. Intramuscular Injections

Intramuscular injections can also be used to produce anesthesia. Paraldehyde is an anesthetic agent, also used as a preanesthetic agent, which can be given by deep intramuscular injection. A dosage of 1 ml/kg body weight will produce anesthesia within 20–30 minutes. It has an advantage in that, unlike pentobarbital, no increased dosage is needed to produce light anesthesia following repeated injections. It is critical that injections be given deep intra-

muscularly as subcutaneous injections will produce ulcers and sloughing of the skin (63).

Another drug combination used intramuscularly is fentanyl and droperidol, a potent neuroleptoanalgesic agent. Each milliliter of the drug combination contains 20 mg of droperidol, a tranquilizer, and 0.4 mg of fentanyl, an analgesic. The usual dosage is 0.22 ml/kg or 1 ml/10 lb body weight. This dosage will produce a certain degree of tranquilization and anesthesia. It is reported that, even when fully under the influence of the drug combination, most rabbits assume a normal sitting position and many are capable of limited locomotion. However, the animals do not respond to auditory stimuli and are apparently unresponsive to painful stimuli (78). At all times the rabbits will respond vigorously to a light tap on the nose; therefore, it is suggested that this method be used with some discretion until more is known about its effects on the rabbit.

4. Oral and Rectal Anesthesia

Oral and rectal anesthesia is seldom performed. However, the administration of paraldehyde by gastric intubation at a dose of 1.5 ml/kg, mixed 1:7 with tap water, will produce satisfactory anesthesia in rabbits (77). The induction time varies, but it will produce moderate to deep narcosis within 30 minutes. Anesthetics are seldom given rectally, except for demonstration purposes. One such agent used is "Equi-Thesin" in doses varying from 0.5 ml/kg to 3.0 ml/kg body weight. The extremely variable results obtained by this method suggest that one of the other routes of administration should be used.

5. Muscle Relaxants

On occasion, muscle relaxation may be needed to perform a particular surgical procedure. Curariform drugs are used in this capacity, but may have an additive effect when used in conjunction with anesthetic agents. Of the more commonly used curariform drugs, D-tubocurarine is long lasting; decamethonium and gallamine triethiodide are of intermediate length; and succinylcholine is of short duration. The dose of these drugs is 0.1–0, 2 mg/kg body weight except for gallamine which requires an increased dosage of about 12.5% (67). It is important to stress that the muscle-relaxing agents are not anesthetics and are used only in conjunction with an acceptable anesthetic to produce greater muscle relaxations.

F. Emergencies and Postanesthetic Care

Because metabolic processes have been slowed during surgery and there has been a loss of body heat, it is essential that the rabbit be placed in a warm environment. The rabbit should be placed on its side in a cage which has a covered floor, and on occasion a heat lamp may be indicated. If the surgical wound is large enough to require bandaging, it is preferable to spray the wound with one of the plastic sprays or to cover the wound with collodion. Cloth bandages become soiled very quickly which may predispose the wound to secondary infections.

The most common emergency associated with anesthesia in the rabbit is respiratory distress. If an inhalation anesthetic has been used, the tongue should be pulled forward so that it does not block the trachea, and systematic pressure and release should be applied to the thorax and abdomen. If the animal is on a closed circuit machine and oxygen is available, the anesthetic should be cut off completely and the mixing bag should be filled with oxygen. Then artificial respiration is applied until normal breathing resumes. If a longer acting anesthetic has been used, it may be necessary to perform a tracheostomy and attach the animal to a respirator. This will only be effective as long as the heart continues to beat. None of the stimulants are very effective against depression caused by an overdose of an anesthetic drug. Picrotoxin is considered the most effective drug antagonist to barbiturate depression and should be given in a dosage of 3 mg/kg body weight (5). Pentylenetetrazol also can be used at a dosage of 0.1 mg/kg body weight. Nikethamide is considered more effective against the volatile anesthetics and is used in a dosage of 0.03 mg/kg body weight. Ephedrine (0.06 mg/kg body weight) and methamphetamine (0.07 mg/kg body weight) may be used to enhance the effectiveness of the aforementioned stimulants. However, caution should be used as these amounts are only approximate, the actual dosage depending on the depth of depression (60). It also has been reported that rabbits tolerate large doses of atropine (25) which can be useful in controlling bronchial secretions.

VI. SPECIALIZED RESEARCH TECHNIQUES

A. Flushing of Ova

Many experiments are designed to make use of the relative ease with which the female rabbit conceives. Flushing of ova is a technique often used to obtain fertilized ova at a particular cell stage. For example, one-cell fertilized eggs can usually be obtained 21 hours postcoitus and two-cell eggs at 24 hours postcoitus. One technique for acquiring fertilized eggs is as follows: (a) Determine the stage at which the fertilized egg is needed and sacrifice the doe at that point using one of the euthanasia techniques described in Section VII. The exact breeding time and a good knowledge of early embryonic development are essential. (b) Immediately following death of the female, open the abdominal

cavity and remove the uteri intact. (c) Cannulate the fimbriated end of the fallopian tube with a short length of fine polyethylene tubing attached to a 50-ml syringe containing a modified Krebs solution (52). Inject 5–10 ml of the fluid into the uterine horn and gently massage it. Then flush the eggs into a sterile centrifuge tube or beaker (1). However, if for some reason the female rabbit is needed for future studies, this technique can be performed without sacrificing the doe. (a) Perform a laparotomy to expose the uterine horns using either methoxyflurane, halothane, or a spinal anesthetic. (2) Cannulate the fallopian tubes with polyethylene tubing attached to a 50-ml syringe containing the modified Krebs solution. Inject a few milliliters into the fallopian tube and massage this fluid from the tubouterine junction toward the uterine horn. Gently clamp off this horn hear the cervix to prevent the escape of fluids. (3) Insert a glass cannula through a small longitudinal incision near the apex of the horn and hold the uterine wall closely around the rim of the cannula to prevent leakage. (4) Gently massage the fluids from the distended length of the horn via the cannula into a watch glass. (5) Close the uterine horn utilizing a Cushing suture and do a routine closure of the abdominal wall (30, 30a, 31, 82).

B. Transplantation of Ova

Once ova have been obtained, they should be maintained in a sterile Krebs solution at body temperature until transfer can be initiated. A recipient doe is selected and anesthetized and a laparotomy performed. A Pasteur pipette is then introduced into the lumen of a uterine horn through a puncture wound near the tubouterine junction. Ova are released from the pipette and transplantation is complete. The abdominal wall is closed as described in Section VI,E (30, 30a, 31).

It is important that a healthy doe be selected as the recipient because elevated uterine temperatures have been shown to produce high embryonic deaths (36). The rabbit has also been utilized for the transfer of fertilized ova of another species from one part of the world to another. For example, highly bred cattle ova have been transferred, via the rabbit, to other geographical locations where the ova were removed and replaced in scrub cattle.

C. Artificial Insemination

Most references on artificial insemination cite the work of Walton (83) as the basic technique for semen collection. However, since this original work, Walton (84) and Grove (28) have described improved designs for an artificial vagina.

Their basic technique is to construct an artificial vagina from a rubber hose about $3\frac{1}{2}$ inches long, $1\frac{1}{2}$ inches in diameter, and $\frac{1}{8}$ inch wall thickness. It is usually lined with a condom which is attached at one end to a collecting vial. Several of these units can be constructed and placed in a warming oven at about 45°C for easy use. Other investigators have designed more intricate artificial vaginas using a warm water jacket and air pressure (28, 84, 86). These can be utilized if the previously described instrument fails.

The method of collecting semen with the use of an artificial vagina, varies. One method is to use a second buck as the dummy and hold the artificial vagina between the hind legs of the dummy buck (28). The dummy buck technique has its advantages since some bucks ejaculate quickly. However, if a male is difficult to train, a doe may be used. The artificial vagina is held between her hind legs and the penis is directed into the artificial vagina when he mounts (28, 86). Another method is to place a rabbit skin in a cage with females to acquire a feminine odor. The artificial vagina is grasped in the hand. Then the rabbit skin is placed over the hand and arm and presented to the male. He will usually mount the arm and serve the artificial vagina held in the hand, which actually makes collection quite simple. Since many bucks are apprehensive, do not expect all of them immediately to make use of the artificial vagina.

The actual technique of inseminating the doe with semen varies between individual investigators. However, a method is to give a subcutaneous injection of 50 IU of pregnant mare serum (PMS) 48 hours before the proposed time of insemination. At the time of insemination, each doe should be given an additional 50 IU of human chorionic gonadotropin (HCG) followed by inseminating 0.25 ml of dilute semen containing at least 5 million motile spermatozoa into the cranial aspect of the vagina (69, 70).

Other techniques which may be useful include a means for quantitatively estimating the sperm producing capabilities of the male. This is done by making accurate measurements of the testis in the live animal and correlating this information with data obtained by testicular biopsy (2, 47, 64, 65). This information, along with semen analysis, can be of considerable value in studying male infertility. Others have claimed that motility, viability, and fertility of an inseminate can be improved by mixing semen samples from different males (6, 31).

D. Microcirculation Techniques

During the twentieth century, living tissues have been studied by numerous investigators who employed either a tissue culture technique or microscopic observation of the tissue within the living animal. The former method, although used extensively, may allow for growth changes which greatly diverge from that which occurs in the living host. Therefore, the technique of placing a chamber within the ear of a rabbit to study microcirculation has become a common practice.

The actual construction of an ear chamber is quite complicated. Although a detailed description of construction has been reported (71, 72), it is recommended that one be purchased commercially. The placement of the chamber in the ear varies a little with each particular investigation; however, most investigators use a rather basic technique which is as follows: (a) Place the rabbit on a restraining board and prepare the ear for aseptic surgery. The region of the ear selected for study should include a main artery on a flat surface, and it should be near the tip of the ear for ease in studying with a microscope. (b) Anesthetize the rabbit's ear with 2% xylocaine in the region of the proposed ear chamber. (c) Excise a square of the external skin slightly smaller in size than the chamber and thereby expose the main artery and the subcutaneous tissues adjacent to it. (d) Longitudinally incise the arterial sheath surrounding the main artery and nerve to expose the vessel. (e) After the space has been cleared, control hemorrhages by pinching off or tying the bleeders; then slip the chamber into position. Fill the chamber with normal saline solution and make sure the main artery is placed between the chamber walls. (f) Fasten the chamber into position with nonabsorbable suture materials. Then turn the ear over and cut away a square of tissue composed of internal skin and cartilage, leaving the chamber entirely transparent. (g) At the end of the operation, make an incision in the ear, beginning at its lateral edge and extending one inch in a direction transverse to its long axis. This incision should lie somewhere between the base of the ear and the chamber region and is designed to overcome the normal curling of the ear. The rabbit must be immobilized for observation in a stock and the ear mounted on a special stage. Illumination is obtained from a 60-watt frosted, electric light bulb, whose rays are passed through a round flask of distilled water containing enough copper acetate to give it a faint greenish-blue tinge. If the skin retracts far enough to expose the edges of the celluloid chamber, liquid parlodion can be spread over this defect and will prolong the life of the chamber. For a more complete description of the technique, the reader is referred to the works of Sandison (72) and Ebert *et al.* (18).

E. Experimental Surgical Procedures

Experimental surgical procedures require that the surgeon be thoroughly familiar with the anatomy involved and that aseptic techniques be employed.

1. Decerebration

Decerebration is a specialized technique utilized in many neurological studies. Most attempts result in lesions larger than planned, high operative mortality, and difficult interpretation of histological results. A recent technique was described which used high frequency current applied by means of two convergent electrodes to sever the brain stem at the midcollicular level. This procedure is relatively simple and has been used with routine success on cats, rabbits, and guinea pigs. For details about the equipment and the technique used, the reader is referred to the works of Koller (51). Decerebration ensures a complete absence of stimulatory effects, a factor which may be of significance when carrying out subsequent physiological or pharmacological studies in acute, as well as in chronic experiments.

2. Perfusion of Cerebral Ventricles

Perfusion of cerebral ventricles of the conscious rabbit has been accomplished using two guide tubes directed through the cranium towards the lateral ventricles and a curved guide tube directed toward the cisterna magna. By this method the cerebrospinal fluid system can be tapped which allows for ventriculocisternal perfusion of the fully conscious animal (59).

3. Cannulation of the Common Bile Duct

Cannulation of the common bile duct is a standard research procedure in the rabbit. The animal should be fasted for 14 hours prior to the surgery and then anesthetized with an acceptable general anesthetic. The entire ventrolateral aspect of the abdomen and thorax is prepared for surgery. The animal is positioned in a modified left lateral position, with the hind limbs secured as for a supine position and with the forelegs secured to the left side of the table. The surgical area is then draped, and, using the superior mammary vein as a guide, a right lateral incision (approximately 7 cm in length) is made parallel to the mammary vein. To expose the gallbladder, the fascia and fat of the right ventral lobes of the liver are bluntly dissected free. The area is then packed with moist sterile sponges while the catheter is prepared. The catheter, 3.0 mm o.d., is a rubber tube prepared by slightly bevelling the end to be inserted. The distal end of the gallbladder is incised, partially drained, and the catheter inserted to a point about 2 mm from the neck leading to the cystic duct. When a portion of the bile remains in the gallbladder, there is less chance of it twisting or folding over on itself. The catheter is then secured in place by looping two sutures around the free end of the gallbladder. The next step is to isolate the common bile duct, a procedure which requires skill and patience, because the vena cava is closely associated with the duct. Once the duct has been isolated, a length of silk suture is used to ligate it. The catheter can be considered patent if there is a 16- and 20-cm rise of bile in the catheter over a 15-minute period. A routine closure of the abdominal wall is used and the catheter is taped over the dorsal external aspect of the animal. A 125-ml polyethylene bottle is often taped on the side opposite the externalized catheter to provide a means for continuous collection of bile. This method has been shown to work over a period of several days with few complications (9, 57).

4. Hysterectomy

Hysterectomy is a surgical procedure which has been employed in reproductive studies for many years. To minimize problems, an acceptable routine should be followed. Properly prepare the ventral midline for surgery; use a general anesthesia (Section V); and place the animal on a reversal tilt board so that the hind limbs of the rabbit are elevated. This displaces the digestive organs in a cranial direction, which allows for easier access to the genital tract. Make a 5-cm skin incision just cranial to the pubic symphysis and expose the viscera and peritoneum. Identify the urinary bladder and observe the uterine horns as they lie on either side of the bladder. Use a retractor or blunt tenaculum, and hook one of the horns to pull it gently out of the abdominal incision. By applying traction, the ovary can be located; however, care should be taken as these tissues tear easily. Place two hemostats on the pedicle cranial to the ovary and tie a ligature in front of the more cranial hemostat. Use a scalpel and sever the ovary and horn from the pedicle. Repeat this procedure on the opposite ovary and its horn. Using gentle traction on the uterine horns, locate the body of the uterus and place two additional hemostats across the uterine body. Secure a ligature around the body of the uterus and then sever the uterus between the two caudal hemostats to remove the ovaries, uterine horns, and uterine body intact. Remove all hemostats and check for bleeding. Then close the body wall including the peritoneum with a continuous suture of catgut. Skin closure is usually made with nonabsorbable interrupted or mattress sutures (45).

5. Splenectomy

Splenectomy is another surgical procedure which is often associated with a specialized research protocol. The animal should be anesthetized with a general anesthetic (Section V) and the ventral midline prepared for surgery. An incision is made along the linea alba, extending from the caudal border of the rib cage to the midabdominal level. The spleen is found in the upper left quadrant of the abdominal cavity and is fixed in this position by its mesentery and by four or more blood vessels. Each blood vessel should be double ligated and then severed between the ligations. Splenic adhesions are broken down by blunt dissection and the spleen removed. It is important that all ligatured vessels be checked for bleeding. Use a routine closure of the abdominal cavity suturing peritoneum, muscle, and skin (42).

6. Adrenalectomy

Adrenalectomy requires complete excision of the adrenal glands which is preferable to simple denervation and other indirect methods of medullary ablation (87). This technique is rendered difficult by the adherence of the right adrenal to the posterior aspect of the vena cava close to the entrance of the right renal vein. The animal must be put under a general anesthesia as any movement may result in the tearing of blood vessels during the search for and removal of the glands. A surgical field is prepared on a ventral midline from the xiphisternum to the umbilicus. Following a cranial midline incision, the large intestine is packed to the left, thus exposing the upper abdominal vena cava, right lobe of the liver, the pancreas, and the portal vein. The posterior peritoneum which is bound to the vena cava is incised parallel to the liver lobe margin and stripped caudally to allow mobilization of the right renal vein and vena cava in the vicinity of the right adrenal. A fine suture is used to ligate the right lumboadrenal vein after which it is severed. This stump is retracted and allows exposure of the right adrenal gland. In dissecting out the gland, great care should be taken not to cut into the vena cava as the margins of this gland actually join with the adventitia of the vena cava. The left adrenal is located at the lower border of the pancreas in the angle between the vena cava and the left renal vein. To remove this gland, the left adrenal vein is exposed and severed between two ligatures. The ligatures are left long and used to apply traction to aid in elevating the adrenal away from its bed of fat. For ease in removal, the splanchnic nerve will have to be severed for both the right and left adrenals. Following removal, a routine abdominal closure is used.

Postoperatively, it is recommended that 2 mg cortisone acetate and 2 mg deoxycorticosterone acetate be given intramuscularly on the first day, followed daily by 1 mg cortisone acetate and 1.5 mg deoxycorticosterone acetate. The animal should be allowed free access to tap water, fresh cabbage, and pelleted rabbit feed. Without this supportive therapy, adrenalectomized rabbits can be expected to die within 48 hours of surgery (43, 87).

7. Hypophysectomy

Hypophysectomy in the rabbit has been described by numerous investigators (20, 22, 37, 44, 62, 75). Of these descriptions the most widely used method is that of Firor (22), and the reader is referred to this article for details. In brief, these are the directions which should be followed: (a) Use a reliable general anesthetic. (b) Elevate the animal's head and turn the head and thorax to one side. (c) Enlarge the palpebral fissure with a lateral incision starting just below the lateral canthus and with a medial incision extending 1 cm from the medial canthus. (d) Retract the lower lid and divide the palpebral conjunctiva to expose the ligament bridging the posterior half of the zygomatic arch. (e) Remove this ligament and some of the underlying fat to provide adequate exposure within the orbit. (f) Incise the tissues attached to the medial end of the supraorbital process. Insert a spatula into the deepest part of the orbit and bring it forward to compress the orbital sinus, which is a venous sinus covering the posterior half of the eye. The eye is then removed from the orbit for a short distance and any hemorrhage can be controlled with a gauze pack applied for a few minutes. (g) With the eye held out of the orbit, remove areolar tissue and fat covering the fifth cranial nerve as it emerges along the medial edge of the internal pterygoid muscle. (h) Use a straight forceps to puncture the thin lateral wall of the sella turcica. This bone is broken rather easily by exerting a slight pressure with forceps. (i) Remove the pituitary gland by applying suction through a beveled glass tube inserted into the sella. It is important to have the bevel of the tube facing away from the pituitary stalk, or infundibulum, or the suction applied to the base of the brain may injure or kill the animal. (j) Withdraw the glass tube, lift out the spatula, and replace the eye within the orbit. Incision lines are then closed and the animal is allowed to recover. With practice the operation can be completed within a 15–20-minute period.

8. Bone Marrow Biopsy Techniques

Bone marrow biopsy techniques have been described by several investigators (19, 79, 80). These methods all allow for the collection of bone marrow; however, most of them will not provide a sterile specimen. Therefore, the method of choice seems to be that of Powsner and Fly (66), and a brief outline of that procedure is as follows. (a) Anesthetize the animal and prepare the lateral surface of the thigh for surgery. (b) Thoroughly wash the animal with a bactericidal detergent from the neck down. (c) Place the rabbit on the surgery table on its side with the prepared limb in a tightly flexed position. The shaved skin is then treated with tincture of iodine and draped with sterile towels. (d) Make a 10-cm incision over the femur between the vastus intermedius and biceps femoris. Continue the dissection through the connective tissue and fascia of the muscle masses to reach the femur. The exposed shaft of the femur is scraped with a small periosteal elevator and swabbed with tincture of iodine. (e) Make two small holes about 4–5 cm apart with a surgical drill. Insert hypodermic needles with a reground dull point into these holes. It is important that the drill and needle sizes be matched before surgery is begun. (f) Attach a syringe containing a balanced salt solution to one of the needles and an empty syringe to the other. By combined gentle pressure on the filled syringe and aspiration of the empty syringe, the marrow between the two needles is readily washed into the empty syringe. Usually about 5–7 ml of washing solution is adequate. Following aspiration, the needles are removed and the wound is closed with a nonabsorbable suture.

VII. EUTHANASIA

The term "euthanasia" implies humane and painless death. Whenever a rabbit is put to death to alleviate suffering or as part of the experimental design, it is the responsibility of the research investigator to see that true euthanasia is accomplished. There are several factors that must be considered in selecting the method used. The procedure must not be painful to the rabbit; it should cause only a minimum of animal apprehension and fear; and it should produce a relatively rapid death. The method used should not interfere with the collection and interpretation of the experimental data. The emotional effect of the procedure on the investigator or other observers should also receive consideration. If painless, this may not be of importance to investigators, but the emotional impact the procedure could have on members of humane societies, antivivisectionists, and other groups who are not as familiar with biological systems should be considered (10).

Ether and chloroform have been used by many researchers in attempts to produce euthanasia. The rabbit is forced to inhale the vapors until death results, but not always euthanasia. The vapors of these agents are irritating to mucous membranes and also produce considerable excitement in the rabbit before there is loss of consciousness. Inhalation of carbon monoxide produces a rapid painless death and does not produce terminal convulsions. The pure form of carbon monoxide should be used as gasoline engine exhaust gases usually contain contaminants which may be noxious and cause pain. The inhalation of carbon dioxide gas in concentrations above 40% produces a rapid onset of anesthesia without pain. The gas then produces an esthetically acceptable euthanasia. Carbon dioxide is the best inhalant agent discussed for true euthanasia. It is also safe for investigators to use if handled in a well-ventilated room (10).

Intravenous administration is generally the preferred route for injectable agents. This can be accomplished while producing a minimum of animal discomfort because a lethal concentration of the agent can be rapidly introduced directly into the blood stream and thus to the respiratory center in the brain. Pentobarbital is the most commonly used barbiturate for intravenous euthanasia in rabbits and is recommended for this use if it will not interfere with experimental results (10).

Many anesthetic agents (Section V) are available that will produce euthanasia if administered in excessive doses. Anesthetics will depress synaptic excitability and cause varying degrees of central nervous system depression. If these agents are given in lethal doses, the eventual cause of death is respiratory failure or an alteration of normal cardiac activity and resultant anoxia. This anoxia occurs long after the animal is unconscious so that rabbit excitement, fear, and convulsive activity do not take place.

Curariform drugs have been considered for euthanasia. Curare, succinylcholine, and other agents that produce only a skeletal muscle paralysis can cause anoxia and death. The rabbit is fully conscious during the initial stages of the anoxia so that this class of drugs should never be considered as the agent to use for euthanasia (10).

There are many other chemical and physical methods that have been used to euthanize rabbits. They will not be discussed in this section because of undesirable properties or actions, or because of esthetic considerations. The reader is returned to the Report of the AVMA Panel on Euthanasia for their recommendations.

VIII. NECROPSY PROCEDURES

Postmortem examinations are an excellent aid in rabbit colony preventative medicine programs. All rabbits that die during a research project should be immediately necropsied and the cause of death determined. For a necropsy to be meaningful, the individual must know the appearance of normal tissues and organs before he is capable of recognizing the abnormal or diseased.

The following is a suggested list of the recommended equipment to have available before the dissection is begun:

1. Autopsy knife
2. Scalpel and blades
3. Toothed forceps
4. Rat-toothed forceps
5. Scissors (blunt-blunt and blunt-sharp)
6. Necropsy scissors (heavy)
7. Probe (1 mm diameter)
8. Metric rule
9. Rib cutters
10. Balance for organ weights
11. Pans with fixative (10% formalin)
12. String for tying vessels
13. Bacteriological swabs
14. Sterile glass containers for fluid collection
15. Syringe with needle (10 ml)
16. Bone saw

A. Necropsy Dissection

The following is one example of an organized procedure that can be followed to accomplish a necropsy dissection. It is advisable to have an assistant available to help with the dissection and a person to record the findings.

1. External Examination

Inspect and palpate the entire body.

a. SKIN. Note discolorations, postmortem lividity, hemorrhages, wounds, ulcers, tumors, scars, external parasites, fur condition, and soiling.

b. HEAD. Examine eyes and nose, inside of mouth including teeth, buccal mucosa and tongue, ears for wounds or foreign materials, and palpate skull for abnormal characteristics.

c. LYMPH NODES. Palpate superficial nodes to detect enlargements.

d. GENITALIA. Palpate and examine the external genital organs for abnormalities.

2. Incise Skin

Starting at the anterior section of the sternum, make an incision along the midline posteriorly to the symphysis. To obtain maximal exposure of the abdominal cavity, cut transversely through the subcutaneous tissue and muscles on each side of the lower abdomen. The skin should then be reflected laterally and the subcutis examined. The abdominal muscles can be incised and reflected laterally. The scapular attachments should be severed to facilitate future manipulations.

3. Open Thorax

Cut through the costal cartilages medial to the costochondral junctions, beginning at the second rib and extending posteriorly. Avoid going too deep so that one avoids cutting into the lungs. Cut the sternum from the diaphragm.

a. EXAMINE THE THORACIC CAVITY. Lift the sternum and examine the pleural cavities. Next smears, cultures, and pleural fluid samples are collected. Remove the breast plate by cutting through the cartilage of the first rib and severing any remaining attachments. Examine lymph nodes and lungs. Pass the gloved hand around the lungs, and if adhesions are encountered, they should be broken. Open the pericardial sac and examine for fluid, adhesions, or hemorrhage. The thoracic duct is examined by lifting the right lung and reflecting it to the left. Blunt dissection may be required to locate the duct.

b. EXAMINE THE ABDOMINAL ORGANS. The abdominal cavity is explored with a gloved hand. The position and size of the spleen, liver, kidneys, stomach, and cecum should be noted. Omental adhesion should be detected. Observe and palpate other structures including the diaphragm, female genital organs, and urinary bladder.

4. Remove Thoracic Organs

a. THE HEART. Remove the heart by lifting it and cutting the venae cavae and pulmonary veins. Sever the aorta and pulmonary artery thus freeing the heart.

b. THE LUNGS. Remove the lungs by cutting the bronchus and other hilar structures on each side. If required, the heart, lungs, and trachea can be removed as a unit.

5. Remove Abdominal Organs

The spleen is elevated and the hilar vessels are severed. The esophagus is then cut at the point where it joins the stomach. The stomach is gently lifted and removed and the remaining gastrointestinal tract is removed as one unit. A knife is used to sever the mesentery from the intestines as they are removed. Care must be taken to isolate and remove the pancreas from the small intestine. The vasculature of the liver is severed and the organ removed.

6. Remove Urogenital System

Cut behind the left kidney and adrenal to free the structures from the retroperitoneal tissues. Open the aorta and probe the renal arteries. Lift the left kidney and note the course of the ureter. Repeat this procedure for the right kidney. Then remove the urinary bladder. The genital organs are then examined and removed from the carcass.

7. Remove Neck Organs

Make an incision from the mandible to the sternum and reflect the skin laterally. Examine and isolate vessels in the neck region. To remove the tongue, larynx, and associated structures, cut through the floor of the mouth starting at the point of the chin and cut to each side along the mandibles. Grasp the tongue with toothed forceps and pull caudally. Continue the manual retraction while dissecting the throat structure and progressing posterior until the desired portions of the esophagus and trachea are removed.

The organs have now been removed for examination and preservation. If required, there are procedures for the removal and examination of skeletal structures, including the vertebral column and structures of the head (4).

B. Specimen Collection for Histopathology

Histopathological examination of tissues is often the single most important factor in reaching a definitive diagnosis or the only means of confirming some tentative diagnoses. Ideally every research worker should have at his disposal facilities capable of performing histopathological slide preparation and professional personnel capable of interpretation. Tissues for examination should be preserved as quickly as possible following death of the animal. Ten percent neutral buffered formalin is a commonly used fixative which quickly penetrates and preserves and also to some degree hardens the tissue. The specimen taken for examination should be representative of the lesion and should include some normal appearing tissue. To assure adequate fixation, the tissue should be cut into sections 6–10 mm thick.

Acknowledgment

We wish to thank Peg Kornacker for her invaluable assistance in the preparation of this manuscript. Special thanks is also given to Dr. Robert Szot and Kingsley Langenberg for helpful comments on the content of the material presented.

REFERENCES

1. Alliston, C. W., Howarth, B., Jr., and Ulberg, L. C. (1965). Embryonic mortality following culture *in vitro* of one and two-cell rabbit eggs at elevated temperatures. *J. Reprod. Fert.* **9**, 337–341.

2. Amann, R. P., and Almquist, J. O. (1962). Reproductive capacity of dairy bulls. VIII. Direct and indirect measurement of testicular sperm production. *J. Dairy Sci.* **45**, 774.

3. Appleman, R. M. (1966). A device for restraining rabbits. *Lab. Anim. Care* **16**, 300–303.

4. Baker, R. D. (1967). "Postmortem Examination, Specific Methods and Procedures," pp. 5–44. Saunders, Philadelphia, Pennsylvania.

5. Barlow, O. W. (1935). Relative efficiency of series of analeptics as antidotes to sublethal and lethal dosages of pentobarbital, chloral hydrate, and tribromoethanol (Avertin). *J. Pharmacol. Exp. Ther.* **55**, 1.

6. Beatty, R. A., and Napier, R. A. N. (1960). Genetics of gametes. II. Strain differences in characteristics of rabbit spermatozoa. *Proc. Roy. Soc. Edinburgh, Sect. B* **68**, 17.

7. Bieter, R. N., Cunningham, R. W., Lenz, O., and McNearny, J. J. (1936). Threshold anesthetic and lethal concentrations of certain spinal anesthetics in the rabbit. *J. Pharmacol. Exp. Ther.* **57**, 221–224.

8. Bito, L. Z. (1969). Animal restrainers for unanesthetized cats and rabbits. *Lab. Anim. Care* **19**, 244–246.

9. Boegli, R. G., and Hall, I. H. (1969). A surgical external biliary fistula for the total collection of bile from rabbits. *Lab. Anim. Care* **19**, 657–658.

10. Breazile, J. E., and Kitchell, R. L. (1969). Euthanasia for laboratory animals. *Fed. Proc., Fed. Amer. Soc. Exp. Biol.* **28**, 1577–1579.

11. Bree, M. M., and Cohen, B. J. (1965). Effects of urethane anesthesia on blood and blood vessels in rabbits. *Lab. Anim. Care* **15**, 254–259.

12. Chen, G., and Bohner, B. (1968). Surgical anesthesia in the rabbit with 2-(ethylamino) - 2 - (2-thienyl) cyclohexanone-HCL (CI-634) and chloral hydrate. *Amer. J. Vet. Res.* **29**, 869–875.

13. Cook, R., and Dorman, R. G. (1969). Anesthesia of germ-free rabbits and rats with halothane. *Lab. Anim. Care* **3**, 101–106.

14. Croft, P. G. (1964). *In* "Small Animal Anesthesia" (O. Graham-Jones, ed.), pp. 99–102. Pergamon, Oxford.

15. Davis, D. G., and Dingman, R. O. (1969). Cord anesthesia in rabbits. *J. Surg. Res.* **9**, 383–386.

16. Di Pasquale, G., and Campbell, W. A. (1966). A gag for gastric intubation of rabbits. *Lab. Anim. Care* **16**, 294–295.

17. Dolowy, W. C., and Hesse, A. L. (1959). Chlorpromazine premedication with pentobarbital anesthesia in the rabbit. *J. Amer. Vet. Med. Ass.* **134**, 183.

18. Ebert, R. H., Florey, H. W., and Pullinger, B. D. (1939). A modification of a Sandison-Clark chamber for observation of transparent tissue in the rabbit's ear. *J. Pathol. Bacteriol.* **48**, 79.

19. Erslev, A. J., and Hughes, J. R. (1960). The influence of environment on iron incorporation and mitotic division in a suspension of normal bone marrow. *Brit. J. Haematol.* **6**, 414.

20. Fee, A. R., and Parkes, A. S. (1929). Studies on ovulation. I. The relation of the anterior pituitary body to ovulation in the rabbit. *J. Physiol. (London)* **67**, 383–388.

21. Field, E. J. (1957). Anesthesia in rabbits. *J. Anim. Tech. Ass.* **8**, 47.

22. Firor, W. M. (1933). Hypophysectomy in pregnant rabbits. *Amer. J. Physiol.* **104**, 204–215.

23. Gardner, A. F. (1964). The development of general anesthesia in the albino rabbit for surgical procedures. *Lab. Anim. Care* **14**, 214–225.

24. Gibson, J. E., and Becker, B. A. (1967). The administration of drugs to one-day old animals. *Lab. Anim. Care* **17**, 524–527.

25. Godeaux, J., and Tonnesen, M. (1949). Investigations into atropine metabolism in the animal organism. *Acta Pharmacol. Toxicol.* **5**, 95.

26. Goodman, L. S., and Gilman, A., eds. (1965). "The Pharmacological Basis of Therapeutics," pp. 94 and 163–174. Collier-Macmillan Canada Limited, Toronto.

27. Grice, H. C. (1964). Methods for obtaining blood and for intravenous injections in laboratory animals. *Lab. Anim. Care* **14**, 483–493.

28. Grove, Von D. (1963). An artificial vagina for the sampling of semen in rabbits. *Deut. Tieraerztl. Wochenschr.* **70**, 492–494. (Ger. art.)

29. Gruber, R. P., and Amato, J. J. (1970). Hypnosis for rabbit surgery. *Lab. Anim. Care* **20**, 741–742.

30. Hancock, J. L., and Hovell, G. J. R. (1961). Transfer of sheep ova. *J. Reprod. Fert.* **2**, 295–306.

30a. Hancock, J. L., and Hovell, G. J. R. (1962). Egg transfer in the sow. *J. Reprod. Fert.* **4**, 195–201.

31. Hess, E. A., Ludwick, T., Richard, H. C., and Ely, F. (1954). Some influences of mixed ejaculates upon bovine fertility. *J. Dairy Sci.* **37**, 649.

32. Hodesson, S., Rich, S. T., Washington, J. O., and Apt, L. (1965). Anesthesia of the rabbit with Equi-Thesin following the administration of preanesthetics. *Lab. Anim. Care* **15, 336.**

33. Hoge, R. S., Hodesson, S., Snow, I. B., and Wood, A. I. (1969). Intubation technique and methoxyflurane administration in rabbits. *Lab. Anim. Care* **19**, 593–595.

34. Hoppe, P. C., Laird, C. W., and Fox, R. R. (1969). A simple technique for bleeding the rabbit ear vein. *Lab. Anim. Care* **19**, 524–525.

35. Hoskins, P. H., ed. (1959). *In* "Canine Medicine," 2nd ed., pp. 44–48. Amer. Vet. Publ., Santa Barbara, California.

36. Howarth, B., Alliston, C. W., and Ulberg, L. C. (1965). Importance of uterine environment on rabbit sperm prior to fertilization. *J. Anim. Sci.* **24**, 1027–1032.

37. Jacobson, D., and Westman, A. (1940). A parapharyngeal method of hypophysectomy in rabbits. *Acta Physiol. Scand.* **1**, 71–78.

38. Kaplan, H. M. (1962). "The Rabbit in Experimental Physiology," p. 5. Scholar's Library, New York.

39. Kaplan, H. M. (1962). "The Rabbit in Experimental Physiology," pp. 9–10. Scholar's Library, New York.

40. Kaplan, H. M. (1962). "The Rabbit In Experimental Physiology," pp. 10–11. Scholar's Library, New York.

41. Kaplan, H. M. (1962). "The Rabbit in Experimental Physiology," p. 11. Scholar's Library, New York.

42. Kaplan, H. M. (1962). "The Rabbit in Experimental Physiology," pp. 61–63. Scholar's Library, New York.

43. Kaplan, H. M. (1962). "The Rabbit in Experimental Physiology," pp. 113–114. Scholar's Library, New York.

44. Kaplan, H. M. (1962). "The Rabbit in Experimental Physiology," pp. 117–118. Scholar's Library, New York.

45. Kaplan, H. M. (1962). "The Rabbit in Experimental Physiology," pp. 119–120. Scholar's Library, New York.

46. Kaplan, H. M., and O'Brien, D. J. (1962). Comparative evaluation of anesthetic properties of six barbiturates and of Viadril in the rabbit. *Proc. Anim. Care Panel* **12**, 1.

47. Kennelly, J. J. and Foote, R. H. (1964). Sampling boar testes to study spermatogenesis quantitatively and to predict sperm production. *J. Anim. Sci.* **23**, 160.

48. Kent, G. M. (1971). General anesthesia in rabbits using methoxyflurane, nitrous oxide, and oxygen. *Lab. Anim. Sci.* **21**, 256–257.

49. Klemm, W. R. (1965). Drug potentiation of hypnotic restraint of

rabbits, as indicated by behavior and brain electrical activity. *Lab. Anim. Care* **15**, 163–167.

50. Knize, D. M., Weatherley-White, R. C. A., Geisterfer, D. J., and Paton, B. C. (1969). Restraint of rabbits during prolonged administration of intravenous fluids. *Lab. Anim. Care* **19**, 394–397.

51. Koller, E. A. (1969). A technique for standard decerebration by high frequency coagulation. *Brain Res.* **14**, 549–552.

52. Lardy, H. A., and Phillips, P. H. (1943). Effect of pH and certain electrolytes on the metabolism of ejaculated spermatozoa. *Amer. J. Physiol.* **138**, 741.

53. Ling, H. W. (1957). Anesthesia in rabbits. *J. Anim. Tech. Ass.* **8**, 58.

54. Luduena, F. P. (1957). Experimental spinal anesthesia. *Arch. Int. Pharmacodyn. Ther.* **109**, 143.

55. Lumb, W. F. (1963). "Small Animal Anesthesia," p. 284, Lea & Febiger, Philadelphia, Pennsylvania.

56. McCormick, M. J., and Ashworth, M. A. (1971). Acepromazine and methoxyflurane anesthesia of immature New Zealand White rabbits. *Lab. Anim. Sci.* **21**, 220–223.

57. Markowitz, J., Archibald, J., and Downie, H. G. (1964). "Experimental Surgery," pp. 563–570. Williams & Wilkins, Baltimore, Maryland.

58. Marston, J. H., Rand, G., and Chang, M. C. (1965). The care, handling and anesthesia of the snowshoe hare (*Lepus americanus*). *Lab. Anim. Care* **15**, 325–327.

59. Moir, A. T. B., and Dow, R. C. (1970). A simple method allowing perfusion of cerebral ventricles of the conscious rabbit. *J. Appl. Physiol.* **28**, 528–529.

60. Murdock, H. R., Jr. (1969). Anesthesia in the rabbit. *Fed. Proc., Fed. Amer. Soc. Exp. Biol.* **28**, 1510–1516.

61. Nelson, N. S., and Hoar, R. M. (1969). A small animal balling gun for oral administration of experimental compounds. *Lab. Anim. Care* **19**, 871–872.

62. Norgren, A. (1966). The parapharyngeal method of hypophysectomy in the rabbit. *Acta Univ. Lund., Sect. 2* **32**, 1–16.

63. Pandeya, N. K., and Lemon, H. M. (1965). Experiments in albino rabbits. *Lab. Anim. Care* **15**, 304–306.

64. Paufler, S. K., and Foote, R. H. (1969). Semen quality and testicular function in rabbits following repeated testicular biopsy and unilateral castration. *Fert. Steril.* **20**, 618–625.

65. Paufler, S. K., Ven Vleck, L. D., and Foote, R. H. (1969). Estimation of testicular size in the live rabbit. *Int. J. Fert.* **14**, 188–191.

66. Powsner, E. R., and Fly, M. N. (1962). Aseptic aspiration of bone marrow from the living rabbit. *J. Appl. Physiol.* **17**, 1021–1022.

67. Randall, L. O. (1951). Synthetic curare-like agents and their antagonists. *Ann. N. Y. Acad. Sci.* **54**, 460.

68. Rapson, W. S., and Jones, T. C. (1964). Restraint of rabbits by hypnosis. *Lab. Anim. Care* **14**, 131–133.

69. Rich, R. D., and Alliston, C. W. (1970). Influence of programmed circadian temperature changes on the reproductive performance of rabbits acclimated to two different temperatures. *J. Anim. Sci.* **30**, 960–965.

70. Roche, J. F., Dzuik, P. J., and Lodge, J. R. (1968). Competition

between fresh and aged spermatozoa in fertilizing rabbit eggs. *J. Reprod. Fert.* **16**, 155.

71. Sanders, A. G., Dodson, L. F., and Florey, H. W. (1954). An improved method for the production of tubercles in a chamber in a rabbit's ear. *Brit. J. Exp. Pathol.* **35**, 331.

72. Sandison, J. C. (1928). The transparent chamber of the rabbit's ear, giving a complete description of improved technic of construction and introduction, and general account of growth and behavior of living cells and tissues as seen with the microscope. *Amer. J. Anat.* **41**, 447–470.

73. Schildt, B. E., and Schildt, E. E. (1962). Thialisobumal (baytinal) as an intravenous anesthetic for rabbits. *Acta Pharmacol. Toxicol.* **19**, 377.

74. Sholkoff, S. D., Glickman, M. G., and Powell, M. R. (1969). Restraint of small animals for radiopharmaceutical studies. *Lab. Anim. Care* **19**, 662–663.

75. Smith, P. E., and Shite, W. E. (1931). The effect of hypophysectomy on ovulation and corpus luteum formation in the rabbit. *J. Amer. Med. Ass.* **97**, 1861–1863.

76. Sollmann, T. (1957). "A Manual of Pharmacology," p. 677. Saunders, Philadelphia, Pennsylvania.

77. Stone, H. H., and Prijot, E. L. (1955). The effect of a barbiturate and paraldehyde on acqueous humor dynamics in rabbits. *Arch. Ophthalmol.* [N. S.] **54**, 834.

78. Strack, L. E., and Kaplan, H. M. (1968). Fentanyl and droperidol for surgical anesthesia of rabbits. *J. Amer. Vet. Med. Ass.* **153**, 822–825.

79. Sundberg, R. D., and Hodgson, R. E. (1949). Aspiration of bone marrow in laboratory animals. *Blood* **4**, 557.

80. Thomas, E. D. (1955). *In vitro* studies of erythropoiesis. I. The effect of normal serum on heme synthesis and oxygen consumption by bone marrow. *Blood* **10**, 600.

81. UFAW, ed. (1967). "The UFAW Handbook on The Care and Management of Laboratory Animals," 3rd ed., pp. 114–117. Livingstone, Edinburgh.

82. Venge, O. (1952). A method for continuous chromosome control of growing rabbits. *Nature* (*London*) **169**, 590.

83. Walton, A. (1933). "The Technique of Artificial Insemination," introductory chapter. Imperial Bureau of Animal Genetics. Oliver & Boyd, Edinburgh.

84. Walton, A. (1958). Improvement in the design of an artificial vagina for the rabbit. *J. Physiol.* (*London*) **143**, 26P–28P.

85. Watson, S. C., and Cowie, A. T. (1966). A simple closed-circuit apparatus for cyclopropane and halothane anesthesia of the rabbit. *Lab. Anim. Care* **16**, 515–519.

86. White, I. G. (1955). The collection of rabbit semen. *Aust. J. Exp. Biol.* **33**, 367–369.

87. White, S. W. (1966). Adrenalectomy in the rabbit. *Aust. J. Exp. Biol. Med. Sci.* **44**, 447–450.

88. Wright, J. G. (1966). "Wright's Veterinary Anesthesia and Analgesia," 6th ed., pp. 193–196. Williams & Wilkins, Baltimore, Maryland.

CHAPTER 5

The Fetus in Experimental Teratology

Howard A. Hartman

I. INTRODUCTION

In the past decade, increased emphasis has been placed on the use of the fetus of various laboratory animal species to detect the presence and/or nature of teratogenic potential of chemical entities prior to their use in human trials to evaluate therapeutic effectiveness. Similar investigations are required to evaluate entities being developed as food additives, pesticides, etc. (54, 61, 81, 82, 96, 119, 132, 151, 234, 303, 311, 401, 414, 415, 428).

The usefulness of animal testing prior to evaluation in humans is frequently questioned with regard to the ever present hazard of interspecies extrapolation of data (56, 223, 228). Yet, only brief reflection upon the gravity of the thalidomide catastrophy is required to reinforce the need to foster and apply continually improving technique to all aspects of pathological embryology. In this emphasis, care must be taken to counter ultraconservative attitudes which might deny development of effective pharmaceuticals and useful chemical products (96, 254, 312, 330).

The user of fetal laboratory animals is usually confronted by the lack of a single source of information dealing with the embryology and fetal pathology peculiar to a particular species or strain. This chapter represents an attempt to collect and assemble appropriate data derived from recent and current reports concerned with the rabbit fetus. The majority of these reports deal with the evaluation of rabbit fetal development following performance of specific experimental procedures on the doe and/or fetus. From these citations, tabulations have been arranged to depict the pathological changes and anatomical variations reported in the fetus. Attention was also directed to a review of embryological phenomena. Although fragmentary, an array of experimental situations created for the pregnant doe or performed on the fetus are presented in tabular form. In addition to the latter, consideration was given to aspects of experimental design applicable to teratological investigation.

These considerations were chosen to provide accessability to relatively obscure information and satisfy the continued need for periodical review and correlation of existing information (177, 426). For those unfamiliar with experimental teratology, it is hoped this chapter will serve as a general review and guide to more detailed reference material. For those currently involved in the field, the tabulations of pathological phenomena noted in rabbit fetuses will be of particular value since few laboratories have yet individually assembled a comparable collection of material. Also, the tables summarizing documented experimental results may add to the design of future experiments.

The basic objective was devoted to enhance the perspective of the biological research worker's understanding of the species with which he must work. Maintaining one's awareness of advantages and limitations related to experi-

mental use of any species and interpreting the data obtained through its use requires little comment.

"Teratology" by current definition encompasses those aspects of embryology and pathology concerned with the study of abnormal development and congenital malformations (109). The term was coined in 1832 by Geoffrey Saint-Hilares (343), making tangible a science evolving from concern over the nature and cause of the monstrosities and various types of defective development seen in children and animals at birth. Legends, reports, and subsequent reviews dealing with these phenomena represent a vast body of literature (36, 343, 418). The degree to which one may wish to broaden the definition of "defect" or malformation has fostered much controversy over terminology and "acceptable ranges" of normal variation, etc. (137, 226, 261, 417, 437). It is conceivable and indeed a practice by some to stress the inclusion of clinical, microscopic, and biochemical abnormalities (224, 267). Unfortunately, most data currently available have been derived from only descriptive anatomical observations. Comparatively little information has been presented concerning abnormal histology and far less about disturbed fetal systemic, biochemical, or cytological phenomena (223, 228). These areas exist as extremely challenging research frontiers in biology (203).

As the science of pathological embryology (175, 177), teratology offers common research areas to the anatomist, embryologist, biochemist, toxicologist, pathologist, physician, and veterinarian. As might be expected, this mixed disciplinary involvement possesses inherent problems, such that "embryologists need to know more pathology and pathologists more embryology," etc. (426). The bias unique to the disciplines involved in teratology must be countered by tolerance and deliberate efforts to recognize, utilize, and supplement the potential of each toward the solution of the multifactoral aspects inherent in the expression of birth defects. While the single cause concept is diminishing, the need to detect the individual factors is gaining more emphasis (138). Hopefully, from this will evolve uniformity in evaluation techniques and a more penetrating analytical approach to explain the complex interrelationships uncovered through research. A coordinated interdisciplinary approach will inevitably bridge many existing gaps toward gaining insights if not answers to the conditions in question (31, 62, 160, 179, 180, 312, 316, 420).

Recently in teratology there has been a rapid development of a toxicologically oriented subgroup (45, 312). Its goal, as that of others, is the prevention of environmentally induced fetal disease in humans; particularly those of chemical origin. Their challenge is the development and improvement of preliminary testing methods of all types to detect or predict the possible teratogenic potential of new chemical entities being developed as drugs, pesticides, food additives, etc. (54, 62, 339). Thus, they face an immediate practical problem of great magnitude, namely the applied problem of evaluating large numbers of compounds for possible teratogenic risk. As such, their activities are frequently preliminary to more academic endeavors of exploring pathogenetic mechanisms.

II. SUMMARIZED CYCLIC AND EMBRYOLOGICAL PHENOMENA

A. Maternal Reproductive Tract

1. Anatomical Features

The rabbit uterus exists in complete duplication. While fused externally near the vagina, two separate uterocervico-vaginal canals exist. A corpus of the uterus is absent. The cervices are not plugged at any time by a mucus seal (114, 335).

The uterine vascular system is composed of longitudinal and circular vascular channels which communicate frequently. There are essentially two uterine blood flows, one to the uterus, not associated with placentae, and the other to the placentae (335). During pregnancy the maternal uterine and fetal circulations cannot be effectively separated from each other. They must be considered as a complex, essentially single unit, each acting on and/or dependent on the other.

At about 20–22 days of pregnancy, a reduced maternal uterine blood flow occurs. It is due to vascular compression caused by uterine stretching to accomodate spherical enlargement of the conceptuses. Also, the concommitant acceleration of fetal growth causes additional stretching. Spontaneous fetal deaths have been noted to increase at this point in time, and may be related to transient fetal anoxia (79).

Uterine blood flow must also be considered relative to the evolving stages of placental development (335). The developing discoid placentae can only be accommodated to a certain degree by the uterine circulation. This either restricts placental development or causes local nutritive deficit and prenatal death (7). The hemodynamics of placental blood flow is considered as a direct contributing factor to fetal growth (262). Hence the changing variety of structural and functional placental factors directly influence litter development.

2. Maternal Cyclical Phenomena

Rabbits are usually in a condition of estrus. However, short diestral periods do occur during which they refuse the male. While estrus cannot be determined by vaginal smear, 4- to 6-day long cyclic cellular phenomena have been detected in the cytology of the vaginal content. The only external sign of estrus is that of a congested dark-colored vaginal mucosa and a slight swelling of the lower end of the vagina and external genitalia. Some question exists concerning the reliability of these signs. The female remains in heat up to 36 days during which time 7-to 10-day long cycles of follicular maturation occur. After this, follicles become atretic. Coital stimulation induces further growth with rupture of the follicles after 10 hours (114–116, 198).

Puberty is dependent on the breed, state of nutrition, and time of birth. It takes place at approximately 5½ months for fall-born and 8½ months for spring-born animals. In general the smaller breeds of rabbit, e.g., Polish, reach puberty earlier than do the larger breeds, e.g., New Zealand White.

Ovulation is a neuroendocrine reflex, usually dependent on coital stimulus. Manual stimulation of the external genitalia, mounting by other females, and diverse experimental procedures can induce ovulation. Seasonal effects on ovulation have been reported (127). Even though pregnant females will mate, ovulation does not occur. Ovulation is thought to be induced by release of gonadotropin (LH) from the anterior pituitary.

In the ovary, interstitial cells are prominent and contain eosinophilic and lipoid inclusions in the cytoplasm (143, 197). This phenomenon is unique to the rabbit and is also referred to as the "glande interstitielle." The mature follicle at coitus is about 1.5 mm in diameter, relatively flat, and bluish tinged. Following coitus, the follicle protrudes slightly, vascularity increases, and macula pellucida (clear area) forms prior to ovulation. One polyovular follicle is noted per 200 follicles (28). Eight days following ovulation the luteal cells reach full development. In pseudopregnancy, luteal cell regression occurs by 18 days, and the corpora lutea become a chalky yellow color. In pregnancy, the corpora lutea persist, and finally regress during lactation.

Pseudopregnancy occurs in response to coitus or suitable physical stimuli such as thread or scalpel injury to the uterine mucosa. It is characterized by persistent functional corpora lutea lasting for about two weeks, during which time they induce histological changes in the uterus and mammary glands (116). At 16 to 19 days, habits common at delivery such as lactation and nest building occur followed by receptivity to mating and ovulation (115). These phenomena require progesterone from the corpora lutea. Pseudopregnant females can also be induced by intravenous (IV) injection of 17 IU of human chorionic gonadotropin (HCG) (Gonogen) (266).

Gestation is approximately 31–32 days in length with a range of 30–35 (28, 83, 115). During pregnancy the vagina enlarges markedly and regresses following parturition. Pregnancy can be determined by abdominal palpation as early as 9 days of gestation (392). In the cervix, the number of clear gobletlike cells increases in pregnancy and pseudopregnancy. Duration of gestation is not seemingly affected by season, age, and number of previous pregnancies. Poor physical condition is thought to lengthen it (430).

Conception rate variations have been recorded, the trend being to reduction from higher spring rates to lower rates in the fall (377).

Implantation and placental development occur initially on days 6–7 when the abembryonic pole of the blastocyst develops a sticky substance thought to be related to the different metabolic patterns existing between the embryonic and abembryonic poles of the blastocyst. At attachment, the endometrial endothelium breaks down. Blastocyst attachment is thought to depend on mechanisms related to the myometrial muscle, adhesives and inversion (183). At eight days the placenta or gestation sac is 1 cm in diameter. The placenta is made up of two swellings on the mesometrial side arising from placental folds of the uterus. It is of simple discoid shape and superficially or centrally oriented in the uterine lumen.

After implantation on day 7, nutrients are initially provided by the yolk sac (19, 104, 105). The nonvascular portion of the yolk sac begins to degenerate by day 10, but the vascular portion persists. During days 10–17, establishment and development of the true chorioallantoic placenta occurs (79).

Histologically the placenta is hemoendothelial in character from day 22 to term, consisting of a virtual intermingling of maternal and fetal blood separated only by the fetal endothelium of the chorionic blood vessels. Initially at day 8 the developing rabbit placenta is endothelial—chorial but ultimately becomes hemoendothelial as pregnancy progresses (19, 25).

Early workers considered histological differences in maternal fetal placental relationships to be indicative of significant physiological differences (19). It must be recognized that although typical histological features of placental structure have been defined as unique for various species, the differences are not sharply defined within one placenta or at various stages of pregnancy. Structural changes take place through the life span of the placenta: The morphology is not uniformly constant. In reality the placenta is subject to more standard variation than any other mammalian organ, never stabilizing structurally or functionally (284, 327).

First postpartum estrus can occur soon after parturition, even in lactating animals. Successful implantation occurs if the nursing litter size is small, perhaps one or two pups, otherwise the new embryos die in the blastocyst stage. This is thought to be relative to the greater degree of involution taking place after a larger litter (115).

B. Products of Conception—Prenatal Development

Comparatively few comprehensive treatises on rabbit embryology exist. (117, 157, 277). Fortunately texts and references are available, which, while dealing with other species, offer useful information (25, 260, 349, 350, 426, 431, 440–442). The following references deal with limited aspects of rabbit embryology:

Embryology and Physiology
  Preimplantation embryology (20, 171)
  Fetal physiology (1, 15)
  Fetal temperature (195)
  Initial breathing stimulus (154)
  Developmental pharmacology (108)
  Antibody transfer (52)
Fetal Membranes and Placenta
  Placental physiology (35, 335)
  Placental transfer (190)
  Fetal membrane morphogenesis (284)
  Fetal membranes and fluid exchange (315)
Organ and Tissue Development
  Lung (13, 236)
  Brain (129)
  Choroid plexus (395)
  Occular development (334)
  Skeleton (39, 59)
  Lymphoid tissue (5, 24, 206, 332)
  Heart (376)
  Fetal circulation (34)
  Ductus arteriosus closure (205)
  Renal development (145, 173)
  External ear (87)
  Pancreatic islets (46)

In timing prenatal anatomical development, it has been suggested that the beginning of the primitive streak (day 9) is a more accurate reference point than the time of conception since growth rates before and after primitive streak formation vary greatly. Using the onset of primitive streak formation as the starting point, prenatal growth data has been expressed in other species as straight line relationships between logarithms of the weight and age (260). Also to be considered are methods based on external features such as crown–rump measurement and the appearance of specific anatomical features, for example, the palpebral fissure, trunk papillae, digital development, ear flaps, neural tube, etc. (117). Obviously with any of these techniques considerable experience has to be gained in using the technique to assess the feasibility and reliability of the data obtained experimentally. In addition to anatomical development, attention has been directed to sequences in physiological development which has demonstrated interesting parallels among many species such as heart beat, hemoglobin, liver glycogen, lung surfactant, sodium/potassium levels, body fats, arterial pressure, etc. (15). Thus, in addition to morphological change, gradients of physiological susceptibility must be considered when evaluating teratogenic potential (426).

1. Ova and Blastocyst—Histological Characteristics

As the ova enters the proximal third of the oviduct, it is surrounded by the zona pellucida acquired while in the ovary. The corona radiata is shed in the first third of the oviduct. Retraction of these cells from the vitelline membrane is accompanied by cytoplasmic degeneration (444). The ovum diameter is 123 μm $\pm$ 1.9 μm without and 188 μm $\pm$ 2.0 μm with the zona pellucida. The eggs are essentially naked in the middle third of the oviduct. In the last third of the oviduct, the ova are surrounded by a thick layer of albumen, (20, 28, 197).

Ova have a fertile life of approximately 6 hours postovulation. Fertilization of ova can occur while they are in or as they leave the liquor folliculi at the ovarian end of the oviduct. Once the albuminous layer is acquired, they can no longer be fertilized (28).

Sufficient numbers of sperm for fertilization are present in the oviduct five hours after mating (168). Thus coitus induced ovulation in the rabbit insures that fertilization of nearly all ova is initiated within 2–3 hours after their arrival into the fallopian tube (30).

Within 2 hours postovulation most ova have travelled half the oviduct length. Following that time, the ova are delayed at the ampullary isthmic junction for about 46 hours. Progress through the isthmus is slow, the ova reaching the uterine lumen about 72 hours postovulation.

The rapid initial descent of ova has been related to strong muscular contraction and abovarian ciliary activity (169).

b. LITTER SIZE. Litter size varies with the strain or breed. It tends to increase until the doe is about three years of age. The sex ratio approximates 53% males: 47% females (28).

a. INVERSION OF THE GERM LAYERS. In the rabbit vesicle, the embryo remains superficial, whereas in most rodents an inversion occurs with the embryo inverted into the vesicle (197).

2. Embryonic and Fetal Development

The embryonic phase includes the period during which characteristic organs are being formed into definitive structure. It is followed by the fetal period which is characterized by maturation of these organ systems (259, 376).

A variety of tabular depictions of prenatal development exists for man (307, 378, 386, 388) and other species (307, 432).

Table I summarizes the characteristics of rabbit development at various times of gestation.

The following tabulation summarizes the chronological sequences of preembryonic conceptus development (11, 29, 94, 171, 189, 240):

TIMETABLE OF EARLY ZYGOTE AND BLASTOCYST DEVELOPMENT

Pregnancy time (approx.)		
Hours	Days	Event
0	0	Coitus
22–26		2-Cell cleavage state
30–34		4-cell cleavage state
38–42		8-Cell cleavage state
46–52		16-Cell cleavage state
41–75		Morula state
60		Slight changes in peripheral cells destined to become trophoblasts
70–80		Blastocyst reaches uterus (depends on length of oviduct)
75		Appearance of blastodermic vesicle
		Cleft appears to separate the primitive mass cells from trophoblastic cells
90		Enlargement of blastocyst and increased number of cells in trophoblast
	5 (Entire diameter 1.3–2.2 mm)	Blastocyst embryonic disc, indefinite outline, 1–2 cell layers thick
	(Disc diameter 0.52–0.90 mm)	Round disc, smooth outline about 3 cell layers thick
	6 (Entire diameter 2.2–4.8 mm)	Disc elongation and growth in posterior direction with slight thickening at anterior end
	(Disc, diameter 0.92–1.35 mm)	Beginning of primitive streak formation with cell condensation on midline
		Primitive streak present in the posterior part of the disc
	7 early (Disc diameter 0.01–1.33 mm)	Mesoderm begins to grow fan-shaped from posterior end of disc

Fetal weight increases rapidly during the last 10 days of pregnancy. At day 20 the fetus weighs less than 5 gm, thereafter it gains about 2 gm per day until day 22, 4 gm per day until day 24, and until day 30 about 5 gm per day (247). One day prior to term fetal weight is approximately 40 gm. In depth considerations of human fetal body and organ weights versus gestational age offer useful insights (299, 380).

Placental weight increases gradually until the twenty-fourth day of pregnancy, thereafter remaining relatively constant. At day 20 the placenta is about one-half the weight of the fetus. The placenta gains about 0.5 gm per day until day 26 (249).

Amniotic fluid decreases in volume rapidly after the twenty-sixth day until term. At day 20 the fluid is about equal to the fetal weight (249).

TABLE I

STAGES OF EMBRYOLOGICAL DEVELOPMENT

Stage[a]	No. of somites[a]	Length[a] (mm)	Day of pregnancy[a]	Stage[b]	Main commencing features[b]
1–7 Cleavage and Bastula	—		0–3.5		
8–11 Gastrula	—		3.5–6.5		
13–17 Neurula	5–8				
	8–9		7.75	1	1–4 Pairs of somites, neural tube open throughout
			8.25	2	Heart formed, neural tube closed to hindbrain
	10–16		8.5	3	Anterior neuropore formation
	16–23		9.0	4	Torsion of embryo, closure of anterior neuropore
18–24 Tailbud embryo		3.2–4.4	9.5	5	Appearance of mandibular and hyoid arches
			9.75		
25 Complete embryo	29		10		
26–33 Metamorphosis	32		10.5	6	Appearance of 3rd and 4th branchial arches
	36–39		11	7	Formation of anterior limb buds
		4.4–6.3	11.5	8	Formation of nasal pits
			12		
			12.5	9	Division of forelimb buds
		8.5±	13	10	Appearance of tubercles on mandibular and hyoid arches
		10.2±	14	11	Ear flap formed, pigment appears in eye
34 Fetal I		12.5±	15	12	Papilla appears over eyes, vibrissary papillae in four rows
		14.8±	16		
		18.0±	16.5	13	Appearance of membranous eyelids, separation of forelimb digits begin
		21.3±	17	14	Complete separation of forelimb digits
			17.5		
		25.0	18	15	First appearance of papillae on limbs
			18.75		
			19	16	Palpebral fissure closes
35 Fetal II	29		20	17	Claws formed
			21		
			24		
			28		
36 Fetal III; ends at birth			32		

[a] Data cited in ref. 376, 431.

[b] Data cited in ref. 117.

3. Prenatal Mortality

Mortality *in utero* is determined by comparing numbers of corpora lutea, numbers of implantation sites, and numbers of fetuses present late in pregnancy. It is subject to error due to inaccurate counting of corpora lutea and implant sites (6). A more reliable method is to examine for uterine swellings by laparotomy shortly after implantation (days 8–10), and subsequent necropsy examination at term (days 28 to 30) (6, 183). Some evidence exists to relate preimplantation loss to chromosome defects in aging ova (30).

Postimplantation embryonic survival depends on inheritance, age of doe, exposure to unknown blood-borne substances, age of gametes at fertilization, and hormonal requirements for maintenance of pregnancy. Survival is also considered to be positively related to the number of implantations, in that, survival decreases above and below an optimal number per uterine horn (183).

Prenatal death classifications should include estimated time of pregnancy and/or anatomical features of the uterus or fetus (7) as follows: *Early*—before 10 days, uterine swellings of 10 mm or less at 10 days; *middle*—10 to 17 days, establishment of true chorioallantoic placenta; and *late*—17 to 24 days, well-established placenta or dead fetus.

Prenatal mortality is recognized as variable and a significant feature to consider in evaluating experimental data (159). Little information exists concerning either the frequency of prenatal mortality in untreated does or the periods of pregnancy during which mortality occurs. In one report a preimplantation ova loss of 9.7% is reported and accompanied by a 7% loss immediately after implantation, 12% between days 8 and 17, and 6% between days 17 and 23 of pregnancy (6).

III. SUMMARIZED LABORATORY TECHNIQUES FOR OBSERVATION AND EVALUATION

The following comments are appropriate regardless of the species being considered.

1. Husbandry

Like other small laboratory animals, the rabbit offers the obvious conveniences of easy handling and maintenance for short periods in the laboratory (10, 65, 301). Unlike the small rodents, establishing and maintaining a trouble-free rabbit breeding colony by the user seldom seems to be a solution to one of the chief practical problems—procurement of both healthy and uniform test animals. Of concern are standards of nutrition and sanitation to which successive animal groups were exposed. Maintenance of colony health is paramount to success in the conduct of one's experiment. The experiences of frustrating delays and losses from a variety of respiratory and intestinal diseases or nutritional imbalance are not difficult to elicit from the rabbit user. Acclimatization in the users laboratory is helpful to combat problems of infectious disease. For teratological investigation it is highly desirable to obtain animals with a uniform genetic background. It is unfortunate if collections of rabbits from various sources are used since this introduces the unpredictable effects of dissimilar genetic pools on the experiment.

It is desirable to maintain a uniform environmental situation for the test animals during the experiment including approximately 12 hours of controlled light daily (131). A disadvantage related to the doe is the inaccuracy inherent in recording body weight and food consumption data—parameters found useful with other species as monitors for adverse or toxicological effects related to one's experimental procedure.

Of particular advantage are the facts that the rabbit does not spontaneously ovulate and can be artificially inseminated. Both allow accurate pregnancy timing. The fetuses are large and easy to examine (147).

2. Mating

a. SIRE.   For conventional studies either natural mating or artificial insemination techniques may be used. The first requisite for either is the availability of a suitable number of proven bucks.

In natural mating, it is customary to take the female to the male's cage, observe the mating process, and remove the female. The rate of successful mating is diminished considerably if the male is taken to the female's cage (10). One variation is the double mating technique, wherein the female is placed first with one buck for observed mating and then immediately removed to the second buck for an-other mating. The entire process usually takes less than 10 minutes per female and results in a conception rate of 80 to 90%. In either case, it is desirable to select does for receptivity based on the presence of a congested vaginal mucosa (10). Selection of the doe can also be based on receptivity for vasectomized bucks (131).

With either natural or artificial insemination techniques, consideration must be given to semen characteristics and quality (17). While it has been shown that seminal fluid is needed for maintaining viability in saline suspensions, pooling semen samples from different males is not adverse to sperm viability. Correlations with sperm morphology (staining, head caplessness, and length) have been made to percentage of ova fertilized and mean litter size. These characteristics of sperm morphology have been cited as indicators of potential fertility (300). Spermatozoa retain their fertilizing capacity for 20–30 hours (147, 285, 424).

Collection of rabbit semen for artificial insemination can be accomplished using an artificial vagina (53, 147, 286, 301, 399). Staining techniques for microscopy (44) and electronic equipment for counting of spermatozoa exist. In the latter a major problem encountered is the presence of irrelevant granular contaminants which cannot be discriminated against by the electronic counter (237). A cyclical variation in sperm count of unknown cause has been described (107).

The principal steps in the artificial insemination technique include (147): (a) collection in a warm (45°C) lined artificial vagina using a teaser female; (b) use of males experienced in natural breeding; (c) sperm motility and numbers evaluated by hemocytometer reading; (d) dilution to 12 million sperm per mm^3; motility evaluations before use of sample; (f) insemination performed with restrained female head down, vagina up (vertical position): and (g) a nonbreakable pipette is used to deliver about 0.25 mm of semen (3 million motile sperm) beyond the urethral orifice when resistance to insertion is detected. A small volume of concentrated sperm is desirable.

Intraperitoneal insemination is also possible in rabbits (8, 9, 182).

b. DAM.   In any normal female rabbit population a substantial portion can be expected not to mate at any given time. This creates a problem in executing studies requiring large numbers of pregnant animals (131).

Various techniques have been used to compensate for this, such as physical stimulation of the vagina with glass rods, artificial insemination, selecting does for receptivity based on vaginal congestion, use of various hormones in combination with natural or artificial physical stimuli, etc. From these, the intravenous use of purified pituitary luteinizing hormone (PLH—Armour) (2.5 mg/2–3 kg doe) has been found to produce ovulation in both receptive and nonreceptive does (131). Induced ovulation occurs chiefly between 10 and 13 hours postinjection, at a time similar to that occurring naturally (193). Such superovulation is accompanied by the inherent risks of uterine overcrowding and prenatal mortality (6, 7).

The need for accuracy in timing of mating and estimating the times of ovulation the onset and stage of pregnancy are extremely important for both scientific and practical reasons. Ova are fertilizable for 6–8 hours postovulation and sperm capable of effecting fertilization for about 20–30 hours after ejaculation (112). Thus, errors in timing alone related to mating technique might introduce a reduced pregnancy rate.

Adherence to routine scheduled collection and preparation of semen, use of pituitary luteinizing hormone, or conventional natural mating procedures allows an accurate estimate of the stages of pregnancy which exist when specific dosages or techniques are applied during an experiment.

Accurate individual animal pregnancy timing and preterm maternal sacrifice timing are critical when examining the fetal specimen. Consideration of each observation should be made versus the actual age of the fetus. It is well known that rapid changes of preterm fetal maturation (83) such as the degree of ossification (325) and fetal size occur during short time periods late in pregnancy.

Timing is, therefore, important in determining the schedules for each pregnant dam for preterm Caesarian delivery of the fetuses. This minimizes the time-related variables of size and skeletal maturation from complicating the experimental data.

In the rabbit, consideration is given to the 24-hour fetal survival rate. As fetuses are delivered earlier than one day preterm, the overnight mortality rate increases introducing additional problems for interpretation.

In an equally important but practical sense, scheduled breeding including terminal necropsy to avoid Caesarian delivery and subsequent detailed fetal examination on weekends and holidays fosters uniformity in the examination techniques rendered. Scheduled during the work week, adequate numbers of skilled personnel are available.

3. Preimplantation Ova and Blastocyst Techniques

The suitability of the rabbit for ova or blastocyst collection and transfer techniques offers an extremely useful approach to experimental embryology (7, 57, 93, 94, 189, 242, 266, 313). The rabbit ova is unique from most species in that it develops an albuminous coat in the oviduct which presumably makes it more tolerant to various *in vitro* and *in vivo* manipulations (170, 184). In many other species the fertilized ovum is less adaptive to collection, *in utero* culture and transfer, etc., apparently due to the absence of this coating. These techniques have been advocated for evaluating the effects of maternal administration of

chemical agents on the preimplantation zygote (11, 12, 93, 123). The usefulness of this technique depends a great deal on the careful regulation of dose, stage of pregnancy, and collection timing. Large numbers of blastocysts must be examined to compensate for the wide ranges of morphological variability existing between and within litters (255). It is considered that the rapidly multiplying cells of the embryonic disc at that period are quite susceptible to the action of the external agents (11).

Preimplantation fertilized zygotes or ova collected from sacrificed donor females can be fixed and stained immediately for histological evaluation, stored *in vitro* pending transplantation to recipient females, or cultured *in vitro* prior to fixation and further examination. Also, by quick-freezing of the entire uterus after killing of the female, assays of radioactivity content can be accomplished on frozen dissected blastocysts (123).

Most ova or zygotes are found in the oviducts up to 70 hours postcopulation. Flushing the freshly dissected oviduct with 10–15 ml sterile physiological saline or Locke's fluid via the fimbra of the oviduct is recommended for collection. After 70 hours, since most ova will be located *in utero*, reverse flushing of the uterine horn with the oviducts removed at the uterine junction is satisfactory (171). The flushings can be examined microscopically and dimensions of the blastocyst and zone pellucida determined using an ocular micrometer (266).

Perhaps the best approach to fixation depends on the user's laboratory evaluation of the available variety of processing techniques preliminary to performing the actual study (11, 62, 171, 278, 308).

In vitro holding and culture techniques allow preparation of specimens for examination or storage prior to transfer to recipient females or fixation. The suitability of any one medium for this purpose depends somewhat on the experimental design (11, 93, 94, 184, 313). Induction of pseudopregnancy has been used as a culture mechanism for blastocysts (266).

Methylene blue (0.05 mg/ml) is a nontoxic vital stain which facilitates examinations of viable ova (and sperm) in vitro (94).

Ova or blastocyst transfer can be accomplished using aseptic laboratory technique with direct instillation of the material into the uterine horn (7, 11) or via flank laparotomy with introduction at the fimbria (266). For successful transfer of zygotes, the stages of development must closely approximate the stage of uterine development in the recipient. Zygotes slightly older tend to survive better than those slightly younger than the recipient's uterine development stage (184).

In the rabbit, refractile ova can be detected in the ovarian end of the uterine horns at 4 days 6 hours. By 5 days 22 hours dilations of the uterine lumen exist. Early gestation swellings are noted at 6 days 22 hours. Mesometrial opacity appears between 7 days 22 hours and 8 days 2 hours. A convenient technique for the detection of these early phenomena consists of bleaching the uterine tracts in hydrogen peroxide (H_2O_2) and clearing through benzol to benzyl benzoate and viewing them with oblique lighting and a dissection microscope. This material is also suitable for further processing for microscopic evaluation of the uterus and trophoblast (284).

4. Fetal Examination Techniques

The usual teratological experiment includes three types of postmortem examinations of the fetus collected before delivery or at birth. These are an external examination, soft tissue examination of brain, thoracic and abdominal viscera, and a skeletal examination. All three types must be performed with uniform levels of diligence and precision. Appropriate anatomical (84, 85) and embryological (117, 157) references should be available for consultation. Criteria should be established and adhered to with reference to the limits of morphological variation deemed "normal" (312). It is also helpful to keep the same personnel involved in any one experiment for each examination procedure for uniformity in data collection. In the long run, documentation and tabulation of all findings regardless of how insignificant they might appear affords one the opportunity of a more effective evaluation. Simply, it is easier to ascribe a lack of significance to findings than to regret a lack of diligence or attention to detail toward specimens no longer available for examination.

The data obtained in a routine experiment from the three general types of examination pose a problem for collection and extraction for review. In addition to the morphological data obtained through the fetal postmortem examinations, it is also important to evaluate the following parameters for each litter: number of resorption sites, number of corpora lutea in each ovary, number of live and dead fetuses, fetal weight, sex, crown–rump measurement (distance from base of tail to anterior curvature of skull with fetus lying free on one side), and transumbilical distance (the fetal diameter at the level of the umbilical cord). Record sheets for these data obviously will vary in format, yet should be concise, comprehensive, and versatile in design.

a. EXTERNAL EXAMINATION. This examination is essentially repeated three times. It is first performed during removal of the fetuses from the uterus, when identifying them and placing them in a 24-hour incubator chamber. Second, during humane killing after the 24-hour incubation period and, third, at the time of evisceration and skinning of the carcass. During the initial exam, close attention is paid to bilateral symmetry of limbs, digits, eyes, ears, and the general body and head conformation. At this time, fetal weight, crown–rump measurement and transumbilical dis-

tance are recorded. Depending on the nature of the experimental procedures, additional examinations might be considered such as anogenital distance measurement (352). During the examination, pathological changes and suspicious anatomical variations should be compared with littermates and litters being examined from other (control) groups. Should features be detected which can be better evaluated by microscopic examination, either a part or the entire fetus can be preserved in formalin or Bouin's solution, appropriately decalcified, and processed.

The 24-hour incubation period allows an assessment to be made of respiratory activity, urine flow, behavior, and group viability. Humane killing is performed using deep ether anesthesia or excessive CO_2. Routinely the majority of fetuses are preserved in 95% alcohol and the remaining fetuses in Bouin's fluid.

b. SOFT TISSUE OR VISCERAL EXAMINATION. Since the majority of fetuses are destined for further processing and skeletal examination, this examination can be performed using a two-phase technique. For those preserved in 95% alcohol destined for skeletal examination, the skin and thoracic and abdominal viscera must be removed prior to skeletal staining. If evisceration is done carefully a reasonable assessment can be made of the components in spite of disturbed orientation due to evisceration. At this time, the fetal sex is recorded. The second phase involves selecting a certain population of fetuses and preserving them in Bouin's fluid for about 2 weeks to allow for sufficient decalcification prior to dissection. These fetuses can then be dissected in a manner deemed appropriate whether it be by serial cross section of the entire carcass at 3-mm intervals (37, 429) or by using conventional dissection technique following a ventral midline incision of the thorax and abdomen. It is also feasible to mount the specimens in thin plastic bags for storage and additional examination (369). The use of a dissecting stereo microscope or a magnifying lens is helpful for either phase. While not always practical, in these dissection or evisceration techniques it is extremely effective if each litter is examined simultaneously, i.e., Bouin's preserved littermates are examined at the same time as those fetuses destined for skeletal processing are eviscerated. Thus, findings from either procedure may prompt review of the littermates to determine the incidence and degree of variation present, etc. Bouin's fixed specimens are also suitable for histological evaluation either as specific tissues or as full cross sections of carcass and viscera. Preparation of whole body sagittal parafin sections is also feasible (218).

c. SKELETAL EXAMINATION (Alizarin Red S Techniques). The ease with which the fetal skeleton can be processed and visualized allows careful evaluation to be made of its development, the features observed serving as indicators of possible previous insult of this system during organogenesis. Few articles deal in detail with fetal skeletal development in rabbits. Several comprehensive studies are reported for the mouse (215, 350) and rat (389, 414, 436). These studies are useful since they provide a needed perspective when considering observations made in rabbits.

The need for accurate pregnancy timing in reference to skeletal examination has been mentioned previously under Mating Techniques.

To facilitate processing, modified compartmented plastic ice cube trays or commercially available plastic fishing tackle boxes are useful for handling several fetuses simultaneously. Either type should be modified by drilling the necessary drainage holes.

The soft tissues of the skinned 95% alcohol fixed fetuses are cleared with potassium hydroxide and then stained with alizarin red S to demonstrate calcified bone. The requirements of this technique are clarity, specimen flexibility, long keeping qualities, and uniformity in reaction through processing (86). Alizarin red S has also been used *in vivo* to demonstrate bone growth postnatally (207). This technique may have some application *in utero*.

Following early descriptions (100, 371), a wide variety of modifications have been suggested (214, 331, 352, 370, 384), most of them claiming a shorter processing time requirement. A satisfactory technique follows: (1) 95% alcohol fixation (minimum 3–4 days) of fetuses prior to skinning and followed by 3–4 additional days fixation in alcohol. Paw and tail skinning are not necessary; (2) clearing in 2% KOH for 24–30 hours until slight clearing is evident. Total immersion in this solution is imperative; (3) fetal staining for approximately 16 hours, the actual time controlled by visual inspection. Once adequate reactions are visible in the limbs and spinal vertebrae the reaction is considered complete. Care must be exercised that the KOH concentration is not too great or duration too long to avoid cartilage disintegration. The staining solutions consist of 2% KOH (500 cc) and distilled water (4000 cc); alizarin red S (15 cc of a 0.5% aqueous solution); (4) dehydration performed for 12–18 hours in 70% alcohol; and (5) examination and storage; glycerol is used as the medium for examination and subsequent storage of the fetal specimens. Glycerol should be replaced frequently to avoid alcohol contamination which will cause a brown discoloration to occur in the cleared fetal soft tissue.

d. Cartilage Staining. By modifying the skeletal staining procedure and including toluidine blue stain, some estimate of the cartilaginous skeletal characteristics can be ascertained (60).

e. Radiographic Examination. Conventional radiographic technique can be used to examine the fetal skeletons without resorting to chemical clearing and staining processes. This allows more facility in examining the soft tissues by detailed dissection and histological technique (308). Radiographs are limiting in that they do not provide a three-dimensional model as do the alizarin preparations, requiring additional positioning to achieve that perspective. A dental radiographic unit might be usefully employed for fetal radiography.

f. Fetal Specimen Storage. The use of plastic bags and containers of various descriptions has facilitated convenient short-term or permanent storage of fetal specimens (26, 121, 369). Reusable tape labels which can be easily removed from an initial storage container and reapplied to the plastic containers reduce error and time required for specimen identification.

5. Placental Examination Techniques

The placenta is a junction of interdependence between two circulatory systems, thus the fetus and placenta cannot be separated during pregnancy. As yet, little appreciation exists concerning pathological change in experimental animal placental tissue and its association with the occurrence of fetal malformation. The need to explore techniques for evaluating the placenta in experimental studies is obvious (47, 426). In humans such correlations have been investigated (178).

A comprehensive discussion and a variety of collection and fixation techniques for preparation of histological materials have been published (283).

6. Data Handling

The data generated in a typical 4-group study which might consist of 100 pregnant does and close to 1000 fetuses all subjected to multiple (individual and collective) examinations is voluminous. The use of computer technology has facilitated its collection and manipulation (199). Also helpful are the variety of indices which have been used by various authors. Some are listed as illustrations. The applicability of an index obviously must be determined by the needs of the experiment and the data available on which it will be used. The references cited represent either the original source, or useful demonstration.

$$\% \text{ Fertility} = \frac{(\text{no. pregnant females} \times 100)}{\text{no. mated}} \quad (410)$$

$$\% \text{ Gestation} = \frac{(\text{no. live litters} \times 100)}{\text{no. pregnancies}} \quad (410)$$

$$\% \text{ Resorption} = \frac{(\text{no. resorption sites and dead fetuses} \times 100)}{\text{no. implant sites}} \quad (372)$$

$$\% \text{ Malformation} = \frac{(\text{no. malformed fetuses} \times 100)}{\text{no. implant sites}} \quad (372)$$

$$\% \text{ Retardation} = \frac{(\text{no. fetuses with retardations} \times 100)}{\text{no. implant sites}}$$

$$(372)$$

Abnormality score $(A) = \dfrac{(C - N)}{(C)} = 1 - \dfrac{(N)}{(C)}$ (372)

C = average no. implant sites in controls
N = no. normal fetuses from each mother at term

Surface to body weight ratio (372)

Teratogenic ratio $= \dfrac{\text{maternal toxic dose}}{\text{fetal toxic dose}}$ (150)

Animal teratogenic dose versus human therapeutic dose (150, 348)
  (Chances of safety considered to increase as animal dose becomes greater than human dose.)

Teratogenic index $= \dfrac{\text{teratogenic dose in animals}}{\text{teratogenic dose in humans}}$ (62)

Average effective dose = produces pharmacological effects in 50% of animals (62)

Average teratogenic dose = producing congenital malformations in embryo (62)

Minimal LD 100 = minimal dose to produce 100% litter resorptions (287)

Teratogenic range = dose levels in which abnormalities are produced (287)

Noneffective range = no abnormalities produced (287)

7. Statistical Analysis

Statistical evaluation is unusually perplexing in that the date generated in teratological investigation arises from many parameters at the same and different points in time if one considers the timetable of embryological development and the pathogenesis of malformations. The parameters used and detected observations must be carefully considered individually and collectively when selecting the appropriate testing procedures. At present, statistical treatments appear conspicuous by their absence—conspicuous in the sense that not only does a vast armamentarium of mathematical and statistical tools exist, but even more so in the presence of an ever growing and generally available computer technology to execute them (268, 385).

8. Historical Control Data

As laboratory experiments are conducted, it frequently becomes of value to maintain an accumulating tabulation of experimental control data or references for current studies. Such a data collection system would indicate the incidence of various types of fetal malformations, variations in development, fetal weights and dimensions, resorption rates, etc., considered versus gestational age. Unfortunately, source and stock or strain variation limit the usefulness of such data for rabbit experimentation.

IV. FETAL PATHOLOGY AND ANATOMICAL VARIATION

A. General Comments

"Developmental Pathology deserves to be recognized as a portion of pathology and biology in which much has been accomplished" (177, 179). While considering the following tabular presentations of cited naturally occurring and induced phenomena noted in rabbit fetuses (Tables II–XIV) the reader is encouraged to "extrapolate in reverse," that is, from humans to animals. In no other species does more extensive documentation exist concerning prenatal disease than for humans (e.g., 280, 347, 396). These data offer a significant resource to evaluating the nature and pathogenesis of fetal disease in animals. Such extrapolation in reverse is hardly an unfair consideration since the research worker is continually challenged to extrapolate the relevance of animal data to humans.

Human prenatal pathology is the subject of texts (e.g., 333, 426) and varied citations far too voluminous to detail. General references relating to nonhuman prenatal mammalian pathology are also available (296, 433, 435).

Little information exists concerning the peculiarities of rabbit fetal response to various etiological agents. Thus, information derived from experimental studies using other species must be considered when evaluating the nature of rabbit fetal disease.

The many disciplines engaged in experimental teratology are gradually broadening their orientation to include a necessary attention to the metabolic and genetic aspects of teratogenesis in addition to the morphological. Thus, it has been suggested that the term congenital malformation be applied to gross structural defects present at birth and that the term congenital anomalies be used to denote microscopic malformations and abnormal physiological or metabolic phenomena (421).

Present accomplishments, though impressive, are only preliminary toward understanding pathogenetic mechanisms of fetal and neonatal disease. It is to these problems that the experimental animal species offer great research potential. Their utilization in studying pathogenetic mechanisms has hardly begun. Deliberations summarizing aspects of current knowledge dealing with the pathogenesis of fetal disease are presented in detail elsewhere (176, 241, 269, 427).

Once an animal sensitivity to teratogenic activity can be predicted, it becomes a tool or an acceptable model to manipulate, evaluate, and hopefully clarify the multitude of factors inherent in the pathogenesis of a particular fetal disease (31, 175). Anyone of these may prove in humans to be of value in preventing or modifying the occurrence of similar fetal or neonatal pathology (420).

Considering pathogenic mechanisms involved in the expression of congenital malformations, one of the most frustrating factors confronting experimental design is the virtually impenetrable, *in utero* seclusion of the injured conceptus from practical close scrutiny. The initially important, yet poorly understood aspects of pathogenesis take place following *in utero* insult prior to the recognition of the aberrant development noted days or weeks later at term. Thus, the defects noted serve as indications or markers of earlier teratogenic effects. The complexity of the inter-related biochemical and physiological alterations presumed to be occuring in association with or preceding them should be investigated, yet these can only be speculated on at this time.

The primary injury is generally thought to be one which disturbs the delicate equilibria of cell growth resulting in cell death and/or altered cell growth, with subsequent effects being related to the lack of cell differentiation or retarded development in specific parts or accompanied by alterations in mutually dependent organs or postnatal functional disorders (212, 259, 427). These phenomena can only be explored by careful study during ontogenesis (224, 269, 406).

Both the basic insult and the period of particular organ or tissue susceptability to injury are considered to occur in a brief period usually during the period of most rapid growth and differentiation (176). If effects occur, the phenomena occurring thereafter either allow fetal survival with malformation or result in fetal death within the remaining *in utero* period. In addition to studying the adverse effects noted in developing tissues, alterations in related anatomical sites and upset functional dependence of other structures must be considered. A delineation of the periods of prenatal development during which effects occur in the conceptus has been summarized as: segmentation damage; blastopathy, occurring before nine days; kyematopathies, entocyst disc lesions or lesions in organs before placentation; embryopathies, lesions induced during the development phase of the embryo; and foetopathies, lesions induced during the development phase of the fetus (241).

B. Observations—Tabular Presentations of Cited Naturally Occurring and Experimentally Induced Phenomena in the Rabbit Fetus

It is well recognized that a malformation can arise through different mechanisms and be caused by different etiological agents (175). The interrelationships of these factors are far from clear in animals and less so in humans.

The following tabular presentations (Tables II—XIV) simultaneously depict both pathological lesions and anatomical variations noted in the rabbit fetus at birth or immediately prior to delivery as reported by numerous investigators. No attempt was made to separate the two types, since the anatomical variants must be viewed in the context of the experimental data to which they belong to determine if their frequency or pattern of occurence is abnormal.

The goal of these tabulations was merely to assemble published data for general reference. Critical analysis was impractical as such efforts would be hazardous and require complete accessibility to data, specimens, etc.—all obviously impossible. The tabulations are presented to develop a better appreciation of the types of abnormal development noted in rabbits. They are not intended to eliminate the need for diligent study of concurrent experimental controls in the user's experiment.

A review of these tabulations allows one to conclude that, as one would expect, the rabbit fetus seems prone to develop an array of naturally occurring and induced malformations similar to those recorded for other species including humans.

The observations have been categorized as "naturally occurring in controls" or as being associated with dams subjected to an "experimental situation." This association is not meant to infer a cause unless the data available justifies such a conclusion.

For those noted in the "naturally occurring" category, it is recognized that there is in all likelihood a true "cause," however, in most situations seldom are sufficient data available to relate their occurrences to a specific etiological agent or situation (136, 175, 224, 268, 312, 426).

Some of the experimental data presented provide strong evidence that the pathological entities were indeed induced. Unfortunately, the data reported are usually preliminary to the performance of the required definitive studies to repeat and to explore the pathogenesis of the reported abnormal development phenomena.

It must also be recognized that in the "experimental situations" column, not all observations cited can be considered as deliberately induced. One must anticipate the natural occurrence of spontaneous phenomena in addition to those induced by the experimental treatments (212, 224). Also, for any observation, the data was usually derived from examining 2–3 experimental groups compared to a single control group, thus the "experimental condition" column represents a data pool at least 2–3 times greater than that of the "naturally occurring" data column.

Of the data reviewed concerning fetal pathology, most pertain to the skeleton—probably because it is the easiest to observe and quantitate. In evaluating these skeletal data, one must consider variants in the terminal stages of ossification which occur just prior to birth. Sampling as early as day 29 rather than later as on day 31 of pregnancy will include many variations in ossification unique to certain bones. Inaccurate pregnancy timing fosters the inclusion of these variables in the experimental data.

Far less pathological data exist for soft tissues probably

because alterations are more difficult to detect and quantitate; often requiring tedious histological evaluation for accuracy.

The anatomical arrangement for tabulation was arbitrary and selected for convenience because little choice exists (426). To compensate for lack of space and limited availability of illustrations, the presence of illustrations in a cited article is so indicated. Data was excluded when its association with the control or experimental group categories was unclear.

The terminology used was that of the cited author. No attempt was made to consolidate data which appeared to be similar in diagnosis, even though less appropriate terms were used.

A problem frequently encountered was a general lack of thorough description of the anomalous fetuses, including consideration of accompanying phenomena.

It is indeed unfortunate that nonspecific terms, such as "malformation, abnormalities, and deformities," were used by many authors rather than precise terminology. Further, seldom were such findings described adequately if at all. The net result being poor utility of the data reported.

Space will not permit detailed consideration of appropriate terminology for pathological embryology. A vast nomenclature has been developed for humans and for the most part is appropriate for other mammalian species. Considerations of specific and modifying terminologies and useful listings are frequent (e.g., 21, 25, 201, 212, 221, 224, 296, 400, 423, 426, 437).

In presenting these tabulations as a collection of noted "lesions and observations," a lesion is defined as a pathological discontinuity (109), and while all lesions are observations, not all observations are necessarily lesions—rather they may be anticipated anatomical variations peculiar to the embryological development of the organ or

species. Indeed, in one's experiment, simultaneous analysis of lesions and observations must be performed when attempting to detect possible teratogenic effect. It is important to define the pattern of response to an experimental condition be it manifested as altered embryological development or as more obvious pathological alterations. Alterations of metabolic processes, though difficult to detact, are an area attracting considerable attention (130, 437).

Thus it is of great importance that lesions and observations not be considered only separately, additionally, they should be evaluated as to their occurrence in a pattern or syndrome of lesions—many occurring at low frequency, but collectively depicting a response to an experimental condition (176, 222). To be considered also is the frequency with which lesions occurring in one system occur simultaneously with those of another system (147). Obviously, consideration must be given to the significance of the malformation to the individual fetus (311). It is unfortunate that such correlations or comments are seldom presented, if indeed performed. The data are merely tabulated.

In this chapter attention is focused only on neonatal pathology. It must not be overlooked that the *in utero* period represents only one brief part of the total life span and that one's perspective on defective development should include the expression of the postnatal developmental morphological alterations, functional disturbances, and the presence of clinically silent lesions detected only at death (147, 228, 296). For most laboratory animal species such postnatal and postmortem pathological data are seldom available. It is not difficult to envision the usefulness of correlations of early postnatal development and disease in animals with *in utero* disease (120) or induced injury within the scope of currently performed experimentation.

TABLE II

ANATOMICAL REGIONS, FETAL MEMBRANES, AND FLUIDS[a]

Organ/Tissue	Lesion/Observation[b]	Control incidence[c] (inapparent etiology)	Stock or strain	Ref.	Incidence[c] (treated animals)	Experimental[d] condition	Dose[e] (mg/kg)	Time[f] (day of gestation)
			Abnormal Anatomical Regions					
Head								
General								
	Acephaly[b]	1/?	N.Z. White	33	—	—	—	—
	Acephaly	1/1114	N.Z. White	83	—	—	—	—
	Acephaly	—	Dutch Belted	88	?	Inherited spina bifida	—	—
	Acephaly	—	Mixed breeds	233	1/200	Cyclopamine	150–300	6–9
	Acephaly	—	Mixed breeds	233	1/50	Jervine	300	7
	Acephaly	1/225	Albino	329	—	—	—	—
	Acephaly	1/8500	N.Z. White	311	—	—	—	—
	Acephaly	3/2640	Mixed breeds	361	—	—	—	—
	Dome shaped	—	N.Z. White	111	1/88	Thalidomide	150	8–16

Table II (*continued*)

Organ/Tissue	Lesion/Observation[b]	Control incidence[c] (inapparent etiology)	Stock or strain	Ref.	Incidence[c] (treated animals)	Experimental[d] condition	Dose[e] (mg/kg)	Time[f] (day of gestation)
	Supraoptic proboscis	1/?	g	390	—	—	—	—
	Cebocephaly[b]	0/16	Mixed breeds	233	3/47	Cyclopamine	22–45	6–9
	Cebocephaly[b]	—	Mixed breeds	233	5/200	Cyclopamine	150–300	6–9
	Cebocephaly	?	N.Z. White	311	—	—	—	—
	Rhinencephaly	—	N.Z. White	33	?	Thalidomide	100	5–18
	Rhinencephaly	—	White Danish	244	?	Thalidomide	100	1–20
	Rhinencephaly	1/5700	N.Z. White	311	—	—	—	—
Face								
	Malformed	—	Not stated	17	?	Insulin	?	?
	Craniofacial hypoplasia	—	Not stated	101	?	Oxygen deficiency	—	7–15
	Bull dog face	1/219	Albino	329	—	—	—	—
Nose								
	Deformity[b]	0/16	Mixed breeds	232	20/47	Cyclopamine	22–45	6–9
	Deformity[b]	—	Mixed breeds	233	23/200	Cyclopamine	150–300	6–9
	Arrhinia	—	N.Z. White	216	1/102	Laparotomy and atmosphere	10 mm	7–10
	Nasal horn[b]	—	Not stated	296	?	Inheritance	—	—
Jaw								
	Cheilognathopalatoschisis	—	Mixed breeds	142	16/331	Cyclophosphamide	30	6–14
Lips								
	Deformity[b]	—	Mixed breeds	233	1/200	Cyclopamine	150–300	6–9
	Deformity	—	Himalayan	248	4/177	Thalidomide	50–450	6–18
	Cleft	—	Mixed breeds	142	?	Cyclophosphamide	30	6–14
	Cleft[b]	1/16	Not stated	232	—	—	—	—
	Hare	—	Mixed breeds	150	?/50	Thalidomide	150	7–17
	Hare	1/17,000	N.Z. White	311	—	—	—	—
	Hare	3/2640	Mixed breeds	361	—	—	—	—
Eyelids								
	Absent	—	N.Z. White	282	1/?	U 11,100	1–5	4–6
	Open	—	White Danish	244	?	Thalidomide	100	1–20
	Open	1/17,000	N.Z. White	311	—	—	—	—
	Open	—	Jap. Albino	394	1/92	Norethisterone mestranol	1–10	8–20
	Open	31/2640	Mixed breeds	361	—	—	—	—
	Open	—	Not stated	393	?	Vincristine	?	7–8
	Entropion	—	Rex	288	?	Inheritance	—	—
Ears								
	Size	—	Not stated	66	?	Inheritance	—	—
	Form	—	—	296	?	Inheritance	—	—
	Malformed	1/2640	Mixed breeds	361	—	—	—	—
	anomalies[b]	—	Not stated	87	?	Inheritance	—	—
	Displaced[b]	1/219	Albino	329	—	—	—	—
	Fleshy	—	Not stated	322	—	Inherited achondroplasia	—	—
	Small	—	Mixed breeds	357	6/109	Thalidomide	500	6–11
	Absent	1/1114	N.Z. White	83	—	—	—	—
	Absent	—	Not stated	356	?	Inheritance	—	—
Jaw								
	Agenesis	1/17,000	N.Z. White	311	—	—	—	—
Limbs, feet and digits (*see* skeletal system)								
Thorax								
	Thoracoschisis	—	Mixed breeds	357	1/109	Thalidomide	500	6–11
	Thoracogastroschisis	1/17,000	N.Z. White	311	—	—	—	—
	Fluid	—	Not stated	322	?	Inherited achondroplasia	—	—
Diaphragm								
	Hernia	—	Dutch Belted	88	?	Inherited spina bifida	—	—
	Hernia	—	N.Z. White	111	5/88	Thalidomide	150	8–16
	Hernia	—	Mixed breeds	150	?/50	Thalidomide	150	7–17
	Hernia	—	White Danish	244	?	Thalidomide	100	1–20
	Hernia	—	Himalayan	248	1/108	Thalidomide	50–450	8–16
	Hernia	—	Chinchilla	381	?	Inheritance	—	—
	Hernia	—	Several strains	408	3/906	Thalidomide	150	12.5–13.5

Table II (*continued*)

Organ/Tissue	Lesion/Observation[b]	Control incidence[c] (inapparent etiology)	Stock or strain	Ref.	Incidence[c] (treated animals)	Experimental[d] condition	Dose[e] (mg/kg)	Time[f] (day of gestation)
	Hiatus hernia[b]	—	N.Z. White	325	4/121	Thalidomide	100–300	7.5–13.5
	Hiatus hernia	—	Hybrid	405	7/120	Thalidomide	150	7.5–13.5
	Hiatus hernia[b]	—	Several strains	408	68/906	Thalidomide	150	12.5–13.5
Abdomen								
	Wall defect	—	Jap. Albino	394	12/247	Chlormadinone acetate	1–10	8–20
	Protuberant	—	Not stated	322	?	Inherited achondroplasia	—	—
	Celosomia	—	Not stated	72	1/83	Dimethyl sulfoxide	5000	6–14
	Coelosomia[b]	—	Not stated	402	?	WR–1339	25–400	?
	Failure of wall closure	—	White Danish	244	?	Thalidomide	100	1–20
	Evertration of viscera[b]	—	N.Z. White	14	2/20	L-Asparaginase	50	8–9
	Evertration of viscera	—	N.Z. White	33	?	Thalidomide	100	5–18
	Evertration of viscera	—	Mixed breeds	233	1/200	Cyclopamine	150–300	6–9
	Evertration of viscera	—	Mixed breeds	233	1/50	Jervine	300	7
	Hernia	—	Mixed breeds	76	1/27	Radiation	250 r	8
	Hernia – lateral	—	White Danish	244	?	Thalidomide	100	1–20
	Hernia	—	White Danish	245	2/13	Thalidomide	60	1–20
	Hernia	—	White Danish	245	3/62	Phenylbutazone	30	1–20
	Hernia	1/1400	N.Z. White	311	—	—	—	—
	Hernia	12/2640	Mixed breeds	361	—	—	—	—
	Omphalocele	—	Mixed breeds	142	4/331	Cyclophosphamide	30	6–14
	Omphalocele	—	Mixed breeds	150	?/50	Thalidomide	150	7–17
	Omphalocele	1/62	N.Z. White	325	1/121	Thalidomide	100–300	7.5–13.5
	Omphalocele	—	Hybrid	405	1/120	Thalidomide	150	7.5–13.5
	Omphalocele	—	Several strains	408	?	Thalidomide	150	12.5–13.5
	Fluid	—	Not stated	322	?	Inherited achondroplasia	—	—
Tail								
	Absent[b]	—	N. Z. White	14	1/18	L-Asparaginase	50	8–9
	Absent[b]	—	N.Z. White	102	?	Thalidomide	150	8–16
	Absent	—	Mixed breeds	233	1/200	Cyclopamine	150–300	6–9
	Absent	—	White Danish	244	?	Thalidomide	100	1–20
	Absent	—	White Danish	245	2/13	Thalidomide	60	1–20
	Absent	—	White Danish	245	1/62	Phenylbutazone	30	1–20
	Absent	1/49	Albino	329	—	—	—	—
	Absent	—	Mixed breeds	357	1/109	Thalidomide	500	6–11
	Absent	—	Dutch Belted	365	42/42	6-Aminonicotinamide	3	9
	Short[b]	—	Mixed breeds	76	34/345	Radiation	250–400 r	2–22
	Short[b]	—	N.Z. White	102	?	Thalidomide	150	8–16
	Short	—	N.Z. White	216	1/104	Laparotomy and atmosphere	15 min	7–10
	Short	—	White Danish	245	1/62	Monophenylbutazone	60–150	1–20
	Short	—	White Danish	245	6/13	Thalidomide	60	1–20
	Short	—	Himalayan	248	7/177	Thalidomide	50–450	8–16
	Short	3/8000	N.Z. White	311	—	—	—	—
	Short	—	N.Z.	373	1/68	Moquizone	20–60	7–17
	Short	1/118	Jap. Albino	394	—	—	—	—
	Short	8/186	Mixed breeds	357	26/109	Thalidomide	500	6–11
	Rudimentary	—	Mixed breeds	150	?/50	Thalidomide	150	7–17
	Rudimentary	—	N.Z. White	216	1/106	Laparotomy and atmosphere	Momentary	7–10
	Stubbed	3/2640	Mixed breeds	361	—	—	—	—
	Fleshy	—	Unknown	322	?	Inherited achondroplasia	—	—
	Abnormalities	—	N.Z.	372	20/370	Thalidomide (oral)	25–200	8–12
	Abnormalities	—	N.Z.	372	9/319	Thalidomide (IV)	5	7–10
	Malformed	—	Unknown	200	2/98	Thalidomide	500	3–16
	Kinky	2/186	Mixed breeds	357	29/109	Thalidomide	500	6–11
External genitalia								
	Absent	—	N.Z. White	216	1/102	Laparotomy and atmosphere	10 min	7–10
	Malformed	—	N.Z. White	102	3/56	Thalidomide	150	8–11

Table II (*continued*)

Organ/Tissue	Lesion/Observation[b]	Control incidence[c] (inapparent etiology)	Stock or strain	Ref.	Incidence[c] (treated animals)	Experimental[d] condition	Dose[e] (mg/kg)	Time[f] (day of gestation)
	Feminization	—	Unknown	118	?	Cyproterone acetate	6–100	13–24
Twins								
	Incomplete[b]	—	Not stated	200	1/98	Thalidomide	500	3–16
	Conjoined	—	Not stated	73	4/252	Inheritance	—	—
	Posterior duplication[b]	?	Not stated	296	—	—	—	—

Fetal Membranes and Fluids

Fetal membranes								
General								
	Bacterial colonies	—	Copenhagen	314	—	*Listeria monocytogenes*	—	—
Yolk sac								
	Bacterial colonies[b]	—	White	238	—	*Listeria monocytogenes*	—	20
Placenta	Weight	—	Not stated	249	—	Normal development	—	—
Placenta	Weight	—	N.Z. White	186	—	Uterine overcrowding	—	2–32
Placenta	Crowding	?	N.Z. White	186	?	Superovulation	—	—
Placenta	Fusion	?	N.Z. White	186	?	Superovulation	—	—
	Placentitis	—	White	238	—	*Listeria monocytogenes*	—	20
	Reduced vasculature[b]	—	N.Z. White and Dutch Belted	243	—	Vitamin A deficiency	—	—
Amnionic fluid								
	Volume	—	Not stated	249	—	Normal development	—	—
	Hydramnion	—	N.Z. White	83	1/2821	Miscellaneous tests	—	—
	Bacterial colonies[b]	—	White	238	—	*Listeria monocytogenes*	—	20

[a] Reported in nonexperimental "control" fetuses or in those from dams exposed to various experimental conditions.

[b] Illustration in reference.

[c] Incidence: no. of affected fetuses/no. of fetuses examined.

[d] The natural occurrence of phenomena in experimental fetuses cannot be overlooked. Comments as to whether the observation was spontaneous or induced or its relation to the experimental treatment were usually absent from the majority of articles reviewed.

[e] Dose: Summarized depiction of dosage range or as otherwise indicated (usually in mg/kg/day).

[f] Time: Period of days administered during gestation.

[g] Article not available for review; original source cited.

? = Incidence not specified yet observation or lesion noted.

TABLE III

Modified Growth and Development and Integument and Muscle[a]

Organ/Tissue	Lesion/Observation	Control incidence[b] (inapparent etiology)	Stock or strain	Ref.	Incidence[b] (treated animals)	Experimental[c] condition	Dose[d] (mg/kg)	Time[e] (day of gestation)
				Modified Growth and Development				
Body size								
	General	—	Mixed breeds	67	?	Inheritance	—	—
	General	—	Not stated	172	?	Inheritance	—	—
	General	?	g	70	?	Inheritance	—	—
	Altered	—	N.Z. White/hybrid	328	?	Inheritance	—	—
	Atrophy	—	g	191	?	Inheritance	—	—
	Atrophy	—	g	363	51/80	Inheritance	—	—
	Racial	—	Mixed breeds	68	?	Inheritance	—	—
	Racial	—	Mixed breeds	71	?	Inheritance	—	—
	Reduced	—	N.Z. White	90	?	Inheritance	—	—
	Reduced	?	N.Z. White	187	?	Maternal feed restriction	—	—
	Reduced	—	N.Z. White	188	?	Maternal feed restriction	—	—
	Reduced	—	Not stated	246	?	Hereditary dwarfism	—	—

Table III (*continued*)

Organ/ Tissue	Lesion/Observation	Control incidence[b] (inapparent etiology)	Stock or strain	Ref.	Incidence[b] (treated animals)	Experimental[c] condition	Dose[d] (mg/kg)	Time[e] (day of gestation)
	Reduced	—	N.Z. White/Dutch Belted	243	?	Vitamin A deficiency	—	?
	Reduced[f]	—	N.Z. White	375	?	Inherited achondroplasia	—	—
	Reduced	—	Dutch Belted	194	?	Vitamin A deficiency	—	—
	Small[f]	—	Not stated	324	?	Inherited osteopetrosis	—	—
	Microsomy	—	N.Z. White	216	1/102	Laparatomy and atmosphere	10 min	7–10
	Runting	—	N.Z. White	79	?	Adrenaline	0.159	22
	Runting	—	N.Z. White	143	?	Marihuana	150–500	10
	Stunting	—	N.Z. White	282	2/?	U-11, 100A	1–5	4–6
	Dwarfism (achondroplasia)	1/1114	N.Z. White	83	1/2821	Misallaneous tests	—	—
	Dwarfism (achondroplasia)	—	Mixed breeds	71	?	Inheritance	—	—
	Dwarfism	—	g	342 291	?	Inheritance	—	—
Body weight								
	General	—	Mixed	66	?	Inheritance	—	—
	Reduced	—	N.Z. White	4	?	Ethamoxytryphetol	25	21–26
	Reduced	—	N.Z. White	113	?	—	—	—
	Reduced	—	N.Z. Red	144	?	Emotional stress (noise)	0.159	22
In utero growth								
	Retardation	—	N.Z. White	79	?	Adrenaline	—	14–24

Integument

Organ/ Tissue	Lesion/Observation	Control incidence[b] (inapparent etiology)	Stock or strain	Ref.	Incidence[b] (treated animals)	Experimental[c] condition	Dose[d] (mg/kg)	Time[e] (day of gestation)
Skin								
	Subcutaneous hemorrhage	—	N.Z.	122	?/43	1-Phthalimidobutane	150	7–12
	Subcutaneous hemorrhage	—	White Danish	244	?	Thalidomide	100	1–20
	Subcutaneous hemorrhage	—	Mixed breeds	341	?	Imipramine	5–30	1–20
	Edema	24/2640	Mixed breeds	361	—	—	—	—
	Hydrops	?	g	298	?	Erythroblastosis	—	—
	Hydrops	—	g	291	?	Inheritance	—	—
	Hematoma[f]	—	N.Z. White	83	5/2821	Miscellaneous tests	—	—
	Hemangioma[f]	—	N.Z. White	102	?	Thalidomide	150	8–16
	Color	—	Several strains	66	?	Inheritance	—	—
	Albinism	—	Several strains	296	?	Inheritance	—	—
	Hypotrichosis	—	Several strains	96	?	Inheritance	—	—
Fat								
	Xanthophyllosis	—	g	326	?	Inheritance	—	—
Hair and nails								
	Keratosis	—	Not stated	296	?	Inheritance	—	—
	Shortening	—	Rex	288	?	Inheritance	—	—
	Furless	—	g	69	?	Inheritance	—	—
	Furless	—	g	292	?	Inheritance	—	—
	Wirehair	—	Not stated	356	?	Inheritance	—	—
Muscles								
	Contracture–lethal	—	Not stated	356	?	Inheritance	—	—
	Defects	—	Not stated	257	40%?	Thalidomide	150	8–16
	Softness	—	Not stated	322	?	Inherited achondroplasia	—	—
	Thigh foramen	—	N.Z. White	83	1/2821	Miscellaneous tests	—	—
	Abdominal–defective	—	N.Z. White	33	?	Thalidomide	100	5–18
	Abdominal–contracture	11/2640	Mixed breeds	361	40%	—	—	—
Ligaments								
	Defect	—	Not stated	257	40%?	Thalidomide	150	8–16

[a] Reported in nonexperimental "control" fetuses or in those from dams exposed to various experimental conditions.

[b] Incidence: no. of affected fetuses/ no. of fetuses examined.

[c] The natural occurrence of phenomena in experimental fetuses cannot be overlooked. Comments as to whether the observation was spontaneous or induced or its relation to the experimental treatment were usually absent from the majority of articles reviewed.

[d] Dose: Summarized depiction of dosage range or as otherwise indicated (usually in mg/kg/day).

[e] Time: Period of days administered during gestation.

[f] Illustration in reference.

[g] Article not available for review; original source cited.

? = Incidence not specified yet observation or lesion noted.

TABLE IV

SKELETAL SYSTEM: MISCELLANEOUS, SKULL, AND STERNUM[a]

Organ/ Tissue	Lesion/Observation[f]	Control incidence[b] (inapparent etiology)	Stock or strain	Ref.	Incidence[b] (treated animals)	Experimental[c] condition	Dose[d] (mg/kg)	Time[e] (day of gestation)
			Skeletal System					
Miscellaneous	Delay in ossification	—	N.Z. White	410	?	Freon aerosol	?	?
	Delay in ossification	—	N.Z. White	410	?	Isoproterenol Aerosol	0.15–0.45	6–16
	Ossification schedule	—	Mixed	89	?	Inheritance	—	—
	Inhibited	—	N.Z. White	368	?	Food restriction	—	—
	Retarded development	—	N.Z. White	4	?	Ethamoxytriphetol	25	21–26
	Osteopetrosis[f]	—	Not stated	318	?	Inherited achondroplasia	—	—
	Osteopetrosis[f]	—	Not stated	319	?	Inherited osteopetrosis	—	—
	Osteopetrosis[f]	—	Not stated	317	?	Inherited osteopetrosis	—	—
	Osteopetrosis[f]	—	Not stated	324	?	Inherited osteopetrosis	—	—
	Distal endochondrial cartilage growth	—	Not stated	322	?	Inherited achondroplasia	—	—
	Decrease bone formation	—	Not stated	322	?	Inherited achondroplasia	—	—
	Acromegalia	—	Not stated	208	?	Inheritance	—	—
	Trabecular abnormalities[f]	—	N.Z. White	2	?	Maternal estrogen	—	—
	Abnormal salt deposition	—	N.Z. White	2	?	Maternal estrogen	—	—
Skull General	Form and variations	—	Not stated	391	?	Inheritance	—	—
	Defects	23/377	Mixed breeds	258	14/46	Thalidomide	300	8–16
	Defects	—	Mixed breeds	258	14/40	Phenobarbital	50	8–16
	Abnormal	—	Not stated	398	3/10	Sperm Storage	4°C	84 hr.
	Assymetrical	—	Mixed breeds	357	1/109	Thalidomide	500	8–11
	Ossification–inhibited	—	Mixed breeds	257	?	Phenobarbital	50	8–15
	Ossification–reduced	19/8000	N.Z. White	311	—	—	—	—
	Ossification–deficiency	—	Not stated	322	?	Inherited achondroplasia	—	—
	Stages of development	?	Mixed breeds	59	—	—	—	—
	Incomplete development	22/1620	Dutch Belted	265	1/88	Methadone HCl	20/40	6–18
	Retarded development	9/125	N.Z.	372	5/370	Thalidomide (oral)	25–200	8–12
	Retarded development	—	N.Z.	372	13/319	Thalidomide (IV)	5	7–10
	Cranial bulge[f]	1/219	Albino	329	—	—	—	—
	Cranial bulge[f]	—	Mixed breeds	341	?	Imipramine	7–30	1–20
	Cephalodysplasia[f]	?	Not stated	74	—	Inheritance	—	—
	Complete or partial absence[f]	—	N.Z. White	111	9/88	Thalidomide	150	8–16
	Acrania	1/17,000	N.Z. White	311	—	—	—	—
	Craniosynostosis[f]	?	Mixed breeds	164	?	Inheritance	—	—
	Craniosynostosis	—	g	166	?	Inheritance	—	—
	Craniosynostosis	—	N.Z. White	281	1/19	Colcemide	0.2	12,15
	Premature fusion							
	Scaphocephaly	—	Mixed breeds	164	?	Inheritance	—	—
	Trigonocephaly[f]	—	Mixed breeds	164	?	Inheritance	—	—
	Plagiocephaly[f]	—	Mixed breeds	164	?	Inheritance	—	—
	Oxycephaly[f]	—	Mixed breeds	164	?	Inheritance	—	—
	Oxycephaly	—	Not stated	166	?	Inheritance	—	—
	Incomplete fusion	—	N.Z. White/ Dutch Belted	235	1/145	Folpet	75	6–16
	Wide central fontanelle	—	N.Z. White	111	2/18	Thalidomide	150	8–16
	Cranioschisis[f]	—	Not stated	74	?	Inbreeding	—	—
	Cranioschisis	—	Dutch Belted	88	?	Inherited spina bifida	—	—
	Cranioschisis	—	N.Z.	122	?/48	Thalidomide	150	7–12
	Cranioschisis	—	N.Z.	122	1/30	3-Nitrothalidomide	150	7–12
	Cranioschisis	—	N.Z. White	281	1/?	Vinblastine	0.3	1
	Cranioschisis	—	N.Z. White	282	1/324	Clomephene	7.5–15.	1
	Cranioschisis	—	N.Z. White	282	1/?	U-11,100	1–5	4–6
	Cranioschisis	1/222	Albino	329	—	—	—	—

Table IV (*continued*)

Organ/ Tissue	Lesion/Observation[f]	Control incidence[b] (inapparent etiology)	Stock or strain	Ref.	Incidence[b] (treated animals)	Experimental[c] condition	Dose[d] (mg/kg)	Time[e] (day of gestation)
	Cranioschisis	1/3400	N.Z. White	311	—	—	—	—
	Cranioschisis	6/2640	Mixed breeds	361	—	—	—	—
	Craniorachischises	1/17000	N.Z. White	311	—	—	—	—
Nasal bones	Assymetrical	—	N.Z. White	357	1/109	Thalidomide	500	6–11
	Defects	—	N.Z. White	33	?	Thalidomide	100	5–18
	Imperfect closure	0/62	N.Z. White	325	1/121	Thalidomide	100–300	7.5–13.5
	Imperfect closure	8/186	Mixed breeds	357	42/109	Thalidomide	500	6–11
	Imperfect nasal, frontal, parietal	—	N.Z. White	111	1/18	Thalidomide	150	8–16
Parietal bones	Imperfect	—	N.Z. White	111	3/18	Thalidomide	150	8–16
	Extra suture	—	Hybrid	200	2/57	Thalidomide	500	3–16
	Abnormal suture	—	Not stated	398	?	Sperm storage	4–20°C	12–84 hr
Supraoc-cipital bone	Underdeveloped	—	N.Z. White	111	3/18	Thalidomide	150	8–16
Frontal bones	Membranous	—	N.Z. White	111	3/18	Thalidomide	150	8–16
	Bossed	—	Mixed breeds	357	2/109	Thalidomide	500	6–11
	Spine[f]	—	Dutch Belted	365	22/42	6-Aminonicotinamide	1–3	9–12
Suture	Reversed	—	Not stated	166	?/500	Inherited–recessive	—	—
	Fused	—	Not stated	166	?/500	Inherited–recessive	—	—
Suture bones	Abnormal[f]	—	Dutch Belted	88	?	Inherited spina bifida	—	—
	Frontal and parietal	—	N.Z. White	83	?/2821	Miscellanous tests	—	—
		258/8000	N.Z. White	311	—	—	—	—
	Parietal	1/186	Mixed breeds	357	6/109	Thalidomide	500	6–11
Accessory bone		—	Not stated	166	?/500	Inherited–recessive	—	—
Jaw	Defects	—	N.Z. White	282	1/324	Clomiphene	7.5–15	1
	Brachygnathia	1/8500	N.Z. White	311	—	—	—	—
	Brachygnathia	—	Chinchilla	381	?	Inheritance	—	—
	Agnathia[f]	1/8500	N.Z. White	311	—	—	—	—
	Agnathia	—	Mixed breeds	142	1/331	Cyclophosphamide	30	6–14
Maxilla	Absent[f]	—	N.Z. White	83	1/2821	Miscellaneous tests	—	—
	Deformity	—	N.Z. White	216	1/102	Laparotomy and atmosphere	—	—
	Bending[f]	—	Not stated	74	?	Inheritance	—	—
	Brachygnathia[f]	—	Japanese	292	?	Inheritance	—	—
Mandible	Dysostosis[f]	—	Not stated	74	?	Inheritance	—	—
	Agnathus[f]	—	N.Z. White	83	1/2821	Miscellaneous tests	—	—
	Agnathus[f]	1/8500	N.Z. White	311	—	—	—	—
	Otocephalus	?	Chinchilla	270	—	—	—	—
	Otocephalus	—	Mixed breeds	357	1/109	Thalidomide	500	6–11
	Otocephalus	—	Pourgogne	345	?	AY 9944	50	1–13
Teeth	Irregular calcification	?	N.Z. White	210	—	—	—	—
	Absent	—	g	290	?	Inheritance	—	—
	Absent–maxillary	—	White Danish	245	1/13	Thalidomide	60	1–20
	Incisors small	—	Not stated	324	?	Inherited osteopetrosis	—	—
	Incisors small	—	Not stated	322	?	Inherited achondroplasia	—	—

Table IV (*continued*)

Organ/ Tissue	Lesion/Observation[f]	Control incidence[b] (inapparent etiology)	Stock or strain	Ref.	Incidence[b] (treated animals)	Experimental[c] condition	Dose[d] (mg/kg)	Time[e] (day of gestation)
	Incisors small	—	White Danish	245	1/13	Thalidomide	60	1–20
	Supernumerary	—	g	290	?	Inheritance	—	—
Hyoid								
	Aplasia	—	Himalayan	248	1/137	Thalidomide	50–450	8–16
Palate								
	Defects	—	N.Z. White	282	1/324	Clomiphene	7.5–15.	1
	Cleft[f]	—	N.Z. White	33	?	Thalidomide	100	5–18
	Cleft	2/1114	N.Z. White	83	—	—	—	—
	Cleft[f]	—	Mixed breeds	83	?	Radiation	250–400	2–22
	Cleft[f]	—	N.Z. White	111	5/88	Thalidomide	150	8–16
	Cleft	—	Mixed breeds	142	73/331	Cyclophosphamide	30	6–14
	Cleft	—	Not stated	150	2/52	Thalidomide	150	7–17
	Cleft	—	Several strains	158	2/59	Thalidomide	20–200	6–18
	Cleft	—	N.Z. White	216	1/104	Laparotomy and atmosphere	15 min	7–10
	Cleft	1/3400	N.Z. White	311	—	—	—	—
	Cleft	2/1725	Himalayan	248	13/177	Thalidomide	50–450	8–16
	Cleft	—	Not stated	322	?	Inherited achondroplasia	—	—
	Cleft[f]	—	Mixed breeds	341	?	Imipramine	5–30	1–20
	Cleft[f]	—	Dutch Belted	365	21/42	6-Aminonicotinamide	1–3	9–12
	Cleft	—	N.Z.	372	?	Thalidomide	25–200	8–12
	Cleft	—	Chinchilla	381	?	Inheritance	—	—
	Cleft	—	N.Z.	373	1/62	Moquizone	20	7–17
	Cleft	—	Jap. Albino	394	14/247	Chlormadinone acetate	1–10	8–20
	Cleft	—	N.Z. White	413	12/92	Cortisone	10–25	13.5–16.5
	Cleft	—	N.Z. White	413	31/167	Triamcinolone	0.1–1.0	13.5–16.5
	Cleft	—	N.Z. White	413	28/81	Dexamethasone	0.1–4.0	13.5–16.5
	Cleft	—	N.Z. White	413	9/78	Prednisolone	1–8	13.5–16.5
	Cleft	?	Dutch Belted	443	—	—	—	—
Pelvis								
Pubis								
	Abnormal[f]	—	Dutch Belted	365	14/42	6 Aminonicotinamide	1–3	9–12
	Absent	—	N.Z. White	111	3/18	Thalidomide	150	8–16
	Aplasia	—	Himalayan	248	2/137	Thalidomide	50–450	8–16
	Hypoplasia[f]	—	Not stated	296	?	Inheritance	—	—
Sternum								
	Malformations[f]	—	Hybrid	174	?	Maternal plasma loss	—	9
	Malformations	10/51	N.Z. White	33	38/72	Trazodone	100	5–18
	Malformations	—	N.Z. White	33	15/57	Thalidomide	100	5–18
	Malformations	—	Dutch Belted	88	?	Inherited spina bifida	—	—
	Malformations	—	Mixed breeds	257	?	Thalidomide	300	8–15
	Malformations	—	Mixed breeds	257	?	Phenobarbital	50	8–15
	Malformations	—	Mixed breeds	357	6/109	Thalidomide	500	6–11
	Malformations	—	N.Z.	372	8/370	Thalidomide (oral)	25–200	8–12
	Malformations	—	N.Z.	372	13/319	Thalidomide (IV)	5	7–10
	Malformations	—	White Danish	245	2/62	Phenylbutazone	30	1–20
	Malformations	—	White Danish	245	1/13	Thalidomide	60	1–20
	Malformations[f]	7/56	Dutch Belted	365	34/88	6-Aminonicotinamide	1–3	9–12
	Defect	56/377	Mixed breeds	258	18/46	Thalidomide	300	8–16
	Defect	—	Mixed breeds	258	13/40	Phenobarbital	50	8–16
	Irregularities	—	Hybrid	200	2/98	Thalidomide	500	3–16
	Irregularities	—	Not stated	251	?	Ethyl Alcohol	10	8–16
	Length variations	—	N.Z. White/ hybrid	328	?	Inheritance	—	—
	Enlarged	—	Chinchilla	381	?	Inheritance	—	—
	Cleft	—	Hybrid	405	1/120	Thalidomide	150	7.5–13.5

Table IV (*continued*)

Organ/ Tissue	Lesion/Observation[f]	Control incidence[b] (inapparent etiology)	Stock or strain	Ref.	Incidence[b] (treated animals)	Experimental[c] condition	Dose[d] (mg/kg)	Time[e] (day of gestation)
	Lack of fusion	—	Hybrid	200	2/98	Thalidomide	500	3–16
	Defective ossification	2/119	N.Z.	51	2/139	Methaqualone	100	1–29
	Defective ossifications	—	N.Z. White	83	Frequent/ 2821	Miscellaneous tests	—	—
	Defective ossification-5th	Frequent	N.Z. White	83	Frequent	Miscellaneous tests	—	—

[a] Reported in nonexperimental "control" fetuses or in those from dams exposed to various experimental conditions.

[b] Incidence: no. of affected fetuses/no. of fetuses examined.

[c] The natural occurrence of phenomena in experimental fetuses cannot be overlooked. Comments as to whether the observation was spontaneous or induced or its relation to the experimental treatment were usually absent from the majority of articles reviewed.

[d] Dose: Summarized depiction of dosage range or as otherwise indicated (usually in mg/kg/day).

[e] Time: Period of days administered during gestation.

[f] Illustrations in reference.

[g] Article not available for review; original source cited.

? = Incidence not specified yet observation or lesion noted.

TABLE V
SKELETAL SYSTEM: STERNEBRAE AND RIBS[a]

Organ/ Tissue	Lesion/Observation	Control incidence[b] (inapparent etiology)	Stock or strain	Ref.	Incidence[b] (treated animals)	Experimental[c] condition	Dose[d] (mg/kg)	Time[e] (day of gestation)
			Skeletal System					
Sternebrae								
	Absent	21/249	Dutch Belted	146	—	—	—	—
	Absent	—	Hybrid	200	3/57	Thalidomide	500	3–16
	Absent or reduced	40/186	Mixed breeds	357	3/109	Thalidomide	500	6–11
	Extra	?	Not stated	162	—	—	—	—
	Extra	2/64	Hybrid	200	—	—	—	—
	Extra (7th segment)	—	N.Z. White		?	Inheritance	—	—
			Hybrid	328				
	Extra and fused	2/249	Dutch Belted	146	—	—	—	—
	Extra and fused	20/1696	Dutch Belted	265	11/139	Methadone HCl	20–40	6–18
	Synostosis[f]	1/1114	N.Z. White	83	—	—	—	—
	Synostosis	—	Not stated	74	139/352	Inbreeding	—	—
	Synostosis	3/249	Dutch Belted	146	—	—	—	—
	Synostosis	—	Hybrid	200	?	Thalidomide	500	3–16
	Synostosis[f]	2/1456	Himalayan	248	17/137	Thalidomide	50–450	8–16
	Synostosis	4/186	Mixed breeds	357	8/109	Thalidomide	500	6–11
	Synotosis—incomplete	1/64	Hybrid	200	—	—	—	—
	Synotosis—centers 1–5	—	N.Z. White	111	1/18	Thalidomide	150	8–16
	Synostosis–and bipartite	1/249	Dutch Belted	146	—	—	—	—
	Atrophic	32/249	Dutch Belted	146	—	—	—	—
	Atrophic and absent	1/249	Dutch Belted	146	—	—	—	—
	Bipartite	—	Hybrid	200	2/57	Thalidomide	500	3–16
	Bipartite (bilaterally)	13/186	Mixed breeds	357	3/109	Thalidomide	500	6–11
	Bipartite (dorsoventral)	—	Mixed breeds	357	6/109	Thalidomide	500	6–11
	Bipartite[f]	2/2640	Mixed breeds	361	—	—	—	—
	Bipartite and atrophic	1/249	Dutch Belted	146	—	—	—	—
	Defective (5th)	—	N.Z. White	111	5/18	Thalidomide	150	8–16
	Defective (5th and 6th)	—	N.Z. White	111	2/18	Thalidomide	150	8–16
	Miscellaneous alterations	258/8000	N.Z. White	311	—	—	—	—

Table V (*continued*)

Organ/Tissue	Lesion/Observation	Control incidence[b] (inapparent etiology)	Stock or strain	Ref.	Incidence[b] (treated animals)	Experimental[c] condition	Dose[d] (mg/kg)	Time[e] (day of gestation)
	Hypoplastic[f]	—	Not stated	174	?	Maternal plasma loss	—	—
Xiphi-sternum	Sternal bars misaligned	12/1696	Dutch Belted	265	1/88	Methadone HCl	20–40	6–18
	Wide	1/64	Hybrid	200	—	—	—	—
	Bifurcation	1/8500	N.Z. White	311	—	—	—	—
Ribs	Abnormal	—	Mixed breeds	357	1/109	Thalidomide	500	6–11
	Malformed	—	Not stated	150	?	Maternal plasma loss	—	9
	Malformed	—	N.Z. White	230	1/126	Methamphetamine	1–5	12–30
	Malformed	6/344	Mixed breeds	258	6/46	Thalidomide	300	8–16
	Abnormalities[f]	—	Dutch Belted	88	?	Inherited spina bifida	—	—
	Abnormalities[f]	—	Hybrid	200	1/57	Thalidomide	500	3–8
	Abnormalities[f]	—	Not stated	252	?	Thalidomide	100	8–16
	Abnormalities	—	N.Z.	372	11/370	Thalidomide (oral)	25–200	8–12
	Abnormalities	—	N.Z.	372	18/319	Thalidomide (IV)	15	7–10
	Irregular	—	Mixed breeds	357	6/109	Thalidomide	500	6–11
	Defective[f]	—	Mixed breeds	74	5/352	Inbreeding	—	—
	Aplasia	7/51	N.Z. White	33	32/57	Thalidomide	100	5–18
	Aplasia	—	N.Z. White	33	29/72	Trazodone	100	5–18
	Hypoplasia	5/51	N.Z. White	33	12/57	Thalidomide	100	5–18
	Hypoplasia	—	N.Z. White	33	23/72	Trazodone	100	5–18
	Short	1/64	Hybrid	200	—	—	—	—
	Cage irregular	—	Not stated	251	?	Ethyl alcohol	10	8–16
	Abnormal flexion	—	Not stated	341	?	Imipramine	5–30	1–20
	Clubbed	1/2640	Mixed breeds	361	—	—	—	—
	Chondrodystrophy	—	Not stated	295	?	Inherited Pelger anomaly	—	—
	Synostosis[f]	—	Not stated	174	?	Maternal plasma loss	—	—
	Fusion[f]	—	N.Z.	51	1/371	Methaqualone	100–400	1–29
	Fusion[f]	—	N.Z. White	33	3/57	Thalidomide	100	5–18
	Fusion[f]	1/1114	N.Z. White	83	—	—	—	—
	Fusion[f]	?	Not stated	74		Inheritance	—	—
	Fusion[f]	—	Not stated	252	?	Thalidomide	100	8–16
	Fusion	—	White Danish	245	1/88	Phenylbutazone	60–150	1–20
	Fusion	103/8000	N.Z. White	311	—	—	—	—
	Fusion	1/64	N.Z.	373	—	—	—	—
	Fusion—8th and 9th	—	N.Z. White	111	1/18	Thalidomide	150	8–16
	Bipartite	57/249	Dutch Belted	146	—	—	—	—
	Bipartite	103/8000	N.Z. White	311	—	—	—	—
	Branched[f]	—	Not stated	74	?	Inbreeding Experiment	—	—
	Branched	1/64	Hybrid	200	—	—	—	—
	Branched or fused	—	Dutch Belted	365	26/88	6-Aminonicotinamide	1–3	9–12
	Variable number	—	Mixed breeds	39	?	Inheritance	—	—
	Variable number	—	Mixed breeds	162	—	—	—	—
	Decreased number	—	Mixed breeds	161	?	Inheritance	—	—
	Increased number	—	Mixed breeds	161	?	Inheritance	—	—
	Supernumerary	26/56	Dutch Belted	365	31/88	6-Aminonicotinamide	1–3	9–12
	Supernumerary	—	Not stated	354	?	Inheritance	—	—
	Supernumerary	—	Not stated	353	?	Inheritance	—	—
	Extra—right	17/249	Dutch Belted	146	—	—	—	—
	Extra—left	18/249	Dutch Belted	146	—	—	—	—
	12 Pairs	54%	N.Z. White	311	—	—	—	—
	13 Pairs	Frequent	N.Z. White	83	frequent	Miscellaneous test	—	—
	13 Pairs[f]	—	N.Z. White	111	9/18	Thalidomide	150	8–16
	13 Pairs	?	Hybrid	163	—	—	—	—
	13 Pairs	482/1696	Dutch Belted	265	76/139	Methadone HCl	20–40	6–18

Table V (*continued*)

Organ/ Tissue	Lesion/Observation	Control incidence[b] (inapparent etiology)	Stock or strain	Ref.	Incidence[b] (treated animals)	Experimental[c] condition	Dose[d] (mg/kg)	Time[e] (day of gestation)
	13 Pairs	46%	N.Z. White	311	—	—	—	—
	Small 13th ribs	—	N.Z. White	111	3/18	Thalidomide	150	8–16
	Absent 13th	—	Mixed breeds	258	?/40	Phenobarbital	50	8–16

[a] Reported in nonexperimental "control" fetuses or in those from dams exposed to various experimental conditions.

[b] Incidence: no. of affected fetuses / no. of fetuses examined.

[c] The natural occurrence of phenomena in experimental fetuses cannot be overlooked. Comments as to whether the observation was spontaneous or induced or its relation to the experimental treatment were usually absent from the majority of articles reviewed.

[d] Dose: Summarized depiction of dosage range or as otherwise indicated (usually in mg/kg/day).

[e] Time: Period of days administered during gestation.

[f] Illustrations in reference.

[g] Article not available for review; original source cited.

? Incidence not specified yet observation or lesion noted.

TABLE VI
SKELETAL SYSTEM: VERTEBRAE[a]

Organ/ Tissue	Lesion/Observation	Control incidence[b] (inapparent etiology)	Stock or strain	Ref.	Incidence[b] (treated animals)	Experimental[c] condition	Dose[d] (mg/kg)	Time[e] (day of gestation)
			Skeletal System					
Vertebrae								
	Malformations	—	Hybrid	150	?	Maternal plasma loss	—	9
	Malformed	—	N.Z. White	230	1/126	Methamphetamine	1.5	12–30
	Lumbar	—	N.Z. White	33	1/57	Thalidomide	100	5–18
	Lumbar	—	N.Z. White	33	1/72	Trazodone	100	5–18
	Defects[f]	85/342	Not stated	75	—	—	—	—
	Defects	4/344	Mixed breeds	258	?	Thalidomide	300	8–16
	Abnormalities[f]	—	Not stated	74	?	Inbreeding	—	—
	Craniorachischisis— see Skull							
	Assymetrical (lumbar)	—	Not stated	252	?	Thalidomide	100	?
	Disorientation[f]	4/1456	Himalayan	248	26/137	Thalidomide	50–450	8–16
	Variations	—	Mixed breeds	161	?	Inheritance	—	—
	Variations—numerical	?	Not stated	358	—	—	—	—
	Aplastic	—	Not stated	252	?	Thalidomide	100	8–11
	Retarded development[f]	—	Not stated	74	?	Inbreeding	—	—
	Retarded development	2/125	N.Z.	372	—	—	—	—
	Regression–caudal	—	Pourgogne	345	13/47	AY 9944	50	1–13
	Fusion[f]	—	Dutch Belted	88	?	Inherited spina bifida	—	—
	Fusion, bipartite hemicental	103/8000	N.Z. White	311	—	—	—	—
	Fusion–caudal	—	Hybrid	200	8/98	Thalidomide	500	3–16
	Fusion	—	N.Z. White	439	?	WU-385	150	6–11
	Fracture[f]	?	N.Z. White	210	—	—	—	—
	Extra (presacral)[f]	?	Not stated	353	—	—	—	—
	Spinous processes variation	?/105	Not stated	162	—	—	—	—
	Spinous processes variation	?	Not stated	355	—	—	—	—
	Spina bifida[f]	—	N.Z. White	14	1/18	L-Asparaginase	50	8–9
	Spina bifida	—	Mixed breeds	76	?	Radiation	250–400r	2–22
	Spina bifida[f]	?	Dutch Belted	88	—	—	—	—

Table VI (*continued*)

Organ/ Tissue	Lesion/Observation	Control incidence[b] (inapparent etiology)	Stock or strain	Ref.	Incidence[b] (treated animals)	Experimental[c] condition	Dose[d] (mg/kg)	Time[e] (day of gestation)
	Spina bifida	—	Hybrid	200	1/57	Thalidomide	500	3–8
	Spina bifida	—	N.Z. White	360	?	Inheritance	—	—
	Spina bifida	1/1725	Himalayan	248	—	—	—	—
	Spina bifida	—	N.Z. White	281	3/?	Colcemide	0.5–5.	9–26
	Spina bifida	—	N.Z. White	282	1/?	U-11, 555A	10–20	9–12
	Spina bifida[f]	?	Not stated	296	—	—	—	—
	Spina bifida	3/586	N.Z.	305	2/703	Disodium etidronate	25–50	2–16
	Spina bifida	1/3400	N.Z. White	311	—	—	—	—
	Spina bifida	—	Pourgogne	345	2/47	AY 9944	50	1–13
	Spina bifida[f]	—	Mixed breeds	341	?	Imipramine	5–30	1–20
	Spina bifida	—	Mixed breeds	357	1/109	Thalidomide	500	6–11
	Spina bifida	3/2640	Mixed breeds	361	—	—	—	—
	Spina bifida	—	N.Z. White	372	1/370	Thalidomide (oral)	25–200	8–12
	Spina bifida	—	N.Z. White	372	5/319	Thalidomide (IV)	10	7–10
	Spina bifida	1/118	Jap. Albino	394	1/92	Norethisterone mestranol	1–10	8–20
	Scoliosis	?	N.Z. White	311	—	—	—	—
	Scoliosis[f]	30/2640	Mixed breeds	361	—	—	—	—
	Scoliosis[f]	—	Dutch Belted	365	24/88	6-Aminonicotinamide	1–3	9–12
	Scoliosis[f]	41/?	N.Z. White	360	—	—	—	—
Coccygeal	Crooked spine	—	Mixed breed	233	1/136	Cyclopamine	150–300	6–9
	Abnormalities	—	N.Z.	372	20/370	Thalidomide (oral)	25–200	8–12
	Abnormalities	—	N.Z.	372	9/319	Thalidomide (IV)	7–10	10
	Not ossified	—	Hybrid	200	1/98	Thalidomide	500	3–1
	Fusion	—	N.Z. White	111	5/18	Thalidomide	150	8–16
	Fusion	5/186	Mixed breeds	357	33/109	Thalidomide	500	6–11
	Bipartite	—	Mixed breeds	357	6/109	Thalidomide	500	6–11
	Hemicentric	7/186	Mixed breeds	357	7/109	Thalidomide	500	6–11
	Assymetrical	2/186	Mixed breeds	357	24/109	Thalidomide	500	6–11
	Variations	?	Hybrid	405	?	Thalidomide	150	7.5–13.5
Neural arches	Abnormalities[f]	—	Dutch Belted	88	?	Inherited spina bifida	—	—
	Abnormalities[f]	2/56	Dutch Belted	365	28/88	6-Aminonicotinamide	1–3	9–12
	Defect	—	Not stated	74	15/352	Inheritance	—	—
	Fused[f]	—	Not stated	74	?	Inheritance	—	—
	Fused	—	Mixed breeds	357	1/109	Thalidomide	500	6–11
	Assymetrical	—	Mixed breeds	357	1/109	Thalidomide	500	6–11
	Enlarged	—	Not stated	74	63/352	Inheritance	—	—

[a] Reported in nonexperimental "control" fetuses or in those from dams exposed to various experimental conditions.

[b] Incidence: no. of affected fetuses/no. of fetuses examined.

[c] The natural occurrence of phenomena in experimental fetuses cannot be overlooked. Comments as to whether the observation was spontaneous or induced or its relation to the experimental treatment were usually absent from the majority of articles reviewed.

[d] Dose: Summarized depiction of dosage range or as otherwise indicated (usually in mg/kg/day).

[e] Time: Period of days administered during gestation.

[f] Illustrations in reference.

? = Incidence not specified yet observation or lesion noted.

TABLE VII
SKELETAL SYSTEM: LIMBS–MISCELLANEOUS AND DIGITS[a]

Organ/Tissue	Lesion/Observation	Control incidence[b] (inapparent etiology)	Stock or strain	Ref.	Incidence[b] (treated animals)	Experimental[c] condition	Dose[d] (mg/kg)	Time[e] (day of gestation)
			Skeletal System					
Limbs								
(miscellaneous)								
Forelimbs and/								
or hindlimbs								
	Malformations[f]	?	N.Z. White	83	?	Miscellaneous tests	—	—
	Malformations	—	N.Z.	122	?/43	Thalidomide	150	7–12
	Malformations	—	White Danish	244	?/46	Thalidomide	100	1–20
	Malformations (forelimb)	—	N.Z.	122	1/30	4-Phthalimidobutyramide	150	7–12
	Malformations (forelimbs)	—	N.Z.	122	3/43	Thalidomide	150	7–12
	Malformations (forelimbs)	—	N.Z.	122	1/59	Phthalimidobenzene	150	7–12
	Malformations (forelimbs)	—	White Danish	245	2/13	Thalidomide	60	1–20
	Malformations (forelimbs)	—	Not stated	271	?	6-Mercaptopurine	?	10–16
	Malformations (forelimbs)	—	Not stated	128	8/17	Thalidomide	50	7–15
	Malformations (forepaws)	—	White Danish	244	?	Thalidomide	100	1–20
	Malformations (hindpaws)	—	White Danish	244	?	Thalidomide	100	1–20
	Abnormalities[f]	—	N.Z. White	111	30/88	Thalidomide	150	8–16
	Abnormalities	—	Hybrid	200	?	Thalidomide	500	3–8
	Abnormalities[f]	7/56	Dutch Belted	365	30/88	6-Aminonicotinamide	1–3	9–12
	Abnormalities[f]	—	N.Z.	372	56/370	Thalidomide (oral)	25–200	8–12
	Abnormalities[f]	—	N.Z.	372	41/319	Thalidomide (IV)	5	7–10
	Abnormalities (forelimbs)	7/2640	Mixed breeds	361	—	—	—	—
	Defects	—	N.Z. White	282	1/324	Clomiphene	7.5–15	1
	Defects	—	Mixed breeds	258	14/46	Thalidomide	300	8–16
	Deformed[f]	—	N.Z. White	83	1/2821	Miscelleanous tests	—	—
	Deformed[f]	—	N.Z. White/ Dutch Belted	103	?	Thalidomide	150	8–16
	Deformed	—	White Danish	245	6/13	Thalidomide	60	1–20
	Deformed (hindlimbs)[f]	—	N.Z. White	102	?	Thalidomide	150	8–16
	Deformed	16/587	Dutch Belted	352	—	—	—	—
	Deformed (hindlimbs)[f]	—	Hybrid	405	59/133	Thalidomide	150	7.5–13.5
	Abnormal development	?	N.Z. White	2	—	MER-25 (estrogen decrease)	—	25–30
	Abnormal development[f]	?	N.Z. White	90	—	—	—	—
	Ossification patterns	—	Not stated	89	?	Inbreeding	—	—
	Achondroplasia[f]	?	Mixed breeds	58	—	—	—	—
	Achondroplasia[f]	1/1114	N.Z. White	83	1/2821	Miscellaneous tests	—	—
	Achondroplasia	—	Not stated	323	?	Inheritance	—	—
	Achondroplasia[f]	—	Not stated	322	?	Inheritance	—	—
	Chondrodystrophy	—	Not stated	295	?	Inherited Pelger anomaly	—	—
	Chondrodystrophy–lethal	—	Not stated	58	?	Inheritance	—	—
	Dysmelia	—	N.Z. White	325	51/121	Thalidomide	150–300	7.5–13.5
	Dysmelia	—	Hybrid	405	76/133	Thalidomide	150	7.5–13.5
	Amelia	1/17,000	N.Z. White	311	—	—	—	—
	Amelia	—	N.Z. White	281	13/13	6-Mercaptopurine	5–20	12–14
	Absent (forelimb)[f]	—	N.Z. White	102	?	Thalidomide	150	8–16
	Hemimelia[f]	?	N.Z. White	83	1/2821	Miscellaneous tests	—	—
	Hemimelia[f]	1/17,000	N.Z. White	311	—	—	—	—
	Phocomelia (forelimb)	—	N.Z. White	102	1/56	Thalidomide	150	8–16
	Phocomelia (forelimb)	—	Mixed breeds	357	1/109	Thalidomide	500	6–11
	Micromelia[f]	—	N.Z. White	111	Occasional	Thalidomide	150	8–16
	Short (forelimb)[f]	—	N.Z. White	83	1/2821	Miscellaneous tests	—	—
	Short (forelimb)[f]	—	N.Z. White	102	?	Thalidomide	150	8–16
	Short (forelimb)	1/105	N.Z. White/ Dutch Belted	235	—	—	—	—
	Short	—	Not stated	295	?	Inherited Pelger anomaly	—	—
	Short (hindlimb)	?	N.Z. White	329	—	—	—	—

Table VII (*continued*)

Organ/ Tissue	Lesion/Observation	Control incidence[b] (inapparent etiology)	Stock or strain	Ref.	Incidence[b] (treated animals)	Experimental[c] condition	Dose[d] (mg/kg)	Time[e] (day of gestation)
	Varus deformity (forelimb)[f]	—	Hybrid	209	9/49	Thalidomide	500	8–15
	Varus deformity[f]	—	N.Z. White	102	?	Thalidomide	150	8–16
	Varus deformity	2/1620	Dutch Belted	265	1/139	Methadone HCl	20–40	6–18
	Splayfoot	—	N.Z. White	210	—	Inheritance	?	?
	Micropus	—	N.Z. White	216	1/109	Laparotomy and atmosphere	5 min	7–10
	Distal foreleg curvature[f]	—	Several strains	320	?	Inheritance	—	—
	Arthrogryposis	—	N.Z.	139	?	Inheritance	—	—
	Arthrogryposis (forelimbs)[f]	—	N.Z. White	14	1/20	L-Asparaginase	50	8–9
	Arthrogryposis (forelimbs)[f]	—	N.Z. White	102	?	Thalidomide	150	8–16
	Arthrogryposis (forelimbs)	1/1700	N.Z. White	311	—	—	—	—
	Arthrogryposis (forelimbs)	1/105	N.Z. White/ Dutch Belted	235	—	—	—	—
	Arthrogryposis (forelimbs)	—	N.Z. White	306	?/132	Thalidomide	150	7–11
	Abnormal angle	—	Hybrid	200	2/57	Thalidomide	500	3–16
	Abnormal angle	—	Mixed breeds	341	?	Imipramine	5–30	1–20
	Contracture	1/118	Jap. Albino	394	17/247	Chlormadenone acetate	1–10	8–20
	Clubbing[f]	—	N.Z. White	102	?	Thalidomide	150	8–16
	Clubbing	2/292	Mixed breeds	150	1/201	Meclozine HCi	25–175	7–17
	Clubbing	—	Mixed breeds	150	?/50	Thalidomide	150	7–17
	Clubbing	1/105	N.Z./Dutch Belted	235	33/104	Thalidomide	75	6–16
	Clubbing	—	Himalayan	248	6/177	Thalidomide	50–450	8–16
	Clubbing	—	Mixed breeds	257	50%	Thalidomide	300	8–15
	Clubbing	—	Mixed breeds	258	23/46	Thalidomide	300	8–16
	Clubbing	—	Mixed breeds	357	33/109	Thalidomide	500	6–11
	Clubbing	?	Dutch Belted	443	—	—	—	—
	Talipes equinovarus	—	Dutch Belted	365	24/48	6-Aminonicotinamide	1–3	9–12
	Talipes volutus[f]	—	N.Z. White	139	2/120	Thalidomide	50–300	4–16
	Talipes varus[f]	—	N.Z. White	102	?	Thalidomide	150	8–16
	Talipes varus[f]	—	N.Z. White	139	14/120	Thalidomide	50–300	4–16
	Talipes varus[f]	—	Hybrid	209	2/49	Thalidomide	500	8–15
	Malrotation	—	N.Z. White	33	?	Thalidomide	100	5–18
	Malrotation (forelimb)	?/?	N.Z. White	51	1/171	Methaqualone	100–400	1–29
	Malrotation (hindlimb)[f]	—	Not stated	74	?	Inbreeding	—	—
	Medial or radial rotation	—	N.Z. White	306	?/132	Thalidomide	150	7–11
	Twisted	—	N.Z. White	306	?/132	Thalidomide	150	7–11
	Inversion	—	N.Z. White	306	?/132	Thalidomide	150	7–11
	Limb flexure[f]	26/8000	N.Z. White	311	—	—	—	—
	Limb flexure	—	Himalayan	248	32/177	Thalidomide	50–450	6–18
	Limb flexure (hindlimb)	1/49	Albino	103	—	—	—	—
	Limb flexure (forelimb)	—	Dutch Belted	352	4/180	Ethynodiol diacetate	0.01–0.5	0–28
	Limb flexure (forelimb)	—	Dutch Belted	352	1/137	Mestranol	0.001–0.25	0–28
	Polydactylism	—	Mixed breeds	150	?/50	Thalidomide		7–17
	Polydactylism[f]	—	Hybrid	209	5/49	Thalidomide	500	8–15
	Polydactylism	—	Not stated	252	?	Thalidomide	100	8–16
	Supernumerary digits	—	N.Z. White	111	4/88	Thalidomide	150	8–16
	Supernumerary defects	—	N.Z. White	325	6/121	Thalidomide	150–300	7.5–13.5
	Supernumerary defects[f]	—	Hybrid	405	17/120	Thalidomide	150	7.5–13.5
	Supernumerary defects[f]	—	N.Z. White/ hybrid	406	?	Thalidomide	?	?
	Hyperphalangism (forelimbs)	?/?	N.Z. White	51	1/171	Methaqualone	100–400	1–29
	Ectrodactyly	—	N.Z.	139	?	Thalidomide	50–300	4–16
	Ectrodactyly[f]	1/17,000	N.Z. White	311	—	—	—	—
	Ectrodactyly and club feet	—	Not stated	346	49/158	Thalidomide	150–250	6–14
	Ectrodactyly	—	Not stated	393	?	Hydroxyurea	?	11
	Ectrodactyly (forelimbs)[f]	—	Dutch Belted	365	3/42	6-Aminonicotinamide	1–3	9–12

Table VII (*continued*)

Organ/ Tissue	Lesion/Observation	Control incidence[b] (inapparent etiology)	Stock or strain	Ref.	Incidence[b] (treated animals)	Experimental[c] condition	Dose[d] (mg/kg)	Time[e] (day of gestation)
	Oligodactylia (forelimbs)	—	Mixed breeds	142	92/331	Cyclophosphamide	30	6–14
	Oligodactylia (forelimbs)	1/118	Jap. Albino	394	2/247	Chlormadenone	1–10	8–20
	Adactylia	—	N.Z. White	33	?/57	Thalidomide	100	5–18
	Adactylia (forelimbs)	1/51	N.Z. White	33	?/57	Thalidomide	100	5–18
	Adactylia (hindlimbs)	—	Mixed breeds	142	?	Cyclophophamide	30	6–14
	Absent digits	—	N.Z. White	111	14/88	Thalidomide	150	8–16
	Absent digits	—	Hybrid	200	5/57	Thalidomide	500	3–16
	Absent digits	—	White Danish	245	3/13	Thalidomide	60	1–20
	Absent digits	—	Chinchilla	381	?	Inheritance	—	—
	Absent (forelimbs)							
	1st	—	Mixed breeds	357	11/109	Thalidomide	500	6–11
	Pollex	—	N.Z. White	51	1/171	Methaqualone	100–400	1–29
	Pollex[f]	—	Not stated	128	?/17	Thalidomide	50	7–15
	Pollex[f]	—	N.Z. White	325	6/121	Thalidomide	150–300	7.5–13.5
	Pollex	1/2000	N.Z. White	311	—	—	—	—
	5th Digit	—	Mixed breeds	357	1/109	Thalidomide	500	6–11
	3rd Digit[f]	1/51	N.Z. White	33	—	—	—	—
	Absent (hindlimbs)	—	Not stated	128	?/17	Thalidomide	50	7–15
	Reduction deformities[f]	—	N.Z. White	325	50/121	Thalidomide	150–300	7.5–13.5
	Reduction deformities[f]	—	Hybrid	405	76/133	Thalidomide	50	7.5–13.5
	Reduction deformities[f]	—	N.Z. White/ hybrid	406	?	Thalidomide	?	?
	Reduced forelimb[f]	—	N.Z. White	83	1/2821	Miscellaneous tests	—	—
	Reduced forelimb	—	Mixed breeds	357	30/109	Thalidomide	500	6–11
	Hypophalangy	—	Himalayan	248	2/137	Thalidomide	50–450	8–16
	Brachydactyly	—	N.Z.	139	?	Thalidomide	50–300	4–16
	Brachydactyly	—	Mixed breeds	142	74/331	Cyclophosphamide	30	6–14
	Brachydactyly[f]	—	English	167	?	Inheritance	—	—
	Brachydactyly	—	N.Z. White	216	1/104	Laparotomy and atmosphere	15 min	7–10
	Brachydactyly	—	g	165	?	Inheritance	—	—
	Brachydactyly	1/17,000	N.Z. White	311	—	—	—	—
	Microdactyly	—	N.Z. White	33	?	Thalidomide	100	5–18
	Syndactyly	—	N.Z. White	83	1/2821	Miscellaneous tests	—	—
	Syndactyly	—	Not stated	274	?	Vitamin A excess	?	?
	Syndactyly	—	Dutch Belted	365	3/42	6-Aminonicotinamide	3	9
	Syndactyly	—	Not stated	393	?	Hydroxyurea	?	11
	Arachnodactyly	—	N.Z.	139	?	Thalidomide	50–300	4–16
Digits (fore-limbs and/or hindlimbs)								
	Platypodia	—	N.Z.	139	1/50	Thalidomide	50–300	4–16
	Pes valgus	—	N.Z.	139	?	Thalidomide	50–300	4–16
	Malformed[f]	—	Dutch Belted	365	22/42	6-Aminonicotinamide	1–3	9–12
	Abnormal development	—	Mixed breeds	142	?	Cyclophosphamide	30	6–14
	Enlarged (forelimbs)	1/1114	N.Z. White	83	—	—	—	—
	Pedunculated (forelimb)[f]	—	N.Z. White	325	8/121	Thalidomide	150–300	7.5–13.5
	Fused (forelimb)	—	Mixed breeds	76	58/345	Radiation	250–400 r	2–22
	Bipartite (forelimb)	—	Mixed breeds	357	1/109	Thalidomide	500	6–11
	Not ossified (forelimb)	1/1114	N.Z. White	83	1/2821	Miscellaneous tests	—	—
	Not ossified (forelimb)	86/8000	N.Z. White	311	—	—	—	—
	Deficient development	—	N.Z. White	51	1/171	Methaqualone	100–400	1–29
	Thumb malposition[f]	—	Himalayan	248	5/177	Thalidomide	50–450	8–16
	Thumb absent	—	N.Z. White	306	?/132	Thalidomide	150	7–11
	Thumb malformed	—	N.Z. White	306	?/132	Thalidomide	150	7–11
	Absent	—	N.Z. White	83	1/2821	Miscellaneous tests	—	—
	Absent	—	Mixed breeds	357	30/109	Thalidomide	500	6–11
	Aplasia[f]	—	Himalayan	248	23/177	Thalidomide	50–450	8–16

Table VII (*continued*)

Organ/ Tissue	Lesion/Observation	Control incidence[b] (inapparent etiology)	Stock or strain	Ref.	Incidence[b] (treated animals)	Experimental[c] condition	Dose[d] (mg/kg)	Time[e] (day of gestation)
	Reduced[f]	—	Mixed breeds	357	30/109	Thalidomide	500	6–11
	Triphalangia[f]	1/1114	N.Z. White	83	—	—	—	—
	Triphalangia[f]	0/50	N.Z. White	325	3/96	Thalidomide	100–300	7.5–13.5
	Pendulous dolichodactyly	—	N.Z.	139	?	Thalidomide	50–300	4–16
	Mummified[f]	—	N.Z. White	325	?	Thalidomide	150–300	7.5–13.5

[a] Reported in nonexperimental "control" fetuses or in those from dams exposed to various experimental conditions.

[b] Incidence: no. of affected fetuses/no. of fetuses examined.

[c] The natural occurrence of phenomena in experimental fetuses cannot be overlooked. Comments as to whether the observation was spontaneous or induced or its relation to the experimental treatment were usually absent from the majority of articles reviewed.

[d] Dose: Summarized depiction of dosage range or as otherwise indicated (usually in mg/kg/day).

[e] Time: Period of days administered during gestation.

[f] Illustrations in reference.

[g] Article not available for review; original source cited.

? = Incidence not specified yet observation or lesion noted.

TABLE VIII
Skeletal System: Forelimbs and Hindlimbs[a]

Organ/ Tissue	Lesion/Observation[f]	Control incidence[b] (inapparent etiology)	Stock or strain	Ref.	Incidence[b] (treated animals)	Experimental[c] condition	Dose[d] (mg/kg)	Time[e] (day of gestation)
			Skeletal System					
Forelimbs								
Clavicle								
	Abnormality	—	Himalayan	248	1/137	Thalidomide	50–450	6–18
Scapula								
	Malformed[f]	—	Dutch Belted	365	2/88	6-Aminonicotinamide	1–3	9–12
	Malformed[f]	—	N.Z. White	90	?	Inherited achondroplasia	—	—
	Crumpled spine	—	Mixed breeds	357	1/109	Thalidomide	500	6–11
Humerus								
	Shortened	—	N.Z. White	83	1/2821	Miscellaneous tests	—	—
	Shortened[f]	—	N.Z. White	102	?	Thalidomide	150	8–16
	Short	—	Chinchilla	381	?	Inheritance	—	—
	Absent	—	Mixed breeds	257	?	Thalidomide	300	8–16
	Absent	—	Mixed breeds	258	14/46	Thalidomide	300	8–16
Radius								
	Folded	—	Not stated	128	8/17	Thalidomide	50	7–15
	Short[f]	—	N.Z. White	111	1/18	Thalidomide	150	8–16
	Short[f]	—	N.Z. White	102	?	Thalidomide	150	8–16
	Short	—	N.Z. White	379	?	Thalidomide	150	8–16
	Short	—	Mixed breeds	357	2/109	Thalidomide	500	6–11
	Short	—	Chinchilla	381	?	Inheritance	—	—
	Absent[f]	—	N.Z. White	83	1/2821	Miscellaneous tests	—	—
	Absent	2/292	Mixed breeds	150	—	—	—	—
	Absent	—	Dutch Belted	384	?/110	Thalidomide	25–200	8–15
	Agenesis	1/17,000	N.Z. White	311	—	—	—	—
	Hypoplasia[f]	—	Hybrid	209	?	Thalidomide	500	8–15
	Hypoplasia[f]	—	Himalayan	248	8/137	Thalidomide	50–450	8–16
	Hypoplasia	—	N.Z. White	379	?/18	Thalidomide	150	8–16
Ulna								
	Folding	—	Not stated	128	8/17	Thalidomide	50	7–15

Talble VIII (*continued*)

Organ/ Tissue	Lesion/Observation[f]	Control incidence[b] (inapparent etiology)	Stock or strain	Ref.	Incidence[b] (treated animals)	Experimental[c] condition	Dose[d] (mg/kg)	Time[e] (day of gestation)
	Absent	—	N.Z. White/ Dutch Belted	103	?	Thalidomide	150	8–16
	Short	—	Chinchilla	381	?	Inheritance	—	—
	Hypoplasia[f]	—	Himalayan	248	1/137	Thalidomide	50–450	8–16
	Hypoplasia	—	N.Z. White	379	?/18	Thalidomide	150	8–16
Meta-carpus	Flexion[f]	5/1114	N.Z. White	83	6/2821	Miscellaneous tests	—	—
	Contracture	1/118	Jap. Albino	394	19/247	Chlormadinone acetate	1–10	8–20
	Dropped wrist	—	Hybrid	200	4/57	Thalidomide	500	3–16
	Synostosis	—	Himalayan	248	2/137	Thalidomide	50–450	8–16
	1st Reduced	—	Mixed breeds	357	2/109	Thalidomide	500	6–11
	1st Absent	—	Hybrid	200	3/57	Thalidomide	500	6–11
Hindlimbs Femur								
	Bent	—	Mixed breeds	258	?/46	Thalidomide	300	8–16
	Bent	—	Mixed breeds	257	?	Thalidomide	300	8–15
	Defect	—	N.Z. White	379	1/18	Thalidomide	150	8–16
	Abnormal shape	—	Dutch Belted	365	1/42	6-Aminonicotinamide	3	9
	Retarded development	—	N.Z. White	4	?	Ethamoxytriphetol	25	21–26
	Decreased collagen synthesis	—	N.Z. White	3	?	Ethamoxytriphetol	25	21–26
	Short[f]	—	N.Z. White	111	1/18	Thalidomide	150	8–16
	Short	—	Mixed breeds	258	14/46	Thalidomide	300	8–16
	Short	—	Chinchilla	381	?	Inheritance	—	—
	Short	—	Hybrid	405	3/133	Thalidomide	150	7.5–13.5
Tibia								
	Abnormalities	—	Hybrid	200	1/57	Thalidomide	500	3–16
	Abnormalities[f]	—	N.Z. White	379	?/18	Thalidomide	150	8–16
	Defects	—	Hybrid	405	59/133	Thalidomide	150	7.5–13.5
	Bent	—	N.Z. White	111	1/18	Thalidomide	150	8–16
	Absent	—	N.Z. White/ Dutch Belted	103	?	Thalidomide	150	8–16
	Absent	—	Mixed breeds	257	?	Thalidomide	300	8–15
	Absent	—	Mixed breeds	258	14/46	Thalidomide	300	8–16
	Absent	1/219	Albino	329	—	—	—	—
	Absent	—	Mixed breeds	357	5/109	Thalidomide	500	6–11
	Aplasia	—	N.Z. White	33	1/57	Thalidomide	100	5–18
	Aplasia[f]	—	Himalayan	248	10/137	Thalidomide	50–450	8–16
	Short[f]	—	N.Z. White	102	?	Thalidomide	150	8–16
	Short	—	Hybrid	200	1/57	Thalidomide	500	3–16
	Hypoplasia[f]	—	Hybrid	209	?	Thalidomide	500	8–15
	Hypoplasia[f]	—	Not stated	252	?	Thalidomide	100	8–16
	Extra collar of bone	1/64	Hybrid	200	—	—	—	—
Fibula								
	Abnormalities	—	Hybrid	200	1/57	Thalidomide	500	3–16
	Malformed	—	Hybrid	405	?	Thalidomide	150	7.5–13.5
	Deformed[f]	—	N.Z. White	379	?/18	Thalidomide	150	8–16
	Absent[f]	—	N.Z. White	111	4/18	Thalidomide	150	8–16
	Absent[f]	—	N.Z. White	102	?	Thalidomide	150	8–16
	Absent	—	Mixed breeds	257	?	Thalidomide	300	8–15
	Absent	—	Mixed breeds	258	14/46	Thalidomide	300	8–16
	Absent	1/219	Albino	329	—	—	—	—
	Short[f]	—	N.Z. White	102	?	Thalidomide	150	8–16
	Hypoplasia[f]	—	Hybrid	209	?	Thalidomide	500	8–15
	Hypoplasia[f]	—	Not stated	252	?	Thalidomide	100	8–16
	Hypoplasia[f]	—	Himalayan	248	4/137	Thalidomide	50–450	8–16
	Hyperplasia	—	N.Z. White	33	1/57	Thalidomide	100	5–18
	Enlarged	—	Hybrid	200	1/57	Thalidomide	500	3–16

Table VIII (*continued*)

Organ/ Tissue	Lesion/Observation[f]	Control incidence[b] (inapparent etiology)	Stock or strain	Ref.	Incidence[b] (treated animals)	Experimental[c] condition	Dose[d] (mg/kg)	Time[e] (day of gestation)
Tarsus- metatarsus Talus								
	Synostosis	—	Himalayan	248	2/137	Thalidomide	50–450	8–16
	Aplasia	—	Himalayan	248	5/137	Thalidomide	50–450	8–16
	Absence, hypoplasia	—	Dutch Belted	365	?	6-Aminonicotinamide	3 mg/kg	12
Astrag- alus Pollex	Aplasia	—	N.Z. White	33	1/57	Thalidomide	100	5–18
	Agenesis	1/2000	N.Z. White	311	—	—	—	—
Meta- arsus	Shortened[f]	—	Hybrid	405	?	Thalidomide	150	7.5–13.5

[a] Reported in nonexperimental "control" fetuses or in those from dams exposed to various experimental conditions.

[b] Incidence: No. of affected fetuses/No. of fetuses examined.

[c] The natural occurrence of phenomena in experimental fetuses cannot be overlooked. Comments as to whether the observation was spontaneous or induced or its relation to the experimental treatment were usually absent from the majority of articles reviewed.

[d] Dose: Summarized depiction of dosage range or as otherwise indicated (usually in mg/kg/day).

[e] Time: Period of days administered during gestation.

[f] Illustrations in reference.

[g] Article not available for review; original source cited.

? = Incidence not specified yet observations or lesion noted.

TABLE IX

NERVOUS SYSTEM: BRAIN AND PERIPHERAL NERVES[a]

Organ/ Tissue	Lesion/Observation[f]	Control incidence[b] (inapparent etiology)	Stock or strain	Ref.	Incidence[b] (treated animals)	Experimental[c] condition	Dose[d] (mg/kg)	Time[e] (day of gestation)
			Nervous System					
Brain								
	Anencephaly[f]	1/1114	N.Z. White	83	1/2821	Miscellaneous tests	—	—
	Anencephaly	—	Mixed	142	1/331	Cyclophosphamide	30	6–14
	Anencephaly	—	N.Z. White	216	1/109	Laparotomy and atmosphere	5 min	7–10
	Anencephaly	—	N.Z. White	282	1/?	U 11,100A	1–5	4–6
	Anencephaly	1/3400	N.Z. White	311	—	—	—	—
	Anencephaly[f]	—	Not stated	403	?	Thalidomide	125–250	6–14
	Anencephaly	—	Not stated	346	12/158	Thalidomide	150–250	6–14
	Arrhinencephaly	1/40,000	Not stated	296	—	—	—	—
	Arrhinencephaly	—	N.Z.	305	1/703	Disodium etidronate	25–500	2–16
	Arrhinencephaly[f]	1/17,000	N.Z. White	311	—	—	—	—
	Microcephaly	—	Not stated	77	?	Insulin	?	?
	Microcephaly	1/2640	Mixed breeds	361	3/112	Dexamethasone	?	?
	Exencephaly	—	Mixed breeds	76	?	Radiation	250–400 r	2–22
	Exencephaly	—	N.Z. White	230	6/126	Methanphetamine	1.5	12–30
	Exencephaly	?	Not stated	251	?	Ethyl alcohol	5	8–16
	Exencephaly	—	Himalayan	248	1/177	Thalidomide	50–450	6–18
	Exencephaly	1/4300	N.Z. White	311	4/247	—	—	—
	Exencephaly	–	Jap. Albino	394	?	Chlormadinone acetate	1–10	8–20
	Encephalocele	1/292	Mixed breeds	150	?/50	Thalidomide	150	7–17
	Encephalocele	—	Mixed breeds	257	25%?	Thalidomide	300	8–15
	Encephalocele	—	N.Z. White/ Dutch Belted	258	14/46	Thalidomide	300	8–16
	Encephalocele	—	Pourgogni	345	?	AY9944	50	1–13
	Encephalocele[f]	—	Mixed breeds	341	?	Imipramine	5–30	1–20

Table IX (*continued*)

Organ/ Tissue	Lesion/Observation[f]	Control incidence[b] (inapparent etiology)	Stock or strain	Ref.	Incidence[b] (treated animals)	Experimental[c] condition	Dose[d] (mg/kg)	Time[e] (day of gestation)
	Encephalocele[f]	—	Not stated	403	?	Thalidomide	125–250	6–14
	Hydrocephalus	—	N.Z. White	14	1/18	L-Asparaginase	50	8–9
	Hydrocephalus	—	Not stated	74	6/352	Inheritance	—	—
	Hydrocephalus	—	N.Z. White	111	4/88	Thalidomide	150	8–16
	Hydrocephalus	—	Not stated	128	4/17	Thalidomide	50	7–15
	Hydrocephalus	—	Dutch Belted	194	?	Vitamin A deficiency	?	6–38 wk
	Hydrocephalus	—	Mixed breeds	233	20/200	Cyclopamine	150–300	6–9
	Hydrocephalus[f]	0/16	Mixed breeds	233	3/47	Cyclopamine	22–45	6–9
	Hydrocephalus	—	Mixed breeds	233	5/50	Jervine	300	7
	Hydrocephalus	5/1333	Himalayan	248	—	—	—	—
	Hydrocephalus[f]	—	Not stated	272	16/16	Vitamin A deficiency	—	1–30
	Hydrocephalus	—	N.Z. White	281	1/?	Vinblastine	0.3	3
	Hydrocephalus	—	N.Z. White	282	1/324	Clomiphene	4–5	1
	Hydrocephalus[f]	—	Albino	273	47/51	Vitamin A deficiency	—	14–38 wk
	Hydrocephalus[f]	—	Mixed breeds	276	82/141	Vitamin A deficiency	—	12–28 wk
	Hydrocephalus	3/696	Dutch Belted	265	1/139	Methadone HCl	20–40	6–18
	Hydrocephalus	1/586	N.Z.	305	3/703	Disodium etidronate	25–500	2–16
	Hydrocephalus	0/62	N.Z. White	325	3/121	Thalidomide	100–300	7.5–13.5
	Hydrocephalus	1/8500	N.Z. White	311	—	—	—	—
	Hydrocephalus	—	Not stated	324	?	Inherited osteopetrosis	—	—
	Hydrocephalus	18/2640	Mixed breeds	361	—	—	—	—
	Hydrocephalus	2/104	Dutch Belted	365	—	—	—	—
	Hydrocephalus	—	N.Z.	372	?	Thalidomide	25–200	8–12
	Hydrocephalus	13/408	Dutch Belted	352	—	—	—	—
	Hydrocephalus	—	g	293	?	Inheritance	—	—
	Hydrocephalus	5/162	Hybrid	405	6/120	Thalidomide	150	7.5–13.5
	Hydrocephalus	?	Dutch Belted	443	—	—	—	—
Cerebellum	Displacement	—	Not stated	273	21/51	Vitamin A deficiency	—	14–38 wk
Medulla oblongata	Displacement	—	Not stated	273	21/51	Vitamin A deficiency	—	14–38 wk
Spinal cord	Syringomyelia	—	g	289	?	Inheritance	—	—
	Syringomyelia	—	Not stated	356	?	Inheritance	—	—
	Spina bifida (see Skeletal System, Vertebral Column)							
Meninges	Meningoencephalocele[f]	—	N.Z. White	102	?	Thalidomide	150	8–16
	Meningocele	—	N.Z.	372	?	Thalidomide	25–200	8–12
Spinal fluid	Increased pressure	—	Not stated	275	?	Maternal hypovitaminosis A	—	—
	Increased pressure	—	Mixed breeds	276	?	Maternal hypovitaminosis A	—	12–28 wk
	Overproduction	—	N.Z. White	111	1/88	Thalidomide	15	8–16
Optic nerve	Atrophy	—	Several strains	134	?	Inheritance	—	—

[a] Reported in nonexperimental "control" fetuses or in those from dams exposed to various experimental conditions.

[b] Incidence: no. of affected fetuses/no. of fetuses examined.

[c] The natural occurrence of phenomena in experimental fetuses cannot be overlooked. Comments as to whether the observation was spontaneous or induced or its relation to the experimental treatment were usually absent from the majority of articles reviewed.

[d] Dose: Summarized depiction of dosage range or as otherwise indicated (usually in mg/kg/day).

[e] Time: Period of days administered during gestation.

[f] Illustrations in reference.

[g] Article not available for review; original source cited.

? = Incidence not specified yet observation or lesion noted.

TABLE X
Nervous System: Eye and Respiratory System[a]

Organ/ Tissue	Lesion/Observation	Control incidence[b] (inapparent etiology)	Stock or strain	Ref.	Incidence[b] (treated animals)	Experimental[c] condition	Dose[d] (mg/kg)	Time[e] (day of gestation)
Eye General						Nervous System		
	Abnormalities	—	N.Z. White	243	?	Vitamin A deficiency	—	—
	Anomalies[f]	—	Not Stated	181	?	Lens antisera	?	9–20
	Defects	—	N.Z. White	282	1/324	Clomiphene	7.5–15.	1
	Defects	—	N.Z. White	372	?	Thalidomide	25–200	8–12
	Posterior defects	—	Not stated	181	?	Lens antisera	?	9–20
	Hemorrhage	58/8000	N.Z. White	311	—	—	—	—
	Rudimentary eyes	—	Not stated	200	1/98	Thalidomide	300–500	3–16
	Exopthalmus	—	Not stated	322	?	Inherited osteopetrosis	—	—
	Exopthalmus	—	Not stated	393	?	Vincristine	?	7–8
	Micropthalmia	—	Dutch Belted	88	?	Inherited spina bifida	—	—
	Micropthalmia	—	Mixed breeds	142	1/331	Cyclophosphamide	30	6–14
	Micropthalmia	—	Mixed breeds	150	?/50	Thalidomide	150	7–17
	Micropthalmia	—	Not stated	181	?	Lens antisera	?	9–20
	Micropthalmia	—	Dutch Belted	235	1/145	Folpet	75	6–16
	Micropthalmia	—	N.Z. White	230	2/126	Methamphetamine	1.5	12–30
	Micropthalmia	—	N.Z. White	325	3/121	Thalidomide	150–300	7.5–13.5
	Micropthalmia	1/8000	N.Z. White	311	—	—	—	—
	Micropthalmia	—	Mixed breeds	357	3/109	Thalidomide	500	6–11
	Micropthalmia[f]	—	Dutch Belted	365	70/88	6-Aminonicotinamide	1–3	9–12
	Micropthalmia	—	Not stated	403	?	Actinomycin D	50–70 γ	?
	Micropthalmia	—	Hybrid	405	4/120	Thalidomide	150	7.5–13.5
	Anophthalmia	—	N.Z. White	111	4/88	Thalidomide	150	8–16
	Anophthalmia	—	N.Z. White	33	?	Thalidomide	100	5–18
	Anophthalmia	0/62	N.Z. White	325	3/121	Thalidomide	100–300	7.5–13.5
	Anophthalmia	—	N.Z. White	305	1/703	Disodium etidronate	25–500	2–16
	Anophthalmia	1/219	Albino	329	—	—	—	—
	Anophthalmia	—	Not stated	403	?	Actinomycin D	50–70 r	?
	Buphthalmia[f]	—	Several strains	134	?	Inheritance	—	—
	Coloboma	—	g	204	?	Inheritance	—	—
	Coloboma	1/586	N.Z.	305	—	—	—	—
	Coloboma	—	Not stated	435	?	Actinomycin D	?	?
	Glaucoma	—	Not stated	181	?	Lens antisera	?	?
	Glaucoma	—	g	411	?	Inheritance	—	—
	Hydrophthalmus	—	g	135	?	Inheritance	—	—
	Cyclopea[f]	—	N.Z. White	83	1/2821	Miscellaneous tests	—	—
	Cyclopea	—	N.Z. White	111	1/88	Thalidomide	150	8–16
	Cyclopea	—	N.Z. White	122	1/61	Phthalimidobenzene	150	7–12
	Cyclopea[f]	0/16	Mixed breeds	232	6/47	Cyclopamine	22, 33, 45	6–9
	Cyclopea	—	Mixed breeds	233	2/50	Jervine	300	7
	Cyclopea[f]	—	Mixed breeds	233	23/200	Cyclopamine	150–300	6–9
	Cyclopea	—	N.Z. White	230	5/126	Methamphetamine	1.5	12–30
	Cyclopea	?	g	270	—	—	—	—
	Cyclopea[f]	1/5700	N.Z. White	311	—	—	—	—
	Cyclopea	—	Pourgogne	345	?	AY 9944	50	1–13
	Cyclopea	—	Jap. Albino	394	1/247	Chloradenone acetate	1–10	8–20
Cornea	Cornification	—	Several strains	134	?	Inheritance	—	—
Iris	Coloboma	—	Mixed breeds	357	2/109	Thalidomide	500	8–11
Lens	Herniated	1/156	N.Z.	305	—	—	—	—
	Cataracts	—	Not stated	181	?	Lens antigen	?	8–17
	Cataracts	—	Not stated	181	?	Lens antisera	?	8–17
	Cataracts	—	g	297	?	Inheritance	—	—

Table X (*continued*)

Organ/ Tissue	Lesion/Observation	Control incidence[b] (inapparent etiology)	Stock or strain	Ref.	Incidence[b] (treated animals)	Experimental[c] condition	Dose[d] (mg/kg)	Time[e] (day of gestation)
Retina								
	Rossettes	—	g	309	?	X-Irradiation	?	?
	Folded	1/586	N.Z.	305	2/703	Disodium etidronate	25–100	2–16
				Respiratory System				
Nostril								
	Absent	—	White Danish	244	?	Thalidomide	100	1–20
	Closed[f]	—	N.Z.	372	1/125	Thalidomide	25–200	8–12
	Single	—	White Danish	244	?	Thalidomide	100	1–20
	Malformed[f]	—	N.Z. White	102	?	Thalidomide	150	8–16
	Abnormal	—	N.Z. White	281	1/?	6-Azauridine	500	13–17
	Deformed[f]	—	Mixed breeds	232	?	Cyclopamine	22–45	6–9
Larynx								
	Choanal stenosis	?	g	291	?	Inheritance	—	—
	Cleft[f]	—	Mixed breeds	408	?	Thalidomide	150	7.5–13.5
Lungs								
	Small	—	Dutch Belted	88	?	Inherited spina bifida	—	—
	Small[f]	—	Mixed breeds	357	3/109	Thalidomide	150	6–11.5
	Lobulation–abnormal	0/62	N.Z. White	325	Occas/121	Thalidomide	100–300	7.5–13.5
	Defective	—	Hybrid	405	3/120	Thalidomide	150	7.5–13.5
	Bilobed right side	—	N.Z. White	14	1/20	L-Asparaginase	50	8–9
	Agenesis intermediate lobe	36/8000	N.Z. White	311	—	—	—	—
	Absent lobe	—	N.Z. White	51	?	Methaqualone	100	8–16
	Absent lobe	—	Not stated	128	1/17	Thalidomide	50	7–15
	Edema	—	Not stated	98	?	Fluid inhalation	—	28
	Pneumonia[f]	—	Not stated	38	?	*Streptococcus foecalis*	—	26–29
	Pneumonia[f]	—	Not stated	38	?	*Escherichia coli*	—	26–29
	Increased weight[f]	—	Not stated	16	?/116	Prenatal hypoxia	—	—
	Increased weight[f]	—	N.Z. White	63	?/18	Tracheal ligation	—	25–30
	Alveolar collapse	?	Not stated	16	?	Prenatal hypoxia	—	28–30

[a] Reported in nonexperimental "control" fetuses or in those from dams exposed to various experimental conditions.

[b] Incidence: no. of affected fetuses/no. of fetuses examined.

[c] The natural occurrence of phenomena in experimental fetuses cannot be overlooked. Comments as to whether the observation was spontaneous or induced or its relation to the experimental treatment were usually absent from the majority of articles reviewed.

[d] Dose: Summarized depiction of dosage range or as otherwise indicated (usually in mg/kg/day).

[e] Time: Period of days administered during gestation.

[f] Illustrations in reference.

[g] Article not available for review; original source cited.

? = Incidence not specified yet observation or lesion noted.

TABLE XI

CARDIOVASCULAR SYSTEM[a]

Organ/ Tissue	Lesion/Observation	Control incidence[b] (inapparent etiology)	Stock or strain	Ref.	Incidence[b] (treated animals)	Experimental[c] condition	Dose[d] (mg/kg)	Time[e] (day of gestation)
				Cardiovascular System				
Heart								
	General Malformation	1/1333	Himalayan	248	1/108	Thalidomide	50–450	8–16
	abnormalities	—	Dutch Belted	88	?	Inherited spina bifida	—	—
	Cardiomegaly	—	Mixed breeds	407	?/757	Thalidomide	150	7.5–13.5

Table XI (*continued*)

Organ/ Tissue	Lesion/Observation	Control incidence[b] (inapparent etiology)	Stock or strain	Ref.	Incidence[b] (treated animals)	Experimental[c] condition	Dose[d] (mg/kg)	Time[e] (day of gestation)
	Ectopia cordis	1/4300	N.Z. White	311	—	—	—	—
	Ectopic	—	Not stated	77	?	Insulin	?	?
	Cor biloculare	1/586	N.Z. White	305	—	—	—	—
	Dextrocardia	1/1700	N.Z. White	311	—	—	—	—
	Dextrocardia[f]	—	Mixed breeds	407	1/757	Thalidomide	150	7.5–13.5
	Physiological phenomena	?	Not stated	99	—	—	—	—
Septum								
	Septal defects	—	Not stated	78	?	Cortisone acetate	0.4–4	11–14.5
	Septal defects[f]	—	Mixed breeds	407	12/757	Thalidomide	150	7.5–13.5
	Ventricular defects	—	Not stated	78	?	Cortisone acetate	0.4–4	11–14.5
	Ventricular defects[f]	—	N.Z. White	405	12/120	Thalidomide	150	7.5–13.5
	Ventricular defects[f]	—	Mixed breeds	407	17/757	Thalidomide	150	7.5–13.5
Auricles								
	Small	26/377	N.Z. White/ Dutch Belted	258	25/34	Tolbutamide	200	8–16
	Small	—	N.Z. White/ Dutch Belted	258	6/46	Thalidomide	300	8–16
	Small	—	N.Z. White/ Dutch Belted	258	12/55	Acetylsalicylic acid	200	8–16
	Small left	Occas/62	N.Z. White	325	Occasional/ 121	Thalidomide	150–300	7.5–13.5
	Enlargement	—	N.Z. White	257	20% of ?	Thalidomide	300	8–15
Valves and cusps								
	Deletion[f]	—	Mixed breeds	407	5/757	Thalidomide	150	7.5–13.5
	Supernumerary[f]	—	Mixed breeds	407	6/757	Thalidomide	150	7.5–13.5
	Disproportion[f]	—	Mixed breeds	407	2/757	Thalidomide	150	7.5–13.5
	Nonseparation[f]	—	Mixed breeds	407	1/757	Thalidomide	150	7.5–13.5
	Pulmonary valve atresia[f]	—	Mixed breeds	407	19/757	Thalidomide	150	7.5–13.5
Vasculature								
General								
	Hypoplasia	—	Mixed breeds	407	16/757	Thalidomide	150	7.5–13.5
	Hypoplasia[f]	—	N.Z. White	405	12/120	Thalidomide	150	7.5–13.5
	Vascular abnormalities[f]	—	Dutch Belted	88	?	Inherited spina bifida	—	—
	Rotational anomalies[f]	—	Mixed breeds	407	?/757	Thalidomide	150	7.5–13.5
	Transportation[f]	—	Mixed breeds	407	7/757	Thalidomide	150	7.5–13.5
Jugular vein								
	Absent	—	Mixed breeds	357	1/109	Thalidomide	500	6–11
Subclavian artery								
	Anomaly	—	N.Z. White	405	3/120	Thalidomide	150	7.5–13.5
	Anomaly[f]	—	Mixed breeds	407	1/757	Thalidomide	150	7.5–13.5
Coronary artery								
	Abnormal pattern[f]	—	Mixed breeds	407	?	Thalidomide	150	7.5–13.5
Aorta								
	Developmental variation[f]	—	Mixed breeds	362	?	Inheritance	—	—
	Anomalies	—	N.Z. White	257	9–21%?	Thalidomide	300	8–15
	Coarctation	—	Mixed breeds	407	4/757	Thalidomide	150	7.5–13.5
	Coarctation	—	N.Z. White	305	1/703	Disodium etidronate	25–500	2–16
	Stenosis	—	N.Z. White	439	?	WU-385	150	6–11
	Supravalvular stenosis[f]	—/26	N.Z. White	140	14/34	Vitamin D excess	—	—
Aortic arches								
Anomalies								
	Defects	—	Not stated	346	23/148	Thalidomide		
	Right descending aorta						150–250	6–14
	4th missing	5/377	N.Z. White/ Dutch Belted	258	14/46	Thalidomide	300	8–16
	Common pulmonary trunk	—	N.Z. White/ Dutch Belted	258	9/40	Phenobarbital	50	8–16

Table XI (*continued*)

Organ/ Tissue	Lesion/Observation	Control incidence[b] (inapparent etiology)	Stock or strain	Ref.	Incidence[b] (treated animals)	Experimental[c] condition	Dose[d] (mg/kg)	Time[e] (day of gestation)
Vessels								
	Stenosis	3/586	N.Z.	305	—	—	—	—
	Interrupted	—	N.Z. White	405	1/120	Thalidomide	150	7.5–13.5
	Interrupted left	—	Mixed breeds	407	4/757	Thalidomide	150	7.5–13.5
	Broad	—	Mixed breeds	357	4/109	Thalidomide	150	6–11
	Broad	—	N.Z. White/ Dutch Belted	258	18%?	Acetylsalicylic acid	200	8–16
	Right	—	Mixed breeds	407	5/757	Thalidomide	150	7.5–13.5
	Fusion with pulmonary arch	1/3400	N.Z. White	311	—	—	—	—
	Fusion with pulmonary arch	—	Mixed breeds	357	2/109	Thalidomide	150	6–11
Pulmonary artery								
	Behind aorta	—	Mixed breeds	357	1/109	Thalidomide	500	6–11
	Absent	1/186	Mixed breeds	357	1/ -	—	—	—
	Absent[f]	—	Mixed breeds	407	3/757	Thalidomide	150	7.5–13.5
	Hypoplasia[f]	—	Mixed breeds	407	10/757	Thalidomide	150	7.5–13.5
	Hypoplasia	—	N.Z. White	405	12/120	Thalidomide	150	7.5–13.5
	Hypertrophy	—	Mixed breeds	407	5/757	Thalidomide	150	7.5–13.5
Ductus arteriosus								
	Closure[f]	?	Not stated	205	—	—	—	—
	Absent	—	Mixed breeds	407	2/757	Thalidomide	150	7.5–13.5
Right ductus arteriosus								
	Accessory[f]	—	Mixed breeds	407	1/757	Thalidomide	150	7.5–13.5
Bronchial artery								
	Abnormal pattern[f]	—	Mixed breeds	407	?	Thalidomide	150	7.5–13.5
Vena cava								
	Displaced[f]	—	Mixed breeds	357	1/109	Thalidomide	500	6–11
	Double	—	Mixed breeds	357	1/109	Thalidomide	500	6–11
Posterior vena cava								
	Abnormal pattern[f]	—	Not stated	263	?	Inheritance	—	—
Iliolumbar arteries								
	Assymetry[f]	—	Mixed breeds	364	?	Inheritance	—	—
Renal vein								
	Double	—	Mixed breeds	357	1/109	Thalidomide	500	6–11
Limb bud vessels								
	Degeneration[f]	—	Not stated	211	?	Inherited brachydactylism	—	—

[a] Reported in nonexperimental "control" fetuses or in those from dams exposed to various experimental conditions.

[b] Incidence: no. of affected fetuses/no. of fetuses examined.

[c] The natural occurrence of phenomena in experimental fetuses cannot be overlooked. Comments as to whether the observation was spontaneous or induced or its relation to the experimental treatment were usually absent from the majority of articles reviewed.

[d] Dose: Summarized depiction of dosage range or as otherwise indicated (usually in mg/kg/day).

[e] Time: Period of days administered during gestation.

[f] Illustration in reference.

[g] Article not available for review; original source cited.

? = Incidence not specified yet observation or lesion noted.

TABLE XII
URINARY AND REPRODUCTIVE SYSTEMS[a]

Organ/Tissue	Lesion/Observation[f]	Control incidence[b] (inapparent etiology)	Stock or strain	Ref.	Incidence[b] (treated animals)	Experimental[c] condition	Dose[d] (mg/kg)	Time[e] (day of gestation)
			Urinary System					
Kidney								
	Absent	—	N.Z. White	51	?	Thalidomide	500	8–13
	Absent	—	Dutch Belted	88	?	Inherited spina bifida	—	—
	Absent	—	N.Z. White	111	8/88	Thalidomide	150	8–16
	Absent	—	Mixed breeds	150	?/50	Thalidomide	150	7–17
	Absent–unilateral[f]	—	Hybrid	209	2/49	Thalidomide	500	8–15
	Absent	—	White Danish	245	1/3	Thalidomide	60	1–20
	Absent or reduced[f]	—	Mixed breeds	357	14/109	Thalidomide	500	6–11
	Absent	—	N.Z. White	439	?	WU-385	150	6–11
	Aplasia	—	Himalayan	248	3/108	Thalidomide	50–450	8–16
	Aplasia	1/62	N.Z. White	325	3/121	Thalidomide	100–300	7.5–13.5
	Agenesis–unilateral	—	N.Z. White Danish	244	?	Thalidomide	100	1–20
	Agenesis–unilateral	1/6000	N.Z. White	311	—	—	—	—
	Agenesis–unilateral	—	Hybrid	405	9/120	Thalidomide	150	7.5–13.5
	Agenesis–unilateral	—	g	95	?	Inheritance		
	Agenesis–unilateral	—	Not stated	346	32/158	Thalidomide	150–250	6–14
	Hypoplasia[f]	—	Hybrid	209	1/49	Thalidomide	500	8–15
	Hypoplasia	2/143	N.Z. White	230	—	—	—	—
	Hypoplasia	—	Himalayan	248	5/108	Thalidomide	50–450	8–16
	Hypoplasia	2/48	Dutch	365	5/48	6-Aminonicotinamide	1–3	9, 12
	Hypoplasia	—	Hybrid	405	1/120	Thalidomide	150	7.5–13.5
	Abnormal	—	White Danish	244	?	Thalidomide	100	1–20
	Deformed	—	N.Z. White	111	1/88	Thalidomide	150	8–16
	Malformations	—	Not stated	344	?	Thalidomide	?	6–14
	Ectopic	2/143	N.Z. White	230	—	—	—	—
	Ectopic	1/8500	N.Z. White	311	—	—	—	—
	Pelvic–unilateral	1/153	N.Z. White	51	1/171	Methaqualone	100	8–16
	Fusion	—	N.Z. White	14	1/20	L-Asparaginase	50	8–9
	Fusion	—	N.Z. White	139	2/120	Thalidomide	50–300	5–30
	Fusion	1/4300	N.Z. White	311	—	—	—	—
	Fusion–right angle	—	Mixed breeds	357	1/109	Thalidomide	500	6–11
	Horseshoe	1/4300	N.Z. White	311	—	—	—	—
	Hydronephrosis	1/1620	Dutch Belted	265	1/139	Methadone HCl	20–40	6–18
	Hydronephrosis	3/586	N.Z.	305	1/703	Disodium etidronate	25–500	2–16
	Hydronephrosis	1/4300	N.Z. White	311	—	—	—	—
Ureter								
	Absent	—	Mixed breeds	150	?/50	Thalidomide	150	7–17
	Absent	—	Hybrid	209	2/49	Thalidomide	500	8–15
	Absent	—	Hybrid	405	9/120	Thalidomide	150	7.5–13.5
	Absent right	1/586	N.Z.	305	—	—	—	—
	Absent–unilateral	1/6000	N.Z. White	311	—	—	—	—
	Displacement	1/3000	N.Z. White	311	—	—	—	—
	Displacement–behind vena cava	—	Mixed breeds	357	2/109	Thalidomide	500	6–11
	Hydroureter	—	Dutch Belted	88	?	Inherited spina bifida	—	—
Bladder								
	Small	—	Mixed breeds	357	1/109	Thalidomide	500	6–11
	Extroversion[f]	—	N.Z. White	14	1/20	L-Asparaginase	50	8–9
	Blood filled	—	White Danish	244	?	Thalidomide	100	1–20
			Reproductive System					
Male Gonads								
	Cryptorchism	1/586	N.Z.	305	—	—	—	—
	Atrophy	2/586	N.Z.	305	1/703	Disodium etidronate	25–500	2–16

Table XII (*continued*)

Organ/ Tissue	Lesion/Observation[f]	Control incidence[b] (inapparent etiology)	Stock or strain	Ref.	Incidence[b] (treated animals)	Experimental[c] condition	Dose[d] (mg/kg)	Time[e] (day of gestation)
Prostate								
	Suppressed development	—	Not stated	118	?	Cryproterone acetate	0.6–100	13–24
Penis								
	Hypoplasia	—	Not stated	220	?	Fetal castration	—	4–22
	Diphally[f]	—	g	291	—	—	—	—
	Hypospadia	—	g	291	?	Inheritance	—	—
Female Gonads						—		
	Absent	—	Hybrid	209	2/49	Thalidomide	500	8–15
Uterus								
	Absent horn	—	Hybrid	209	2/49	Thalidomide	500	8–15
	Interrupted[f]	1/2000	N.Z. White	311	—	—	—	—
External genitalia (see Abnormal Anatomical Regions)								
Accessory glands and structures (male and female)								
Secondary sex organs								
	Absent	—	Dutch Belted	88	?	Inherited spina bifida	—	—
Accessory glands								
	Failure to develop	—	Not stated	118	?	Cyproterone acetate	0.6–100	13–24
Cowpers glands								
	Suppressed development	—	Not stated	118	?	Cyproterone acetate	0.6–100	13–24
Vesicular gland								
	Suppressed development	—	Not stated	118	?	Cyproterone acetate	0.6–100	13–24
Wolfian ducts								
	Hypoplasia	—	Not stated	220	?	Fetal castration	—	21–22
	Regression	—	Not stated	118	?	Cyproterone acetate	0.6–100	13–24
	Persistence	—	Not stated	118	?	Cyproterone acetate	0.6–100	13–24
Mullerian ducts								
	Regression	—	Not stated	118	?	Cyproterone acetate	0.6–100	13–24

[a] Reported in nonexperimental "control" fetuses or in those from dams exposed to various experimental conditions.

[b] Incidence: no. of affected fetuses/no. of fetuses examined.

[c] The natural occurrence of phenomena in experimental fetuses cannot be overlooked. Comments as to whether the observation was spontaneous or induced or its relation to the experimental treatment were usually absent from the majority of articles reviewed.

[d] Dose: Summarized depiction of dosage range or as otherwise indicated (usually in mg/kg/day).

[e] Time: Period of days administered during gestation.

[f] Illustration in reference.

[g] Article not available for review; original source cited.

? = Incidence not specified yet observation or lesion noted.

TABLE XIII
DIGESTIVE SYSTEM[a]

Organ/ Tissue	Lesion/Observation	Control incidence (inapparent etiology)	Stock or strain	Ref.	Incidence[b] (treated animals)	Experimental condition[c]	Dose[d] (mg/kg)	Time[e] (day of gestation)
			Digestive System					
Tongue								
	Protruding		Not stated	322		Inherited achondroplasia	—	—
	Macroglossia	1/1725	Himalayan	248	—	—	—	—

Table XIII (*continued*)

Organ/ Tissue	Lesion/Observation	Control incidence (inapparent etiology)	Stock or strain	Ref.	Incidence[b] (treated animals)	Experimental condition[c]	Dose[d] (mg/kg)	Time[e] (day of gestation)
Esophagus								
	Tracheoesophageal fistula	—	N.Z. White	325	2/121	Thalidomide	100–300	7.5–13.5
	Tracheoesophageal fistula[f]	—	Several strains	408	27/906	Thalidomide	150	12.5–13.5
	Megaoesophagus	—	N.Z. White	325	3/121	Thalidomide	100–300	7.5–13.5
	Megaoesophagus	—	Hybrid	405	7/121	Thalidomide	150	7.5–13.5
	Megaoesophagus [f]	—	Several strains	408	68?/906	Thalidomide	150	12.5–13.5
	Bacterial colonies[f]	—	White	238	—	*Listeria monocytogenes*	?	20 hr
Stomach								
	Thoracogastroschisis	1/1400	N.Z. White	311	—	—	—	—
	Gastroschisis	—	N.Z. White	33	?	Thalidomide	100	5–18
	Gastroschisis	—	N.Z. White	83	2/2821	Miscellaneous tests	—	—
	Gastroschisis	1/1725	N.Z. White	111	3/88	Thalidomide	150	8–16
	Gastroschisis	—	N.Z. White	216	1/109	Lapartomy and atmosphere	5 min	7–10
	Gastroschisis	—	Himalayan	248	3/177	Thalidomide	50–450	8–16
	Gastroschisis	—	N.Z. White	281	1/?	Colcemide	0.5–5	2–5
	Gastroschisis	—	N.Z. White	281	1/?	Vinblastine	0.3	5
	Gastroschisis	—	N.Z. White	282	1/?	Parahydroxypropio-phenone	10–130	1–4
	Gastroschisis	—	N.Z. White	282	1/324	Clomiphene	7.5–15	1
	Gastroschisis	—	N.Z. White	282	1/?	U-11, 100	1–5	2–4
	Gastroschisis	—	N.Z. White	305	1/703	Disodium etidronate	25–500	2–16
	Gastroschisis	1/1400	N.Z. White	311	—	—	—	—
	Gastroschisis	1/2640	Mixed breeds	361	—	—	—	—
	Ectopic	—	White Danish	244	?	Thalidomide	100	1–20
	Displaced[f]	—	Mixed breeds	357	11/109	Thalidomide	500	6–11
	Rudimentary	—	Mixed breeds	150	?/50	Thalidomide	150	7–17
	Small[f]	—	Mixed breeds	357	5/109	Thalidomide	500	6–11
	Enlarged	—	White Danish	245	1/13	Thalidomide	60	1–20
Intestine								
	Herniated	1/143	N.Z. White	230	1/126	Methamphetamine	1.5	12–30
Jejunum								
	Atresia[f]	—	N.Z. White	325	2/121	Thalidomide	100–300	7.5–13.5
	Atresia	—	Several strains	408	2/906	Thalidomide	150	12.5–13.5
	Diverticulum	—	N.Z. White	51	1/171	Methaqualone	100	8–16
Colon								
	Diverticula	—	Mixed breeds	357	7/109	Thalidomide	500	6–11
	Diverticula[f]	—	Serveral strains	408	6/906	Thalidomide	150	12.5–13.5
	Diverticula	—	N.Z. White	439	?	WU-385	150	6–11
	Sacculate[f]	—	Mixed breeds	357	15/109	Thalidomide	150	6–11
	Succulate[f]	—	Several strains	408	8/906	Thalidomide	150	12.5–13.5
	Megacolon	—	N.Z. White	439	?	WU-385	150	6–11
	Dilated	—	White Danish	245	6/13	Thalidomide	60	1–20
	Ruptured	—	Mixed breeds	357	1/109	Thalidomide	150	6–11
Rectum								
	Rectourethral fistula[f]	—	Hybrid	405	7/120	Thalidomide	150	7.5–13.5
	Rectourethral fistula[f]	—	Several strains	408	?	Thalidomide	150	12.5–13.5
	Rectoanal fistula[f]	—	Several strains	408	2/906	Thalidomide	150	12.5–13.5
	Atresia[f]	—	N.Z. White	325	2/121	Thalidomide	100–300	7.5–13.5
	Atresia	—	Hybrid	405	13/120	Thalidomide	150	7.5–13.5
	Atresia	—	Several strains	408	32/906	Thalidomide	150	12.5–13.5
	Obstruction	—	Hybrid	405	7/120	Thalidomide	150	7.5–13.5
Anus								
	Atresia	—	N.Z. White	83	1/2821	Miscellaneous tests	—	—
	Imperforate	—	N.Z. White	216	1/104	Laparotomy and atomos-phere	15 min	7–10
Cloaca								
	Persistent	—	Mixed breeds	357	1/109	Thalidomide	500	6–11

Table XIII (*continued*)

Organ/ Tissue	Lesion/Observation	Control incidence (inapparent etiology)	Stock or strain	Ref.	Incidence[b] (treated animals)	Experimental condition[c]	Dose[d] (mg/kg)	Time[e] (day of gestation)
Liver								
	Enlargement[f]	—	Not stated	322	?	Inherited achondroplasia	—	—
	Hyperfissuring	0/62	N.Z. White	325	Occasional/121	Thalidomide	100–300	7.5–13.5
	Hyperfissuring	—	Several strains	408	Infrequent	Thalidomide	150	12.5–13.5
	Absent left posterior lobe	—	Mixed breeds	357	1/109	Thalidomide	500	6–11
	Reduced weight	?	N.Z.	187	?	Maternal feed restriction	—	?
	Glycogen increase	?	Not stated	374	—	—	—	—
Gallbladder								
	Aplasia	4/1333	Himalayan	248	4/108	Thalidomide	50–450	8–16
	Agenesis	—	Mixed breeds	150	?/50	Thalidomide	150	7–17
	Agenesis	72/8000	N.Z. White	311	—	—	—	—
	Hypoplasia	7/153	N.Z. White	51	14/171	Methaqualone	100	8–16
	Hypoplasia	Frequent/62	N.Z. White	325	Frequent/121	Thalidomide	100–300	7.5–13.5
	Reduction	72/8000	N.Z. White	311	—	—	—	—
	Variation	—	Several strains	408	Numerous	Thalidomide	150	12.5–13.5
	Variation	?	N.Z. White	359	—	—	—	—
	Lobulation	Variable	Hybrid	405	Variable	Thalidomide	150	7.5–13.5
	Bifurication	72/8000	N.Z. White	311	—	—	—	—
	Double	—	Hybrid	200	?	Thalidomide	500	3–8
	Double	—	Mixed breeds	357	3/109	Thalidomide	500	6–11
	Duplicate	2/1333	Himalayan	248	21/108	Thalidomide	50–450	8–16

[a] Reported in nonexperimental "control" fetuses or in those from dams exposed to various experimental conditions.

[b] Incidence: no. of affected fetuses/no. of fetuses examined.

[c] The natural occurrence of phenomena in experimental fetuses cannot be overlooked. Comments as to whether the observation was spontaneous or induced or its relation to the experimental treatment were usually absent from the majority of articles reviewed.

[d] Dose: Summarized depiction of dosage range or as otherwise indicated (usually in mg/kg/day).

[e] Time: Period of days administered during gestation.

[f] Illustration in reference.

? = Incidence not specified yet observation or lesion noted.

TABLE XIV

HEMATOPOIETIC AND ENDOCRINE SYSTEMS[a]

Organ/ Tissue	Lesion/Observation	Control incidence (inapparent etiology)	Stock or strain	Ref.	Incidence[b] (treated animals)	Experimental[c] condition	Dose[d] (mg/kg)	Time[e] (day of gestation)
			Hematopoietic System					
General								
	Pelger anomaly	—	Not stated	296	?	Inheritance	—	—
	Erythroblastosis	[g]		294	?	Inheritance	—	—
Spleen								
	Hypertrophy	—	Not stated	40	?	Decapitation *in utero*	—	22
	Hyperplasia	—	Not stated	43	?	Decapitation	—	20–22
	Enlargement[f]	—	Not stated	322	?	Inherited achondroplasia	—	—
	Atrophy[f]	—	Mixed breeds	202	?/178	Propiomazine hydrochloride	2–4	pp. 1–3
	Small	—	White Danish	245	1/13	Thalidomide	60	1–20
	Lymphoid depletion[f]	—	Not stated	322	?	Inherited achondroplasia	—	—
	Abnormalities	—	Dutch Belted	88	?	Inherited spina bifida	—	—
Lymph nodes								
	Atrophy[f]	—	Mixed breeds	202	?/178	Propiomazine HCl	2–4	pp. 1–3
	Lymphoid depletion[f]	—	Not stated	322	?	Inherited achondroplasia	—	—

Table XIV (*continued*)

Organ/ Tissue	Lesion/Observation	Control incidence (inapparent etiology)	Stock or strain	Ref.	Incidence[b] (treated animals)	Experimental[c] condition	Dose[d] (mg/kg)	Time[e] (day of gestation)
Thymus								
	Atrophy[f]	—	Mixed breeds	202	?/178	Propiomazine HCl	2–4	pp. 1–3
	Enlargement	—	Not stated	41	?	Decapitation	—	20–22
	Enlargement	—	Not stated	43	?	Decapitation	—	20–22
Blood								
	Composition	—	g	64	?	Inheritance	—	—
	Immature erythrocytic forms[f]	—	Not stated	322	?	Inherited achondroplasia	—	—
	Suppressed leukocyte maturation[f]	—	Not stated	295	?	Inherited Pelger–anomaly	—	—
	Elevated triglycerides and cholesterol	—	Not stated	42	?	Hypophysectomy (by decapitation)	—	—
Liver								
	Reduced hematopoiesis[f]	—	Not stated	322	?	Inherited achondroplasia	—	—
			Endocrine System					
Adrenal								
	Absent	—	N.Z. White	111	8/81	Thalidomide	150	8–16
	Absent	—	White Danish	245	1/13	Thalidomide	60	1–20
	Hypoplasia	—	Not stated	41	?	Surgical decapitation (apituitarianism)	—	20–22
	Hypoplasia	—	Not stated	43	?	Surgical decapitation (apituitarianism)	—	20–22
	Hypoplasia	—	Not stated	40	?	Surgical decapitation (apituitarianism)	—	20–22
Parathyroid								
	Hyperplasia	—	Not stated	319	?	Inherited osteopetrosis	—	—
Pancreas (islets)								
	Beta cell hyperfunction[f]	—	N.Z. White	422	?	Alloxan diabetes	—	—
Pituitary								
	Absence	—	Not stated	43	?	Surgical decapitation	—	21

[a] Reported in nonexperimental "control" fetuses or in those from dams exposed to various experimental conditions.

[b] Incidence: no. of affected fetuses/no. of fetuses examined.

[c] The natural occurrence of phenomena in experimental fetuses cannot be overlooked. Comments as to whether the observation was spontaneous or induced or its relation to the experimental treatment were usually absent from the majority of articles reviewed.

[d] Dose: Summarized depiction of dosage range or as otherwise indicated (usually in mg/kg/day).

[e] Time: Period of days administered during gestation. pp = days postpartum.

[f] Illustration in reference.

[g] Article not available for review; original source cited.

? = Incidence not specified yet observation or lesion noted.

V. TERATOLOGICAL INVESTIGATION

Pathological prenatal development is thought to result from three general factors; inheritance, environment, or interaction between the two. Thus, there is perhaps no one teratogen, but rather an interaction of activity involving several factors. At present, no predictable relationships have been detected between teratogenesis and chemical structure or pharmacological activity (62, 180, 341). Further complicating a description of a teratogen is the concept that any drug (or situation) at the right dose in an embryo of proper species will be effective in causing disturbances in embryonic development (228). If one follows this premise, it is then only a matter of time until suitable conditions are arranged to demonstrate effects of virtually any situation or teratogenic agent on the developing fetus.

Successful teratogens tend to exhibit dose related effects. At low exposure adverse effects are absent and, as exposure is increased, both the frequency of affected animals and severity of damage usually increase, eventually reaching a lethal level (180).

The incidence levels and patterns of malformations produced become different from the anticipated levels of spontaneously occurring phenomena (147, 226). The anomalies induced by a teratogen are not considered to be different from the types of expected to occur spontaneously. The expression of teratogenic activity is thought to be a result of a disturbed equilibrium between developing cells, which initiates or allows aberrant sequences or pathways of development to occur (311).

Depending on the time of exposure and the stage of embryological development, significant differences in effect can be expected.

A. Factors to Consider in Experimental Design

The degree to which any or all of the following variables had been considered in published reports was frequently not apparent, complicating evaluation of the experimental data and interpretations rendered. Supposedly, with any agent the right combination of factors will cause disturbed embryonic development (229). Experimental design is the subject of several general references (31, 49, 304, 341, 429, 433).

1. Animal Species

Unfortunately, there is no priority basis for selecting the species most likely to respond to a potential teratogen (96, 228). Species are used empirically since the action of most experimental agents is as yet unpredictable, thus no species is more advantageous to use than another (62). To compensate for unknown species peculiarities, it is considered desirable to use several species to screen for teratogenic potential (62, 228, 312, 339, 404). Such animal studies are preliminary. Once effects are detected, reproducibility in the same and other species should be evaluated and the pathogenesis studied (341).

2. Species Metabolism

Definition of maternal and fetal metabolic characteristics and their interrelationships requires continued exploration. Availability of such data is extremely important for development of protocols to provide more sophisticated teratogenic testing (56).

The need to consider and compare patterns of chemical absorption, distribution, metabolic fate, and excretion not only between man and animals but also between dam and fetus is obvious. The variations of normal or altered maternal metabolism have the potential of altering eventual effects on the fetus (62). The agent might be rendered innocuous or its metabolite found to be of greater potential risk to the fetus (155, 231, 341). Age differences in metabolism and tolerance to toxic effects of chemicals are recognized to exist between dam and neonate (133, 152, 196). Little has been done to explore such differences between fetus and neonate or fetus and dam.

Differences or similarities between a test species and man may alter the intended protocol or negate use of that species entirely. Regardless of the results of metabolic comparisons with man, it is important to determine if, when, and to what material the animal fetus was actually exposed. Theoretically, one could better determine if the experimental design (route, time dose level, etc.) offered sufficient fetal challenge and hence data sufficient for interpretation.

3. Dose Period—Gestational Age and Duration of Treatment

Accurate timing of pregnancy is necessary for correlations to be made between gestational age and duration of treatment (117). As controversy exists over nomenclature concerning day 0 (the first 24 hours postcopulation versus day 1, etc.), it is important to clearly specify the experimental timing used (225).

The stages of conceptus development can be characterized by a variety of criteria, among them differential rates and times of development of the various tissues and/ or organ systems. Each progresses on a different timetable of development within which they have periods of differing susceptibility to environmental agents or stresses (25, 224). Thus, the effects produced by a given agent will vary depending on the actual time of administration (62). The period of greatest sensitivity to teratogenic activity in rabbits is during the sixth to eighteenth day of gestation (81).

In addition to determining whether something crosses the placenta, correlation must be made between dosage time and the timetables of normal embryological and placental development to provide a frame of reference on which to evaluate the nature of the observed malformation (190, 224, 283, 426). Our comprehension of the ever-changing anatomical and metabolic maternal–fetal equilibria occurring during prenatal development has barely begun. Since attempts to produce teratogenic insult to the fetus by an agent which is partially metabolized by the dam (and thus of an unknown chemical nature), it is most difficult to characterize the specific challenge to the fetus. We must be able to determine when, with how much, and to what we have exposed the fetus. An agent administered early and theoretically long enough to act on various stages of embryological organogenesis runs the risk of maternal metabolic alteration. As such it might be then negated or if metabolically altered result in a syndrome of multiple deformities due to the interference with several critical developmental stages. On the other hand, while a single brief dose reduces risk of maternal metabolic conversion, the chance selection of a suitable time period for organ sensitivity versus dose level and unknown teratogenic potential is probably of equal or greater frustration (224).

Once teratogenic activity is detected during routine testing, it is feasible to repeat with segmented dose periods to be more precise in detecting the critical stages and sequences of organ development affected (81).

4. Dose Selection

In conjunction with the gestational age challenge, the proper dose level must be chosen in the effort to detect teratogenic activity. This dose need not be deleterious to the dam (427). Unless sufficient challenge can be demonstrated to the maternal and/or fetal test system, data produced are subject to serious question as to their relevance (150, 339).

In attempting to detect effects on fetal development it is desirable to be near yet below doses which induce embryo lethality (62), but sufficiently high to include the theoretical range for teratogenic activity if present for the agent. It must be remembered that fetal death does not necessarily mean teratogenesis has occurred, rather it may represent a separate drug action (62, 223).

Even if available, variations in experimental design and differences in animal stocks render literature data of lesser value in selecting appropriate dose levels. Data regarding subacute toxicity and/or LD_{50} for the species being used may be of value, however, as the anticipated treatment period is 10–13 days and involves pregnant animals; even these data are not directly applicable.

A useful approach to compensate for this lack of information is that of abbreviated dose selection studies (312). These might involve 6 to 8 dose levels in pregnant animals dosed for the time anticipated for the complete study, including necropsy of the pregnant females in midpregnancy (rabbits, day 18). It has been suggested that dosage be developed based on surface body weight relationships (372). The goals are detection of maternal toxicity as symptoms and/or death and evidence of embryo lethality (resorption). From this data adverse effect levels in both dam and fetuses can be compared and some estimate of the possible teratogenic range made (433). To achieve the "optimal teratogenic dose level," dose levels in the subsequent evaluations should be adjusted to avoid maternal toxicity, hopefully to prevent introduction of additional complicating features in the fetal environment (235) and be at or below the level producing increased or total fetal resorption, this range usually being narrow (224, 312).

The concept of teratogenic risk is developed by relating the differences between general maternal toxic level and embryo lethal level, i.e., teratogenic ratio. The broader the differences, the greater the possibility for teratogenic activity because the fetus is being affected independently of the dam at a lower dose than the dam (339, 403). This suggests that selective embryopathy probably exists (343). If maternal toxicity and fetal toxicity levels are similar, teratogenic risk is low (51).

That such variations of toxicity can exist between dam and fetus is clearly suggested by the wide differences in toxicity known to exist between adult and neonate animals (78). The selective effects of thalidomide in rabbit and human fetuses versus lack of effect in the dams are classic examples.

By using these pilot studies a plan for the final complete evaluation of effects on the fetus can be designed. In these animal studies the pattern and frequency of malformations noted should differ from the anticipated spontaneous rate and pattern if teratogenic activity is present (147), and this should be dose related (180) in that the severity and frequency increases with the increasing dose level.

5. Route of Administration

Many routes can be chosen, including inhalation (409), to administer materials, to the doe, hence presumably to the fetus. Direct administration to the fetus or adjacent fluids, uterine lumen, etc., requires some degree of surgery introducing an element of risk for maintenance of pregnancy or fetal injury.

While administration via feed offers the obvious convenience of simplicity and minimizes repreated excitement and physical trauma to the dam, it is grossly inaccurate and not to be recommended. Gastric intubation is more desirable since it insures accurate delivery of calculated amounts of material to each animal.

The route of choice is also governed by knowledge of the behavior of the agent within the test species, i.e., the effect of the test species on the agent, via gastric juice, metabolism, etc. An example of this is the conversion by rabbit stomach acid of teratogenic cyclopamine to the nonteratogenic veratrine (232). Via the oral route, absorption delays insult to fetus, whereas intraperitoneal (IP) and IV routes are rapid. For example, by changing from an oral route to an intravenous route, the thalidomide effect is enhanced in rats (372). The immediate gastric availability of intubated material is also different from the amount received in 24 hours when mixed in feed.

6. Extrapolation of Animal Data to Humans

Teratological testing procedures are in reality subject to the classic working rules of toxicological investigation (45, 62, 96, 252, 254). The well-known hazards of extrapolation among species, including man, are equally appropriate for teratological data. In addition, the presence of the maternal organism between the agent and fetus further complicates the situation.

One of the chief goals of experimental teratology is predicting the inherent risk of human exposure to a particular agent or chemical. Another is the desire for determining the relationship of various causative factors to the pathogenetic mechanisms of abnormal development. For the latter, the availability of various laboratory animal species and emphasis for continual sophistication of laboratory techniques offer inspiration toward development of prospective and retrospective studies in other species (47, 56). As to the former, it is well recognized that man still remains the ultimate species for investigation (18, 62, 279, 296, 341). Because of this, extreme caution must be exercised to avoid exposure to pregnant females unless specifically indicated (279). Animal research data must be considered as a source of insight to understanding phenomena occurring in man, not as a reliable source of specific answers.

While reports of positive effects in animals appear frequently (227), it is in reality the vast body of negative or lack of effect data which must be viewed with caution! Often these conclusions are based on obviously inadequate numbers, beyond that, questions can be raised concerning lack of thoroughness in experimentation. The lack of animal effect introduces an unavoidable, subconscious tendency to extrpolate this to represent a theoretical lack of risk in man (54). In presentation of such reports, the burden of proof is great to furnish distinct evidence that an effective challenge situation existed for the conceptus (32). Demonstration that sufficient challenge did not exist should automatically preclude presentation of data as being negative. To assemble such proof requires an intimate knowledge of the agent and species' metabolic interrelationships, an extremely costly and tedious research effort in its own right.

Defining the maternal fetal relationship in a teratogen-induced syndrome in laboratory animals may well give clues to similar relationships or phenomena in humans. The similarity of actual animal lesions being insignificant to the demonstration of activity.

It is well recognized that an undercurrent or background rate of "spontaneous" malformation is common to all mammalian species. Such lesions occurring in the experimental treatment groups introduce serious complications in interpretation (339, 404, 443). For the most part, they are inadvertently grouped or included in the syndrome being caused by or related to the particular experimental treatment. Seldom is the naturally occurring rate of malformation in treated animals mentioned. Their existence should be recognized. While it is difficult or impossible to sort spontaneous or naturally occurring from induced, it must be remembered that by experimental design in an experiment with 1 control and 2–3 treatment groups, chance alone favors the occurrence of malformations at low incidence in the treated groups.

The relevance of animal data to humans will always be a source of controversy (54, 56, 228). At present, little choice exists; as has been said, "animal models are as good as we shall ever have concerning means of studying the pathology of development in man" (180, 426). It is unlikely that man will submit to becoming a test species, however, some studies have been performed in cases of legalized abortions (412).

For the future, it is hoped that improved development and interpretation of the animal data will allow a more accurate basis for critical appraisal of potential hazards to humans, the ultimate species (56, 62), animal testing being recognized as an initial step (223).

B. Factors Altering Prenatal Development in Rabbits

The expression of abnormal embryological development is governed by the complex interaction of two general factors, heredity and environment. The fact that the genotype can influence a particular species' response to these agents is recognized and has been demonstrated in several species (47, 155, 175, 235, 351, 421, 427). The rabbit has not been studied to a great extent concerning this aspect of teratogenesis, since there is a general lack of availability of inbred strains.

The rabbit has been advocated for experimented teratology because it possesses a less inbred, more heterogeneous genetic constitution—hence, likely to be more sensitive and phenotypically expressive to the teratogenic activity of unknown materials (155). On the other hand, the relevance of this sensitivity is criticized with the thought that with the genetic uniformity of smaller rodents, expression of such activity is more meaningful (312, 351). Certainly it can be appreciated that experimentally induced effects as influenced by specific genetic susceptibility are more difficult to investigate in rabbits than in rats and mice.

A variety of inherited phenomena have been detected and studied in the rabbit (e.g., 338). The results of some of these are summarized in Table XV and pathological changes, some directly associated with these conditions, are presented in Table II through XIV.

A wide variety of conditions have been evaluated in many species including man and found capable of producing cogenital malformations naturally or experimentally. While they have been classified variously (137, 175, 259, 427), when considering the experimental animal species the following scheme is suggested:

Causes of Altered Prenatal Development

A. Maternal origin (naturally occuring phenomena)
1. Physical
2. Inheritance
3. Metabolic disorders
4. Nutritional imbalances
5. Infections
B. Environmental experimental agents or stresses
1. Physical
2. Chemicals, drugs, and hormones
3. Infections

By slightly different arrangement, Tables XV–XVII summarize results of specific experimental attempts to alter embryological development in rabbits. These experiments were meant to detect sensitivity of the embryo or fetus to the conditions mentioned. Citations include only those dealing with abnormal embryological and fetal development. The tables were arbitrarily arranged where possible by intended use as drugs (92) or chemicals, etc. References dealing with effects on preimplantation zygotes (e.g., 11, 94, 124, 156, 255, 256, 278, 313) were omitted.

The study of congenital malformations at birth in any species represents only one aspect of phenomena occurring from conception to postnatal death. The stages of the pre-implantation zygote, postorganogenic prebirth fetal maturation, and postnatal development and the clinically silent malformations detected at death have received scant attention in most species except humans (147).

Listings of suspected or known teratogens and fetal reactions to experimental treatments have been presented, often comparing several species (22, 32, 55, 62, 80, 106, 312, 341).

The conditions listed in the following tabulation have been described as teratogenic in rabbits (See Tables XV to XVII).

Acetylsalicylic acid	Mer 25 (ethamoxytriphetol)
Chlormadinone acetate	Methamphetamine
Colecemid	Oxygen deficiency
Cyclizine	Phenobarbital
Cyclopamine	Thalidomide
Imiprimine	Prednisone
Insulin	Vitamin A deficiency
Jervine	Vitamin D excess
6-Mercaptopurine	X-Irradiation

The situations listed below have, thus far, been associated with adverse effects on rabbit fetal development. (See Tables XV–XVII.)

Adrenaline	Laser beams
Caffeine	L-Asparaginase
Captan	Listeria monocytogenes
Colchicine	Marijuana
Cyclophosphamide	Maternal estrogen
Difoltan	Methotrexate
Estrogen	Stress (noise)
Feed restriction	Testosterone
Iodine	Thymectomy (fetal)

Of the successful teratogens thalidomide has been recommended as a reference test standard in experimental animal teratology studies in many species to predict the sensitivity of the animals used to a known teratogen. Indeed, a large number of reports have demonstrated its teratogenic effect in rabbits and the lack of effect in response to thalidomide metabolites (Table XVII). Thus far, no one has achieved a method of correlating chemical structure and/or pharmacogenic effects with teratogenic effect in any species (62, 180, 341). With this in mind, it is difficult to understand the significance of the additional data generated animals from test standards in reference to an unknown chemical entity in this type of experiment. While the experience gained using such reference standards is excellent as a training procedure, it seems to have doubtful value in interpreting the data obtained from other unknown chemical substances in similar experiments. Perhaps that expenditure of effort would be of greater value if applied to basic research or to more thorough experimental design related to evaluating the unknown test compounds.

C. Tabular Presentations of Summarized Results as Cited in Reported Experimental Studies

TABLE XV

THE DAM: PHYSICAL, METABOLIC, NUTRITIONAL IMBALANCES, AND INFECTIOUS DISEASES AFFECTING FETAL DEVELOPMENT[a]

Experimental condition	Stock or strain	Ref.	Route	Time[b] (day of gestation)	Dose[c] (mg/kg)	Sample size (no. dams)	Effect lack (non-teratological)	Increased resorption rate	Altered growth or development	General pathological effects on systems
Physical Phenomena (Dam)										
Emotional stress (noise)	N. Z. Red	144	—	1–30	—	?			Decreased fetal weight	
X-Irradiation	Not stated	309	?	?	?	?				(+) Nervous
Surgical laparotomy	N. Z. White	216	Oral	7–10	—	49	+			
Metabolic Phenomena (Dam)										
Maternal estrogen	N. Z. White	2	—	—	—	6			Bone	(+) Skeletal
Alloxan diabetes	N. Z. White	422	—	—	—	9				(+) Endocrine

Table XV (*continued*)

Experimental condition	Stock or strain	Ref.	Route	Time[b] (day of gestation)	Dose[c] (mg/kg)	Sample size (no. dams)	Effect lack (non-teratological)	Increased resorption rate	Altered growth or development	General pathological effects on systems
Hyperglycemia (maternal)	N. Z. White	422	—	—	—	9				(+) Endocrine
Maternal plasma loss	Hybrid	174	—	9	—	?				(+) Skeletal
Lactation	N. Z./Dutch Belted	183	—	—	—	?		+		
Advanced age–parity	N. Z./Dutch Belted	183	—	—	—	?	+			
Aged sperm	N. Z. White	397	—	—	—	?	+			
Nutritional Imbalances (Dam)										
Vitamin A deficiency	Not stated	272	—	1–30	—	16				(+) Nervous
Vitamin A deficiency	Dutch Belted	194	—	?	—	?			Decreased size	(+) Nervous
Vitamin A deficiency	Mixed	276	—	12–28 wk[d]	—	31				(+) Nervous
Vitamin A deficiency	Albino	273	—	14–38 wk[d]	—	11				(+) Nervous
Vitamin A deficiency	Dutch/N. Z. White	243	—	?	—	42		+	+	(+) Nervous
Feed restriction	N. Z.	187	—	1–30	—	?			Decreased size	
Infectious Diseases (Dam)										
Listeria monocytogenes	Copenhagen	314	?	?	?	1			Fetal death	
Listeria monocytogenes	White	238	IV	20	100,000 organisms	16			Placental lesion	
Inherited Phenomena (fetus)										
Body size	Miscellaneous	67	—	—	—	?			Decrease size	
Body size	Miscellaneous	71	—	—	—	?			Decrease size	
Dwarfism	Not stated	246	—	—	—	?			+	
Skeletal conditions										
Osteopetrosis	Not stated	318	—	—	—	?				(+) Integument
Osteopetrosis	Not stated	317	—	—	—	?				(+) Integument
Osteopetrosis	Not stated	324	—	—	—	?			Decrease size	(+) Miscellaneous, (+) integument
Osteopetrosis	Not stated	319	—	—	—	?				(+) Integument, (?) endocrine
Achondroplasia	Miscellaneous	58	—	—	—	?			Early postnatal death	(+) Integument
Achondroplasia	N. Z. White	90	—	—	—	60			Skeleton	(+) Integument
Achondroplasia	Not stated	323	—	—	—	?				(+) Integument
Achondroplasia	Not stated	322	—	—	—	?				(+) Integument, (+) hematological
Achondroplasia	Not stated	375	—	—	—	46			Achondroplasia	(+) Integument
Chondrodystrophy	Not stated	87	—	—	—	?			Dachs rabbit	(+) Integument
Osseous pattern–limbs	Miscellaneous	89	—	—	—	?				(+) Integument
Rib variation	Miscellaneous	162	—	—	—	?				(+) Integument
Rib pattern	Not stated	39	—	—	—	?				(+) Integument
Brachydactylia	English	167	—	—	—	?				(+) Integument
Brachydactylia	?	211	—	—	—	?				(+) Integument, (+) skeletal
Digtal foreleg curvation	Beveren, Dutch	321	—	—	—	?				(+) Integument, (+) hematological (+) Integument
Pelger anomaly	Not stated	295	—	—	—	260				(+) hematological
Buphthalmia	N. Z. White	134	—	—	—	?				(+) Miscellaneous

Table XV (*continued*)

Experimental condition	Stock or strain	Ref.	Route	Time[b] (day of gestation)	Dose[c] (mg/kg)	Sample size (no. dams)	Effect lack (non-teratological)	Increased resorption rate	Altered growth or development	General pathological effects on systems
Spontaneous mal-formation	Miscellaneous	361	—	—	—	?				(+) Integument
										(+) Integument
Sternal development	N. Z. White/ hybrid	328	–	—	—	?				(+) Integument
Asymmetrical vascular pattern	Miscellaneous	364	—	—	—	?				(+) Skeletal
Aoritic arch variation	Miscellaneous	362	—	—	—	?				(+) Skeletal
Anomalies	Not stated	74	—	—	—	?				(+) Integument

[a] Based on cited author's summarized interpretation of experimental data.
[b] Time: Periods in excess of 32 days include duration of pregnancy. Shorter periods refer to times during gestation.
[c] Dose: Summarized depiction of dosage range or as otherwise indicated (usually in mg/kg/day).
[d] Period preceding and including gestation.
? = Details not readily apparent.
+ = presence and/or occurrence.

TABLE XVI
Inherited and Environmental Effects on Fetal Development[a]

Experimental condition	Stock or strain	Ref.	Route[b]	Time[c] (day of gestation)	Dose[d] (mg/kg)	Sample size (no. dams)	Effect lack (non-terato-logical)	Increased resorption rate	Altered growth or development	General pathological effects on systems
Physical Phenomena (fetus)										
Fluid inhalation										
Saline	Not stated	98	—	29–30	—	?				(+) Respiratory
Amnionic fluid	Not stated	98	—	29–30	—	?				(+) Respiratory
Anoxia cardio- vascular res.	Not stated	99	—	24–29	—	?				(+) Physiological
Hypoxia prenatal	Not stated	16	—	28–32	—	?				(+) Reproductive system
Oxygen deficiency	Not stated	101	—	7–15	—	?				(+) Nervous system; (+) skeletal
Umbilical cord clamping	Not stated	110	—	28–31	—	?				(+) Nervous
Irradiation										
Cobalt	Mixed	76	—	2–22	50–400 r	59				(+) Skeletal
Laser beams	Not stated	94	—	22–50 hr	?	?			+	
Surgery										
Tracheal ligation	N.Z. White	63	—	23–30	—	8				(+) Respiratory
Decapitation	Not stated	42	—	—	—	?				(+) Hematological
Decapitation	Not stated	41	—	21	—	?				(+) Hematological, (+) endocrine
Decapitation	Not stated	43	—	21	—	?				(+) Hematological, (+) endocrine
Decapitation	Not stated	40	—	22	—	?				(+) Hematological, (+) endocrine
Thymectomy	Not stated	23	—	?	—	?			Retarded	(+) Hematological
Castration	Not stated	426	—	21–22	—	?				(+) Reproductive

Table XVI (*continued*)

Experimental condition	Stock or strain	Ref.	Route[b]	Time[c] (day of gestation)	Dose[d] (mg/kg)	Sample size (no. dams)	Results[a]			
							Effect lack (non-teratological)	Adverse effects		
								Increased resorption rate	Altered growth or development	General pathological effects on systems
Overcrowding (uterine)	N.Z. White	186	—	2–32	—	?		+		
Overcrowding (uterine)	N.Z. White	8	—	2–32	—	?		+		
Undercrowding (uterine)	N.Z. White	183	—	2–32	—			+		
Infectious Agents (fetus)										
Viruses										
Rubella	N.Z. White	308	Intranasal	3–14	10^5 $TCID_{50}$	13	+			
Herpesvirus hominus	N.Z. White	308	Intranasal	10–14	10^6 $TCID_{50}$	4	+			
Bacteria										
Streptococcus foecalis	Not stated	38	Amnion	26–29	?	6				(+) Respiratory
Escherichia coli	Not stated	38	Amnion	26–29	?	6				(+) Respiratory
Brucella suis	Not stated	264	?	14–20	?	?	+			
Chemicals (fetus)										
Miscellaneous										
Iodine	N.Z. White/ Dutch Belted	27	Oral	28–32	250–1000	pm ?			Poor/ survival	
DMSO	Not stated	72	Oral and SQ	6–14	5 q	10	+			(+) Respiratory
Carbon monoxide	N.Z. White	91	Inhalation	4,12,29	?	6				(+) Hematological
Lens antigen	Not stated	181	?	8–17	2.5–4.0	?				(+) Nervous
Lens antisera	Not stated	181	IV	8–17	?	?				(+) Nervous
Alkaloid Q	Mixed	236	Oral	7	300	8	+			
Trypan Blue	N.Z. White	308	IM	7–10	300	3		+		
Freon 114/12	N.Z. White	410	Inhalation	−5–5	?	12	+			(±) Skeletal
6 Aminonico-tinamide	Dutch Belted	365	P	9, 12	1–3	37				(+) Skeletal
Moquizone	N.Z.	373	Oral	7–17	20–60	43	+			
Ethyl alcohol	Not stated	251	Oral	8–16	5,14	?				(+) Skeletal
Veratrum californicum	N.Z. White	217	Oral	3–24	50% diet	8	+			
Cylopamine	Mixed	232	Oral	9–15	22–45	8	+			
Cylopamine	Mixed	232	Oral	6–9	22–45	4				(+) Nervous
Cylopamine	Mixed	233	Oral	6–16	150–300	31				(+) Nervous
Jervine	Mixed	233	Oral	7	250–300	7				(+) Nervous
Pesticides										
Imidan	N.Z. White	124	Oral	7–12	35	5	+			
Formothion	N.Z. White	239	Oral	6–18	6–30	20	+			
Thiometon	N.Z. White	239	Oral	6–18	1,5	20	+			
Herbicides										
Carbaryl	N.Z. White	337	Oral	5–15	100–200	8	+			
Diazinon	N.Z. White	337	Oral	5–15	7,30	11	+			
Antiinflammatory										
Phenylbutazone	Dutch Belted	366	Oral	6–18	50	11	+			
Sodium meclofenamate	Dutch Belted	366	Oral	6–18	3.5	11	+			
Anticalculus										
Disodium Etidronate	N.Z.	305	Oral	2–16	25–500	49	+			

Table XVI (*continued*)

Experimental condition	Stock or strain	Ref.	Route[b]	Time[c] (day of gestation)	Dose[d] (mg/kg)	Sample size (no. dams)	Effect lack (non-teratological)	Increased resorption rate	Altered growth or development	General pathological effects on systems
Antifertility										
Ethamoxytriphetol (Mer-25)	N.Z. White	4	IM	21–26	25	6				(+) Skeletal
Ethamoxytriphetol	N.Z. White	3	Oral	21–26	25	12				(+) Skeletal
Ergocornine methanesulfonate	N.Z. White	281	SQ	3–4	1.5–2.5	5	+			
Norehisterone + mestranol	Jap. Albino	394	Oral	8–20	1,3,10	12	+			
Vitamins										
Hypervitaminosis	Not stated	7	?	0–5	60–300,000	?	+			
B$_{12}$ analogs	Not stated	7	?	−10–30	0.1–0.5	?	+			
Vitamin D excess	N.Z. White	140	IM		1.5 million total	?				Cardiovascular
Diuretics										
Caffeine and nicotine	Not stated	251	SO	6–14	100/10	?			Retarded	
Phthalamidine	N.Z. White	124	Oral	7–12	150	6	+			
Hypotensive agents										
Veratramine	Mixed	233	Oral	7	75	19	+			
Rubijervine	Mixed	233	Oral	7	300	8	+			
Methyldopa	Albino	329	Oral	8–16	50–200	24	+			
Sympathetic stimulants (adrenergic)										
Adrenaline (see Hormones)										
Noradrenaline (see Hormones)										
Isoproterenol	N.Z. White	410	Inhal.	6–16	0.15, 0.45	24	+			(±) Skeletal
Isoproterenol	N.Z. White	433	Oral	7–16	16–50	49	+			
Metaproterenol	N.Z. White	433	Oral	7–16	16–50	44	+			
Parasympathetic depressants (anticholinergic)										
Smooth muscle relaxants										
Dicyclomine HCl	Not stated	147	Oral	9–16	10–100	?	+	±		
Antihistamines										
Meclozine HCl	Miscellaneous	150	Oral	7–17	25–175	28	+			
Doxylamine succinate	Not stated	147	Oral	9.16	10–100	?	+	±		
Dimenhydrinate	N.Z. White/ Dutch Belted	258	Oral	8–16	100	6	+			
Diphenhydramine HCl	Dutch Belted	367	Oral	6–18	3–15	36	+			
Ganglionic agents										
Hexamethonium	Not stated	99	IV	28–29	24–29	?				(+) Cardiovascular
Acetylsalicylic acid	N.Z. White	113	Oral	6–12	250	12			Reduced size	
Acetylsalicylic acid	Not stated	251	Oral	7–16	100–300	?	+			
Acetylsalicylic acid	N.Z. White	257	Oral	8–15	200	?				(+) Cardiovascular

Table XVI (*continued*)

Experimental condition	Stock or strain	Ref.	Route[b]	Time[c] (day of gestation)	Dose[d] (mg/kg)	Sample size (no. dams)	Effect lack (non-terato-logical)	Increased resorp-tion rate	Altered growth or develop-ment	General pathological effects on systems
								Results[a]		
									Adverse effects	
Acetylsalicylic acid	N.Z. White/ Dutch Belted	258	Oral	8–16	200	6				(+) Cardiovascular
Acetylsalicylic acid	Dutch belted	366	Oral	6–18	200–250	12	+			
Sodium salicylate	Not stated	213	SQ	30	500–1000	9	?			
Aminopyrine	Not stated	251	Oral	6–16	10–90	?	+			
Phenylbutazone	White Danish	245	SQ	1–20	30,60	12	+			
Monophenyl-butazone	White Danish	245	SQ	1–20	60,150	12	+			
Barbiturates										
Pentobarbital	N.Z. White	216	IV	7–10	40	?		±		
Phenobarbital	N.Z. White	257	Oral	8–15	50	?		+		(+) Cardiovascular
Phenobarbital	N.Z. White/ Dutch Belted	258	Oral	8–16	50	7		+		(+) Cardiovascular (+) Skeletal
Psychotomimetic Agents										
Lysergide LSD	N.Z. White	125	Oral	4–12	0.02–0.1	14	+			
Marihuana	Not stated	143	Q	7–10	130–500	68		+	Runting	(+) Nervous, (+) skeletal
Trazodone	N.Z. White	33	Oral	5–18	100–210	10	+			
Stimulants										
Metham-phetamine	N.Z. White	230	IV	12–30	1.5	36				(+) Skeletal, (+) cardio-vascular, (+) digestive
Imiprimine	White Danish	244	SQ	3–20	15–25	10	+		Fetal	(+) Nervous,
Imipramine	Miscellaneous	341	SQ	1–20	5–30	12		+	toxicity	(±) cardio-vascular
Imipramine	Not stated	192	?	?	30	?				(+) Nervous, (±) skeletal
Glutaramide derivatives										
Glutethimide	N.Z. White/ Dutch Belted	258	Oral	8–16	150	6				
Succinimido-glutorimide	N.Z./White/ Chinchilla	122	Oral	7–12	150	3	+			
Thalidomide (see Table XVII)										
Methaqualone	N. Z. White/ California	51	Oral	8–16	100–500	42	+			
Methaqualone	N. Z. White/ Dutch Belted	258	Oral	8–16	100	4				(±) Skeletal
Tranquilizers										
Haloperidol	Not stated	49	?	?	?	?	+			
HORMONES										
Adrenal cortex										
Betamethasone	N. Z. White/ Dutch Belted	413	IM	13–21	1–4	24				(±) Skeletal
Dexamethasone	N. Z. White/ Dutch Belted	413	IM	13–21	0.1–4.0	?				(+) Skeletal
Cortisone	N. Z. White/ Dutch Belted	413	IM	13–21	10–25	?				(?) Skeletal

Table XVI (*continued*)

Experimental condition	Stock or strain	Ref.	Route[b]	Time[c] (day of gestation)	Dose[d] (mg/kg)	Sample size (no. dams)	Effect lack (non-teratological)	Increased resorption rate	Altered growth or development	General pathological effects on systems
Triamcinolone	N. Z. White/ Dutch Belted	413	IM	13–21	0.01–5.0	?				(+) Skeletal
Prednisalone	N. Z. White/ Dutch Belted	413	IM	13–21	1–8	?				(+) Skeletal
Methylpredni- solone	N. Z. White/ Dutch Belted	413	IM	13–21	0.1–4	7	+			
Adrenal medulla										
Adrenaline	Not stated	99	IV	24–29	2–3 μg	?				(+) Cardiovascular
Adrenaline	N. Z. White	79	IV, IM	22	0.159	30			Runting	
Adrenaline	White	425	IV	22	0.159	2			Runting	
Adrenaline	Not stated	110	IV	28–31	10–25 μg	?				(+) Physiological
Noradrenaline	Not stated	110	IV	28–31	10–25 μg	?				(+) Physiological
Noradrenaline	Not stated	99	IV	24–29	0.2–3 μg	?				(+) Cardiovascular
Estrogenic steriods										
Estrogens	Not stated	93	?	18 hr	?	?			?	
Androgenic steriods										
Testosterone	Not stated	93	?	18 hr	?	?			?	
Antiandrogens										
Cyproterone acetate	Not stated	118	IV	13–24	0.625–100	?				(+) Reproductive
Progesterones										
Chlormadinone acetate	Jap. Albino	394	Oral	8–20	1,3,10	30				(+) Nervous, (+) skeletal
Progesterone	Not stated	93	?	18 hr	?	?			+	
Δ-6,6-Chloro- 17α-acetoxy- progesterone	Not stated	75	Oral	8–?	0.2–1.5	?	+			
Anti-diabetics										
Tolbutamide	N. Z. White	247	IV, IM, SQ	7–14	125	40	+			
Tolbutamide	N. Z. White/ Dutch Belted	258	Oral	8–16	200	5		+		(+) Reproductive
Insulin	Not stated	77	?	?	?	?				(+) Nervous, (+) skeletal, (+) cardiovas- cular
Estrogen and Anti- estrogens										
Clomiphene	N. Z. White	253	Intraamnionic	10–12	25–50	54			Death	
Clomiphene	N. Z. White	282	SQ	1–24	7.5–15	32	+			
Parahydroxy- phoprophene	N. Z. White	282	SQ	1–4	10–130	15	+			
U-11,100A	N. Z. White	282	?	1–17	0.5–5	25	+			
U-11,555A	N. Z. White	282	SQ	1,18	10–20	30	+			
ORF 3858	N. Z. White	282	?	1–25	0.2–1	22	+			
Ethnynodial diacetate	Not stated	352	Oral	0–10	1–2	38	+			
Mestranol	Not stated	352	Oral/SQ	0–28	0.05–0.1	34	+			
Folic acid antago- nists										
Methotrexate	N. Z. White	219	IV	10	9.6	?			Lethal	(+) Miscellaneous, (+) nervous
Antimitotic activity										
Cyclophos- phamide	Mixed	142	IV	6–14	30	61			Reduced fetal weight	(+) Nervous, (+) skeletal
Colcemide	N. Z. White	281	SQ	9–15	0.1–5	38			Lethal	

Table XVI (*continued*)

Experimental condition	Stock or strain	Ref.	Route[b]	Time[c] (day of gestation)	Dose[d] (mg/kg)	Sample size (no. dams)	Effect lack (non-teratological)	Increased resorption rate	Altered growth or development	General pathological effects on systems
Vinblastine	N. Z. White	281	IV	−4,1,5	0.3	3				(+) Nervous, (+) digestive
Desacetylamino-colchicine	N. Z. White	281	SQ	9–15	0.1–5	?			Lethal	
L-Asparaginase (*E. coli*)	N. Z. White	14	IV	8–9	50 IU	4		+	Decreased body size	(+) Nervous, (+) skeletal
Purine Antagonists										
5-Fluorouracil	N. Z. White	281	?	4–13	5–10	10	+			
BW-57-323H	N. Z. White	281	IP	12–14	5–20	4				(+) Skeletal
6-Azauridine	N. Z. White	281	?	13–17	500	1			Reduced litter size	
6-Mercapto-purine	Not stated	271	?	6–16	?	?				(+) Nervous, (+) skeletal
Antiviral										
Adamontine HCl	N. Z. White	242	Oral	7–10	100	?	+			
Benedectin	Not stated	147	Oral	9–16	1,2,10	?	+			
Fungicides										
Captan	N. Z. White	235	Oral	6–16	18–75	18		+	Decreased size	
Captan	Dutch Belted	235	Oral	6–16	75	6	+			
Captan	N. Z. White	124	Oral	7–12	80	4	+			
Phthalton	N. Z. White	124	Oral	7–12	80	6	+			
Difolatan	N. Z. White	235	Oral	6–16	18–150	24		+	Decreased size	
Difolatan	Dutch Belted	235	Oral	6–16	75	10	+			
Folpet	Dutch Belted	235	Oral	6–16	75	9	+			
Folpet	N. Z. White	235	Oral	6–16	18–75	17		+	Decreased size	
Antibacterial										
Sulfadiazine	Not stated	251	Oral	2–11	10–500	?	+			
Sulphamoprine	N. Z. White	310	Oral	?	0.05–1	22	+			

[a] Based on cited author's summarized interpretation of experimental data.
[b] SQ = subcutaneous; IM = intramuscular; inhal. = inhalation; IP = intraperitoneal; IV = intravenous.
[c] Time: Period in excess of 32 days represents times prior to and including duration of pregnancy. Shorter periods refer to times during gestation.
[d] Dose: Summarized depiction of dosage range or as otherwise indicated (usually in mg/kg/day).
? = Details not readily apparent.

TABLE XVII

EFFECTS OF THALIDOMIDE, ITS METABOLITES, AND RELATED COMPOUNDS ON FETAL DEVELOPMENT

Experimental condition	Stock or strain	Ref.	Route[b]	Time[c] (day of gestation)	Dose[d] (mg/kg)	Sample size (no. dams)	Effect lack (nontetralogical)	Increased resorption rate	Altered growth or development	General pathological effects on systems
				Thalidomide						
Thalidomide	Hybrid/N.Z. White	406	Oral	?	?	68				(+) Skeletal
Thalidomide	White Danish	245	SQ	1–20	60	8				(+) Skeletal
Thalidomide	White Danish	244	SQ	1–20	100	14				(+) Skeletal (+) urinary

Table XVII (*continued*)

Experimental condition	Stock or strain	Ref.	Route[b]	Time[c] (day of gestation)	Dose[d] (mg/kg)	Sample size (no. dams)	Effect lack (non-terato-logical)	Increased resorp-tion rate	Altered growth or develop-ment	General pathological effects on systems
Thalidomide	Hybrid	200	Oral	3–16	300–500	53				(+) Skeletal, (+) digestive
Thalidomide	N.Z. White	139	Oral	4–16	50–300	19				(+) Skeletal (+) urinary
Thalidomide	N.Z. White	33	Oral	5–18	100	10				(+) Skeletal
Thalidomide	Mixed	357	Oral	6–11	500	30?				(+) Skeletal, (+) cardiovascular, (+) digestive (+) urinary
Thalidomide	Dutch Belted	235	Oral	6–16	75	7				(+) Nervous
Thalidomide	N.Z. White	235	Oral	6–16	75	10				(+) Nervous
Thalidomide	N.Z. White Chinchilla	122	Oral	7–12	150	10				(+) Skeletal
Thalidomide	N.Z. White	124	Oral	7–12	150	5				(+) Skeletal
Thalidomide	Not stated	128	Oral	7–15	50	5				(+) Nervous (+) skeletal
Thalidomide	Miscellaneous	150	Oral	7–17	150	15				(+) Nervous (+) skeletal (+) digestive, (+) urinary
Thalidomide	Hybrid	405	Oral	7.5–13.5	150	23				(+) Nervous, (+) skeletal, (+) cardio-vascular, (+) respiratory, (+) digestive, (+) urinary
Thalidomide	Miscellaneous	407	Oral	7.5–13.5	150	100?				(+) Cardiovascular
Thalidomide	N.Z. White	325	Oral	7.5–13.5	150, 300	22				(+) Nervous, (+) skeletal, (+) respiratory, (+) digestive, (+) Cardiovascular. (+) urinary
Thalidomide	Not stated	344	?	6–14	?	?				
Thalidomide	Not stated	346	Oral	6–14	15–250	35			Fetal death	(+) Nervous (+) skeletal, (+) cardio-vascular, (+) urinary
Thalidomide	N.Z. White	372	IV	7–10	10	12				(+) Skeletal
Thalidomide	N.Z. White	372	Oral	8–12	25–200	46				(+) Skeletal
Thalidomide	N.Z. White	51	Oral	8–13	500	7		+		
Thalidomide	N.Z. White	257	Oral	8–15	300	?		+		(+) Skeletal, (+) cardio-vascular,
Thalidomide	Hybrid	209	Oral	8–14	500	10				(+) Skeletal, (+) urinary, (+) reproductive
Thalidomide	Himalayan	248	Oral	6–18	50–450	50				(+) Skeletal, (+) urinary
Thalidomide	N.Z. White	379	Oral	8–16	150	4				(+) Skeletal
Thalidomide	N.Z. White	311	Oral	8–16	150	15		+		(+) Skeletal (+) digestive
Thalidomide	N.Z. White/Dutch Belted	258	Oral	8–16	300	6		+		(+) Skeletal, (+) cardio-vascular
Thalidomide	N.Z. White	306	Oral	7–11	150	15		+		(+) Skeletal
Thalidomide	6 Strains	408	Oral	12.5–13.5	150	?				(+) Digestive

Table XVII (*continued*)

Experimental condition	Stock or strain	Ref.	Route[b]	Time[c] (day of gestation)	Dose[d] (mg/kg)	Sample size (no. dams)	Effect lack (non-teratological)	Increased resorption rate	Altered growth or development	General pathological effects on systems
										Results[a] / Adverse effects

Thalidomide Metabolities and Related Compounds

Experimental condition	Stock or strain	Ref.	Route[b]	Time[c] (day of gestation)	Dose[d] (mg/kg)	Sample size (no. dams)	Effect lack (non-teratological)	Increased resorption rate	Altered growth or development	General pathological effects on systems
Phthalimide	N.Z. White/ Chinchilla	122	Oral	7–12	150	3	+			
Phthalimide	Dutch Belted	235	Oral	6–16	75	10	±			
N-Carbethoxyphthalimide	N.Z. White	438	Oral	6–11	150	4	+			
1-Phthalimidobutane	N.Z. White/ Chinchilla	122	Oral	7–12	150	5	+			
2-Phthalimidobutane	N.Z. White/ Chinchilla	122	Oral	7–12	150	2	+			
Phthalimidobenzene	N.Z. White/ Chinchilla	122	Oral	7–12	105	7	+			
3-Phthalimidopyridine	N.Z. White/ Chinchilla	122	Oral	7–12	150	4	+			
4-Phthalimidobutyramide	N.Z. White/ Chinchilla	122	Oral	7–12	150	2	+			
α-Phthalimidoaspar-timide	N.Z. White/ Chinchilla	122	Oral	7–12	150	4	+			
Hexahydrothalidomide	N.Z. White/ Chinchilla	122	Oral	7–12	150	3	+			
N-Methoxythalidomide	N.Z. White	439	Oral	6–11	150	7		+		
3-Nitrothalidomide	N.Z. White/ Chinchilla	122	Oral	7–12	150	4	+			
α-Aminoglutarimide	N.Z. White/ Chinchilla	122	Oral	7–12	150	4	+			
Methyl-4-phthalimido-DL-glutaramate	N.Z. White	439	Oral	6–11	150	5				(+) Skeletal, (+) cardiovascular, (+) urinary
Glutamic acid imide·HCl	Not stated	141	Oral	6–14	100	7	+			(+) Digestive
N-(O-Carboxybenzoyl)-DL-isoglutamine	Not stated	141	Oral	6–14	100	5	+			
2-Phthalimido-DL-gluturic diamide	N. Z. White	439	Oral	6–11	150	8		+		
Phthaloyl-DL-isoglutamine	Hybrid	200	Oral	0–5	100–200	6	+			
N-Phthalyl-DL-isoglutamine	Not stated	141	Oral	6–14	100	15	+			
2-Phthalimidoglutaric acid anhydride	N.Z. White/ Chinchilla	122	Oral	7–12	150	6	+			
N-(O-Carboxybenzoyl)-DL-glutamic acid	Not stated	141	Oral	6–14	100	9	+			
DL-O-Carboxybenzoyl-glutamic acid	Hybrid	200	Oral	0–5	200—400	4	+			
N-phthalyl-DL-glutamic acid	Not stated	141	Oral	6–14	100	9	+			
N-phthalyl-DL-glutamine	Not stated	141	Oral	6–14	100	8	+			
N-(O-carboxybenzoyl)-DL-isoglutamine	Not stated	141	Oral	6–14	100	6	+			

[a]Based on cited author's summarized interpretation of experimental data.

[b]SQ = subcutaneous and IV = intravenous.

[c]Time: Periods in excess of 32 days include duration of pregnancy. Shorter periods refer to time during gestation.

[d]Dose: Summarized depiction of dosage range or as otherwise indicated (usually in mg/kg//day).

? = Details not readily apparent.

+ = presence of or occurrence.

VI. CONCLUSIONS

In this chapter, the fetal rabbit has been considered in regard to its past and current use in the field of experimental teratology. Also, an outline has been presented detailing the occurrence of naturally occurring and induced pathological phenomena. An obvious question is ever present. Is the rabbit fetus a suitable test species in this area of research? In answer to this, it can be said that the use of the rabbit fetus as an instrument for teratological investigation is not without limitation and need for further understanding. However, it seems that enough is known about this species to justify its experimental use.

In experimented teratology, one of the paramount deliberations is not which species is used but how to achieve its effective use. The solution to this largely depends on the rationale unique to the investigator's experimental design. Obviously, achieving a realistic balance between the practical and the ideal approach is an ever present problem in design.

Current efforts to detect teratogenic activity have evolved a variety of animal screening procedures focused primarily on the detection of environmentally induced abnormal *in utero* fetal development as judged by morphological examination. Little emphasis has been placed on detecting effects resulting from *in utero* insult occurring in the postnatal period. This approach continues to be the subject of much criticism as being inadequate and misleading.

Regardless of the discipline involved, the field of teratology as it is currently recognized is a young science, barely 40–50 years old with its most significant development occurring in the past 10–15 years. Considering this, the current testing approaches, while admittedly open to criticism, do represent steps toward developing a better understanding of pathological embryology. With regard to the use of animal testing methods, among the next steps which should be contemplated are in-depth evaluations of pathogenetic mechanisms in abnormal development and further evaluation of prenatal and early postnatal development, including the maturation of the fetal and neonatal physiological and metabolic systems. It is hoped that as more effective methods are established, those less appropriate will be discontinued.

Frontiers for research in teratology are many and obvious. The technology available to each of the many disciplines participating in this field is awesome and undergoing continued sophistication. In spite of this, significant achievement seems only possible and proportional to the degree with which an integrated multidisciplinary approach is encouraged and patiently applied.

Acknowledgments

I wish to recognize the patient assistance of Miss Patricia Garry and Miss Catherine North in the preparation of this manuscript and the encouragement of the Management of Sandoz-Wander, Inc. to pursue this undertaking.

REFERENCES

1. Abdul-Karim, R. W. (1968). Fetal physiology. *Obstet. Surv. (Baltimore)* **23**, 713–745.
2. Abdul-Karim, R. W., and Marshall L. D. (1969). The influence of ethamoxytriphetol on the collagen and calcium contents of the femurs of rabbit fetuses. *Toxicol. Appl. Pharmacol.* **15**, 185–188.
3. Abdul-Karim, R. W., Prior, J. T., and Nesbitt, R. E. L. (1968). Influence of maternal estrogen on fetal bone development in the rabbit. *Obstet. Gynecol.* **31**, 346–353.
4. Adbul-Karim, R. W., Shelley, T. F., Marshall, L. D., and Rizk, P. T. (1970). Influence of ethamoxytriphetol on the fetal bone development in the rabbit. *J. Reprod. Med.* **4**, 137–140.
5. Ackerman, G. A. (1966). The origin of the lymphocytes in the appendix and tonsil iliaca of the embryonic and neonatal rabbit. *Anat. Rec.* **154**, 21–39.
6. Adams, C. E. (1960). Prenatal mortality in the rabbit *Oryctolagus cuniculus*. *J. Reprod. Fert.* **1**, 36.
7. Adams, C. E. (1962). Studies on prenatal mortality in the rabbit, *Oryctolagus cuniculus*: The effect of transferring varying numbers of eggs. *J. Endocrinol.* **24**, 471–490.
8. Adams, C. E. (1968). Fertilizing capacity of rabbit spermatozoa deposited in the vagina, fallopian tubes or peritoneal cavity. *Congr. Int. Reprod. Anim. Insem. Artif., 6th, 1900* Vol. 1, pp. 31–33.
9. Adams, C. E. (1969). Intraperitoneal insemination in the rabbit. *J. Reprod. Fert.* **18**, 333–339.
10. Adams, C. E., Aitken, F. C., and Worden, A. N. (1966). The rabbit. *In* "The UFAW Handbook on the Care and Management of Laboratory Animals" (W. Lane-Petter, ed.), 3rd ed., pp. 396–448. Williams & Williams, Baltimore, Maryland.
11. Adams, C. E., Hay, M. F., and Lutwak-Mann, C. (1961). The action of various agents upon the rabbit embryo. *J. Embryol. Exp. Morphol.* **9**, 468–491.
12. Adams, C. E., and Lutwak-Mann, C. (1960). The effect of certain agents upon the early rabbit embryo. *J. Endocrinol.* **20**, 1–2.
13. Adams, F. H., Enhorning, G., and Norman, A. (1966). Surface properties of lung extracts. II. Comparison of fetal and adult rabbits. *Acta Physiol. Scand.* **68**, 28–36.
14. Adamson, R. H. (1970). Evaluation of the embryotoxic activity of L-asparaginase. *Arch. Int. Pharmacodyn. Ther.* **186**, 310–320.
15. Adolf, E. F. (1970). Physiological stages in the development of mammals. *Growth* **34**, 113–124.
16. Aherne, W., and Dawkins, M. J. (1964). The removal of fluid from the pulmonary airways after birth in the rabbit, and the effect on this of prematurity and pre-natal hypoxia. *Biol. Neonatorum* **7**, 214–229.
17. Amann, R. P., and Lambiase, J. T., Jr. (1967). The male rabbit. I. Changes in semen characteristics and sperm output between puberty and one year of age. *J. Reprod. Fert.* **14**, 329–332.
18. American Medical Association. (1965). Animal drug teratogenicity: Its application to man. *J. Amer. Med. Ass.* **194**, 1007–1008.
19. Amoroso, E. C. (1962). Placentation. *In* "Marshall's Physiology of Reproduction" (A. S. Parkes, ed.), 4th ed., Vol. 2, pp. 127–311. Longmans, Green, New York.
20. Anderson, E., Condon, W., and Sharp, D. (1970). A study of oogenesis and early embryogenesis in the rabbit, *Oryctolagus cuni-*

culus, with special reference to the structural changes of mitochondria. *J. Morphol.* **130**, 67–92.

21. Anderson, W. A. D. (1966). "Pathology." Mosby, St. Louis, Missouri.

22. Apgar, V. (1964). Drugs in pregnancy. *J. Amer. Med. Ass.* **190**, 840–841.

23. Archer, O. K., Sutherland, D. E., and Good, R. A. (1963). Appendix of the rabbit; a homologue of the bursa in the chicken. *Nature (London)* **200**, 337–339.

24. Archer, O. K., Sutherland, D. E., and Good, R. A. (1964). The developmental biology of lymphoid tissue in the rabbit. Consideration of the role of thymus and appendix. *Lab. Invest.* **13**, 259–271.

25. Arey, L. B. (1954). "Developmental Anatomy." Saunders, Philadelphia, Pennsylvania.

26. Armed Forces Institute of Pathology. (1960). "Manual of Histologic and Special Staining Techniques," 2nd ed. McGraw-Hill, New York.

27. Arrington, L. R., Taylor, R. N., Jr., and Ammerman, C. B. (1965). Effects of excess dietary iodine upon rabbits, hamsters, rats and swine. *J. Nutr.* **87**, 394–398.

28. Asdell. S. A. (1964). "Patterns of Mammalian Reproduction," 2nd ed. Cornell Univ. Press, Ithaca, New York.

29. Austin, C. R. (1961). "The Mammalian Egg." Blackwell, Oxford.

30. Austin, C. R. (1967). Chromosome deterioration in ageing eggs of the rabbit. *Nature (London)* **213**, 1018–1019.

31. Axelrod, L. R. (1970). Drugs and nonhuman primate teratogenesis. *In* "Fetal Homeostasis" (R. Wynn, ed.), Vol. 4, pp. 217–230. Appleton, New York.

.32. Baker, J. B. E. (1960). The effects of drugs on the foetus. *Pharmacol. Rev.* **12**, 37–90.

33. Barcellona, P. S. (1970). Investigations on the possible effects of Trazodone in rats and rabbits. *Boll. Chim. Farm.* **109**, 323–332.

34. Barclay, A. E., Franklin, K. J., and Prichard, M. M. L. (1945). "The Foetal Circulation and Cardiovascular System, and the Changes That They Undergo at Birth." Thomas, Springfield, Illinois.

35. Barron, D. H., and Battaglia, F. C. (1956). The oxygen concentration gradient between the plasmas in the maternal and fetal capillaries of the placenta of the rabbit. *Yale J. Biol. Med.* **28**, 197–207.

36. Barrow, M. V. (1971). A brief history of teratology to the early 20th century. *Teratology* **4**, 119–129.

37. Barrow, M. V., and Taylor, W. J. (1969). A rapid method for detecting malformations in rat fetuses. *J. Morphol.* **127**, 291–306.

38. Barter, R. (1953). The histopathology of congenital pneumonia, a clinical and experimental study. *J. Pathol. Bacteriol.* **66**, 407–415.

39. Baumgartner, I. M., and Sawin, P. B. (1943). Familial variation in the pattern of rib ossification in the rabbit. *Anat. Rec.* **86**, 473–487.

40. Bearn, J. G. (1967). The influence of the foetal adrenals on the development of the spleen of the rabbit foetus. *Acta Anat.* **68**, 97–101.

41. Bearn, J. G. (1967). Role of the fetal pituitary and adrenal glands in the development of the fetal thymus of the rabbit. *Endocrinology* **80**, 979–982.

42. Bearn, J. G. (1968). The thymus and the pituitary-adrenal axis in anencephaly. *Brit. J. Exp. Pathol.* **44**, 136–144.

43. Bearn, J. G., Antonis, A., and Pilkington, T. (1967). Foetal and maternal plasma triglyceride and cholesterol levels in the rabbit after foetal hypophysectomy by decapitation *in utero*. *J. Endocrinol.* **37**, No. 4, 479–480.

44. Beatty, R. A. (1957). Nigrosin-eosin staining of rabbit spermatozoa and the fertility of semen. *Proc. Roy. Soc. Edinburgh* **67**, 1–31.

45. Becker, B. A. (1969). Teratology in toxicology. *Toxicol. Appl. Pharmacol.* **15** (editorial).

46. Bencosme, S. A. (1955). The histogenesis and cytology of the pancreatic islets in the rabbit. *Amer. J. Anat.* **96**, 103–151.

47. Benirschke, K. (1969). Models for cytogenetics and embryology. *Fed. Proc., Fed. Amer. Soc. Exp. Biol.* **28**, 170–178.

48. Bergstrom, R. M. (1967). Neurophysiological effects of teratogenic agents in ontogeny. *Acta Neurol. Scand.* **43**, 37–42.

49. Bertelli, A., and Donati, L., eds. (1969). "Teratology." Excerpta Med. Found., Amsterdam.

50. Bertelli, A., Polani, P. E., Spector, R., Seller, M. J., Tuchmann-Duplessis, H., and Mercier-Parot L. (1968). Retentissement d'un neuroleptique, l'Halopéridol, sur la gestation et le développement prénatal des rongeurs. Resultats de trois groupes de recherches. *Arzneim.-Forsch.* **18**, 1420–1424.

51. Bough, R. G., Gurd, M. R., Hall, J. E., and Lessel, B. (1963). Effect of methaqualone hydrochloride in pregnant rabbits and rats. *Nature (London)* **200**, 656–657.

52. Brambell, F. W. R. Hemmings, W. A., Henderson, M., Parry, H. J., and Rowlands, W. T. (1949) The route of antibodies passing from the maternal circulation to the foetal circulation in rabbits. *Proc. Roy. Soc., Ser. Bis.* **138**, 195–204.

53. Bredderman, P. J., Foote, R. H., and Yassen, A. M. (1964). An improved artificial vagina for collecting rabbit semen. *J. Reprod. Fert.* **7**, 401–403.

54. Brent, R. L. (1964). Drug testing in animals for teratogenic effects: Thalidomide in the pregnant rat. *Med. Progr.* **64**, 762–770.

55. Brent, R. L. (1967). Medicolegal aspects of teratology. *J. Pediat.* **71**, 288–298.

56. Brent, R. L. (1972). Protecting the public from teratogenic and mutagenic hazards. *J. Clin. Pharmacol. New Drugs* **12**, 61–70.

57. Brinster, R. L. (1970). Culture of two-cell rabbit embryos to morulae. *J. Reprod. Fert.* **21**, 17–22.

58. Brown, W. H., and Pearce, L. (1945). Hereditary achondroplasia in the rabbit. I. Physical appearance and general feature. *J. Exp. Med.* **82**, 241–260.

59. Bruce, J. A., (1941). Time and order of appearance of ossification centers and their development in the skull of the rabbit. *Amer. J. Anat.* **68**, 41–67.

60. Burdi, A. R. (1965). Toluidine blue-alizarin red S staining of cartilage and bone in whole-mount skeletons *in vitro*. *Stain Technol.* **40**, 45–48.

61. Cahal, D. A. (1965). Drug embryopathies, preventive measures—the British point of view. *In* "Embryopathic Activity of Drugs" (J. Robson, F. Sullivan, and R. Smith, eds.), pp. 279–288. Little, Brown, Boston, Massachusetts.

62. Cahen, R. L. (1964). Evaluation of the teratogenicity of drugs. *Clin. Pharmacol. Exp. Ther.* **5**, 480–514.

63. Carmel, J. A., Friedman, F., and Adams, F. H. (1965). Fetal tracheal ligation and lung development. *Amer. J. Dis. Child.* **109**, 452–456.

64. Casey, A. E., Rosahn, P. D., Hu, C. K., and Pearce, L. (1934). Hereditary variations in the blood cytology of normal rabbits. *Science* **79**, 189–190.

65. Cass, J. S., Campbell, I. R., and Lange, L. (1960). A guide to production, care and use of laboratory animals. *Fed. Proc., Fed. Amer. Soc. Exp. Biol.*, Suppl. No. 6.

66. Castle, W. E. (1929). A further study of size inheritance in rabbits, with special reference to the existence of genes for size characters. *J. Exp. Zool.* **53**, 421–453.

67. Castle, W. E. (1932). Growth rates and racial size in rabbits and birds. *Science* **76**, 259–260.

68. Castle, W. E. (1933). The furless rabbit. *J. Hered.* **24**, 81–86.

69. Castle, W. E. (1940). "Mammalian Genetics." Harvard Univ. Press, Cambridge, Massachusetts.

70. Castle, W. E., and Gregory P. W. (1929). The embryological basis of size inheritance in the rabbit. *J. Morphol. Physiol.* **48**, 81–103.

71. Castle, W. E., Walter, W. E., Mullenix, R. C., and Cobb, S. (1909). Studies of inheritance in rabbits. *Carnegie Inst. Wash. Publ.* **114**, pp. 81–103.

72. Caujolle, F. M., Caujolle, D. H., and Cros, S. B. (1967). Limits of

toxic and teratogenic tolerance of dimethyl sulfoxide. *Ann. N.Y. Acad. Sci.* **141**, 110–126.

73. Chai, C. K., and Crary, D. D. (1947) Conjoined twinning in rabbits. *Teratology* **4**, 433–444.

74. Chai, C. K., and Degenhardt, K. H. (1962). Developmental anomalies in inbred rabbits. *J. Hered.* **53**, 174–182.

75. Chambron, Y., Touret, J. L., and Depagne, A. (1967). Teratogenic study of delta 6, 6-chloro-17-alpha-acetoxyprogesterone on the fetuses of castrated or intact doe rabbits (Fre). *Ann. Endocrinol.* **28**, 333–342.

76. Chang, M. C., Hunt, D. M., and Harvey, E. B. (1963). Effects of radiocobalt irradiation of pregnant rabbits on the development of fetuses. *Anat. Rec.* **145**, 455–466.

77. Chomette, G. (1955). Entwicklungsstörungen nach insulinshock bein trächtigen kaninchen. *Beitr. Pathol. Anat. All. Pathol.* **115**, 439–451.

78. Clavert, J., Buck, P., and Rumpler, Y. (1965). Cardio-vascular malformations produced in embryos by injecting pregnant rabbits with cortisones. *Therapie* **20**, 1579–1584.

79. Cliff, M. M., and Reynolds, S. R. M. (1959). A dose-stress response of adrenaline affecting fetuses at a critical time in pregnant rabbit. *Anat. Rec.* **134**, 379–384.

80. Cohlan, S. Q. (1963). Teratogenic agents and cogenital malformations. *J. Pediat.* **63**, 650–659.

81. Cook, M. J., and Fairweather, F. A. (1968). Methods used in teratogenic testing. *Lab Anim.* **2**, 219–228.

82. Cook, M. J., Fairweather, F. A., and Hardwick, M. (1969). Further thoughts on teratogenic testing. *In* "Teratology" (A. Bertelli and L. Donati, eds.), pp. 34–42. Excerpta Med. Found., Amsterdam, Holland.

83. Cozens, D. D. (1965). Abnormalities of the external form and of the skeleton in the New Zealand White rabbit. *Food Cosmet. Toxicol.* **3**, 695–700.

84. Crabb, E. D. (1931). "Principles of Functional Anatomy of the Rabbit." McGraw-Hill (Blakiston), New York.

85. Craigie, E. H. (1957). "Bensley's Practical Anatomy of the Rabbit." Univ. of Toronto Press, Toronto.

86. Crary, D. D. (1962). Modified benzyl alcohol clearing of alizarin-stained specimens without loss of flexibility. *Stain Technol.* **37**, 124–125.

87. Crary, D. D. (1964). Development of the external ear in the dacks rabbit. *Anat. Rec.* **150**, 441–447.

88. Crary, D. D., Fox, R. R., and Sawin, P. B. (1966). Spina bifida in the rabbit. *J. Hered.* **57**, 236–243.

89. Crary, D. D., and Sawin, P. B. (1949). Morphogenetic studies in the rabbit. VI. Genetic factors influencing the ossification pattern of the limbs. *Genetics* **34**, 508–523.

90. Crary, D. D., and Sawin, P. B. (1952). A second recessive achondroplasia in the domestic rabbit. *J. Hered.* **43**, 255–259.

91. Curtis, G. W., Algeri, E. J., McBay, A. J., and Ford, R. (1955). The transplacental diffusion of carbon monoxide. *AMA Arch. Pathol.* **59**, 677.

92. Cutting, W. C. (1967). "Handbook of Pharmacology. The Actions and Uses of Drugs," 3rd ed. Meredith, New York.

93. Daniel, J. C. (1964). Some effects of steroids on cleavage of rabbit eggs *in vitro. Endocrinology* **75**, 706–710.

94. Daniel, J. C., and Takahashi, K. (1965). Selective laser destruction of rabbit blastomeres and continued cleavage of survivors *in vitro. Exp. Cell Res.* **39**, 475–482.

95. Da Rosa, F. M. (1943). Renal agenesis, a new mutation in the rabbit. *Rev. Med. Vet. (Lisboa)* **38**, 349–363.

96. Davey, D. G. (1965). The study of the toxicity of a potential drug—basic principles. *Proc. Eur. Soc. Study Drug Toxicity* **6**, Suppl., 1–13.

97. David, L. T. (1932). The external expression and comparative

dermal histology of hereditary hairlessness in mammals. *Z. Zellforsch. Mikvosk. Anat.* **14**, 616–719.

98. Davis, J. A., and Stafford, A. (1964). Respiratory distress in new born rabbits. *Biol. Neonatorum* **2**, 129–140.

99. Dawes, G. S., Handler, J. J., and Mott, J. C. (1957). Some cardiovascular responses in foetal new-born and adult rabbits. *J. Physiol. (London)* **139**, 123–136.

100. Dawson, A. B. (1926). A note on the staining of the skeleton of cleared specimens with alizarin red S. *Stain Technol.* **1**, 123–124.

101. Degenhardt, K. H. (1960). Cranio-facial dysplasia induced by oxygen deficiency in rabbits. *Biol. Neonatorum* **2**, 93–104.

102. Dekker, A., and Mehrizia, A. (1964). The use of thalidomide as teratogenic agent in rabbits. *Bull. Johns Hopkins Hosp.* **115**, 223–230.

103. Delahunt, C. S., Lassen, L. J., and Rieser, N. (1966). Some comparative teratogenic studies with thalidomide. *Proc. Eur. Soc. Study Drug Toxicity* **7**, 229–240.

104. Deren, J. J., Padykula, H. A., and Wilson, T. H. (1966). Development of structure and function in the mammalian yolk sac. II. Vitamin B_{12} uptake by rabbit yolk sacs. *Develop. Biol.* **13**, 344–369.

105. Deren, J. J., Padykula, H. A., and Wilson, T. H. (1966). Development of structure and function in the mammalian yolk sac. III. Amino acid transport by rabbit yolk sac. *Develop. Biol.* **13**, 370–384.

106. Dipaola, J. A., and Kotin, P. (1966). Teratogenesis—oncogenesis: A study of possible relationships. *Arch. Pathol.* **81**, 3–23.

107. Doggett, V. C., (1956). Periodicity in the fecundity of male rabbits. *Amer. J. Physiol.* **187**, 445–450.

108. Done, A. K. (1964). Developmental pharmacology. *Clin. Pharmacol. Ther.* **5**, 432–479.

109. Dorland (1965). "Illustrated Medical Dictionary," 24th ed. Saunders, Philadelphia, Pennsylvania.

110. Dornhorst, A. C., and Young, I. M. (1952). Action of adrenaline and nor-adrenaline on the placental and foetal circulations in the rabbit and guinea-pig. *J. Physiol. (London)* **118**, 282–288.

111. Drobeck, H. P., Coulston, I., and Cornelius, D. (1965). Effects of thalidomide on fetal development in rabbits and on establishment of pregnancy in monkeys. *Toxicol. Appl. Pharmacol.* **7**, 165–178.

112. Dukelow, W. R., Chernoff, H. N., and Williams, W. L. (1967). Fertilization life of the rabbit ovum relative to sperm capacitation. *Amer. J. Physiol.* **213**, 1397–1400.

113. Earley, P. A., and Hayden, J. (1964). Effect of acetylsalicylic acid on foetal rabbits. *Lancet* **1**, 763.

114. Eckstein, P., and Zuckerman, S. (1962). Morphology of the reproductive tract. *In* "Marshall's Physiology of Reproduction" (A. S. Parkes, ed.), 4th ed., Vol. 1, Part 1, pp. 43–155. Longmans, Green, New York.

115. Eckstein, P., and Zuckerman, S. (1962). The oestrus cycle in the mammalia. *In* "Marshall's Physiology of Reproduction" (A. S. Parkes, ed.), 7th ed., Vol. 1, Part 1, pp. 226–396. Longmans, Green, New York.

116. Eckstein, P., and Zuckerman, S. (1962). Changes in the accessory reproductive organs of the non-pregnant female. *In* "Marshall's Physiology of Reproduction" (A. S. Parkes, ed.), 4th ed., Vol. 1, Part 1, pp. 543–654. Longmans, Green, New York.

117. Edwards, J. A. (1968). The external development of the rabbit embryo. *Advan. Teratol.* **3**, 239–262.

118. Elger, W. (1966). The role of fetal androgens in sex differentiation in rabbits and their distinction from other hormonal and somatic factors by means of a powerful antiandrogen. *Arch. Anat. Microsc. Morphol. Exp.* **55**, Suppl., 657–743.

119. Ellenhorn, M. J. (1964). The Food and Drug Administration and the prevention of drug embryopathy. *J. New Drugs* **4**, 12–20.

120. Emery, J. L. (1967). Evidence from bone growth that most of the infants dying in the neonatal period had been ill before birth. *Acta Paediat. Scand., Suppl.* **172**, 55–59.

121. Evans, H. E. (1948). Clearing and staining small vertebrates, *in toto*, for demonstration, ossification. *Turtox News* **26**, 42–47.

122. Fabro, S., Schumacher, H., Smith, R. L., and Williams, R. T. (1964). Teratogenic activity of thalidomide and related compounds. *Life Sci.* **3**, 987–992.

123. Fabro, S., Schumacher, H., Smith, R. L., and Williams, R. T. (1964). Identification of thalidomide in rabbit blastocysts. *Nature (London)* **201**, 1125–1126.

124. Fabro, S., and Sieber, S. M. (1969). Caffeine and nicotine penetrate the pre-implantation blastocyst. *Nature (London)* **223**, 410–411.

125. Fabro, S., Sieber, S. M., Sato, U., and Pergament, E. (1968). Is lysergide a teratogen? *Lancet* **1**, 639–640.

126. Fabro, S., Smith, R. L., and Williams, R. T. (1966). Embryotoxic activity of some pesticides and drugs related to phthalimide. *Food Cosmet. Toxicol.* **3**, 587–590.

127. Farrell, G., Powers, D., and Otani, T. (1968). Inhibition of ovulation in the rabbit: Seasonal variation and the effects of indoles. *Endocrinology* **83**, 599–603.

128. Felisati, D. (1962). Thalidomide and congenital abnormalities. *Lancet* **2**, 724–725.

129. Fernandez, V. (1969). An autoradiographic study of the development of the anterior thalamic group and limbic cortex in the rabbit. *J. Comp. Neurol.* **136**, 423–52.

130. Fishbein, M., ed. (1963). "Birth Defects." Lippincott, Philadelphia, Pennsylvania.

131. Foote, R. H., Hafs, H. D., Staples, R. E., Grégoire, A. T., and Bratton, R. W. (1963). Ovulation rates and litter sizes in sexually receptive and non-receptive artificially insemminated rabbits given varying dosages of luteinizing hormone. *J. Reprod. Fert.* **5**, 59–66.

132. Forsberg, U., and Grant, C. A. (1971). Current practices of pharmaceutical companies for detecting and reporting foetal anomalies obtained in animal tests. *Proc. Eur. Study Drug Toxicity* **12**, 352–356.

133. Fouts, J. R., and Adamson, R. H. (1959). Drug metabolism in the newborn rabbit. *Science* **129**, 897–898.

134. Fox, R. R., Crary, D. D., Babino, E. J., Jr., and Sheppard, L. B. (1969). Buphthalmia in the rabbit. *J. Hered.* **60**, 206–212.

135. Franceschetti, A. (1930). Die Vererburg von Augenleiden. *Kurzes Handb. Ophthalmol* **1**, 631–855.

136. Fraser, F. C., Kalter, H., Walker, B. E., and Fainstate, T. D. (1954). The experimental production of cleft palate with cortisone and other hormones. *J. Cell. Comp. Physiol.* **43**, 237–259.

137. Fraser, F. C. (1959). Causes of congenital malformations in human beings. *J. Chronic Dis.* **10**, 99–110.

138. Fraser, F. C. (1971). Developmental thresholds and teratogenetics. *Fed. Proc., Fed. Amer. Soc. Exp. Biol.* **30**, 100–101.

139. Fratta, I. D., Sigg, E. B., and Maiorana, K. (1965). Teratogenic effects of thalidomide in rabbits, rats, hamsters, and mice. *Toxicol. Appl. Pharmacol.* **7**, 268–286.

140. Friedman, W. F., and Roberts, W. C. (1966). Vitamin D and the supravalvar aortic stenosis syndrome. *Circulation* **34**, 77–86.

141. Fritz, H. (1966). Failure of thalidomide metabolites to produce malformation in the rabbit embryos. *J. Reprod. Fert.* **11**, 157–159.

142. Fritz, H., and Hess, R. (1971). Effects of cyclophosphamide on embryonic development in the rabbit. *Agents Actions* **2**, 83–86.

143. Geber, W. F., and Anderson, T. A. (1967). Abnormal fetal growth in the albino rat and rabbit induced by maternal stress. *Biol. Neonatorum* **11**, 209–215.

144. Geber, W. F., and Schramm, L. C. (1969). Effect of marihuana extract on fetal hamsters and rabbits. *Toxicol. Appl. Pharmacol.* **14**, 276–282.

145. Gersh, I. (1937). The correlation of structure and function in the developing mesonephros and metanephros. *Contrib. Embryol. Carnegie Inst.* **153**, 35–57.

146. Gibson, J. P., Staples, R. E., Larson, E. J., Kuhn, W. L., Holtkamp, D. E., and Newberne, J. W. (1968). Teratology and reproduction studies with antinauseant. *Toxicol. Appl. Pharmacol.* **12**, 293–294.

147. Gibson, J. P., Staples, R. E., and Newberne, J. W. (1966). Use of the rabbit in teratogenicity studies. *Toxicol. Appl. Pharmacol.* **9**, 398–408.

148. Gilchrist, F. G. (1968). "A Survey of Embryology." McGraw-Hill, New York.

149. Giordana, A. (1969). Human pathology and experimental teratology. *In* "Teratology" (A. Bertelli and L. Donati, eds.), pp. 205–210. Excerpta Med. Found., Amsterdam.

150. Giurgea, M., and Puigdevall, J. (1966). Experimental teratology with meclozine. *Med. Pharmacol. Exp.* **15**, 375–388.

151. Goldenthal, E. I. (1968). Current views on safety evaluation of drugs. *FDA (Food Drug Admin.) Pap.* **2**, 13–18.

152. Goldenthal, E. I. (1971). A compilation of LD_{50} values in newborn and adult animals. *Toxicol. Appl. Pharmacol.* **18**, 185–207.

153. Goldstein, I., and Wexler, D. (1931). Rosette formation in the eyes of irradiated human embryos. *Arch. Ophthalmol.* [N.S.] **5**, 591–600.

154. Goodlin, R. C. (1965). Fetal incubator studies. III. Factors associated with breathing in fetal rabbits. *Biol. Neonatorum* **8**, 274–280.

155. Gottsche, G. H. (1971). Voraussetzungen und empfehlungen für das teratologische experiment. *Arzneim.-Forsch.* **21**, 2169–2172.

156. Gottschewski, G. H. M. (1964). Mammalian blastopathies due to drugs. *Nature (London)* **201**, 1232–1233.

157. Gottschewski, G. H. M., and Zimmermann, W. (1971). Die Embryonalentwicklung des hauskaninchens. "Neuerscheinung." Verlag M. & H. Schaper, Hannover.

158. Grauwiler, J. (1969). Variations in physiological reproduction data and frequency of spontaneous malformations in teratological studies with rats and rabbits. *In* "Teratology" (A. Bertelli and L. Danati, eds.), pp. 129–135. Excerpta Med. Found., Amsterdam.

159. Grauwiler, J. (1969). Variations in physiological reproduction data and frequency of spontaneous malformations in teratological studies with rats and rabbits. *In* "Teratology" (A. Bertelli and L. Donati, eds.), pp. 129–135. Excerpta Med. Found, Amsterdam.

160. Grauwiler, J. (1971). An industrial teratologists' comments. *Teratology* **4**, 479–480.

161. Green, E. L. (1938). The inheritance of costal and vertebral variations in the rabbit. *Genetics* **23**, 149.

162. Green, E. L. (1940). The inheritance of a rib variation in the rabbit. *Anat. Rec.* **74**, 47–60.

163. Green, E. L., and Sawin, P. B. (1938). The anatomy and genetics of some skeletal variations in the rabbit. *Anat. Rec.* **70**, 32.

164. Greene, H., and Brown, W. (1932). Hereditary variations in the skull of the rabbit. *Science* **76**, 421–422.

165. Green, H., and Saxton, J. A. (1939). Hereditary brachydactylia and allied abnormalities in the rabbit. *J. Exp. Med.* **69**, 301–314.

166. Greene, H. S. N. (1933). Oxycephaly and allied conditions in man and in the rabbit. *J. Exp. Med.* **57**, 967–976.

167. Greene, H. S. N. (1935). Hereditary brachydactylia and allied abnormalities in the rabbit. *Science* **81**, 405–407.

168. Greenwald, G. S. (1956). Sperm transport in the reproductive tract of the female rabbit. *Science* **124**, 586.

169. Greenwald, G. S. (1961). A study of the transport of ova through the rabbit oviduct. *Fert. Steril.* **12**, 80–95.

170. Greenwald, G. S. (1962). The role of the Mucin layer in development of the rabbit blastocyst. *Anat. Rec.* **142**, 407–415.

171. Gregory, P. W. (1930). The early embryology of the rabbit. *Embryology* **125**, 141–168.

172. Gregory, P. W., and Castle, W. E. (1931). Further studies on the embryological basis of size inheritance in the rabbit. *J. Exp. Zool.* **59**, 199–211.

173. Grillo, T. A. (1964). The occurrence of insulin in the pancreas of foetuses of some rodents. *J. Endocrinol.* **31**, 67–73.

174. Grote, W. (1969). Embryonale skeletmissbildungen nach plasma-entzug bei graviden kaninchen. *Z. Anat. Entwicklungs gesch.* **129**, 346–352.

175. Gruenwald, P. (1947). Mechanisms of abnormal development. *Arch. Pathol.*, **44**, 398–436.

176. Gruenwald, P. (1947). Mechanisms of abnormal development. *Arch. Pathol.* **44**, 495–559.

177. Gruenwald, P. (1947). Mechanisms of abnormal development. *Arch. Pathol.* **44**, 648–664.

178. Gruenwald, P. (1964). Examination of the placenta by the pathologist. *Arch. Pathol.* **77**, 41–46.

179. Gruenwald, P. (1966). The place of teratology in present-day biology and pathology. *Arch. Pathol.* **81**, 1–2.

180. Grunberg, H. (1963). "The Pathology of Development." Wiley, New York.

181. Guyer, M. F., and Smith, E. A. (1924). Further studies on inheritance of eye defects induced in rabbits. *J. Exp. Zool.* **38**, 449–474.

182. Hadek, R. (1958). Intraperitoneal insemination of rabbit doe. *Proc. Soc. Exp. Biol. Med.* **99**, 39–40.

183. Hafez, E. S. E. (1961). Embryonic survival in relation to number and size of implantation swellings in the rabbit *Oryctolagus cuniculus. Proc. Soc. Exp. Biol. Med.* **107**, 680–684.

184. Hafez, E. S. E. (1961). Storage of rabbit ova in gelled media at 10°C. *J. Reprod. Fert.* **2**, 163–178.

185. Hafez, E. S. E. (1964). The effects of overcrowding *in utero* on implantation and fetal development in the rabbit. *J. Exp. Zool.* **156**, 269–287.

186. Hafez, E. S. E. (1968). Some maternal factors causing postimplantation mortality in the rabbit. *Congr. Int. Reprod. Anim. Insem. Artif., 6th,* vol. 1, pp. 425–427.

187. Hafez, E. S. E., Gollnick, P. D., and Moustafa, L. A. (1967). Effect of maternal feed intake on body composition of neonatal rabbits before and after withholding of feed. *Amer. J. Vet. Res.* **28**, 1837–1841.

188. Hafez, E. S. E., Lindsay, D. R., and Moustafa, L. A. (1967). Effect of feed intake of pregnant rabbits on nutritional reserves of neonates. *Amer. J. Vet. Res.* **28**, 1153–1158.

189. Hafez, E. S. E., and Rojakoski, E. (1964). Growth and survival of blastocysts in the domestic rabbit. *J. Reprod. Fert.* **7**, 229–240.

190. Hagerman, D. D., and Villee, C. A. (1960). Transport functions of the placenta. *Physiol. Rev.* **40**, 313–330.

191. Hammond, J. (1928). Die Kontrolle der fruchtbarkeit bei tieren. *Zuechtungskunde* 3, 523–547.

192. Harper, K. H., Palmer, A. K., and Davies, R. E. (1965). Effect of imipramine upon the pregnancy of laboratory animals. *Arzneim.-Forsch.* **15**, 1218–1221.

193. Harper M. J. K. (1961). The time of ovulation in the rabbit following the injection of luteinizing hormone. *J. Endocrinol.* **22**, 147–152.

194. Harrington, D. D., and Newberne, P. M. (1970). Correlation of maternal blood levels of vitamin A at conception and the incidence of hydrocephalus in newborn rabbits: An experimental animal model. *Lab. Anim. Care* **20**, 675–680.

195. Hart, F. M., and Faber, J. J. (1965). Fetal and maternal temperatures in rabbits. *J. Appl. Physiol.* **20**, 737–741.

196. Hart, L. G., Adamson, R. H., Dixon, R. L., and Fouts, J. R. (1962). Stimulation of hepatic microsomal drug metabolism in the newborn and fetal rabbit. *J. Pharmacol. Exp. Ther.* **137**, 103–106.

197. Hartman, C. G. (1925). On some characters of taxonomic value appertaining to the egg and the ovary of rabbits. *J. Mammal.* **6**, 114–121.

198. Hartman C. G. (1945). The mating of mammals. *Ann. N. Y. Acad. Sci.* **46**, 23–44.

199. Hartman, H. A., Hrab, R., and Carty, P. (1970). A versatile data processing system developed for animal reproduction and teratology data. *Toxicol. Appl. Pharmacol.* **17**, 291–292.

200. Hay, M. F. (1964). Effects of thalidomide on pregnancy in the rabbit. *J. Reprod. Fert.* **8**, 59–76.

201. Hay, S., and Towascia, S. (1968). "A Classification of Congenital Malformations." U.S. Department of Health, Education and Welfare, Public Health Service, Dental Health Center, San Francisco, California.

202. Henson, E. C., Ball, D. A., and Brunson, J. G. (1968). Effects of propiomazine hydrochloride on newborn rabbits. *Arch. Pathol.* **85**, 357–365.

203. Herrman, H. (1963) Chemical causes of birth defects. *In* "Birth Defects" (M. Fishbein, ed.), pp. 181–191. Lippincott, Philadelphia, Pennsylvania.

204. Hippel, E. von (1903). Embryologische Untersuchungen über die Entslehungsweise der typischen angeborenen Spaltbildungen (Colobome des Augapfels. *Arch. Ophthalmol.* **55**, 507–548.

205. Hörnblad, P. Y. (1969). Embryological observations of the ductus arteriosus in the guinea-pig, rabbit, rat, and mouse. Studies on closure of the ductus arteriosus. IV. *Acta Physiol. Scand.* **76**, 49–57.

206. Hostetler, J. R., and Ackerman, G. A. (1969). Lymphopoiesis and lymph node histogenesis in the embryonic and neonatal rabbit. *Amer. J. Anat.* **124**, 57–75.

207. Hoyte, D. A. N. (1960). Alizarin as an indicator of bone growth. *J. Anat.* **94**, 432–442.

208. Hu, C. K., and Greene, H. S. N. (1935). A lethal mutation with stigmata of an acromegalic disorder. *Science* **81**, 25–26.

209. Ingalls, T. H., Curley, F. J., and Zappasodi, P. (1964). Thalidomide embryopathy in hybrid rabbits. *N. Engl. J. Med.* **271**, 441–444.

210. Ingham, B., Brentnall, D. W., Woollam, D. H. M., and Millen, J. W. (1965). Occurrence of tetany in litters of N. Z. White rabbits associated with spontaneous vertebral fractures and limb malformations. *Brit. Med. J.* **2**, 32.

211. Inman, O. R. (1941). Embryology of hereditary brachydactyly in the rabbit. *Anat. Rec.* **79**, 483–505.

212. Innes, J. R. M., and Saunders, L. Z. (1962). "Comparative Neuropathology." Academic Press, New York.

213. Jackson, A. V. (1948). Toxic effects of salicylate on the foetus and mother. *J. Pathol. Bacteriol.* **60**, 587–593.

214. Jensh, R. P., and Brent, R. L. (1966). Rapid schedules for KOH clearing and alizarin red S staining of fetal rat bone. *Stain Technol.* **41**, 179–183.

215. Johnson, L. (1933). The time and order of appearance of ossification centers in the albino mouse. *Amer. J. Anat.* **52**, 241–271.

216. Johnson, W. E. (1971). Fetal loss from anesthesia and surgical trauma in the rabbit. *Toxicol. Appl. Pharmacol.* **18**, 773–779.

217. Johnson, W. E., and Martin, A. R. (1969). Nonteratogenicity of *Veratrum californicum* in rabbits. *J. Pharm. Sci.* **58**, 1165–1166.

218. Jones, S. R., Stair, E. L., Gleiser, C. A., and Bridges, C. H. (1971). Use of whole-body sagittal paraffin sections of infant mice for immunoflourescent and histopathologic studies. *Amer. J. Vet. Res.* **32**, 1137–1142.

219. Jordan, R. L., Terapane, J. F., and Schumacher, H. J. (1970). Studies on the teratogenicity of methotrexate in rabbits. *Teratology* **3**, 203.

220. Jost, A. (1948). Le contrôle hormonal de la différentiation du sexe. *Biol. Rev. Cambridge Phil. Soc.* **23**, 201–236.

221. Jubb, K. V. F., and Kennedy, P. C. (1970). "Pathology of Domestic Animals," 2nd ed., 2 vols. Academic Press, New York.

222. Kallen, B., and Winberg, J. (1969). Multiple malformations studied with a national register of malformations. *Pediactrics* **44**, 410–417.

223. Kalter, H. (1967). Drug induced embryopathy. *Clin. Pharmacol. Ther.* **2**, 123–129.

224. Kalter, H. (1968). "Teratology of the Central Nervous System." Univ. of Chicago Press, Chicago, Illinois.

225. Kalter, H. (1968). How should times during pregnancy be called in teratology? *Teratology* **1**, 231–234.

226. Kalter, H. (1970). Editorial: The difference between mutagenesis

and teratogenesis, or how to tell the players from the spectators. *Teratology* 3, 221–222.

227. Kalter, H., and Warkany, J. (1959). Experimental production of congenital malformations in mammals by metabolic procedure. *Physiol. Rev.* 39, 69–115.

228. Karnofsky, D. A. (1964). Drugs as teratogens in animals and man. *Annu. Rev. Pharmacol.* 5, 447–472.

229. Karnofsky, D. A. (1965). Mechanisms of action of certain growth inhibiting drugs. *In* "Teratology, Principles and Techniques" (J. G. Wilson and J. Warkany, ed.), pp. 185–213. Univ. of Chicago Press, Chicago, Illinois.

230. Kasirsky, G., and Tansy, M. F. (1971). Teratogenic effects of methamphetamine in mice and rabbit. *Teratology* 4, 131–134.

231. Keberle, H., Schmid, K., Faigle, J. W., Fritz, H., and Loustalot, P. (1966). Uber die penetration von korperfremden stoffen in den jungen wirbeltieskeim. *Bull. Schweiz. Akad. Med. Wiss.* 22, 134–152.

232. Keeler, R. F. (1970). Teratogenic compounds of *Veratrum californicum* (Durand). X. Cyclopia in rabbits produced by cyclopamine. *Teratology* 3, 175–180.

233. Keeler, R. F. (1971). Teratogenic compounds of *Veratrum californicum* (Durand). XI. Gestational chronology and compound specificity in rabbits (35453). *Exp. Biol. Med.* 136, 1174–1179.

234. Kelsey, F. O. (1965). Drug embryopathies, preventive measures—the American point of view. *In* "Embryopathic Activity of Drugs" (J. Robson, F. Sullivan, and R. Smith, eds.), pp. 261–278. Little, Brown, Boston, Massachusetts.

235. Kennedy, G., Fancher, O. E., and Calandra, J. C. (1968). An investigation of the teratogenic potential of Captan, Folpet and Difolatan. *Toxicol. Appl. Pharmacol.* 13, 420–430.

236. Kikkawa, Y., Motoyama, E. K., and Gluck, L. (1968). Study of the lungs of fetal and newborn rabbits. *Amer. J. Pathol.* 52, 177–209.

237. Kilhström, J. E., and Fjellström, D. (1967). Automatic counting of spermatozoa in rabbit semen. *J. Reprod. Fert.* 14, 155–157.

238. King, C., and Olsen, C. (1967). Uterine-amnionic pathway of infecting the rabbit fetus with *Listeria monocytogenes. Amer. J. Vet. Res.* 28, 1555–1567.

239. Klotzschc, C. (1970). Teratologische und embryotoxische untersuchungen mit formothion und thiometon. *Pharm. Acta Helv.* 45, 434–440.

240. Kodituwakku, G. E., and Hafez, E. S. E. (1969). Blastocyst size in Dutch-belted and New Zealand rabbits. *J. Reprod. Fert.* 19, 187–190.

241. Kreybig, T. von (1969). The critical sensitivity of the developmental phase and the organotropic action of different teratogenic agents, receptors of morphogeneses in the mammalian embryo. *In* "Teratology" (A. Bertelli and L. Donati, ed), pp. 152–159. Excepta Med. Found., Amsterdam.

242. Lamar, J. K., Calhoun, F. J., and Darr, A. G. (1970). Effects of amantadine hydrochloride on cleavage and embryonic development in the rat and rabbit. *Toxicol. Appl. Pharmacol.* 17, 272.

243. Lamming, G. E., Salisburg, G. W., Hays, R. L., and Kendall, K. A. (1954). The effect of incipient vitamin A deficiency on reproduction in the rabbit. II. Embryonic and fetal development. *J. Nutr.* 52, 227–239.

244. Larsen, V. (1963). The teratogenic effects of thalidomide, imipramine HCl, and imipramine-*N*-oxide HCl on white Danish rabbits. *Acta Pharmacol. Toxicol.* 20, 186–200.

245. Larsen, V., and Bredahl, E. (1966). The embryotoxic effect on rabbits of monophenylbutazone, (Monazan (R)) compared with phenylbutazone and thalidomide. *Acta Pharmacol. Toxicol.* 24, 443–445.

246. Latimer, H. B., and Sawin, P. B. (1955). Morphogenetic studies of the rabbit. XIII. The influence of the dwarf gene upon organ size and variability in race X. *Anat. Rec.* 123, 447–466.

247. Lazarus, S. S., and Volk, B. W. (1963). Absense of teratogenic effect of tolbutamide in rabbits. *J. Clin. Endocrinol. Metab.* 23, 597–599.

248. Lehman, H., and Niggeschulze, A. (1971). The teratologic effects of thalidomide in Himalayan rabbits. *Toxicol. Appl. Pharmacol.* 18, 208–219.

249. Lell, W. A. (1931). The relation of the volume of the amniotic fluid to the weight of the fetus at different stages of pregnancy in the rabbit. *Anat. Rec.* 51, 119–124.

250. Little, W. A. (1966). Drugs in pregnancy. *Amer. J. Nurs.* 66, 1303–1307.

251. Loosli, R., Loustalot, P., Schalch, W. R., Sievers, K., and Stenger, E. G. (1964). Joint study in teratogenicity research. Preliminary communication. *Proc. Eur. Soc. Study Drug Toxicity* 4, 214–217.

252. Loosli, R., and Theiss, E. (1964). Methodik und Problematic der medikamintos—experimtellin teratogenese. *Teratogenesis; Symp. Schweiz. Akad. Med. Wiss. (Basel)* pp. 398–416.

253. Lopez-Escobar, G., and Fridhandler, L. (1969). Studies of clomiphene effects on rabbit embryo development and biosynthetic activity. *Fert. Steril.* 20, 697–714.

254. Lukens, M. M. (1971) Clinical perspectives concerning teratological information. *Amer. J. Pharm. Educ.* 35, 776–791.

255. Lutwak-Mann, C., and Hay, M. F. (1962). Effect on the early embryo of agents administered to the mother. *Brit. Med. J.* 2, 944–946.

256. Lutwak-Mann, C., Hay, M. F., and New, D. A. T. (1969). Action of various agents on rabbit blastocysts *in vivo* and *in vitro. J. Reprod. Fert.* 18, 235–257.

257. McColl. J. D. (1966). Teratogenicity studies. *Appl. Ther.* 8, 48–52.

258. McColl, J. D., Robinson, S., and Globus, M. (1967). Effect of some therapeutic agents on the rabbit fetus. *Toxicol. Appl. Pharmacol.* 10, 244–252.

259. McCutcheon, R. S. (1969). Teratology. *Essays Toxicol.* 1, 61–82.

260. MacDowell, E. C., Allen, E., and MacDowell, C. G. (1927). The prenatal growth of the mouse. *J. Gen. Physiol.* 11, 57–70.

261. McIntosh, R. (1959). The problem of congenital malformations. *J. Chronic Dis.* 10, 139–151.

262. McLaren, A., and Michie, D. (1960). Control of prenatal growth in mammals. *Nature (London)* 187, 363–365.

263. McNutt, C. W., and Sawin, P. B. (1943). Hereditary variations in the vena cava inferior of the rabbit. *Amer. J. Anat.* 72, 259–289.

264. Manresa, M. (1932). Inheritance of resistance and susceptibility to infectious abortion. *J. Infec. Dis.* 51, 30–71.

265. Markham, J. K., Emmerson, J. L., and Owen, N. J. (1971). Teratogenicity studies of methadone HCl in rats and rabbits. *Nature (London)* 233, 342–343.

266. Mauer, R. E., Hafez, E. S. E., Ehlers, M. H., and King, J. R. (1968). Culture of two cell rabbit eggs in chemically defined media. *Exp. Cell. Res.* 52, 293–300.

267. Mellin, G. W. (1963). The frequency of birth defects. *In* "Birth Defects" (M. Fishbein, ed.), pp. 1–17. Lippincott, Philadelphia, Pennsylvania.

268. Mellin, G. W., and Katzenstein, M. (1964). Increased incidence of malformations—chance or change? *J. Amer. Med. Ass.* 187, 570–573.

269. Menkes, B., Sandor, S., and Ilies, A. (1970). Cell death in teratogeneses. *In* "Fetal Homeostasis" (R. Wynn, ed.), Vol. IV, pp. 170–215. Appleton, New York.

270. Menschaw, G. B. (1934). Fattori letali nel coniglio cincilla. *Riv. Coniglicolt.* 6, 8–9.

271. Mercier Parot, L., and Tuchmann-Duplessis, H. (1967). Obtention de malformations des membres par la 6—mercaptopurine chez trois espèces: lapin, rat et souris. *C. R. Soc. Biol.* 161, 762–768.

272. Millen J. W. (1962). Thalidomide and limb deformities. *Lancet* 2, 599–600.

273. Millen, J. W., and Dickson, A. D. (1957). The effect of vitamin A upon the cerebrospinal fluid pressures of young rabbits suffering from hydroceohalus due to maternal hypovitaminosis A. *Brit. J. Nutr.* 11, 440–446.

274. Millen, J. W., and Woollam, D. H. M. (1956). The effect of duration of vitamin A deficiency in female rabbits upon the incidence of hydrocephalus in their young. *J. Neurol. Neurosurg. Psychiat.* [N.S.] **19**, 17–20.

275. Millen, J. W., Woollam, D. H. M., and Lamming, G. E. (1953). Hydrocephalus associated with deficiency of vitamin A. *Lancet* **2**, 1234–1236.

276. Millen, J. W., Woollam, D. H. M., and Lamming, G. E. (1954). Congenital hydrocephalus due to experimental hypovitaminosis A. *Lancet* **2**, 679–683.

277. Minot, C., and Taylor, F. (1905). Normal plots of the development of the rabbit (*Lepus cuniculus*). *In* "Normentafeln zur Entwicklunggeschichte der Wirkeltiere" (F. Keibel, ed.). Fischer, Jena.

278. Moog, F., and Lutwak-Mann, C. (1958). Observations on rabbit blastocysts prepared as flat mounts. *J. Embryol. Exp. Morphol.* **67**, 57–67.

279. Moore, K. L. (1963). The vulnerable embryo. Causes of malformation in man. *Manitoba Med. Rev.* **43**, 306–319.

280. Morison, J. E. (1963). "Foetal and Neonatal Pathology." Butterworth, London.

281. Morris, J. M., van Wagenen, G., Hurteau, G. D., Johnston, D. W., and Carlsen, R. A. (1967). Compounds interfering with ovum implantation and development. I. Alkaloids and antimetabolites. *Fert. Steril.* **18**, 7–17.

282. Morris, J. M., van Wagenen, G., McCann, T., and Jacob, D. (1967). Compounds interfering with ovum implantation and development, II. Synthetic estrogens and antiestrogens. *Fert. Steril.* **18**, 18–34.

283. Mossman, H. W. (1926). Rabbit placenta and the problem of placental transmission. *Amer. J. Anat.* **37**, 433–497.

284. Mossman, H. W. (1937). Comparative morphogenesis of the fetal membranes and accessory uterine structures. *Contrib. Embryol. Carnegie Inst.* **158**, 133–246.

285. Mroueh, A., and Mastroianni, L., Jr. (1966). Insemination via the intraperitoneal route in rabbits. *Fert. Steril.* **17**, 76–82.

286. Murdoch, R. N., and White, I. G. (1967). The metabolism of labelled glucose by rabbit spermatozoa after incubation *in utero*. *J. Reprod. Fert.* **14**, 213–223.

287. Murphy, M. L. (1960). Teratogenic effects of tumor-inhibiting chemicals in the foetal rat. "*Ciba Found. Symp. on Congenital Malfunction* (G. E. W. Wolstenholme and C. M. O'connor, eds.), pp. 78–107. Little, Brown, Boston.

288. Nachtsheim, H. (1929). Das rexkaninchen und seine genetik. *Z. Vererbungslehre* **52**, 1–52.

289. Nachtsheim, H. (1931). Über eine erbliche Nervenkrankheit (Syringomyelic) beim Kaninchen. *Z. Pelztier Rauchwarenk.* **3**, 254–259.

290. Nachtsheim, H. (1936). Erbliche zahnanomalien bein Kaninchen. *Zuechtungskunde* **11**, 273–287.

291. Nachtsheim, H. (1937). Erbpathologie des Kaninchen. *Erbarzt (Leipzig)* **4**, 25–30, and 50–55.

292. Nachtsheim, H. (1937). Erbpathologische untersuchungen am Kaninchen. *Z. Indukt. Abstamm. Vererbungsl.* **73**, 463–466.

293. Nachtsheim, H. (1939). Erbleiden des nervensystem bei Säugetieren. In "Handbook der Erbiologie des Menschen" (G. Just, ed.), Vol. 5, pp. 1–55. Springer, Berlin.

294. Nachtsheim, H. (1947). Eine erbliche fetale erythroblastose beim tier und ihre Beziehungen zu den gruppenfaktoren des blutes. *Klin. Wochenschr.* **24/25**, 590–592.

295. Nachtsheim, H. (1950). The Pelger-anomaly in man and rabbit. A Mendelian character of the nuclei of the leucocytes. *J. Hered.* **41**, 131–137.

296. Nachtsheim, H. (1958). Erbpathologie der Nagetiere. *In* "Pathologie der Laboratoriumstiere" (P. Cahrs, R. Jaffe, and H. Meessen, eds.), Vol. II, pp. 310–452. Springer-Verlag, Berlin and New York.

297. Nachtsheim, H., and Gürich, H. (1939). Erbleiden des Kaninchenauges. I. Erbliche Nahtbändchentrübung der linse mit nachfolgenden Kernstar. *Z. Konstitutions lehre* **23**, 463–483.

298. Nachtsheim, H., and Klein, H. (1948). Hydrops congenitus universalis bein Kaninchen, eine erbliche fetale erythroblastose. *Abh. Deut. Akad. Wiss. Berlin* No. 5.

299. Naeye, R. L., and Kelly, J. A. (1966). Judgment of fetal age. III. The pathologist's evaluation. *Pediat. Clin. N. Amer.* **13**, 849–862.

300. Napier, R. A. N. (1961). Fertility in the male rabbit. I. Sensitivity of spermatozoa to handling techniques. *J. Reprod. Fert.* **2**, 246–259.

301. Napier, R. A. N. (1963). Rabbits. *In* "Animals for Research" (W. Lane-Petter, ed.), pp. 323–364. Academic Press, New York.

302. National Institutes of Health. (1967). "Status of Research in Pharmacology and Toxicology." Report by the Pharmacology and Toxicology Training Committee, pp. 30–46. Nat. Inst. Health, Washington, D.C.

303. Nelson, N., Coon, J. M., Friedman, L., Gosselin, R. E., Kurin, C. M., Loomis, T. A., Schulick, P., Whittenberger, J. L., and Wilson, J. G. (1970). Food and Drug Administration advisory committee on protocols for safety evaluations. Panel on reproduction report on reproduction studies in safety evaluations of food additives and pesticide residues. *Toxicol. Appl. Pharmacol.* **16**, 264–296.

304. Nishimura, H. M., and Miller, J. R. (1968). "Methods for Teratological studies in Experimental Animals and Man." *Igaku Shoin Ltd.* Medical Examination Publishing Co., Inc., Flushing, New York.

305. Nolen, G. A., and Buehler, E. V. (1971). The effects of disodium etidronate reproductive functions and embryogeny of albino rats and New Zealand rabbits. *Toxicol. Appl. Pharmacol.* **18**, 548–561.

306. Nudleman, K. L., and Travill, A. A. (1971). A morphological and histochemical study of thalidomide-induced upper limb malformations in rabbit fetuses. *Teratology* **4**, 409–426.

307. Otis, E. M., and Brent, R. (1954). Equivalent ages in mouse and human embryos. *Anat. Rec.* **120**, 33–63.

308. Oxford, J. S., and Sutton, R. N. P. (1968). The effect of rubella and herpesvirus hominis on the pre-and post-implantation stages of pregnancy in laboratory animals. *J. Embryol. Exp. Morphol.* **20**, 285–294.

309. Pagenstecher (1916). Strahlwirkung auf das wascheende auge: Experimentelle untersvehungen uhr die entsteburg der netzhaut rossetten. *Versamml. Ophthalmol. Ges.* **40**, 447.

310. Paget, G. E., and Thorpe, E. (1964). A teratogenic effect of a sulphonamide in experimental animals. *Brit. J. Pharmacol.* **23**, 305–312.

311. Palmer, A. K. (1968). Spontaneous malformations of the New Zealand White rabbit: The background to safety evaluation tests. *Lab. Anim.* **2**, 195–206.

312. Palmer, A. K. (1969). The relationship between screening tests for drug safety and other terotological investigations. *In* "Teratology" (A. Bertelli and L. Donati, eds.), pp. 55–72. Excerpta Med. Found., Amsterdam.

313. Palmer, W. M., and Fridhandler, L. (1968). Effects of growth-inhibiting antibiotics on macromolecule biosynthesis in pre-implantation rabbit conceptus. *Fert. Steril.* **19**, 273–285.

314. Paterson, J. S. (1940). A case of naturally occurring listerellosis in an adult rabbit. *J. Pathol. Bacteriol.* **51**, 441–442.

315. Paul, W., Enns, T., Reynolds, S. R. M., and Chinard, F. P. (1946). Sites of water exchange between the maternal system and the amniotic fluid of rabbits. *J. Clin. Invest.* **35**, 634–640.

316. Pauling, L. (1963). Our hope for the future. *In* "Birth Defects" (M. Fishbein, ed.), pp. 164–170. Lippincott, Philadelphia, Pennsylvania.

317. Pearce, L. (1948). Hereditary osteopetrosis of the rabbit. II. X-Ray, hematologic, and chemical observations. *J. Exp. Med.* **88**, 597–620.

318. Pearce, L. (1950). Hereditary osteopetrosis of the rabbit. III. Pathologic observations; skeletal abnormalities. *J. Exp. Med.* **92**, 591–600.

319. Pearce, L. (1950). Hereditary osteopetrosis of the rabbit. IV. Pathologic observations; general features. *J. Exp. Med.* **92**, 601–624.

320. Pearce, L. (1960). Hereditary distal foreleg curvature in the rabbit. I. Manifestations and course of the bowing deformity: Genetic studies. *J. Exp. Med.* **3**, 801–822.

321. Pearce, L. (1960). Hereditary distal foreleg curvature in the rabbit. II. Genetic and pathological aspects. *J. Exp. Med.* **3**, 823–829.

322. Pearce, L., and Brown, W. H. (1945). Hereditary achondroplasia in the rabbit. II. Pathological aspects. *J. Exp. Med.* **82**, 261–280.

323. Pearce, L., and Brown, W. H. (1945). Hereditary achondroplasia in the rabbit. III. Genetic aspects; general considerations. *J. Exp. Med.* **82**, 281–295.

324. Pearce, L., and Brown, W. H. (1948). Hereditary osteopetrosis of the rabbit. I. General features and course of disease; genetic aspects. *J. Exp. Med.* **88**, 579–596.

325. Pearn, J. H., and Vickers, T. H. (1965). The rabbit thalidomide embryopathy. *Brit. J. Exp. Pathol.* **47**, 186–192.

326. Pease, M. (1928). Yellow fat in rabbits. A linked character? *Z. Vererbungslehre, Suppl.* **2**, 1153–1156.

327. Pecile, A., and Finzi, C., ed., (1969). "The Foeto-Placental Unit." Excerpta Med. Found., Amsterdam.

328. Peck, E. D., and Sawin, P. B. (1950). Morphogenetic studies of the rabbit. *J. Exp. Zool.* **114**, 335–357.

329. Peck, H. M., Mattis, P. A., and Zawoiski, E. J. (1965). The evaluation of drugs for their effects on reproduction and fetal development. *Proc. Symp. Drug Induced Dis.* pp. 19–29. Excerpta Med. Found. Int. Cong. Series, p. 85.

330. Peck, H. M. (1963). The preclinical evaluation of drugs for evidence of teratogenic activity. *Amer. Pharm. Ass. Conf.*

331. Peltzer, M. A., and Schardein, J. L. (1966). A convenient method for processing fetuses for skeletal staining. *Stain Technol.* **41**, 300–302.

332. Perey, D. Y. E., and Good, R. A. (1968). Experimental arrest and induction of lymphoid development in intestinal lymphoepithelial tissues of rabbits. *Lab. Invest.* **18**, 15–26.

333. Potter, E. L. (1962). "Pathology of the Fetus and the Infant," 2nd ed. Yearbook Publ., Chicago, Illinois.

334. Prince, J. H. (1964). "The Rabbit in Eye Research." Thomas, Springfield, Illinois.

335. Reynolds, S. R. M. (1962). Maternal blood flow in the uterus and placenta. *In* "Handbook of Physiology" (Amer. Physiol. Soc., J. Field, ed.), Sect. 2, Vol. I. Chapter 45, pp. 1585–1618. Williams and Wilkins, Baltimore, Maryland.

336. Robens, J. F. (1968). Teratogenic effects of Carbaryl and other pesticides in the hamster, the rabbit, and the guinea pig. *Toxicol. Appl. Pharmacol.* **12**, 294.

337. Robens, J. F. (1969). Teratologic studies of Carbaryl, Diazinon, Norea, disulfiram and thiram in small laboratory animals. *Toxicol. Appl. Pharmacol.* **15**, 152–163.

338. Robinson, R. (1958). Genetic studies of the rabbit. *Bibliogr. Genet.* **17**, 229–558.

339. Robson, J. M. (1963). The problem of teratogenicity. *Practioner* **191**, 136–142.

340. Robson, J. M., and Sullivan, F. M. (1963). The production of foetal abnormalities in rabbits by imipramine. *Lancet* **1**, 638–639.

341. Robson, J. M., Sullivan, F. M., and Smith, C. L., eds. (1965). "Embryopathic Activity of Drugs." Little, Brown, Boston, Massachusetts.

342. Rosahn, P. D., and Greene, H. S. M. (1935). Birth weight criterion of dwarfism in the rabbit. *Proc. Soc. Exp. Biol. Med.* **32**, 1580–1583.

343. Rosselli, G. S. (1969). A hundred years and more of experimental teratology. *In* "Teratology" (A. Bertelli and L. Donati, eds.), pp. 1—7. Excerpta Med. Found., Amsterdam.

344. Roux, C., Aubry, M. M., and Dupuis, R. (1969). Action tetratogène d'un inhibiteur de la synthèse du cholesterol, le AY 9944, sur différentes aspèces animales. *C. R. Soc. Biol.* **163**, 327–332.

345. Roux, C., Cahen, R., and Dupuis, R. (1965). Visceral malformations caused by thalidomide in the rabbit. *C.R. Soc. Biol.* **159**, 1059–1063.

346. Roux, C., Emerit, I., and Taillemite, J. L. (1971). Chromosomal breakage and teratogenesis. *Teratology* **4**, 303–316.

347. Rubin, A. (1967). "Handbook of Congenital Malformations." Saunders, Philadelphia, Pennsylvania.

348. Ruffalo, P. R., and Ferm, V. H. (1965). The embryocidal and teratogenic effects of 5-bromodeoxyuridine in the pregnant hamster. *Lab. Invest.* **14**, 1547–1553.

349. Rugh, R. (1964). "Vertebrate Embryology." Harcourt, New York.

350. Rugh, R. (1968). "The Mouse, its Reproduction and Development." Burgess, Minneapolis, Minnesota.

351. Runner, M. N. (1954). Inheritance of susceptibility to congenital deformity—embryonic instability. *J. Nat. Cancer Inst.* **151**, 637–649.

352. Saunders, F. J., and Elto, R. L. (1967). Effects of ethynnodiol diacetate and mestranol in rats and rabbits, on conception, on the outcome of pregnancy and on the offspring. *Toxicol. Appl. Pharmacol.* **11**, 229–244.

353. Sawin, P. B. (1937). Preliminary studies of hereditary variation in the axial skeleton of the rabbit. *Anat. Rec.* **69**, 407–427.

354. Sawin, P. B. (1945). Morphogenetic studies of the rabbit. I. Regional specificity of hereditary factors affecting homoeotic variations in the axial skeleton. *J. Exp. Zool.* **100**, 301–329.

355. Sawin, P. B. (1946). Morphogenetic studies of the rabbit. III. Skeletal variations resulting from the interaction of gene determined growth forces. *Anat. Rec.* **96**, 183–200.

356. Sawin, P. B. (1955). Recent genetics of the domestic rabbit. *Advan. Genet.* **7**, 183–226.

357. Sawin, P. B., and Crary, D. D. (1951). Morphogenetic variations in the rabbit. X. Racial variations in the gall bladder. *Anat. Rec.* **110**, 573–590.

358. Sawin, P. B., and Crary, D. D. (1955). Hereditary scoliosis in the rabbit. *Anat Rec.* **121**, 449–450.

359. Sawin, P. B., and Crary, D. D. (1964). Genetics of skeletal deformities in the domestic rabbit (*Orytolagus cuniculus*). *Clin. Orthop. Relat. Res.* **33**, 71–90.

360. Sawin, P. B., Crary, D. D., Fox, R. R., and Wuest, H. M. (1965). Thalidomide malformation and genetic background in the rabbit. *Experientia* **21**, 672–678.

361. Sawin, P. B., and Edmonds, H. W. (1949). Morphogenetic studies of the rabbit. VII. Aortic arch variations in relation to regionally specific growth differences. *Anat. Rec.* **105**, 377–397.

362. Sawin, P. B., and Gadbois, D. S. (1947). Genetic influences upon the sex ratio on the rabbit. *Genetics* **32**, 286–302.

363. Sawin, P. B., and Hull, I. B. (1946). Morphogenetic studies of the rabbit. II. Evidence of regionally specific hereditary factors influencing the extent of the lumbar region. *J. Morphol.* **78**, 1–26.

364. Sawin, P. B., and Nace, M. A. G. (1948). Morphogenetic studies of the rabbit. V. Inheritance of an asymmetrical vascular pattern. *J. Morphol.* **82**, 331–354.

365. Schardein, J. L., Blatz, A. T., Woosley, E. T., and Kaump, D. H. (1969). Reproduction studies on sodium meclofenamate in comparison to aspirin and phenylbutazone. *Toxicol. Appl. Pharmacol.* **15**, 46–55.

366. Schardein, J. L., Hentz, D. L., Petrere, J. A., and Kurtz, S. M. (1971). Teratogensis studies with diphenhydramine HCl. *Toxicol. Appl. Pharmacol.* **18**, 971–976.

367. Schardein, J. L., Woosley, E. T., Peltzer, M. A., and Kaump, D. H. (1967). Congenital malformations induced by 6-aminonicotinamide in rabbit kits. *Exp. Mol. Pathol.* **6**, 335–346.

368. Schneider, M., and Adar, U. (1964). Effect of inanition of rabbit growth cartilage plates. *Arch. Pathol.* **78**, 149–156.

369. Schnell, V., and Newberne, J. W. (1965). Improved method for processing gross fetal sections of rabbits and rats. *Amer. J. Clin. Pathol.* **44**, 702–703.

370. Schnell, V., and Newberne, J. W. (1970). Accelerated clearing and staining of teratologic specimens by heat and light. *Teratology* **3**, 345.

371. Schultz, O. (1897). Über Herstellung und Conservirung durchsightigen Embryonen zum Stadium der Skeletbildung. *Verh. Anat. Ges. Jena, Anat. Anz.* **13**, 3–5.

372. Schumacher, H., Blake, D. A., Gurian, J. M., and Gillette, J. R. (1968). A comparison of the teratogenic activity of thalidomide in rabbits and rats. *J. Pharmacol. Exp. Ther.* **160**, 189–200.

373. Setnikar, I., and Magistretti, M. J. (1970). Maternal and fetal toxicity of moquizone. *Arzneim.-Forsch.* **20**, 1559–1561.

374. Shelley, H. J. (1961). Glycogen reserves and their changes at birth and in anoxia. *Brit. Med. Bull.* **17**, 137–143.

375. Shepard, T. H., Fry, L. R., and Moffett, B. C. (1969). Microscopic studies of achondroplastic rabbit cartilage. *Teratology* **2**, 13–22.

376. Sissman, N. J. (1970). Development landmarks in cardiac morphogenesis: Comparative chronology. *Amer. J. Cardiol.* **25**, 141–148.

377. Sittmann, D. B., Rollins, W. C., Sittmann, K., and Casady, R. B. (1964). Seasonal variation in reproductive traits of New Zealand White rabbits. *J. Reprod. Fert.* **8**, 29–37.

378. Sledge, C. B. (1966). Some morphologic and experimental aspects of limb development. *Clin. Orthop.* **44**, 241–264.

379. Somers, G. F. (1962). Thalidomide and congenital abnormalities. *Lancet* **1**, 912.

380. Spencer, R. P., and Coulombe, M. J. (1964). Observation on fetal weight and gestational age. *Growth* **28**, 243–247.

381. Staemmler, M., Helm, F., and Kiel, H. (1964). Congenital malformations of skeleton in rabbits. *Med. Exp.* **10**, 22–26.

382. Staples, R. E., Holtkamp, D. E., and Dorsey, E. (1966). Influence of body weight upon corpus luteum formation and maintenance of pregnancy in the rabbit. *J. Reprod. Fert.* **12**, 221–224.

383. Staples, R. E., Holtkamp, D. E., and Warkany, J. (1963). Effect of parental treatment with thalidomide on fetal development of rats and rabbits. *3rd Annu. Meet. Teratol. Soc.*

384. Staples, R. E., and Schnell, V. L. (1964). Refinements in rapid clearing technique in the KOH-alizarin red S method for fetal bone. *Stain Technol.* **39**, 62–63.

385. Sterling, T. D. (1971). Difficulty in evaluating toxicity and teratogenicity of 2, 4, 5,-T from existing animal experiments. *Science* **174**, 1358–1359.

386. Streeter, G. (1942). Developmental horizons in human embryos. Description of age group XI, 13 to 20 somites, and age group XII, 21 to 29 somites. *Contrib. Embryol. Carnegie. Inst.* **30**, 211–245.

387. Streeter, G. (1945). Developmental horizons in human embryos. Description of age group XIII, embryos about 4 or 5 millimeters long, and age group XIV, period of indentation of the lens vesicle. *Contrib. Embryol. Carnegie Inst.* **31**, 27–63.

388. Streeter, G. (1948). Developmental horizons in human embryos. Sescription of age groups XV, XVI, XVII, and XVIII, being the third issue of a survey of the Carnegie collection. *Contrib. Embryol. Carnegie Inst.* **32**, 133–203.

389. Strong, R. M. (1925). The order, time and rate of ossification of the albino rat skeleton. *Amer. J. Anat.* **36**, 313–355.

390. Stupka, W. (1931). Uber die Bauverhaltmisse des Gehirns eiver Zyklpischen ziege. *Arb. Neurol. Inst. (Inst. Anat. Physiol. Zentralnerrensyst.) Univ. Wien.* **33**, 315–394.

391. Suchalla, H. (1943). Variabilitat und Erblichkeit von Schaldelmerkmalen bei Zwerg-und Riesenrassen; dargestellt an Hermelin-und Widderkaninchen. *Z. Morphol. Anthropol.* **15**, 274–333.

392. Suitor, A. E. (1946). Palpating domestic rabbits to determine pregnancy. *U.S., Dep. Agri. Bull.* **245**.

393. Szabo, K. T., and Kang, J. Y. (1969). Comparative teratogenic studies with various theraputic agents in mice and rabbits. *Teratology* **2**, 270.

394. Takano, K., Yamamura, H., and Suzuki, M. (1966). Teratogenic effect of chlormadinone acetate in mice and rabbits. *Proc. Soc. Exp. Biol. Med.* **121**, 455–457.

395. Takekoshi, S. (1965). Effects of Urethan on the teratogenic action of hypervitaminosis A. *Gumma J. Med. Sci.* **14**, 210–212.

396. Ten Cate, G. (1962). "Teratology," Abstracts and Titles of Papers on Congenital Malformations collected from the 1962 Medical Literature. Excerpta Med. Found., Amsterdam.

397. Tesh, J. M. (1969). Effects of ageing of rabbit spermatozoa *in utero* on fertilization and prenatal development. *J. Reprod. Fert.* **20**, 299–306.

398. Tesh, J. M. (1971). Teratogenic studies and artificial insemination: Effects of semen storage on fertility and pre-natal development in the rabbit. *Proc. Eur. Soc. Study Drug Toxicity* **12**, 337–341.

399. Tesh, S. A., and Tesh, J. M. (1971). Artificial insemination in the rabbit and its use in routine teratogenic studies. *Proc. Eur. Soc. Study Drug Toxicity* **12**, 332–336.

400. Thompson, E. T., and Hayden, A. C. (1961). "Standard Nomenclature of Diseases, and Operations." McGraw-Hill, New York.

401. Tuchmann-Duplessis, H. (1965). Design and interpretation of teratogenic tests *In* "Embryopathic Activity of Drugs" (J. Robson, F. Sullivan, and R. Smith, eds.), pp. 56–87. Little, Brown, Boston, Massachusetts.

402. Tuchmann-Duplessis, H., and Mercier-Parot, L. (1964). Abortions and malformations due to the effect of an agent inducing hyperlipemia and hypercholesterolemia. *Bull. Acad. Nat. Med., Paris* **148**, 392–398.

403. Tuchmann-Duplessis, H., and Mercier-Parot, L. (1964). Repercussions des neuroleptiques et des antitumoraux sur le developpement prenatal. *Bull. Schweiz. Akad. Med. Wiss.* **20**, 490–526.

404. Tuchmann-Duplessis, H., and Mercier-Parot, L. (1964). Apropos of teratogenic tests. Spontaneous malformations in rabbits. *C. R. Soc. Biol.* **158**, 666–670.

405. Vickers, T. H. (1967). The thalidomide embryopathy in hybrid rabbits. *Brit. J. Exp. Pathol.* **48**, 107–117.

406. Vickers, T. H. (1967). Concerning the morphogenesis of thalidomide dysmelia in rabbits. *Brit. J. Exp. Pathol.* **48**, 579–591.

407. Vickers, T. H. (1968). The cardiovascular malformations in the rabbit thalidomide embryopathy. *Brit. J. Exp. Pathol.* **49**, 179–196.

408. Vickers, T. H. (1970). The alimentary tract malformations in the rabbit thalidomide embryopathy. *Brit. J. Exp. Pathol.* **51**, 286–297.

409. Vogin, E. E., Goldhamer, R. E., and Carson, S. (1968). Teratologic study in rats and rabbits exposed to an isoproterenol aersol nibair. *Toxicol. Appl. Pharmacol.* **12**, 294–295.

410. Vogin, E. E., Goldhamer, R. E., Scheimberg, J., and Carson, S. (1970). Teratology studies in rats and rabbits exposed to an isoproterenol aresol. *Toxicol. Appl. Pharmacol.* **16**, 374–381.

411. Vogt, A. (1919). Vererbter hydrophthalmus beim Kanichen. *Klin. Monatsbl. Augenheilk.* **63**, 233.

412. von Kobyletzki, D. (1971). New aspects for teratology research: Pharmacokinetics, prospective information and legalized abortions. *Proc. Eur. Soc. Study Drug Toxicity* **12**, 342–346.

413. Walker, B. E. (1967). Induction of cleft palate in rabbits by several glucocorticoids. *Proc. Soc. Exp. Biol. Med.* **125**, 1281–1284.

414. Walker, D. G., and Wirtschater, Z. T. (1957). "The Genesis of the Rat Skeleton—A Laboratory Atlas." Thomas, Springfield, Illinois.

415. Ward, C. O. (1968). Screening methods in teratology. *Drug. Intel.* **2**, 324–327.

416. Ward Orsini, M. (1962). Study of ovo-implantation in the hamster, rat, mouse, guinea-pig and rabbit in cleared uterine tracts. *J. Reprod. Fert.* **3**, 288–293.

417. Warkany, J. (1963). Pathology and experimental teratology. *Arch. Pathol.* **75**, 579–581.
418. Warkany, J. (1963). Birth defects through the ages. *In* "Birth Defects" (M. Fishbein, ed.), pp. 18–24. Lippincott, Philadelphia, Pennsylvania.
419. Warkany, J., Chairman, (1963). Report: Conference on Prenatal Effects of Drugs. Commission on Drug Safety, Chicago, Illinois.
420. Warkany, J. (1970). Trends in teratologic research. Epilogue to the third international conference on cogenital malformations. *Teratology* **3**, 89–92.
421. Warkany, J., and Kalter, H. (1961). Congenital malformations. *N. Engl. J. Med.* **265**, 993–1001.
422. Wellmann, K. F., Volk, B. W., Lazarus, S. S., and Brancato, P. (1969). Pancreatic B cell morphology and insulin content of normal and alloxan-diabetic rabbits and their offspring. *Diabetes* **18**, 138–145.
423. Wells, A. H. (1965). "Systematized Nomenclature of Pathology." College of American Pathologists, Chicago, Illinois.
424. White, I. G. (1955). The collection of rabbit semen. *Aust. J. Exp. Biol.* **33**, 367–370.
425. Wier, K. (1959). Effect on the weight of fetuses and fetal lymphoid organs of adrenaline given to rabbit at a critical period of pregnancy: Observations on spontaneous and induced runting. *Anat. Rec.* **153**, 373–376.
426. Willis, R. A. (1962). "The Borderland of Embryology and Pathology," 2nd ed. Butterworth, London.
427. Wilson, J. G. (1959). Experimental studies on congenital malformations. *J. Chronic Dis.* **10**, 111–130.
428. Wilson, J. G., Tuchman-Duplessis H., and Peck, H. M. (1967). Principles for the testing of drugs for teratogenicity. *World Health Organ., Tech. Rep. Ser.* **364**, 5–18.
429. Wilson, J. G., and Warkany, J. (1965). "Teratology, Principles and Techniques." Univ. of Chicago Press, Chicago, Illinois.
430. Wing, F. (1945). The gestation period of the rabbit and factors associated with it. Master's Thesis, Brown University.
431. Witschi, E. (1956). "Development of Vertebrates." Saunders, Philadelphia, Pennsylvania.
432. Witschi, E. (1962). Growth including reproduction and development. *In* "Biological Handbooks." Fed. Amer. Soc. Exp. Biol., Washington, D.C.
433. Wolstenholme, G. E. W., and O'Connor, C. M., eds. (1960). "Congenital Malformations," Ciba Found. Symp. Little, Brown, Boston, Massachusetts.
434. Woodard, M. W., Woodard, G., Hollingsworth, R. L., and Scott, R. L. (1971). Fetal rabbit ductus arteriosus assessed in a teratological study on Isopreterenol and Metaproterenol. *Toxicol. Appl. Pharmacol.* **18**, 231–234.
435. Woollam, D. H., ed. (1966–1969). "Advances in Teratology," vols. 1–4. Academic Press, New York.
436. Wright, H. V. (1958). Prenatal development of skeleton in Long Evans rats *Anat. Rec.* **130**, 659–670.
437. Wright, S. W. (1963). The borderline between normal and abnormal. *In* "Birth Defects" (M. Fishbein, ed.), pp. 211–218. Lippincott, Philadelphia, Pennsylvania.
438. Wuest, H. M., and Fox, R. R. (1968). The relationship between teratogeny and structure in the thalidomide field. *Experientia* **24**, 993–994.
439. Wuest, H. M., Fox, R. R., and Crary, D. D. (1969). Thalidomide: Lack of teratogenic action of N-carbethoxyphthalimide (WU 374) in New Zealand rabbit. *Teratology* **2**, 273.
440. Wynn, R. M., ed. (1965). *"Fetal Homeostasis,"* Vol. 1. N. Y. Acad. Sci. Interdisciplinary Commun. Program, New York.
441. Wynn, R. M., ed. (1967). *"Fetal Homeostasis,"* Vol. 2. N. Y. Acad. Sci. Interdisciplinary Commun. Program, New York.
442. Wynn, R. M., ed. (1968). *"Fetal Homeostasis."* Vol. 3. N. Y. Acad. Sci. Interdisciplinary Commun. Program, New York.
443. Yeary, R. A., (1964). Teratogenic agents in man and in animals. *Lancet,* **1**, 831.
444. Zamboni, L., and Mastroianni, L., Jr. (1966). Electron microscopic studies on rabbit ova. I. The follicular oocyte. *J. Ultastruct. Res.* **14**, 95–117.

CHAPTER 6

Specialized Research Applications:
I. Arteriosclerosis Research

Thomas B. Clarkson, Noel D. M. Lehner, and Bill C. Bullock

I. INTRODUCTION AND SCOPE OF THE REVIEW

Because atherosclerosis of human beings is so prevalent and its clinical consequences so serious there has been a major research effort to better understand its pathogenesis and thereby provide a more rational approach to prophylaxis and therapy. The kinds of experimental approaches that can be made on human patients are limited because of difficulty in control, the slow rate of lesion development, and the inability to make quantitative antemortem determinations of the extent and severity of atherosclerotic lesions. For these reasons characterization of appropriate animal models has been an important segment of atherosclerosis research. Studies on the pathogenesis of atherosclerosis with use of animal models have progressed remarkably and the relevancy of observations made on animal models to the human lesions is becoming of increasing interest to research workers.

The purpose of this communication is to familiarize the reader with some of the characteristics of rabbits as models of atherosclerosis. Our review of the literature is by no means complete, rather we have elected to refer to published observations that might influence the research worker's selection of the rabbit as a model and the methods he might select for its study.

II. HISTORY OF THE USE OF RABBITS IN ARTERIOSCLEROSIS RESEARCH

Rabbits were the first animal model of atherosclerosis and have been used experimentally in such studies for 65 years. The earliest investigations of experimentally induced atherosclerosis of rabbits were those of Ignatowski (36) who found that diets of milk, meat, and eggs were capable of producing lipid-containing intimal lesions of arteries having some resemblance to fatty streaks of human beings. Ignatowski erroneously believed the lesions he saw were due to the high level of animal protein in the rabbits' diets. Later studies by a number of workers demonstrated that the atherogenic effect of meat, milk, and eggs was in the fat portion of the diet (3–5). It is now generally accepted that the atherogenic component of the diet was the cholesterol.

Until the mid-1950's rabbits were the most frequently used animal model for atherosclerosis research; however, in recent times they have been viewed by comparative pathologists with some disfavor. The basis for the disfavor stems from the fact that rabbits are by nature herbivous and have a quite different whole-body cholesterol metabolism from man. In addition, the pathological characteristics of the lesions are dissimilar to those seen in man. While the most extensive aortic atherosclerosis of man is seen in the abdominal aorta, the disease in rabbits has a striking predilection for the aortic arch and the thoracic aorta. Microscopically, the lesions of rabbits resemble thick fatty streaks of man, rarely developing into complicated lesions with fibrosis, ulceration, and thrombosis. The coronary artery lesions of rabbits occur primarily in the small intramyocardial branches with lesions of the large proximal arteries being rare.

Most regimens to induce atherosclerosis in the rabbit include feeding a diet high in fat and cholesterol; this causes extreme serum cholesterol levels in the rabbit, levels almost never seen in human beings. Associated with the extreme hypercholesterolemia are lipid storage lesions of all the visceral organs resulting in a syndrome in the rabbit which has more the appearance of a lipid storage disease than atherosclerosis as it is seen in higher primates.

III. NATURALLY OCCURRING ARTERIOSCLEROSIS OF RABBITS

The naturally occurring arterial lesions of rabbits are described in detail elsewhere in this volume (Chapter 15) and will not be discussed in depth here. Investigators using rabbits should be aware of these predominantly medial lesions and that the prevalence of these lesions varies considerably among rabbits from different sources (60, 61). Zeek (67) was able to selectively breed rabbits for presence or absence of these arterial lesions. Bragdon (10) has described what appears to be a milk factor that affects the prevalence of aortic lesions in young rabbits. There is a tendency for investigators to order "New Zealand White" or "Dutch Belted" rabbits from the most convenient source. If continuing projects are planned some careful background work on the prevalence of "spontaneous" lesions and the response of animals to the proposed experimental manipulation is indicated.

IV. DIET-INDUCED ATHEROSCLEROSIS OF RABBITS

A. Composition of Diets

The usual dietary regimen for the induction of atherosclerosis in the rabbit is a diet of 1–3% cholesterol and 4–8% fat added to commercial rabbit chow. The relationship between the amount of cholesterol fed and the severity of induced atherosclerosis is obscure. Scebat et al. (59) reported that when rabbits were fed 0.25 gm cholesterol per day the amount of atherosclerosis that developed was practically as severe as when 1 gm of cholesterol per day was fed, even though the level of serum cholesterol was only half as high. The lack of relationship between cholesterol dose and the severity of disease points up some very important questions about the process.

The amount of added fat which is coadministered with the cholesterol and whether the cholesterol is dissolved or suspended in the fat is of importance in determining the amount of disease (Table I). Feeding rabbits cholesterol without added fat will usually cause atherosclerosis which is more severe than that of rabbits fed diets containing cholesterol plus fat (37–39). This phenomenon is thought to be due to the mobilization of endogenous fat. Endogenous fat is mobilized when cholesterol is fed without added dietary fat, and is more saturated than the fat usually incorporated in the diatary regimen.

The type of fat which is fed along with the cholesterol supplement is a determinant in the atherogenicity of the dietary cholesterol (Table II). In general, this is related to the saturation of the fat and is said to vary inversely with the iodine number. The principal work with rabbits concerning the saturation of the fat supplement has been done by Kritchevsky et al. (37, 40, 41) and Vles et al. (63).

Whether or not the fat is heated prior to its inclusion in the diet is also a variable in determining the amount of atherosclerosis produced in the rabbit (Table III). Kritchevsky et al. (43, 44) and Kritchevsky (37) have shown that most edible fats are hydrolyzed and release fatty acids

TABLE I
EFFECT OF THE CHOLESTEROL VEHICLE IN DIET-INDUCED ATHEROSCLEROSIS OF RABBITS[a]

Dietary fat and cholesterol	Serum total cholesterol (mg/100 ml)	Extent of atherosclerosis[b]
None + 2% cholesterol	1640	3.05
Corn oil (6%) + 2% cholesterol-dissolved	2262	2.35
Corn oil (6%) + 2% cholesterol-suspended	2100	1.78

[a] After Kritchevsky (37).
[b] Expressed as the mean gross grade on a scale of 0 being none and 5 being the most severe, thoracic aorta only.

TABLE II
THE EFFECT OF THE TYPE OF DIETARY FAT ON DIET-INDUCED ATHEROSCLEROSIS OF RABBITS[a]

Diet	Serum total cholesterol (mg/100 ml)	Extent of atherosclerosis[b]
No fat + 2% cholesterol	1214	2.3
Coconut oil + 2% cholesterol	2827	2.5
Lard + 2% cholesterol	2245	2.0
Corn oil + 2% cholesterol	1908	1.3
Hydrogenated corn oil + 2% cholesterol	1984	1.7

[a] After Kritchevsky (37).
[b] Expressed as the mean gross grade on a scale of 0 being none and 5 being the most severe, thoracic aorta only.

TABLE III
EFFECT OF PREHEATING OF DIETARY FAT ON DIET-INDUCED ATHEROSCLEROSIS OF RABBITS[a]

Diet	Serum total cholesterol (mg/100 ml)	Extent of atherosclerosis[b]
Cholesterol 2% + corn oil-suspension	2203	1.2
Cholesterol 2% + corn oil-suspension preheated	3534	2.1

[a] After Kritchevsky (37).
[b] Expressed as the mean gross grade on a scale of 0 being none and 5 being the most severe, thoracic aorta only.

to some extent when heated in the air. The addition of these fatty acids enhances the atherogenic effect of a cholesterol fat diet.

The source of dietary carbohydrate has been shown to affect the severity of atherosclerosis of rabbits on cholesterol-fat diets. Cholesterol-lactose-fed rabbits develop significantly more severe atherosclerosis than cholesterol-sucrose-fed rabbits (64, 65), for example.

The type and amount of protein in the cholesterol-fat diet seems to represent still another variable in determining the amount of atherosclerosis that will be induced, however, some carefully done studies have failed to show an effect (48). In general, it could be stated that in cholesterol-induced atherosclerosis of the rabbit, cholesterol and fat are more atherogenic when the level of protein in the diet is low than when it is high, and less atherogenic when coadministered with protein of vegetable origin than with protein animal origin. A number of studies seem to establish that casein is itself an atherogenic protein and is additive to the atherogenic effect of the cholesterol in the diet.

Perhaps the most useful characteristic of rabbits as animal models of atherosclerosis is their response to dietary saturated fat added to synthetic diets. Donomae et al. (20) reported on the production of severe coronary artery disease in rabbits by feeding lanolin added to okara. Okara is the refuse left in making soybean curd (contains about 85% protein and 2% fat). Each rabbit was given the basic okara diet each day to which 5–7 gm of an equal parts mixture of lanolin and cottonseed oil had been added. Lanolin contains 15% mixed sterols about two-thirds of which is cholesterol. The rabbits developed extensive aortic and coronary artery atherosclerosis. As compared with the usual cholesterol-induced atherosclerosis the lesions of the coronary arteries occurred more often in the larger branches, were more fibrotic, and were associated with myocardial infarction.

Stormby and Wigand (62) extended these observations and fed adult male rabbits for seven months a semisynthetic diet without added cholesterol but containing 8% hydrogenated coconut oil. The serum total cholesterol concentration of these animals ranged from 435–1310 mg/100 ml. The lesions in these animals resembled those of human beings and were quite severe. Some investigators have speculated that endogenously derived cholesterol is less atherogenic than that of exogenous origin. The coconut oil-induced lesions provide clear evidence that this is not true.

A somewhat surprising finding has been that when the saturated fats are added to commercial rabbit chow rather than to a synthetic or semisynthetic diet, hypercholesterolemia and atherosclerosis do not develop (42). These same authors have investigated the mechanism of this "protective" action of rabbit chow. In their experiments they attempted to find out if the protective material in the rabbit chow was in the lipid or in some nonlipid fraction. Their conclusion was that complete rabbit chow was required to overcome the hypercholesterolemia and atherogenic effects of the hydrogenated coconut oil. Numerous other species exhibit the "chow effect" when fed cholesterol-containing diets and this phenomenon deserves further study.

Cookson et al. (14) have shown that the hypercholesterolemia which follows the oral administration of cholesterol to rabbits can be prevented by feeding a diet consisting mainly of alfalfa. Horlick *et al.* (34) provided evidence that the alfalfa effect was due to interference with the absorption of cholesterol from the gastrointestinal tract. Later, Cookson and Feddoroff (15) reported on the results of experiments intended to determine the relationship between the amount of cholesterol administered and the amount of alfalfa necessary to prevent hypercholesterolemia. They found that the ingestion of approximately 300 gm of alfalfa per week was adequate to prevent hypercholesterolemia in rabbits fed up to 0.6 gm of cholesterol per day. From the data derived in this experiment the authors felt that the main effect of the alfalfa was an interference in cholesterol absorption. This interference with absorption was not due to β-sitosterol.

B. Selection of Animals

The effect of age on the susceptibility of rabbits to diet-induced atherosclerosis is not clear. Harman (31) has indicated that young rabbits are more susceptible to cholesterol-induced atherosclerosis than are older rabbits. In contrast, however, Pollak (54) provided data which indicated that one-year-old rabbits were more susceptible to diet-induced atherosclerosis than were young animals of weaning age. We would conclude that the predominant evidence suggests that older animals are more susceptible than weaning age animals.

The sex of the rabbits may be another variable affecting the results of experiments on atherosclerosis. Female rabbits have higher serum cholesterol concentrations than do males and their hypercholesterolemic response to cholesterol-containing diets is more pronounced (25, 56). One group reported that female rabbits are more susceptible to diet-induced atherosclerosis than are males, although a significant sex difference has not been observed by all investigators (56).

No data are available on breed differences in susceptibility to diet-induced atherosclerosis. It should be restated, however, that those breeds with the highest frequency of naturally occurring medial lesions tend to have somewhat more atherosclerosis. The size of the animal should be considered, however, when selecting the breed of the rabbit for atherosclerosis experiments. If an investigator requires large samples of blood, for example, for lipoprotein studies, a large breed should be selected (see Table III, Chapter 1). One of the largest commonly available breeds is the Flemish which have adult body weights of 10 to 14 lb. The next largest breed is the New Zealand White rabbit, adults of which weigh 9–12 lb. If, on the other hand, an investigator is feeding a very expensive diet or evaluating a drug that is in short supply he may wish to use a small breed. The smallest breeds that are commonly available in the United States are the Polish (2–3 lb) and the Dutch (4–5 lb).

C. Plasma Cholesterol Transport in the Rabbit

All major lipids in the plasma, including cholesterol, are made soluble in this aqueous medium by being complexed with proteins. These giant molecules called lipoproteins have been investigated by various means including ultracentrifugation, electrophoresis, chromatography, immunology, and chemical composition. These methods examine different aspects of the macromolecular structure of these molecules and have resulted in various classifications of these substances. The atherogenicity of blood lipids appears not only to be a function of the kinds of lipid and their concentrations, but also the kinds of lipoprotein complexes in which they are transported. Abnormalities in the concentrations of these substances have therefore become considered as lipoprotein disorders and not necessarily just lipid abnormalities (9). At least five hyperlipoproteinemic disorders have been identified in man (26).

Early studies using rabbits contributed to the evolving hypothesis on lipoprotein abnormalities and atherosclerosis. Low density serum lipoproteins of rabbits fed laboratory chow diets when analyzed by analytical ultra-centrifugation appear as a single component with flotation rates of 5 to 8 Svedberg units (S_f)(1 S_f = 10^{-13} cm/sec/dyne/gm). These lipoproteins contain about 30% cholesterol and have a hydrated density of 1.03 gm/ml. When rabbits are fed cholesterol the initial increase in the serum cholesterol concentration is in this previously existing lipoprotein spectrum and this fraction may increase as much as fourfold. Further increases in the serum cholesterol concentration are associated with the appearance of a new series of cholesterol-containing lipoproteins, most of which have flotation rates of S_f 10–30 and hydrated densities of 1.01 gm/ml or less. Lipoproteins with S_f values greater than 8 do not appear in the serum until the cholesterol concentration reaches 200–250 mg/dl (29).

The type of hyperlipoproteinemia induced in cholesterol-fed rabbits can be altered experimentally and may have marked effects on the development of atherosclerosis. Cholesterol-fed alloxan diabetic rabbits develop extreme serum cholesterol concentrations of 2000 mg/dl or greater. A great proportion of the serum cholesterol of these rabbits is present in lipoproteins with flotation rates of S_f 100 or greater and they appear to have a metabolic block in the conversion of these lipoproteins to those of the S_f 10–30 class (52). In spite of the marked hypercholesterolemia, the alloxan diabetic rabbits do not develop much atherosclerosis (22, 45). In contrast, normal rabbits fed cholesterol have lower serum cholesterol concentrations, mostly in lipoproteins with flotation rates of S_f 10–30 and become severely atherosclerotic. It is these lipoproteins with flotation rates of S_f 10–30 which have been found to be significantly and positively correlated with the development of atherosclerosis in the rabbit (29).

D. Cholesterol Metabolism in the Rabbit

Homeostasis of body cholesterol is dependent on a balance between cholesterol intake and cholesterol outgo. The regulation of body cholesterol metabolism is a complicated interrelation of cholesterol absorption, synthesis, degradation, and excretion (19). Rabbits are particularly sensitive to dietary cholesterol and develop extreme serum concentrations when fed this substance. The marked hypercholesterolemia induced in rabbits by cholesterol feeding results from the absorption of great quantities of dietary cholesterol without compensatory increases in cholesterol degradation and excretion.

Noncholesterol-fed rabbits synthesize and excrete about 100 mg of cholesterol per day and maintain serum cholesterol concentrations of 100 mg/dl or less (32, 47). It is estimated that rabbits fed 1.6% cholesterol absorb 77% of that ingested, or about 0.25 gm/kg of body weight/day (13). The liver and intestine are the two principal sources of endogenously synthesized cholesterol and hepatic cholesterol synthesis is regulated by a negative feedback mechanism to absorbed dietary cholesterol. Even though hepatic cholesterol synthesis is inhibited in cholesterol-fed rabbits (30), this is not sufficient to compensate for the massive influx of absorbed dietary cholesterol.

The absorbed dietary cholesterol may be excreted as such, be degraded to bile acids and be excreted, or enter the miscible pools of body cholesterol. Little is known of the effects of cholesterol feeding on cholesterol excretion per se in the rabbit, but the excretion of bile acids, the major catabolite of body cholesterol, is not increased (32, 47). As little as 0.2 gm of cholesterol/kg of body weight given as a single oral does may induce a 2- to 3-fold increase in serum cholesterol concentration of rabbits. Similarly, feeding 2% cholesterol may induce a hypercholesterolemia of 650 mg/dl within 72 hours, while feeding cholesterol for extended periods results in hypercholesterolemia up to 2000 mg/dl (27). It is apparent that the rabbit is unable to degrade or excrete sufficient amounts of cholesterol to compensate for that which is absorbed. With the development of marked hypercholesterolemia, cholesterol is deposited in many organs and tissues and in several months takes on the appearance of a lipid storage disease (55).

E. Lesions of Diet-Induced Atherosclerosis

The lesions produced in the cardiovascular system of rabbits by cholesterol-fat feeding have been well described

(21, 23, 24). In contrast to man, the primary site of aortic lesion formation is in the thoracic aortic arch rather than in the abdominal aorta. After a one-month period of cholesterol-fat feeding, aortic lesions become microscopically demonstrable and are characterized by the accumulation of foam cells resting on the internal elastic membrane underneath an intact endothelial surface. By two months of feeding, the lesions progress from fatty streaks to gross yellow plaques that are raised above the intimal surface. These lesions are located principally in the aortic arch and at the origins of the intercostal arteries. Microscopically, the lesion progresses to a point where there is both intra- and extracellular lipid and slight "degenerative" changes in the adjacent media. After three to four months, there is a progressive but small increase in the fibrous and smooth muscle elements of the plaque. With continuous feeding of cholesterol and fat for periods up to one year, there are usually no complications such as calcification, ulceration, and thrombus formation. Occasionally there will be calcification in a necrotic area of an intimal plaque (Fig. 1), however, medial calcification, not necessarily related to a plaque is more common.

Lesions occur in the coronary arteries of the cholesterol-

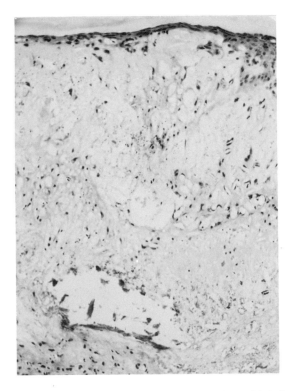

Fig. 1. A lesion from the thoracic aorta of a rabbit fed 0.5% cholesterol in a diet containing 10% lard for about six months. The thickened intima contained foam cells and extracellular lipid. A portion of the media is visible at the lower left corner. There is a thin fibromuscular "cap" at upper right and a calcified area of necrotic material just above the media. All photomicrographs are of paraffin sections stained with hematoxylin and eosin. ×140.

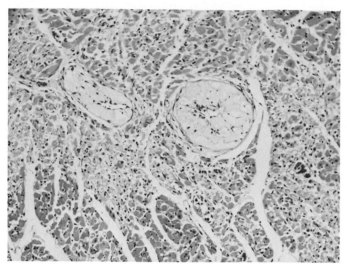

Fig. 2. Typical coronary artery lesions from a cholesterol-fat-fed rabbit. The small intramyocardial arteries are nearly occluded by foam cells. The original media is not discernible at this magnification. There is some patchy myocytolysis in the adjacent myocardium. ×126.

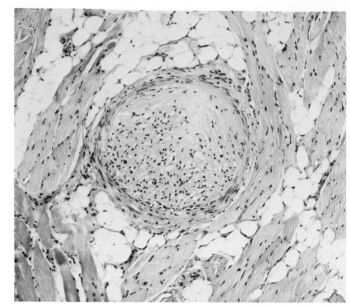

Fig. 4. A branch of the lingual artery appears to be completely stenosed. ×126.

fat-fed rabbits quite early. By one month of feeding there is a homogeneous accumulation of lipid in the intimal space beneath an intact endothelium. Prior *et al.* (55) point out that the distribution of these lesions in rabbits is quite different from man. In contradistinction to man, the small intramyocardial branches are affected (Fig. 2) and the larger coronary arteries are relatively uninvolved (Fig. 3). As the cholesterol-fat regimen is continued, the entire lumens of the intramyocardial arteries are transformed into

homogeneous lipid masses. Lesions of tongue arteries have been striking in cholesterol-fed rabbits (Fig. 4).

In human atherosclerosis the pulmonary artery is usually spared. In the cholesterol-fat-fed rabbit the pulmonary arteries and veins show marked luminal stenosis by masses of lipid-filled histiocytes (Fig. 5).

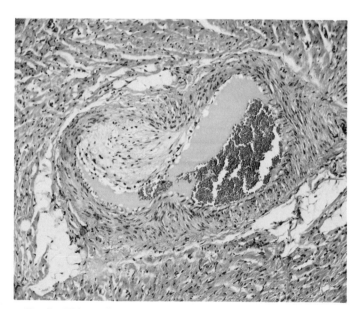

Fig. 3. This medium-sized coronary artery has a foam cell lesion at a point of branching. Lesions in this size or larger coronary arteries are less common than in those shown in Fig. 2. ×118.

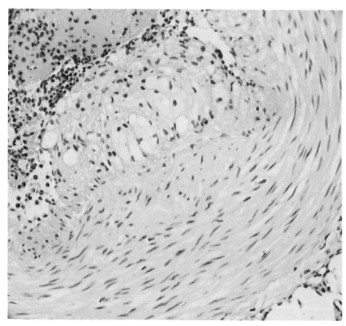

Fig. 5. This pulmonary artery has a lesion typical of those of cholesterol-fed rabbits. The lesion is composed of foam cells and even the endothelium is vacuolated due to lipid having been removed during processing. ×210.

The effect of cholesterol-fat feeding on the cerebral arteries of the rabbit has received very little attention. Prior *et al.* (55) were unable to produce lesions in studies lasting up to six months. Pollak (53) was unable to produce cerebral atherosclerosis in the rabbit with a variety of techniques. Using diets of milk and egg yolk, Altschul (2) produced some changes in the capillaries but not in the arteries of the cerebral circulation of rabbits that were fed for periods up to 234 days. Foam cells were found in the vascular tissue of the choroid plexus in all of these cases and in the capillaries of the suprachiasmatic regions of five of seventeen animals. In four of the cases, the ependymal lining of the third ventricle was distorted. The leptomeninges showed occasional foam cell aggregation.

F. Arterial Metabolism of Rabbits

Rabbits have been used extensively for studies on changes in arterial metabolism with progressing atherosclerosis. Using radiolabeled cholesterol, Biggs and Kritchevsky (6) showed in 1951 that cholesterol enters the arterial wall from plasma in both normal and hypercholesterolemic rabbits. The increase in aortic cholesterol and cholesterol ester of cholesterol-fed rabbits appears to be from increased influx from plasma (49). More recent data clearly indicates that at least a portion of the cholesterol ester is derived from *in situ* esterification of free cholesterol.

During the past 20 years there has been considerable interest in whether arteries contribute through synthesis to cholesterol accumulated in atherosclerotic lesions. Numerous studies have been done on a variety of species. It now seems clear that a very small amount of cholesterol may be derived from *de novo* synthesis, however, its contribution to lesion cholesterol is quite small. As concerns rabbits it is doubtful that this species synthesizes cholesterol *in situ* in arteries. The most conclusive study of this question was reported by Whereat (66) who failed to demonstrate cholesterol synthesis from either acetate or mevalonate in preparations of intima of either normal or atherosclerotic rabbit aorta.

Atherosclerotic rabbit arteries like those from other animals and human beings have enhanced rates of fatty acid synthesis and enhanced rates of esterification of newly synthesized fatty acid to cholesterol. The exact mechanism of this enhanced cholesterol esterification is not known. Abdulla *et al.* (1) reported that lecithin:cholesterol acyltransferase activity in arterial intima media preparations was increased during atherogenesis. Experiments at our center, however, have been unable to confirm this observation (58). It appears that the anatomical site of the cholesterol esterification is in the foam cells of the lesion. Day and Tume (18) reported that isolated foam cells from rabbit atherosclerotic aorta actively incorporated labeled oleic acid into cholesterol ester.

Aortic tissue has the capability to hydrolyze cholesterol esters (8, 17, 35, 51). At least one study provided evidence that there is a decrease in aortic cholesterol ester hydrolysis during atherogenesis. There is no information suggesting that there is a degradation of cholesterol in arterial tissue.

The careful work of Zilversmit and co-workers has clearly shown that the origin of the increased phospholipid of atherosclerotic lesions can be accounted for by *de novo* synthesis within the artery (69). Further studies by Zilversmit and McCandless (68) provided evidence that the increased arterial phospholipid synthesis was independent of elevated serum cholesterol concentration.

G. Experimental Attempts to Induce Complicated Atherosclerosis

As stated earlier one of the principal criticisms of rabbits as animal models of atherosclerosis concerns the lack of complications of their lesions. Considerable progress has been made in recent years to develop methods that consistently produce complicated lesions.

Constantinides (12) made a significant contribution by describing the production of complicated aortic atherosclerotic plaques in rabbits by the intermittent feeding of cholesterol-fat diets. Two hundred and fifty rabbits were fed alternately cholesterol-fat and normal diets for two- or three-month periods. The alternate feeding periods were repeated for up to two years. He reported that the prevalence of the various features of complications was as follows: "capsule-gruel formation" 90%, calcification 80%, medial breakdown of slight to moderate degree 80%, capillarization 25%, hemorrhage 15%, necrosis and complete breakdown of the media 10%, and ulceration 2%. Atherosclerosis of the basilar artery was seen in several rabbits of this series.

One of the most exciting accomplishments in the experimental production of complicated rabbit atherosclerosis has been described by Minick and co-workers (46). These authors reported on the synergy of allergic injury to arteries combined with feeding cholesterol-containing diets to rabbits. The combined regimen resulted in a change both in the quality and distribution of the atherosclerotic lesions. The coronary artery lesions of the rabbits subjected to allergic injury occurred more frequently in the large arteries and were more cellular (Table IV). The lesions produced by their method bear a striking resemblance to human lesions.

Cooper and Gutstein (16) induced calcific aortic atherosclerosis of rabbits by combining the feeding of cholesterol and Pitressin treatment. The morphology of the lesions produced by this regimen again more closely resembled the human disease than those produced by atherogenic diet alone.

TABLE IV

THE EFFECT OF ALLERGIC INJURY TO ARTERIES AND DIETARY CHOLESTEROL ON CORONARY ATHEROSCLEROTIC LESIONS OF RABBITS[a]

Lesion characteristic[b]	Dietary cholesterol (%)	Treatment control diet and horse serum injection (%)	0.5% Dietary cholesterol and horse serum injection (%)
Large coronary arteries affected	0	22	29
Small coronary arteries affected	96	42	38
Predominantly foam cell lesion	100	0	4
Lipid containing-proliferative lesion	0	0	90

[a] After Minick et al. (46).
[b] Expressed as percent of all lesions observed.

H. Lesions in Other Organs among Rabbits Fed Cholesterol-Containing Diets

One of the frequent objections raised to the use of rabbits in experiments involving dietary cholesterol is that cholesterol and other lipids accumulate earlier and to a relatively greater extent in tissues other than arteries. This was noticed by some of the earlier investigators and was described in detail by Duff in 1935 (21). Most authors have chosen to emphasize the similarities between human atherosclerosis and the experimental arterial lesion in rabbits and largely ignore some of the striking differences. Prior et al. (55) have also described the lesions in tissues other than arteries. Our own observations are similar to those of Duff (21) and of

Prior (55). When rabbits are fed "atherogenic" diets containing cholesterol, lipid usually accumulates in a number of tissues, notably the liver, adrenal glands, spleen, and bone marrow. These lipid accumulations occur before lipid-containing arterial lesions are found. If cholesterol is fed continually for several months, lipid accumulates in practically every tissue including parts of the eye, skin (Fig. 6), submucosa of the gastrointestinal tract, liver (Fig. 7), interstitum of the kidney, choroid plexus (Fig. 8), heart valves (Fig. 9), the splenic capsule, and in practically every area of inflammation regardless of initial cause. Some of these changes are particularly striking. The adrenal glands are frequently so lipid-laden that they will float in water. The renal glomeruli sometimes become a nearly solid mass of foam cells (glomerular lipidosis). Other glomeruli have a thickened mesangium with or without foam cells. The latter lesion may have been preceded by a proliferative response (Fig. 10).

The extent of accumulation of cholesterol in various tissues of rabbits fed a diet containing 2% cholesterol for nine months has been studied by Ho and Taylor (33). A summary of their observations is presented in Table V.

A careful study has been reported by Parker and Odland (50) on xanthoma of cholesterol-fed rabbits. After about a month of cholesterol feeding lipid begins to accumulate in skin around small vessels. By 3–4 months lipid-filled macrophages increase in number and occupy the upper half of the dermis. These authors suggest that the same pathogenic mechanisms exist in early xanthoma and atheroma forma-

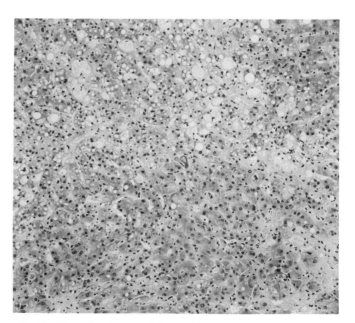

Fig. 7.   Fatty liver from a cholesterol-fat-fed rabbit. Some hepatic cord cells have several vacuoles in the cytoplasm while others have a large single vacuole which has displaced the nucleus to one side. There are a few necrotic cord cells. ×119.

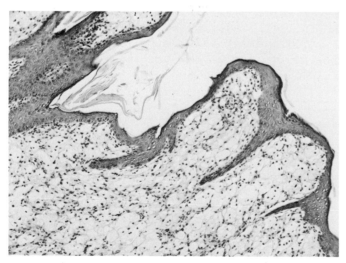

Fig. 6.   The dermis of this cutaneous xanthoma contains many large foamy macrophages under an irregular epithelial covering. ×216.

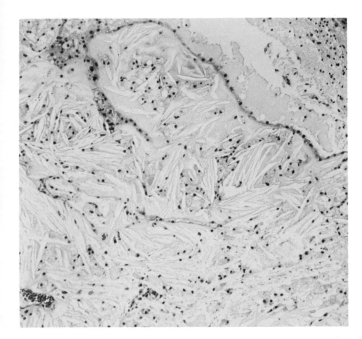

Fig. 8. The choroid plexus of a cholesterol-fed rabbit is distorted by the accumulation of foamy macrophages and extracellular lipid. Sterol clefts are numerous. ×133.

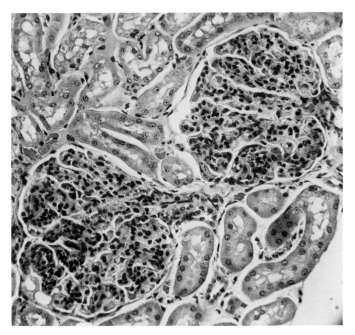

Fig. 10. These glomeruli appear to have an increased number of nuclei (see text). ×210.

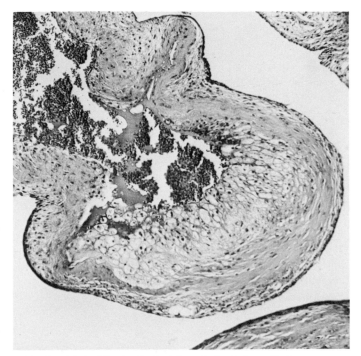

Fig. 9. A left AV valve leaflet is irregularly thickened by foamy macrophages. ×133.

TABLE V
Tissue Cholesterol Content of Rabbits Fed Control or Cholesterol-Containing Diets[a]

Organ	Control diet group[b]	Cholesterol-fed group[b]
Adrenal	19	34
Brain	9.5	9
Lung	2.5	8
Spleen	1.7	16
Kidney	1.75	6
Testis	1.6	4.5
Small intestine	1.4	4.25
Colon	1.35	5
Liver	1.1	12
Aorta	0.6	11
Pancreas	0.7	1.75
Heart	0.75	1.8
Skin	0.3	10.5
Muscle	0.4	1.4
Fat	0.1	0.5

[a] After Ho and Taylor (33).
[b] Expressed as the mean grams of cholesterol per 100 gm of dry tissue.

tion and thus skin could be used as a model system for studies on atherosclerosis.

Roscoe and Vogel (57) described the changes in the eyes of cholesterol-fed rabbits. Large increases occurred in the cholesterol content of the cornea, iris, and ciliary body. In this study the authors examined the relationships between increased aortic cholesterol and the cholesterol content of cornea and iris. There was a poor correlation between corneal cholesterol and aortic atherosclerosis but a strong positive correlation between iridic and aortic cholesterol ($r = 0.72$).

V. REGRESSION OF CHOLESTEROL-INDUCED ATHEROSCLEROSIS OF RABBITS

Advanced cholesterol-induced atherosclerosis of rabbits does not appear to regress after the diet is changed back to normal while the lipid content of early lesions tends to disappear with time. Bortz (7) has presented evidence to suggest that fatty streaks produced in rabbits after a few weeks of cholesterol feeding regress considerably in 300 days. Friedman and Byers (28) and Connor *et al.* (11) have concluded that the atherosclerosis of rabbits, produced by much longer periods of cholesterol feeding is irreversible. Data from our own laboratories support the observation of irreversibility in the rabbit model.

REFERENCES

1. Abdulla, Y. H., Orton, C. C., and Adams, C. W. M. (1968). Cholesterol esterification by transacylation in human and experimental atheromatous lesions. *J. Atheroscler. Res.* **8**, 967–973.
2. Altschul, R. (1950). "Selected Studies on Arteriosclerosis." Thomas, Springfield, Illinois.
3. Anitschkow, N. (1914). Über die Atherosclerose der Aorta beim Kaninchen und über deren Entstehungsbedingungen. *Beitr. Pathol. Anat. Allg. Pathol.* **59**, 308–348.
4. Anitschkow, N. (1933). Experimental arteriosclerosis in animals. *In* "Arteriosclerosis; A Survey of the Problem" (E. V. Cowdry, ed.), Chapter 10. Macmillan, New York.
5. Anitschkow, N., and Chalatow, S. S. (1913). Über experimenteller Cholesterinsteatose und ihre Bedeutung für die Entstehung einiger pathologischer Prozesse. *Zentralbl. Allg. Pathol. Pathol. Anat.* **24**, 1–9.
6. Biggs, M. W., and Kritchevsky, D. (1951). Observations with radioactive hydrogen (^{3}H) in experimental atherosclerosis. *Circulation* **4**, 34–42.
7. Bortz, W. M. (1968). Reversibility of atherosclerosis in cholesterol-fed rabbits. *Circ. Res.* **22**, 135–139.
8. Bowyer, D. E., Howard, A. N., Gresham, G. A., Bates, D., and Palmer, B. V. (1968). Aortic perfusion in experimental animals: A system for study of lipid synthesis and accumulation. *Progr. Biochem. Pharmacol.* **4**, 235–243.
9. Boyd, B. S., Noble, F. P., and Schettler, F. G. (1969). Plasma lipids and lipoproteins. *In* "Atherosclerosis—Pathology, Physiology, Aetiology, Diagnosis and Clinical Management" (F. G. Schettler and G. S. Boyd, eds.), p. 531. Elsevier, Amsterdam.
10. Bragdon, J. H. (1952). Spontaneous atherosclerosis in the rabbit. *Circulation* **5**, 641–646.
11. Connor, W. E. Armstrong, M. L., Jackson, C. S., and Ali, A. M. (1966). Persistence of cholesterol-4C^{14} in atherosclerotic aortas of animals treated with a diet high in polyunsaturated fat. *J. Clin. Invest.* **45**, 997.
12. Constantinides, P. (1961). Production of experimental atherosclerosis in animals. *J. Atheroscler. Res.* **1**, 374–385.
13. Cook, R. P., and Thomson, R. O. (1951). The absorption of fat and cholesterol in the rat, guinea pig, and rabbit. *Quart. J. Exp. Physiol. Cog. Med. Sci.* **36**, 61–74.
14. Cookson, F. B., Altschul, R., and Fedoroff, S. (1967). The effects of alfalfa on serum cholesterol-induced atherosclerosis in rabbits. *J. Atheroscler. Res.* **7**, 69–81.
15. Cookson, F. B., and Fedoroff, S. (1968). Quantitative relationships between administered cholesterol and alfalfa required to prevent hypercholesterolaemia in rabbits. *Brit. J. Exp. Pathol.* **49**, 348–355.
16. Cooper, J., and Gutstein, W. H. (1966). Calcific aortic atherosclerosis of the rabbit after cholesterol and Pitressin treatment. *J. Atheroscler. Res.* **6**, 75–86.
17. Day, A. J., and Gould-Hurst, P. R. S. (1966). Cholesterol esterase activity of normal and atherosclerotic rabbit aorta. *Biochim. Biophys. Acta* **116**, 169–171.
18. Day, A. J., and Tume, R. K. (1969). *In vitro* incorporation of ^{14}C-labelled oleic acid into combined lipid by foam cells isolated from rabbit atheromatous lesions. *J. Atheroscler. Res.* **9**, 141–149.
19. Dietschy, J. M., and Wilson, J. D. (1970). Regulation of cholesterol metabolism. *N. Engl. J. Med.* **282**, 1128–1138, 1179–1183, and 1241–1249.
20. Donomae, I., Matsumoto, Y., Kokubu, T., and Koide, R. (1957). Production of coronary heart disease in the rabbit by lanolin feeding. *Circ. Res.* **5**, 645–649.
21. Duff, G. L. (1935). Experimental cholesterol arteriosclerosis and its relationship to human arteriosclerosis. *Arch. Pathol.* **20**, 81–123 and 259–304.
22. Duff, G. L., and McMillan, G. C. (1949). The effect of alloxan diabetes on experimental cholesterol atherosclerosis in the rabbit. I. The inhibition of experimental cholesterol atherosclerosis in alloxan diabetes. II. The effect of alloxan diabetes on the regression of experimental cholesterol atherosclerosis. *J. Exp. Med.* **89**, 611–630.
23. Duff, G. L., and McMillan, G. C. (1951). Pathology of atherosclerosis. *Amer. J. Med.* **11**, 92–108.
24. Duff, G. L. McMillan, G. C., and Ritchie, A. C. (1957). The morphology of early atherosclerotic lesions of the aorta demonstrated by the surface technique in rabbits fed cholesterol. Together with a description of the anatomy of the rabbit's aorta and the "spontaneous" lesions which occur in it. *Amer. J. Pathol.* **33**, 845–873.
25. Fillios, L. C., and Mann, G. V. (1956). The importance of sex in the variability of the cholesteremic response of rabbits fed cholesterol. *Circ. Res.* **4**, 406–412.
26. Fredrickson, D. S., Levy, R. I., and Lees, R. S. (1967). Fat transport in lipoproteins—an integrated approach to mechanisms and disorders. *N. Engl. J. Med.* **278**, 34–44, 94–103, 148–156, 215–224, and 273–281.
27. Friedman, M., and Byers, S. O. (1954). Pathogenesis of dietary induced hypercholesterolemia in the rabbit. *Amer. J. Physiol.* **179**, 201.
28. Friedman, M., and Byers, S. O. (1963). Observations concerning the evolution of atherosclerosis in the rabbit after cessation of cholesterol feeding. *Amer. J. Pathol.* **43**, 349.
29. Gofman, J. W., Lindgren, F., Elliot, H., Mantz, W., Hewitt, J., Strisower, B., and Herring, V. (1950). The role of lipids and lipoproteins in atherosclerosis. *Science* **3**, 166–171 and 186.
30. Gould, R. G. (1951). Lipid metabolism and atherosclerosis. *Amer. J. Med.* **11**, 209–227.
31. Harman, D. (1962). Atherosclerosis—effect of rate of growth. *Circ. Res.* **10**, 851.
32. Hellström, K. (1965). On the bile acid and neutral fecal steroid excretion in man and rabbits following cholesterol feeding: Bile acids and steroids 150. *Acta Physiol. Scand.* **63**, 21–35.
33. Ho, K., and Taylor, C. B. (1968). Comparative studies on tissue cholesterol. *Arch. Pathol.* **86**, 585–596.
34. Horlick, L., Cookson, F. B., and Fedoroff, S. (1967). Effect of alfalfa feeding on the excretion of fecal neutral sterols in the rabbit. *Circulation* **35** and **36**, Suppl. II, 18.
35. Howard, C. F., Jr., and Portman, O. W. (1966). Hydrolysis of cholesteryl linoleate by a high speed supernatant preparation of rat and monkey aorta. *Biochim. Biophys. Acta* **125**, 623–626.
36. Ignatowski, A. I. (1908). Influence of animal food on the organism of rabbits. *S. -Peterb. Izv. Imp. Voyenno-Med. Akad.* **16**, 154–176.
37. Kritchevsky, D. (1970). Role of cholesterol vehicle in experimental atherosclerosis. *Amer. J. Clin. Nutr.* **23**, 1105–1110.

38. Kritchevsky, D., Langan, J., Markowitz, J., Berry, J. F., and Turner, D. A. (1961). Cholesterol vehicle in experimental atherosclerosis. III. Effects of absence or presence of fatty vehicle. *J. Amer. Oil Chem. Soc.* **38**, 74.

39. Kritchevsky, D., Moyer, A. W., Tesar, W. C., Logan, J. B., Brown, R. A., Davies, M. C., and Cox, H. R. (1954). Effect of cholesterol vehicle in experimental atherosclerosis. *Amer. J. Physiol.* **178**, 30.

40. Kritchevsky, D., Moyer, A. W., Tesar, W. C., McCandless, R. F. J., Logan, J. B., Brown, R. A., and Englert, M. E. (1956). Cholesterol vehicle in experimental atherosclerosis. II. Influence of unsaturation. *Amer. J. Physiol.* **185**, 279.

41. Kritchevsky, D., and Tepper, S. A. (1965). Cholesterol vehicle in experimental atherosclerosis. VII. Influence of naturally occurring saturated fats. *Med. Pharmacol. Exp.* **12**, 315.

42. Kritchevsky, D., and Tepper, S. A. (1968). Experimental atherosclerosis in rabbits fed cholesterol-free diets: Influence of chow components. *J. Atheroscler. Res.* **8**, 357–369.

43. Kritchevsky, D., Tepper, S. A., and Langan, J. (1962). Influence of short-term heating on composition of edible fats. *J. Nutr.* **77**, 127.

44. Kritchevsky, D., Tepper, S. A., and Langan, J. (1962). Cholesterol vehicle in experimental atherosclerosis. IV. Influence of heated fat and fatty acids. *J. Atheroscler. Res.* **2**, 115.

45. McGill, H. C., Jr., and Holman, R. L. (1949). The influence of alloxan diabetes on cholesterol atheromatosis in the rabbit. *Proc. Soc. Exp. Biol. Med.* **72**, 72–75.

46. Minick, C. R., Murphy, G. E., and Campbell, W. G. Jr. (1966). Experimental induction of athero-arteriosclerosis by the synergy of allergic injury to arteries and lipid-rich diet. *J. Exp. Med.* **124**, 635–652.

47. Mosbach, E. H., Halpern, E., and Brunder, J. (1956). Sterol metabolism in the rabbit. *Fed. Proc., Fed. Amer. Soc. Exp. Biol.* **15**, 525.

48. Munro, H. N., Steel, M. H., and Forbes, W. (1965). Effect of dietary protein level on deposition of cholesterol in the tissues of the cholesterol-fed rabbit. *Brit. J. Exp. Pathol.* **49**, 489–496.

49. Newman, H. A. I., and Zilversmit, D. B. (1962). Quantitative aspects of cholesterol flux in rabbit atheromatosis. *J. Biol. Chem.* **237**, 2078.

50. Parker, F., and Oldand, G. F. (1968). Experimental xanthoma. A correlative biochemical, histologic, histochemical, and electron microscopic study. *Amer. J. Pathol.* **53**, 537–564.

51. Patelski, J., Bowyer, D. E., Howard, A. N., and Gresham, G. A. (1968). Changes in phospholipase A, lipase and cholesterol esterase activity in the aorta in experimental atherosclerosis in rabbit and rat. *J. Atheroscler. Res.* **8**, 221–228.

52. Pierce, F. T. (1952). The relationship of serum lipoproteins to atherosclerosis in cholesterol-fed alloxanized rabbit. Circulation **5**, 401–407.

53. Pollak, O. J. (1945). Attempts to produce cerebral atherosclerosis. *Arch. Pathol.* **39**, 16–21.

54. Pollak, O. J. (1947). Age and weight as factors in the development of experimental cholesterol atherosclerosis in rabbits. *Arch. Pathol.* **43**, 387.

55. Prior, J. T., Kuntz, D. M., and Ziegler, D. D. (1961). The hypercholesterolemic rabbit. *Arch. Pathol.* **71**, 672–684.

56. Rona, G., Chappel, C. I., and Gaudry, R. (1959). Aggravation of cholesterol atherosclerosis in rabbits by free unsaturated fatty acids. *Can. J. Biochem. Physiol.* **37**, 479–483.

57. Roscoe, H. C., and Vogel, A. W. (1968). Lipid changes in the eye concomitant with the development of atherosclerosis in the aorta in the rabbit. *Circ. Res.* **33**, 633–643.

58. St. Clair, R. W. Personal communication.

59. Scebat, L., Renais, J., and Lenegre, J. (1964). *In* "Lipid Pharmacology" (R. Paoletti, ed.), p. 127. Academic Press, New York.

60. Schenk, E. A. Gaman, E., and Feigenbaum, A. S. (1966). Spontaneous aortic lesions in rabbits. I. Morphologic characteristics. *Circ. Res.* **19**, 80–88.

61. Schenk, E. A., Gaman, E., and Feigenbaum, A. S. (1966). Spontaneous aortic lesions in rabbits. II. Relationship to experimental atherosclerosis. *Circ. Res.* **19**, 89–95.

62. Stormby, N. G., and Wigand, G. (1963). Morphological changes in hypercholesterolemic rabbits given saturated fat without supplementary cholesterol. *J. Atheroscler. Res.* **3**, 103–120.

63. Vles, R. D., Buller, J., Gottenbos, J. J., and Thomasson, H. J. (1964). Influence of type of dietary fat on cholesterol-induced atherosclerosis in the rabbit. *J. Atheroscler. Res.* **4**, 170.

64. Wells, W. W., and Anderson, S. C. (1959). The increased severity of atherosclerosis in rabbits on a lactose-containing diet. *J. Nutr.* **68**, 541.

65. Wells, W. W., Quan-ma, R., Cook, C. R., and Anderson, S. C. (1962). Lactose diets and cholesterol metabolism. II. Effect of dietary cholesterol, succinylsulfathiazole and mode of feeding on atherogenesis in the rabbit. *J. Nutr.* **76**, 41–47.

66. Whereat, A. F. (1964). Lipid biosynthesis in aortic intima from normal and cholesterol-fed rabbits. *J. Atheroscler. Res.* **4**, 272–282.

67. Zeek, P. M. (1933). Familial factors in arteriosclerosis in rabbits. *Arch. Pathol.* **16**, 302.

68. Zilversmit, D. B., and McCandless, E. L. (1959). Independence of arterial phospholipid synthesis from alterations in blood lipids. *J. Lipid Res.* **1**, 118.

69. Zilversmit, D. B., Shore, M. L., and Ackerman, R. F. (1954). The origin of aortic phospholipid in rabbit atheromatosis. *Circulation* **9**, 581.

CHAPTER 7

Specialized Research Applications: II. Serological Genetics

Carl Cohen and Robert G. Tissot

The rabbit is the animal of choice for most immunological studies on the humoral response to antigens. It is not necessary in this volume to enumerate the methods that have been used to produce antibody in the rabbit to a variety of antigens or to describe the techniques for measuring the antibody response both *in vivo* and *in vitro*. A method book such as that by Chase and Williams (72) or Campbell *et al.* (7) should be consulted for such information. A list of antigens, the specific regimen for producing antibody, and the kind of test used for antibody measurement have been published (24).

The specific purpose in this chapter is to discuss the genetic systems of the rabbit that are detectable by serological reactions. These systems include the antigenic systems of cells and body fluids and the genetic systems identified in the formation of antibodies. Emphasis in this chapter will be on the cellular antigens.

I. THE BLOOD GROUP ANTIGENS

A. Methods of Isoantibody Production

A blood group system is dependent on the occurrence of at least two alternative forms of erythrocyte antigens in the population and also on the availability of a reagent to discriminate between alternative forms.

The ABO blood group antigens in man, discovered by

Landsteiner in 1900, stimulated investigators to search for equivalent systems in animals, both domestic and laboratory species. Because of the nature of the ABO system, that is, because there are two antibodies normally present in man, Lansteiner was able to divide humans into four groups. The first studies outside of man, therefore, involved a search for naturally occurring antibodies in the serum of the species under investigation. The usual procedure was to bleed the animals, mix the sera of one individual with the cells of the others, and hope that some combinations would give agglutination reactions.

The rabbit is one of the species in which no significant level of natural isoagglutinins appear; and, although there have been reports from time to time of natural antibodies, these antibodies have not proved useful in regularly detecting antigens under clear-cut genetic control.

At about the time that Landsteiner's search for normally occurring antibody was going on, Ehrlich and Morganroth were carrying out a series of experiments in which blood of one goat was injected into other goats and the recipient's serum was then tested against the donor to see whether an antibody had been induced. The development of knowledge of the rabbit blood group system has depended upon this process of cross-immunization within the species and the production, separation, and analysis of reliable serologically distinct typing reagents.

Until some antibodies were produced that could be used to discriminate between two or more rabbits, the donor–recipient pairing was done at random. That is, a quantity of red cells or whole blood from a donor was injected into a recipient without knowledge of the antigenic types of either animal. With the development of typing sera it became possible to carry out immunizations in which there were known antigenic differences between the donor and the recipient.

A number of methods have been used to enhance antibody response because, in general, the isoimmunization procedures are not as effective in inducing a response as are the immunizations across species lines. In some of the very first meaningful studies on rabbit blood groups, Levine and Landsteiner (54, 55) used cells that were lysed in distilled water and then introduced by intravenous inoculation. Fleischer (36) used as injectants, in addition to rabbit red cells, such substances as killed typhoid bacilli and human serum from patients with measles. As methods of enhancing the antibody response were developed they were used in isoimmunization studies. We (12) found that isoantibody production was greatly enhanced by preinjection of the recipient rabbits with an intradermal injection of the bacillus of Calmette and Guerin (BCG). A systematic test of the BCG procedure yielded the data that in the BCG-treated group, 26 animals of 46 undergoing isoimmunization regimens gave useful isoimmune sera, whereas only three of the 27 untreated animals responded. These iso-

immunizations were done in random donor–recipient pairs, and it was from this BCG procedure that we obtained the antisera to the seven different antigens that began our own study of rabbit red cell antigens.

This BCG treatment is currently under investigation as a means of enhancing tumor immunity.

Our present procedure (29), a procedure that we have used for several years, is based in part on our BCG results and the finding of Anderson (1) that good typing sera could be obtained by incorporating rabbit red cells in Freund's adjuvant (see ref. 7). In this procedure 2 ml of a suspension consisting of one volume of 50% rabbit red cells mixed with three volumes of Freund's complete adjuvant are introduced subcutaneously into multiple sites on the back of the rabbit on two occasions with a one-week interval between injections. These subcutaneous injections are followed by nine intravenous injections of 2 ml each of a glycerol-red cell mixture given over a three-week period. The glycerol-red cell mixture is used because it enables us to prepare a relatively large batch of cells, which are then frozen at $-20°C$ in several aliquots; each aliquot is thawed when needed for injection. The procedure for preparation of the glycerol-citrate mixture follows the method described by Chaplin (11). The cells prepared in this way are well preserved, cell lysis is minimal, and there are no untoward reactions in the recipient.

In skin transplantation studies we (30) found that a very effective adjuvant action occurred for production of anti-red cell antibodies when the incompatible cells were introduced under and/or around an ear skin graft that was destined to be rejected. The graft itself frequently induced the formation of anti-red cell antibody (*vide infra*).

B. Preparation of the Typing Reagents

When an effective antibody is found, as shown by agglutination of the donor's cells, tests must be made against a panel of known cell types and absorption procedures must be carried out in order to produce a monospecific typing reagent. The absorption procedure is based on the fact that because the antigen-antibody complex is relatively stable, centrifugation of a mixture of red cells and serum leads to the sedimentation of the cells and anything absorbed onto the cells; the soluble, not absorbed, serum components remain in the supernatant fluid. The antiserum containing antibodies of more than one specificity is mixed with red cells that lack only one specific antigen, namely, the antigen corresponding to the single antibody to be prepared as a monospecific typing reagent. The test of the whole serum against the cell panel allows us to select from the panel the cells that have the specific antigen array which can be used for absorption. It may be necessary to use successive absorptions by one or more than one cell type to exhaust the antibodies that are to be eliminated in order to make the monospecific serum.

As each new specific antibody is recognized, prepared, and tested against a population of animals the numbers of unidentified antibodies produced by isoimmunization will decrease and selection of suitable absorbing cells to prepare monospecific typing sera becomes a relatively uncomplicated matter. A serum must, of course, be recognized as monospecific before any genetic hypotheses as to mode of inheritance or relationship to a previously recognized blood group system can be made.

C. Genetic Analysis of Blood Group Data

When a monospecific serum is tested against a population, the pattern of reactions gives insight into the nature of inheritance of the antigen. The genetic influences come only after the establishment of the immunological acceptability of the serum based on its reliability as a monospecific typing reagent.

The test of populations usually includes the test of family groups and the pattern of reactions with these groups suggests an hypothesis as to the mode of inheritance from a limited number of hypothesis including the following:

1. The antigen is inherited as a codominant trait and appears whenever the gene is present.

2. The antigen is inherited as a recessive trait and appears only in the homozygote.

3. The gone controlling the antigen is a new allelle at a previously recognized locus.

4. The antigen is controlled by a previously recognized allele and is part of a complex of antigens controlled by a "complex" locus.

5. The mode of inheritance is not uncomplicated, and the antigen may be controlled by an interaction between two alleles or two or more loci.

6. The pattern of reaction indicates no reasonable genetic hypothesis for the presence of the antigen; it may be environmental in origin or indeed it may be an artifact of an inconsistent typing serum or a polyspecific serum.

In any discussions of serological systems it is necessary to distinguish between the serological trait—the antigen—and the gene controlling the presence or absence of the antigen. The term allele is the name of a gene and is used in the description of the genotype; the antigen and the blood type are phenotypic descriptions; the antibody is the specific detector of the antigen. There may not necessarily be a 1:1 association between the genotype and phenotype.

D. Development of the Rabbit Blood Group Systems

The significant studies of rabbit blood groups which yielded information that is interpretable today began with the work of Levine and Landsteiner (54, 55) and Fischer and Klinkhardt (34, 35) in 1929. These investigators used the process of isoimmunization to produce antibody and showed clearly that the blood cells of rabbits may contain antigens for which no normally occurring agglutinins occur. Cameron and Snyder (6) and Castle and Keeler (8) used Levine and Landsteiners's two sera and the human ABO system as a model and found that with these sera, called anti-H_1 and anti-H_2, all rabbits could be classified as H_1, H_2, H_1H_2, or O. Their general conclusion was that the two recognized red cell antigens of the rabbit were controlled by a series of three alleles giving rise to six different genotypes. Fischer and Klinkhardt's sera, called anti-K_1 and K_2, were identical to those of Levine and Landsteiner except that the symbols used were different.

An unusual aspect of the early studies of rabbit blood groups is that they were carried out by investigators who were major figures of their time. Landsteiner was the discoverer of the ABO system in man, and with Levine, of the MN and P systems. Levine was the first to recognize that the Rh system, discovered by Landsteiner and Weiner, was significant in erythroblastosis. Castle was one of the earliest mammalian geneticists in the country, and he and his students were the stimuli for much of the mammalian genetics of today. The rabbit, then, had attractive features for blood group studies and the rabbit still has a role as a model in immunological genetics.

The antibodies anti-H_1 and -H_2 and -K_1 and -K_2 are the only ones that have come down to the present in an identifiable state. Although other antigens and antibodies were reported, they cannot be identified or compared with those we now have. For example, Fischer (34) found three more agglutinogens, called K_3, K_4, and K_5. He reported that the K_3 and K_4 antigens were not confined to the erythrocyte but also occurred in tissues other then blood. Marcussen (58) reported another antigen and postulated the discovery of others. Knopfmacher (53) reported an antigen called H_3, independent of H_1 and H_2, but we do not know how it fits into our present knowledge.

Up to this time the studies of the rabbit erythrocyte antigens were oriented primarily toward verifying the fact that genetic systems exist in the rabbit and they can serve as models for problems involving blood groups in man. The discovery of the Rh system in man and its role in erythroblastosis encouraged additional investigations in the rabbit. The Rh system has unusual genetic complexity and, in addition, the variation in the nature of antibody plays a unique role in the manifestation of the disease. Nachtsheim (66) reported a naturally occurring gene-controlled hydrops fetalis in the rabbit which was thought to be the result of a maternal-fetal incompatibility for red cells antigens. The animal stocks involved were made available to Coombs' laboratory in Cambridge, where an effort was mounted to investigate the phenomenon. Heard *et al.* (41, 42) could not corroborate Nachtsheim's finding although they tried deliberate immunizations to exaggerate the antigenic differences between the fetus and the mother. They were, however, able to produce typing sera to detect three antigens controlled by three allelic genes. This work was continued in Cambridge by Joysey (48).

Again, at about the same time that the Cambridge group was investigating the rabbit blood groups, the question of normally occurring antibody came up; Oswald (67) reported that 20% of 405 rabbits showed the presence of normal isoantibodies. This finding has not been corroborated generally, although recently Padma (68) reported normal antibody detectable in 10 out of 18 rabbits under study. The antibodies were detected as relatively weak agglutinins in undiluted sera.

Kellner and Hedal (50, 51) demonstrated that the rabbit could serve as a model for the Rh system, at least in terms of a maternal-fetal incompatibility. They prepared antibodies called anti-G and anti-g. These appeared to be controlled by a pair of alleles, and all rabbits had one or both of these antigens. On the basis of the distribution of the alleles in the population, Kellner assumed that these antigens were not the same as the H_1 and H_2 of the earlier American investigators.

The renewed interest in rabbit blood groups was evidenced by the fact that by the time Kellner and Hedal reported their work (1951), three other separate, systematic investigations were under way on rabbit blood groups. Anderson (2, 3) was studying isoimmunization and the hydrops problem, Joysey (48) was continuing the work initiated by Heard *et al.*, and we were studying the genetics of isoimmunization.

Later Dahr and his co-workers (31, 32), using the animal stocks that Nachtsheim had produced, became involved in the hydrops problem. Ivanyi (43–47) also became interested in the rabbit hydrops problem and branched out into transplantation problems in the rabbit. Additional work in blood group systems in the rabbit was reported from the Pasteur Institute by Eyquem and his associates (33, 56), by Yokoyama (74), and by Nelken (66a).

It is unfortunate that many of the people who became interested in rabbit blood group systems started without referring to previous accomplishments and were unaware that they had recognized the same antigens first reported by Levine and Landsteiner and Fischer and Klinkhardt.

The investigation of hydrops fetalis and transfusion reactions represents the major portion of the studies in the rabbit. Although a large number of individuals spent some effort on the studies, they all found the same system: the H_1H_2 or K_1K_2 systems of the earliest workers. The descriptions of hydrops fetalis and mode of induction as published in 1951 by Kellner and Hedal are unchallenged and, in a sense, remain the classics on the hydrops problem. A further result of all the work is an array of wildly varied

nomenclature, all applied to the same system. The nomenclature problem in the literature was clarified several years ago (17). The consideration of what is genotype or phenotype was not a preoccupation of most of the investigators; the potential for complexity of the one most commonly found rabbit system has gone far beyond the capabilities of the nomenclatures presented by most of the investigators.

E. Present Status of the Red Cell Antigens in the Rabbit

Although our work started without any known sera on hand, our animals were derived from some of the animals used in studies by Castle. The first two antibodies studied in depth appeared to be controlled by alleles; and since some animals were negative to both of the typing sera we assumed that these antigens were controlled by a locus with at least three alleles. We had no doubt that we were dealing with the classical H_1, H_2 system. After comparing sera with Kellner, we found that his antigens G and g were also the same as H_1 and H_2. Unfortunately, because the animal population he used did not have the same gene frequencies as Castle's animals, he had not recognized the fact that he had rediscovered the original Levine and Landsteiner antibodies. To avoid further confusion we adopted a nomenclature (13, 14) acknowledging the existence of the early work and Kellner's more recent contribution which we thought would be a common model for fetal-maternal incompatibility. Thus, the gene symbol *Hg* was devised and the alleles that we had recognized were named *Hg*^{A} and *Hg*^{F}. Shortly thereafter, when an antibody detecting the third allele was prepared, we had a system of three codominant alleles. We were able to compare our sera with those of Anderson and Joysey, and, by a stroke of good fortune, we were able to test some of Fischer's original serum. These sera all fit into a pattern; that is, Anderson, Fischer, Castle, and Kellner had antisera to two antigens controlled by two alleles of a three-allele system. Both in Cambridge and in our laboratory antibody was prepared to detect the products of each of the alleles we now call *Hg*^{A}, *Hg*^{D}, and *Hg*^{F}.

The antigens are not equal in their ability to induce isoantibody following deliberate immunization regimens (17). Indeed, the probability of inducing anti-D in an animal with antigen F is of the order of 4% (18, 21). This low probability is probably the reason why so many investigators failed to produce this antibody and consequently were unable to do more elegant genetic analysis of their rabbits.

Table I shows the present status of the *Hg* system, which has continued to grow more interesting with time. It resembles in its complexity the Rh and HL-A systems of man, the H-2 system of the mouse and the B system of cattle since it now has four alleles and 12 antigens in 10 combina-

TABLE I
THE RABBIT BLOOD GROUP SYSTEMS

Locus	Alleles	Typing sera	Number of recognized phenotypes
Hg	Hg^A, Hg^D, Hg^F, Hg^N	Anti-A,-D,-F,-I,-J,-K,-N,-P,-R,-T,-V,-W(12)	10
Hb	Hb^B, Hb^M	Anti-B,-B(2)	3
Hc	Hc^C, Hc^L	Anti-C,-L(2)	3
He	He, he	Anti-E(1)	2
Hh	Hh, hh	Anti-H(1)	2
Hq	Hq^Q, hq^S	Anti-Q,-S(2)	3

tions. Our knowledge of the system evolved from three alleles and three antigens in the following manner.

An antiserum was prepared by isoimmunization as part of a continuing study of the genetics of isoimmunization; analysis indicated that the antiserum had two antibodies directed at antigens controlled by the Hg system. One of these clearly was anti-D, an expected specificity, and the second antibody was isolated by absorption of the serum with type D cells and type A cells. This serum, called anti-I (16), when tested against a panel, reacted only with cells that were heterozygous for the antigens A and D. Population studies and breeding experiments showed conclusively that this antibody detected an antigen that was present only in the heterozygote Hg^A/Hg^D and was not present in any other type. Although the rare occurrence in offspring of cellular antigens that do not occur in either parent had been reported before, this was the first report of its occurrence as a heterozygote at a known locus and the first report of its occurrence in the mammal.

The next antigen in the Hg system for which an antibody was discovered was another interaction antigen (22), antigen J, which occurred in all animals having the allele Hg^D and also in animals heterozygous Hg^A/Hg^F. Antigen J occurs as an interaction antigen in one case, Hg^A/Hg^F, and as a typical direct antigen in the animal having the Hg^D allele. This second interaction antigen indicates that heterozygous effects may not be uncommon and that some insights into the gene–antigen relationship can be derived from the study of these particular antigens.

In the same serum from which anti-J was prepared another specificity was found that reacted with all cells of blood type D or blood type F but did not react with cells of animals homozygous Hg^A/Hg^A. This antibody, called anti-K, detected an antigen that was controlled independently by the genes Hg^D and Hg^F. This addition to the series of antigens in the Hg system did not require the hypothesis of a new allele. Thus the locus which appeared to support the simple one gene-one antigen hypothesis has grown into a complex system that begins to challenge the meaning of the terms locus and gene.

In time two new antibodies were found, anti-P and anti-R, which detected additional antigens of the Hg system. These were very much like antigen K in their mode of inheritance in that each was controlled, independently, by two alleles. Antigen P was not present in an animal lacking either the gene Hg^A or Hg^F; that is, the homozygote Hg^D/Hg^D lacked antigen P. Similarly the antigen R was present in animals with the allele Hg^A or Hg^D. The antigens K, P, and R are not interaction antigens, but are indicative of similarity of structure in portions of the cistrons of the Hg alleles.

As the antigenic systems controlled by the Hg locus became more complex, we were in fact studying aspects of the gene–antigen relationship because we had but three alleles and were detecting the various antigens in combinations as expressed by each allele or by pairs of alleles. The finding of a fourth allele at the Hg locus evolved from the discovery of a family of rabbits in our colony which failed to react with anti-A, anti-D, or anti-F. On the basis of a three allele system, the cells of each of these animals should have been agglutinated by at least one of these three typing reagents. This family of animals was derived from a cross involving an animal introduced into our closed colony after its establishment. The parents and grandparents appeared to have antigen D in earlier tests but on retest the parents failed to react with the more recently prepared anti-D which was used to test the offspring. Analysis of various anti-D typing sera following a series of absorptions disclosed that in reality there were two antibodies present in some anti-D sera; one of the antibodies reacted with all rabbits which had been classified as having the Hg^D allele and also reacted with cells from the unusual litter. The antibody we originally called anti-D reacted with all Hg^D animals but did not react with the odd family.

We therefore hypothesized that the Hg^D locus controlled an additional antigen, now called N, and some animals had antigen N but not the antigen we called D. The introduced animal brought into the colony a different allele which had a strong resemblance to the Hg^D allele but was not precisely the same. We called this new allele Hg^N and tested the animals which appeared to be homozygous for the Hg^N allele with all other reagents which react with cells from animals having the Hg^D allele in homozygous or heterozygous state. The reactions showed that the allele Hg^N gave rise to all antigens detected in Hg^D individuals except the antigen D. Even the antigen I, the interaction antigen, appeared in the Hg^A/Hg^N heterozygote. The Hg^N allele may have arisen from Hg^D by a mechanism of unequal crossover or by mutation, but is certainly closely related in origin to the Hg^D allele.

Since the products of the allele Hg^N so closely resemble those of the Hg^D allele, the continuing development of the Hg system will not include consideration of the Hg^N allele. Whenever a condition for complexity includes the allele

Hg^D it should be implicit that allele Hg^N may be substituted for it. Table I includes all four alleles and their combinations.

An addition to the Hg system is the antigen T, another interaction antigen. The antibody, anti-T, detects an antigen present in every case where the allele Hg^A is present in a heterozygous animal. Table I shows that three out of the ten classes of individuals have antigen T. This is an antigen resulting from the interaction of one allele with any other allele in the allelic series.

Another antigen, similar in mode of gene control to T, has also been found; this antigen, called V, appears only when the gene Hg^D is present in combination with any allele other than Hg^D.

The most recent antibody which we have prepared, called anti-W, detects an antigen present when the allele Hg^A is present in a double dose. That is, it appears to fit the definition of a recessive antigen. The antigen W is in sharp contrast to antigen T in that the latter is present only when the gene is present in a single dose and the former antigen appears only when the gene Hg^A is in the homozygous state, that is, in a double dose.

The evolution of our knowledge of the Hg locus can be seen in the arrangement of Table II, which shows all known antisera used to test the ten known genotypes. There are 12 different antibodies so far in this system, which when first recognized in 1929 was considered as a model for the simplicity implicit in a one gene-one enzyme era. The complexity in the system has made its comprehension difficult, but in many ways it has some of the characteristics of the Rh blood group and HLA tissue compatibility antigens of man and the H-2 antigens of the mouse and can now be considered an excellent model for complex systems. The mechanism of the translation of the gene to give rise to the complex of antigens is not simple and certainly in-

dicates that we do not have a simple one gene-one antigen relationship. We feel that to digress into the arguments involving the gene–antigen relationship is not appropriate in this volume.

In addition to the Hg system there are five other blood group systems in the rabbit. All the blood group systems are shown in Table II. It can be seen that in addition to the Hg system, other systems exist where the exact genotypes in heterozygotes can be determined directly; the Hb system, the Hc system and the Hq system each appear to have two codominant genes responsible for the antigens.

In all of our studies on the blood group systems we had ample opportunity to discover antigens indicative of complexity in any of the blood group systems; aside from the Hg system nothing unusual has appeared. We therefore have concentrated our efforts on attempts to understand the Hg system as a model for the study of gene action and all the immunogenetic consequences that are involved.

F. Rabbit Blood Group Systems as Models for Study of Immunological Problems

1. Prevention of RH Immunization

We have already mentioned that the study of allelic interaction antigens may give insight into the nature of the gene–antigen relationship. This is of particular importance in the Rh system because there are many antigens involved and there has been considerable controversy as to the nature of the antigens and their genetic control. The Hg system of the rabbit has been looked on as having many of the properties necessary for the understanding of the Rh blood groups but the rabbit has not become a popular model because it is much less convenient than the mouse.

TABLE II
THE EVOLUTION OF OUR KNOWLEDGE OF THE RABBIT Hg BLOOD GROUP SYSTEM

Alleles	Genotypes[a]	Original typing reagents			Interaction antigen reagents		New allele detector	Antigens of multiple origin reagents			New interaction reagents			Phenotypes (type)
		anti-A	-D	-F	-I	-J	(-N)	-K	-P	-R	-T	-V	-W	
Hg^A	A/A	+	−	−	−	−	−	−	+	+	−	−	+	APRW
Hg^D	A/D	+	+	−	+	+	+	+	+	+	+	+	−	ADIJKNRTV
Hg^F	A/F	+	−	+	−	+	−	+	+	+	+	−	−	AFJKPRT
	D/D	−	+	−	−	+	+	+	−	+	−	−	−	DJKNR
	D/F	−	+	+	−	+	+	+	+	+	−	+	−	DFJKNPRV
	F/F	−	−	+	−	−	−	+	+	−	−	−	−	FKP
New allele Hg^N	N/N	−	−	−	−	+	+	+	−	+	−	−	−	JKNR
	A/N	+	−	−	+	+	+	+	+	+	+	+	−	AIJKNRTV
	F/N	−	−	+	−	+	+	+	+	+	−	+	−	FKNPRV
	D/N	−	+	−	−	+	+	+	−	+	−	−	−	DJKNR

Of much more interest to the students of the human blood group problems is the recognition that the *Hg* system can be used to gain some understanding of the mechanism by which Rh sensitization may be prevented. In particular, the regularity in which anti-A can be produced in non-A rabbits has led to the use of this particular rabbit combination to study the fate of donor red cells after they enter the circulation of an immune individual [see Mollison (65)].

Following our work (26) on the nature of immunization in the presence of circulating antibody a number of investigators, including Pollack (70) and Woodrow (73), have used the rabbit blood group systems for studies on the way in which anti-red cell antibodies may suppress isoimmunization. The protective effect of passively transferred preformed antibody is the basis of the most recent and surprisingly effective method of preventing the erythroblastosis resulting from Rh dependent maternal-fetal incompatibility.

2. Studies of the Genetics of Isoantibody Formation

The complex series of actions leading from the entrance of antigen into an individual to the ultimate appearance of an immune response in the form of a circulating antibody is one of the areas of investigation in immunology today. The isoantigens may not have the same pathway of recognition and information processing as do the heteroantigens. The factors that lead to formation of isoantibody or the failure to produce isoantibody are of concern, again, particularly in regard to models of the human Rh system.

The variations in response to isoantigens which we have studied (16, 21, 23), as well as measures of the rate of reaction of the isoantibodies with red cell antigens (18), are useful models for problems in human isoimmunization where antigenic differences between two individuals (such as the mother–fetus interaction or the donor–recipient combination in transfusion or transplantation) are the basis for unwanted immunological reactions.

3. Tolerance to Red Cell Antigens

Several years ago, when actively acquired immunological tolerance was first described by Medawar it was suggested that tolerance to red cell antigens may play some role in whether or not antibody will be formed by a mother with an incompatible fetus or by a particular transfusion recipient. We have developed a means of inducing tolerance in rabbits to rabbit red cell antigens. The method of inducing tolerance is not difficult (21) and some of the effects on transfusion (28), the anamnestic reaction (20), as well as the means of breaking tolerance (25) still have much to offer as models for studying the factors in isoimmunization.

II. THE TRANSPLANTATION ANTIGENS

A. Review of Animal Models for Transplantation Studies

The genetic factors that control histocompatibility antigens are surely among the most complex control systems in mammals; tissue transplanted between any two random individuals found in nature is universally rejected because no two are alike. The single exception is, of course, identical twins, where a single genome is expressed in two organisms, thus demonstrating that the histocompatibility antigens are indeed genetically determined.

Historically the genetics of transplantation has been studied by using genetically artificial identical twins produced by rigorous inbreeding. The most commonly used laboratory animal in transplantation studies during the past 60 years has been the inbred laboratory mouse. The systematic development of this model system began in the early 1900's with the collaborative experiments of Little and Tyzzer (57), in which the acceptance or rejection of a transplantable tumor was shown to be under the control of many independent genetic factors (later called histocompatibility loci). The success of these early experiments, along with the concomitant development of several inbred strains, quickly led to the development of the inbred mouse as the model system for the isolation and study of the histocompatibility antigens and their genetic control.

In the mouse, and in the few other species adequately studied so far, the ability to accept or reject tissue transplanted from another member of the species (allografts) is controlled by a major histocompatibility locus and many (14 or more) minor histocompatibility loci. Although allelic incompatibility for any single minor histocompatibility locus may be adequate to cause eventual rejection of non-invasive grafts, incompatibilities between a donor and a recipient for several minor loci produce a rejection response of sufficient strength to mimic the vigorous response given by allelic incompatibility at the major histocompatibility locus (37). In the mouse the antigens of the major histocompatibility locus (*H-2*) and some of the minor loci have also been shown to be expressed as blood group antigens; blood typing has therefore been a useful method of identifying histocompatibility alleles of the *H-2* locus. As we review the genetics of transplantation in the rabbit, the use, and perhaps the misuse, of this mouse model system in the design of experiments will become evident.

Very few reported experiments in the rabbit were designed to investigate directly the genetic factors involved in transplantation; however, useful information can often be gleaned from reports of transplantation experiments not specifically designed to produce genetic data. As is often the case with this method of gathering information, no rigorous proof of an hypothesis can be found, but inferences can be drawn from the data, in much the same way that the detective in a mystery novel gathers his information by bits and pieces. When all of these pieces are correctly fitted together the detective is able to solve the problem that is set for him. The geneticist, however, is often confronted with the more difficult problem of fitting all the clues together properly without prior knowledge of what problem he is trying to solve. He has no alternative but to turn to previous experiences in other species used as models for aid in interpreting these clues.

Histocompatibility experiments in the rabbit generally can be fitted into one of three categories: (a) experiments demonstrating the complexity of the histocompatibility system, (b) investigations of the role of blood group loci as histocompatibility loci, and (c) direct attempts at matching or typing for antigens identifying histocompatibility loci.

B. Complexity of the Rabbit Histocompatibility System

The complexity of the rabbit histocompatibility system was evident from Medawar's first experiments in which he demonstrated the specificity of the homograft second-set reaction. Using multiple donor–recipient combinations Medawar found that at least seven independent genetic factors were required to explain the heterogeneity found in his noninbred rabbit population (62, 63).

Klassen and Milgrom (52) have also provided data that give further insight into the complexity of the rabbit histocompatibility system, although their experiments were designed to determine the role of humoral antibody in the rejection of renal allografts. Using a very sensitive mixed agglutination technique to detect humoral antibody, they found that four of twenty-seven rabbits produced circulating antibody after the rejection of a primary renal allograft. Those recipients that did not produce detectable antibody were regrafted with a second kidney, either from the original donor or from an unrelated donor. Three recipients out of five (60%) produced antibody when the second kidney graft was not from the original donor. These data indicate that the probability of eliciting humoral antibody after second-set rejection is independent of the source of the second renal allograft in random bred rabbits.

The number of histocompatibility loci in the rabbit has been estimated directly by Chai (9) using a mathematical model developed by Chai and Chiang (10). When this model was applied to the various inbred lines of rabbits of the Roscoe B. Jackson Laboratory Colony, the estimates for the number of loci increased as inbreeding progressed; for a hypothetical gene frequency $p = 0.50$ at generation 0, the data from the fifth to seventh inbreeding generations were compatible with 22–23 loci, data from the eleventh to fourteenth generations were compatible with 56–57 loci, and data from the fifteenth to seventeenth generations

suggested 64 loci. The mathematical model used to produce these estimates does not correct for the selective maintenance of polymorphic systems during inbreeding; however, the segregation of the blood groups of animals from these lines suggests that a selective maintenance of polymorphism may indeed exist. If selection for heterozygotes was occurring, the estimates for the number of loci from the earlier generations are better estimates than those involving later inbreeding generations.

Harrison and Bartlett (40) have also used inbred rabbits from the Jackson Laboratory Colony Y-line in the fourteenth to seventeenth generation of inbreeding with coefficients of inbreeding from 0.7 to 0.9. All 15 untreated inbred animals were still segregating for histocompatibility genes; mean graft survival time for skin grafts exchanged between them was 35 days, while noninbred rabbits used as grafting controls maintained their skin grafts for 11.7 ± 3 days. The Y-line inbred rabbits were probably homozygous for the major histocompatibility locus since animals treated with mild immunosuppression maintained skin allografts from Y-line donors for a significantly prolonged time (71). Therefore, the 35-day rejection time found in the untreated inbred animals reflects only residual allelic segregation at minor histocompatibility loci.

These experiments, although not generally conceived as genetic experiments, provide some insight into the complexity of the control of the transplantation antigens in the rabbit. That there must be great diversity between animals is shown by residual graft rejection between highly inbred rabbits and also by the consistent early rejection of grafts exchanged between unrelated rabbits. However, there must also be a certain degree of similarity as shown by the kidney grafting experiments. The extent of the complexity of the rabbit histocompatibility system, especially the number and complexity of the minor histocompatibility loci, presents an insoluble problem at the present time—a problem that may have to wait for the further development of inbred strains of rabbits to use in meaningful experiments in this area.

C. Blood Groups and Histocompatibility in the Rabbit

Medawar was the first to show that the mouse and the rabbit histocompatibility systems differ when he was unable to demonstrate humoral antibody to erythrocytes following the rejection of skin allografts (64). The role of these blood group antigens as histocompatibility antigens in the rejection of skin allografts has not been consistently established, although the presence of the antigens of the *Hg*, *Hb*, *Hc*, and *Hh* blood group loci have definitely been shown by a variety of techniques to be on tissue cells as well as on erythrocytes (30, 38, 69). Hancock and Mullan (39) and Matej (59) found hemagglutinins directed at the antigens controlled by the *Hg* locus following skin grafting but were unable to demonstrate that the *Hg* locus is also a histocompatibility locus since skin grafts compatible for the *Hg* locus did not have prolonged graft survival times. Kapitchnikov *et al.* found that grafts from blood type *AF* donors on type *F* recipients were rejected faster than grafts from blood type *F* donors on type *AF* recipients; these findings imply that the *Hg* locus is a histocompatibility locus (49).

Cohen *et al.* (27) demonstrated that hemagglutinins to several of the rabbit blood group antigens were produced following skin grafting but that, of all the blood group loci, only the *Hg* blood group locus was important in graft survival time. Skin grafts compatible for antigens *A*, *D*, or *F* were maintained for a mean survival time of 9.3 ± 0.82 days; Hg^D/Hg^D homozygotes (blood type *D*) rejected skin grafts incompatible for antigen *F* with a mean survival time of 6.1 ± 0.99 days, and rejected a graft incompatible for antigen *A* in 6.4 ± 1.42 days. Hg^F/Hg^F homozygotes and Hg^D/Hg^F heterozygotes also rejected antigen *A* incompatible grafts faster than compatible grafts, 6.2 ± 1.16 days. Skin grafts exchanged between animals of all other genetic combinations of antigens controlled by the *Hg* locus were maintained with a mean survival time not significantly different from that for compatible grafts.

When all of the blood group experiments are put together as a group they demonstrate that, although the erythrocyte antigens are definitely on cells of other tissues, none of the rabbit blood group loci function as the major histocompatibility locus. Significant prolongation of grafts in blood group compatible animals did not consistently occur. However, the *Hg* locus must either function as a minor histocompatibility locus or be closely linked to a minor histocompatibility locus, since in some rabbit colonies it is strongly correlated with accelerated skin graft rejection in certain donor–recipient combinations.

D. Histocompatibility Matching and the Major Histocompatibility Locus in the Rabbit

Only a few investigators have attempted to use the rabbit in direct histocompatibility testing. Matsukura *et al.* (61) used rejection time of second set skin grafts to predict more compatible donor–recipient pairs. This test involved grafting from strain A to strain B; and, after rejection of the first graft, a second graft from the original donor and grafts from a closely related rabbit (i.e., from the same strain or breeder) and an unrelated rabbit were put on as second set grafts. The order of second set rejection was predictable in 8 out of 10 cases. This test does not lead to the identification of genetic loci but is useful only as a matching test.

Berg and Dausset (4) used a leukocyte intradermal test to predict compatibility between donor and recipient, or at least to eliminate known incompatible donors. Their test involved a panel of ten immune rabbits, each rabbit immunized with peritoneal exudate cells from a single donor. These immune rabbits were then injected with leukocytes from each potential donor and recipient to be typed; the degree of inflammation indicated antigens in common between the test rabbit and the immunizing rabbit. The number of incompatibilities between potential donors and recipients could then be estimated by comparing the pattern of positive and negative reaction of the donor and recipient against these ten rabbits. Classifying rabbits with 0, 1, or 2 incompatibilities by the intradermal leukocyte test had an average graft survival of 9.25 days, while those with 4, 5, or 6 incompatibilities had an average graft survival time of 7.5 days.

The first attempt at actual cell typing for histocompatibility and the identification of a histocompatibility locus in the rabbit was reported by Black in 1967 (5). After Black immunized a panel of rabbits with spleen cells to produce cytotoxic antisera he tested these supposedly independent antisera against a panel of unrelated cells. Comparisons using Fisher's 2×2 test showed that some antisera were significantly positively correlated, while others were significantly negatively correlated. These results suggested that all of his cytotoxic antisera detected antigens controlled by a single locus and that one group of antisera detected antigens of one allele, *RLC-la*, while a second group of antisera detected antigenic products of the alternative allele, *RLC-lb*. Animals could then be classified as la/la, la/lb. Those animals which were found to possess neither allele *RLC-la* or *RLC-lb* were classified as 1x. On 13 of the 14 recipients tested with multiple donors the longest surviving graft was of a phenotype compatible with the host for this locus. Since matching for the antigens controlled by alleles at a single histocompatibility locus has a significant effect on graft survival time, this locus must be a major histocompatibility locus. Matej (60) has recently repeated Black's experiments by using skin grafts to induce the formation of cytotoxic antibody. Matej was able to determine three antigenic types, 1, 2, and 3 and was able to demonstrate that compatibility or incompatibility for these antigenic types consistently affected graft survival time.

Our own work on the major histocompatibility locus, now called the *RL-A* locus, exploited a colony of rabbits from which the variability usually encountered in random-bred rabbits has been removed by intensive inbreeding. More than 50 cytotoxic antisera have been produced by skin graft exchange within our colony and all detect antigenic products of a single locus. Seven complex alleles of this locus have been isolated so far, and these alleles appear to be quite similar in complexity to the *H-2* locus of the mouse; i.e., several specificities are shared by some alleles, and other alleles have different specificities in common (see

TABLE III
ALLELES OF THE RLA LOCUS THAT HAVE BEEN IDENTIFIED
BY TYPING KNOWN HOMOZYGOTES WITHIN THE INBRED LINES
OF THE COHEN COLONY

Typing sera	Family line	Alleles						
		$RL\text{-}A^a$	$RL\text{-}A^b$	$RL\text{-}A^c$	$RL\text{-}A^d$	$RL\text{-}A^e$	$RL\text{-}A^f$	$RL\text{-}A^g$
R-1	Sh1	+	+	+	+	−	+	−
R-2	Sh13	+	+	−	−	−	−	−
R-3	Fa5	−	+	+	+	+	+	+
R-4	Sh5	+	+	+	+	+	−	+
R-5	Sh6	−	−	+	−	+	+	−
R-6	Sh13	+	+	−	−	+	−	+
R-7	Sh13	+	+	−	+	−	−	+
R-8	Sh1	+	+	−	−	+	−	+
R-9	Sh6	−		−	+			−
R-10	Fa5	+	−	−	−	−	−	−

Table III). There can be little doubt that this locus is the major histocompatibility locus since a significant number of *RL-A* compatible sibling skin grafts (over 50%) were maintained for more than 12 days, while incompatible grafts were all sloughed before the twelfth day postgrafting. Within our colony it is possible to match animals for alleles as well as for antigenic groups. Significant prolongation of skin graft survival time was induced with relatively mild immunosuppression (6 mg/kg/day-6-mercaptopurine for 10 days) in *RL-A* compatible animals; in incompatible animals there was no prolongation of graft survival time, even with twice the dose of 6-mercaptopurine (71).

It is assumed that the locus detected in the experiments of Black, Matej, and our laboratory is the same locus, the *RL-A* locus, since all three groups have reported a major effect on graft survival time. It also appears that all cytotoxic antisera are directed at antigens of this *RL-A* locus.

Thus, the rabbit histocompatibility system offers a significant new model for transplantation genetics, for unlike the inbred mouse, the major histocompatibility locus of the rabbit is not also a blood group locus. The *RL-A* locus does rival the *H-2* in complexity and in importance to the homograft reaction, and many homologies must also exist between the minor loci of the mouse and rabbit systems, although the minor loci of the rabbit may never be investigated to the same degree as those of the inbred mouse. One significant development in the rabbit model is the ability to produce by genetic manipulation *RL-A* identical unrelated rabbits to be used for studying some of the problems encountered in human transplantation.

REFERENCES

1. Anderson, J. R. (1955). Isoimmunization of rabbits by means of Freund-McDermott adjuvants. *Brit. J. Exp. Pathol.* **36**, 137–142.

2. Anderson, J. R. (1955). Blood groups in rabbits. *Brit. J. Haematol.* **1**, 378–385.
3. Anderson, J. R. (1956). The experimental production of erythroblastosis foetalis in rabbits. *Brit. J. Haematol.* **2**, 44–60.
4. Berg, P., and Dausset, J. (1964). Survival-time of skin grafts in rabbits compared with results obtained from an investigation of leucocyte intradermal tests. *Nature (London)* **202**, 1354–1355.
5. Black, L (1967). Histocompatibility testing in the rabbit. *Transplantation* **5**, 390–409.
6. Cameron, R. D., and Snyder, L. H. (1933). The inheritance of isohemagglutinogens in rabbits. *Ohio. J. Sci.* **33**, 50–54.
7. Campbell, D. H., Garvey, J. S., Cremer, N. E., and Susdorf, D. H. (1970). "Methods in Immunology," 2nd ed. Benjamin, New York.
8. Castle, W. E., and Keeler, C. D. (1933). Blood group inheritance in the rabbit. *Proc. Nat. Acad. Sci. U.S.* **19**, 92–98.
9. Chai, C. K. (1968). The effect of inbreeding in rabbits. Skin transplantation. *Transplantation* **6**, 689–693.
10. Chai, C. K., and Chiang, S. M. (1963). A method of estimating the number of histocompatibility loci in a sib-mating mouse population. *Genetics* **48**, 1153–1161.
11. Chaplin, H., Jr., and Mollison, P. L. (1953). Improved storage of red cells at −20°C. *Lancet* **1**, 215–218.
12. Cohen, C. (1954). Stimulatory effect of BCG on isoantibody production in the rabbit. *Proc. Soc. Exp. Biol. Med.* **85**, 375–377.
13. Cohen, C. (1955). Blood group factors in the rabbit. *J. Immunol.* **74**, 432–438.
14. Cohen, C. (1955). Blood group factors in the rabbit. II. The inheritance of six factors. *Genetics* **40**, 770–780.
15. Cohen, C. (1956). Occurrence of the three red blood cell antigens in rabbit as the result of the interaction of two genes. *Science* **123**, 935.
16. Cohen, C. (1957). The study of blood group factors in the rabbit as an approach to some problems relating to human blood groups and transfusion. *Bibl. Haematol. (Basel)* **7**, 165–170.
17. Cohen, C. (1958). On blood groups and confusion in the rabbit. *Transplant. Bull.* **5**, 21–23.
18. Cohen, C. (1958). Influences upon the agglutinability of rabbit erythrocytes in the presence of isoantibody: Genetic factors. *J. Immunol.* **80**, 73–76.
19. Cohen, C. (1958). Factors influencing isoimmunization in the rabbit. I. The effect of early postnatal immunization on the specific isoantibody response. *Proc. Soc. Exp. Biol. Med.* **99**, 607–610.
20. Cohen, C. (1959). Anamnestic reactions in rabbits showing unresponsiveness to erythrocytic isoantigens. *Transplant. Bull.* **6**, 426–428.
21. Cohen, C. (1960). Genetic influences on isoimmunization. *J. Immunol.* **85**, 144–147.
22. Cohen, C. (1960). A second example of a rabbit red blood cell antigen resulting from the interaction of two genes. *J. Immunol.* **84**, 501–506.
23. Cohen, C. (1962). Blood groups in rabbits. *Ann. N. Y. Acad. Sci.* **97**, 26–36.
24. Cohen, C. (1966). Methods in immunization. *In* "Handbook of Biochemistry and Biophysics" (H. C. Damm, ed.), pp. 567–589. World Publ., Cleveland, Ohio.
25. Cohen, C., and Allton, W. H. (1962). The tolerant state to blood group isoantigens in the rabbit. *Fed. Proc., Fed. Amer. Soc. Exp. Biol.* **21**, 41.
26. Cohen, C., and Allton, W. H. (1962). Iso-immunization in the rabbit with antibody-coated erythrocytes. *Nature (London)* **193**, 990.
27. Cohen, C., DePalma, R. G., Colberg, J. E., Tissot, R. G., and Hubay, C. A. (1964). The relationship between blood groups and histocompatibility in the rabbit. *Ann. N. Y. Acad. Sci.* **120**, 356–361.
28. Cohen, C., and Gurney, H. C. (1960). Rate of disappearance of Cr⁵¹ tagged erythrocytes in rabbits rendered unresponsive to isoantigens. *Fed. Proc., Fed. Amer. Soc. Exp. Biol.* **19**, 197.
29. Cohen, C., and Tissot, R. G. (1965). Blood groups in the rabbit. Two additional isoantibodies and the red cell antigens they identify. *J. Immunol.* **95**, 148–155.

30. Colberg, J. E., Ribak, B., and Cohen, C. (1969). The role of tissue cells in anti-A hemagglutinin formation in rabbit skin allografts. *Transplantation* **8**, 582–594.

31. Dahr, P., and Fischer, K. (1955). Über Experimentelle blutgruppenmassig bedingte Erythroblastose beim Kaninchen. *Ther. Ber.* **27**, 212–215.

32. Dahr, P., Fischer, K., and Kindler, M. (1955). Serologische Befunde bei Kaninchen—Erythroblastose. II. Weitere Ergebnisse bei Versuchen, Kranke Neugeborene zu gewinnen und neue Antikorper zu erzeugen. *Z. Hyg.* **141**, 91–102.

33. Eyquem, A., and Podliachouk, L. (1954). Les groupes sanguins des chats. *Ann. Inst. Pasteur, Paris* **87**, 91–94.

34. Fischer, W. (1935). Ueber blutgruppeneigenschaften beim Kaninchen. *Z. Immunitaetsforsch.* **86**, 97–129.

35. Fischer, W., and Klinhart, G. (1929). Ueber Isohämagglutination und Isohämolyse beim Kaninchen. *Staats. Inst. Exp. Ther. Frankfurt am Main* **22**, 31–39.

36. Fleischer, L. (1927). Studien uber die Hamagglutination bei Tier und Mensch. *Z. Immunitaets. Porsch. Exp. Ther.* **49**, 121–138.

37. Graff, R. J., Silvers, W. K., Billingham, R. E., Hildemann, W. H., and Snell, G. D. (1966). The cumulative effect of histocompatibility antigens. *Transplantation* **4**, 605–617.

38. Grothaus, E. A., and Cohen, C. (1971). Isoantigens common to erythrocytes and other tissue cells in the rabbit. *Transplantation* **11**, 122–127.

39. Hancock, D. M., and Mullan, F. A. (1962). The appearance of hemagglutinins after homografting in rabbits. *Ann. N. Y. Acad. Sci.* **99**, 534–541.

40. Harrison, H. N., and Bartlett, J. A. (1970). Immunosuppression across weak histocompatibility barriers. I. 6-Mercaptopurine in inbred rabbits: Graft survival and toxicity. *Transplantation* **10**, 358–360.

41. Heard, D. H. (1955). The recognition of four red cell antigen-antibody systems in the rabbit. *J. Hyg.* **53**, 398–407.

42. Heard, D. H., Hinde, I. T., and Mynors, L. S. (1949). An experimental study of haemolytic disease of the newborn due to isoimmunization of pregnancy. I. An attempt to produce the syndrome in the rabbit. *J. Hyg.* **47**, 119–130.

43. Ivanyi, P., Czambelova, A., Dornetzhuber, V., and Ujhelyiova, M. (1961). Immunological tolerance in rabbits. II. The influence of intraembryonal injections of group incompatible blood on the formation of immune isoagglutinins. *Folia Biol. (Prague)* **7**, 337–341.

44. Ivanyi, P., Hodzova, O., Brozman, M., and Ujhelyiova, M. (1963). Damage to the females during experimental erythroblastosis foetalis in rabbits. *Folia Biol. (Prague)* **9**, 433–439.

45. Ivanyi, P., and Ivanyi, D. (1962). The influence of the degree of relationship on the induction of tolerance in newborn rabbits. *Mech. Immunol. Tolerance, Proc. Symp., 1961*, pp. 165–171.

46. Ivanyi, P., Tomaskova, M., Smetana, K., and Dobrkovska, A. (1956). Experimental foetal erythroblastosis in rabbits. *Cesk. Biol.* **5**, 309–316.

47. Ivanyi, P., Tomaskova, M., Soukup, F., Smetana, K., and Ivanyi, J. (1958). Experimental erythroblastosis in rabbits. II. *Cesk. Biol.* **7**, 201—209.

48. Joysey, V. C. (1955). A study of the blood groups of the rabbit, with reference to the inheritance of three antigens, and the agglutinability of the red cells carrying them. *J. Exp. Zool.* **32**, 440–450.

49. Kapitchnikov, M. M., Ballantyne, D. L., Jr., and Stetson, C. A. (1962). Immunological reactions to skin homotransplantation in rabbits and rats. *Ann. N. Y. Acad. Sci.* **99**, 497–503.

50. Kellner, A., and Hedal, E. F. (1953). Experimental erythroblastosis fetalis in rabbits. I. Characterization of a pair of allelic blood group factors and their specific immune isoantibodies. *J. Exp. Med.* **97**, 33–49.

51. Kellner, A., and Hedal, E. F. (1953). Experimental erythroblastosis fetalis in rabbits. II. The passage of blood group antigens and their specific isoantibodies across the placenta. *J. Exp. Med.* **97**, 51–60.

52. Klassen, J., and Milgrom, F. (1969). The role of humoral antibodies in the rejection of renal homografts in rabbits. *Transplantation* **8**, 566–575.

53. Knopfmacher, H. P. (1942). A study of four antigenic components of rabbits' erythrocytes. *J. Immunol.* **44**, 121–128.

54. Levine, P., and Landsteiner, K. (1929). On immune isoagglutinins in rabbits. I. *J. Immunol.* **17**, 559–565.

55. Levine, P., and Landsteiner, K. (1931). On immune isoagglutinins in rabbits. II. *J. Immunol.* **21**, 513–515.

56. Lille-Szyszkowicz, M. M., and Eyquem, A. (1959). Les groupes sanguins des lapins. *Ann. Inst. Pasteur, Paris* **96**, 184–195.

57. Little, C. C., and Tyzzer, E. E. (1916). Further studies on inheritance of susceptibility to a transplantable tumor of Japanese Waltzing mice. *J. Med. Res.* **33**, 393–425.

58. Marcussen, P. V. (1936). On group differentiation in the rabbit, with special regard to the specificity of the iso-immune sera. *Z. Immunitactsforsch. Exp. Ther.* **89**, 453–477.

59. Matej, H. (1970). Relationship of rabbit erythrocyte antigens to survival of skin allografts. *Arch. Immunol. Ther. Exp.* **18**, 166–172.

60. Matej, H. (1970). Studies on histocompatibility in skin allotransplantation in rabbits. *Arch. Immunol. Ther. Exp.* **18**, 315–326.

61. Matsukura, J., Mery, A. M., Arniel, J. L., and Mathe, G. (1963). Investigation on a test of histocompatibility for allogeneic grafts. II. A study on rabbits. *Transplantation* **1**, 61–64.

62. Medawar, P. B. (1944). The behavior and fate of skin autografts and skin homografts in rabbits. *J. Anat.* **78**, 176–199.

63. Medawar, P. B. (1945). A second study of the behavior and fate of skin homografts in rabbits. *J. Anat.* **79**, 157–176.

64. Medawar, P. B. (1946). Immunity to homologous grafted skin. II. The relationship between the antigens of blood and skin. *Brit. J. Exp. Path.*, **27**, 15—24.

65. Mollison, P. L. (1967). "Blood Transfusion in Clinical Medicine," 4th ed. Davis, Philadelphia, Pennsylvania.

66. Nachtsheim, H. (1947). Eine Erbliche Fetale Erythroblastose beim Tier und ihre beziehungen zu den Gruppenfaktoren des Blutes. *Klin. Wochenschr.* **37/38**, 590–592.

66a. Nelken, D., Burnbaum, D., and Gurevitch, J. (1960). Blood groups in rabbits. *Nature (London)* **186**, 321–322.

67. Oswald, K. (1949). Uber die Moglichkeit eines Blutgruppensystems by Kaninchen. *Z. Immunol.* **106**, 364–368.

68. Padma, M. C. (1968). Some features of the isoantigens of cattle and rabbit spermatazoa. D.Sc. Thesis, University of Mysore.

69. Pious, D. A., and Mills, S. E. (1963). Killing of cultured rabbit fibroblasts with isoimmune serum. *Science* **142**, 52–53.

70. Pollack, W., Gorman, J. G., Hager, H. J., Freda, V. J., and Tripodi, D. (1968). Antibody-mediated immune suppression to the Rh factor: Animal models suggesting mechanism of action. *Transfusion* **8**, 134–145.

71. Tissot, R. G., and Cohen, C. (1972). Histocompatibility in the rabbit. Identification of the major locus. *Tissue Antigens* **2**, 267–279.

72. Williams, C. A., and Chase, M. W., eds. (1967). "Methods in Immunology and Immunochemistry." Academic Press, New York.

73. Woodrow, J. C. (1970). Rh immunization and its prevention. *Ser. Haematol.* **3**, 5–151.

74. Yokoyama, M., Takeuchi, M., Nakada, K., and Cohen, C. (1958). The study of blood group factors in the rabbit. *J. Jap. Soc. Blood Transf.* **5**, 161—166.

CHAPTER 8

Gnotobiology

Henry L. Foster

I. INTRODUCTION

Even though work on germfree systems and gnotobiotic animals dates back to before the turn of the last century (30), gnotobiology is a relatively new science. Its distinctness as a science has been made possible only through the recent availability of the necessary technology. The terms gnotobiology and gnotobiotic were coined in the late nineteen-fifties (24), in an effort to provide a more exact meaning than the loose and sometimes controversial word germfree. Gnotobiotic is derived from the Greek words gnosis and bios and means known life (gnosis, knowledge; bios, life). The implication of this term is that gnotobiotic animals would, of necessity, have to be maintained in a totally controlled environment such as an isolator system to be truly definable. Outside of the confines of an isolator system (no matter what precautions may be taken against contamination by personal or the environment) it is unlikely that true gnotobiotic conditions can be engendered.

A. History of Gnotobiotic Technology

Gnotobiology has undergone very dramatic changes since the advent of the Trexler flexible film isolator (51). Prior to this time workers used rigid isolators constructed of stainless steel or stainless steel and glass (19,41). The cost of these units was 20 to 30 times that of the plastic, flexible film units. They occupied generous amounts of floor space, since their weight and size prohibited their stacking one above the other. For a period after the advent of plastic isolators, some of the early investigators questioned the microbiological integrity of these systems, theorizing that more security would be afforded by a rigid steel wall than by a barrier of 15 to 20 ml of vinyl plastic. Time and experience have demonstrated that flexible film isolators are as reliable, and possibly even more so, as the earlier rigid models. Because of greatly reduced cost and flexibility of design, there are indeed many more plastic, flexible film units in use today than there are of any other type. Costs have been so reduced by design and material changes that some units are even considered disposable. (See Fig. 1.) As Foster (15) has shown, it is currently possible to economically provide nutrients for gnotobiotic animals in a prepackaged state on a large-scale basis.

Complete descriptions of the development of various types of germfree isolators, i.e., plastic, stainless steel, remote-controlled, etc., can be found in the report of the 1958 conference, Germfree Vertebrates: Present Status (42).

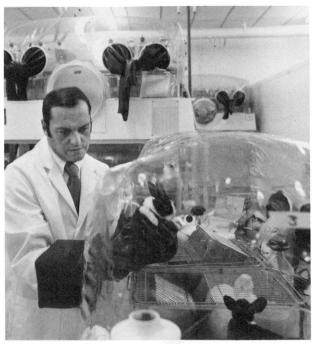

Fig. 1.   Large flexible isolator for maintaining rabbits.

B. Early Workers in Gnotobiology

An examination of the literature reveals that early studies involving gnotobiotic rabbits (56), were conducted by investigators who had successfully hand reared, maintained, and reproduced axenic (Gr.—a, without; xenos, strangers; i.e., free of all detectable microorganisms) mice and rats. This work was a natural extension of derivation and feeding methods already proved successful (2) and of attempts to render laboratory animal colonies free of pathogenic organisms by more conventional techniques (29). Obvious modifications were required, however, because of the dissimilarity of animal size and because of fundamental nutritional differences in the rabbit when compared to mice or rats. Initial studies were directed toward the development of Caesarean delivery techniques, hand-rearing instrumentation and methodology, and nutrition. Later work involved the determination of various anatomical and physiological measurements; studies utilizing the rabbit freed of its natural flora as well as axenic animals were undertaken (21,31). Further studies reported successful reproduction in the axenic state (36), verifying that the rabbit too, along with the rat and mouse, could be rendered axenic and subsequently maintained successfully through the reproductive cycle.

The gnotobiologist works principally with axenic animals and sometimes with the "associated gnotobiote," an axenic animal intentionally associated with one or more defined and recoverable microorganisms within the confines of an isolator system. He is the purist in a sense, since his interests are directed toward the employment of the animal in its gnotobiotic state, with the view to developing data on the gnotobiote or to the use of the animal as a control within the design of his experiment. Probably the greatest value of gnotobiotic technology is in the rederivation of animal populations free of their horizontally transmitted infections. For mice and rats there are numerous reports (18, 24, 27, 34, 41 of the initial axenic derivation and transfer of these species into closed, sterile isolator systems and their subsequent association with defined bacterial flora (45). Ultimately such animals were used as seed stock for new colonies which were maintained within a barrier facility (where the entrance of organisms pathogenic to the species could, hopefully, be prevented). Evidence exists (13) that this concept can be successful when exacting gnotobiotic techniques are utilized in the derivation phase and when contamination control is rigidly enforced during the mass production stage within a barrier facility. Sterilization techniques must be employed—techniques that render all materials coming in contact with the animals, such as feed (16), bedding, air, caging, and personnel (14a) free of pathogens.

C. The Importance of Gnotobiotic Rabbits

The literature contains reports of modifications of the basic gnotobiotic technique in which rabbits were derived by Caesarean section and placed almost immediately into a clean barrier facility, bypassing the axenic-isolator phase (46). This method was used for the purpose of freeing the rabbit from the specific pathogens causing the greatest disturbance to biomedical research. Collins et al. (9) have reported on experiments to liberate rabbits from the pathogenic load of the snuffles syndrome and coccidiosis. While there have been reports where certain diseases have been eliminated through hysterectomy and subsequent aseptic Caesarean birth followed by neonate transfer into a clean barrier system, many unsolved problems still exist. Without exception, these reports note the existence of nonspecific gastrointestinal disturbances, causing varying degrees of mortality. These deaths seem to be most prevalent in 8 to 12-week-old rabbits but are not necessarily confined to this age group. Workers studying the intestinal flora have demonstrated certain unique characteristics of the rabbit intestinal tract, such as very low pH in the stomach and the absence of coliforms, particularly *Escherichia coli* (17, 47). It is believed by many researchers that particular groups of anaerobes vital to the digestion of the high cellulose diet consumed by rabbits play an important role in maintaining a balanced floral distribution. Possibly these are in part responsible for restricting and even crowding out the coliforms. Although controlled experiments have not conclusively proved this hypothesis, a body of circumstantial evidence lends credibility to the theory. Work in this laboratory has demonstrated that colony stability follows immediately after the oral introduction into rabbits of cecal contents of guinea pigs containing fusobacteria, bacteroides, and possibly other yet-to-be-identified anaerobes. However, after repeated inoculations, the benefits seemed to lessen and eventually ceased.

D. Common Diseases Affecting Research

It is difficult to describe the normal complementation of rabbit flora, since it is almost impossible to define the word "normal" as it applies to this or any other species. To have some base for an appreciation of the gnotobiotic derivation of *Oryctolagus cuniculus* (the domestic rabbit), a practical and realistic approach would be to enumerate some of the more common and obvious diseases, parasites, and pathological conditions which can, hopefully, be bypassed by initial Caesarean delivery. These include snuffles (Pasteurellosis), hepatic and intestinal coccidiosis, mucoid enteritis, and the common external and internal parasitic infestations. As an example, Poole et al. (37) surveyed the disease status

of conventional rabbits procured from 10 commercial sources. The results indicated that 5% had diarrhea, 4% were suffering from respiratory distress, 14% had tapeworm cysts, 23% ear mites, 27% were infected with *Pasteurella multocida*, 33% were infested with intestinal nematodes, and 50% had coccidia. One goal of gnotobiotic technology would be to establish breeding colonies from which these diseases had been excluded by Caesarean derivation.

II. THE CAESARIAN CONCEPT

All the early studies in germfree or gnotobiotic research have clearly pointed out that the placenta acts as a highly refined filter. It is capable of preventing the vertical transmission of bacterial diseases (and probably most viral diseases for which we are able to test) as well as most parasitic and protozoan infections. Certain nematodes in their larval stage are capable of penetrating the placenta. They enter the circulatory system of the fetus, migrate through the lungs, and ultimately reach the gastrointestinal system after being swallowed. For this reason, fecal examinations should be performed and anthelmintics administered on all adults prior to mating. Ascariasis, as a practical matter, is principally a problem in the gnotobiotic derivation of canines.

A more important consideration with respect to vertical transmission in the rabbit is *Encephalitozoon cuniculi*, a small protozoan previously suspected of transversing the placenta, which has now been found to pass vertically from doe to fetus. Hunt et al. (22) reported finding encephalitozoons in serial brain sections of germfree animals submitted for examination. Since it is believed that this infection is also transmitted horizontally, possibly through the urine, it would be advantageous to perform an *in vivo* test on pregnant does prior to the Caesarean delivery of their offspring. Pakes et al. (32) reported an immunological skin test that appears to be most encouraging. When this method is further developed, rabbits can be screened prior to mating and again during pregnancy.

Does are mated, using the standard technique of carrying them to the bucks, where copulation is observed. On the thirtieth or thirty-first day of gestation the ventral abdomen is shaved from the genitalia to the sternum using electric clippers. To preclude premature kindling and for the convenience of the technical staff, progesterone can be administered at a dose level of 31 mg. Although this technique to forestall natural birth during the night is not 100% effective, it is sufficiently reliable to have become a routine procedure in many laboratories.

Two basic methods are employed when deriving animals by Caesarean section. The hysterotomy technique reported by Gustafsson (18) is a one-step procedure under complete asepsis. The hysterectomy technique (12) is conducted in two steps—one outside the sterile system, followed by completion within the sterile system.

A. Hysterotomy

A surgical flexible film isolator is made ready by incorporating the necessary surgical instruments and supplies before sterilization. The isolator configuration depends on individual laboratory preference but usually includes at least two sets of surgical sleeves and gloves, an entry port for supplies, and intake and exhaust filters.

Utilizing the hysterotomy technique, the shaved and disinfected abdomen of the anesthetized pregnant doe is brought in contact with and cemented to a window in the floor of the isolator. The surgeon, with his arms and hands encompassed by the rubber sleeves and gloves inside the sterile isolator, cuts through the plastic window into the abdominal cavity of the doe with an electric cautery or a scalpel. This creates direct sterile access from the isolator interior to the abdominal cavity and gravid uterus of the female. The uterus is incised and the fetuses removed and freed of their fetal membranes. The neonates are wiped clean with sterile, distilled water and transferred into an adjoining isolator rearing unit via a sterile connector sleeve. The uterus, placentae, and membranes are dropped back into the doe's abdominal cavity, after which she is generally euthanized, although surgical repair can be accomplished. A 12-inch ring is sometimes incorporated into the isolator floor in preparation for another procedure, permitting reclosure of the surgical opening by stretching a new mylar membrane over it. In this way, multiple hysterotomy procedures can be performed, while maintaining the sterility of the surgical unit.

B. Hysterectomy

This technique is a more popular one and has been used routinely for the Caesarean delivery of many species of animals. The pregnant, shaved doe is either injected with a local anesthetic along the ventral median line (56), given general anesthesia with pentobarbital (55) or ether, or sacrificed at the onset by cervical fracture. Cook and Dorman (10) have recently reported a simple and effective method of anesthetizing germfree rabbits with halothane for the collection of heart blood samples. It is possible that this technique could be applied to Caesarean sectioning. In the hundreds of hysterectomies performed in our laboratories, ether anesthesia has been utilized because of the maximum control available over the depth of anesthesia. See Figs. 2 to 7. Using clean, but not necessarily aseptic,

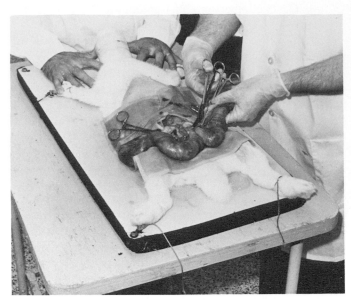

Fig. 2. Extra-isolator hysterectomy phase.

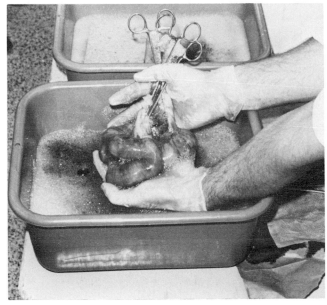

Fig. 3. Immersion of gravid uterus in germicidal bath prior to isolator entry.

techniques, a hysterectomy is performed and the gravid uterus (with surgical clamps attached to the cervical portion and to the uterine cornuae) removed. This extra-isolator phase is best performed close to, but not within the confines of, a sterile surgical isolator system. At this point, the doe can either be repaired surgically or euthanized by an over-administration of anesthesia. The intact uterus is immersed in one or more germicides for a minute or two. Iodides, quaternary ammoniums, and chlorines have all proved successful. A flexible, or rigid tubular, attachment from the

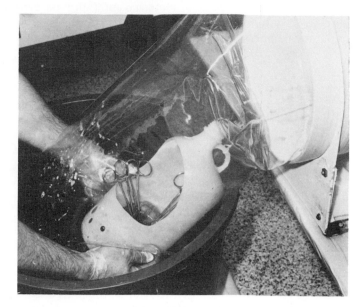

Fig. 4. Gravid uterus within plastic container guided from beneath surface of germicide into transfer tube connected to isolator.

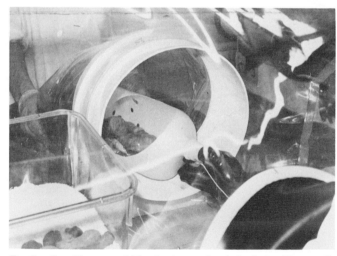

Fig. 5. Gravid uterus within plastic container being hoisted into sterile isolator system.

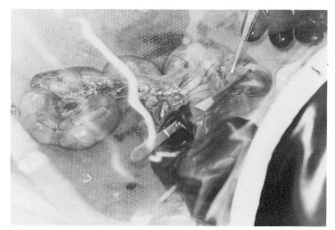

Fig. 6. Caesarean section being performed within the confines of the isolator chamber.

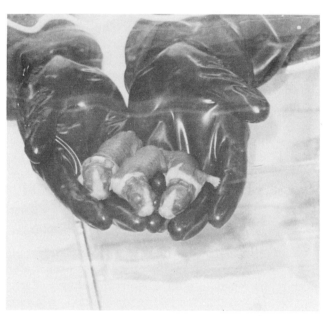

Fig. 7. Dutch Belted neonates 15 minutes post-Caesarean.

isolator projects beneath the surface of the germicide, so that after the initial immersion, the gravid uterus can be passed into the mouth of the tube beneath the germicide surface. It is then usually transferred into the isolator by the surgical technician, guided on the outside by another technician. Many different devices have been improvised for this step; but a practical apparatus consisting of a 200-ml plastic jug (one side removed, perforations for drainage in the bottom, and attached to a hoisting cord) works quite well and can be readily made in most laboratories. Once within the sterile confines of the surgical isolator, the fetuses are freed of their membranes and maternal attachments, are washed clean, and massaged to stimulate respiration. Successive procedures can be conducted with sterility maintained. The uterine transport tube must *always* remain beneath the surface of the germicide, or the sterile integrity of the surgical unit will be lost. A plastic cage or box with taut mylar held in place across the top by a large rubber band can act as the surgical table within the isolator. On this table the uterus is incised and the fetuses removed and freed of their fetal membranes. The neonates are washed clean with sterile distilled water. The mylar is punctured and the uterus and fetal membranes are dropped into the cage. A new mylar membrane is affixed across the top of the cage in preparation for another procedure. The entire operation from initial incision to birth of the last fetus should take approximately 8 minutes.

III. HAND REARING

After the Caesareans have been completed, the litters are carefully examined and the runts are removed. Two methods are available for carrying the neonates to weaning. If lactating gnotobiotic rabbits are available, foster nursing of the newborns is a relatively simple procedure. Since it is unlikely that axenic lactating does are available in most cases, preparation for hand rearing follows. The literature describes the evolution of hand-rearing technology from its beginnings to its present day status. Early workers, including Pleasants (34), followed procedures successfully utilized with mice and rats. Controlled feeding devices, consisting of a pipette, a flexible tube with regulatory clamp, and a latex nipple, constituted a workable system (36). Multiple feedings of a semisynthetic diet around the clock permitted the successful weaning of axenic rabbits. Davey (11) has described the hand feeding of gnotobiotic rats and guinea pigs, a process similar to that for axenic rabbits.

In 1956 Hills and McDonald (20) hand reared rabbits feeding from a doll's bottle on a mixture of cow's milk with casilan (Glaxo) at 4-hour intervals. Bernard (3) used the same diet in 1962 with a twice daily feeding regime. Coates and O'Donoghue (8), however, have noted a bovine milk allergy in infant germfree rabbits, accompanied by degenerating mast leukocytes and suggestions of changes in the "milk spots" of the omentum. Davis' observation of the doe's feeding habits in nature constituted the basis for the once to twice a day feedings.

Many modifications of these earlier reports greatly simplified the procedure and permitted successful results. Broadfoot (5) recognized that in the wild state the lactating doe suckled her young only once daily; she successfully followed this schedule after Caesarean delivery. Zarrow *et al.* (59) showed that the restriction of suckling in young rabbits to once each day appears to depend on the mother and not on the pups.

Improvements in hand-feeding apparatus for germfree rats and rabbits and improvements in milk formulas for axenic rabbits have been reported by Zimmermann and Wostmann (60). They note that these improvements greatly increased rabbit survival.

Before actual hand rearing commences, it is recommended that sexing of neonates allows the hand feeding process to be conducted at a ratio compatible with production colony rules: 2 to 10. Utilizing the method described by Fox and Crary (16a), neonates are sexed into conveniently sized groups and weighed. Our experience with some 2200 Caesarean-born rabbits indicates the highest survival rate occurs when the birth weight is about 50 gm; this, however, does not predicate a lack of success at 40 gm and over. It is not uncommon in larger litters to find neonatal weights of even less than 40 gm, but in nature these

young receive the benefit of maternal milk and maternal care—a situation which is impossible to duplicate precisely under artificial conditions.

Rearing Isolator

Once the Caesareans have been performed, the isolator interior is swabbed and cultures are taken from the newborns to confirm microbiological sterility. The neonates are not fed or handled during the ensuing 24 hours; this permits a report of preliminary findings on the microbial status. The young are moved either immediately after birth into the rearing unit or are transferred the following day. The hand-rearing unit usually consists of multiple pairs of gloves, two or three, to permit simultaneous feedings by 2 or 3 technicians. The isolator is approximately 5 feet long by 2 feet high and 2 feet deep. Heating is provided by means of thermostatically controlled heating pads located between the plastic floor of the isolator and its wood base enclosure. Others have temperature control devices, such as water baths, inside the isolator. A temperature of 39°C is maintained to approximate the nest temperature of the doe in the wild.

Wostmann and Pleasants (56) in 1959 described their technique of hand feeding a steam sterilized milk formula, L-449G-El (Table I). A 10-ml glass tube having a 10-ml rubber bulb on one end and a latex nipple on the other was used.

This diet was autoclaved for 20 minutes at 121°C (15 lb of pressure) after being ampuled. Sterilized in separate ampules, the following supplements were added to 160 ml of basic diet after sterilization: (a) 3 ml of a 8% $CaCO_3$ suspension (0.24 gm $CaCO_3$ or 0.1 gm Ca); (b) 1 ml each of salts solutions A and B (see Table IA); (c) 3 ml of a solu-

TABLE I

COMPOSITION AND PREPARATION OF MILK DIET L-449G-El[a]

Basic Diet:	
Milk, homogenized–vitamin D	25 ml
Cream, coffee (18% butterfat)	75 ml
Vi-Syneral + E + K[b]	0.2 ml
Mixed tocopherols (1 gm = 253 IU)	3.3 ml
DL-Methionine	0.2 gm
1-Tryptophan	0.06 gm
1-Cystine	0.1 gm
Choline H$_2$·citrate	0.1 gm
Dialyzed, lyophilized skim milk	6.0 gm

[a] After Wostmann and Pleasants (56).

[b] The 0.2 ml of Vi-Syneral fortified with vitamin E and vitamin K provides: vitamin A 1667 U.S.P. units; vitamin D 333 U.S.P. units; ascorbic acid 16.7 mg; thiamine HCl 0.33 mg; riboflavin 0.133 mg; pyridoxine HCl 0.1 mg; niacinamide 1.67; pantothenic acid 0.67 mg; mixed tocopherols (1 gm = 253 IU) 3.3 mg; menadione 0.33 mg.

TABLE IA
CONTENTS OF SALT SOLUTIONS USED TO SUPPLEMENT
STERILIZED DIET

Solution A (1 ml)			
176 mg KH$_2$PO$_4$			
180 mg Na$_2$HPO$_4$			
400 mg KI			
Solution B (1 ml)			
14 mg MgSO$_4$		2.4 mg	CuCl$_2$
4 mg MnCl$_2$·4H$_2$O		5.2 mg	ZnSO$_4$·H$_2$O
24 mg Ferric ammonium		0.8 mg	CoCl$_2$·6H$_2$O
citrate			

tion supplying thiamine-HCl, 3 mg; riboflavin, 1 mg; pyridoxine HCl, 1 mg; calcium-panthothenate, 12.5 mg; niacinamide, 2.5 mg; biotin, 0.025 mg; folic acid, 0.25 mg; isoinositol, 100 mg; choline H$_2$ citrate, 100 mg; cyanocobalamin, 0.025 mg; and (d) 5 ml of 1% calcium ascorbate dihydrate (5 mg Ca, 45 mg vitamin C).

The rabbits are nipple fed for 3 weeks at 2- to 3-hour intervals and then offered the same formula in dishes. Water is fed ad lib.

Semisolid diet (Table II) is consumed in small quantities until the milk formula is stopped at 4 weeks. Pleasants reports weight gains of about one-half as fast as naturally fed stock rabbits. Most of the losses are a result of food aspiration. He states further that rabbits in earlier experiments showed abnormally fatty livers during the early post-weaning period. This condition was corrected with his newer diet, L-449G-E1. The semisolid diet used was L-445 (Table II). The composition of the postweaning solid diet, L-461B-E1, is given in Table III.

Hair loss at the end of the hand-feeding period might be taken to imply that the milk formula is not totally adequate in the germfree environment. Even though the sterile diet, L-461B-E1, gives as good growth as does the comparable nonsterilized Purina Rabbit Chow when fed to conventional rabbits, there is still question of its suitability for germfree animals.

In all experiments on the hand rearing and subsequent

TABLE II
COMPOSITION OF SEMISOLID DIET L-445[a,b]

Ingredients	Amounts
Rolled oats, Quaker	10 gm
Ground Purina Lab Chow	10 gm
Dextrose	1 gm
NaCl	0.5 gm
Dry yeast, Fleischmans	1.5 gm
Water	100 ml

[a] After Wostmann and Pleasants (56).
[b] Supplemented within the germfree unit with 1 ml per rabbit per day of an aqueous solution containing 1% ascorbic acid and 0.05% thiamine HCl.

TABLE III
COMPOSITION OF SOLID DIET L-461B-E1[a]

Constituent	Amount/100 gm
Purina Rabbit Chow, ground	62.16 mg
Casein, Sheffield new process	7.0 mg
Ground bran	30.0 mg
BC mix-45	
Ascorbic acid	35.0 mg
Isoinositol	35.0 mg
Riboflavin	0.7 mg
Calcium pantothenate	3.5 mg
Niacinamide	3.5 mg
Pyridoxine hydrochloride	0.7 mg
Biotin	0.0175 mg
Folic acid	0.7 mg
Thiamine hydrochloride	1.75 mg
B$_{12}$, 0.1% in Mannitol	0.7 mg
Choline chloride	70.0 mg
Starch, corn	198.43 mg
Ladek-3	
Vitamin A concentrate, natural ester form	280 IU
Vitamin D, Delsterol	35 IU
Mixed tocopherols (1 gm = 253 IU)	52.5 mg
Menadione	3.5 mg
Corn oil	700 mg

[a] After Wostmann and Pleasants (56).

germfree maintenance of rabbits without exception, cecal enlargement was prominent. Wostmann and Pleasants (56), during this same period, report irregular weight gains, muscular wasting, and nephrotic signs. Their later work utilized L-462, a proven formula for germfree rats and one of basically higher protein content. It consists of a cereal mixture containing soybean meal.

In 1962 Pleasants and Wostmann (35), reported a milk formula which considerably improved the formula used earlier where survival to weaning time was approximately 23%. This new diet, 466-E3 (Table IV), contained milk protein concentrated by centrifugation to minimize protein changes, making it less susceptible to coagulation during sterilization. In addition, a nipple clamp was fastened to the glass tube and a spring pressure clamp to the nipple. This retarded the intake from a few seconds to about half a minute.

An improved solid diet 473-E5 (Table V), plus cecal ligation, improves postweaning survival and permits reproduction in the axenic state. This solid diet is mixed with water and autoclaved as a soft mush after a preliminary vacuum of 20 inches has been drawn. This is maintained for 10 minutes, broken by steam kept free-flowing for 5 minutes and followed by a holding period of 25 minutes at 121°C.

Between 1963 and 1965, Broadfoot (5) developed hand-rearing techniques suitable to the production of gnotobiotic animals. Her work, reported in 1969, avoided some of the difficulties encountered by others. Her equipment and diet

TABLE IV
COMPOSITION OF MILK DIET 466-E3 FOR INFANT RABBITS[a]

Combined in milk ampule	
Centrifuged skim milk concentrate	43.0 ml
Coffee cream (18% butterfat)	57.0 ml
Specially fortified Vi-Syneral[b]	0.2 ml
DL-Methionine	0.14 gm
L-Tryptophan	0.05 gm
Supplements to be ampuled and autoclaved separately	
B-mix 103[c]	3 ml
Salts 15A[d]	1 ml
Salts 15B[e]	1 ml
CaCO$_3$, 8% suspension in H$_2$O	3 ml
Calcium ascorbate·2H$_2$O	
1% in triple-distilled H$_2$O	5 ml

[a] After Pleasants *et al.* (36).

[b] To Vi-Syneral as purchased are added (per 0.2 ml); 0.33 mg Vitamin K$_3$ (menadione), and 3.3 mg mixed tocopherols (253 IU vitamin E per gm).

[c] B-mix 103 supplies in 3 ml triple-distilled H$_2$O: 3 mg thiamine HCl; 1 mg riboflavin; 1 mg pyridoxine HCl; 12.5 mg calcium pantothenate; 2.5 mg niacinamide; 0.25 mg folic acid; 0.025 mg biotin; 0.025 mg cyanocobalamin; 100 mg choline dihydrogen citrate; 100 mg iso-inositol.

[d] Salts 15A supplies in 1 ml distilled H$_2$O; 180 mg Na$_2$HPO$_4$; 176 mg KH$_2$PO$_4$; 0.4 mg KI.

[e] Salts 15B supplies in 1 ml distilled H$_2$O: 24 mg ferric ammonium citrate, U.S.P. XIII; 14 mg MgSO$_4$; 4 mg MnCl$_2$·4H$_2$O; 5.2 mg ZnSO$_4$·H$_2$O; 2.4 mg CuCl$_2$; 0.8 mg CoCl$_2$·6H$_2$O.

formulation were the basic reasons for her increased success.

Teats with different sized apertures, ranging from 0.009 to 0.013 inch, were used. They were manufactured using an aluminum alloy former with a stainless steel wire pin inserted in the end. This gives a small tube, the end of which is cut off to provide the size of aperture desired. Latex (Veedips, Ltd.) is used for the teat material and is set after oven-dipping at 80°C. There are many materials and material combinations which enable one to achieve varying degrees of flexibility and resiliancy.

The milk reservoir consists of a 41-ml flat-bottomed bottle, 95 mm long by 35 mm wide, with a vertical spout opening, 10 mm in diameter. There is a ring at the mouth which holds the teat in place.

The diet is a bovine colostrum base collected less than 24 hours after parturition stored in 500 ml cartons at −40°C. Prior to use, it is thawed for 36 to 48 hours and put through a valve-type homogenizer at 2500 psi at 50°C, after which it is vacuum spin freeze-dried, packed in Kilner jars, sterilized at 2.5 megarads, and stored at 4°C until reconstituted as formula RD 16 at 30°C. [Formula for RD 16: water, 100 ml and colostrum, 35 gm; Albevite (Crookes) × 2 recommended quantity.* After reconstitution with a laboratory homogenizer, the liquid diet is bottled in 100-ml medical flats and stored until use.

* The Crookes Laboratories, Ltd., Basingstoke, Hunts, England.

TABLE V
COMPOSITION OF SOLID DIET 473-E5[a]

Major ingredients (gm/100 gm dry diet)	
Whole wheat flour	31.5
Yellow corn meal	9.7
Rolled oats	21.0
Whole milk powder	10.5
Lactalbumin	10.5
Technical casein	5.3
Alfalfa meal	2.0
Liver powder	2.0
B-mix 75[b]	0.3
Ladek 3+[c]	2.0
Distilled H$_2$O[d]	30.0
Minerals (mg/100 gm dry diet)	
K acetate[d]	900
Mg acetate·4H$_2$O[d]	950
MgSO$_4$	450
CaCO$_3$	1100
Ferric citrate·5H$_2$O	450
NaCl	550
KI	4.5
MnSO$_4$·H$_2$O	57.7
CuSO$_4$·5H$_2$O	22.4
AlK(SO$_4$)$_2$·12H$_2$O	4.5
Na$_2$B$_4$O$_7$·10H$_2$O	3.0
ZnSO$_4$·H$_2$O	3.8
CoCl$_2$·6H$_2$O	3.0
NaF	1.5
MoO$_3$	3.0

[a] After Pleasants *et al.* (36).

[b] 50 mg isoinositol was added to 0.25 gm B-mix 75 especially for these diets. 0.25 gm B-mix 75 contains: 3.0 mg thiamine HCl; 1.5 mg riboflavin; 2.5 mg nicotinamide; 2.5 mg nicotinic acid; 15.0 mg calcium pantothenate; 100 mg choline Cl; 1.0 mg pyridoxine HCl; 0.2 mg pyridoxamine di HCl; 0.05 mg biotin; 0.5 mg folic acid; 2.5 mg *p*-aminobenzoic acid; 12.5 mg triturate of vitamin B$_{12}$ in mannitol (0.1% B$_{12}$); 108.8 mg cornstarch carrier.

[c] 2.0 gm Ladek 3+ contains: 4 mg (800 IU) vitamin A concentrate, natural ester, 200,000 IU/gm; 0.5 mg (100 IU) vitamin D$_2$ (Dawsterol) concentrate, 200,000 IU/gm; 10 mg DL-α-tocopheryl acetate; 150 mg (38 IU) vitamin E, mixed tocopherols, 253 IU/gm; 10 mg vitamin K$_3$ (menadione); 1825 mg corn oil carrier.

[d] Because of their deliquescence, the Mg and K acetates were added to the 30 ml of water which was mixed with the diet as the final step.

The rabbits are fed 5 ml of this diet on the first day and are gradually increased up to 20 to 25 ml per day in a single feeding. The rabbits are offered irradiated hay at about 2 weeks of age and irradiated conventional rabbit pellets at about 18 days. Fresh pellets are provided daily ad lib until weaning.

In 1957, Blount's analysis of bovine colostrum showed 3.5% fat and 17% protein (4). The fat content is almost identical to that of bovine milk, 3.63% mean. The major benefit, therefore, in colostrum is its digestibility and its high quality of protein. An analysis of rabbit milk shows a mean of 17.6% fat and an average of 14.3% protein. Apparently, added fat of rabbit milk does not accrue any particular benefit to the rabbit, since the highly nutritious protein in the cow's colostrum appears to fill all the needs of the young rabbit.

Paterson and Cook (33), in 1969, reported on their experiences in the hand rearing of Caesarean-derived rabbits. They were deriving rabbits for a pathogen-free animal colony and additionally performed Caesarean operations to provide germfree stocks. Their diet was based on homogenized, pasteurized, and freeze-dried cow's colostrum. The resultant powder, fortified with vitamins and minerals and packed in 400-gm quantities in glass jars suitable for fruit preserving, is then sterilized by γ-irradiation (4.0 megarads).

When the formula is to be used for the rearing of barrier facility animals rather than for gnotobiotic rearing in a sterile isolator system, antibiotics are added and γ-irradiation is accomplished at 2.5 megarads. This formula is given in Table VI.

The fortified colostrum powder is prepared for feeding as follows. Forty grams of the powder, 10 ml sunflower oil, and 100 ml of sterile water at 60°C are mixed and mechanically homogenized. The homogenizer is fitted with a fine mesh stainless steel sieve to remove particles that might occlude the teat orifice of the artificial feeder. Neonates are fed once daily, receiving 5 ml initially which increased gradually to approximately 40 ml at 2 weeks.

A problem of brittle bones was overcome by the inclusion of calcium glycerophosphate in the diet. The liquid formula

TABLE VI
DIET FOR HAND REARING RABBITS[a]

Colostral powder[b]	380 gm
Albevite[c]	4 gm
DL-Methionine	1.4 gm
L-Tryptophan	0.5 gm
Choline	0.6 gm
Issinosital	1.0 gm
Ascorbic acid	0.25 gm
Ferric ammonium citrate	0.24 gm
Thiamine	20 mg
Cyanocobalamin	2 mg
Biotin	10 mg
Folic acid	5 mg
Penbritin powder[d, e]	10 gm
Framomycin[e, f]	2 sachets
Calcium glycerophosphate	28 gm
$MnSO_4 \cdot 5H_2O$	0.39 gm

[a] After Paterson and Cook (33).

[b] Prepared from cows' colostrum collected within 24 hours of parturition and freeze-dried after homogenization.

[c] Albevite—a water-miscible vitamin and mineral powder. The Crookes Laboratories Ltd., Basingstoke, Hants, England.

[d] Penbritin Veterinary Powder contains 100 mg of ampicillin per 3.4 gm. Beecham Research Laboratories, Brentford, Middlesex, England.

[e] These antibiotics are used when feeding pathogen-free rabbits.

[f] Framomycin packs of 100 sachets each contain 250 mg of framycetin sulfate. The Crookes Laboratories Ltd., Basingstoke, Hants, England.

is continued until the thirty-fifth day; the rabbits are gradually weaned onto a solid diet. Lobund diet 478 is utilized to feed the germfree weaned rabbits.

In 1971 Appel *et al.* (1) from Germany reported on production performed between 1968 and 1970, during which some 270 hysterectomies were performed and some 900 young rabbits raised. These workers found that with modifications of formulas (which utilized a base of cow's milk) they encountered difficulties in actual formulation and also in sterilization. Animal mortality ranged from 30 to 90%. Reasoning that the best formula would probably be milk from lactating does, they developed an efficient breast pump which enabled them to collect about 40 ml per milking; this procedure was repeated twice weekly. The milk was autoclaved at 115°C for no longer than 15 minutes to prevent carmelization. The neonates were fed 24 hours after hysterectomy-Caesarean birth. Young rabbits were fed twice daily until 3 days of age and then once daily.

Their equipment consisted of a 50-ml plastic bottle with a small opening cut in the bottom, so that the flow could be controlled by placing a finger over the opening. Two orifices in the rubber teat allowed sufficient flow; the milk was strained prior to use to remove lumps which may have resulted from sterilization.

The autoclaved milk shows no significant changes in nutrient value and will keep safely for 4 months in a refrigerator. In their experiments, Appel and his coworkers reported a loss of only 8.5% from birth to 28 days, compared to an 8.9% loss in naturally reared rabbits. This work, however, was not done in a germfree milieu, since the rabbits were associated with a defined flora one day after birth within an isolator system. The rabbits were maintained in the gnotobiotic isolators throughout the hand-rearing period. Owen and his colleagues (31) have published their findings on the characterization and measuremrnt of isoalloxazines in the ceca of conventional rabbits kept on an adequate diet, conventional rabbits maintained on the same diet from vitamin B_2 has been removed, and germfree rabbits fed an adequate diet. Riboflavin (up to 12 mg), unaccompanied by any of its degradation products, was found in the cecum of the germfree animals and 8–9 μg per ml were found in their urine. Hydroxymethylflavine was found in the cecum of conventional rabbits. The germfree cecal contents averaged 22.4% of body weight, while, for conventional rabbits, this figure was only 10.5%.

In our laboratories we employ both a slightly different technique and a different philosophy than those already described. Rabbits are not kept axenic through weaning nor are they associated at birth. In nature the intestinal tract of the rabbit remains relatively free of microorganisms for the first two weeks of life. Canas-Rodriguez and Smith (6) in 1966 described the unique enzymatic action of the rabbit's stomach, which in combination with the high fat content of rabbit's milk, renders the stomach and subsequently the small intestine sterile. They postulate that the presence of free fatty acids, especially *n*-decanoic acid and *n*-octanoic acid, in the stomachs of young rabbits is responsible for this sterility. Interestingly, the sterile intestinal contents demonstrate a lack of antimicrobial activity, irrespective of pH. This indicates the absorption of the active principle in the small intestine. With this knowledge, we maintain complete sterility for 14 days. The sterile homogenized milk formulation is kept without refrigeration within the confines of the axenic environment of the isolator in sealed bottles. The 100-ml sealed glass bottles are sprayed with standard peracetic acid techniques and transferred into the isolater. The hysterectomy-delivered rabbits remain unattended for 24 hours in a sterile handrearing unit. Plastic 50-ml bottles with latex nipples are used for the once-a-day feedings which occur in the forenoon (Fig. 8). Swabs from the rabbits are taken once or twice a week to ascertain sterility. At 14 days the young rabbits are associated with a bacterial mixture consisting of the modified Schaedler flora (45) containing in addition some additional aerobes and anaerobes retrieved from conventional guinea pigs and rabbits. Sterilized prefortified commercial rabbit formula is provided ab lib at 21 days. Nipple feeding is discontinued entirely at the twenty-eighth day, and the young rabbits are supplied with water and sterile pellets ad lib. For detailed information on diet formulas, diet sterilization, and diet storage for germfree rabbits, as well as other animals, the reader is referred to a recent work of Reddy, Wostmann, and Pleasants (39).

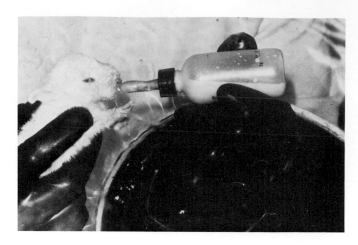

Fig. 8. Hand feeding gnotobiotic 18-day-old rabbit within a flexible film isolator system.

IV. INTESTINAL FLORA

An understanding of the rabbit flora presents some unique challenges and provides some problems yet to be fully solved. Smith (48) in 1965 clearly demonstrated certain characteristics of the gastrointestinal tract which are unique to this species. *Escherichia coli, Clostridium welchii, Lactobacillus,* and yeasts were not recovered from the stomach, small intestine, cecum, or feces. However, *Bacteroides* were found throughout the gastrointestingal tract. This is somewhat contrary to his findings in the mouse, rat, and hamster. The guinea pig showed no *E. coli* or *C. welchii. Bacteroides* could be recovered only in the cecum and feces. In pigs and calves, positive dilatation reactions were produced in ligated segments of intestine by the injection of bacteria-free fluids prepared from soft agar cultures of strains of *E. coli* that had been isolated from animals with diarrhea. Dilatation has also been produced when live bacterial suspensions have been injected into ligated intestines (50).

Smith, in this same work, pointed out the very low pH (1.9) of the rabbit's stomach contents when compared with other species. This very acid condition creates a bactericidal environment—a fact which should be kept in mind when the oral administration of microorganisms to gnotobiotes is contemplated. It is interesting to note that no other antibacterial mechanisms appears to exist in adults, for when the acidity of the stomach contents was adjusted to pH 7.0, they supported bacterial growth *in vitro.* The almost complete absence of living organisms from the stomachs and small intestines of suckling rabbits has been demonstrated to be due primarily to antimicrobial substances produced in the stomach contents by enzymic action on the chloroform-soluble fraction of rabbit milk (49). It thus appears from Smith's work and later reports of the same year, Smith (47), that *Bacteroides* form the bulk of the fecal flora of breeding does. Streptococci are usually present in much smaller numbers, while other organisms are present in very small numbers and only occasionally. These very interesting findings, taken together with the lack of more detailed information in the literature and coupled with the small number of colonies built from germfree stocks, suggest that there are many answers yet to be found.

An interesting phenomenon involving some 250 Caesarean-delivered, gnotobiotic rabbits appeared over an eight-month period. These rabbits were hand reared in sterile isolators and maintained in an axenic state for two weeks prior to their association with a stabilized flora obtained from a Caesarean-derived guinea pig colony maintained in a barrier facility. The rabbits appeared to thrive and to reproduce in the isolator environment without mortality (except for two nonspecific, explainable deaths). In rabbits reared in the identical manner but maintained for 18 months outside of an isolator, but within barrier facilities there were sporadic cases of acute diarrhea followed by deaths in 5 to 10 hours. These deaths occurred in the absence of known rabbit pathogens and without specific gross or histological lesions. Studies are underway to determine the exact nature of the isolator-rabbit flora, both qualitatively and quantitatively, with the hope that differences can be demonstrated between the rabbits maintained exclusively in isolators with gnotobiotic techniques and the colony maintained outside of isolators.

Work conducted in our laboratories has concentrated on the cultivation of organisms which are clearly seen in stained fecal smears. One might classify some of these as fusobacteria. We believe that combinations of fastidious aneaerobes are required for the population of the gut tract of the gnotobiote before the animal can cope successfully with conventional or even barrier colony environments. There have been many reports of enteritis in which coliforms were recovered at death or from moribund animals. In the healthy state, coliforms should never be found; if any occur, they should be in very low concentrations.

We are also studying the order of association of certain microorganisms with the gnotobiote, since it is possible that irreversible qualitative and quantitative bacterial relationships may be established.

Those pioneers of gnotobiology who have attempted the derivation of gnotobiotic rabbits for utilization as seed stock in disease-free colonies have taken one of three approaches. Firstly, the Caesarean-delivered neonate can simply be exposed to the ambient atmosphere where it will pick up organisms from personnel and from the environment. A second approach has been the addition of a microflora successfully employed in mice and rats, after which one continues along the lines of the first approach. Lastly, and ideally, one can attempt to recover, in pure culture, organisms from conventional rabbits, subsequently reintroducing these after determining that the flora is devoid of pathogens.

V. STERILITY TESTING

It is of the utmost importance that frequent and continuing bacteriological testing of animals and isolators be conducted. These examinations should begin immediately after caesarean section. Because the task of preparing isolator systems and formulating and hand-feeding milk formulas is an arduous one, it is an urgent matter to learn of an accidental contamination as soon as it occurs. If the project dictates that the gnotobiote be axenic, the presence of a microorganism causes suspension of the work. If, however, the project is one in which the neonates will subsequently be associated with a defined microflora, it is possible that the contaminant, if in pure culture, would be admissible, thus permitting continuance of the feeding program.

Routine microbiology, as described by Wagner (54) in 1959 and as practiced in quality gnotobiotic facilities, is carried out. This practice usually consists of stained wet mount direct smears of excreta and the cultivation of fecal material at 22°, 37°, and 55°C, utilizing a variety of different media. There are many modifications of this basic technique.

Usually, contaminations can be traced to their origin, although the break in the system or procedures occasionally remains elusive. Mechanical contamination can occur through glove puncture, a break in the plastic film, or an undetected puncture in the filter system. Incomplete sterilization of supplies and poor technique during introduction of supplies are also common causes of contamination. It is therefore essential that personnel be properly trained in these procedures and that rigid adherence be paid to proper practice.

VI. GNOTOBIOTIC ISOLATORS AND THEIR MAINTENANCE

The basic unit for maintaining gnotobiotes is the isolater system. Since rabbits require fairly large living areas, plastic film, vinyl isolators offer the maximum flexibility with respect to size and configuration. In our laboratories we utilize an isolator 6 feet in length by 36 inches in height and width. Two wire mesh cages (mesh size 1 inch by 1 inch) are supported on pipe racks 18 inches above the floor of the isolator so that animal droppings can be removed readily and so that the maximum floor area can be made available for supplies. An 18-inch Fiberglas port is positioned in the center of one side with a pair of neoprene 15-ml sleeves on each side. This maximizes area for the entrance of supplies and enables two technicians to work simultaneously. With rabbits, particularly adults, some of the routine operations (such as collecting fecal samples) requires two technicians. The rabbits are supplied sterile water in 1000-ml flasks with stainless steel sipper tubes inserted into one-hole rubber stoppers. The number of bottles furnished is governed by the number of rabbits present in each cage. To minimize glove entries, it is advisable to provide sufficient water for a 24-hour period. Automatic watering devices based on heat sterilization and/or final mechanical filtration systems are worthy of consideration, but the ever-present risk of a central failure and a consequent large-scale contamination makes the conservative gnotobiologist wary of any purported labor-saving advantages.

Diets

Weaned axenic rabbits grow and reproduce on Lobund formula L-478 (38) and on L-478-E1, which contains added cellulose powder (nonnutritive fiber, General Biochemicals, Inc.). Gnotobiotes with defined flora, maintained either in isolator systems or in barrier rooms, do quite well on pre-fortified rabbit food designed to withstand autoclaving at 121°C for 25 minutes. The food can be sterilized in 50-lb paper bags if the porosity of the bag permits the flow of steam. The chamber should be evacuated with high vacuum equipment and the vacuum broken with steam while the pump continues to operate. This insures complete steam saturation and penetration. Foster *et al.* (16) in 1964 described a pasteurization process which is applicable to rabbit diets, even though the work was concerned with mouse and rat formulas. The length of time and temperature can vary according to the pellet hardness and size.

VII. MICROBIOLOGICAL MONITORING AND HISTOLOGICAL EVALUATION

After the hard-rearing and weaning phases have been completed, the ultimate use of the rabbits will determine the monitoring procedures required. If the plans are to raise axenic rabbits for special research projects, monitoring becomes relatively simple, for the true axenic is easily monitored. However, if the progeny are to be used for seed stock of defined animal colonies, the task is one of different proportion. Weisbroth and Scher (55) described monitoring techniques for the commonly occurring pathogens and parasites of rabbits. The major diseases appear to be pasteurellosis (sniffles), coccidiosis, and enteritis. The indirect fluorescent antibody technique described appears to be a practical and efficient diagnostic tool for *Pasteurella multocida*. Fecal examination for coccidial oocysts and helminth ova are routine, straightforward procedures. Detecting external parasites is also readily performed with standard techniques. However, early diagnosis for the elusive enteric conditions, mucoid enteritis and nonspecific acute enteritis, does not exist, since their primary etiologies are not known. Thus one must define the initial flora and monitor it for shifts in bacterial population and for the occurrence of additional organisms.

Histological examination is important, since it may reveal subtle nutritional deficiencies or near deficiencies. It is definitive and reliable method for the diagnosis of encephalitozoonosis. Unless histology is performed and until a diagnostic test is totally perfected, the disease will remain occult.

IX. RESEARCH UTILIZATION

Utilization of axenic gnotobiotes has been fairly limited, principally because of the high cost of maintaining an animal the size of the rabbit in the gnotobiotic state and

secondarily, because of its relatively poor reproductive performance (14). An excellent summary of investigations employing germfree vertebrates can be found in the recent publication of Coates (7).

Yoshida *et al.* (57) conducted studies to determine the efficiency of digestion in germfree and conventional rabbits. They placed collars on the rabbits to prevent coprophagy while feeding an autoclaved diet with added cellulose. After analyzing fecal excretion to determine nutrient digestibility, they suggested that the intestinal flora play an important role in carbohydrate digestion.

Reddy *et al.* (38) employed the axenic rabbit to study iron and copper utilization. They concluded that germfree rabbits fed diet L-478 maintained normal levels of circulating iron yet stored less iron in the liver, spleen, and kidney than do conventional rabbits. There appeared to be no difference between germfree and conventional rabbits in the concentration of content of copper in the liver, spleen, and kidney, with the exception of total kidney and spleen copper. This last difference, however, is possibly attributable to the organ weight difference in germfree and conventional rabbits.

A. Seed Stock for Disease-Free Colonies

All diseases of known infectious etiology can be bypassed by Caesarean intervention, with the exception of vertically transmitted encephalitozoonosis. Gnotobiotic rabbits can successfully be hand reared and subsequently transferred to clean environments as seed stock for disease-free colonies. Although the classic snuffles syndrome, coccidiosis, and both external and intestinal parasites can be eliminated, there appears to be an unsolved enteritis of unknown etiology in both Caesarean-derived and conventional stocks. Notwithstanding this, a cleaner and healthier rabbit can be obtained as a nucleus stock for healthy rabbit colonies. Experimenters are better off when they use a well-defined animal, since so many variables occasioned by the horizontally transmitted infections can be circumvented.

B. Research on Rabbit Diseases

Gnotobiote or defined, barrier-reared rabbits provide excellent working substrates for the study of the pathogenesis of known rabbit diseases and for investigation of those of unknown etiology. Unfortunately, large quantities of gnotobiotes do not exist, nor are large populations of rabbits with defined flora available. It is evident to the biomedical research community that long-term studies, as for example toxicology investigations and teratological research, require rabbits of a defined microbial status. Such animals must also be able to survive the stresses of experimentation and be free of latent infection which can ultimately confuse experimental data.

REFERENCES

1. Appel, K.-R., Busse, H., Schulz, K.-D., and Werner, W. (1971). Beitrag zur handaufzucht von gnotobiotischen und SPF-kaninchen. *Z. Versuchstierk.* **13**, 282–290.

2. Bender, L. (1925). Spontaneous central nervous system lesions in the laboratory rabbit. *Amer. J. Pathol.* **1**, 653–656.

3. Bernard, E. (1962). Methods and problems concerned with hand-rearing rabbits. *J. Anim. Tech. Ass.* **13**, 35–40.

4. Blount, W. P. (1957). "Rabbits' Ailments." "Fur and Feather" Ltd., Idle, Bradford.

5. Broadfoot, J. (1969). Hand rearing rabbits. *J. Inst. Anim. Tech.* **20**, 91–99.

6. Canas-Rodriguez, A., and Smith, W. H. (1966). The identification of the antimicrobial factors of the stomach contents of suckling rabbits. *Biochem. J.* **100**, 79–82.

7. Coates, M. E., ed. (1968). "The Germ-Free Animal in Research." Academic Press, New York.

8. Coates, M. E., and O'Donoghue, P. N. (1967). Milk allergy in infant germfree rabbits. *Nature* (*London*) **213**, 307–308.

9. Collins, G. R., Scher, S., and Bond, E. (1963). The establishment and maintenance of a specific-pathogen-free rabbit colony. *Lab. Anim. Care* **13**, 544.

10. Cook, R., and Dorman, R. G. (1969). Anaesthesia of germfree rabbits and rats with halothane. *Lab. Anim.* **3**, 101–106.

11. Davey, D. G. (1959). Establishing and maintaining a colony of specific pathogen free mice, rats, and guinea pigs. *Lab. Anim. Cent. Collect. Pap.* **8**, 17–34.

12. Foster, H. L. (1959). A procedure for obtaining nucleus stock for a pathogen-free animal colony. *Proc. Anim. Care Panel* **9**, 135–142.

13. Foster, H. L. (1959). Housing of disease-free vertebrates. *Ann. N. Y. Acad. Sci.* **78**, 80–88.

14. Foster, H. L. (1962). Establishment and operation of S. P. F. colonies. *In* "The Problems of Laboratory Animal Disease" (R. J. C. Harris, ed.), pp. 249–259. Academic Press, New York.

14a. Foster, H. L. (1963). Specific pathogen-free animals. *In* "Animals for Research" (W. Lane-Petter, ed.), pp. 109–138. Academic Press, New York.

15. Foster, H. L. (1968). A canned sterile source of water and feed for the breeding and maintenance of gnotobiotic mice. *In* "Advances in Germfree Research and Gnotobiology" (M. Miyakawa and T. D. Luckey, eds.), pp. 20–29. CRC Press, Cleveland, Ohio.

16. Foster, H. L., Black, C. L., and Pfau, E. S. (1964). A pasteurization process for pelleted diets. *Lab. Anim. Care* **17**, 400–405.

16a. Fox, R. R., and Crary, D. D. (1972). A simple technique for the sexing of newborn rabbits. *Lab. Anim. Sci.* **22** (4), 556–558.

17. Glantz, P. J. (1968). Identification of unclassified *Escherichia coli* strains. *Appl. Microbiol.* **16**, 417–418.

18. Gustafsson, B. (1948). Germfree rearing of rats. *Acta Pathol. Microbiol. Scand., Suppl.* **73**, 1–130.

19. Gustafsson, B. E. (1959). Lightweight stainless steel systems for rearing germfree animals. *Ann. N. Y. Acad. Sci.* **78**, 17–28.

20. Hills, D. M., and McDonald, I. (1956). Hand rearing of rabbits. *Nature* (*London*) **178**, 704–706.

21. Hofmann, A. F., Mosback, E. H., and Sweeley, C. C. (1969). Bile acid conposition of bile from germfree rabbits. *Biochim. Biophys. Acta* **176**, 204–207.

22. Hunt, R. D., King, N. W., and Foster, H. L. (1972). Encephalito-zoonosis; Evidence for vertical transmission. *J. Infec. Dis.* **126**(2), 212–214.

23. Koller, L. D. (1969). Spontaneous *Nosema cuniculi* infection in laboratory rabbits. *J. Amer. Vet. Med. Ass.* **155**, 1108–1114.

24. Luckey, T. D. (1963). "Germfree Life and Gnotobiology." Academic Press, New York.

25. Mack, R. (1962). Disorders of the digestive tract of domesticated rabbits. *Vet. Bull.* **32**, 191–199.

26. Malherbe, H., and Munday, V. (1958). *Encephalitozoon cuniculi* infection of laboratory rabbits and mice in South Africa. *J. S. Afr. Vet. Med. Ass.* **29**, 241–246.

27. Miyakawa, M. (1952). Germfree rearing of animals. *Igaku No Ayumi* **16**, 137–146.

28. Nelson, J. B. (1951). Studies on endemic pneumonia of the albino rat. IV. Development of a rat colony free from respiratory infections. *J. Exp. Med.* **94**, 377–386.

29. Nelson, J. B., and Collins, G. R. (1961). The establishment and maintenance of a specific pathogen-free colony of swiss mice. *Proc. Anim. Care Panel* **11**, 65–72.

30. Nuttall, G. H. F., and Thierfelder, H. (1897). Thierisches leben ohne bakterien im verdauungskanal. III. Mittheilung. Versuche an Huhnern. *Hoppe-Seyler's Z. Physiol. Chem.* **23**, 231–235.

31. Owen, E. C., West, D. W., and Coates, M. E. (1970). Metabolism of riboflavine in germfree and conventional rabbits. *Brit. J. Nutr.* **24**, 259–267.

32. Pakes, S. P., Shadduck, J. A., and Olsen, R. G. (1972). A diagnostic skin test for encephalitozoonosis in rabbits. *Lab. Anim. Sci.* **22**(6) 870–877.

33. Paterson, J. S., and Cook, R. (1969). Production and use of pathogen-free animals.' *In* "The I.A.T. Manual of Laboratory Animal Practice and Techniques." 2nd ed. Crosby Lockwood, London.

34. Pleasants, J. R. (1959). Rearing germfree Caesarean-born rats, mice and rabbits through weaning. *Ann. N. Y. Acad. Sci.* **78**, 116–126.

35. Pleasants, J. R., and Wostmann, B. S. (1962). Rearing germfree rats on chemically defined antigen low diets. *Proc. Indiana Acad. Sci.* **72**, 87.

36. Pleasants, J. R., Wostmann, B. S., and Zimmermann, D. R. (1964). Improved hand rearing methods for small rodents. *Lab. Anim. Care* **14**, 37–47.

37. Poole, C. M., Keenan, W. G., Tolle, D. V., Fritz, T. E., Brennan, P. C., Simkins, R. C., and Flynn, R. J. (1967). Disease status of commercially produced rabbits. *U.S., At. Energy Comm., Argonne Nat. Lab.,* **ANL-7409**, 219–220.

38. Reddy, B. S., Pleasants, J. R., Zimmermann, D. R., and Wostmann, B. S. (1965). Iron and copper utilization in rabbits as affected by diet and germfree status. *J. Nutr.* **87**, 189–196.

39. Reddy, B. S., Wostmann, B. S., and Pleasants, J. R. (1968). Nutritionally adequate diets for germfree animals. *In* "The Germ-Free Animal in Research" (M. E. Coates, ed.), pp. 87–111. Academic Press, New York.

40. Reyniers, J. A. (1949). Some observations on rearing laboratory vertebrates germfree. *Proc. N. Y. State Ass. Pub. Health Lab.* **28**, 60–69.

41. Reyniers, J. A. (1959). Design and operation of apparatus for rearing germfree animals. *Ann. N. Y. Acad. Sci.* **78**, 1–400.

42. Reyniers, J. A., ed. (1959). Germfree vertebrates: Present status. *Ann. N. Y. Acad. Sci.* **78**, 1–400.

43. Richter, C. B., and Hendren, R. L. (1969). The pathology and epidemiology of acute enteritis in captive cottontail rabbits. *Pathol. Vet.* **6**, 159–175.

44. Robinson, J. J. (1954). Common infectious diseases of laboratory rabbits questionably attributed to *Encephalitozoon cuniculi.* *AMA Arch. Pathol.* **58**, 71–84.

45. Schaedler, R. W., Dubos, R., and Costello, R. (1965). Association of germfree mice with bacteria isolated from normal mice. *J. Exp. Med.* **122**, 77–82.

46. Scher, S., Collins, G. R., and Weisbroth, S. H. (1969). The establishment of a specific pathogen free rabbit breeding colony. I. Procedures for establishment and maintenance. *Lab. Anim. Care* **19**, 610–616.

47. Smith, H. W. (1965). Observations on the flora of the alimentary tract of animals and factors affecting its composition. *J. Pathol. Bacteriol.* **89**, 95.

48. Smith, H. W. (1965). The development of the flora of the alimentary tract in young animals. *J. Pathol. Bacteriol.* **90**, 495–513.

49. Smith, H. W. (1966). The antimicrobial activity of the stomach contents of suckling rabbits. *J. Pathol. Bacteriol.* **91**, 1–9.

50. Smith, H. W., and Halls, S. (1967). Studies on *Escherichia coli* enterotoxin. *J. Pathol. Bacteriol.* **93**, 531–543.

51. Trexler, P. C. (1959). The use of plastics in the design of isolator systems. *Ann. N. Y. Acad. Sci.* **78**, 29–36.

52. Trexler, P. C., and Reynolds, L. I. (1957). Flexible film apparatus for the rearing and use of germfree animals. *Appl. Microbiol.* **5**, 406–412.

53. Tufts, J. M. (1966). Unidentified rabbit problem—hemorrhagic colitis. *Lab. Anim. Dig.* **2**, 16–17.

54. Wagner, M. (1959). Determination of germfree status. *Ann. N. Y. Acad. Sci.* **78**, 89–101.

55. Weisbroth, S. H., and Scher, S. (1969). The establishment of a specific-pathogen-free rabbit breeding colony. II. Monitoring for disease and health statistics. *Lab. Anim. Care* **19**, 795–799.

56. Wostmann, B. S., and Pleasants, J. R. (1959). Rearing of germfree rabbits. *Proc. Anim. Care Panel* **9**, 47–54.

57. Yoshida, T., Pleasants, J. R., Reddy, B. S., and Wostmann, B. S. (1968). Efficiency of digestion in germ-free and conventional rabbits. *Brit. J. Nutr.* **22**, 723–737.

58. Yuill, T. M., and Hanson, R. P. (1965). Coliform enteritis of cottontail rabbits. *J. Bacteriol.* **89**, 1–8.

59. Zarrow, M. X., Denenberg, V. H., and Anderson, C. O. (1965). Rabbit: *Frequency of suckling in the pup. Science* **150**, 1835–1836.

60. Zimmermann, D. R., and Wostmann, B. S. (1963). Hand-feeding of suckling rodents. *Lab. Anim. Care* **13**, 582–587.

CHAPTER 9

Bacterial Diseases

Ronald E. Flatt

I. INTRODUCTION

The diseases caused by bacteria are among the most common of the naturally occurring diseases of rabbits. This group of diseases has been responsible for considerable economic loss to rabbit breeders and investigators using rabbits as research animals. Some diseases of historical interest which are not common, such as tuberculosis and tularemia, are included in this chapter to assist those with special interest in these diseases.

II. PASTEURELLOSIS

The term pasteurellosis will be used in this chapter to refer to diseases caused by *Pasteurella multocida*. Rabbits are very susceptible to infection with *P. multocida* and a number of clinical forms occur. Among these forms are snuffles, enzootic pneumonia, otitis media, conjunctivitis, pyometra, orchitis, abscesses, as well as generalized septicemias. Since there are several clinical forms of this disease, the term pasteurellosis is used only in reference to the entire group of diseases. The various clinical forms will be discussed individually.

A. Snuffles

1. History

A point of confusion has arisen concerning the use of the term snuffles. Some have broadened the original meaning to include not only rhinitis, for which it was intended, but also pneumonia, otitis media, and conjunctivitis. In this chapter, the term snuffles will be used to refer to rhinitis and paranasal sinusitis characterized by a serous, mucus, or mucopurulent nasal exudate.

In 1920, Ferry and Hoskins (59) studied the etiology of snuffles. They concluded that snuffles was caused by any one of several bacterial organisms; however *Bordetella bronchiseptica* was considered to be responsible in the majority of cases (they also considered this organism to be the cause of distemper in dogs). They found that *Pasteurella multocida* and *Staphylococcus aureus* were important etiological factors since they were found in a large number of cases. Hoskins reported similar findings in 1920 (110). He also indicated that the clinical disease may subside only to recur later. This is an important factor when control of the disease is attempted. When rabbits with snuffles were killed and necropsied, Hoskins found the nasal sinuses to be the only site of lesions. He found congestion of the mucous membranes and mucopurulent exudate.

In 1923, McCartney and Olitsky (144) reported that snuffles consisted of chronic inflammation of both the nasal passages and the paranasal sinuses. They found lesions in 10 to 25% of clinically normal rabbits and indicated that these carriers would develop clinical disease when their resistance was lowered due to stresses such as chilling or experimentation. The results of bacterial cultures yielded the same three organisms mentioned previously; however, they concluded that these organisms did not play a primary role in causing snuffles since they were unable to reproduce the disease experimentally.

The series of papers by Webster (244–250), Smith and Webster (213), and Smith (212) is the most definitive work on pasteurellosis in rabbits. It remains as a classic to the present day. They defined the disease, described its clinical appearance, and noted that there was a relationship between snuffles and the occasional finding of chronic abscesses, pleuropneumonia, otitis media, meningitis, and septicemias in their rabbit colony.

Of 100 randomly selected rabbits in Webster's colony, 58 had snuffles and *P. multocida* was isolated from 55 of the 58 affected rabbits. Webster also determined that this organism was present in the nasal cavity of some clinically normal rabbits and that 60 to 70% of the rabbits in their entire colony harbored this organism. Webster concluded that *P. multocida* was the predominant organism in the nasal passage of rabbits affected with snuffles and that *B. bronchiseptica* could be found commonly in the nasal cavity of either normal or affected rabbits. After making frequent nasal bacteriological examinations and clinical observations for snuffles, Webster also concluded that *P. multocida* was present in the nasal cavity at some time prior to the appearance of the clinical disease and that the organism remained present throughout its clinical course. When the

clinical disease subsided, the numbers of *P. multocida* isolated from the nasal cavity was diminished or there was complete disappearance of the organism. Attempts to reproduce snuffles by introducing *P. multocida* into the nasal cavity of rabbits yielded mixed results. Some rabbits became carriers only, some developed snuffles, and some developed snuffles and pneumonia. Dr. Webster attributed the difference in response to resistance of the individual rabbits and the virulence of the strain of organism used.

2. Etiology

Pasteurella multocida, the cause of snuffles, is a gram-negative, nonspore-forming, short, bipolar rod.

This organism, (known previously as *Bacterium lepiseptica*, *Bacillus lepiseptica*, *Pasteurella lepiseptica*, and *P. septica*) may form smooth, rough, or mucoid colonies when grown on artificial media (36). An isolate may undergo disassociation, that is, change from one type to another when grown on artificial media. The mucoid type is thought to be most pathogenic. Recent isolates from clinical material are usually of the mucoid variety.

A hemagglutination test was developed by Carter (36) to differentiate several serological types of *P. multocida*. He established 5 antigenic groups by this method and named them A, B, C, D, and E. Serotypes A and D were most commonly isolated from rabbits by Carter (37) while Hagen (101) reported types B and C to be most common in his rabbits.

Although *Staphylococcus aureas* and *Bordetella bronchiseptica* are commonly isolated from the nasal sinuses of both healthy rabbits and those with snuffles, the work of Webster (245–248) and Smith (212) still stands unchallenged in showing that *P. multocida* is the etiological agent of snuffles.

3. Incidence

There is little precise information concerning the incidence of snuffles; however, it is known to be one of the most commonly observed diseases in domestic rabbits. Reports vary from about 20% to 70% of rabbits affected within individual colonies (102, 212, 246). Webster (244) reported a seasonal influence on the incidence of snuffles with peaks of higher incidence in the fall and spring and the lowest incidence in the summer.

Since many rabbits asymptomatically carry the etiological agent in the nasal cavity, it is thought that some form of stress debilitates the host, allowing the bacteria to multiply, thus initiating episodes of overt clinical disease. The stressing agents or conditions usually are not identified; however, the stress of experimentation, inclement weather, pregnancy, concurrent disease, or their combination may be logical sources of stress to suspect (247).

4. Epidemiology

Hagen (98) indicates that *P. multocida* may be spread from the dam to the offspring via the respiratory route shortly after the time of birth. The introduction and spread of *P. multocida* within a rabbit colony may occur when new stock are brought into the colony. The absence of clinical disease in carrier rabbits is disarming and may allow the introduction of infected rabbits into a colony. The organism may pass rapidly through a colony of susceptible rabbits and produce high mortality through septicemia and pneumonia while other rabbits develop snuffles, otitis media, and the other clinical manifestations previously mentioned. It appears that the etiological agent spreads by direct contact as well as by airborne means.

5. Clinical Signs

Snuffles is characterized clinically by a serious, mucus, or muncopurulent nasal discharge (Fig. 1) (210, 244, 245).

Fig. 1. Nasal exudate in rabbit with snuffles.

Nasal exudation usually stimulates the rabbit to rub the external nares with the medical aspect of the front legs. The exudate causes wetness and matting of the fur in this location and indicates nasal discharge even when the nares themselves may appear dry. Additionally, abnormal respiratory sounds are made. These include sneezing, coughing, and a snuffling sound from which the disease derives its name. It seems likely that the bacterial organism becomes airborne through these respiratory mechanisms. Little work has been done to further characterize this disease. The rectal temperature, hematological changes, and respiratory rates usually are not measured and recorded since the diagnosis is so readily made on clinical signs alone.

6. Pathology

The pathological changes depend on the duration of disease. As the disease changes from acute to chronic, the exudate changes from serous, to mucus, and to mucopurulent. The exudate may cause inflammation in the skin surrounding the nares. Exudate is present in the nasal and paranasal sinuses (Fig. 2). The mucosa lining these cavities may be reddened and edematous or in the more chronic stages it may be normal in color and mildly to moderately thickened. Microscopic changes also vary with the duration of the disease. There may be congestion of the submucosal blood vessels and edema of the submucosa. Heterophils are present within the submucosa. The mucosal epithelium may contain many goblet cells in the subacute to chronic stages, and it may be eroded in some areas. The lumen of the nasal cavity usually contains numerous heterophils in various stages of degeneration mixed with mucus and bacteria (Fig. 3).

7. Diagnosis

The tentative diagnosis of snuffles is made on the basis of a nasal discharge, especially if the exudate is mucus to mucopurulent. The isolation and identification of *P. multocida* by bacterial examination of the nasal exudate is essential to making a definitve diagnosis. It seems logical that other bacteria such as *Staphylococcus* can cause chronic inflammation and nasal discharge in rabbits; however, there is little information available concerning other causes of inflammation of the nasal and paranasal sinuses. Indirect immunofluoresence with specificity for *P. multocida* has been employed as a screening test both for antigens (*P. multocida* in nasal swabs) and antibodies in rabbit sera (251a).

8. Control and Prevention

The establishment of breeding colonies free of *P. multocida* has provided the surest method of preventing diseases

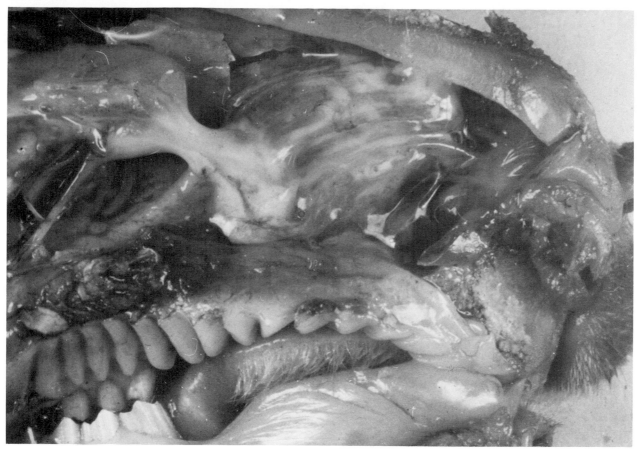

Fig. 2. Sagittal section through head of a rabbit with snuffles. Mucopurulent exudate is adhering to nasal turbinates.

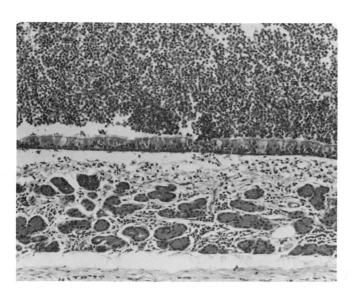

Fig. 3. Nasal sinus filled with heterophils.

caused by this organism. Colonies of this type were initially established by selecting breeding stock free of clinical signs of snuffles and free of *P. multocida* as determined by successive negative bacterial examinations of the nasal cavity (251). Clean facilities were used and frequent bacterial monitoring was performed. Griffin (94) also reported the establishment of a breeding colony of rabbits free of *P. multocida* using the same techniques. More recently, Caesarean derivation and hand rearing of breeding stock has been used to establish breeding colonies free of *P. multocida* and many other pathogens (202) (See Chapter 8). This is the preferred method of prevention where closed colonies are maintained. In colonies in which new stock are continually introduced (open colonies), exposure to *P. multocida* may be minimized through elimination of clinically ill animals and elimination of carriers as determined by nasal bacterial examination.

Alexander *et al.* (3) reported the attempted control of snuffles and pneumonia in a large rabbit colony through the use of a *P. multocida* vaccine. Killed organisms were given intravenously and the death rate due to respiratory disease was reduced. The effectiveness of vaccination in controlling diseases caused by *P. multocida* is in doubt and this technique is infrequently used.

Antibiotics are used to treat snuffles. A combination of 400,000 units of penicillin and 1/2 gm of streptomycin has been recommended to treat individual rabbits. Sulfaquinoxaline has also been shown to be effective in treating snuffles and pneumonia when used at a rate of 225 gm per ton of feed. Similarly, furazolidone is used at a rate of 50 gm per ton of feed (101, 102). With antibiotic treatment there may be remission of clinical signs, however, removal of antibiotics may be accompanied by a recurrence of the disease. In breeding colonies supplying research institutions, the control of the disease through the use of antibiotics seems ill-advised since the investigator using the rabbits may desire antibiotic-free feed or unknowingly change the diet to antibiotic-free feed. This change in diet coupled with shipping, tattooing, and experimental manipulation will create optimum conditions for activation of latent disease.

B. Enzootic Pneumonia

1. History

Acute deaths from enzootic pneumonia have been commonly observed for many years by both rabbit producers and those using rabbits in teaching and research programs. This disease usually takes the form of an acute fibrinopurulent pneumonia and pleuritis often with terminal septicemia (3, 58, 83, 98, 99, 173, 199, 210, 212, 215, 249). A subclinical form of this disease is also recognized in rabbits (61).

2. Etiology

Dr. Theobald Smith, renowned for his work in linking the transmission of *Babesia bigemina* (the etiological agent of Texas cattle fever) to the tick *Boophilus annulatus*, described a fatal septicemia with fibrinous pleuritis in rabbits in 1887. The organism he isolated and described is compatible with the present day descriptions of *P. multocida* (215). Following his report, there were many others describing acute deaths, pneumonia, pleuritis, and septicemia in rabbits with some detailed comments on the isolation and characterization of the causative bacterial organisms (13, 58, 83, 94, 98, 126, 127, 143, 199, 240). As in the case with snuffles, several organisms have been proposed to be the etiological agent.

In 1925, Webster clearly demonstrated that pneumonia in rabbits was associated with *P. multocida* (249). The organisms were isolated from naturally occurring cases,

identified, and used to experimentally reproduce the disease. Pneumonia also has been reproduced with *P. multocida* unintentionally by investigators studying other forms of pasteurellosis (snuffles and abscesses). Two bacterial organisms, *Bordetella bronchiseptica* and *Staphylococcus aureus*, can be isolated frequently from normal rabbit lungs. In pneumonic rabbit lungs, these 2 organisms plus a third, *P. multocida*, are consistently present but only the latter will induce the disease experimentally (63).

3. Incidence

Colony mortality rates from pneumonia are reported to vary from less than 5% to over 50% (3). Mortality rates probably are not a true reflection of the incidence of pneumonia but they do indicate the disease is common and may be severe. Hagen (98) reported that 35 (4%) of 857 young rabbits being studied developed respiratory infection and died. Lund found pneumonia in 23% of 3210 rabbits that died. The majority of cases of pneumonia occurred in rabbits 4 to 8 weeks old; however, he indicated that 53% of the deaths in mature animals were caused by pneumonia. In a similar study, Ostler found that 27% of 200 rabbits from his colony that died had pneumonia (173). The majority of the cases with pneumonia were 6 to 14 weeks of age.

In another study, 8- to 10-week-old apparently healthy rabbits were slaughtered for human consumption and their lungs examined for gross evidence of pneumonia (61). Approximately 20% of the 3967 rabbits examined had evidence of pneumonia; however, the incidence varied from less that 5% to nearly 50% in different batches. It was also observed that there was a significant difference in the incidence of pneumonia in rabbits from different geographical areas.

4. Epizootiology

The spread of *P. multocida* as the cause of pneumonia is considered to be the same as that described for snuffles. It is very likely that nasal infection precedes the other clinical forms of pasteurellosis and spread of the bacterial agent then may occur by various routes (Fig. 4).

Pasteurella multocida may enter the lungs through the trachea or via the blood vascular route. Experimental evidence offers support for each of these 2 routes, and it is likely that both occur naturally (63, 199, 249). The distribution of the gross lesions suggests that the intratracheal route may be most common.

5. Clinical Signs

The clinical features of pneumonia in rabbits are rarely observed in the naturally occurring disease. Domestic rabbits usually have little opportunity to exercise and this

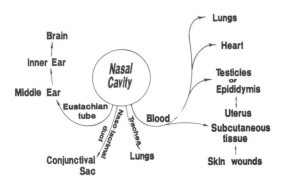

Fig. 4. Diagram of possible routes of spread of *Pasteurella multocida* within a rabbit.

may allow consolidation of the majority of the lung parenchyma without evidence of respiratory distress.

Usually the first indications of disease are anorexia and depression. Very commonly a rabbit is found dead that was apparently well the day before.

6. Pathology

a. GROSS. The gross features of enzootic pneumonia vary according to the duration and severity of the disease. The lesions may involve any portion of the lungs; however, the anteroventral areas are most consistently affected. The gross lesions have been divided into 4 categories including consolidation, atelectasis, abscesses, and small nodular gray foci (61). The disease is initiated by an acute inflammatory reaction recognized by consolidation (Fig. 5). There may be hemorrhage within the lung parenchyma and fibrin may cover the pleural surface. As resolution occurs, atelectasis becomes apparent, and if the

pneumonia was severe, abscesses surrounded by fibrous tissue may be present. Cavitation of abscesses or entire lobes is a common terminal stage of chronicity.

b. MICROSCOPIC. The microscopic changes vary in a similar manner to the gross lesions. The consolidated lesions vary from purulent to fibrinopurulent bronchopneumonia (Fig. 6). There may be hemorrhage, fibrin, and necrosis in the more severe cases. There is simple atelectasis in those lesions recognized grossly as atelectasis. The alveolar walls may be mildly thickened and macrophages are commonly present in the alveoli. Lymphoid nodules are common around blood vessels and adjacent to the large airways.

7. Diagnosis

Because clinical signs are usually unrecognized the diagnosis of enzootic pneumonia depends on postmortem examination and bacterial culture.

8. Control and Prevention

The treatment and prevention of enzootic pneumonia is the same as that described under snuffles.

C. Otitis Media

1. History

Otitis media has been known to occur in domestic rabbits for many years. It is sometimes referred to as wry neck in the older literature because of the predominant clinical finding of torticollis (207, 232). As early as 1925 the relation-

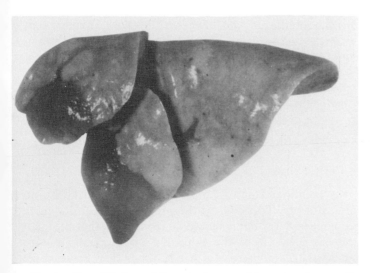

Fig. 5. Consolidation of the anterior-ventral lung area in enzootic pneumonia.

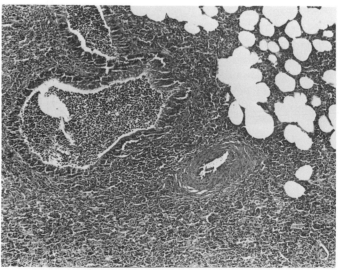

Fig. 6. Pulmonary consolidation in enzootic pneumonia.

ship of otitis media to respiratory infection in rabbits was reported. Webster (244) found that otitis media was related to snuffles and other clinical manifestations of disease caused by *P. multocida*. The information she reported on the clinical appearance, pathology, and etiology is as meaningful today as when it was published. Their observations have been repeated; however, little new information has been added since their paper.

2. Etiology

Otitis media is most commonly caused by *P. multocida*. In one reported series of cases of otitis media, *P. multocida* was isolated from 88 to 91 (97%) affected rabbits (65). In another series of cases, *P. multocida* was isolated from 20 of 25 (80%) affected rabbits (20). *Staphylococcus* spp., *Bordetella* spp., and other bacteria may also be found occasionally either with *P. multocida* or in its absence (65, 136, 213). The intranasal inoculation of susceptible rabbits with *P. multocida* may cause otitis media indistinguishable from naturally occurring cases (213).

Ear mites have long been thought to be a cause of otitis media in rabbits (207, 210, 232). It does not seem unreasonable that the extensive otitis externa caused by *Psoroptes cuniculi* (see Chapter 12) could extend through the tympanic membrane and lead to secondary bacterial infection of the middle ear. Whether or not this actually occurs is questionable. Fox *et al.* (65) was unable to correlate the presence of ear mites with otitis media in the 195 cases they observed nor did Smith and Webster (213) find ear mites as an etiological agent in their studies. The association of otitis externa caused by ear mites with otitis media has not been verified by clinical or postmortem examination and specific evidence is lacking to incriminate ear mites as an initiating agent of otitis media. In some cases of torticollis, the tympanic cavities appear normal on postmortem examination and bacterial cultures yield nothing of significance. It seems obvious from the clinical signs that the inner ear or the brain is affected; however, the cause is unknown. Some cases of torticollis in which otitis media is not present are thought to be related to a chronic granulomatous encephalitis caused by *Encephalitizoon cuniculi* (75). (See Chapter 11.) Some cases are as yet unexplained.

3. Incidence

Otitis media is known to occur commonly but there is little exact information on incidence. Fox *et al.* (65) reported that approximately 2% of their large rabbit colony developed torticollis. Generally the middle ears of rabbits are not examined unless the rabbit had torticollis; therefore, the only information available is the incidence of torticollis. The incidence may, of course, be expected to vary depending on many factors such as the incidence of

infection of *P. multocida* within the colony, stress, etc. Smith and Webster (213) reported examining the middle ears of 102 consecutively necropsied rabbits. They found that 33 (32%) of the rabbits had suppurative exudate in one or both middle ears. They also indicated that 5% of their own stock rabbits developed torticollis; however, 45% of their rabbits necropsied at the termination of various research projects had suppurative exudate in the middle ears. It seems clear that the incidence of otitis media might be quite high if all cases were were recognized; however, only those cases which subsequently cause inner ear involvement and torticollis are detected clinically and the incidence of these cases is usually less than 5%. There does not appear to be any age or sex predisposition to otitis media; however, a genetic predisposition may occur in some strains of rabbits (65).

4. Epidemiology

Otitis media occurs in colonies of rabbits which have *P. multocida* endemic within them. It is logical then that otitis media has been associated with several other clinical manifestations of *P. multocida* in the rabbit including subcutaneous abscesses, snuffles, pneumonia, septicemia, and conjunctivitis (65, 94, 212, 213, 244).

Rabbits may become infected with *P. multocida* by direct contact with infected rabbits, by contaminated fomites (cages, feeders, etc.), and perhaps by the airborne route. Once acquired the rabbit may become a carrier or develop one of the several clinical manifestations. It is generally believed that an upper respiratory infection (which may or may not be clinically apparent) precedes the otitis media and *P. multocida* ascends the eustachian tube to the middle ear and causes infection there. The infection may then spread to affect the inner ear as well as the meninges and brain parenchyma in some cases. It is very likely that stress plays an important role both in the spread of the organism into the middle ear, the inner ear, and brain. In one recent case, the author observed the development of torticollis in a doe on the day she kindled. The torticollis became progressively worse over the next two days and she died on the third day. She had a unilateral otitis media from which *P. multocida* was isolated and suppurative meningitis and encephalitis were present. It appeared that the stress of pregnancy triggered the spread of an existing infection.

5. Clinical Signs

As mentioned above, otitis media may exist with no clinical signs. Torticollis is the primary clinical finding in recognized cases but it should be recognized that the torticollis is a result of the extension of infection into the inner ear (otitis interna) or brain rather than from simple otitis media. The torticollis may be seen in various degrees (Fig. 7). In the more severe cases the rabbit may roll over

Fig. 7. Torticollis as a result of otitis interna.

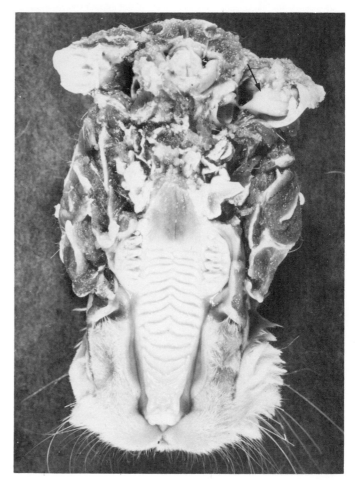

Fig. 8. Suppurative exudate in tympanic cavity (arrow). Same rabbit as in Fig. 6.

in the direction of the head tilt until it rests against the side of the enclosure. In severe cases the rabbit may be unable to eat and drink adequately and weight loss and dehydration may be present. Incoordination and other nervous signs may occur when the meninges and brain are affected by extension of the infection.

6. Pathology

The gross lesions associated with otitis media consist primarily of a white, creamy exudate in one or both tympanic cavities (Fig. 8). Smith and Webster (213) indicate that there is an initial reddening of the tympanic membrane and the lining of the tympanic cavity; however, this is not usually observed because it occurs early in the course of the disease. Suppurative exudate fills the tympanic cavity (Fig. 9), the lining epithelium may contain many goblet cells, and lymphocytes and plasma cells infiltrate the submucosa. Occasionally the tympanic membrane ruptures and suppurative exudate is discharged into the external ear canal. Extension of middle and inner ear infection to the brain results in suppurative meningoencephalitis (Fig. 10). Rabbits with otitis interna may also show considerable weight loss as indicated above.

7. Diagnosis

A tentative diagnosis of otitis media and interna is made when torticollis is observed. Confirmation of the diagnosis

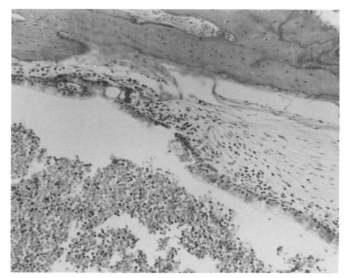

Fig. 9. Tympanic cavity filled with degenerating heterophils.

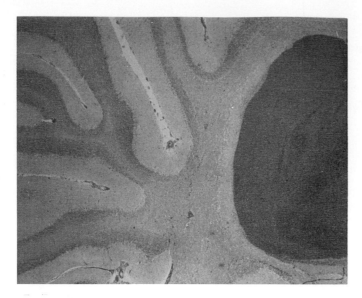

Fig. 10. Suppurative encephalitis in cerebellum as a result of extension of middle and inner ear infections.

is usually made by demonstrating a suppurative exudate in one or both tympanic cavities at postmortem examination followed by bacterial examination of the exudate.

8. Control and Prevention

There is little information available pertaining directly to the prevention of otitis media, however, the information given in the discussion on snuffles applies here as well as to the other diseases caused by *P. multocida*.

Fox (65) reported treatment of 7 cases of torticollis with penicillin although the dosage was not indicated. Only 1 of the 7 cases responded sufficiently that its head could be held normally, and in this case the tympanic membrane had ruptured. Tympanocentesis and antibiotic therapy, therefore, were suggested as a means of therapy to be considered. Generally antibiotics alone are felt to be of little value.

D. Genital Infection

1. General

Genital infections in rabbits include metritis or pyometra in the doe and orchitis and epididymitis in the buck (60, 103, 112, 128, 151, 210, 232). *Pasteurella multocida* is consistently isolated from affected organs, often in pure culture. Genital infections are not frequent but they are observed with some regularity. This disease occurs in adults and young adults and is observed more often in does and less frequently in bucks.

Venereal transmission occurs when infected bucks breed uninfected does or vice versa. It is not known whether the etiological agent is inseminated into does with the ejaculate

at breeding or whether contact with the infected prepuce serves as the method of transmission. It seems likely that both of these methods serve in the venereal transmission of *P. multocida*. An infected buck may serve to spread this genital infection widely throughout a breeding colony with disastrous results. Accurate breeding records are valuable in an epidemiological study. Then records may serve to identify which buck bred a particular doe which was found to have pyometra. Conversely, if a buck is found to have genital infection of this type, the does bred by the buck may be identified and examined.

The possibility of septicemic spread of *P. multocida* to the genital tract has not been considered thus far in the few papers dealing with this topic. Septicemias are known to occur commonly with *P. multocida* and localization of the agent in the genital system is a possibility. A similar mode of genital infection is known to occur in rams infected with *Brucella ovis*, in bulls with *B. abortus*, and in dogs infected with *B. canis*.

2. Clinical Signs

Rabbits with acute and subacute infection are seldom observed, but does may have a serous, mucus, or mucopurulent vaginal discharge. Of course, if septicemia occurs as part of the acute infection, the outcome may be fatal. With chronic infections, there are usually no overt clinical signs. However, does may fail to conceive even following several breedings and there may be a mucopurulent vaginal discharge. Bucks may have a low conception rate and some does bred by him may develop a vaginal discharge or die acutely. Bucks may also have one or both testicles enlarged and firm.

3. Pathology

One or both uteri may be slightly to greatly dilated (Fig. 11). In the more acutely affected uterus, dilatation is slight and a watery gray exudate is present in the lumen. The uterus which is chronically affected may be greatly dilated, and the uterine wall thin and light tan. The color is related to the presence of thick cream-colored exudate filling the lumen (Fig. 12). In some cases only the anterior portion of the uterus contains exudate. The suppurative exudate may adhere to the endometrium of the uterus and microscopically the epithelium may be ulcerated with polymorphonuclear leukocytes infiltrating into the underlying lamina propria (Fig. 13).

The nature of the lesions in the genitalia of the buck are not clear. Enlarged testicles with abscesses have been described (Fig. 14); however, since no microscopic descriptions have been included the exact location of the abscesses is uncertain. The epididymis rather than the testis may be the site of initial infection as is observed in *Brucella* infections in rams, bulls, and dogs referred to previously.

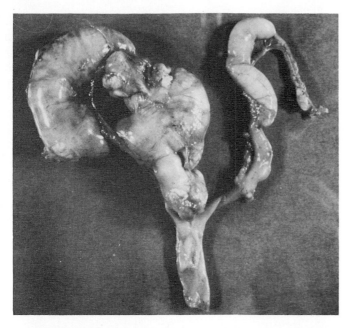

Fig. 11. Genital tract with both uteri dilated with suppurative exudate.

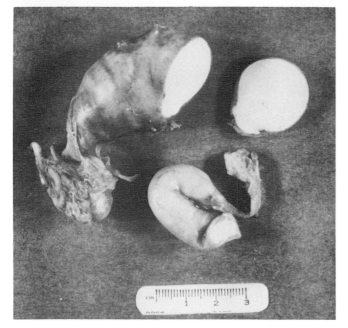

Fig. 12. Same uteri as in Fig. 11. Suppurative exudate filling lumen.

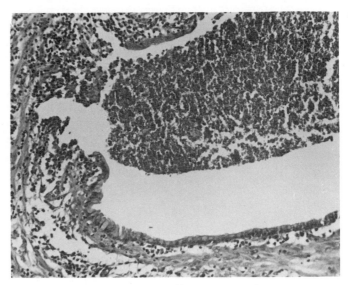

Fig. 13. Uterus with pyometra. The lumen is filled with suppurative exudate, the epithelium is ulcerated, and the lamina propria infiltrated with heterophils. (Courtesy of *Laboratory Animal Science*.)

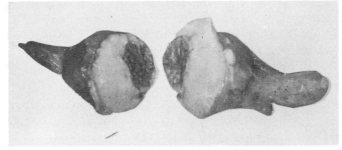

Fig. 14. Testicle containing suppurative exudate.

4. Diagnosis

Diagnosis is usually made through postmortem examination of culls from a breeding colony. The typical gross and microscopic lesions must be accompanied by isolation and identification of *P. multocida* from the affected organs.

5. Control and Prevention

Treatment usually is not attempted since diagnosis is seldom made antemortem. In pet rabbits castration or ovariohysterectomy coupled with antibiotic therapy may be attempted but the author knows of no such treatment being attempted.

Prevention again is emphasized rather than treatment. Regular examination of the external genitalia of breeding stock and regular examination of breeding records is valuable in determining the presence of disease. Bucks with low rates of conception and does that consistently fail to conceive or develop vaginal discharge after breeding should be culled. Bacterial cultures of the vagina or prepuce of suspected or new breeding stock would be valuable in early diagnosis. The establishment of a closed, *Pastuerella*-free colony is desirable although not always practical.

E. Abscesses

1. History

Subcutaneous abscesses have been described in rabbits from colonies experiencing epidemics of enzootic pneumonia and snuffles (51, 52, 94, 127, 136, 210, 244). The bacterial organisms cultured were thought to be identical to those causing snuffles and pneumonia. When the same organisms were given subcutaneously, abscesses similar to those occurring naturally were produced. When given intranasally or intratracheally, fibrinous pneumonia was produced similar to that occurring in other rabbits of the colony. Abscesses also occur in many sites not obvious by clinical examination of affected rabbits. These sites include the lungs (following pneumonia), brain, heart (Fig. 15), muscles, testicles, and conceivably any organ or tissue of the body.

2. Etiology

Bacterial cultures of these lesions very commonly yield *P. multocida* in pure culture. See Section II, A, 2 for discussion of this organism.

3. Epidemiology

The bacterial agent may reach a suitable site to form abscesses by (a) external wound contamination such as bites, scratches from other rabbits, or sharp wires in the cage; (b) septicemic spread from a distant site of infection in the body; or (c) by direct extension of the infection such as from the inner ear to the brain.

4. Clinical Signs

Subcutaneous swellings that vary in size are associated with abscesses in that site; however, clinical signs usually are not associated with abscesses of internal organs. As with the other forms of pasteurellosis, septicemia and death may follow the development of abscesses.

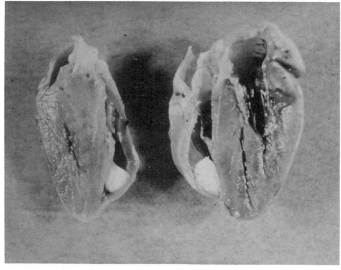

Fig. 15.   Heart with abscess in right ventricle.

5. Pathology

Abscesses caused by *P. multocida* are not unique in their gross or microscopic appearance. They usually contain white to tan, thick, creamy exudate, and they may be surrounded by a fibrous capsule depending on the duration of the disease.

6. Diagnosis

The presence of subcutaneous abscesses can be determined by the presence of a focal swelling that may have a soft center. Aspiration of the contents of the swelling will assist in differentiating abscesses from tumors or parasitic lesions such as coenurus cysts. The aspirated exudate from subcutaneous abscesses may be cultured to determine the specific etiological agent. Abscesses in internal organs are usually recognized at necropsy rather than antemortem.

7. Control and Prevention

Subcutaneous abscesses may be lanced, drained, and treated with topical antibiotics or Lugol's solution. Systemic antibiotics such as penicillin and streptomycin (see treatment of snuffles) should accompany the draining of the abscess. Elimination of the affected rabbit from the colony should be considered as an alternative to treatment in order to prevent further spread of the organism and contamination of equipment and facilities.

F. Conjunctivitis

1. History

Conjunctivitis caused by *P. multocida* has received little attention in papers dealing with pasteurellosis (22, 136, 173, 210). The incidence of this clinical form of the disease is unknown; however, it is quite common. Lesbouyries (136) suggests that the bacteria may gain entrance into the conjunctival sac via the nasolacrimal duct and cause a subacute to chronic conjunctivitis. Both young and mature rabbits may be affected as with the other clinical forms of pasteurellosis but the young may be more commonly affected.

2. Clinical Signs

The eyelids may be moderately swollen and glued shut from exudates and the conjunctiva reddened. A serious, mucus, and finally a mucopurulent exudate occurs. The inflammation may become chronic and the reddening and swelling subside, however, the exudate may continue resulting in epiphora (Fig. 16).

3. Diagnosis

Bacterial cultures of the conjunctival sac are needed to identify the causative organism and to determine sensitivity.

Fig. 16. Chronic conjunctivitis resulting in mucopurulent exudate at medial canthus of eye.

Obviously, there are causes of conjunctivitis other than *P. multocida*; however, no definitive studies have been reported.

4. Control and Prevention

Treatment consists of antibiotic ophthalmic ointment containing such antibiotics as penicillin or chloramphenicol. The likelihood of *P. multocida* persisting in the nasal sinuses must not be forgotten and recurrence of the ocular infection may occur. Prevention of this form of pasteurellosis is the same as that described in the section on snuffles.

G. Septicemia

1. History

Septicemia with *P. multocida* may be a sequella to any of the other clinical forms of the pasteurellosis or it may occur previous to their development. A review of this form of the disease would entail a rehearsal of the history, etiology, incidence, and epidemiology given in the section on snuffles, enzootic pneumonia, otitis media, genital infections, and abscesses. The reader is referred to these sections for additional information. Septicemia commonly is associated with snuffles, pneumonia, and pleuritis (126, 210, 215, 244, 249), while it is less commonly reported in association with some of the other clinical forms (52, 151).

2. Clinical Signs

Clinical signs usually are not observed since the affected animal dies rapidly during this episode. If other forms of pasteurellosis are present, the clinical signs associated with them may be observed.

3. Pathology

As with other animals dying peracutely, there are few if any lesions to be observed grossly or microscopically. There may be congestion of the abdominal and thoracic organs and hemorrhages beneath serous membranes and subcutaneously.

4. Diagnosis

Diagnosis depends on bacterial cultures and identification of the causative organism in blood and various parenchymal organs.

5. Control and Prevention

Since septicemia is not recognized in the antemortem state, treatment is not attempted. Control depends on eliminating *P. multocida* or minimizing exposure to it and preventing stress (see discussion under snuffles).

III. TULAREMIA

A. General

1. History

Tularemia is a naturally occurring acute febrile septicemic disease of numerous vertebrates caused by *Francisella tularensis*. The disease was first described as a disease similar to plague in ground squirrels (*Citellus beecheyi*) from 9 counties in California (145).

The disease in ground squirrels was characterized by enlarged, congested livers and spleens containing numerous areas of focal necrosis. Caseous necrosis was a prominent feature observed in lymph nodes. The disease was transmitted to guinea pigs, mice, rabbits, monkeys, and gophers. McCoy indicated that the disease might be transmitted by the ground squirrel flea, *Ceratophyllus acutus*, and then wrote the prophetic statement, "We do not know whether the organism causing this disease is pathogenic for man, but judging from the large number of species that are susceptible, we are inclined to suspect that man might contract the disease." McCoy and Chapin (146) reported the isolation of the responsible bacteria and names it *Bacterium tularense* after Tulare County, California. The prophecy of human susceptibility was soon fulfilled when, in 1914, Wherry and Lamb reported the first recognized human infections (252, 253). Two cases of severe ulcerative conjunctivitis accompanied by fever, lymphadenitis, and prostration were recognized and the agent isolated from the conjunctiva. The organism isolated produced conjunctival ulceration, septicemia, and death in guinea pigs and rabbits, and these animals had gross lesions in the liver and spleen like those described in ground squirrels. One of these first human cases was a 21-year-old man who dressed meat for a restaurant and presumably he handled infected rabbits (66). The other case was in the wife of an Indiana farmer. She had dressed rabbits for a meal. Rabbits were reported to be dying in the area of their Indiana farm, and Wherry and Lamb (252) found typical gross lesions and isolated the causative organism from 2 wild rabbits found dead 4 miles from the patient's farm. This was the first demonstration of tularemia in cottontail rabbits (*Sylvilagus floridanus*). McCoy and Chapin reported that the serum of both Dr. Chapin and a laboratory attendant contained complement fixing and agglutination antibodies. Although the positive results were puzzling to the investigators at the time, both positive individuals had been extensively involved in handling and examining infected rodents. In addition, Dr. Chapin had a serious febrile illness that caused him to miss work for 28 days previous to the test. It seems nearly certain that these were the first two unrecognized human cases of tularemia identified serologically (66). In his interesting review of tularemia, Francis (66) tells of many human cases of tularemia and relates individual histories that frequently include handling cottontail rabbits. He tells of a man working in a Washington D.C. meat market who went to a physician and indicated that he had "rabbit fever", a disease well known to the men of the market. A diagnosis of tularemia was made based on clinical signs and a 1:320 agglutination titer in his serum. Francis inspected livers of wild rabbits killed in Tennessee and imported for sale at a Washington D.C. market. Of 914 livers inspected in one month, 7 had gross lesions typical of tularemia. Guinea pigs inoculated with affected liver preparations died in about 5 days with typical lesions and *F. tularensis* was isolated. In 1911 Pearce described 6 human cases of deerfly fever, so named because infection was postulated to be caused by the bite of the deerfly, *Chrysops discalis* (66). It was not immediately clear that the disease known as deerfly fever was identical to tularemia and that the deerfly was transmitting the bacterial agent. Eight years after the description of deerfly fever, Francis (66) established its identity with tularemia by isolating the agent from 7 human cases and 17 jackrabbits. He then named the disease tularemia and indicated that cottontail rabbits, jackrabbits, and ground squirrels were likely to be the main reservoirs of infection.

In addition to the ground squirrel flea (*Ceratophyllus acutus*) and deerfly (*Chrysops discalis*) already mentioned, Parker (178) showed that wood ticks (*Dermacentor andersoni*) could carry and transmit the agent of tularemia and that it played an important role in transmitting tularemia in Montana. Similarly, Green (93) presented evidence that the eastern wood tick, *D. variabilis*, could carry the agent of tularemia. In 1926 Francis and Moore (67) reported that a disease, known as Ohara's Disease in Japan, was identified with tularemia. Dr. Ohara, a physician in Japan, reported a number of human cases of disease characterized by fever, lymphadenopathy, and cutaneous lesions. The disease

seemed to be related to a particular district and was known to local villagers for about 20 years. Typical of Dr. Ohara's cases was that of a mother and her 2 sons who acquired the disease after skinning and dressing a wild rabbit. Dr. Ohara did not know the cause of this disease but he became convinced that the disease was caused by contact with the uncooked meat of infected rabbits. Epizootics of a fatal disease were known to occur periodically in the local rabbit population. When the opportunity presented itself, Dr. Ohara's wife offered to help her husband prove his theory. She allowed some blood and tissue fluids of a dead rabbit found in the affected district to rubbed on the back of her left hand. After 20 minutes exposure she washed her hands with soap and water. After several days she became quite ill with fever, enlarged painful left axillary lymph nodes, and headaches. She was confined to bed and finally underwent surgery to remove several axillary lymph nodes. Apparently she recovered. Francis and Moore (67) isolated *Francisella tularensis* from one of Dr. Ohara's patients and they showed positive agglutination titers in the serum of some of his patients.

2. Etiology

Tularemia is caused by a nonsporeforming gram-negative, aerobic, pleomorphic bacteria currently named *Francisella tularensis*. The organism was initially named *Bacterium tularense* by McCoy and Chapin in 1912 (146). In 1957, the seventh edition of "Bergey's Manual of Determinative Bacteriology" listed this organism as *Pasteurella tularensis*. The assignment of this organism to the genus *Pasteurella* was not well accepted and *Francisella tularensis* is the currently accepted name for the etiological agent of tularemia.

3. Incidence

Francisella tularensis infects many vertebrates (32, 55, 93), but rodents and lagomorphs especially are involved in epizootics. There are no meaningful statistics on the incidence of either naturally occurring infection or disease in these animals. Occasionally fatal epizootics occur in domestic and wild animals, especially rodents and rabbits, causing dramatic reduction in their numbers (23, 66, 118, 146, 149, 182, 204, 252, 256). Human infection has been related to some of these epizootics and there are some data on the incidence in humans. In 1941, Belding and Merrill (14) reported an outbreak of tularemia in humans related to contact with cottontail rabbits imported from Missouri and Arkansas into a hunting club in Massachusetts. They indicated that all 48 states and Washington D.C. reported cases; however, most cases were reported in Illinois, Ohio, and Virginia. Although 1939 was the peak year for human cases in the United States, human cases still occur with some regularity. There were 2594 cases including 22 deaths reported from 1960 to 1968. In the spring of 1968 and 1969 an epidemic of tularemia occurred in Vermont (25, 256). Seventy-two people were infected and all had handled hides or carcasses of muskrats being trapped along several related streams. The clinical signs consisted of fever, cutaneous lesions, and lymphadenopathy in most infected people, although 20% had no recognized symptoms. This epidemic was thought to be related to a concurrent epizootic in the muskrats being trapped. *Francisella tularemia* was isolated from water and mud along one of the streams being trapped, as well as from 4 of 78 (5%) muskrats examined subsequent to the reported outbreak. Most diagnoses in humans were made serologically, with agglutination titers of 1:160 or greater. Five of 12 apparently healthy muskrats examined serologically had serum agglutination titers of 1:1280 or greater. Bacterial cultures of these animals were negative for *F. tularensis*. Borg *et al.* (23) reported an epidemic of tularemia in the varying hare (*Lepus timidus*) in 1967 in Sweden which was associated with a concurrent epidemic in the human population. It was estimated that a fifth to a tenth of the population of hares was wiped out. They observed that 75% of the 211 hares submitted for examination were lactating females. No other reports suggest any sex or age predispotition. Borg *et al.* (23) was puzzled by this unexpected observation but speculated that the disease was transmitted primarily by mosquitoes and that the mammary glands of the suckling females may provide more exposed skin surface for mosquitoes to bite. The stress of pregnancy and lactation may be an added factor.

Brooks and Buchanan (25) point out the existence of a seasonal difference in the occurrence of cases in the eastern versus the western United States. West of the Mississippi River most human cases are recorded during the summer months. These cases are mostly contracted by the bites of ticks and other blood-sucking ectoparasites and with the increased outdoor activity during the summer there is greater opportunity for exposure. In the eastern United States, except New England and the mid-Atlantic States where cases occur sporadically throughout the year, most cases are seen in the winter and are related to direct contact between man and rabbit that occurs in the skinning and dressing of rabbits during hunting season.

4. Epidemiology

Francisella tularensis is a versatile organism in regard to mechanisms of transmission. The mechanisms used by the organism are reviewed by Reilly (189) and include the following:

a. Transmission by blood-sucking arthropods including mites, ticks, flies, midges, fleas, mosquitoes, and lice. Transmission by these arthropods may be either mechanical by contaminated mouth parts or biological with the organism proliferating within this vector and transmitted by bite or contamination of the hosts's skin by excreta. Since tularemia is a septicemic disease and the affected animal often depressed and lethargic, transmission by blood-sucking arthropods is greatly facilitated. Transmission by this mechanism may be from animal to animal or animal to man.

b. Transmission by direct contact with infected vertebrates is the most common mechanism whereby man is infected. It is estimated that 90% of the human cases in the United States is related to direct contact with rabbits, primarily cottontail rabbits but also jackrabbits and to a very small extent the snowshoe hare (*Lepis americanus*). Direct contact serves to transmit tularemia from animals to man, but there is one unusual case reported of transmission from man to man by this mechanism. A mother contracted the disease after accidentally sticking herself with a needle while opening a cutaneous lesion of tularemia on her son (66). It has often been suggested that *F. tularensis* can penetrate the intact skin. That it can is suggested by the fact that many people acquire the disease by handling infected animal carcasses and they develop no cutaneous lesions and had no known skin wounds at the time of exposure. In the Vermont outbreak affecting 72 people who handled affected muskrats, 22 had no cutaneous lesions associated with the disease (25, 256).

c. Transmission may occur by ingestion. Ingestion of infected carcasses may effect the transmission of tularemia from animal to animal and ingestion of insufficiently cooked meat may cause infection in man. Water of streams may be contaminated with *F. tularensis* from carcasses of infected rodents. Ingestion of the water by other vertebrates including man may transmit the disease (118, 189, 256).

d. Transmission may rarely occur by inhalation of feces, contaminated dust (32), or by inhalation of organisms in aerosols created while skinning rabbits (256).

In summary, transmission from animal to animal or animal to man may occur through the cutaneous, pulmonary, or gastrointestinal route. Direct contact with infected carcasses constitutes the greatest hazard for man in acquiring the disease, with bites by blood-sucking arthropods also playing an important role.

5. Public Health Significance

Tularemia is a public health hazard as illustrated by the two recent epidemics, one in Vermont in 1968 and 1969 (25, 256) and another in Sweden in 1967 (23). See Chapter 18 for additional information.

B. Clinical Signs

Clinical signs are seldom observed since affected animals are generally found dead. Hares and cottontail rabbits show apathy, anorexia, and ataxia before becoming moribund and dying (23, 198). Affected rodents and rabbits are sometimes easily caught and this may lead to human exposure.

C. Pathology

The lesions present in the rabbits and hares are somewhat variable and depend on the duration of the disease. In the septicemia form, few lesions are present. Usually the spleen is enlarged, dark red, and may contain pinpoint white foci. Similarly, the liver may be congested and contain numerous pinpoint white foci. The lungs may be congested and contain patchy areas of consolidation (23, 198, 252, 255). The bone marrow may also contain areas of focal necrosis.

Microscopically, there are focal areas of necrosis and congestion in the liver (Fig. 17), spleen, lungs, and bone marrow. Thrombi in the small blood vessels of affected organs are thought to contribute to the foci of necrosis.

D. Diagnosis

Tularemia in rabbits is usually diagnosed at postmortem by the appropriate (but not specific) gross and microscopic

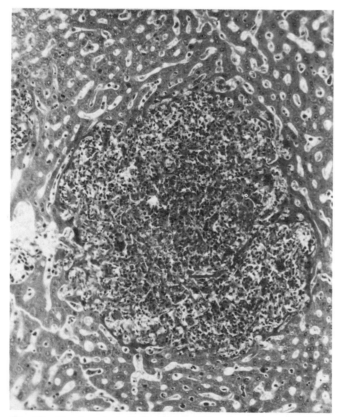

Fig. 17. Focal necrosis in the liver resulting from tularemia. (Tissue section provided courtesy of Veterinary Pathology Division, Armed Forces Institute of Pathology.)

lesions and the isolation of *F. tularensis.* When tularemia is suspected, it is common to inoculate guinea pigs intraperitoneally with affected organs. Guinea pigs are expected to die in 5 to 10 days with lesions similar to those in rabbits plus enlarged lymph nodes containing foci of caseous necrosis. The agent may then be isolated from guinea pig tissues. Serum agglutination tests, although not commonly done on rabbits, have been used in various animals and man as a common antemortem diagnostic test. Other diagnostic tests are occasionally used in man and may be applied to animal cases (198).

E. Control and Prevention

Treatment of tularemia in rabbits is not reported since antemortem diagnosis is rarely made and wildlife populations are not available for treatment. Streptomycin has been accepted as an effective treatment for man and presumably would be effective in animals.

The control of tularemia is equally difficult to apply to wildlife; however, the control of the disease in man can be effective to some measure. Sportsmen may diminish their liklihood of contracting the disease by wearing rubber gloves when skinning and dressing lagomorphs, avoiding ticks and flies in enzootic areas, avoiding drinking from streams in enzootic areas, and cooking meat thoroughly (198). Eradication of rodent population in limited areas may be effective in diminishing the spread of the disease and contact with other animals and man. In Russia, a live vaccine is used in man and this vaccine might be used in high risk populations such as those people handling carcasses.

IV. YERSINIOSIS (PSEUDOTUBERCULOSIS)

A. General

Yersiniosis is usually a chronic debilitating disease affecting numerous animals including rodents, lagomorphs, and man. It is caused by *Yersinia pseudotuberculosis.*

1. History

There is surprisingly little information concerning this disease in rabbits in the current literature presumably because it is not commonly seen.

In 1883 Malassez and Vignal published the first description of yersiniosis and as a result the etiological agent is commonly referred to as the bacillus of Malassez and Vignal. According to Seifried (208), who reviewed this disease in rabbits, the first description of yersiniosis in rabbits was reported in 1885 by Eberth. He described nodules in the internal organs of rabbits and labeled the causative organism the bacillus of pseudotuberculosis. Seifried (208) cites several reports of yersiniosis in rabbits and guinea pigs between 1890 and 1911. Sporadic reports of yersiniosis in rabbits have been made more recently including those by Pallaske (175), Olt (171), Moretti (158), and Genov (77). Truche (226) reviewed yersiniosis in rabbits and other species.

2. Etiology

The bacterial organism that causes yersiniosis has been known by several names including *Bacterium pseudotuberculosis rodentium, Corynebacterium pseudotuberculosis, Corynebacterium rodentium*, and *Pasteurella pseudotuberculosis*. Currently the organism is classified as *Yersinia pseudotuberculosis*. The organism is gram-negative, motile, and pleomorphic with coccoid to bacillary morphology. There are five serotypes of *Y. pseudotuberculosis* based on somatic antigen differences. Types I and II are most common in animals as well as man (118a).

3. Incidence

Pseudotuberculosis is much more common in rodents, especially guinea pigs, than it is in rabbits. The disease is not commonly seen in domestic rabbits in the United States although it is reported to be common in the United Kingdom (1). In Mair's review of pseudotuberculosis in wild animals, he indicates that the incidence of this disease in the European hare varies from 13 to 17% (153). In addition to guinea pigs and rabbits, many other species are affected by this organism including cattle, horses, sheep, pigs, goats, foxes, chinchillas, birds, monkeys, and man (29).

4. Epidemiology

Yersiniosis is thought to be transmitted most often by ingestion of the organism in contaminated feed or water (22, 29, 49, 56). The organism then may produce lesions in the digestive tract and it may be passed in the feces. Fecal contamination of vegetation is thought to play a primary role in the transmission of the organism. The organism has been isolated from the feces of 9 of 25 affected guinea pigs as well as from the feces of 1 of 5 human cases (49). The organism was also cultured from the feces of guinea pigs from a colony in which the organism was thought to be repeatedly reintroduced on green feed contaminated with excreta from infected wild pigeons (180). After entry into the body the organism may affect the digestive tract and then pass via the lymphatics to the mesenteric lymph nodes. Subsequent bacteremia may occur with the liver, spleen, and lungs being the most common sites affected (22, 208). Infection of the tonsils may occur followed by hematogen-ous spread to other sites. In addition to infection by contaminated feed, animals may be infected by direct contact with other infected animals and rodents are generally considered to be reservoirs of the disease.

5. Public Health Significance

Yersinia pseudotuberculosis causes mesenteric lymphadenitis, appendicitis, and septicemia in man (111). See Chapter 18.

B. Clinical Signs

Clinical signs are often absent but when present may include diarrhea, enlargement of mesenteric lymph nodes, and emaciation (56, 136, 158, 171, 208, 226). In the rarely recognized septicemic form, there may be the nonspecific signs usually associated with septicemias including fever, depression, dyspnea, and death (56, 226).

C. Pathology

1. Gross

The digestive tract may be free of gross changes; however, enlargement and necrosis have been observed in Peyer's patches of the small intestine (56, 158, 175, 208). There may be caseous necrosis in the cecum and especially in the appendix (22, 136, 175, 208). The tonsils also may contain foci of necrosis (175, 208). The mesenteric lymph nodes are most consistently affected. They may be enlarged several times and contain large areas of caseous necrosis (Fig. 18). Large areas of caseous necrosis also may be present in the liver and spleen. Occasionally the lungs and kidneys are similarly affected. In addition to the focal areas of necrosis, the spleen is enlarged. The bronchial lymph nodes may be enlarged and contain focal areas of necrosis. Caseous necrosis less commonly may be seen in the vagina and uterus (perhaps spread venereally), lymph nodes of the body and extremities, bones and joints, and in the heart (208).

2. Microscopic

The microscopic lesions in the affected organs consist of caseous necrosis surrounded by macrophages, epithelial cells, fibrous tissue, and lymphocytes. Occasionally giant cells are present. These changes are typical of chronic granulomatous inflammation.

D. Diagnosis

Diagnosis is generally made at necropsy and depends on finding typical gross and microscopic lesions and the

Fig. 18. Liver with attached mesentery containing large areas of caseous necrotic material. (Photo courtesy of Division of Comparative Medicine, The University of Florida.)

isolation of the causative organism. One complicating feature is that in many chronic caseous nodules, bacterial cultures are negative. In clinical cases, the feces may be cultured (49, 180). Serological tests, agglutination and hemagglutination, can be used but they may cause some confusion because of cross-reaction with *Salmonella*, *Brucella*, and *Yersinia pestis* (29, 118a, 143a).

E. Control and Prevention

Treatment in animals is seldom attempted. The diagnosis is seldom made antemortem, and when it is, removal of the affected animals from the colony is desirable. Elimination of affected animals and sanitation or sterilization of contaminated cages and equipment is recommended when the diagnosis is made. Attempts should be made to determine and eliminate the source of infection. The introduction of new stock is a common method of introducing the disease into a colony. The rapid hemagglutination test may be helpful in recognizing and eliminating carriers. Paterson and Cook (180) report the repeated reintroduction of *Y.*

pseudotuberculosis on contaminated greens being fed to their guinea pigs. Similar contamination of feed, water, or equipment should be considered and avoided when possible.

Streptomycin has been shown to have a beneficial effect in treating experimentally infected guinea pigs (234) and under special circumstances treatment with this antibiotic may be indicated in affected rabbits. Tetracyclines and chloramphenicol may be of value based on *in vitro* tests. Bacterins and avirulent cultures have been used to immunize guinea pigs against yersiniosis (29, 105a, 154a, 234). Again, under some circumstances this technique could be of value to prevent the disease in rabbits.

V. NECROBACILLOSIS

A. General

Necrobacillosis, also known in rabbits as Schmorl's disease, is a sporadic condition characterized by necrosis, ulceration, and abscessation of the skin and subcutaneous tissue, especially in the region of the face, head, and neck. It is caused by the bacterium, *Fusobacterium necrophorum*.

1. History

The organism causing necrobacillosis was first described by Leoffler in 1884 from cases of calf diphtheria (21, 205, 208). The disease was first described in rabbits by Schmorl (205) in 1891 as an infectious disease characterized by progressive necrosis of the skin of the face with rarely occurring bacteremias which lead to necrosis in visceral and thoracic organs. He was the first to isolate the etiological agent, which he called *Streptothrix cuniculi*. He experimentally infected many animals and found rabbits and mice to be quite susceptible, whereas the dog, cat, guinea pig, pigeon, and hen were more resistant to infection. In 1908 Basset (10) also described necrobacillosis in rabbits and showed a variation in the virulence of the organism isolated compared with some common rabbit bacterial pathogens.

Beattie (12) reported an epidemic causing the death of 40 rabbits. The disease he observed was morphologically similar to Schmorl's bacillus, but rather than nonmotile and strictly anaerobic as Schmorl described, it was motile and aerobic. I have observed two cases of progressive cutaneous ulceration in rabbits, with abscessation and necrosis on the face and neck from which *Corynebacterium pyogenes* was isolated (Fig. 19).

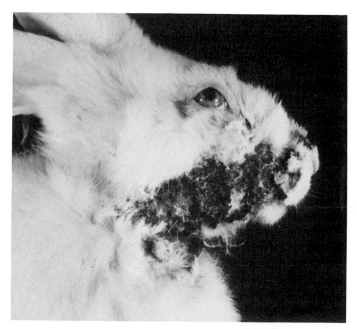

Fig. 19. Necrosis and ulceration of the skin of the face and neck caused by *Corynebacterium pyogenes*.

2. Etiology

The bacterial agent which is responsible for causing necrobacillosis is *Fusobacterium necrophorum*. There are many synonyms, including *Bacillus diphtheriae vitulorum*, *Streptothrix cuniculi*, *Corynebacterium necrophorum*, *Cladothrix cuniculi*, *Bacterium necrophorum*, *Spherophorus necrophorus*, *Actinomyces necrophorus* and *Fusiformis necrophorus* (30). The numerous synonyms suggest the disagreement in classifying this nonsporeforming, gram-negative, anaerobic organism. Morphologically, the organism is pleomorphic and usually occurs in long filaments but coccoid forms occur commonly also. This organism is usually considered to be a secondary invader rather than a primary pathogen (106). It is known to affect many domestic and wild animals following tissue damage caused by various bacterial, viral, or trauma-inducing agents. Schmorl and others have isolated and described this organism from affected rabbits (205, 208), and it was been shown to reproduce the clinical syndrome when inoculated subcutaneously into rabbits (106, 205). The chronic cutaneous lesions, especially those that have ulcerated, contain numerous types of organisms in addition to *F. necrophorum*.

3. Incidence

Necrobacillosis is a sporadic disease with no known breed, age, or sex predisposition. The etiological agent is thought to be present in the digestive tract of rabbits and many other animals. The agent and disease are known to have wide geographical distribution. Carnivorous animals are generally resistant to infection (21, 30, 205).

4. Epidemiology

Infections generally occur when animals are kept in filthy conditions, such as when there are feces under foot. When there are scratches, bite wounds, or skin wounds from other sources under conditions where they can be readily contaminated with feces containing the etiological agent, then the disease is likely to occur (88). *Fusobacterium necrophorum* is considered to be quite sensitive to exposure to aerobic conditions and will not live long when exposed openly to the air. When freshly defecated feces contaminate a wound in the foot or face, the organisms may find suitable anaerobic conditions and initiate the disease. Schmorl (205) observed that the lesions commonly began on the lips and gums and it seems likely that the practice of coprophagy by rabbits leads to contamination of wounds on the lips and in the mouth with fresh feces.

The disease usually begins as an acute inflammatory reaction in the subcutaneous tissue of the skin and necrosis and ulceration follow. The organism occasionally is spread hematogenously and as a result, necrotic lesions may occur in the thoracic and abdominal viscera (1, 56, 136, 205).

Necrobacillosis is not known to be contagious (106), but may be spread by contamination of the environment.

5. Public Health Significance

Fusobacterium necrophorum will affect man but is not a common cause of disease (21, 30). It has been found in ulcerative lesions of the colon and in abscesses of the liver and lung.

B. Clinical Signs

The most consistent clinical signs are swelling and necrosis with abscessation and ulceration involving the lips, skin of the face, head, neck, as well as metacrapal and metatarsal areas and the plantar surface of the feet (1, 56, 88, 136, 205, 208). Abscesses are commonly associated with the angle of the jaw and intramuscular abscesses at this site are observed (12, 56). The necrotizing lesions tend to be progressive and may exist for weeks to months. The open lesions usually have a foul odor. Affected rabbits may have an elevated temperature, they usually lose weight, and eventually become emaciated (1, 12, 57, 88, 136, 205). Lesions may begin in the oral mucosa or gingiva and when this occurs, the rabbit often is unwilling to eat and weight loss is rapid.

C. Pathology

The lesions of necrobacillosis are essentially those described above. The microscopic lesions consist of caseous necrosis of the skin and subcutaneous tissue with ulceration. Chronic active inflammation occurs adjacent to the necrotic areas including fibrosis and infiltration with heterophils and lymphocytes. On some occasions a bacteremia occurs and necrotic lesions may occur in the lungs, liver, and kidneys as well as other organs (1, 56, 205, 208).

D. Diagnosis

A tentative diagnosis can be made on the basis of typical clinical signs but bacterial cultures are needed for confirmation.

E. Control and Prevention

Control of necrobacillosis should include removal or isolation of affected rabbits, disinfection of cages and accessories, and improvement of management conditions if they are contributing to produce this disease.

Treatment of affected rabbits is not commonly attempted because of the desire to remove the infected animals from the colony and minimize contamination of the cages and accessories. Treatment by opening and expressing the necrotic cutaneous nodules and topical application of iodine has been recommended (88, 136). The effectiveness of sulfanilamide has been shown experimentally in rabbits when treatment is started early (106). Sulfamethazine, sulfapyridine, sulfathiazole, and sulfadimidine all have been used successfully in treating necrobacillosis in domestic animals (21, 30). Treatment with chloramphenicol, chlortetracycline, and oxytetracycline is recommended for infection in man (161).

Prevention of necrobacillosis is primarily directed toward prevention of cutaneous wounds and preventing contamination of wounds by feces. Cages and accessories should be examined for sharp wires or surfaces that could produce cutaneous wounds. When two or more rabbits are housed in the same enclosure, care should be taken to have compatible groups to minimize fighting. The cleanliness of the cages and accessories should be assured by regular sanitation and disinfection.

VI. SALMONELLOSIS

A. General

Salmonellosis is an uncommon disease of rabbits characterized by septicemia and rapid death with diarrhea and abortion commonly observed but not constantly present. The disease in rabbits is most commonly caused by *Salmonella typhimurium* and *S. enteritidis*.

1. History

Karsten, in 1927, described an outbreak of salmonellosis which caused the death of many hares (120). The lesions were characterized by inflammation and hemorrhage in the large intestine, serofibrinous exudate in the pleural and peritoneal cavities, focal necrosis in the liver, and swelling of the spleen. He grew the paratyphoid organism in pure cultures from the liver, kidney, and spleen. Olney (1928) (169) described an unusual outbreak of salmonellosis in rabbits which had been fed infertile hen's eggs mixed in their grain mash. About 125 rabbits died in a 2-day period, and *S. pullorum* was isolated from the heart, liver, and spleen. Duthie and Mitchell (57) reported the death of 6 of 15 young rabbits from which they isolated *S. enteritidis*. In 1936 Garofolo (76) reported a severe outbreak of salmonellosis affecting only pregnant female rabbits. Numerous abortions occurred and the does frequently died following abortion. Pregnant females which had not aborted also died and *S. enteritidis* was isolated. Over 100 of 700 rabbits died in an explosive outbreak of salmonellosis reported by Grasovsky (86) in 1939. Most rabbits died acutely of septicemia with anorexia, diarrhea, prostration, and death progressing rapidly. Scattered hemorrhages, splenic enlargement, numerous focal areas of necrosis in the liver, and mucosal ulceration in the cecum and colon characterized the disease and *S. typhimurium* was isolated. Morel (157) isolated *S. pullorum* from a rabbit originating from a colony in which unexplained mortality had been observed for over a year. There have been several reports of the isolation of *S. enteritidis* and *S. typhimurium* from laboratory rabbits in India (74, 123, 186, 237). Iyer and Uppal (114) described anorexia, fever, diarrhea, and rapid deaths in rabbits. Most rabbits had widespread vascular congestion, numerous petechial hemorrhages, with swelling of the liver, spleen, and mesenteric lymph nodes. There were coalescing focal areas of necrosis in Peyer's patches and a few rabbits had focal necrosis of the liver. *Salmonella enteritidis* was isolated. Ghosh and Chatterjee (78) reported the sudden unexpected deaths of 4 rabbits in which clinical signs were not observed. Necropsy revealed hemorrhagic enteritis and *S. typhimurium* was isolated (22, 131, 173, 185, 208). There have been several reviews of salmonellosis in rabbits including that of Habermann and Fletcher (97a) who also reported isolating *S. typhimurium* from rabbits and indicated that rabbits appeared to be more resistant to salmonellosis than other common laboratory animals.

2. Etiology

Salmonellosis in rabbits is most commonly caused by *S. typhimurium* and *S. enteritidis* but *S. pullorum* was also isolated in 2 separate outbreaks of disease (157, 169). Like other members of the *Salmonella* group, these organisms are gram-negative, aerobic, nonspore-forming, nonlactose-fermenting rods. The various species are quite similar and antigenic differences are used to classify them.

3. Incidence

The recognition of salmonellosis in rabbits in the United States is not common; however, recent reports indicate the disease may be common in India (74, 114, 123, 186, 237). One outbreak was reported in rabbits from Hungary (221). When outbreaks occur they tend to be explosive with high morbidity and high mortality. Several reports included data on mortality including death of 6 of 15 (40%) rabbits (57), 109 of 700 (16%) rabbits (86), and 40% of 22 rabbits (186). Mortality is highest in the young and the pregnant although rabbits of all ages may be affected. The fact that young rabbits and pregnant rabbits are more susceptible suggests the important role of stress in the pathogenesis of the disease.

4. Epidemiology

The epidemiology of salmonellosis has been extensively studied in numerous animals and man. The epidemiology of this disease in rabbits has not been studied specifically; however, there is no reason to doubt that both epidemiology and pathogenesis of the disease in rabbits are like those in other animals. The bacterial organisms are discharged in the feces of carriers and clinically ill animals. There is a wide host range for *S. typhimurium* and *S. enteritidis* including mammals, reptiles, and birds, and an infected animal may spread the organisms to many others. Contamination of feed, water, bedding, and cages as well as direct contact with an infected animal or animal caretaker may lead to infection of susceptible animals. Wild rodents are often suspected of being a source of infection. After ingestion of the organisms in a susceptible host, they may be passed in the feces temporarily. The organisms enter the cytoplasm of cells in the lymphoid tissue of the digestive tract as well as the mesenteric lymph nodes. After multiplication in these sites an initial septicemia occurs followed by removal of the organism by cells of the reticuloendothelial system which are especially numerous in the liver, spleen, and lymphoid tissue. Fever, anorexia, and depression may be associated with this bacteremic phase. The disease may subside at this stage and the host recover or become a carrier. There also may be another proliferation period for the organisms and a second, more severe, bacteremia may occur. Infection of the gallbladder may occur, and the organism may multiply here and be discharged with the bile into the digestive tract. Acute signs of illness with diarrhea, septicemia, and death may be observed coincident with the second bacteremia. The acute clinical signs in both the primary and secondary bacteremia are probably associated with the effects of endotoxin resulting from the release and fragmentation of the gram-negative organism in the bloodstream. A chronic state of inapparent infection may be established with the host delicately balanced between health and disease; if the balance of power tips in favor of the organism and they proliferate rapidly, a generalized bacteremia occurs, and acute clinical disease develops.

5. Public Health Significance

The species of *Salmonella* that are pathogenic for animals will also cause disease in man. In addition, there is a species that appears to be especially adapted to man. This organism, *S. typhosa*, produces typhoid fever in man. (See Chapter 18.) In 1948 Gualandi (96a) reported the experience of a family that had a severe digestive upset following the ingestion of contaminated rabbit meat, *Salmonella typhimurium* was isolated from the feces and vomitus of a member of the affected family. In 1971, *S. typhimurium* and *S. sofia* were found in the livers and intestines of rabbits being slaughtered for human consumption (160).

Since most members of the genus *Salmonella* are not host-specific and are pathogenic for man, there is constant need to guard against the transmission of salmonellosis from animals to man. Likewise, there is concern in preventing the infection of disease-free laboratory animals from animal caretakers.

B. Clinical Signs

The clinical signs of salmonellosis may be overlooked in peracute cases and rabbits may simply be found dead. In acute cases the signs are nonspecific and consist of anorexia, depression, and fever, Diarrhea may be present but this is not a consistent finding. Abortions and muco-purulent vaginal discharge have been observed in affected pregnant rabbits, and this may be the major clinical observation. Death frequently follows abortion; however, if the female recovers following an abortion she may not be able to bear young again.

C. Pathology

The lesions vary with the duration of the disease. When death occurs during the peracute phase of the disease, the lesions associated with septicemia are present. There is vascular congestion in most organs and petechial hemorrhages are common on the surface of abdominal and thoracic organs and other serosal surfaces. Serous to serosanguinous fluid may be present in the thoracic and abdominal cavities. In the acute stage, classically there are pinpoint focal areas of necrosis in the liver, the spleen is enlarged and congested, and there are swollen lymphoid nodules visible through the serosal surface of the intestinal wall. The largest of these areas, the Peyer's patches, may contain focal areas of necrosis and there may be ulceration of the mucosa over these areas. Occasionally there is also congestion and hemorrhage in the mucosa of the intestine. The submucosa may be edematous and contain fibrin and polymorpho-nuclear leukocytes. The mesenteric lymph nodes may be enlarged and edematous. A suppurative metritis with ulceration of the mucosal surface may be present in pregnant rabbits or those which have aborted.

D. Diagnosis

Diagnosis of salmonellosis depends on the isolation and identification of the specific agent or the demonstration of the rise in titer of specific circulating antibodies. The isolation of the etiological agent is most commonly accomplished by culturing blood, liver, spleen, and other organs taken at postmortem examination. In the live animal, blood and feces may be cultured if salmonellosis is suspected; however, negative results should not be interpreted as conclusive evidence of freedom from infection. Morgan

(159) suggests that the fresh stool specimen should be streaked immediately on selective media such as SS agar of sodium desoxycholate citrate. In addition it should be streaked on nonselective media such as EMB or Mac-Conkey's and inoculated into an enrichment medium such as tetrathionate or selenite broth which allows the pathogens to multiply while inhibiting coliforms. After incubation in the enrichment medium for 8 to 12 hours, it is streaked on selective and nonselective media. Colonies suspected of being *Salmonella* are subcultured and identified by biochemical and agglutinate tests with specific absorbed sera.

E. Control and Prevention

Treatment of active infection rarely has been attempted in rabbits and little is known about the drugs and dosages that are effective in this host. Serious questions should be raised about the wisdom of treating rabbits known to have the disease because of the possibility of survivors becoming inapparent carriers. If treatment is deemed to be the best course of action then *in vitro* sensitivity of the specific organism could be used in selecting the most effective drug. Sulfonamides, penicillin, streptomycin, and chloramphenicol have all been used with variable results in man (159). The intracellular location of the organism in reticuloendothelial cells seems to afford it some degree of protection from both antibiotics and host-immune mechanisms.

The control of salmonellosis is essentially one of eliminating potential sources of infection of susceptible animals. Since this disease in rabbits is not common, few specific measures are used to prevent transmission besides the eternally needed good sanitation practices. In areas of higher incidence, special precautions can be instituted to detect and eliminate carrier animals and eliminate vehicles of transmission. Carriers are detected by repeated fecal cultures and serological tests. There are numerous potential opportunities for transmission of *Salmonella* to laboratory animals such as infected birds or rodents defecating in feed or bedding during or after its manufacture, transportation, or storage. It may be necessary to check each product with which susceptible animals have contact to be assured they are not contaminated. It also may be necessary to screen animal technicians to identify carriers.

VII. TYZZER'S DISEASE

A. General

Tyzzer's disease is characterized by profuse diarrhea, dehydration, and rapid death. The etiological agent is *Bacillus piliformis*, a filamentous bacterial organism.

1. History

In 1917 Tyzzer (231) described an epizootic disease in Japanese waltzing mice he was using in a tumor transplantation study. The disease, which now bears his name, was characterized by diarrhea, high mortality, and focal areas of necrosis in the liver. Attempts to grow the etiological agent on artificial media failed; however, the agent was observed in bundles in the cytoplasm of hepatic parenchymal cells and intestinal epithelial cells. Tyzzer described the bacterial organisms and gave them the name *Bacillus piliformis*. Tyzzer's disease has been described in mouse colonies from time to time since the original description but it was not until 1965 that the first report of Tyzzer's disease in rabbits was published. Allen *et al.* (4) described 2 epizootic outbreaks of the disease in a rabbit production colony. Watery diarrhea principally affecting weanling and preweanling rabbits was observed. Nearly all of the clinically affected rabbits died in 12–48 hours, but rarely one would survive. Occasionally the dams of affected litters also would become ill and die. All 3 breeds of rabbits present were affected and monthly losses in preweanling rabbits varied from 3 to 6% in some rabbit rooms to as high as 50% in others. The lesions they observed were similar to those described in mice by Tyzzer and included multiple focal areas of necrosis in the liver as well as extensive necrosis of the mucosa and submucosa of the distal ileum, cecum, and proximal colon. Some rabbits also had areas of necrosis in the heart. Periodic acid-Schiff (PAS) and Warthin-Starry methods of staining were most effective in demonstrating the organisms which were filamentous gram-negative and occurred in intracytoplasmic bundles in hepatocytes, smooth and cardiac muscle, and in intestinal epithelium. Attempts to isolate the organism were unsuccessful. In 1971, there were 2 additional reports of Tyzzer's disease in rabbits. Cutlip *et al.* (48) reported an epizootic primarily affecting rabbits 6 to 7 weeks old. The disease was characterized by diarrhea, listlessness, anorexia, dehydration, and death 1 to 3 days after onset of illness. Approximately 135 of 450 (30%) rabbits died in a 4-month period and 90% to 95% of the clinically affected rabbits died. The lesions consisted of necrotic enteritis involving the terminal ileum, cecum, and proximal colon. The liver contained numerous focal areas of necrosis and the myocardium also contained necrotic areas in some cases. Rabbits surviving the clinical episode were stunted and an intestinal stenosis developed due to chronic inflammation and fibrosis at the sites of earlier necrosis. Organisms typical of *B. piliformis* were observed in affected tissues; however, all attempts at isolation failed. Fluctuating Iowa temperatures reaching as high as 95°F were thought to be the precipitating factor in these rabbits. They were moved into air-conditioned quarters and the epizootic promptly ended. Van Kruiningen and Blodgett (235) reported diarrhea and death in 60 of 112 rabbits which were 10 to 12 weeks old as well as death in a few younger and older rabbits. Gross and microscopic lesions were typical of those described before and the organisms were demonstrated in intestinal epithelium, hepatocytes, and smooth muscle cells. The rabbits were treated with oxytetracycline in the drinking water and the epizootic stopped in 36 hours. Since the description of Tyzzer's disease in rabbits in 1965, the disease has become more widely known and as a result it has been observed in the rat, hamster, gerbil, cat, rhesus monkey, and muskrat in addition, of course, to mice (72, 201).

In 1966, Tufts (227) described an unidentified disease of rabbits which seems likely to be Tyzzer's disease. Several outbreaks involving rabbits from all sections of the United States were observed. The rabbits developed profuse diarrhea and died in 24 to 48 hours. The mortality was high. The lesions consisted of reddening and hemorrhage in the colon and focal necrosis in the liver. Attempts to isolate the etiological agent were unsuccessful. Also among the ranks of unidentified disease is that described by Richter and Hendren (192) in cottontail rabbits. The rabbits had been captured and brought into the laboratory where over 50% died during a 4-year period. Severe diarrhea and dehydration occurred and most rabbits died within 24 hours. The lesions consisted of hemorrhages, edema, and necrosis of the mucosa affecting primarily the cecum but also the terminal ileum and colon. Ultrastructural studies of this disease were conducted but the cause of the disease remains unknown (26). Although the diagnosis cannot be made on the basis of their description, the possibility of this disease being identical to Tyzzer's disease seems quite likely.

2. Etiology

The organism causing Tyzzer's disease has been poorly studied because of the difficulty in isolating it. Most investigators have been unable to grow it on either artificial or living media although there is an unconfirmed report of its growth on artificial media (119). Rights *et al.* (193) reported growing the agent in mouse embryo cells but the organism was not pathogenic for mice after the second passage. Craigie (46) isolated *B. piliformis* from the liver of affected mice by growing it in the yolk sac of embryonated hens' eggs. Ganaway *et al.* (73) have isolated *B. piliformis* in embryonated eggs from affected rabbits. The organism reproduced disease in weanling rabbits in the thirty-second egg passage and the organisms were reisolated from the liver of affected rabbits. They reported that initial attempts to isolate *B. piliformis* from the liver by egg inoculation failed repeatedly because of overgrowth with enteric bacteria. When sulfaquinoxaline was added to the drinking water of experimentally infected animals, *B. piliformis* could be isolated in eggs with some consistency. The drug had an inhibitory effect on most of the enteric bacteria which were finding their way to the liver; however, it had no inhibiting effect on *B. piliformis*. *Bacillus piliformis* is a filamentous, pleomorphic, gram-negative, sporeforming, motile bacterium which apparently requires the intracytoplasmic environment to reproduce.

3. Incidence

Tyzzer's disease is not well known among rabbit producers and the author has observed outbreaks of Tyzzer's disease in rabbit colonies in which the owner has diagnosed mucoid enteritis. It is likely that *B. piliformis* is widespread in rabbits throughout the United States and perhaps other countries. As the disease becomes better known, it is likely that it will be recognized more frequently. The disease has a high morbidity and high mortality in affected colonies. Cutlip *et al.* (48) reported losing about 135 to 450 rabbits and Van Kruiningen and Blodgett (235) reported the loss of 60 of 112 rabbits which were 10–12 weeks old. Six- to 12-week-old rabbits are most commonly affected; however, preweanling and adult rabbits may contract the disease. Stress is a key factor in triggering the clinical disease. Tyzzer (231) was transplanting tumors in mice which were overcrowded and in an unclean environment. Allen *et al.* (4) used cortisone inoculations in rabbits exposed to contaminated bedding to experimentally reproduce the disease. Cutlip *et al.* (48) reported exposure to temperatures as high as 95°F in their affected rabbits and termination of the epizootic when the rabbits were moved into air-conditioned quarters. Allen *et al.* (4) also reported that no breed preference was observed; however, there may be some degree of sex predisposition since dams of young litters would occasionally become ill and die, whereas adult males were not observed to be affected. The matter of stress might explain the slight sex predisposition since females with suckling young may be under greater stress than adult males.

4. Epidemiology

Bacillus piliformis is passed in the feces of infected animals and infection occurs by ingestion (4, 231). Infection of the epithelium of the small intestine, cecum, and colon occurs and usually multiplication is slow, tissue damage is minimal, and clinical disease is not present (231). When the animal is stressed, such as by overheating, overcrowding, or experimental manipulation, the organisms multiply rapidly and cause necrosis of the intestinal mucosa and underlying tissue. The organism enter the portal circulation, pass to the liver as well as other organs, and clinical disease then occurs. Fecal contamination of feed, water, bedding material, etc., is the only recognized natural method of transmission; however, transmission has not been studied in detail.

5. Public Health Significance

The infection of man with *B. piliformis* is not known to occur. The wide variety of species that the organism is capable of infecting, including a case in a nonhuman primate, would lead the author to believe that infection in man will be recognized in the future.

B. Clinical Signs

The clinical disease usually is very acute and is characterized by profuse watery diarrhea and fecal staining of the hindquarters. Affected rabbits are depressed, they do not eat and rapidly become dehydrated. Death usually occurs in 12–48 hours after the onset of clinical symptoms. In the few rabbits that survive the acute clinical episode, there may be a depressed appetite and stunting of growth.

C. Pathology

1. Gross

Dehydration and fecal staining of the carcass may be extensive. The serosal surface of the cecum and occasionally the distal ileum and proximal colon are reddened, and subserosal petechial hemorrhages frequently are present (Fig. 20). The wall of the cecum is edematous and thickened. The lumen of the cecum and colon may contain brown watery fecal material and the mucosal surface of the cecum is reddened and rough and has a fine granular appearance. Fecal material may adhere tightly to some portions of the mucosal surface. Similar but usually less severe changes are present in the ileum and colon near their

Fig. 20. Hemorrhages on the serosal surface of the cecum of a rabbit with Tyzzer's disease.

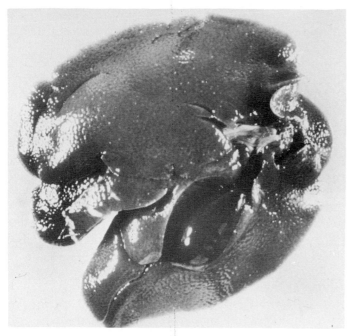

Fig. 21.   Liver with focal areas of necrosis, Tyzzer's disease.

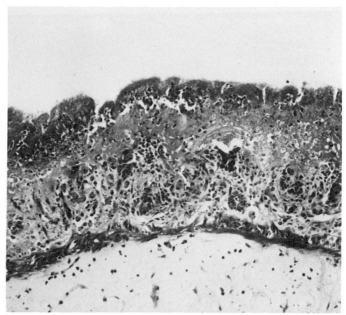

Fig. 22.   Cecum with extensive mucosal necrosis, bacterial growth on denuded surface, and submucosal edema in a case of Tyzzer's disease.

junction with the cecum. In more chronic cases there may be stenosis of the intestion due to fibrosis at the sites of extensive necrosis. The liver frequently contains numerous pinpoint pale foci throughout the parenchyma (Fig. 21). Occasionally pale streaks 0.5 to 2 mm wide and 4 to 8 mm long or pale foci are present in the myocardium.

2. Microscopic

There is extensive necrosis of the mucosal epithelium in the cecum and occasionally in the distal ileum and proximal portion of the colon (Fig. 22). The mucosa is denuded and there is extensive bacterial growth on the denuded surface. The necrosis occasionally extends into the submucosa and tunica muscularis. There is extensive edema of the submucosa and inflammatory cells are present only in small numbers. Red blood cells in the subserosal connective tissue may be present. Giemsa, Periodic acid-Schiff, or Warthin-Starry stains are used to demonstrate the filamentous clumps of organisms in the cytoplasm of viable epithelial cells and occasionally in smooth muscle cells. In more chronic cases there is fibrosis and chronic inflammation and the organism is not present. The liver contains numerous focal areas of necrosis which are frequently adjacent to portal areas (Fig. 23). These areas contain cellular debris and a few polymorphonuclear leukocytes which often are degenerating. At the periphery of the necrotic foci, occasional viable hepatocytes contain bundles of *B. piliformis* within their cytoplasm (Fig. 24). Special stains, as mentioned above, are usually required to demonstrate these

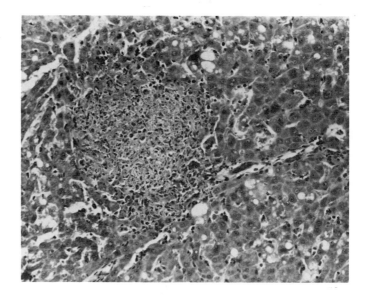

Fig. 23.   Focal area of necrosis in the liver of a rabbit with Tyzzer's disease.

organisms. In hepatic lesions which have existed for a longer duration, there may be fibrosis, multinucleated giant cells and occasionally mineralization. No bacilli are present. The pale streaks observed grossly in the myocardium consist of linear areas of necrosis and *B. piliformis* can also be demonstrated in myocardial fibers adjacent to these areas. Mineralization and giant cells are likewise seen in the healing stages of this lesion.

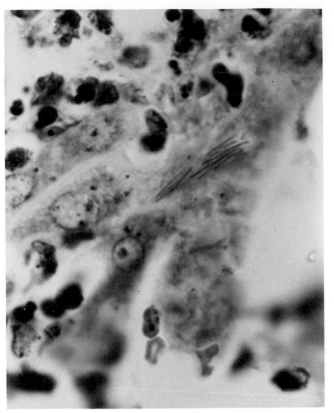

Fig. 24. *Bacillus piliformis* growing in hepatic cells adjacent to a focal area of necrosis, Tyzzer's disease (stained with Giemsa stain).

D. Diagnosis

The gross lesions of Tyzzer's disease are typical but should not be considered pathognomonic. The demonstration of *B. piliformis* within the cytoplasm of affected tissues is the basis of diagnosis just as it was when Tyzzer first described the disease in 1917. Immunofluorescence may be used to assist in the task of finding and identifying the causative organisms (201), but this technique is of more value in experimental work than in diagnostic service. Likewise, complement fixation tests and agar gel diffusion tests have been described for studies in mice and used to screen the sera of retired breeders to determine if a colony has had exposure to the organism (69). Isolation of *B. piliformis* by yolk sac inoculation of embryonated eggs is a valuable technique. However, it is unlikely to replace demonstration of typical intracellular organisms as the basis for diagnosis because it is less reliable, more difficult, and more time-consuming. As cultural techniques are refined and become more reliable, they should replace the current method used to diagnose Tyzzer's disease.

E. Control and Prevention

Treatment of clinically ill rabbits is seldom attempted and no treatment is known to be effective. In one clinical outbreak oxytetraycline was used in the drinking water and the epizootic stopped in 36 hours (235). Variable results have been reported concerning the effect of antibiotics in experimentally infected mice; however, tetracycline and penicillin have been found to be beneficial in the hands of some investigators (46, 72, 119). Ganaway *et al.* (73) tested the ability of several antibiotics to inhibit the growth of *B. piliformis* in embryonated eggs. None of the antibiotics tested were completely effective in inhibiting growth; however, erthromycin, penicillin, streptomycin, and chlortetracycline caused incomplete inhibition of growth. Chloramphenicol and sulfamethazine cause no inhibition of growth. The brief clinical course, the intracellular location of the agent, and the formation of spores tend to make antibiotic therapy ineffective and suggests that emphasis should be placed on prevention.

Prevention of Tyzzer's disease is poorly understood. Attempts have been made to protect mice by immunization with several products including a formalized vaccine prepared from diseased mouse liver (70), sublethal dose of *B. piliformis* grown in eggs (47), and hyperimmune mouse or rabbit serum (71). These products did provide some degree of protection on an experimental basis but thus far there is no practical method to vaccinate mice or other species.

The prevention of Tyzzer's disease in rabbits might be possible in the future by obtaining Caesarean-derived breeding stock and maintaining a closed, uninfected colony. Caesarean-derived rabbits are not readily available at the present time and this method may not be practical for many producers.

There are no known methods of eliminating infection from an infected colony. The prophylactic use of antibiotics during times of stress may be of value in colonies known to be infected. The prevention or spread of the organism in an infected colony is another important factor in preventing the disease. The isolation or elimination of sick rabbits and good sanitation practices are of obvious value but there is little specific information on methods to prevent the spread of this organism. Fecal contamination of the environment is known to be the method of transmission to other rabbits. What role the spores play besides resisting unfavorable environment is not known. It is not known if the spores are transmitted by the airborne route. Additionally, minimizing or eliminating the causes of stress that trigger the disease is needed. Elimination of some types of stress is impossible or not practical. Such things as weaning or experimental manipulation are known to be stressing and may initiate the clinical disease. Irradiation, cortisone administration, overcrowding, transporting, high environmental temperature, etc., all have served to initiate the disease in mice or rabbits. Ganaway *et al.* (72) indicated that the oral administration of sulfonamides to rabbits should also be added to this list. They gave sulfaquinoxaline to 6 rabbits born and raised in a room where no deaths from Tyzzer's disease had been observed previously, and within 30 days all had died of

Tyzzer's disease. Other rabbits in the same room remained well. They point out, in retrospect, that the colony of rabbits from which Tyzzer's disease was first seen recently had been treated with sulfonamides to eliminate coccidiosis.

VIII. LISTERIOSIS

A. General

Listeriosis is a septicemic disease of rabbits characterized by sudden deaths or abortions, or both. The disease is caused by *Listeria monocytogenes*, a small, gram-positive bacteria.

1. History

In 1926 Murray *et al.* (164) described an acute fatal disease in young rabbits, most of which were 4 to 12 weeks of age although some were older. He observed 65 cases of this disease in his rabbits as well as 13 cases in guinea pigs. A gram-positive bacillus was isolated from both the rabbits and guinea pigs and the disease was reproduced experimentally in both species. The authors were impressed with the elevated number of large mononuclear cells in the peripheral blood and utilized this facet of the disease in naming the causative organism *Bacterium monocytogenes*. Hülphers (113) described lesions of focal hepatic necrosis in a rabbit and isolated an organism named *Bacillus hepatis*. Some authors have suggested that the disease in rabbits described by Hülphers (113) in 1911 was listeriosis (89). Others have considered the disease described by Hülphers to be necrobacillosis (208). Pirie (183) described what he believed to be a new disease in African rodents (*Tatera lobengulae*). He named the disease "Tiger River Disease" after the region in which the disease was found. He isolated a gram-positive, motile bacillus which he named *Listerella hepatolytica* in honor of Lord Lister. The disease was experimentally reproduced in rabbits, guinea pigs, and rats. Pirie pointed out that his organism was similar to that described by Murray *et al.* (164) and that if the organisms were identical, the name proposed by the latter authors would have preference. Pirie (184) later changed the generic name to *Listeria* because the previously proposed generic name had already been used. In 1940 Paterson (179) described listeriosis in a young adult doe. This doe had been bred but failed to kindle at the expected time. She became ill and was killed and necropsied. The liver was enlarged and contained numerous focal areas of necrosis. The uterus contained 6 dehydrated fetuses enveloped in pus. Pure cultures of *L. monocytogenes* were isolated from the spleen, liver, abdominal cavity, uterine exudate, and from a fetus. Traub's description of listeriosis in Angora rabbits is unique in that he observed torticollis and rolling, and microscopically there was meningoencephalitis involving the brain stem,

the base of the cerebellum, and the cervical spinal cord (223). In 1943 Henricson (107) described a case of listeriosis in a hare, and in 1946 Schoop (206) described the death of 130 of 500 Angora does caused by *L. monocytogenes*. These does became depressed, refused to eat, and died within a few days. The lesions consisted of focal hepatic necrosis, splenic enlargement, and a consistent necrosis and suppuration in the uterus. Abortions as well as retention and decomposition of feti occurred. Carlotto (35) also described a naturally occurring outbreak of listeriosis, and Vetesi and Kemenes (239) described the death of 13 pregnant does due to listeriosis. Several review articles on listeriosis have been published (85, 133, 236), the most extensive of which is that by Gray and Killinger (89).

2. Etiology

Listeriosis is caused by *L. monocytogenes*. The synonyms include *Bacterium hepatis*, *Bacterium monocytogenes*, *Listerella hepatolytica*, and others (89). It is now accepted that this organism affects many species including sheep, goats, cattle, pigs, horses, dogs, cats, rabbits, guinea pigs, chinchillas, man, and many more.

The organism is a small gram-positive, nonsporeforming, motile, nonacid-fast rod with rounded ends. It may be difficult to isolate on artificial media and it may be confused with members of the genus *Corynebacterium*.

Pure cultures of *L. monocytogenes* have been used on numerous occasions to experimentally reproduce disease in rabbits which is similar to the naturally occurring disease (90, 91, 107, 164, 183).

3. Incidence

Listeriosis in rabbits is a sporadic, seldom recognized disease. Stress may play a role in initiating or promoting clinical signs of listeriosis. Murray *et al.* (164) observed that the rabbits involved in their outbreak were young and they were given inadequate amounts of feed which was of poor quality. The stress induced by poor nutrition may have been an important factor in the disease. Pregnancy seems to be an important factor in promoting or perhaps initiating listeriosis. Murray *et al.* (164) indicate that their first isolation of *L. monocytogenes* came from the heart's blood of a pregnant doe. Paterson (179) described a fatal case of listeriosis in a doe which was pregnant, and many of the 130 Angora rabbits which Schoop (206) studied were pregnant. Additionally, Gray *et al.* (90, 91) observed that experimental infection of rabbits orally or by conjunctival instillation caused serious illness, abortion, and some death in pregnant does. In nonpregnant females and in males, there were no signs of disease except at the site of inoculation; however, *L. monocytogenes* could be isolated from the vagina and uterus of nonpregnant females. Pregnancy is a physiologically stressing condition and the stress of pregnancy may make the doe much more susceptible to listeriosis. It should be pointed out, however, that the contents of the pregnant uterus may simply offer a more favorable environment for the organism to grow and thereby induce a more severe disease with stress playing an insignificant role. Gray and Killinger (89) indicated that listeriosis may occur in man following the administration of cortisone or its derivatives. This again suggests the possible importance of stress in initiating the clinical disease.

In cattle and sheep listeriosis is diagnosed much more frequently than in rabbits (85, 92). In these species the disease tends to have a low morbidity but high mortality, and the central nervous system is most consistently affected. In rabbits, on the other hand, there is too little data to make generalizations regarding morbidity and mortality. Schoop (206) reported 130 deaths of 500 rabbits. Murray *et al.* (164) reported 65 deaths in rabbits but did not give the total number at risk. Contrary to cattle and sheep, *L. monocytogenes* seldom affects the central nervous system but spreads via the bloodstream to consistently affect the liver, spleen, and gravid uterus.

4. Epidemiology

The method of natural transmission of *L. monocytogenes* is not well understood. The organism can survive long periods under various conditions (92) and it may be spread by ingestion of feed or water that has been contaminated with the feces of an infected animal. Gray (89) indicated that many domestic animals as well as man may have inapparent infections and shed the organisms in the feces. Stress may diminish the resistance of the carrier and clinical disease then develops. The association of listeriosis in cattle with the ingestion of silage certainly supports the concept that ingestion is an important natural route of infection. Also in support of the concept of infection by the oral route is the isolation of *L. monocytogenes* from the mesenteric lymph node, liver, spleen, and kidney of a sheep with listeric encephalitis. Other routes of infection have been suggested and should be considered. They include infection via the respiratory route, especially in sheep and cattle. In support of this concept is the isolation of *L. monocytogenes* from the nasal cavity of normal as well as clinically ill sheep, and from cattle and man (89, 188). Venereal transmission also has been proposed in man (188).

After entering the body, the bacteria are able to gain entrance into the bloodstream of rabbits and spread to numerous organs as evidenced by the lesions of septicemia, isolation of *L. monocytogenes* from the blood, and focal necrosis in the liver and spleen. It is not known if rabbits may become inapparent carriers as occurs in some domestic animals and man.

5. Public Health Significance

Listeria monocytogenes infects man and has been associated with several clinical forms including meningoence-phalitis, septicemia, habitual abortion, endocarditis, conjunctivitis, and others (89). For additional details see Chapter 18.

B. Clinical Signs

The clinical appearance of listeriosis is quite variable and usually nonspecific. Rabbits may stop eating, become depressed, lose weight, and die or simply die suddenly with no observed clinical illness. Pregnant does may experience abortions and have a blood-tinged vaginal exudate. In the one instance of central nervous system involvement, torticollis, incoordination, and rolling were observed (223).

C. Pathology

1. Gross

There may be few gross lesions in cases of peracute septicemia; however, congestion of the visceral and thoracic organs and a few scattered hemorrhages may be seen. In acute cases there is enlargement and edema of the cervical and mesenteric lymph nodes and clear fluid in the thoracic cavity, pericardial sac, and abdominal cavity. There also may be subcutaneous and pulmonary edema. The liver consistently contains numerous pinpoint pale gray foci throughout the parenchyma. The spleen occasionally contains similar foci. Infarcts in the lungs and necrosis within the adrenal glands also has been described (164). In pregnant does there may be acute metritis with accumulation of suppurative exudate in the lumen. Degenerating fetuses may also be present in the lumen if they were not aborted previous to the death of the doe. The uterine wall may contain necrotic areas and it is usually thickened by the inflammatory process.

2. Microscopic

Focal areas of necrosis in the liver are the most consistently reported microscopic feature of listeriosis in rabbits (Fig. 25). Paterson (179) reports that there is little inflammatory reaction associated with these areas of necrosis but occasional clumps of bacteria are observed in association with the necrotic foci. Similar lesions are occasionally present in the spleen. An intense suppurative inflammation with areas of necrosis may be present in the uterine wall. Many bacteria are present on the surface of fetal membranes and throughout the fetus. Traub (223) reported finding no gross lesions in the rabbits with central nervous system signs; however, there was meningoencephalitis with focal areas of necrosis associated with many polymorphonuclear leukocytes and mononuclear cuffing of nearby blood vessels. These lesions were most prominent in the brain stem, base of the cerebellum, and the anterior cervical spinal cord.

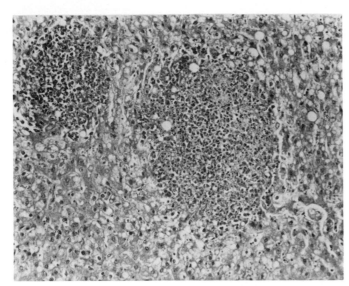

Fig. 25. Focal areas of necrosis in the liver in a case of listeriosis. (Tissue section provided courtesy of Veterinary Pathology Division, Armed Forces Institute of Pathology.)

D. Diagnosis

Diagnosis of listeriosis depends on isolating the etiological agent, *L. monocytogenes*. Neither the clinical signs nor the pathological changes are sufficiently specific to make a diagnosis. Gray (89) states that serological tests are not satisfactory in making a diagnosis of listeriosis and isolation of the organism is the only reliable method currently available. Cultures should be made from blood, cerebrospinal fluid, and vaginal exudates in antemortem cases and from the blood and visceral organs and brain in postmortem examinations.

E. Control and Prevention

Since antemortem diagnosis of listeriosis is rarely made in rabbits, treatment is seldom if ever attempted. Tetracycline or a combination of penicillin and streptomycin have been recommended for treatment of listeriosis in domestic animals and in man.

Since the epidemiology is so poorly understood, there is little information available on prevention of the disease. Attempts to vaccinate animals with both killed and living organisms have not been successful (85, 89).

IX. TUBERCULOSIS

A. General

Tuberculosis is a very rare disease in domestic rabbits characterized by granulomatous inflammation of the lungs, digestive tract, kidneys, liver, spleen, and lymph nodes and nonspecific clinical signs such as emaciation. The disease most commonly has been associated with the bovine type of *Mycobacterium tuberculosis*; however, the avian and human types also produce the disease in rabbits.

1. History

Tuberculosis is reported to have been described first in rabbits by Robert Koch in 1884, only 2 years after he first isolated and cultivated the tubercle bacillus. He described tuberculosis in 8 rabbits and, following him, Strauss reported 1 naturally occurring case in a rabbit in 1895 (44, 45, 105, 136, 208). In 1908, Guérin (97) reported the death of an emaciated rabbit which had nodular lesions throughout the lungs, parietal pleura, pericardium, peritoneum, liver, and kidneys. The lung nodules had caseous centers and fibrous walls and acid-fast organisms were demonstrated in the lesions. Rothe (198) in 1912 reported 5 additional cases of rabbit tuberculosis from a rabbitry of 80 to 90 rabbits. The lungs, bronchial lymph nodes, pleura, and kidneys contained caseous nodules. There were similar nodules in the digestive tract, especially Peyer's patches, the sacculus rotundus, and appendix. The mesenteric lymph nodes contained caseous foci. The bovine type of *M. tuberculosis* was isolated. Rothe thought the lesions suggested infection by inhalation of the organisms. Also in 1912, Raymond (187) reported a naturally occurring case of tuberculosis in a rabbit. There were lesions in both thoracic and abdominal viscera; however, no lesions were recognized in the digestive tract and mesenteric lymph nodes. The bovine type of organism was identified. The kidney lesions were studied histologically by Chrétien (41). Cobbett (43) described two cases of tuberculosis in rabbits caused by the avian tubercle bacillus. These rabbits were raised with poultry which presumably had the disease. Besides visceral lesions similar to those already described, these rabbits had tuberculosis lesions of several joints including the stifle, hock, and elbow joints. In 1923, Bru (27) also described skeletal involvement in a spontaneous case of tuberculosis in a rabbit. In addition to involvement of the lungs, pleura, bronchial and mediastinal lymph nodes, kidneys, liver, and intestinal tract, there were tuberculous lesions of the olecranon, humerus, and lumbar vertebrae. The bovine-type organism was isolated. In 1924, Coulaud (44) described 17 cases of naturally occurring rabbit tuberculosis. Lesions primarily involved the lungs but kidneys, liver, and mesenteric lymph nodes were also involved. The bovine-type of organism was identified. The organism was used to infect rabbits experimentally by ingestion and lesions with a similar morphology and distribution were produced. He concluded that ingestion was the route of natural infection rather than inhalation which was cited as reported by Koch, Strauss, and Rothe (136, 198, 208, 233). Valtis (233) injected virulent *M. tuberculosis* of the bovine type into a doe on the same day she kindled. After one month her 4 young were weaned and placed in individual cages. Three of the 4 young were necropsied and lesions of tuberculosis were in the intestine and mesenteric lymph nodes but the lungs and bronchial lymph nodes were free of disease. Valtis (233) concurred with the conclusion of Coulaud that natural infection occurred via ingestion. Calmette (33) who reviewed naturally occurring tuberculosis in guinea pigs and rabbits, shared their opinion. Numerous additional reports and reviews of spontaneous tuberculosis in rabbits have been published (1, 9, 22, 24, 28, 45, 56, 64, 87, 95, 96, 105, 136, 139, 174, 200, 208, 225).

2. Etiology

Tuberculosis is caused by a slender, rod-shaped, gram-positive, aerobic, acid-fast bacterium of which there are three recognized types, avian, bovine, and human. In the cases in which specific identification of type was made, the bovine type of organism was found most frequently (9, 27, 28, 44, 64, 95, 105, 139, 187, 198, 225); however, the avian (24, 43, 95, 96) and human types (174, 200) were found. The organisms *M. tuberculosis* (*Bacterium tuberculosis*) are rich in lipids which are thought to provoke the granulation tissue reaction which characterizes tuberculosis.

3. Incidence

Tuberculosis is rarely found as a naturally occurring disease in rabbits. This disease in rabbits has always been considered to be uncommon or rare and it appears to be recognized with decreasing frequency because of the control of tuberculosis in man and other animals. In an extensive review of resistance to tuberculosis Lurie (143a) indicated that some inbred stocks of rabbits were more susceptible to experimental infection than others. In resistant rabbits, the infection remained localized while in the susceptible rabbits there was rapid spread of the organisms within the body.

4. Epidemiology

The epidemiology of tuberculosis in man and some economically important animals has been extensively studied. The disease is so uncommon in rabbits and the rabbit being of little economic importance, there is little known about the epidemiology of this disease in rabbits. In man, there are genetic, age, and sex predispositions. The disease in man is usually caused by the human-type organism which is almost exclusively spread by the inhalation of airborne droplets (154). The ulceration of the mucosa of pulmonary airways in infected persons releases organisms into the mucosal secretions which then contaminate the saliva. Coughs and sneezes then provide a mechanism for

the organisms to become airborne and be inhaled by another person. Man may be infected by ingestion, especially by ingestion of the bovine-type organism in unpasteurized milk. Coincident with the widespread pasteurization of milk and the tuberculosis testing and control programs in the bovine, this type of infection has been drastically reduced. Most of the reports of tuberculosis in rabbits involved the bovine-type organism presumably because it was a common practice to feed unpasteurized cow's milk to rabbits to stimulate their growth (22, 87, 105). Brune (28) reported 5 naturally occurring cases in rabbits caused by the bovine-type organism. The owner of the rabbits recently had a cow with tuberculosis.

Two primary modes of infection of rabbits with naturally occurring cases of tuberculosis have been proposed. Inhalation as the means of infection was proposed by Koch, Strauss, and Rothe (136, 198, 208, 233). This conclusion was reached primarily on the basis of distribution of the lesions, that is, there appeared to be primary lung involvement in the opinion of these authors. Coulaud (44) and Valtis (233) rejected this hypothesis in favor of ingestion as the means of transmission. They based their opinion on the ability to produce disease with the same distribution of lesions as the naturally occurring disease by experimentally feeding the organism. Intestinal lesions could be found without lung lesions, and in cases where there were lung but no intestinal lesions, the mesenteric lymph nodes contained lesions and organisms. Pacchioni (174) concurred with their conclusion. The fact that milk was commonly fed to rabbits, the bovine-type organism was most frequently found, and the lesion could be reproduced by ingestion suggests that infection by ingestion was the most common route of infection. Ingested organisms probably penetrate the mucosal epithelium of the digestive tract, where they may be taken up by the cells of the lymphoreticular system and pass via the lymphatics to the mesenteric lymph nodes (136). If numerous organisms are ingested they may localize in the wall of the intestine, multiply within the lymphoreticular cells and cause focal areas of necrosis in the gut wall. The organism may cause caseous necrosis in the mesenteric lymph node, pass via the lymphatics to the blood stream and then to the lungs and other parts of the body. It is speculated that the organism also may travel from the intestinal wall to the liver directly via the portal circulation. The possibility of inhalation of the organism should not be overlooked. Direct infection of the lungs with spread to the bronchial lymph nodes, then to the bloodstream may occur. The couching up and swallowing of the organism may account for digestive tract and mesenteric lymph node involvement.

The fact that the turbercle bacillus may be passed in the feces of infected rabbits has received very little attention. Infected feces may serve as a source of contamination of feed, water, etc., or it may be a source of infected dust that may be inhaled.

5. Public Health Significance

Tuberculosis is of considerable interest to those dealing with public health and since this disease does occur in man and rabbits there is a potential threat to human health (see Chapter 18).

B. Clinical Signs

In most reported cases of tuberculosis in rabbits, few clinical signs were observed. The most consistent findings were anorexia and emaciation. In addition, other clinical findings have occasionally been observed including elevation of body temperature, pallor of mucous membranes due to anemia, diarrhea, skeletal deformities, and ocular lesions (136). Gross enlargement of elbow, stifle, and hock joints was observed (27, 43). In one reported case a severe spondilitis occurred and posterior paralysis was observed (174). In 1943 Bablet and Van Deinse (9) observed 3 cases of tuberculous ocular lesions in experimentally infected rabbits, and Florio and Saurat (64) observed naturally occurring ocular lesions. The affected rabbit had a poor appetite and was emaciated. The right eye was normal in size but the palpebral reflex was absent and there was a severe keratitis and posterior synechia. The iris was discolored, and the crystalline lens was opaque. The bovine-type tubercle bacillus was isolated from the eye as well as other affected tissues.

C. Pathology

1. Gross

The gross lesions usually consist of emaciation of the carcass and light tan to gray, firm nodules in various organs. The nodules vary in size from less than 1 mm to several centimeters. The nodules usually are found in the lungs, visceral and parietal pleura, pericardium, bronchial lymph nodes, mesenteric lymph nodes, kidneys, and liver. They are less commonly found in the spleen. These nodules usually have a caseous center and a fibrous capsule. The nodules in the lungs may coalesce and cavitation similar to that seen in man may occur. The digestive tract is commonly but not consistently affected. The serosal surface of small intestine, cecum, appendix, and large intestine may contain slightly elevated, firm, areas of various sizes. The mucosal surface opposite these sites is ulcerated and there may be caseous necrosis in the wall surrounding the ulcerated site. The ulcerated areas within the digestive tract commonly are located at the sites of lymphoid tissue including Peyer's patches, the sacculus rotundus, appendix, and lymphoid nodules in the wall of the cecum. Other sites also

may be involved. The bronchial and mediastinal lymph nodes frequently are enlarged and contain foci of caseous necrosis. The involvement of the skeleton has already been mentioned. In addition to joint enlargement that is visible in the live animal, there may be involvement of joints (95) and bones which are not obvious until necropsy or until the skeleton has been macerated (27, 96). Bru (27) indicated that after maceration, roughened areas of lysis and production of bone associated with tuberculosis became obvious. He demonstrated advanced involvement of the lumbar vertebrae in a rabbit that clinically had a progressive posterior paralysis. Similar involvement of the lumbar vertebrae with posterior paralysis was reported by Pacchioni (174).

2. Microscopic

The microscopic lesions of tuberculosis in rabbits are not unique. The tubercle bacilli stimulate a chronic inflammation characterized by foci of caseous necrosis surrounded by epithelioid cells and multinucleated giant cells. Fibrous tissue surrounds these nodular lesions. These nodules are typical of those produced in many animals and they are generally referred to as granulomas or tubercles. Brack (24) indicated that he observed a case of tuberculosis in a rabbit with lesions of granulomatous inflammation similar to the microscopic lesions of tuberculosis in carnivores. In his case, which was caused by the avian type tubercle bacillus, there were numerous epithelioid cells which were not organized into discrete granulomas. There was only a small amount of necrosis and multinucleated giant cells were scarce. Mineralization of the lesions in rabbits has not been observed. Acid-fast organisms can be found in impression smears or tissue sections of the lesions. In those animals with skeletal lesions, there was granulomatous inflammation of the bone with caseous necrosis and both production and lysis of bone.

D. Diagnosis

The diagnosis of tuberculosis in rabbits is nearly always made by microbiological examination at postmortem. The demonstration of typical gross and microscopic lesions with the presence of acid-fast organisms has been considered diagnostic in the past; however, the isolation and identification of the organism should be considered an essential part of the diagnosis. The clinical signs are nonspecific and antemortem diagnosis is unlikely. Fecal and salivary cultures could be attempted on suspect live animals if their value warrants this. Tuberculin testing has been attempted but it was not a reliable indication of disease (136, 208).

E. Control and Prevention

Treatment of naturally occurring tuberculosis in rabbits has not been reported, nor does it seem logical because of the possibility of transmission of the disease to man or other animals. The prevention of the disease has had little consideration because of its rare occurrence in rabbits. Thirty to 60 years ago when most of the cases of naturally occurring tuberculosis were reported, it was not uncommon to feed unpasteurized milk to rabbits and to raise rabbits and poultry together in an open yard. These practices resulted in the transmission of tuberculosis from cattle and poultry to rabbits. Pasteurization of milk, separation of various animals, and testing programs to eliminate infected animals all have contributed to the control of tuberculosis.

X. TREPONEMATOSIS

A. General

Treponematosis is a frequently mentioned, seldom recognized disease of domestic and wild rabbits characterized by inflammation of the genitalia, face, and regional lymph nodes. This disease is caused by *Treponema cuniculi*, a spirochete that is similar in many ways to *T. pallidum*, the agent responsible for syphilis in man. This disease has a number of synonyms including rabbit syphilis, vent disease, spirochetosis, cuniculosis, and venereal spirochetosis.

1. History

Ross (195) in 1912 is usually given credit for the first published mention of treponematosis in rabbits. As a footnote, in this humble beginning Ross indicated that parasites similar to those found in syphilis were found in the blood of rabbits which had chancres, buboes, and ulcers on the genitalia, anus, and mouth. Bayon (11) described papules on the penis of a rabbit which had been inoculated with material from the rabbit described in the paper by Ross (195). Spirochetes were demonstrated by dark field microscopy and the author raised the question of the validity of much of syphilis research which had been performed on rabbits not known to be free of this naturally occurring disease. In 1914, Ross (196) commented further on the disease he observed earlier in rabbits and described its passage in rabbits. He indicated that there were lesions in the livers of some experimentally infected rabbits which resembled gummata. Gummata are granulomatous lesions associated with syphilis in man; however, the lesions in rabbits have not been confirmed by the findings of other investigators and it is likely that the lesions described by Ross were caused by the larvae of the tapeworm *Taenia pisiformis*. In 1914 Arzt and Kerl (8) reported the results of examining the rabbits of 5 rabbitries. They found a total of 72 of 267 (27%) rabbits with gross lesions of treponematosis with 2 of the rabbitries free of the disease and 7.5 to 36% of rabbits affected in the other rabbitries. Organisms indistinguishable from *T. pallidum* were demonstrated. Swelling of the inguinal lymph nodes was observed in 4 of the affected rabbits. Schereschewsky (203) described a naturally occurring case with papular and ulcerative lesions in 1920 and in the same year Jacobsthal (115) described the same types of lesions in rabbits and proposed to name the disease paralues cuniculi and suggested the causative organism be named *Spirochaeta paralues cuniculi*. Noguchi (166, 167) found lesions of treponematosis in 11 of 70 (16%) rabbits he examined. He proposed the name *Treponema cuniculi* for the etiological agent and indicated that this agent was morphologically similar to *T. pallidum* although possibly a little thicker. Wassermann tests on infected rabbits were negative. Many investigators wondered if *T. cuniculi* and *T. pallidum* were identical and if *T. cuniculi* was pathogenic for man. Lersey *et al.* (130) reported an interesting situation in which a man who raised rabbits developed lesions typical of syphilis. Lesions like that of syphilis were discovered in one of the patient's rabbits and spirochetes indistinguishable from *T. pallidum* were demonstrated. The question naturally arose as to whether the patient had acquired the disease from the rabbit, the rabbit had acquired the disease from the man, or both had acquired the disease elsewhere. Further investigations revealed that the patient had sexual relations with his girl friend about $2\frac{1}{2}$ months prior to becoming ill and that, although she had no clinical signs of infection, both had strongly positive Wassermann tests. In 1921, Levaditi *et al.* (137, 138) reported finding several naturally occurring cases of treponematosis in rabbits and they described crusty papules and ulcers on the vulva and nose of affected rabbits. They demonstrated that the disease could be transmitted venereally. They described the histological changes in the affected tissues and demonstrated spirochetes which resembled *T. pallidum*. They wondered if the agent in rabbits was transmissible to man and if it could be used to produce a vaccine against syphilis in man similar to the use of vaccinia virus of cattle to immunize against smallpox in man. In an attempt to clarify this question, Nicolau and Levaditi inoculated themselves by scarification on the external surface of the arm. They used an inoculum rich in spirochetes obtained from an infected rabbit. In addition to themselves, they inoculated a male rabbit and a cynomolgus monkey (*Macaca irus*). The skin lesions from scarification healed in 5 days and neither scientist was infected. Both had negative Wassermann reactions before and after inoculation. The monkey likewise did not become infected and the rabbit developed typical lesions on the prepuce which contained numerous spirochetes. Klarenbeek (124) described treponematosis in 5 adult rabbits from Holland and passaged the disease in rabbits. He felt the etiological agent was identical to *T. pallidum*. To the contrary, Warthin *et al.* (242) indicated that anyone acquainted with both *T. cuniculi* and *T. pallidum* should have no trouble differentiating them. They differentiated the 2 organisms on the basis of morphology of the organisms and the histopathology of the lesions induced in rabbits. Their experimentally infected rabbits developed no general immunity, remained negative to the Wassermann test, and showed no evidence of generalization of the infection. Additional descriptions of the naturally occurring disease and the etiological agent have been published published (2, 16, 17, 19, 39, 50, 68, 165) including the observation of the disease in wild rabbits in England (2) and the description of two epidemics of treponematosis in Russia (68).

2. Etiology

Treponematosis in rabbits is caused by *T. cuniculi* which is a slender spiral-shaped bacterium. Other members of this genus include *T. pallidum* (the cause of syphilis in man), *T. pertenue* (the cause of yaws in man), and *T. carateum* (the cause of pinta in man) (228). *Treponema cuniculi* and the other pathogenic members of this genus have not been grown on artificial media, embryonated eggs, or tissue culture, although some nonpathogenic members of the genus have been grown by these means. *Treponema cuniculi* is still considered to be morphologically indistinguishable from *T. pallidum* (228), however, some authors have disagreed. Warthin *et al.* (242) indicated that *T. pallidum* was thinner and its ends were tapered in contrast to *T. cuniculi* which was thicker and had blunt ends. Additionally, *T. pallidum* was more rigid, had tighter turns, and had curved rather than right-angle bends in the middle of the organism. Differentiation between *T. cuniculi* and *T. pallidum* has also been attempted on the basis of pathology and serology with conflicting opinions similar to those concerning their morphological differentiation (19, 214, 229, 230, 242).

3. Incidence

There is some confusion concerning the incidence of treponematosis in rabbits. The clinical disease is not commonly observed; however, when serological tests are performed using serum of apparently healthy rabbits, it is not uncommon to find positive results (42, 125, 176, 229). Small and Newman (211) for instance, found positive results in 6 of 41 rabbits from one colony and 10 of 142 from another colony. Lesions and spirochetes were not found in many serologically positive animals. Turner and Hollander (229) reported that rabbits from commercial breeders often have had prior infection with *T. cuniculi* and that close examination of the skin over the testicles may reveal stellate scars that result from healing of ulcerated areas. Rabbits with scars of this type often had positive agglutination tests. A

positive test may be interpreted to mean a past or present experience with *Treponema* or a false positive.

Within affected colonies, there may be many rabbits with the clinical disease. Arzt and Kerl (8) reported 65 of 189, 3 of 40, and 4 of 11 rabbits with gross lesions in 3 different colonies. Noguchi (166, 167) reported 5 of 50 (10%) and 6 of 20 (30%) rabbits affected with the clinical disease in two separate groups. Adams *et al.* (2) estimated that 20 to 40% of the wild rabbits in England were affected with treponemiasis. Fried and Orlov (68) found that 640 of 5000 (13%) rabbits from a single colony had gross lesions of treponematosis and in the same colony 8 months later they found 704 of 3650 (19%) had macroscopic lesions. They also found that more females than males were affected. About 55% of the females over 1 year of age were affected and 35% of the males of the same age were affected. Likewise, they found both age and breeding activity to be an important factor in the incidence. In rabbits of both sexes that were over 1 year of age and had been mated, 1304 of 2893 (46%) were affected, whereas only 40 of 5800 (0.7%) of the rabbits under 8 months and not mated were affected. They also observed that rabbits housed in a cold barn or outside in the winter season developed more extensive lesions than those living in warm quarters.

4. Epidemiology

The transmission of *T. cuniculi* from rabbit to rabbit may be by genital or extragenital contact. Venereal transmission is thought to be the most common mechanism of transmission (68, 137). Extragenital infection is suggested by the infection of 40 of 5800 nonmated rabbits under 8 months of age and 5 of 285 rabbits which were 6 to 8 weeks of age (68). Natural cases of intrauterine infection have not been observed.

Most authors have considered treponematosis of rabbits to be a disease involving only the skin and mucous membranes without involvement of visceral organs. Fried and Orlov (68) as well as Arzt and Kerl (8) found that there was swelling of the regional lymph nodes in a few of their cases but spirochetes were seldom demonstrated in the glands. It has been shown experimentally that inoculation of rabbits with *T. cuniculi* may result in lymph node infection and that the organisms survive in this tissue for long periods of time (18, 152, 214). Rabbits with lymph node infection may be clinically healthy. The importance of inapparent carriers in the transmission of this disease is unknown; however, it seems likely that stress could cause clinical disease in a carrier and initiate an outbreak of the disease.

Once a rabbit becomes infected the organisms localize and proliferate in the skin and mucosa near mucocutaneous junctions. There is an initial acute stage followed by a chronic inflammatory reaction and there may be focal ulcerations in the skin. The external genitalia are the usual site of initial involvement but lesions may also occur on the chin, lips, face, nostrils, eyelids, ears, and occasionally other sites. The spread of the organism from one body surface to another presumably occurs by autoinfection. Spread of the organism via lymphatics to local lymph nodes occurs but whether a bacteremia follows is not known (68).

5. Public Health Significance

Through the efforts of Levaditi and Nicolau and others (137, 138), *T. cuniculi* is considered nonpathogenic for man.

B. Clinical Signs and Gross Lesions

The earliest signs of disease are reddening and edema and these usually begin on the external genitalia and perianal area. Minute vesicles are rarely observed before there is serum exudation which helps to form a brownish crusty surface over the area. Focal ulceration of the epithelium may occur beneath the crusty surface and hemorrhage may occur from these sites. The lesions are said to be painful (68) and this may facilitate autoinoculation of the face, chin, nose, etc. Lesions on the face may spread peripherally and the hair may fall out at affected sites but quickly grows out during the healing process. The chronic lesions tend to be dry, scaly, slightly raised areas that may easily be overlooked. The inguinal and popliteal lymph nodes may be swollen. The lesions usually resolve spontaneously over a period of several weeks. Affected rabbits are generally alert and appear healthy except for the superficial lesions.

C. Microscopic Lesions

There is acanthosis and hyperkeratosis of the epidermis with rete pegs extending deep into the dermis. The superficial dermis contains numerous lymphocytes and plasma cells and occasional polymorphonuclear leukocytes. In areas of ulceration of the epidermis, polymorphonuclear leukocytes may be numerous in the adjacent dermis. The chronic inflammatory cells do not tend to be oriented around blood vessels as they do when *T. pallidum* is experimentally inoculated into rabbits. Numerous spirochetes can be demonstrated in the epidermis and superficial dermis with the use of Warthin-Starry silver stained tissue sections. The inguinal and popliteal lymph nodes are hyperplastic with large germinal centers containing many immature lymphoreticular cells. Focal accumulations of epithelioid cells may be present.

D. Diagnosis

Diagnosis usually depends on the demonstration of gross lesions and the demonstration of the etiological agent by dark field microscopic examination of scrapings from the lesions. The organisms also may be demonstrated by silver stained smears or tissue sections. Rabbits may be examined serologically using one of several available tests. The Wassermann test, commonly used in the diagnosis of syphilis in man is a complement fixation test (242). When Wassermann developed the test, the antigen used was fetal liver rich in *T. pallidum* and complement-fixing antibodies were present in high titers in patients with syphilis. It was soon discovered that the normal liver could serve just as well as a source of antigen and the antigen commonly used now is a purified lipid extract of beef heart (cardiolipin) to which lecithin and cholesterol are added (242). A number of other serological tests have been developed to detect the same antibody (sometimes called reagin) detected in the Wassermann test. The Venereal Disease Research Laboratory (VDRL) slide test and the rapid plasma reagin (RPR) card test are convenient tests used to detect the presence of the Wassermann antibody. A more recently developed test utilizing the fluorescent antibody techniques and fresh or frozen *T. pallidum* is commonly used. This test is called the fluorescent treponemal antigen (FTA) test. The RPR card test, VDRL slide test, and FTA test as well as the Wassermann complement fixation test have been used to test rabbits. Most of the early reports indicate a negative or variable Wassermann reaction in rabbits shown to be infected with *T. cuniculi* (68, 166, 167, 242). Kolmer and Casselman (125) observed that some of their normal laboratory rabbits gave positive tests and they did not have an explanation for this finding. It has since been shown that there are cross-reacting antibodies between *T. pallidum* and *T. cuniculi* and that *T. cuniculi* will stimulate the production of Wassermann antibody when administered experimentally (152, 230). More recently, the VDRL slide Test, RPR, and FTA tests have been used to screen the sera of rabbits (42, 176, 211). Rabbits found to be positive by serological test are considered as suspects. The demonstration of *Treponema* by dark field microscopy of lymph expressed from cutaneous lesions or from prepucial washings is essential to confirm the diagnosis in these rabbits (211).

E. Control and Prevention

Treponematosis, like syphilis in man, responds to systematic treatment with penicillin (39). Prevention of treponematosis in rabbits has received little attention perhaps because clinical outbreaks are seldom observed and they respond well to treatment. In colonies where infection of breeding stock is especially undesirable, a closed breeding colony is desirable. If it is necessary to bring in additional rabbits, clinical and serological screening should precede admission to the colony.

XI. STAPHYLOCOCCOSIS

A. General

Staphylococcosis is a commonly occurring disease in both domestic and wild rabbits. This disease is caused by *Staphylococcus aureus* and is characterized by fatal septicemia or suppurative inflammation in nearly any organ or site.

1. History

In 1903 Catterina observed an outbreak of acute septicemia and death in rabbits (38). The lesions observed were nonspecific and a *Staphylococcus* spp. was isolated. In 1915 Jacob (116) described subcutaneous abscesses caused by *Staphylococcus* in the area of the head, eyelids, and back of domestic rabbits. Sustmann (220) observed a chronic suppurative alveolar periostitis and osteomyelitis of the maxilla of a rabbit. *Staphylococcus* was isolated. Soituz (216) reported abscesses and inflammation of the mammary glands in rabbits caused by *Staphylococcus*. In 1945, Lesbouyries (135) reviewed staphylococcosis in rabbits. He indicated that *S. aureus*, *S. albus*, and *S. citreus* may be involved in this disease. Septicemias as well as subcutaneous and mammary gland inflammation were described. In 1963, Hagen (100) reported disseminated staphylococcosis in rabbits 3 to 5 days old. Numerous subcutaneous abscesses were present on the lower abdomen, inner aspect of the forelegs, and lower jaw. Abscesses were observed in the heart and lungs during postmortem examination. An α-hemolytic, coagulase positive *S. aureus* was isolated.

2. Etiology

Staphylococci are spherical gram-positive organisms which are quite uniform in size and occur in irregular masses. *Staphylococcus aureus*, which were coagulase-positive and hemolytic were most consistently isolated from affected rabbits (100, 104, 147, 191).

3. Incidence

Little objective data is available concerning the incidence of this disease; however, it is generally conceded to be quite common. Likewise, the disease appears to be common among wild rabbits including jackrabbits (172) and cottontails (15, 40, 147). The septicemic form is most common in young rabbits or rabbits stressed from various causes.

4. Epidemiology

Rabbits may be infected by *S. aureus* with little or no clinical disease (62, 104, 191). The organism may reside in the nasal sinuses or lungs and they may be spread by direct contact or by aerosol. Renquist and Soave reported that

S. aureus of the same phage type as in the nasal cavity of their rabbits was isolated from the nasal cavity of the technician caring for their rabbits. Once infected, a rabbit is likely to develop clinical disease when its resistance is sufficiently decreased. Infection of skin wounds is a common form of acquiring the organism and the result is suppurative inflammation of the skin and subcutaneous abscesses (40, 173). Septicemia may also result from skin infection.

B. Clinical Signs

The clinical signs depend on the site and duration of infection. There may, of course, be no clinical disease related to infection; however, swelling, redness, and induration of the skin, subcutaneous tissue, and mammary glands may occur in acute and subacute infections. In chronic infections, abscesses develop in the subcutaneous tissue and mammary glands, and these appear as firm or fluctuant swellings. In cases of acute septicemia there may be fever, anorexia, depression, and death.

C. Pathology

Suppurative inflammation of the skin and subcutaneous tissue is typically seen in staphylococcosis. The inflammation usually becomes chronic, and thick-walled abscesses containing thick white suppurative exudate develop. In the acute stages of skin infection there is swelling, redness, and induration. Acute to chronic mastitis may be observed with lesions varying from swelling, redness, and induration to chronic abscessation like that seen in the skin.

Septicemia may result in peracute death with only a few nonspecific lesions; however, if the rabbit survives this phase, abscesses may develop in many internal organs. The heart, kidneys, lungs, liver, spleen, epididymis, testis, and joints may be affected (56, 100, 147, 172, 191). Osteitis and osteomyelitis similar to that seen in actinomycosis has been described (220).

D. Diagnosis

Diagnosis of staphylococcosis depends on isolation and identification of the etiological agent.

E. Control and Prevention

Infections with *S. aureus* are usually treated with local or systemic antibiotics. *In vitro* antibiotic sensitivity testing is commonly practiced to predict which antibiotic will be most effective. Cameron (34) vaccinated rabbits with *S. aureus* and found them to be resistant to skin infection when challenged by intradermal inoculation of living organisms.

He indicated that there was no correlation between the degree or resistance and the demonstrated titer of serum antibodies.

XII. MISCELLANEOUS BACTERIAL DISEASES

A number of bacterial diseases are poorly defined, of little importance, or of doubtful existence in the opinion of the author. These diseases have been grouped together and will be discussed briefly in this section.

A. Colibacillosis

Acute outbreaks of diarrhea with a high mortality rate has been observed frequently in rabbits. The etiology is usually undetermined; however, in some instances the etiology has been attributed to *Escherichia coli* or its endotoxin (18, 82, 132, 140, 141, 153a, 238, 243, 257). The work of Matthes (153a) suggests that environmental changes sometimes lead to rapid change in the gut flora. Normally the gut flora is predominantly gram-positive but when conditions favor the rapid shift to a predominantly gram-negative flora, acute diarrhea and death usually result. *Escherichia coli*, especially serotype 0128:K67(B12), was the predominant organism found in young rabbits dying under these conditions. Loliger *et al.* (140–143a, 153, 153a) also found *E. coli*, especially serotype 0128, to be an important cause of diarrhea and acute death in young rabbits. Vetesi (238) described what he called mucoid enteritis in rabbits in Hungary. Over a 7-year period, 156 of 1254 (12%) rabbits necropsied had this disease. Most of the affected rabbits were 1 to 3 months of age. The clinical disease was characterized by diarrhea and salivation with 20 to 30% of the rabbits in some colonies dying from this disease. Occasionally all of the rabbits from a colony died as a result of this condition. Pure cultures of *E. coli* were commonly obtained from the cecum of affected rabbits. The lesions consisted of subserosal and submucosal hemorrhages in the cecum and colon. These organs also were edematous and the lumen was filled with gelatinous mucin. Occasionally lesions were found outside the intestinal tract, and these included focal areas of necrosis in the liver and heart. The disease was regarded by the author as colibacillosis. The resemblance between the clinical signs and lesions of the disease described in this report and those occurring in Tyzzer's disease is striking. Glantz (81) reported the association of *E. coli* belonging to the unclassified serogroup OX1 with fatal cases of diarrhea in rabbits. This same organism was isolated from a few clinically healthy rabbits. The *E. coli* belonging to serogroup OX1 produced a positive reaction when injected into ligated intestinal loops of young rabbits similar to the reaction of *E. coli* strains known to be

pathogenic in man. Weber and Manz (243) reported typing the 0/antigenic groups of 61 strains of *E. coli*. Serogroup 085 was found in 47 of 54 strains isolated from rabbits with diarrhea. Colibacillosis has been reported in European hares (*Lepus europeus*) (238) and in cottontail rabbits (257). Yuill and Hanson (257) reported high mortality in cottontail rabbits kept in outside pens. They associated the initiation of the disease with severe winter weather. They found that they could induce the disease by artificially exposing the rabbits to low temperatures and they observed that 42 of 47 serum samples of normal cottontails contained *E. coli* agglutinins.

The understanding of colibacillosis in rabbits is quite incomplete; however, it appears that this disease may prove to be very important. Part of the disease complex called mucoid enteritis may prove to be colibacillosis.

B. Diseases Associated with *Proteus vulgaris*

This organism is seldom considered to be of importance in producing disease in animals. Lesbouyries (132, 136) reported that this organism could cause subcutaneous abscesses in rabbits similar to those produced by *Pasteurella multocida* and *Staphylococcus aureus*. The abscesses may be numerous and are variable in size. Morel (156) reported the unusual association of *Proteus* with a progressive paralysis in a rabbit. The organism was isolated from the brain and was thought to be responsible for producing an encephalomyelitis.

C. Diseases Associated with *Pseudomonas aeruginosa*

Subcutaneous abscesses and septicemia in rabbits have been associated with *P. aeruginosa* (5, 22, 80, 136, 148, 168, 194, 241). McDonald and Pinheiro (148) reported sporadic outbreaks of fulminating pneumonia and diarrhea in young rabbits. *Pseudomonas aeruginosa* was isolated from the water delivered in an automatic watering system. The watering system was cleaned, disinfected, and intermittent chlorination of the water was instituted using 1.5 to 2 parts per million.

The septicemic disease has a sudden onset with lethargy, fever, nasal and ocular discharged, dyspnea, and death. Diarrhea may also be present. The lesions are nonspecific and include widespread congestion with clear to blood-tinged fluid in the thoracic cavity, pericardial sac, and abdominal cavity. The spleen may be enlarged and hemorrhages may be present on serosal surfaces. Consolidation of the lungs and pulmonary abscesses also may be present (5, 148). Both polmyxin and neomycin have been suggested for treatment of this disease (5, 22).

Hibbs *et al.* (108) isolated *P. aeruginosa* and *Aeromonas hydrophila* in rabbits dying following abdominal surgery.

Hemorrhage and congestion of the cecum were observed postmortem. Through their attempts to reproduce the disease experimentally, they concluded that *A. hydrophila* rather than *P. aeruginosa* was responsible for the naturally occurring disease they observed.

D. Melioidosis

Malleomyces pseudomallei is a bacterial organism similar to *P. aeruginosa* in many of its cultural features. It has been reported on one occasion to cause a fatal septicemic disease within a rabbit colony (136, 219). The clinical signs included nasal and ocular discharge, dyspnea, and death. In those animals that did not die peracutely, miliary foci of necrosis were present in both the thoracic and abdominal viscera. The testes and epididymis frequently contained areas of caseous necrosis. Lymph nodes of the neck and axilla were enlarged and contained caseous nodules. In some of the subacute to chronic cases the lungs were consolidated.

E. Actinobacillosis

In 1961 Arseculeratne (6) described 3 cases of swelling and inflammation in the soft tissue around the tarsal joints of rabbits. The affected joints enlarged over several weeks and the affected animals became emaciated. Two of the 3 died within 3 months of the first appearance of the disease. Granulomatous inflammation characterized the soft tissue swelling around the affected joints. Colonies of gram-negative bacilli were present within this tissue and the colonies were surrounded by a row of eosinophilic clubs. In one affected rabbit there was caseous necrosis in the ipsilateral popliteal lymph node and a 3-mm nodule of granulomatous inflammation in the lung. An *Actinobacillus* was isolated but it was considered to be distint from *A. lignieresi*. The following year the same author reported 2 additional cases of the disease and indicated that in addition to the tarsal joint, the stifle and digits may be affected (7). The responsible organism resembled *A. lignieresi* except that it had a capsule. The name *A. capsulatus* was proposed for this organism.

F. Actinomycosis

Actinomycosis is a rarely recognized disease in domestic rabbits caused by the *Actinomyces* spp. (208, 217). The disease produces osteitis with bone production and lysis similar to the reaction caused by these organisms in other domestic animals. Involvement of mandible, nasal bones, foot and hock joints, and lumbar vertebrae have been observed. Soft tissue lesions including subcutaneous inflammation, boils, and fistulas of the head, neck, and trunk

also have been observed. Granulomatous inflammation with "sulfur granules" characterize the histological pathology.

G. *Hemophilus*

There is a single report of *Hemophilus* spp. being isolated from subcutaneous abscesses in a domestic rabbit (79). The abscesses were firm, contained white suppurative exudate, and were surrounded by a thick fibrous capsule. The gram-negative, nonmotile, short rod that was isolated was identified as a member of the genus *Hemophilus*.

H. Sylvatic Plague

Sylvatic or rodent plague is occasionally found in wild rabbits (121, 122, 136, 170, 208) and is caused by *Yersenia pestis*. Plague has been recognized in the bush rabbit (*Sylvilagus bachmoni*), desert cottontails (*S. auduboni*), the hare (*Lepus californicus*), and the mountain cottontail (*S. nuttalli*) (122).

The rabbit had not been considered of epidemiological importance until an outbreak of human cases was shown to have been acquired from wild rabbits. This outbreak occurred in New Mexico in 1950 (121). Ten years later 2 additional human cases occurred in New Mexico simultaneously with an epizootic of plague in cottontail rabbits.

I. Brucellosis

Brucellosis has rarely been recognized in domestic rabbits; however, it has been reported more frequently in various species of wild rabbits (84, 181, 190, 208, 222, 254). Peres and Granon-Fabre (181) described abortions in 4 domestic rabbits which were raised on a farm with sheep and goats known to have brucellosis. Both skin tests and a serum agglutination test were positive on 2 females that survived following abortion. The lesions in the two that died included pyometra and abscesses in the lungs, liver, spleen, and axillary lymph nodes. *Brucella melitensis* was isolated from both cases. Renoux and Sacquet (190) found that the serum of young rabbits frequently contained naturally occurring antibodies to *Brucella*. In one group of 98 rabbits, 25 had agglutination titers of 1:40 to 1:320. Necropsy of several rabbits with positive agglutination reaction revealed no gross lesions; however, an organism which they identified as *Brucella intermedia* was isolated from 2 rabbits.

Brucellosis is known to occur with some frequency in European hares (*L. europaeus*) (84, 254). In their review of brucellosis in wild mammals, Witter and O'Meara site the existence of this disease in European hares in Switzerland, France, Hungary, Germany Czechoslovakia,

Denmark, and Russia (254). Approximately 10% of the hares submitted for postmortem examination in Switzerland had brucellosis and approximately 8% of a group examined from Hungary had the disease. *Brucella suis*, *B. melitensis*, and *B. abortus* all have been incriminated. Thorpe *et al.* (222) reported brucellosis in black-tailed jackrabbits (*L. californicus*) in Utah, and Goyon (84) reported a case of brucellosis in a European hare. In the latter case, the spleen was enlarged and contained numerous small abscesses. Multiple abscesses were also present in the liver, lungs, and subcutaneous tissue. A Danish type *B. suis* was isolated. The author sites numerous other references to brucellosis in hares.

J. *Streptococcus*

Streptococcus has been associated with acute septicemias particularly involving young rabbits (31, 56, 109, 117, 136, 208). The clinical disease is characterized by anorexia, depression, fever, dyspnea, inconsistent diarrhea, and death. Few lesions are observed at postmortem, and these are nonspecific. Jansen (117) described a peracute septicemia in rabbits and isolated a *Streptococcus* with cultural features he believed to be unique. He names the organism *S. cuniculi*.

K. *Diplococcus*

Acute septicemia and death in rabbits have been associated with *Diplococcus* spp. (56, 129, 136). Like other septicemic diseases the clinical signs and lesions are not specific. The disease is characterized clinically by dyspnea, fever, depression, and death. Profuse diarrhea also has been described. Gross lesions include widespread congestion, clear to blood-tinged fluid in the thoracic and abdominal cavities, and petechial hemorrhages on serosal surfaces. The spleen and lymph nodes may be enlarged. Occasionally pericarditis, myocarditis, and pneumonia are present. Munoz and Holford (163) found naturally occurring antibodies against pneumococcus Type II in 21 of 32 normal rabbits that would protect mice against 10 to 1000 lethal doses of the organism. They concluded that many of their rabbits had a previous experience with this or a similar organism.

L. *Clostridium*

Acute disturbances characterized by anorexia, diarrhea, tympany, and death have been described in rabbits (134, 136, 155, 162, 208). This disease is usually referred to as enterotoxemia. It affects lactating rabbits as well as young rabbits of either sex. Lactating rabbits usually become ill

8 to 30 days after parturition. The cause of this disease was not identified by Moussu (162). Morcos (155) described an acute fatal disease in rabbits 1 to 5 months old. The disease was characterized by anorexia, diarrhea which sometimes contained blood, and tympany. He isolated an organism that he named *Bacillus* (*Clostridium*) *tympani-cuniculi*. Others have considered the agent to be *C. perfringens* or a *C. perfringens*-like organism (134, 136, 208). The lesions include the distension of the abdomen due to the accumulation of gas in the stomach and small intestine. Occasionally ulcers are present on the mucosal surface of the small intestine.

M. Erysipelas

Erysipelothrix rhusiopathiae has been reported to occur in rabbits on only a few occasions. Lucas *et al.* (142) described a bilateral orchitis in a hare from which he isolated this agent. He also sites a similar case in a hare. There is one report of an outbreak of erysipelas in domestic rabbits in which nervous symptoms, torticollis, abortions, and neonatal mortality were observed (224). *Erysipelothrix insidiosa* was isolated in this outbreak.

N. Leptospirosis

Leptospirosis is known to occur in cottontail and swamp rabbits in southeastern United States (150, 197, 209). Few isolations have been made; however, in a serological survey 10 of 48 (21%) cottontail and swamp rabbits had serological titers considered to be positive (209). Focal nephritis was present in 92% of the kidneys examined from the same group of rabbits.

O. *Mycoplasma*

Two strains of *Mycoplasma pulmonis* have been isolated from the nares and oropharynx of 4 New Zealand White rabbits (53). These organisms have been characterized immunologically (54), however no disease has been associated with these organisms.

P. *Chlamydia*

Psittacosis-lymphogranuloma venereum (PLV) agents, or chalmydial agents, have been isolated from both wild and domestic rabbits. Spalatin *et al.* (218) isolated a chlamydial agent from snowshoe hares involved in a fatal epizootic in Saskatchewan, Canada. These hares had enlarged spleens and livers which were congested and contained focal areas of necrosis. Parker *et al.* (177) briefly mentioned the isolation of a chlamydial agent from domestic rabbits experiencing abortions. Flatt and Dungworth (62) isolated a chlamydial agent from the lungs of a domestic rabbit with pneumonia and reproduced pneumonia experimentally by giving the agent intratracheally.

REFERENCES

1. Adams, C. E., Aitken, F. C., and Worden, A. N. (1967). The rabbit. *In* "The UFAW Handbook on the Care and Management of Laboratory Animals" (Staff of UFAW, ed.), 3rd ed., pp. 419–438. Livingstone, Edinburgh.
2. Adams, D. K., Cappell, D. F., and McCluskie, A. S. (1928). Cutaneous spirochaetosis due to *Treponema cuniculi* in British rabbits. *J. Pathol. Bacteriol.* **31**, 157–161.
3. Alexander, M. M., Sawin, P. B., and Roehm, D. A. (1952). Respiratory infection in the rabbit: An enzootic caused by *Pasteurella lepiseptica* and attempts to control it by vaccination. *J. Infec. Dis.* **90**, 30–33.
4. Allen, A. M., Ganaway, J. R., Moore, T. D., and Kinard, R. F. (1965). Tyzzer's disease syndrome in laboratory rabbits. *Amer. J. Pathol.* **46**, 859–882.
5. Alpen, G. R., and Maerz, K. (1969). The incidence of a pathogenic strain of *Pseudomonas* in a rabbit colony. *J. Inst. Anim. Tech.* **20**, 72–74.
6. Arseculeratne, S. N. (1961). A preliminary report on actinobacillosis as a natural infection in laboratory rabbits. *Ceylon Vet. J.* **9**, 5–8.
7. Arseculeratne, S. N. (1962). Actinobacillosis in joints of rabbits. *J. Comp. Pathol.* **72**, 33–39.
8. Arzt, L., and Kerl, W. (1914). Weitere Mitteilungen über Spirochätenbefunde bei Kaninchen. *Wien. Klin. Wochenschr.* **27**, 1053–1055.
9. Bablet, J., and Van Deinse, F. (1943). Trois cas d'infection oculaire specifique au cours de la tuberculose expérimentale par bacilles attenues. *Ann. Inst. Pasteur, Paris* **69**, 45–47.
10. Basset, M. J. (1908). Contribution á l'étude de la "Maladie die Schmorl." *Bull. Soc. Cent. Med. Vet.* **62**, 345–350.
11. Bayon, H. (1913). A new species of *Treponema* found in the genital sores of rabbits. *Brit. Med. J.* **2**, 1159.
12. Beattie, J. M., Yates, A. G., and Donoldson, M. A. (1913). An epidemic disease in rabbits resembling that produced by *B. necrosis* (Schmorl), but caused by an aerobic bacillus. *J. Pathol. Bacterial.* **18**, 34–36.
13. Beck, M. (1893), Der Bacillus der Brustseuche beim Kaninchen. *Z. Hyg. Infektionskr.* **15**, 363–368.
14. Belding, D. L., and Merrill, B. (1941). Tularemia in imported rabbits in Massachusetts *N. Engl. J. Med.* **224**, 1085–1087.
15. Bell, J. F., and Chalgren, W. S. (1943). Some wildlife diseases in the Eastern United States. *J. Wildl. Manage.* **7**, 270–278.
16. Bessemans, A. (1928). La spirochétose spontanée du lapin en Belgique. *C. R. Soc. Biol.* **99**, 331–333.
17. Bessemans, A., and DeGeest, B. (1928). Sur quelques propriétés du *Treponema cuniculi*. *C. R. Soc. Biol.* **99**, 334–335.
18. Bessemans, A., and DeWilde, H. (1937). Nouvelles données sur la pallidoidose receptivité inapparente du lapin, du cobaye, de la souris, du rat et du hamster. *C. R. Soc. Biol.* **126**, 264–266.
19. Bessemans, A., and Van Canneyt, J. (1930). Sur un procédé expérimental de diagnostic différentiel entre *Treponema pallidum* et *Treponema cuniculi* par production de lésions oculaires primaires. *C. R. Soc. Biol.* **102**, 951–954.
20. Betts, A. O. (1952). Respiratory diseases of pigs. V. Some clinical and epidemiological aspects of virus pneumonia of pigs. *Vet Rec.* **64**, 283–288.
21. Beveridge, W. I. B. (1959). Necrobacillosis, foot-rot, etc. (Diseases due to non-sporing anaerobes). *In* 'Infectious Diseases of Animals: Diseases Due to Bacteria" (A. W. Stableforth and I. A. Galloway, eds.), Vol. 2, pp. 397–412. Butterworth, London.
22. Blount, W. P. (1957). "Rabbits' Ailments," pp. 70–86. "Fur and Feathers," Idle, Bradford, England.
23. Borg, K., Hanko, E., Krunajevic, T., Nilsson, N. G., and Nilsson, P. O. (1969). On tularemia in the varying hare (*Lepus timidus L.*). *Nord. Veterinaer med.* **21**, 95–104.
24. Brack, M. (1966). Spontane, durch *Mycobacterium avium* verursachte Tuberkulose beim Hauskaninchen. *Deut. Tierärztl. Wochenschr.* **73**, 317–321.
25. Brooks, G. F., and Buchanan, T. M. (1970). Tularemia in the United States: Epidemiologic aspects in the 1960s and follow-up at the outbreak of tularemia in Vermont. *J. Infec. Dis.* **121**, 357–359.
26. Brown, R. C., Richter, C. B., and Bloomer, M. D. (1969). Ultrastructural pathology of an acute fatal enteritis of captive cottontail rabbits. Search for a viral etiologic agent. *Amer. J. Pathol.* **57**, 93–126.
27. Bru, P. (1923). Tuberculose spontanée du lapin avec lesions d'ostéopériostite. *Rev. Gen. Med. Vet.* **32**, 631–637.
28. Brune, Cachenback, La Grange, and Roger (1935). Sur la tuberculose spontanée du lapin. *Rev. Gen. Med. Vet.* **44**, 17–19.
29. Bruner, W. D., and Gillespie, J. H. (1966). *In* "Hagen's Infectious Diseases of Domestic Animals," 5th ed., pp. 267–269. Cornell Univ. Press (Comstock), Ithaca, New York.
30. Bruner, W. D., and Gillespie, J. H. (1966). *In* 'Hagen Infectious Diseases of Domestic Animals," 5th ed., pp. 386–392. Cornell Univ. Press (Comstock), Ithaca New York.
31. Burnet, E. (1928). Streptocoques de sortie chez le lapin. *C. R. Soc. Biol.* **98**, 440–442.
32. Burroughs, A. L., Holdenried, R., Longanecker, D. S., and Meyer, K. F. (1945). A field study of latent tularemia in rodents with a list of all known naturally infected vertebrates. *J. Infec. Dis.* **76**, 115–119.
33. Calmette, A. (1933). L'infection tuberculeuse spontanée du cobaye et du lapin. *Ann. Inst. Pasteur, Paris* **50**, 148–160.
34. Cameron, C. M. (1971). Evaluation of serological tests as criteria for immunity to staphylococcal skin infection in rabbits. *Onderstepoort J. Vet. Res.* **38**, 99–110.
35. Carlotto, F. (1965). Un foclaio di listeriosi nel coniglio. *Nuova Vet.* **41**, 223–225.
36. Carter, G. R. (1955). Studies on *Pasteurella multocida*. I. A hemagglutination test for the identification of serological types. *Amer. J. Vet. Res.* **16**, 481–484.
37. Carter, G. R. (1967). Pasteurellosis: *Pasteurella multocida* and *Pasteurella hemolytica*. *Advan. Vet. Sci.* **11**, 321–379.
38. Catterina, G. (1903). Ueber eine bewimperte Micrococcusform, welche in einer Septikämie der Kaninchen gefunden werde. *Zentralbl. Bakteriol., Parasitenk. Inteklienskr.*, Abt. 1: Orig. **34**, 108–112.
39. Chapman, M. P. (1947). The use of penicillin in the treatment of spirochetosis (vent disease) of domestic rabbits. *N. Amer. Vet.* **28**, 740–742.
40. Cheatum, E. L. (1941). Lymphadenitis in New York cottontails. *J. Wildl. Manage.* **5**, 304–308.
41. Chrétien, A. (1912). Aut sujet de la tuberculose zoogléique et de la tuberculose bacillaire du rein due lapin. *Hyg. Viande Lait* **6**, 432–434.
42. Clark, J. W., Jr., (1970). Serological tests for syphilis in healthy rabbits. *Brit. J. Vener. Dis.* **46**, 191–197.
43. Cobbett, L. (1913). Two cases of spontaneous tuberculosis tuberculosis in the rabbit caused by the avian tubercle bacillus. *J. Comp. Pathol. Ther.* **26**, 33–45.
44. Coulaud, E. (1924). La tuberculose par contamination naturelle chez le lapin. (Premiere Mémoire). *Ann. Inst. Pasteur, Paris* **38**, 581–597.
45. Coulaud, E. (1931). La tuberculose par contamination naturelle

chez le lapin. (Deuxieme Mémoire). *Ann Inst. Pasteur, Paris* **45**, 424–443.

46. Craigie, J. (1966). *Bacillus piliformis* (Tyzzer) and Tyzzer's disease of the laboratory mouse. I. Propagation of the organism in enbryonated eggs. *Proc. Roy. Soc. Ser. B* **165**, 35–60.

47. Craigie, J. (1966). *Bacillus piliformis* (Tyzzer) and Tyzzer's disease of the laboratory mouse. II. Mouse pathogenicity of *B. piliformis* grown in embryonated eggs. *Proc. Roy. Soc., Ser. B* **165**, 61–77.

48. Cutlip, R. C., Amtower, W. C., Beall, C. W., and Matthews, P. J. (1971). An epizootic of Tyzzer's disease in rabbits. *Lab. Anim. Sci.* **21**, 356–361.

49. Daniels, J. J. H. M. (1961). Enteral infections with *Pasteurella pseudotuberculosis. Brit. Med. J.* **2**, 997.

50. Danila, P., and Stroe, A. (1923). Sur la spirochetose du lapin. *C. R. Soc. Biol.* **88**, 892–894.

51. Davis, D. J. (1913). A bacillus from spontaneous abscesses in rabbits and its relationship to the influenza bacillus. *J. Infec. Dis.* **12**, 42–5.

52. Davis, D. J. (1917). Further observations on subcutaneous abscesses in rabbits. The carrier state and its relation to rabbit septicemia. *J. Infec. Dis.* **21**, 314–321.

53. Deeb, B. J., and Kenny, G. E. (1967). Characterization of *Mycoplasma pulmonis* variants isolated from rabbits. I. Identification and properties of isolates. *J. Bacteriol.* **93**, 1416–1424.

54. Deeb, B. J., and Kenny, G. E. (1967). Characterization of *Mycoplasma pulmonis* variants isolated from rabbits. II. Basis for differentiation of antigenic subtypes. *J. Bacteriol.* **93**, 1425–1429.

55. Dieter, L. V., and Rhodes, B. (1926). Tularemia in wild rats. *J. Infec. Dis.* **38**, 541–546.

56. Dumas, J. (1953). "Les animaux de laboratoire." Editions Médicales Flammarion, Paris.

57. Duthie, R. C., and Mitchell, C. A. (1931). *Salmonella enteritidis* infection in guinea pigs and rabbits. *J. Amer. Vet. Med. Ass.* **78**, 27–41.

58. Ferry, N. S. (1913–1914). Bacteriology and control of acute infections in laboratory animals. *J. Pathol. Bacteriol.* **18**, 445–446.

59. Ferry, N. S., and Hoskins, H. P. (1920). Bacteriology and control of contagious nasal catarrh (snuffles) of rabbits. *J. Lab. Clin. Med.* **5**, 311–318.

60. Flatt, R. E. (1969). Pyometra and uterine adenocarcinoma in a rabbit. *Lab. Anim. Care* **19**, 398–401.

61. Flatt, R. E., and Dungworth, D. L. (1971). Enzootic pneumonia in rabbits: Naturally occurring lesions in lungs of apparently healthy young rabbits. *Amer. J. Vet. Res.* **32**, 621–626.

62. Flatt, R. E., and Dungworth, D. L. (1971). Enzootic pneumonia in rabbits: Microbiology and comparison with lesions experimentally produced by *Pasteurella multocida* and a chlamydial organism. *Amer. J. Vet. Res.* **32**, 627–637.

63. Omitted.

64. Florio, R., and Saurat, P. (1943). Sur en cas d'iritis tuberculeuse chez le lapin, *Rev. Med. Vet.* **94**, 164–167.

65. Fox, R. R., Norberg, R. F., and Myers, D. D. (1971). The relationship of *Pasteurella multocida* to otitis media in the domestic rabbit (*Oryctolagus cuniculus*). *Lab. Anim. Care* **21**, 45–48.

66. Francis, E. (1925). Tularemia. *J. Amer. Med. Ass.* **84**, 1243–1250.

67. Francis, E., and Moore, D. (1926). Identity of Ohara's disease and tularemia. *J. Amer. Med. Ass.* **86**, 1329–1332.

68. Fried, S. M., and Orlov, S. S. (1932). Spontaneous spirochetosis and experimental syphilis in rabbits. *Arch. Dermatol. Syphilol.* **25**, 893–905.

69. Fujiwara, K. (1967). Complement fixation reaction and agar gel double diffusion test in Tyzzer's disease of mice. *Jap. J. Microbiol.* **11**, 103–117.

70. Fujiwara, K., Kurashina, H., Maejima, K., Tajima, Y., Takagaki, Y., and Naiki, M. (1965). Actively induced immune resistance to the experimental Tyzzer's disease of mice. *Jap. J. Exp. Med.* **35**, 259–275.

71. Fujiwara, K., Takahashi, R., Kurashina, H., and Matsunuma, N. (1970). Protective serum antibodies in Tyzzer's disease of mice. *Jap. J. Exp. Med. 39*, 491–504.

72. Ganaway, J. R., Allen, A. M., and Moore, T. D. (1971). Tyzzer's disease. *Amer. J. Pathol.* **64**, 717–732.

73. Ganaway, J. R., Allen, A. M., and Moore, T. D. (1971). Tyzzer's disease of rabbits: Isolation and propagation of *Bacillus piliformis* (Tyzzer) in embryonated eggs. *Infec. and Immunity* **3**, 429–437.

74. Ganguli, S. (1958). *Salmonella* serotypes in India. *Indian J. Med. Res.* **46**, 637–642.

75. Garner, F. M., Innes, J. R. M., and Nelson, D. H. (1967). Murine neuropathology. *In* "Pathology of Laboratory Rats and Mice" (E. Cotchin and F. J. C. Roe, eds.), pp. 295–348. Davis, Philadelphia, Pennsylvania.

76. Garofolo, T. (1936). Enzoozia da b. paratifo b. Breslavia in un allevamento di conigli. *Clin Vet. Milano* **59**, 527–531.

77. Genov, I. (1961). Research into the pseudotuberculosis in rabbits and guinea pigs. *Izv. Vet. Inst. Zaraz. Parazit. Bolesti, Sofia* **2**, 257–268.

78. Ghosh, G. K., and Chatterjee, A. (1960). Salmonellosis in guinea pigs, rabbits and pigeons. *Indian Vet. J.* **37**, 144–148.

79. Gibbons, N. E. (1929). *Hemophilus* sp. and *Neisseria* sp. in skin abscesses in rabbits and guinea pigs. *J. Infec. Dis.* **45**, 288–292.

80. Giorgi, W. (1966). Isolamento de uma amostra de *Pseudomonas aeruginosa* capsulada, de pus de coelho. *Arq. Inst. Biol. (Sao Paulo)* **33**, 49–52.

81. Glantz, P. J. (1970). Unclassified *Escherichia coli* serogroup OXI isolated from fatal diarrhea of rabbits. *Can. J. Comp. Med.* **34**, 47–49.

82. Glantz, P. J. (1971). Serotypes of *Escherichia coli* associated with colibacillosis in neonatal animals. *Ann. N. Y. Acad. Sci.* **176**, 67–79.

83. Glaue, D. (1911). Ueber den Erreger einer Kaninchen-Pleuropneumonic. *Zentralbl. Bakteriol. Parasitenk. Infekximskr., Abt. 1: Orig.* **60**, 176–188.

84. Goyon, M. (1958). Nouveau cas de brucellose du lièvre à *Brucella suis*, type danois. *Rec. Med. Vet.* **134**, 533–541.

85. Graham, R., Levine, N. D., and Morrill, C. C. (1943). Listerellosis in domestic animals. A technical discussion of field and laboratory investigations. *Ill., Agr. Exp. Sta., Bull.* **499**, 1–99.

86. Grasovsky, Y. S. (1939). The occurrence of *Salmonella*, bact. typhi-murium in a disease of rabbits. *Vet. J.* **95**, 294–296.

87. Gray, H. (1925). Tuberculosis in the rabbit and guinea pig. *Vet. J.* **81**, 212–217.

88. Gray, H. (1931). Some of the common diseases of the rabbit. *Vet. Rec.* **11**, 921–927.

89. Gray, M. L., and Killinger, A. H. (1966). *Listeria monocytogenes* and listeric infections. *Bacteriol. Rev.* **30**, 309–382.

90. Gray, M. L., Singh, C., and Thorp, F., Jr., (1955). Abortion, stillbirth, early death of young in rabbits by *Listeria monocytogenes*. I. Ocular instillation. *Proc. Soc. Exp. Biol. Med.* **89**, 163–169.

91. Gray, M. L., Singh, C., and Thorp, F., Jr., (1955). Abortions, stillbirth, early death of young in rabbits by *Listeria monocytogenes*. II. Oral exposure. *Proc. Soc. Exp. Biol. Med.* **89**, 169–175.

92. Gray, M. L., Stafseth, H. J., and Thorp, F., Jr., (1951). A four-year study of listeriosis in Michigan. *J. Amer. Vet. Med. Ass.* **118**, 242–252.

93. Green, R. G. (1913). The occurrence of *Bacterium* tularense in the Eastern wood tick, *Dermacentor variabilis. Amer. J. Hyg.* **14**, 600–613.

94. Griffin, C. A. (1952). Respiratory infection among rabbits. *Proc. Anim. Care Panel* **3**, 3–13.

95. Griffith, A. S. (1928). Tuberculosis of the domesticated species of animals. *J. Comp. Pathol. Ther.* **41**, 109–122.

96. Griffith, A. S. (1932). A study of the BCG strain of tubercle bacillus. *Lancet* **222**, 303–308 and 361–363.

96a. Gualandi, G. (1948). Le carni di coniglio quale fonte di tossinfezione alimentare nell'uomo. *Clin. Vet. Milano* **71**, 313–317.

97. Guérin, G. (1908). Cas de tuberculose spontanée du lapin. *Hyg. Viande Lait* **2**, 7–8.

97a. Habermann, R. T., and Williams, F. P., Jr. (1958). Salmonellosis in laboratory animals. *J. Nat. Cancer Inst.* **20**, 933–947.

98. Hagen, K. W., Jr. (1958). Enzootic pasteurellosis in domestic rabbits. I. Pathology and bacteriology. *J. Amer. Vet. Med. Ass.* **133**, 77–80.

99. Hagen, K. W., Jr. (1959). Chronic respiratory infection in the domestic rabbit. *Proc. Anim. Care Panel* **9**, 55–60.

100. Hagen, K. W., Jr. (1963). Disseminated staphylococcic infection in young domestic rabbits. *J. Amer. Vet. Med. Ass.* **142**, 1421–1422.

101. Hagen, K. W. Jr. (1966). Enzootic pasteurellosis in domestic rabbits. II. Strain types and methods of control. *Lab. Anim. Care* **16**, 487–491.

102. Hagen, K. W., Jr. (1967). Effect of antibiotic-sulfonamide therapy on certain microorganisms in the nasal turbinates of domestic rabbits. *Lab. Anim. Care* **17**, 77–80.

103. Hagen, K. W., Jr. and Lund, E. E. (1962). Common diseases of domestic rabbits. *U.S. Dep. Agr., Agr. Res. Serv. Publ.* **Ars 45–3**.

104. Hajek, V., and Marsalek, E. (1971). A study of staphylococci isolated from the upper respiratory tract of different animal species. IV. Physiologic properties of *Staphylococcus aureus* strains of hare origin. *Zentratbl. Bakteriol., Parasitenk., Infektienskir. Hyg., Abt. I: Orig.* **216**, 168–174.

105. Harkins, M. J., and Saleeby, E. R. (1928). Spontaneous tuberculosis of rabbits. *J. Infec. Dis.* **43**, 554–556.

105a. Helluy, J. R., deLavergne, E., Burdin, J. C., Schmitt, J., and Percebois, G. (1960). Mise en èvidence de l'allergie dans la pseudo-tuberculose expérimentale du lapin par un nouvel antigène. *C. R. Soc. Biol.* **154**, 1259–1261.

106. Hammens, E. S., and Dack, G. M. (1939). The effect of sulfanilamide on experimental infections with *Bacterium necrophorum* in rabbits. *J. Infec. Dis.* **64**, 43–48.

107. Henricson, T. (1943). Ett fall av listerellos hos hare. *Sv. Vet. Tidskr.* **48**, 1–9; Abstr. in *Vet. Bull.* **13**, 312.

108. Hibbs, C. M., Merker, J. W., and Kruckenberg, S. M. (1971). Experimental *Aeromonas hydrophila* infection in rabbits. *Cornell Vet.* **61**, 380–386.

109. Horne, H. (1913). Eine Kaninchenseptikämie (verursacht durch Streptokokken). *Z. Tiermed.* **17**, 49–76.

110. Hoskins, H. P. (1920). Snuffles (contagious nasal catarrh) of rabbits; its etiology and treatment. *J. Amer. Vet. Med. Ass.* **57**, 317–321.

111. Hubbert, W. T., Petenyi, C. W., Glasgow, L. A., Uyeda, C. T., and Creighton, S. A. (1971). *Yersinia pseudotuberculosis* infection in the United States: Septicemia, appendicitis and mesenteric lymphadenitis *Amer. J. Trop. Med. Hyg.* **20**, 679–684.

112. Huebner, R. A. (1938). Pasteurella pyometra in a rabbit. *J. Amer. Vet. Med. Ass.* **93**, 389.

113. Hülphers, G. (1911). Lefvernekros hos kanin orsakad af en ej Förut Beskrifven Bakterie. *Sv. Vet. Tidskr.* **16**, 265–273.

114. Iyer, P. R. K., and Uppal, D. R. (1956). Salmonellosis in rabbits. A comparison with rinderpest. *Indian Vet. J.* **32**, 430–438.

115. Jacobsthal, E. (1920). Untersuchungen über eine Syphilisähnliche Spontanerkrankung des Kaninchens (*Paralues cuniculi*). *Dermatol. Wockenschr.* **71**, 569–571.

116. Jakob, H. (1915). Mitteilungen aus der Klinik für kleine Haustiere der Reichstierarzneischule zu Utrecht (Holland). *Berlin Tieraerztl. Wochenschr.* **31**, 483–485.

117. Jansen, J. (1933). Peracute septicaemie van een konijn veroorzaakt door een onbekende streptococcus (*Streptococcus cuniculi nov. spec.*). *Tijdschr. Diergeneesk.* **60**, 925–928.

118. Jellison, W. L., Kohls, G. M., Butler, W. J., and Weaver, J. A. (1942). Epizootic tularemia in the beaver, *Castor canadensis*, and the contamination of stream water with *Pasteurella tularensis*. *Amer. J. Hyg.* **36**, 168–182.

118a. Joubert, L. (1968). La pseudo-tuberculose, zoonose, d'avenir. *Rev. Med. Vet.* **119**, 311–322.

119. Kanazawa, K., and Imai, A. (1959). Pure culture of the pathogenic agent of Tyzzer's disease of mice. *Nature* (*London*) **184**, 1810–1811.

120. Karsten (1927). Zur Ätiologie der Paratyphuserkrankungen unserer Haustiere auf Grund der im Tierseuchen-institut zu Hannover festgestellten Paratyphusbefunde. *Deut. Tieraerztl. Wochenschr.* **35**, 781–790.

121. Kartman, L. (1960). The role of rabbits in sylvatic plague epidemiology, with special attention to human cases in New Mexico and use of the fluorescent antibody technique for detection of *Pasteurella pestis* in field specimens. *Zoonoses Res.* **1**, 1–27.

122. Kartman, L., Goldenberg, M. I., and Hubbert, W. T. (1966). Recent Observations on the epidemiology of plague in the United States. *Amer. J. Pub. Health* **56**, 1554–1569.

123. Khera, S. S. (1962). Animal salmonellosis in India. *Indian J. Med. Res.* **50**, 569–579.

124. Klarenbeek, A. (1921). Recherches expérimentales avec un spirochete, se trouvant spontanement chez le lapin et ressemblant au *Treponema pallidum*. *Ann. Inst. Pasteur*, Paris **34**, 326–331.

125. Kolmer, J. A., and Casselman, A. J. (1913). Concerning the Wassermann reaction with normal rabbit serum. *J. Med. Res.* **28**, 369–375.

126. Krans, R. (1897). Ueber den Erreger einer Influenza-artigen Kaninchenseuche. *Z. Hyg. Infektionskr.* **24**, 396–402.

127. Kurita, S. (1909). Ueber den Grustseuchebacillus des Kaninchens. *Zentralbl. Bakteriol., Parasitenk. Infektionskr., Abt. 1: Orig.* **49**, 508–510.

128. Kyaw, M. H. (1945). Note on a bacterial cause of sterility in rabbits. *Vet. Rec.* **57**, 502–503.

129. Lanfranchi, A. (1907). Di un'infezione diplococcica a tipo setticoemico di alcuni roditori. *Clin. Vet.* **30**, 269–280.

130. Lersey, P., Dosquet, H., and Kuezynsk, M. (1921). Ein Beitrab zur Kenntnis der "Originären Kaninchensyphilis." *Berlin Klin. Wochenschr.* **58**, 546–548.

131. Lesbouyries (1931). Paratyphose du lapin. *Rec. Med. Vet.* **107**, 257–264.

132. Lesbouyries (1942). Protéose du lapin. *Bull. Acad. Vet.* [N.S.] **15**, 278–280.

133. Lesbouyries (1943). Listériose du lapin. *Rec. Med. Vet.* **119**, 145–150.

134. Lesbouyries and Berthelon (1936). Entéro-toxémie du lapin. *Bull. Acad. Vet.* [N.S.] **89**, 74–82.

135. Lesbouyries, G. (1945). Staphylococcie du lapin du liévre. *Rec. Med. Vet.* **121**, 321–325.

136. Lesbouyries, G. (1963). "Pathologie du Lapin," pp. 140–205. Librairie Maloine, Société Anonyme d'Editions Médicales et Scientifiques, Paris.

137. Levaditi, C., Marie, A., and Isaicu, L., (1921). Recherches sur la spirochetose spontanée du lapin. *C. R. Soc. Biol.* **85**, 51–54.

138. Levaditi, C., Marie, A., and Nicolau, S. (1921). Virulence pour l'homme du spirochète de la spirillose spontanée du lapin. *C. R. Acad. Sci.* **172**, 1542–1543.

139. Lindau, A., and Jensen, K. A. (1931). Über Spontantuberkulose bei Kaninchen und Meerschweinchen. *Acta Pathol. Microbiol. Scand.* **8**, 368–384.

140. Loliger, H. C., Matthes, S., Schubert, H. J., and Hockmann, F. (1969). Die akuten Dysenterien der Jungkaninchen. I. Untersuchungen zur Ätiologie und Pathogenese. *Deut. Tieraerztl. Wochenschr.* **76**, 16–20.

141. Loliger, H. C., Matthes, S., Schubert, H. J., and Hockmann, F. (1969). Die akuten Dysenterien der Jungkaninchen. II. Untersuchungen über Praventive Krankheitsbekampfung. *Deut. Tieraerztl. Wochenschr.* **76**, 38–41.

142. Lucas, A., Chauvrat, J., and Laroche, M. (1960). Infection du lievre a *Erysipelothrix rhusiopathiae*, bacille du rouget du porc. *Rec. Med. Vet.* **136**, 1207–1208.

143. Ludwig, L. (1910). Ueber ein für Kaninchen und Meerschweinchen

Pathogenese, noch nicht beschriebenes Bacterium. *Zentralbl. Bakteriol. Parasitenk. Infektionskr. Abt. 1: Orig.* **54**, 97–105.

143a Lurie, M. B. (1964). "Resistance to Tuberculosis: Experimental Studies in Native and Acquired Defense Mechanisms." Harvard Univ. Press, Cambridge, Massachusetts.

144. McCartney, J. E., and Olitsky, P. K. (1923). Studies on the etiology of snuffles in stock rabbits. Paranasal sinusitis a factor in the interpretation of experimental results. *J. Exp. Med.* **38**, 591–604.

145. McCoy, G. W. (1911). A plague-like disease of rodents. *Pub. Health Bull.* **42**, 53–71.

146. McCoy, G. W., and Chapin, C. W. (1912). Further observations on a plague-like disease of rodents with a preliminary note on the causative agent, *Bacterium tularense. J. Infec. Dis.* **10**, 61–72.

147. McCoy, R. H., and Steenbergen, F. (1969). Staphylococcus epizootic in western Oregon cottontails. *Bull. Wildl. Dis. Ass.* **5**, 11.

148. McDonald, R. A., and Pinheiro, A. F. (1967). Water chlorination controls *Pseudomonas aeruginosa* in a rabbitry. *J. Amer. Vet. Med. Ass.* **151**, 863–864.

149. McGinnes, B. W. (1964). Depletion of a cottontail rabbit population attributed to tularemia. *Wildl. Dis.* **34**, 1–13.

150. McKeever, S., Gorman, G. W., Chapman, J. F., Galton, M. M., and Powers, D. K. (1958). Incidence of leptospirosis in wild mammals from South-western Georgia with report of new hosts for six serotypes of leptospires. *Amer. J. Trop. Med. Hyg.* **7**, 646–655.

151. McKennedy, F. D., and Shillinger, J. E. (1938). Transmission of *Pasteurella cuniculicida* in rabbits by breeding. *J. Amer. Vet. Med. Ass.* **93**, 161–164.

152. McLeod, C., and Turner, T. B. (1946). Studies on the biologic relationship between the causative agents of syphilis, yaws, and venereal spirochetosis of rabbits. II. Comparison of the experimental disease produced in rabbits. *Amer. J. Syph. Gonor., Vener. Dis.* **30**, 455–462.

153. Mair, N. W. (1969). Pseudotuberculosis in free-living wild animals. *In* "Diseases in Free-Living Wild Animals" (A. McDiarmid, ed.), pp. 107–117. Academic Press, New York.

153a. Matthes, S. (1969). Die Darmflora gesunder und dysenteriekranker Jungkaninchen. *Zentralbl. Veterinaer Med., Reihe B* **16**, 563–570.

154. Middlebrook, G. (1972). The mycobacteria. *In* "Bacterial and Mycotic Infections of Man" (R. J. Dubos and J. G. Hirsch, eds.), 4th ed., pp. 490–529. Lippincott, Philadelphia, Pennsylvania.

154a. Mollaret, H. H. (1961). Contribution à l'étude de l'inoculation par voie oculaire de *Pasteurella pseudotuberculosis* (bacilli de Malassez et Vignal). *Ann. Inst. Pasteur, Paris* **100**, 753–764.

155. Morcos, Z. (1932). New disease in young rabbits, infectious tympanitis. *J. Bacteriol.* **23**, 449–454.

156. Morel, P. (1955). Protéose du lapin. *Rec. Med. Vet.* **131**, 96–97.

157. Morel, P. (1958). Infection du lapin par *Salmonella pullorum. Rec. Med. Vet.* **134**, 281–283.

158. Moretti, B. (1938). Ein Beitrag zur Pseudotuberkulose. *Deut. Tieraerztl. Wochenschr.* **46**, 35–37.

159. Morgan, H. R. (1965). The enteric bacteria. *In* "Bacterial and Mycotic Infections of Man" (R. J. Dubos and J. G. Hirsch, eds.), 4th ed., pp. 610–648. Lippincott, Philadelphia, Pennsylvania.

160. Morganti, L., and Ricci Bitti, G. (1971). Sulla presenza di salamonelle in viscera di conigli regularmente macellati. *Nuova Vet.* **47**, 86–88.

161. Morton, H. E. (1972). Bacteroides. *In* "Bacterial and Mycotic Infections of Man" (R. J. Dubos and J. G. Hirsch, eds.), 4th ed., pp. 770–774. Lippincott, Philadelphia, Pennsylvania.

162. Moussu, R. (1927). Sur une maladie indéterminé du lapin. *Bull. Soc. Cent. Med. Vet.* **80**, 251–253.

163. Munoz, J., and Holford, F. E. (1949). The occurrence of antibodies for pneumococcus Type II in the serum of normal rabbits. *J. Immunol.* **63**, 51–69.

164. Murray, E. G. D., Webb, R. A., and Swann, M. B. R. (1926). A disease of rabbits characterized by a large mononuclear leukocytosis, caused by a hitherto undescribed bacillus bacterium monocytogenes (n. sp.). *J. Pathol. Bacteriol.* **40**, 407–439.

165. Neumann, F. (1923). Über das spontane Auftreten von Spirochäten des Pallidatyps bei einem nichtsyphilitischen, isolierten Kaninchen. *Klin. Wochenschr.* **2**, 256–257.

166. Noguchi, H. (1921). A note on the venereal spirochetosis of rabbits. A new technic for staining *Treponema pallidum. J. Amer. Med. Ass.* **77**, 2052–2053.

167. Noguchi, H. (1922). Venereal spirochetosis in American rabbits. *J. Exp. Med.* **35**, 391–407.

168. O'Donoghue, P. N., and Whatley, B. F. (1971). *Pseudomonas aeruginosa* in rabbit fur. *Lab. Anim.* **5**, 251–255.

169. Olney, J. F. (1928). *Salmonella pullorum* infections in rabbits. *J. Amer. Vet. Med. Ass.* **73**, 631–633.

170. Olsen, P. F. (1970). Sylvatic (wild rodent) plague. *In* "Infectious Diseases of Wild Mammals" (J. W. Davis, L. H. Karstad, and D. O. Trainer, eds.), pp. 200–213. Iowa State Univ. Press, Ames.

171. Olt, A. (1937). Ueber das seuchenhafte Auftreten der Rodentiose unter den Hasen Zeitsch. *Z. Infektionskr. Haust.* **52**, 89–107.

172. Osebold, J. W., and Gray, D. M. (1960). Disseminated staphylococcal infections in wild jack rabbits (*Lepus californicus*). *J. Infec. Dis.* **106**, 91–94.

173. Ostler, D. C. (1961). The diseases of broiler rabbits. *Vet. Rec.* **73**, 1237–1252.

174. Pacchioni, G. (1936). Grave spondilite tubercolare in un coniglio contagiatosi naturalmente. *Nuova Vet.* **14**, 121–133.

175. Pallaske, G. (1933). Beitrag zur Patho- und Histogenese der Pseudotuberkulose (*Bact. Pseudotub. rodentium*) der Tier. *Z. Infektionskr. Haust.* **44**, 43–66.

176. Pannu, J. S., Rosenberg, M. A., Israel, C. W., and Smith, J. L. (1967). Incidence of reactive VDRL tests in the normal rabbit. *Brit. J. Vener. Dis.* **43**, 114–116.

177. Parker, H. D., Hawkins, W. W., Jr., and Brenner, E. (1966). Epizootiologic studies on ovine virus abortion. *Amer. J. Vet. Res.* **27**, 869–877.

178. Parker, R. R., Spencer, R. R., and Francis, E. (1924). Tularemia. XI. Tularemia infection in ticks of the species *Dermacentor andersoni* stiles in the Bitterroot Valley, Mont. *Pub. Health Rep.* **39**, 1057–1073.

179. Paterson, J. S. (1940). A case of naturally occurring listerellosis in an adult rabbit. *J. Pathol. Bacteriol.* **51**, 441–442.

180. Paterson, J. S., and Cook, R. (1963). A method for the recovery of *Pasteurella pseudotuberculosis* from faeces. *J. Pathol. Bacteriol.* **85**, 241–242.

181. Peres, G., and Granon-Fabre, P. (1935). La melitococcie chez le lapin. *Rev. Gen. Med. Vet.* **44**, 201–204.

182. Philip, C. B., Jellison, W. L., and Wilkins, H. F. (1935). Epizootic tick-borne tularemia in sheep in Montana. *J. Amer. Vet. Med. Ass.* **86**, 726–744.

183. Pirie, J. H. H. (1927). A new disease of veld rodents. "Tiger River disease." *S. Afr. Inst. Med. Res.* **3**, 163–186.

184. Pirie, J. H. H. (1940). *Listeria*: Change of name for a genus of bacteria. *Nature (London)* **145**, 264.

185. Ratcliffe, H. L. (1945). Infectious diseases of laboratory animals. *Ann. N. Y. Acad. Sci.* **46**, 77–96.

186. Ray, J. P., and Mallick, B. B. (1970). Public health significance of *Salmonella* infections in laboratory animals. *Indian Vet. J.* **47**, 1033–1037.

187. Raymond (1912). Tuberculose du lapin. *Hyg. Viande Lait* **6**, 430–431.

188. Reed, R. W. (1965). Listeria and erysipelothrix. *In* "Bacterial and Mycotic Infections of Man" (R. J. Dubos and J. G. Hirsch, eds.), 4th ed., pp. 752–762. Lippincott, Philadelphia, Pennsylvania.

189. Reilly, J. R. (1970). Tularemia. *In* "Infectious Diseases of Wild Mammals" (J. W. Davis, L. H. Karstad, and D. O. Trainer, eds.), pp. 175–190. Iowa State Univ. Press, Ames.

190. Renoux, G. and Sacquet, E. (1957). Brucellose spontanée du lapin domestique. *Arch. Inst. Pasteur Tunis* **34**, 231–232.

191. Renquist, D., and Soave, O. (1969). Staphylococcal pneumonia in a laboratory rabbit: An epidemiologic follow-up study. *J. Amer. Vet. Med. Ass.* **155**, 1221–1223.

192. Richter, C. B., and Hendren, R. L. (1969). The pathology and epidemiology of acute enteritis in captive cottontail rabbits (*Sylvilagus floridanus*). *Pathol. Vet.* **6**, 159–175.

193. Rights, F. L., Jackson, E. B., and Smadel, J. E. (1947). Observations on Tyzzer's disease in mice. *Amer. J. Pathol.* **23**, 627–635.

194. Rinjard, P., and Homutov, P. (1936). Action pathogène du B. pyocyanique chez le lapin. *Bull. Acad. Vet. Fr.* [N.S.] **9**, 100–108.

195. Ross, E. H. (1912). An intracellular parasite developing into spirochaetes. *Brit. Med. J.* **2**, 1651–1654.

196. Ross, H. (1914). The history of the parasite of syphilis. *Brit. Med. J.* **1**, 341–343.

197. Roth, E. E. (1970). Leptospirosis. *In* "Infectious Diseases of Wild Mammals" (J. W. Davis, L. H. Karstad, and D. O. Trainer, eds.), pp. 293–303. Iowa State Univ. Press, Ames.

198. Rothe, S. (1912). Studien üer spontane Kaninchentuberkulose. *Deut. Med. Wochenschr.* **1**, 642–643.

199. Saelhof, C. C. (1921). The correlation of rabbit pneumonia and human influenzal pneumonia. *J. Infec. Dis.* **28**, 374–380.

200. Saenz, A. (1932). Sur l'infection tuberculeuse spontanée du lapin et du cobaye. *C. R. Soc. Biol.* **109**, 437–439.

201. Savage, N. L., and Lewis, D. H. (1972). Application of immunofluorescence to detection of Tyzzer's disease agent (*Bacillus piliformis*) in experimentally infected mice. *Amer. J. Vet. Res.* **33**, 1007–1011.

202. Scher, S., Collins, G. R., and Weisbroth, S. H. (1969). The establishment of a specific pathogen-free rabbit breeding colony. I. Procedures for establishment and maintenance. *Lab. Anim. Care* **19**, 610–616.

203. Schereschewsky, J. (1920). Geschlochtlich übertragbare originäre Kaninchensyphilis und Chinin-Spirochätotropie *Berlin Klin. Wochenschr.* **57**, 1142–1144.

204. Schlotthauer, C. F., Thompson, L., and Olsen, C., Jr., (1935). Tularemia in wild gray foxes: Report of an epizootic. *J. Infec. Dis.* **56**, 28–30.

205. Schmorl, G. (1891). Ueber ein Pathogenes Fabenbacterium (*Streptothrix cuniculi*). *Deut. Z. Tiermed.* **17**, 375–408.

206. Schoop, G. (1946). Metritis Infectiosa trägender Angorahasinnen. *Deut. Tieraerztl. Wochenschr.* **53**, 42–43.

207. Schwartz, B. (1928). Rabbit parasites and diseases. *U.S., Dep. Agr., Farmers' Bull.* **1568**.

208. Seifried, O. (1937). "Die Krankheiten des Kaninchens," pp. 1–97. Springer-Verlag, Berlin and New York.

209. Shotts, E. G., Jr., Andres, C. L., Sulzer, C., and Greene, E. (1971). Leptospirosis in cottontail and swamp rabbits of the Mississippi Delta. *J. Wild. Dis.* **7**, 115–118.

210. Siegmund, O. H. (1961). Pasteurellosis. *In* "The Merck Veterinary Manual" (O. H. Siegmund, ed.), 2nd ed., p. 1375. Merck & Co., Inc., Rahway, New Jersey.

211. Small, J. D., and Newman, B. (1972). Venereal spirochetosis of rabbits (rabbit syphilis) due to *Treponema cuniculi*: A clinical, serological, and histopathological study. *Lab. Anim. Sci.* **22**, 77–89.

212. Smith, D. T. (1927). Epidemiological studies on respiratory infections of the rabbit. X. A spontaneous epidemic of pneumonia and snuffles caused by *Bacterium lepisepticum* among a stock of rabbits at Saranac Lake, New York. *J. Exp. Med.* **45**, 553–559.

213. Smith, D. T., and Webster, L. T. (1925). Epidemiological studies on respiratory infections of the rabbit. VI. Etiology of otitis media. *J. Exp. Med.* **41**, 275–283.

214. Smith, J. L., and Pesetsky, B. R. (1967). The current status of *Treponema cuniculi*. Review of the literature. *Brit. J. Vener. Dis.* **43**, 117–127.

215. Smith, T. (1887). A contribution to the study of the microbe of rabbit septicemia. *J. Comp. Med. Surg.* **8**, 24–37.

216. Soituz, V. (1930). Sur un staphylocoque adapté la glande mammaire de la lapin. *Rev. Pathol. Comp. Hyg. Gen.* **30**, 381–389.

217. Sorenson, B., and Saliba, A. M. (1961). Actinomicose espontanea em coelhos. *Biologico* **27**, 131–135.

218. Spalatin, J., Fraser, C. E. O., Connell, R., Hanson, R. P., and Berman, D. T. (1966). Agents of psittacosis-lymphogranuloma venereum group isolated from muskrats and snowshoe hares in Saskatchewan. *Can. J. Comp. Med. Vet. Sci.* **30**, 347–363.

219. Stanton, A. T., and Fletcher, W. (1925). Melioidosis and its relation to glanders. *J. Hyg.* **23**, 347–363.

220. Sustmann (1921). Etwas über Kaninchenkrankheiten und deren Behandlung. *Deut. Tieraerztl. Wochenschr.* **29**, 247–249.

221. Szemeredi, G. (1965). Adatok a hazinyulak salmonellosisanak hazai elofordulasahoz. *Magy. Allatorv. Lapja* **20**, 411–413.

222. Thorpe, B. D., Sidwell, R. W., Bushman, J. B., Smart, K. L., and Moyes, R. (1965). Brucellosis in wildlife and livestock of west central Utah. *J. Amer. Vet. Med. Ass.* **146**, 225–232.

223. Traub, E. (1942). Ueber eine mit Listerelle-ähnlichen Bakterien vergeseilschaftete Meningo-Encephalomyelitis der Kaninchen. *Zentralbl. Bakteriol. Parasitenk. Infektienskr., Abt. I: Orig.* **149**, 38–49.

224. Trbic, B., and Stojkovic-Atanackovic, M. (1970). Pojava crvenog vetra na jednoj farmi kunica. *Vet. Glas.* **24**, 395–396.

225. Truche, C., and Saenz, A. (1931). Sur un cas de tuberculose spontanée chez le lapin. *Ann. Inst. Pasteur. Paris* **47**, 472–474.

226. Truche, M. (1938). La pseudo-tuberculose chez les animaux. *Rev. Pathol. Comp. Hyg. Gen.* **38**, 874–883.

227. Tufts, J. M. (1966). Unidentified rabbit problem—hemorrhagic colitis. *Lab. Anin. Dig.* **2**, 16–17.

228. Turner, T. B. (1965). The spirochetes. *In* "Bacterial and Mycotic Infections of Man" (R. J. Dubos and J. G. Hirsch, eds.), 4th ed., pp. 573–609. Lippincott, Philadelphia, Pennsylvania.

229. Turner, T. B., and Hollander, D. H. (1957). Biology of the treponematoses. *World Health Organ. Monogr. Ser.* **35**.

230. Turner, T. B., McLeod, C., and Updyke, E. (1947). Cross immunity in experimental syphilis, yaws, and venereal spirochetosis of rabbits. *Amer. J. Hyg.* **46**, 287–295.

231. Tyzzer, E. E. (1917). A fatal disease of the Japanese waltzing mouse caused by a spore-bearing bacillus (*Bacillus piliformis, n. sp.*) *J. Med. Res.* **37**, 307–338.

232. Vail, E. L., and McKenny, F. D. (1943). Diseases of domestic rabbits. *U.S., Fish Wildl. Serv., Conserv. Bull.* **31**.

233. Valtis, J. (1924). Sur l'infection tuberculeuse spontanée du jeune lapin. *C. R. Soc. Biol.* **91**, 853–854.

234. Van Dorssen, C. A. (1952). Orienteerende Proven over Therapie en Vaccinatie Bij Pseudotuberculosis van Knaagdieren. *Tijdschr. Diergeneesk.* **77**, 235–255.

235. Van Kruiningen, H. J., and Blodgett, S. B. (1971). Tyzzer's disease in a Connecticut rabbitry. *J. Amer. Vet. Med. Ass.* **158**, 1205–1212.

236. Verge, J., and Goret, P. (1941). Les maladies communes à l'homme et aux animaux. La listériose ou listerellose. *Rec. Med. Vet.* **67**, 5–27.

237. Verma, N. S., and Sharma, S. P. (1969). Salmonellosis in laboratory animals. *Indian. Vet. J.* **46**, 1101–1102.

238. Vetesi, F. (1970). A nyúl ún mucoid enteritise (colienteritoxaemiája). *Magy. Allatorv. Lapja* **25**, 465–472.

239. Vetesi, F., and Kemenes, F. (1965). A vemhes hazinyuoak listeriosisa hazankban. *Magy. Allatorv. Lapja* **20**, 405–410.

240. Volk, R. (1902). Ueber eine Kaninchenseuche. *Zentralbl. Bakteriol., Parasitenk. Infektionskr., Abt. 1: Orig.* **31**, 177–182.

241. Wahl, R., and Forgeot, P. (1954). Un cas d'infection naturelle du lapin par *Pseudomonas aeruginosa*. *Bull. Acad. Vet. Fr.* [N.S.] **27**, 147–150.

242. Warthin, A. S., Buffington, E., and Wanstrom, R. C. (1923). A study of rabbit spirochetosis. *J. Infec. Dis.* **32**, 315–332.

243. Weber, V. A., and Manz, J. (1971). Serologische Untersuchungen der O-Antigene von *E. coli*-Stammen isoliert von Kaninchen. *Berlin. Munechen. Tieraerztl. Wochenschr.* **84**, 441–443.

244. Webster, L. T. (1924). The epidemiology of a rabbit respiratory infection. I. Introduction. *J. Exp. Med.* **39**, 837–841.

245. Webster, L. T. (1924). The epidemiology of a rabbit respiratory infection. II. Clinical, pathological, and bacteriological study of snuffles. *J. Exp. Med.* **39**, 843–856.

246. Webster, L. T. (1924). The epidemiology of a rabbit respiratory infection. III. Nasal flora of laboratory rabbits. *J. Exp. Med.* **39**, 857–877.

247. Webster, L. T. (1924). The epidemiology of a rabbit respiratory infection. IV. Susceptibility of rabbits to spontaneous snuffles. *J. Exp. Med.* **40**, 109–116.

248. Webster, L. T. (1924). The epidemiology of a rabbit respiratory infection. V. Experimental snuffles. *J. Exp. Med.* **40**, 117–127.

249. Webster, L. T. (1925). Epidemiological studies on respiratory infections of the rabbit. VII. Pneumonia associated with *Bacterium lepisepticum*. *J. Exp. Med.* **43**, 555–572.

250. Webster, L. T. (1927). Epidemiological studies on respiratory infections of the rabbit. IX. The spread of *Bacterium lepisepticum* infection at a rabbit farm in New City, New Jersey—an epidemio-

logical study. *J. Exp. Med.* **45**, 529–551.

251. Webster, L. T., and Burn, C. G. (1927). *Bacterium lepisepticum* infection. Its mode of spread and control. *J. Exp. Med.* **45**, 911–935.

251a. Weisbroth, S. H., and Scher, S. (1969). The establishment of a specific-pathogen-free rabbit breeding colony. II. Monitoring for disease and health statistics. *Lab. Anim. Care* **19**, 795–799.

252. Wherry, W. B., and Lamb, B. H. (1914). Discovery of *Bacterium tularense* in wild rabbits and the danger of its transfer to man. Preliminary note. *J. Amer. Med. Ass.* **63**, 2041.

253. Wherry, W. B., and Lamb, B. H. (1914). Infection of man with *Bacterium tularense. J. Infec. Dis.* **15**, 331–340.

254. Witter, J. F., and O'Meara, D. C. (1970). Brucellosis. *In* "Infectious Diseases of Wild Mammals" (J. W. Davis, L. H. Karstad, and D. O. Trainer, eds.), pp. 249–255. Iowa State Univ. Press, Ames.

255. Woolley, P. G. (1915). The lesions in experimental infection with *Bacterium tularense. J. Infec. Dis.* **17**, 510–513.

256. Young, L. S., Bicknell, D. S., Archer, B. G., Clinton, J. M., Leavens, L. J., Feeley, J. C., and Brachman, P. S. (1969). Tularemia epidemic: Vermont, 1928. Forty-seven cases linked to contact with muskrats. *N. Engl. J. Med.* **280**, 1253–1260.

257. Yuill, T. M., and Hanson, R. P. (1965). Coliform enteritis of cottontail rabbits. *J. Bacteriol.* **89**, 1–8.

CHAPTER 10

Viral Diseases

C. J. Maré

I. INTRODUCTION

The science of animal virology has been through an era of phenomenal growth during the past two decades. The explosion of knowledge which has occurred during this time can be attributed largely to a single technical advance, namely the application of cell culture techniques to the study of animal viruses. The wide-spread adoption of cell cultures as the host system of choice in diagnostic and experimental virology has resulted in the isolation and characterization of many new viruses of man and animals. It has also led to a more fundamental understanding of the viruses themselves and of virus–host interrelationships.

The recognition of the stable physical and chemical characteristics of the viruses has been followed by several attempts at viral classification based on the fundamental properties of the virus particle rather than on the tissue affinity or the disease-producing characteristics of the agent. Several extremely practical classification schemes have been proposed but as yet general agreement has not been reached on which system should become universally adopted. The state of flux which exists in the area of virus taxonomy and nomenclature encourages confusion and thus every effort will be made in this review to use the viral terminology which is most widely accepted. Wherever practicable the recommendations of the International Committee for the Nomenclature of Viruses (167) will be followed, but well-established colloquial names for viruses will be presented wherever appropriate.

The viral diseases of rabbits will be discussed in a sequence based on the taxonomic groups to which the causative viruses belong rather than on the clinical or pathological syndromes which may be caused by the agents. The sequence in which the viruses are discussed does not imply a descending or ascending degree of epidemiological importance. While the skeleton of the chapter is formed by this sequence of virus families, the subject matter is presented in a form that recognizes the applied nature of this book.

In reviewing the literature on viral diseases of rabbits, one is struck by the paucity of information which exists in this area. Not only have relatively few viruses of rabbits been recognized but with the exception of the myxoma and fibroma viruses and a few papovaviruses, those which are known have as yet been relatively poorly characterized especially in respect of their disease-producing potential. In the past, those viruses which cause dramatic diseases in rabbits have naturally received most attention and it is only recently that the importance of viruses causing subclinical and latent infections has been recognized. Such viruses are included in this chapter since the importance of latent viruses in experimental animals cannot be over emphasized.

Only the naturally occurring viral infections of rabbits will be discussed in this review. Such viruses as the pseudorabies virus, to which the rabbit is highly susceptible under experimental conditions, will not be discussed since natural infection with this virus has not been reported. While the principal emphasis will be on virus infections of domestic rabbits of the genus *Oryctolagus*, naturally occurring infections of other rabbits and hares will also be discussed.

II. DNA VIRUS INFECTIONS

A. Poxvirus Infections

Poxviruses cause several economically important diseases in domestic and wild mammals and birds. Infection with viruses of this group usually results in relatively mild disease syndromes involving the skin of the infected animals, but generalized and often fatal disease may also occur, for example, in smallpox in man and myxomatosis in rabbits.

Close antigenic relationships exist between many poxviruses derived from different animal species and it is still unclear as to whether such closely related viruses should be regarded as distinct entities. In discussing the poxviruses of rabbits those viruses which produce distinct disease syndromes will be discussed as separate entities in spite of the fact that very close antigenic relationships between some of these viruses may exist.

1. Myxoma Virus

a. HISTORY. The disease myxomatosis, now known to be caused by the myxoma virus, was first recognized by Sanarelli in Uruguay in 1896. A number of European rabbits of the genus *Oryctolagus* which he had acquired for antiserum production suddenly developed a highly fatal disease characterized by the development of numerous mucinous skin tumors. He named this disease "infectious myxomatosis of rabbits" and since he could detect no microbial agents, he proposed that the disease could possibly be caused by the newly recognized group of infectious agents known as "filterable viruses" (143). The term "myxoma" was derived from the Greek word "myxo" meaning mucous which describes the appearance of the cutaneous tumours when sectioned.

The virus which caused the first known outbreak of myxomatosis is believed to have originated from the Tapeti or tropical forest rabbit (*Sylvilagus brasiliensis*) in which the virus causes relatively mild disease. Transmission from wild to domestic rabbits probably occurred via vector mosquitoes of the genus *Aedes* (8,43).

Myxomatosis soon spread to other countries of South America where it now occasionally causes sporadic outbreaks of disease in domesticated rabbits. In Chile, the disease is considered enzootic in the wild European rabbit population (43). The disease was first recognized in North America in 1930 when natural outbreaks of a fatal disease of rabbits resembling myxomatosis occurred in several rabbit colonies near San Diego in southern California (87). It is believed that the virus which caused the first outbreaks in California was introduced into the United States from Mexico by importation of infected domestic rabbits (164). The disease is now considered to be enzootic in the western United States where the brush rabbit (*S. bachmani*) has been incriminated as the natural reservoir of the virus (104, 105).

The disease myxomatosis was intentionally introduced into Australia in an effort to control what had become Australia's major animal pest, namely the European rabbit, *Oryctolagus cuniculus*. The virus was first introduced into Australia in 1926, but for more than two decades the agent was used only in experimental studies aimed at determining the feasibility of its use as a rabbit control measure. In 1950 the virus was finally released into the wild rabbit population where, after a somewhat slow start, it became established and spread rapidly decimating the rabbit population of the continent by 1953. The disease is now enzootic in the wild rabbit population of Australia, where it occasionally assumes enzootic proportions when climatic conditions favor vector activity.

Within a decade following release of the myxoma virus into the Australian rabbit population it became evident that through a process of natural selection genetically resistant strains of rabbits had emerged. In these rabbits a virulent strain of myxoma virus caused only 25% mortality as compared to the 90% mortality in nonresistant strains of rabbit (43).

Genetic modification of the myxoma virus was recognized soon after release of the virus into the rabbit population, and by the fourth year markedly attenuated strains of the virus had replaced the virulent virus as the dominant virus strains. The naturally attenuated viruses cause a milder disease of longer duration which favors vector transmission and thus persistence of the virus in nature (42, 46). The evolution of myxomatosis in Australia is the classic example of natural modification of both a virus and its host until a state of equilibrium is reached, a state which allows the continued existence of both.

The introduction of myxomatosis into Europe followed the early successes of the Australian campaign. In 1952, while French officials were considering the desirability of introducing the disease, a private individual acquired the virus and released it onto his own estate in an effort to control the rabbit population. The virus spread rapidly through the countryside and by the end of 1953, myxomatosis had been diagnosed in Belgium, the Netherlands, Germany, Luxembourg, Spain, and England (43, 100).

Myxomatosis is thus now known to be enzootic in rabbits of the genus *Sylvilagus* in both South and North America and in wild rabbits of the genus *Oryctolagus* in South America, Europe, and Australia.

b. ETIOLOGY. Myxomatosis is caused by any one of several strains of the myxoma virus, a member of the poxvirus group. Antigenic differences which can be demonstrated between the different strains of the virus (38, 129) have prompted some to consider the California strains of the virus as distinct from the myxoma virus and the designation "California rabbit fibroma virus" has been used to describe this virus. In the opinion of this author the demonstrable antigenic differences are not sufficient to justify the above distinction and thus in this discussion the California strains of the virus will be discussed as strains of the myxoma virus and not as separate entities.

The myxoma virus is antigenically very closely related to the rabbit fibroma virus as demonstrated by agar-gel diffusion microprecipitation techniques (38). Heat-inactivated myxoma has been reactivated by fibroma virus (12, 37), further demonstrating the close relationship between these two viruses. The Berry-Dedrick phenomenon of poxvirus reactivation was confirmed by Smith (151) who was also able to demonstrate a spectrum of virulence for strains of myxoma and fibroma viruses. The fibroma viruses ranged in virulence from the relatively avirulent 1A strain of the virus through the 0A strain to the virulent Boerlage strain isolated by Shope in 1942 [cited by Smith (151)]. The spectrum of virulence of the myxoma virus was further elucidated by Fenner and Marshall (41) in an exhaustive study involving 92 strains of the virus. They were able to establish a virulence spectrum ranging from strains which cause over 99% mortality in European rabbits to others which

cause less than 30% mortality. The most virulent strains were the Standard Laboratory, Lausanne and California strains while the least virulent were the neuromyxoma and Nottingham strains.

A complete virulence spectrum can thus be seen to exist ranging from the most avirulent fibroma virus to the almost uniformly lethal strains of myxoma virus. It seems plausible that ecological pressures such as those previously described as occurring in Australia could have been responsible for the emergence of many of these strains of viruses. In many instances however, man himself has manipulated the viruses to the point of permanent modification (91, 92).

The chemical and physical characteristics of the myxoma virus and the reasons for its inclusion in the poxvirus group have been adequately presented elsewhere (6, 35, 43). It has been conclusively shown that the myxoma virus is a poxvirus, morphologically indistinguishable from the vaccinia virus.

The myxoma virus is readily propagated on the chorioallantoic membrane of embryonated hens eggs at 35°C and forms distinct pocks (40). Different strains of the virus may cause pocks of varying sizes, the variation being sufficiently distinct to allow tentative strain identification. The South American strains of the virus cause very large pocks, whereas the California strains produce very small focal lesions on the membrane (41).

The virus can also be propagated in cell cultures derived from rabbits and other species; for example, chicken, squirrel, rat, hamster, guinea pig, and human cells (6, 170). Distinct differences in plaque size on rabbit kidney cell cultures can be demonstrated between the South American and California strains of the virus, the former causing much larger plaques (170). The most sensitive method for the isolation of myxoma virus under laboratory conditions is still inoculation of the skin of European rabbits (40).

c. INCIDENCE. Myxomatosis is now enzootic on four continents; namely, South America, North America, Europe, and Australia. In Brazil and Uruguay the virus has long been considered enzootic in the wild rabbits of the genus *Sylvilagus*, particularly *Sylvilagus brasiliensis* (8). A similar situation may exist in Panama and Colombia, but the strains of virus isolated from these countries, while similar in their virulence to the South American strains, are antigenically more closely related to the California strains of the virus (38). In the forested part of Argentina the virus is also enzootic in *Sylvilagus* rabbits but in the southern part of the country and in Chile, the principal reservoir of the virus is the wild European rabbit (43).

The California strains of myxoma virus are enzootic in wild rabbits of the genus *Sylvilagus*, especially the brush rabbit (*S. bachmani*) which serves as the principal source for infection of domestic rabbits (104).

In Australia the myxoma virus has been considered enzootic in wild European rabbits since its introduction into the rabbit population in 1950. Following the introduction of the virus into France in 1952, myxomatosis has become established in most countries of Europe, the wild European rabbit (*Oryctolagus cuniculus*) serving as the prevailing host species.

d. Epidemiology. Very intensive investigations have been done to determine the host range of the myxoma virus since this information was required by the Australian authorities before the decision could be made to release the virus. Numerous species of wild and domestic animals and birds were tested for susceptibility to the virus and ultimately it was shown that under natural conditions the myxoma virus would produce disease in leporids only. Naturally susceptible species are the European rabbit (*O. cuniculus*), the European hare (*Lepus europaeus*), the mountain hare (*L. timidus*), the Tapeti or tropical forest rabbit (*Sylvilagus brasiliensis*), the brush rabbit (*S. bachmani*), and the Eastern cottontail (*S. floridanus*). Under experimental conditions several additional species of the genus *Sylvilagus* could be infected (43).

Transmission of myxomatosis by contact can occur if an infected rabbit excreting virus via ocular discharges or oozing skin lesions is kept in close contact with susceptible rabbits. It has been demonstrated under experimental conditions that the virus can spread from infected to noninfected rabbits in the absence of arthropod vectors and it has been suggested that such contact transmission may also occur under natural conditions in rabbit warrens (113, 114). The respiratory or oral routes of infection do not seem to be significant in the spread of the disease.

The principal natural mode of transmission of the virus is via arthropod vectors, mosquitoes and fleas being most often incriminated. Since transmission of the virus occurs by simple mechanical transport of virus on the mouth parts (Fig. 1) the species of mosquito is not important and thus any mosquito which feeds on rabbits (58), as well as

other biting flies and gnats (112), may serve as vectors of the disease.

The source of virus is usually the superficial layers of the skin especially of the eyelids and in the region of the base of the ears where even surface-feeding arthropods such as mites and lice may pick up the virus and serve as mechanical vectors of myxomatosis. Transmission of myxoma virus via contaminated spines of thistles (*Circium vulgare*) has been described (32, 114). It has also been shown that the claws of predatory birds and carrion-feeders such as buzzards and crows may be contaminated with the virus and in Sweden it has been suggested that such birds may play a role in dissemination of the virus (13).

Arthropod transmission of the myxoma virus in its original habitat, South America, has not been intensively evaluated. The experimental findings of Aragao (8) indicate that under laboratory conditions the mosquitoes, *Aedes aegypti* and *Aedes scapularis*, as well as the cat flea, *Ctenocephalides felis*, may transmit the disease. Circumstantial evidence supports the conclusion that mosquitoes of the genus *Aedes* are the important natural vectors of myxomatosis in South America.

In California, Grodhaus *et al.* (58) have shown experimentally that *Anopheles freeborni*, *Aedes serriensis*, *Aedes aegypti*, *Culiseta incidens*, and *Culex tarsalis* can transmit myxoma virus between brush rabbits (*S. bachmani*). The virus was isolated from wild-caught *An. freeborni* mosquitoes in an area where myxomatosis outbreaks were occurring, further implicating it as a possible vector (104). The available evidence would indicate that the principal vectors of myxomatosis in California are mosquitoes, and that several species may be involved in transmission of the disease.

The principal vectors of myxomatosis in Australia are also mosquitoes, and characteristically the disease has spread along river valleys, the main vector mosquitoes being *Culex annulirostris*, *Anopheles annulipes*, *Culex pipiens australicus*, and several *Aedes* species (43). Several species of blackflies (*Simulium* spp.) have been incriminated as possible vectors in Australia, but only one species (*S. melatun*) has been conclusively shown to serve as a myxoma vector (112). Stickfast fleas of the genus *Echidnophaga* and the cat flea, *Ctenocephalides felis*, have been shown to be capable of transmitting the disease, and Mykytowcyz (113) has shown that the louse (*Haemodispus ventricosus*) and the mite, *Cheyletiella parasitivorax*, are also vectors. None of these arthropods are believed to play a significant role in epizootic myxomatosis in Australia where mosquitoes are unquestionably the major vectors.

In France little definitive investigation into the vector-spectrum of myxomatosis has been done. There is, however, considerable circumstantial evidence supporting the contention that mosquitoes are the principal vectors of summer epizootics of the disease in France. Several spe-

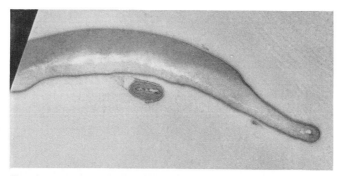

Fig. 1. Electron micrograph of myxoma virus loosely attached to maxillary stylet of *Aedes aegypti* (Courtesy of B. K. Filshie, C.S.I.R.O., Canberra, Australia).

cies of the genus *Anopheles* have been incriminated as possible myxoma vectors. The rabbit flea, *Spilopsyllus cuniculi*, is probably a major vector, especially during winter months where mosquito activity is low (43).

Myxomatosis in Britain is characterized by the virtual absence of the seasonal fluctuations in disease incidence which are observed in Australia, California, and France. A possible explanation for this phenomenon is the fact that in Britain mosquitoes play only a minor vector role, while the rabbit flea (*Spilopsyllus cuniculi*) which is far less influenced by changes in season, is the major vector, (7, 106). The myxoma virus in Britain has not undergone the rapid loss of virulence which has been observed with the Australian and French viruses. While mild strains of virus have emerged in Britain, the predominant strains are still moderately virulent ones (17, 39). The different evolution of the myxoma virus in Britain has been attributed to the fact that it is predominantly flea transmitted. The flea is less seasonal and it is far less mobile than the mosquito. The fact that fleas will move in large numbers from dead animals while moving only occasionally from live ones would seem to act selectively in favor of transmission of virulent virus strains (43).

The flea is also an effective reservoir of virus, possessing a far longer life span than mosquitoes. The life span of active female mosquitoes is usually 2–3 weeks while fleas have been known to feed actively for over a year. It has been demonstrated that the myxoma virus can persist for 105 days in rabbit fleas held in artificial burrows with no rabbit contact (18).

e. PUBLIC HEALTH SIGNIFICANCE.   Man is not susceptible to the myxoma virus.

f. CLINICAL SIGNS.   Considerable differences in the virulence of myxoma virus strains complicate discussion of the clinical disease, as does the fact that different species and strains of rabbits vary considerably in their susceptibility to the myxoma virus. Major emphasis will be given to discussion of the disease produced in Oryctolagus by the major strains of myxoma virus found in South America, Europe, California, and Australia.

g. SYMPTOMS IN *SYLVILAGUS* SPECIES.   Infection of the Tapeti or tropical forest rabbits (*S. brasiliensis*) with the South American virus under natural or experimental conditions usually results only in the development of skin tumors (fibromas) at the site of the virus inoculation or mosquito bite. These nodules usually appear from four to eight days after exposure to the virus and may persist for up to 40 days. Very young rabbits may, however, succumb to generalized disease following infection with myxoma virus (8).

The response of the brush rabbit (*S. bachmani*) to experimental inoculation with both the South American and California strains of the virus was evaluated by Marshall and Regnery (105). Infection with both strains of virus resulted only in the development of local skin tumors, the South American virus causing slightly more prominent lumps than the California virus. In cottontail rabbits (*S. floridanus*, *S. nuttallii*, and *S. audoboni*) small local lesions develop at the site of California virus inoculation but in *S. nuttallii* and *S. audoboni* the South American virus results in a more severe reaction.

The rabbits of the genus *Sylvilagus*, probably the natural host genus of the virus, are thus relatively resistant to infection with the virus when compared to the European rabbit, *O. cuniculus.*

h. SYMPTOMS IN *LEPUS* SPECIES.   The European hare, *Lepus europaeus* has been shown to be highly resistant to the myxoma virus under experimental conditions, and field experience supports this observation. Occasionally, however, individual hares (*L. europaeus* and *L. timidus*) showing mild to severe generalized myxomatosis have been encountered (43).

i. SYMPTOMS IN *ORYCTOLAGUS CUNICULUS.*   Myxoma virus infection in the European rabbit usually results in severe disease with mortality of up to 99%. The clinical disease is, however, largely predicated by the strain of myxoma virus as well as the genetic resistance of the strain of rabbit used (152, 153). The variety of clinical syndromes which can result from infection with various strains of virus in various strains of rabbit have been presented in detail elsewhere (41, 43). The discussion which follows is a summation of the findings of these workers and others (17, 86).

Symptoms which develop following infection with either the South American or California strains of the virus in fully susceptible rabbits will vary depending on the length of time the animal survives. In the peracute form of the disease in which rabbits die within a week of exposure to the virus, slight edema of the eyelids and cerebral depression immediately preceding death may be all that is observed. The peracute form of the disease is most often associated with infection with the California strains of the virus. With the acute form of the disease in which rabbits survive for one to two weeks, more distinct symptoms develop. With the California strains the first symptom observed at six to seven days is usually edema of the eyelids resulting in a "droopy" appearance of the eyes. Inflammation and edema around the anal, genital, oral, and nasal orifices are frequently observed at this time as are skin hemorrhages and convulsions preceding death on the ninth or tenth day. The few rabbits which survive beyond ten days may develop a purulent blepharoconjunctivitis and edema at the base of the ears, symptoms more often associated with the other strains of myxoma virus.

The nodular lesion which develops at the site of inoculation with the California virus is not a clearly defined tumor and under natural conditions the development of myxomatous tumors is not characteristically seen. While nodule development on the ears, head, and legs has been reported (86), other workers have not been able to demonstrate nodule development following inoculation with California strains of the virus (41).

The acute disease which follows inoculation of rabbits with the "standard laboratory strain" of myxoma virus results in a mean survival time of eleven days. This virus, the original South American isolate of Moses (110), is the strain which has been most extensively used to infect Australian rabbits and results in a mortality of 100% in susceptible rabbits.

Three to four days following inoculation or natural infection with this virus, a primary tumor may become evident but usually generalized tumors become evident only on the sixth or seventh day. At this time edema of the eyelids is observed followed by the development of mucopurulent blepharoconjunctivitis often resulting in complete closing of both eyes (Fig. 2). Mucopurulent nasal discharge and pronounced edema of the base of the ears, the perineal region, the external genitalia, and lips are frequently seen. By the tenth day, hard convex lumps may cover much of the body, head, and ears and occasionally also the legs. The lumps are not sharply demarcated but may become markedly congested and ultimately necrotic in rabbits surviving for two weeks. Labored breathing

Fig. 2. Rabbit myxomatosis. Mucopurulent conjunctivitis with complete closing of eye.

is often seen in protracted cases of the disease, but surprisingly, appetite may be maintained until shortly before death. Terminal convulsions frequently precede death which usually occurs eight to fifteen days postinfection. Infection with the less virulent South American and Australian strains of virus results in milder disease with less edema, less nasal and ocular discharge, more clearly demarcated nodule development and low mortality. With the laboratory-attenuated "neuromyxoma" virus (76) a very mild disease with little or no mortality is observed.

The predominant myxoma strains in Europe are the virulent "Lausanne" strain and its naturally attenuated derivatives originating from the first virus introduced into France from Brazil in 1952. The virulent European viruses cause severe disease in rabbits resulting in mortality of up to 100%. Clinically the disease produced by this virus differs in several respects to the disease produced by the "standard laboratory strain." The European disease is characterized by the very rapid proliferation of large, protuberant lumps, usually purple to black in color by the end of a week. The lumps may break open by the tenth day, oozing a serous discharge. Lumps may occur all over the body but are seldom seen on the ears. Infection with this strain of virus also results in very pronounced edema of the face and anal region, as well as seropurulent discharges from the eyes and nose. In general, there is considerable skin congestion associated with infection with this virus.

More attenuated strains of the European virus result in somewhat less dramatic disease with lower mortality. With some of the naturally attenuated British viruses the mortality is considerably decreased and the lumps which develop are flat rather than distinctly convex, in this sense resembling some of the attenuated field strains of virus encountered in Australia (17).

j. PATHOLOGY.   The gross and microscopic pathology of myxomatosis have been comprehensively reviewed (43, 75, 133).

In adult rabbits of the genus *Sylvilagus* the lesions induced by the myxoma virus are usually localized skin tumors resembling the fibromas produced in European rabbits by the rabbit fibroma virus (see later). In hares or young *Sylvilagus* rabbits, fibromatous to myxomatous nodules similar to those seen in acute myxomatosis may be found but as a rule the myxoma virus causes a mild localized infection in these animals also.

k. GROSS LESIONS.   The most prominent gross lesions encountered at necropsy of European rabbits with myxomatosis are the skin tumors (not characteristic of the Californian disease) and the pronounced cutaneous and subcutaneous edema especially in the area of the face and around body orifices (Fig. 3). Skin hemorrhages and subserosal petechiae and ecchymoses may be observed in the stomach and intestines especially following infection with

Fig. 3. Rabbit myxomatosis. Pronounced cutaneous and subcutaneous edema especially of the lips.

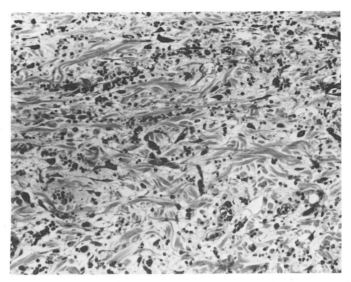

Fig. 4. Rabbit myxomatosis. Characteristic microscopic appearance of core of myxoma nodule. Note homogeneous mucinous material interspersed with inflammatory cells. (Courtesy of Dr. N. F. Cheville, N.A.D.L., Ames, Iowa.)

the California virus. Subepicardial and subendocardial hemorrhages may also occur. No consistent gross changes have been described for the spleen, liver, kidneys, or lungs but various degrees of congestion of these organs will occur depending on the severity of the disease.

l. MICROSCOPIC LESIONS. The lesions in the skin involve the epithelial cells, fibroblasts, and the endothelial cells. Skin lesions may range from predominantly proliferative lesions to essentially degenerative lesions depending on the strain of virus involved. The skin tumors result from initial proliferation of undifferentiated mesenchymal cells which become large stellate cells referred to as "myxoma" cells. These cells lie in a homogeneous matrix of mucinous material interspersed with capillaries and inflammatory cells (Fig. 4). It is because of this mucinous substance that the tumors are referred to as "myxomas" and the disease "myxomatosis."

Endothelial proliferation with narrowing of the lumen and extrusion of stellate "myxoma" cells has been described by Hurst (75) who considers this lesion characteristic of myxomatosis. The necrosis observed in the center of the myxomas may be in part attributed to the luminal occlusion of blood vessels resulting from endothelial proliferation.

The epithelial cells overlying the tumor may appear normal in early tumors, or they may show hyperplasia or degeneration (Fig. 5). Intracytoplasmic inclusions have been observed in various cells types but are especially prominent in the prickle-cell layer (138).

Lesions in other organs, while not consistently present, reflect the generalized nature of myxomatosis. Cellular

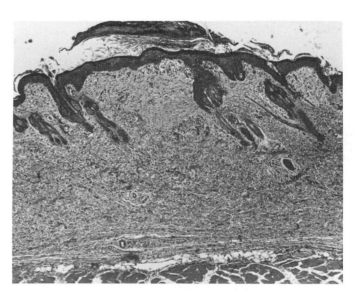

Fig. 5. Rabbit myxomatosis. Microscopic appearance of myxoma lesion. Note intact epithelium overlying tumor. (Courtesy of Dr. N. F. Cheville, N.A.D.L., Ames, Iowa.)

proliferation, a lesion invariably present in the skin, has also been described in the pulmonary alveolar epithelium and in the reticulum cells of lymph nodes and spleen (75). Focal or interstitial hemorrhages may be observed in the skin, kidneys, lymph nodes, testes, heart, stomach, and intestinal walls. Degeneration and necrosis of lymph nodes, pulmonary alveoli, spleen, and central veins of the hepatic lobules are frequently seen. Stellate cells resembling the "myxoma" cells may be encountered in the lymph nodes, the bone marrow, the uterus and ovaries, the testes, and the lung.

m. DIAGNOSIS. Myxomatosis as seen in European rabbits is usually a sufficiently distinct disease syndrome to allow diagnosis on the basis of the clinical-pathological syndrome alone. Infection with the California viruses may be hard to detect, however, due to the frequent absence of skin nodules as well as other symptoms of the disease. Diagnosis should be confirmed by virus isolation. The technique of choice is intracutaneous inoculation of young susceptible rabbits with fresh tissue collected from lesions free from secondary bacterial contamination. Lesions should develop at the site of inoculation within a week. The virus can also be readily isolated by chorioallantoic membrane inoculation of eleven- to thirteen-day-old embryonated eggs followed by incubation at 35°C for four to six days. Distinct focal pocks should develop if the virus is successfully isolated. The South American viruses cause large pocks, the California virus intermediate sized pox, and the fibroma virus minute pocks.

Virus isolation on chicken embryo fibroblast cell cultures or on any one of several other cell types can also be accomplished.

The virus isolated by any one of the above methods should be identified as myxoma virus using one of several serological procedures. The best currently available methods for virus identification are the fluorescent antibody procedure (161, 162), the plaque-neutralization test (170), and the agar-gel diffusion microprecipitation test (38).

Infection of *Sylvilagus* rabbits with myxoma virus clinically resembles fibromatosis and should be differentiated from the latter disease by inoculation of young susceptible rabbits of the genus *Oryctolagus*. If the disease is myxomatosis, severe to fatal disease will ensue whereas fibromatosis will cause only a localized fibroma.

n. CONTROL AND PREVENTION. Control of myxomatosis is of prime importance to rabbit producers in areas where the disease is endemic in the wild rabbit population. In such areas vector control and adequate screening methods aimed at keeping out mosquitoes serve to keep the disease under control.

Introduction of new rabbits into a colony should be through a quarantine procedure in which the new rabbits are held for two weeks in insect-proof quarters. To prevent spread of the disease within a colony all sick rabbits should be isolated until myxomatosis can be excluded as the cause of the disease.

Vaccination has not been widely adopted largely because most of the early inactivated vaccines proved to be of doubtful value. The fibroma virus has been used as a live vaccine for myxomatosis but results have been extremely variable (43).

A live attenuated myxomatosis vaccine (the MSD strain) was developed by McKercher and Saito (101). Inoculation of this virus into susceptible rabbits resulted in a very mild reaction followed by immunity persisting for nine months. Jiran *et al.* (83a) found this virus to be too virulent for use as a vaccine and further attenuated it by serial passage in rabbit kidney cell cultures. The additionally modified virus, designated MSD/B was found to be completely harmless but retained a high degree of immunogenicity.

2. Rabbit (Shope) Fibroma Virus

a. HISTORY. The transmissible tumor-producing agent now known as the rabbit fibroma or Shope fibroma virus was isolated from a cottontail rabbit (*Sylvilagus floridanus*) in 1932 (145). The virus was shown to be transmissible to cottontail rabbits and European rabbits (*Oryctolagus cuniculus*) producing localized fibromata in both species. In his original report Shope described the gross and microscopic lesions of both the natural and experimental disease and in a subsequent paper (146) he identified the causative agent of the disease as a virus. He named the virus the "fibroma" virus and showed that it was antigenically related to the myxoma virus. The similarities between the fibroma and myxoma viruses have subsequently been confirmed by cross-immunity tests (148), ether sensitivity (4, 35), virus-reactivation studies (12), and by microprecipitation procedures (38).

The fibroma virus was initially believed to cause only the localized benign fibromata described by Shope, but it was later shown to cause severe generalized disease in newborn European rabbits (30, 84), and in newborn cottontail rabbits (173).

The disease is historically considered as a benign enzootic disease of wild cottontail rabbits, of little economic significance to commercial rabbit producers or laboratory investigators. However, an epizootic of fibromatosis in a commercial rabbitry resulting in high morbidity and the death of two newborn rabbits was recently reported (84). The disease can thus be a threat to commercial rabbits in areas where it is enzootic in the wild rabbit population and where outdoor husbandry practices would allow contact with arthropod vectors.

b. ETIOLOGY. The fibroma virus is a member of the poxvirus group and is closely related to the myxoma virus (35) and the squirrel and hare fibroma viruses (38). The chemical and physical characteristics of the fibroma virus and the reasons for its inclusion in the poxvirus group have been adequately summarized by Andrewes and Pereira (6).

The virus can be propagated on the chorioallantoic membrane of chicken embryos but characteristic lesions as observed with myxoma virus are not produced (6). The fibroma virus has been propagated in cell cultures derived from rats, man, and guinea pigs (19), and also in rabbit cell cultures derived from cottontail and domestic rabbits (22, 72, 90, 118). Fibroma virus replication in primary or serially cultured rabbit kidney cells resulted in pronounced changes in cell growth and morphology. Such fibroma virus-transformed cells inoculated into the cheek pouch of hamsters resulted in tumor formation (72).

c. INCIDENCE. Since rabbit fibroma virus was first isolated from a cottontail rabbit in New Jersey by Shope in 1932, the disease has been recognized in several other states of the U.S.A. as well as in Canada. Herman *et al.* (67) found that more than 50% of the wild cottontails trapped in the Patuxent Wildlife Refuge in Maryland were positive for fibroma virus or contained fibroma virus antibodies. The virus has also been isolated from cottontails in Wisconsin (173) and Michigan (67). The recent recognition of the disease in Texas (84) indicates that the disease may be more widespread than was formerly believed.

d. EPIDEMIOLOGY. The natural transmission cycle of the fibroma virus has not been clearly elucidated. Experimental evidence has shown, however, that the virus may persist in the epidermis of infected cottontail rabbits for five to ten months (93, 94). This situation would serve to enhance the likelihood of mechanical arthropod transmission. Under experimental conditions it has been shown that several species of mosquitoes as well as triatomes, fleas, and bedbugs can serve as vectors of the fibroma virus (25, 93, 95). Free-flying infected mosquitoes have been shown by Kilham and Dalmat (93), to be capable of infecting wild rabbits with production of leg, ear, and nasal tumors. Experimental and strong circumstantial evidence would thus suggest that the principal transmission mechanism for the fibroma virus is by way of biting arthropods, a situation similar to that described for myxomatosis.

The natural host of the fibroma virus is the Eastern cottontail rabbit (*Sylvilagus floridanus*). Three other species of the genus *Sylvilagus* (*S. bachmani*, *S. nuttallii*, and *S. audobonii*) have been shown by Regnery and Marshall (cited in 43) to be refractory to fibroma virus. The European rabbit is susceptible to the virus (145) as is the snowshoe hare, *Lepus americanus* (172), but the European hare, *Lepus europaeus* is refractory to the virus (43). While the European rabbit can be readily infected, it does not serve as a good source of infection of mosquitoes due to low virus titers in the skin (43).

e. CLINICAL SIGNS. The clinical signs are largely those described by Shope (145, 146, 148). The tumors observed in natural fibromatosis of Eastern cottontail rabbits are almost invariably on the legs or feet, usually one to three tumors occurring on an infected rabbit. Occasionally the tumors may be encountered on the muzzle and around the

eyes of infected rabbits. Tumors may measure up to 7 cm in maximum diameter and are usually in the region of 1–2 cm thick. The tumors are subcutaneous and are not attached to the underlying tissues. The tumors may persist for several months, and in some instances (93) for nearly a year.

Other clinical signs are usually not observed, the tumor-bearing rabbit remaining apparently normal throughout the disease. Metastases from the original tumors do not occur (145).

The clinical signs observed in experimentally infected European rabbits resemble those seen in cottontail rabbits. Regression of tumors is, however, usually more rapid than in the cottontails. Inoculation of newborn European rabbits frequently results in generalized and fatal infections (31). Generalized infections of adult European rabbits have also been described by Hurst (77) but the usual result of infection with the fibroma virus is the development of localized benign tumors. In the only reported natural outbreak of fibromatosis of European rabbits, Joiner *et al.* (84) observed diffuse subcutaneous induration involving the underlying tissues. Hyperemia and edema of the genitalia of infected male and female rabbits was also observed. Visible tumors (Fig. 6) were the prominent symptom in the seven visibly infected adult rabbits, but the twelve of twenty-four suckling rabbits which became visibly ill showed lethargy and loss of condition in addition to the skin lesions.

A spectrum of virulence has been shown by Smith (151) to exist for the fibroma viruses. The least virulent strain which she tested was the 1A strain, a strain modified by human manipulation, while the Boerlage strain isolated by Shope in 1942 was the most virulent. Most of the clinical studies described above have been done with the strains resembling the Boerlage strain.

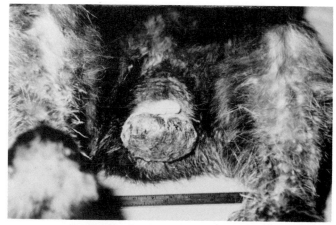

Fig. 6. Perianal fibroma observed in natural case of fibromatosis in female European rabbit. (Courtesy of Dr. G. N. Joiner, Texas A & M University, College Station.)

f. PATHOLOGY. The gross and microscopic pathological changes associated with fibroma virus infection in naturally and experimentally infected cottontail rabbits and in experimentally infected European rabbits have been described by Shope (145), Andrewes (3), Ahlström (1), Fisher (47). Kilham and Fisher (94), Dalmat and Stanton (26), and Youill and Hanson (173). The pathological changes observed in the only reported natural outbreak of fibromatosis of European rabbits were described by Joiner *et al.* (84).

g. GROSS LESIONS. The earliest gross lesion in infected cottontails is slight thickening of the subcutaneous tissue followed by the development of clearly demarcated soft swellings which may be evident on the sixth day postinoculation. As the tumors increase in size they also increase in consistency, usually reaching their maximum size by the twelfth day. The average size of the swellings is approximately 4 × 6 cm with a thickness of about 2 cm. These tumors may persist for months before regressing, leaving the animal essentially normal. Newborn cottontail rabbits, artificially infected with the fibroma virus, may die of generalized fibromatosis. Under natural conditions, however, this type of infection has not been observed.

The gross lesions observed in experimentally infected European rabbits infected with the virus are similar to those observed in cottontails, but regression of the tumors occurs more rapidly in the European rabbit. The gross lesions observed in the natural outbreak reported by Joiner *et al.* (84) have been described earlier.

h. MICROSCOPIC LESIONS. The earliest microscopic lesion in cottontail rabbits is an acute inflammatory reaction followed by localized fibroblastic proliferation accompanied by both mononuclear and polymorphonuclear leukocyte infiltration. Fibroblastic proliferation continues until a distinct tumor is formed consisting of spindle-shaped and polygonal connective tissue cells with abundant cytoplasm. Mitotic figures are not numerous. Many cells may have large intracytoplasmic inclusions characteristic of poxvirus infections. Mononuclear leukocyte cuffing of vessels adjacent to the tumor may be observed and pronounced accumulation of lymphocytes at the base of the tumor is often seen. Degeneration of the epidermis overlying the tumor may result from pressure ischemia. This may be followed by necrosis and sloughing of the epithelium and the tumor. In many instances, however, the tumors regress without sloughing of epithelium. Regression is usually complete within two months after appearance of the tumors.

Andrewes (3) described a strain of fibroma virus which caused a more inflammatory and less proliferative lesion than the fibroma virus strain isolated by Shope (145).

Histologically typical fibromata are observed in European rabbits experimentally infected with fibroma virus.

The lesions differ only slightly from those observed in cottontail rabbits, the principal difference being the absence of the epidermal degeneration often observed in the latter species (145). Ahlström (1) has, however, described the epidermal degeneration in European rabbits also.

The microscopic lesions observed in the natural outbreak of fibromatosis in European rabbits (84) ranged from tumors resembling myxomata (Fig. 7) to typical fibromata. Eosinophilic cytoplasmic inclusions were occasionally observed in tumor cells, and similar inclusions possibly of viral origin, were seen in epithelial cells overlying the tumors. The lesions encountered in adult and newborn rabbits did not differ significantly.

i. DIAGNOSIS. The two diseases from which fibromatosis needs to be differentiated are rabbit papillomatosis and myxomatosis. On clinical basis alone differentiation from papillomatosis should be readily accomplished. The fibromata are essentially flat, subcutaneous, loose, rubbery tumors, whereas papillomata are tumors of the skin, they are heavily keratinized, and project in irregular fashion from the skin surface. Histopathological differentiation is also easily accomplished, especially when fibroma inclusions are observed.

Myxomatosis in cottontail rabbits may resemble fibromatosis and is most easily differentiated by subcutaneous inoculation of young European rabbits with a suspension of tumor material. Fibroma virus should result in the development of local fibromata, whereas the myxoma virus would cause severe generalized and often fatal disease.

Virus isolation in cell cultures or on the chorioallantoic membranes of chicken embryos should be attempted to confirm diagnosis. The virus can then be identified by

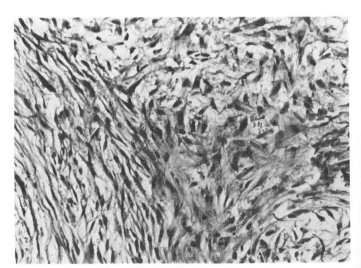

Fig. 7. Myxomatous lesion observed in natural case of fibromatosis of European rabbits. Note the bands of dense connective tissue. (Courtesy of Dr. G. N. Joiner, Texas A & M University, College Station.)

serum-virus-neutralization tests as proposed by Youill (172).

j. CONTROL AND PREVENTION. Since the disease is endemic and of little significance in the cottontail population, no control measures have been developed. Fibromatosis is also not considered to be an important problem in domestic rabbits but since the disease has now been encountered in a rabbitry in an area where the disease is apparently endemic in the wild population, control measures should be developed.

In such areas the obvious method of preventing infection of rabbits held in outdoor rabbitries is vector control. Careful analysis of the role of various possible vectors in natural outbreaks should be made in preparation for the establishment of vector control methods.

3. Hare Fibroma Virus

a. HISTORY. A series of epizootics of a fibromatous disease of European hares (*Lepus europaeus*) occurred in France and Northern Italy, in 1959 (96, 98). The causative agent of the disease was shown by Leinati and his co-workers (99) to be a poxvirus related to the myxoma virus. In retrospect it appears probable that a nodular skin disease of hares designated "hare sarcoma" by Dungern and Coca in Germany in 1909 is the disease now known as hare fibromatosis (29).

b. ETIOLOGY. The causative agent of hare fibromatosis has been shown by electron microscopic examination to be a poxvirus (38). The virus is a member of the myxoma-fibroma subgroup of the poxviruses and has been shown by plaque neutralization and cross-protection tests to be antigenically very closely related to the myxoma, rabbit fibroma, and squirrel fibroma viruses (169). Agar-gel diffusion microprecipitation techniques have been used to demonstrate that hare fibroma virus shares more common antigens with rabbit fibroma virus than with myxoma virus (38).

A substantial degree of cross-protection between the hare fibroma virus and myxoma has been demonstrated. European rabbits immune to myxoma virus are completely refractory to hare fibroma virus, whereas rabbits immunized with hare fibroma virus will develop symptoms when exposed to myxoma virus, but will not die, indicating a degree of protection (170).

c. INCIDENCE. Hare fibromatosis has been recognized only in Europe where under natural conditions it infects the European hare. The European rabbit has been shown to be susceptible to the virus but natural outbreaks of the disease in rabbits have not been recorded.

d. EPIDEMIOLOGY. A definite seasonal incidence of the disease has been reported, the late summer and autumn being the seasons of peak incidence (96, 98). The modes of transmission and interepidemic survival of the virus are not known.

e. CLINICAL SIGNS. In European hares the disease is characterized by the development of numerous skin nodules, up to one inch in diameter, occurring especially on the face, the eyelids, and around the ears. The nodules closely resemble rabbit fibromata (99).

In adult European rabbits the virus causes formation of very small fibromata, but in newborn rabbits large fibromata resembling the lesions of rabbit fibromatosis are produced (43).

f. PATHOLOGY. The gross and microscopic appearance of the lesions of hare fibroma are very similar to the lesions of rabbit fibroma (96, 99).

g. DIAGNOSIS. The disease is usually diagnosed on clinicopathological basis. The diagnosis can be confirmed by virus isolation in rabbit kidney cell cultures or on the chorioallantoic membrane of chicken embryos. Serological characterization of the virus can be achieved using the agar-gel diffusion technique (38).

4. Squirrel Fibroma Virus

The squirrel fibroma virus which causes large fibromata in gray squirrels (*Sciurus carolinensis*) is another poxvirus closely related to the myxoma, rabbit fibroma, and hare fibroma viruses (38). While it is not known to cause natural disease of rabbits, under experimental conditions this virus causes mild nodular skin lesions in European rabbits (43).

5. Rabbit Pox

a. HISTORY. Rabbit pox was first diagnosed at the Rockefeller Institute in New York when a severe, highly fatal disease broke out spontaneously in the rabbit colony in 1932 (50, 51, 139). A smaller outbreak of the disease had occurred in 1930. The disease was characterized by an erythematous rash followed by the development of cutaneous eruptions closely resembling the pocks seen in human infection with variola virus (smallpox). The disease was extremely contagious and caused very high mortality especially in young rabbits and pregnant females. Belgian hares were also susceptible to the disease. The causative agent of the disease was shown to be a poxvirus (123, 139).

A spontaneous outbreak of a similar disease in a laboratory rabbit colony in Holland was described by Jansen, (81) who named the disease "rabbit plague." The disease was highly fatal and differed from the Rockefeller outbreak in that it was not exanthematous, giving rise to the name "pockless" rabbit pox (83). This disease was also shown to be caused by a poxvirus closely related to vaccinia (82).

A second outbreak of the disease was reported in Holland by Verlinde and Wersinck in 1951 (165).

The second epidemic of rabbit pox reported in the United States occurred in New York (20) and was also of the "pockless" type first observed by Jansen In Holland. Another serious epidemic of the disease resulting in mortality of 95% occurred following the introduction of supposedly inactivated rabbit pox virus of the Dutch (Utrecht) strain into the rabbit colony of a New York medical school (20).

b. ETIOLOGY.   The rabbit pox virus is antigenically very closely related to the vaccinia virus (36, 74, 82) and only distantly related to the other poxviruses of leporids, namely, the myxoma, fibroma, and hare fibroma viruses (169). Comparison of the biological properties of the Utrecht (81) and Rockefeller (51) strains of rabbit poxvirus with numerous strains of vaccinia virus revealed that both rabbit pox strains were practically indistinguishable from certain "neurovaccinia" strains (36). The very close antigenic relationship between the rabbit poxvirus and the vaccinia virus taken together with the fact that all reported outbreaks of rabbit pox have occurred in laboratory colonies has led to the suggestion that rabbit pox may be a laboratory "sport" of the vaccinia virus (53, 165). This contention is rejected by other workers (20) who consider the rabbit pox virus to be a distinct entity within the variola subgroup of the poxviruses.

The rabbit pox virus can be readily propagated on the chorioallantoic membrane of eleven- to thirteen-day-old chicken embryos with the development of distinct pocks. The predominant pock type is hemorrhagic (82), but white pock variants have been described (36).

Cell culture propagation of rabbit pox virus has been demonstrated in cells derived from many species of animals (20). Those described by Christensen et al. (20) as being able to support the growth of rabbit poxvirus are HeLa, Chang Liver, L929 (mouse fibroblast), human heart, KB (human epithelial), FL (human amnion), AT (Chinese hamster epithelial), PK-2A (pig kidney), and FAF (Chinese hamster fibroblast).

The Rockefeller strain of rabbit pox virus hemagglutinates chicken erythrocytes but the Utrecht strain and the strain isolated from the first American outbreak of "pockless" rabbit pox do not hemagglutinate (20).

c. INCIDENCE.   The disease is relatively rare, only six outbreaks having been reported since 1930. All the confirmed outbreaks thus far reported have occurred in the United States and Holland, and all have occurred in laboratory colonies.

While no age incidence is evident in infections with this virus, highest mortality occurs in young rabbits and in pregnant females (51).

d. EPIDEMIOLOGY.   No field outbreaks of rabbit pox have been reported. Within infected colonies spread of the disease is extremely rapid and in the outbreak of 1932, even removal and isolation of infected animals failed to prevent the disease from spreading throughout the colony (51). The virus appears to spread via nasal secretions, rich in virus, which usually appear on the third day after infection and which may be inhaled or ingested by susceptible rabbits (11). Recovery from infection does not appear to result in the establishment of a carrier state. It has been shown that recovered rabbits can be safely mated with susceptible rabbits and also that clean colonies can be derived from recovered stock (51).

Arthropods have not been shown to play a role in the epidemiology of rabbit pox infection. The primary sources of infection resulting in the Dutch and United States outbreaks of the disease have not been determined. The spontaneous nature of the outbreaks would lead one to suspect the presence of carrier animals, but the extremely long lag between outbreaks seems to eliminate this proposition.

The origin of the rabbit pox virus and its mode of inter-epidemic survival remain two of the unsolved mysteries of animal virology.

e. PUBLIC HEALTH SIGNIFICANCE.   Although closely related to the vaccinia virus the rabbit pox virus does not infect man (61).

f. CLINICAL SIGNS.   The clinical disease has been described in detail by Greene (51), Bedson and Duckworth (11), and Christensen et al. (20). The nasal cavity is frequently the site of primary infection with the virus which appears to multiply in the nasal mucosa and later in the lymph nodes of the respiratory tract and in the lung and spleen. A febrile response is usually observed two to three days after infection at which time a profuse nasal discharge is commonly seen. Another early symptom which is frequently observed is enlargement and induration of the lymph nodes especially the popliteal and inguinal nodes. The lymphadenitis usually persists throughout the course of the disease.

Skin lesions usually appear about five days after infection, initially as an erythematous or macular rash but later developing into papules which may remain small or which may develop into nodules of up to a centimeter in diameter. These nodules eventually dry up forming superficial crusts. Distribution of the macules and papules is over the entire skin and they may also occur on the mucus membranes of the oral and nasal cavities. Extensive edema of the face and oral cavity is often observed as is focal necrosis of the hard palate and the gums. Hemorrhages in the skin may occur in severe cases of the disease.

Severe orchitis with extensive scrotal edema is frequently seen in male rabbits which may also have papules in the sheath and urethra. Similar lesions may also occur in the vulvae of females, and when extensive edema is present, urine retention may occur in both males and females.

Eye lesions are almost invariably observed and may range from a mild blepharitis to an acute keratitis with corneal ulceration. Purulent conjunctivitis is frequently present.

Death usually occurs seven to ten days after infection, but may occur as early as five days or rabbits may linger for several weeks before dying. Pregnant females infected with the virus usually abort.

The generalized disease syndrome described above represents the findings of Greene, (51) and Bedson and Duckworth (11) with the Rockefeller strain of virus. This strain of virus may occasionally result in peracute disease in which death is preceded only by fever, anorexia, and occasionally blepharitis. This peracute form of the disease is unusual with the Rockefeller strain of the virus, but natural outbreaks of rabbit pox of the so-called "pockless" type (20, 81–83) in which no erythema or other skin lesions are observed, have occurred. In the first Dutch outbreak of the disease some rabbits died within a week after infection showing only anorexia, fever, and listlessness (81). In the first American outbreak of the "pockless" disease, animals died seven to ten days postinfection following a mild fever, conjunctivitis, and sometimes diarrhea (20). Jansen (81–83) as well as Christensen *et al.* (20) have described the occasional presence of a few scattered papules on the lips and tongues of rabbits with "pockless" rabbit pox. Under experimental conditions Bedson and Duckworth (11) showed that both the Utrecht and Rockefeller strains of the virus produce skin lesions.

g. PATHOLOGY. The gross and microscopic pathology of rabbit pox has been described in detail by Greene (52) for the Rockefeller Institute outbreak and by Christensen *et al.* (20) for the first outbreak of "pockless" rabbit pox in the United States.

h. GROSS LESIONS. The most distinctive gross lesions observed at necropsy are the skin lesions which may range in severity from a few localized papules to severe, almost confluent skin lesions with extensive necrosis and hemorrhage. Papular lesions are frequently also found in the mouth, the upper respiratory tract and the spleen, liver, and lungs. Occasional papules may occur in almost any part of the body.

Subcutaneous edema and edema of the mouth and other body openings are commonly seen. Only rabbits with very severe lesions of the mouth appear emaciated at necropsy.

Specific lesions are seldom observed in the gastrointestinal tract, but focal papules in the peritoneum and omentum may be encountered. The liver is usually enlarged, yellowish in color, and numerous small gray nodules may be found throughout the parenchyma. Small focal areas of necrosis may be seen. Small nodules may also be present in the gallbladder.

The spleen is usually moderately enlarged with focal nodules or small areas of focal necrosis occasionally present. The lungs may be riddled with the small whitish nodules and in advanced cases focal areas of necrosis may be encountered.

The testicles, ovaries, and uterus are also frequently riddled with the whitish nodules and marked edema, and necrosis of the testes is frequently seen. Focal abscessation of the uterus is occasionally observed. The focal lesions may be encountered in the lymph nodes, adrenals, thyroid, parathyroid, and rarely in the heart. Specific gross lesions are seldom observed in the central nervous system or the kidneys.

In the "pockless" form of the disease a few pocks in the mouth may be found and occasional skin lesions may be demonstrated by shaving the rabbit. The prominent gross lesions at necropsy are pleuritis, focal necrosis of the liver, enlarged spleen, and edema and hemorrhage of the testes. The small white nodules so abundant in the other form of the disease are occasionally encountered in the lungs and adrenals.

i. MICROSCOPIC LESIONS. The predominant lesion in rabbit pox is the papule or nodule which occurs in the skin and in many other organs of the body. Histologically a typical nodule consists of a central zone of necrosis surrounded by a zone of mononuclear cell infiltration. Edema and occasionally hemorrhage of adjacent tissues is frequently seen. Diffuse lesions with massive mononuclear cell infiltration, necrosis, hemorrhage, and edema are often encountered. Vesiculation and pustule formation, as seen with variola infection in man, is not characteristically seen with rabbit pox infection.

Pronounced swelling of the vascular endothelium is present, and it is believed that the consequent vascular occlusion may contribute to the development of necrotic lesions.

Focal nodular lesions and diffuse areas of pneumonitis are encountered in the lungs. The pneumonitis is characterized by perivascular mononuclear and polymorphonuclear cell infiltration. Focal to extremely extensive necrosis of the lungs may be observed.

Severe congestion of the spleen and marked distension of the sinuses with mononuclear cells is commonly seen. Edema of the malpighian corpuscles and focal to diffuse necrosis is present in many instances.

Lymph nodes are generally greatly enlarged mainly as a result of severe edema. Extensive necrosis of lymph nodes and other lymphoid tissues such as Peyer's patches may occur. Hemorrhage and necrosis of the bone marrow is frequently seen interspersed with areas of mononuclear cell infiltration.

The extensive degeneration and necrosis of liver parenchyma may be focal or diffuse in nature and may involve the whole organ. Focal necrosis with edema is characteristically also seen in the testes as are necrotic foci in the adrenals, uterus, thyroid, thymus, and salivary glands.

The cytoplasmic inclusions characteristically associated with poxvirus infections are seldom encountered in rabbit pox lesions.

j. DIAGNOSIS.   Rabbit pox can be diagnosed on the basis of the clinical signs and the very characteristic gross and microscopic lesions associated with the disease. Confirmatory diagnosis can be made by fluorescent antibody examination of tissues or tissue impression smears (20) or by virus isolation and identification. The virus can be isolated by chorioallantoic membrane inoculation of chicken embryos or by cell culture propagation of the virus on cells derived from rabbits, mice, or any one of several other animal species. The virus may then be identified by the fluorescent antibody procedure, by hemagglutination inhibition (some strains), or by cross-protection tests using vaccinia-immune and susceptible rabbits. Vaccinia-immune rabbits should be highly resistant, whereas severe disease with high mortality should occur in the susceptible rabbits.

k. CONTROL AND PREVENTION.   Since the natural source of virus causing outbreaks has not been determined, control measures to prevent the occurrence of the disease cannot be intelligently developed. In the outbreaks that do occur attempts can be made to prevent spread of virus through the colony by isolation of sick animals but this procedure has proved of little value in practice. Investigators using rabbit poxvirus in experimental systems should take exceptional precautions to prevent the virus from reaching susceptible rabbit populations. In the face of an epidemic in a large colony vaccination with vaccinia virus can be used to protect the uninfected population.

A prophylactic approach which has not been used in rabbit pox but which may prove valuable in the face of epidemics is the use of antiviral drugs. The drug Rifampicin has been shown to be active against poxviruses (124, 157), and the isatin-B derivative of thiosemicarbazone has been extensively used to protect human populations in the face of variola (smallpox) epidemics (125). The success of field trials with the latter drug would certainly warrant its consideration as a prophylactic in rabbit pox epidemics.

B. Herpesvirus Infections

The herpesvirus group is one of the rapidly expanding groups of animal viruses the clinical importance of which is becoming increasingly more obvious. Viruses of this group have long been recognized as the causative agents of respiratory and genital disease in man, cattle, horses, and swine. They are also the recognized cause of other disease syndromes in many species of animals. Recent evidence has linked viruses of this group with neoplastic disease in chickens, frogs, monkeys, rabbits, and possibly man, thus further emphasizing the importance of this group of viruses.

An important characteristic of the herpesviruses is their capacity to cause mild or subclinical disease after which viral persistence in a latent state may ensue. Stresses of various kinds may result in viral recrudescence even after prolonged periods of latency.

The laboratory rabbit is by definition an experimental animal and may frequently be exposed to stresses of various kinds. The possible existence of latent virus infections thus has an important influence on the quality of the experimental animals and possibly on the validity of the findings of the investigator. It is in this context that the two currently recognized herpesviruses of rabbits will be discussed.

1. Herpesvirus cuniculi (*Synonym*: Virus III of rabbits)

a. HISTORY.   Nesburn (115) described the isolation of this virus from primary rabbit- kidney cell cultures and named the virus *Herpesvirus cuniculi*. He acknowledged that this virus isolation could represent a reisolation of virus III of rabbits described by Rivers and Tillett in 1923 (135) but since the original strain of virus is no longer available no direct serological comparisons can be made. However, by comparing the known characteristics of the original virus III with the characteristics of his viral isolate Nesburn presents a very convincing argument in favor of considering these two viral isolates as one and the same virus.

Virus III was isolated during attempts to find the etiologic agent of varicella (chicken pox). Rivers and Tillett (135) showed that the agent, when inoculated into rabbits, produced fever, exanthema, skin vesicles, corneal lesions, and intranuclear inclusions reminiscent of varicella infection, and concluded that they had isolated the causative agent of the disease. In the following year the same authors (136) had to withdraw their claim of having isolated the causative agent of varicella when they found that 4 out of 20 uninoculated rabbits possessed neutralizing antibodies to virus III and also that convalescent sera from patients with varicella infection did not inactivate their viral isolate. In 1924 Miller, Andrewes, and Swift (108) isolated a similar virus from serially passaged normal rabbit testicular tissue while investigating the etiology of rheumatic fever. McCartney isolated the virus while working on scarlet fever in England (163), and the agent has also been isolated by Doerr in Switzerland (2).

b. ETIOLOGY.   Nesburn (115) has shown that his strain of virus (designated 923J) possesses the physical, chemical, cytopathic, histological, and electron microscopic characteristics of a herpesvirus and has proposed that the name *Herpesvirus cuniculi* be adopted for this virus. He has shown that the virus can be readily propagated in primary or established line cells of rabbit origin and will also grow in African grivet monkey kidney cells.

c. INCIDENCE. The prevalence of natural infection with *H. cuniculi* is not known. In their original investigations Rivers and Tillett (136) found that 4 out of 20 stock rabbits possessed antibodies to the virus. Later they showed an incidence of 15% in 200 rabbits which they tested for antibodies. Andrewes (2) in England found 97.9% of 377 experimental rabbits susceptible to the virus. He concluded that the virus was probably endemic in some rabbit colonies and not in others. Topacio and Hyde (163), in Maryland, found that 17% of 76 rabbits which they tested were immune to the virus. They felt that some of the older bucks which were resistant to infection could be serving as carriers of the virus. Nesburn (115) has not been able to reisolate the virus from over 100 batches of primary rabbit kidney cell cultures which he has prepared since his initial isolation of strain 923J of *H. cuniculi*.

d. CLINICAL SIGNS. All the reported isolations of *H. cuniculi* have been made from apparently normal rabbits, and no naturally occurring disease has been attributed to infection with this virus. Under experimental conditions intradermal inoculation of rabbits with the virus results in pronounced erythema at the site of inoculation after 4 to 7 days. The erythema usually disappears within two weeks (136). Occasional generalized reactions with anorexia, diarrhea, emaciation, and temperatures of up to 107° F may be observed and skin vesicles have been reported. Topacio and Hyde (163) reported that intradermal inoculation may occasionally result in papular elevations as well as erythema. They have also shown that corneal scarification with the virus results in swelling and vesiculation of the corneal cells. Intratesticular inoculation of rabbits results in acute orchitis and fever within 3 to 4 days, (2). Repeated intramuscular and subcutaneous inoculation of rabbits with strain 923J of the virus failed to produce symptoms. However, corneal scarification with the virus resulted in the development of a mild punctate keratitis (115).

e. PATHOLOGY. No gross pathological changes have been observed in the internal organs of experimentally infected rabbits. On examination of stained sections of infected testes, skin, or cornea, edema and an intense mononuclear leukocyte infiltration are observed. Large eosinophilic intranuclear inclusions typical of herpesviruses are characteristically found (137). These inclusions may occur in the corneal epithelium, in the interstitial cells of the testicles, and in the endothelial leukocytes of the skin. Pearce (121, 122) has described severe myocarditis with intranuclear inclusions typical of herpesvirus infections in rabbits inoculated intracardially with the virus. *Herpesvirus cuniculi* thus clearly possesses the capacity to produce dis-

ease in the laboratory rabbit. The absence of reported cases of this disease could be attributed to the relative rarity of the condition or to lack of intensive etiological investigation of those cases which do in fact occur.

2. Herpesvirus sylvilagus (*Synonym*: Cottontail virus)

a. HISTORY. This virus was recently isolated by Hinze (68, 70) from primary kidney cell cultures prepared from the kidneys of apparently normal weanling cottontail rabbits (*Sylvilagus floridanus*). The virus was detected in the cells when after 14 days incubation focal areas of cell destruction were observed in the cell monolayers. The virus has since been successfully propagated in kidney cells of both cottontail and domestic rabbits.

b. ETIOLOGY. Hinze has clearly demonstrated (70) that his isolate of the virus designated CHV-1 possesses physical, chemical, and biological properties that justify its inclusion in the herpesvirus group. He has proposed that the name *Herpesvirus sylvilagus* be adopted for his virus.

The virus can be readily propagated in cells of both cottontail and domestic rabbits but failed to grow in cells derived from man, monkeys, hamsters, and mice, nor could it be propagated in chicken embyros. Hinze has shown that this virus possesses no antigenic relationship to *Herpesvirus cuniculi* or to any of the other four mammalian herpesviruses with which he compared it.

c. INCIDENCE. The natural incidence of this virus in wild cottontail rabbit populations has not been determined.

d. CLINICAL SIGNS. Inoculation of cottontail rabbits by the subcutaneous, intradermal, or intraperitoneal route results in chronic infection with persistent low grade viremia believed to persist for the lifetime of the rabbit (71). Rabbits persistently infected with this virus show a pronounced leukocytosis and lymphocytosis, with differential lymphocyte counts of up to 95% compared to 50% to 60% in normal rabbits (69). Repeated attempts to infect domestic New Zealand White rabbits have met with failure. All the available evidence suggests that only rabbits of the genus *Sylvilagus* are susceptible to this virus.

e. PATHOLOGY. The lesions observed in cottontail rabbits infected with *Herpesvirus sylvilagus* are exclusively changes in the lymphoid system (69, 71). Extensive infiltration of various tissues with immature lymphocytes at various stages of development is observed 6 to 8 weeks following experimental inoculation. The intensity of the experimental lymphoproliferative disease varies from benign lymphoid hyperplasia to lesions consistent with a diagnosis of malignant lymphoma.

The etiological role of the herpesvirus in this syndrome has not been conclusively established but considerable evidence supports the contention that the virus is involved in the syndrome as an inducing agent.

Both *Herpesvirus cuniculi* of the domestic rabbit and *Herpesvirus sylvilagus* of cottontail rabbits possess the capacity to persist in their infected hosts in a state of subclinical infection. The presence of such infections, if not recognized, could result in considerable confusion in experimental results especially if the experimental conditions are such that they act to stimulate recrudescence of the latent viruses.

Steps should thus be taken to ensure that rabbit colonies of high repute are free of infection with the known viruses of this group.

A viral agent has recently been isolated from the nasal cavities of European rabbits with upper respiratory disease (131). Preliminary investigations have revealed this virus to be an ether-sensitive DNA virus measuring approximately 100 nm in diameter. On the basis of these and other characteristics the authors have tentatively concluded that the new virus is a herpesvirus. Serological comparisons of this virus with either *H. cuniculi* or *H. sylvilagus* have not yet been reported.

Should further investigation prove this agent to be a herpesvirus, this will be the first indication that herpesviruses are disease-producing agents in rabbits.

C. Papovavirus Infections

The virus family Papovaviridae has been subdivided into two genera, the genus *Papillomavirus* and the genus *Polyomavirus*. The former genus consists entirely of viruses which cause papillomata (warts) of various animal species. Two rabbit viruses belong to the genus *Papillomavirus*, namely the rabbit (Shope) papilloma virus and the rabbit oral papilloma virus. The genus *Polyomavirus* contains one virus of rabbits, the rabbit vacuolating virus.

1. Rabbit (Shope) Papilloma Virus

a. HISTORY. The virus now known as the rabbit or Shope papilloma virus was first recognized as a transmissible agent by Shope and Hurst (150) while attempting to define the etiology of wart-like tumors observed on cottontail rabbits in the midwestern United States. They were able to demonstrate the filterability of the infectious agent, transmitted it to other cottontails as well as domestic rabbits (147), and briefly described the histopathology of the original disease.

Rabbit papillomatosis was, until quite recently, considered to be a natural benign disease of cottontail rabbits only. Spontaneous outbreaks of papillomatosis have, however, recently been reported by Hagen (60), indicating that this disease is indeed of relevance to the rabbit producer.

Soon after discovery of the rabbit papilloma virus it was shown to cause not only the benign papillomata from which it was originally isolated but also malignant neoplasms histologically resembling squamous cell carcinomata (89, 141, 142). Since that time the rabbit papilloma virus has served as a model system in numerous investigations aimed at the elucidation of the role of viruses in the etiology of cancer in man and animals (33, 34).

b. ETIOLOGY. The rabbit (Shope) papilloma virus is the type species of the genus *Papillomavirus* of the Papovaviridae (167) and possesses the characteristic circular DNA genome, cubic symmetry, and other chemical and physical characteristics belonging to viruses of this family (85). A brief summary of these characteristics has been presented by Andrewes and Pereira (6). The rabbit papilloma virus is antigenically distinct from other members of the genus *Papillomavirus*. The rabbit papilloma virus has been propagated in rabbit skin cell cultures (21) and in organ cultures derived from newborn domestic rabbits (28) and in transplants of embryonic rabbit skin (54). Distinct cellular proliferation was observed in these cultures, but increase of free virus could not be demonstrated. Several workers (80, 171) have been able to propagate cottontail rabbit papilloma cells *in vitro* and have readily demonstrated viral antigen by immunofluorescence but not by rabbit inoculation. The virus can be maintained by serial propagation of the skin of cottontail rabbits (150) inoculated intracutaneously or scarified with the virus. Intramuscular inoculation does not result in clinical papillomatosis. Dogs, cats, pigs, goats, rats, mice, and guinea pigs were found to be refractory to the virus. Greene (55) has, however, shown that the embryonic rat skin is susceptible to the papilloma virus, and that typical papillomata may develop following inoculation.

c. INCIDENCE. Shope has observed that rabbit papillomatosis most frequently occurs as a natural disease of cottontail rabbits in the midwestern United States extending from Minnesota and South Dakota in the north through Iowa, Nebraska, Kansas, and Missouri and into Oklahoma and Texas (59, 60). The virus has apparently not become established in the eastern states.

The only natural outbreaks of the disease thus far reported in domestic rabbits were observed in southern California by Hagen (60), who suggested that the virus must be present in the wild cottontail population of the area from which it spread to domestic rabbits via arthropod vectors.

d. EPIDEMIOLOGY. The cottontail rabbit is the natural host of rabbit papillomatosis but the domestic rabbit has also been shown to be susceptible to the virus (147) and natural outbreaks of the disease have been reported in commercial rabbitries (60). The jackrabbit (*Lepus californicus*) has also been shown to be susceptible to the virus (10).

Infection of the skin of domestic rabbits results in the formation of papillomata essentially devoid of infective virus (147). The papillomata can however be serially transferred in rabbits, but infectious virus can seldom be demonstrated (144, 147) and thus infected domestic rabbits are not a source of virus for arthropod vectors. The findings of the Russian worker Nartsissov do not support Shope's early observations. Working with virus obtained from Shope he has been able to serially passage the virus in domestic rabbits and can consistently demonstrate infective virus in the experimentally produced tumors [cited by Gross (59)].

Infection of cottontail rabbits results in the production of lesions containing a high concentration of virus. Various arthropod vectors can readily pick up the virus and transfer it to other susceptible hosts. Contact transmission of papillomatosis may occur but transmission of the virus by the rabbit tick, *Haemaphysalis leporis-palustris*, is probably the most common natural mode of transmission (97). Transmission by mosquitoes and reduviid bugs has been demonstrated under experimental conditions (24) and it has been suggested that in view of the low frequency of incidence of ticks and bugs in commercial rabbitries, the mosquito may in fact be the principal vector between wild cottontail rabbits and domestic rabbits (60). This hypothesis is strengthened by the observation that in the natural cases of the disease in domestic rabbits, lesions were confined to the relatively hairless parts of the body around the eyes, ears, and anus, areas where mosquitoes are apt to bite.

It has also been suggested that nematodes may be involved in the natural transmission of rabbit papillomatosis (130). Under experimental conditions papillomata were produced when papilloma virus plus the larvae of the nematode *Nippostrongylus muris* were applied to rabbit skin, but not by virus or the nematode alone.

e. PUBLIC HEALTH SIGNIFICANCE. Man is not known to be susceptible to the rabbit papilloma virus.

f. CLINICAL SIGNS AND GROSS LESIONS. Papillomatosis of wild cottontail rabbits is characterized by the presence of horny warts usually on the neck, shoulders, or abdomen of the affected rabbit. These warts start off as red raised areas at the site of infection, they grow to become typical papillomata with rough rounded surfaces and may later develop into large keratinized horny growths (150). The virus was initially believed to cause only transient papillomatosis but it was later shown that in naturally infected cottontail rabbits the papillomata may become malignant being replaced by squamous-cell carcinomata (159). This phenomenon was later shown to be a relatively frequent occurrence in both naturally infected and experimentally infected cottontail rabbits when Syverton *et al.* (160) found that 25% of infected cottontails developed carcinomatous lesions following papilloma virus infection. In approximately 35% of naturally infected rabbits, papilloma lesions had disappeared within six months after infection.

The clinical signs observed in natural outbreaks of papillomatosis of domestic rabbits in southern California have been described by Hagen (60). The papillomata most commonly occurred on the eyelids (Fig. 8) and ears (Fig. 9) and were described by Hagen as follows: "The growths were well keratinized and the upper surfaces were irregular and often split. The lower portions of the growth were fleshy to the touch and pinkish in color. The lateral surfaces appeared to be striated. As the lesions became older they increased in size, became more cornified and were hard to the touch. On the cut section, the growth had a pink fleshy center. These growths were easily scratched off by the rabbits or knocked off when the rabbits were handled. Growths removed this way left a free-bleeding surface which usually healed without complications. On two occasions, a second growth appeared at the site of the original lesion."

Experimentally induced papillomata developed more slowly in domestic rabbits than in cottontail rabbits, reached a stationary phase, and then occasionally regressed. Rous and Beard (141, 142) were the first to recognize the malignant potential of the rabbit papilloma virus when they demonstrated that inoculation of papillomatous tissue intramuscularly into domestic rabbits resulted in the development of invasive tumors histologically identifiable as squamous-cell carcinomata. Syverton (158) later showed that 75% of experimentally infected domestic rabbits would develop carcinomata if kept for longer than six months.

g. MICROSCOPIC LESIONS. The benign warts which develop following infection with the rabbit papilloma virus in cottontail and domestic rabbits are typical papillomata. The malignant tumors which may arise from such papillomata are squamous-cell carcinomata. A detailed description of the histopathology of rabbit fibromatosis appears elsewhere in this book.

h. DIAGNOSIS. Rabbit papillomatosis is readily diagnosed clinically by the characteristic skin tumors which never occur in the mouth, and may be confirmed by histopathological examination.

i. CONTROL AND PREVENTION. The endemic natural disease of cottontail rabbits is of little economic significance and thus no prophylactic methods have been developed. In view of the recent evidence that natural infection does occur in domestic rabbits (60), the adoption of control methods may become necessary. In areas where the disease is endemic in the wild rabbits, where arthropod vectors are present, and where outdoor rabbit husbandry is practiced, arthropod control would appear to be one logical approach. Rabbits could also be immunized by two intraperitoneal inoculations with glycerinated rabbit papilloma suspensions as described by Shope (149). Hagen (60) showed that domestic rabbits with experimentally induced papillomata resisted challenge with virus derived from cottontail papillomata.

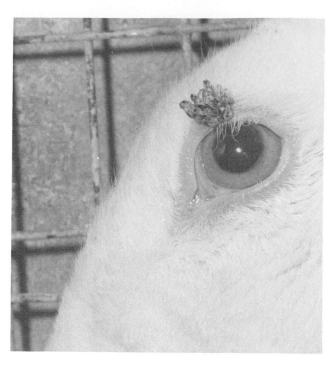

Fig. 8. Papilloma on the eyelid of a naturally infected European rabbit. (Courtesy of Dr. K. W. Hagen).

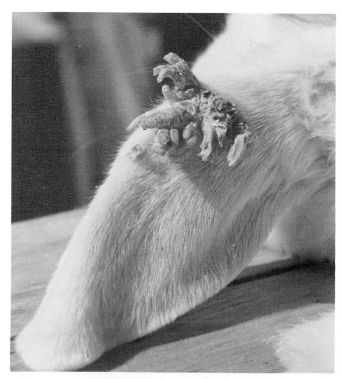

Fig. 9. Large papilloma on ear of naturally infected European rabbit. (Courtesy of Dr. K. W. Hagen).

2. Rabbit Oral Papilloma Virus

a. HISTORY. Oral papillomatosis of rabbits was first recognized as a distinct viral disease of domestic rabbits by Parsons and Kidd (119). They observed a 17% incidence of small papillomata in the mouths of rabbits of several breeds in the New York City area. The lesions were usually confined to the lower surface of the tongue. They were able to demonstrate that the causative virus could be transmitted to both domestic and cottontail rabbits, and by cross-immunity tests and tissue-susceptibility studies demonstrated that the virus was distinct from the rabbit (Shope) papilloma virus.

In subsequent investigations Parsons and Kidd (120) repeated some of their early findings and showed in addition that virus is frequently present in the mouths of rabbits without lesions. They also demonstrated that tattooing or the licking of tar may stimulate lesion development in such carrier rabbits.

A spontaneous outbreak of rabbit oral papillomatosis, once again involving several breeds of rabbits, was recently described from New York State (166).

b. ETIOLOGY. The rabbit oral papilloma virus was partly characterized by Parsons and Kidd (120) and was later placed in the papovavirus group by Melnick (107). It is now classed as a member of the genus *Papillomavirus* of the family Papovaviridae (167). The virus has been shown to be immunologically distinct from the rabbit (Shope) papilloma virus and to infect only leporids (119, 120). Nuclear viral replication characteristic for viruses of the papovirus group has been demonstrated by Richter *et al.* (132) and Rdzok *et al.* (127). The virus has not been successfully propagated outside of the intact susceptible host animal.

c. INCIDENCE. The disease has been described from New York State and Massachusetts only.

d. EPIDEMIOLOGY. The virus probably spreads by direct contact and lesion development may require oral trauma to allow entry of the virus. Excessively coarse food, maloccluded teeth, or other oral irritants may serve as the trigger mechanism (120, 166). Experimentally induced lesions appear from 9 to 38 days postinoculation (119), but the natural incubation period of the disease is not known. Weisbroth and Scher (166) have noted that the disease generally seems to occur in rabbits between the ages of 2 and 18 months and not in younger or older rabbits.

c. PUBLIC HEALTH SIGNIFICANCE. None, only leporids are susceptible.

f. CLINICAL SIGNS AND GROSS LESIONS. Rabbit oral papillomatosis is a benign disease characterized clinically by the appearance of numerous discrete whitish growths on the underside of the tongue (Fig. 10). The early lesions

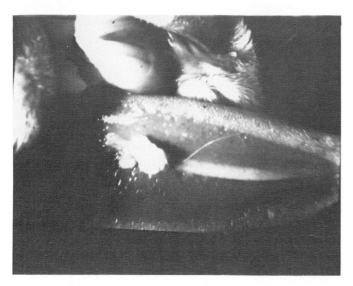

Fig. 10. Oral papilloma on ventral aspect of the tongue of a European rabbit. (Courtesy of Dr. S. H. Weisbroth.)

are sessile, later become rugose or pedunculated and ultimately ulcerate (119, 166). They are seldom more than 5 mm in diameter and 4 mm high and are usually substantially smaller than this. The growths have been known to persist for as long as 145 days but usually they disappear in a matter of weeks (119). Lesions may rarely occur elsewhere in the mouth but never elsewhere on the body.

g. MICROSCOPIC LESIONS. The lesions are microscopically typical papillomata and have been described in detail by Parsons and Kidd (120), Richter *et al.* (132), Rdzok *et al.* (127), and Weisbroth and Scher (166). These lesions are also described in this book in the chapter dealing with tumors of rabbits.

h. DIAGNOSIS. The disease is diagnosed on the basis of the typical lesions occurring only in the mouth in contrast to rabbit papillomatosis in which lesions are not observed in the mouth but only on the skin.

i. IMMUNITY. Rabbits recovering from the disease are resistant to reinfection but remain fully susceptible to the unrelated rabbit papilloma virus.

3. Rabbit Kidney Vacuolating Virus

The rabbit kidney vacuolating (RKV) virus was isolated by Hartley and Rowe (65) in primary rabbit kidney cell cultures from cottontail papillomata collected in Kansas. The virus caused distinct vacuolar cytopathic effects in cell cultures and from this derived its name.

The virus resembles the rabbit (Shope) papilloma virus, but has been shown to be a distinct entity. It does not produce papillomata when inoculated into rabbits and does not

immunize rabbits against the rabbit papilloma virus. The virus is slightly smaller than the viruses of the genus *Papillomavirus* and resembles the polyomavirus in its size, morphology, and DNA composition (23). On the basis of present knowledge this virus has been classed as a member of the genus *Polyomavirus* within the family Papovaviridae (167). A detailed ultrastructural study of the replication of RKV virus shows it to be a typical papovavirus in its replication pattern (16).

Thus far the RKV virus has not been shown to be pathogenic for either domestic or cottontail rabbits or any other animal species. Inoculation of both infant and adult domestic and cottontail rabbits by several routes has resulted in no evidence of disease (65).

Antibodies to RKV virus have been found in wild cottontail rabbits in Kansas and Maryland, but not in domestic rabbits (65).

The RKV virus thus appears to be a fairly common nonpathogenic virus of cottontail rabbits causing only latent infections.

D. Adenovirus Infections

Adenoviruses have been isolated from man and from several species of domestic and wild animals. They are frequently associated with upper respiratory disease in the animals they infect but may cause severe generalized and even fatal disease as exemplified by infectious canine hepatitis. This group of viruses is currently receiving considerable attention because of their proven capacity to induce malignant tumors in several species of laboratory animals, notably baby hamsters.

Adenovirus isolation from rabbits has not been reported, but it has been shown that under experimental conditions rabbits are readily infected with human adenovirus type 5. Such infection results in no clinical response but a persistent viral infection is induced in the rabbit from which virus can be reisolated for as long as one year after exposure (128). Since adenovirus infection in man is common and frequently respiratory in nature, the possibility of infection of rabbits by man should not be overlooked. Such infection may result in the persistence of latent adenovirus which may confuse experimental findings.

III. DISEASES CAUSED BY RNA VIRUSES

A. Picornavirus Infections

The small RNA viruses making up the picornavirus group have been isolated from man and most of his domestic animals. Most picornaviruses inhabit the gastrointestinal tract of the host species, being particularly common in young animals. They seldom cause disease of the intestinal tract but may infect the central nervous system of the host animals causing severe and often fatal disease. Examples of picornavirus infections of the central nervous system are poliomyelitis of man, Teschen disease of swine, and avian encephalomyelitis. The rhinoviruses, a subgroup of the picornavirus group, characteristically infect the upper respiratory tract causing mild to severe respiratory disease. Rhinoviruses are the cause of the "common cold" in man and also cause disease in cattle and horses.

Surprisingly, no members of the picornavirus group have as yet been isolated from rabbits. Since picornaviruses are extremely common inhabitants of the intestinal tract of man and many species of animals it seems unlikely that the rabbit does not harbor viruses belonging to this group. It is the opinion of the author that in time rabbits, too, will be shown to possess a spectrum of picornaviruses resembling those of other species of animals. In outbreaks of respiratory or central nervous system diseases in rabbits the possibility that picornaviruses may be the primary etiologic agents should be considered.

B. Reovirus Infections

The reovirus group is a newly recognized group of viruses containing agents associated with upper respiratory disease and possibly diarrhea in man, chimpanzees, and cattle. No reoviruses have as yet been isolated from rabbits, but naturally occurring antibodies to these viruses have been detected in rabbits (140, 155). No disease has been associated with such natural infection of rabbits.

C. Orthomyxovirus Infections

The orthomyxovirus group is the home of the influenza viruses which infect many species of animals, usually causing disease of the upper respiratory tract. Under natural conditions, influenza viruses cause disease of man, swine, horses, and several species of birds. Many species of laboratory animals are susceptible to the influenza viruses under experimental conditions, but the rabbit appears to be completely refractory to viruses of this group (48, 78).

D. Paramyxovirus Infections

The rabbit synctial virus was isolated by Morris *et al.* (109) from one of thirty-five trapped cottontail rabbits (*S. floridanus*) in Virginia. The only lesions of note in the rabbit from which the virus was isolated were pronounced ascites and hydatid cysts in the liver. The virus was isolated in chicken embryos and on rhesus monkey kidney cells. The syncytial cytopathic effect observed in cell cultures is the basis for the name adopted for this virus.

The virus has been shown to be ether-sensitive, antibiotic-resistant, and 130 to 180 nm in diameter, and it possesses an RNA genome. These characteristics justify only tentative inclusion of this virus in the paramyxovirus group. The virus is serologically distinct from influenza, mumps, parainfluenza-3, SV-5, peromyscus, and respiratory syncytial viruses.

Suckling Swiss mice are highly susceptible to the virus but weaned mice, guinea pigs, European rabbits, and cottontail rabbits developed no symptoms or lesions following inoculation with the virus. Antibodies did, however, develop in these animals. Eight of 25 wild-caught cottontail rabbits had antibodies to the disease but no antibodies could be demonstrated in the sera of seven other species of wild animals trapped in the same area or in human sera, indicating that the virus probably has a narrow host range in nature.

E. Rhabdovirus Infections

The rhabdovirus group is a rapidly growing group of viruses encompassing viruses of mammals, birds, reptiles, arthropods, and possibly plants. The two viruses which cause disease in a variety of mammals and which may cause disease in rabbits are the vesicular stomatitis virus and the rabies virus.

Vesicular stomatitis virus has been shown under experimental conditions to infect rabbits (6), but natural infection of rabbits with this virus has not been reported. In view of the fact that the virus is arthopod transmitted, infection of rabbits in endemic areas (e.g., the southeastern United States) is a distinct possibility.

The rabies virus kills rabbits in much the same way as it kills most other mammals and natural cases of the disease in rabbits, while extremely rare, do occur (57, 168). The appearance of disease involving the central nervous system of wild rabbits living in close association with populations of mustelids and viverrids should arouse suspicion of rabies infection. The rabies virus is reported to be endemic in certain populations of jackrabbits (probably *Lepus townsendi*) in the northwestern United States where they serve as a reservoir for infection of domestic animals (168).

F. Togavirus (Arbovirus) Infections

The togavirus group is the largest group of animal viruses and encompasses principally viruses which are transmitted by arthropods and which replicate both in the arthropod vector and in the mammalian or avian host. Togavirus infections of birds and animals commonly result in disease characterized by central nervous system involvement.

Natural disease of leporids resulting from togavirus infection has not been reported but natural infection with several togaviruses commonly occurs.

The California encephalitis (CEV) subgroup of the toga-viruses includes at least eleven different but closely related viruses some of which may cause serious disease in man.

The first virus of this group to be isolated from leporids was the "snowshoe hare" virus isolated in 1959 from the blood of a sick snowshoe hare (*Lepus americanus*) in western Montana (15). Earlier experiments had shown that CEV antibodies were present in cottontail and jackrabbits in California (62). Subsequent investigations have revealed that the CEV group of togaviruses are prevalent in the snowshoe hare population of North America, antibody levels ranging from 58% to 95% of adult hares tested (73, 116). The virus has also been isolated from the rabbit tick, *Haemaphysalis leporis-palustris* (116).

Antibody studies in Europe reveal that the European hare (*L. europeaus*) and wild European rabbit are both prob-ably involved in the natural maintenance of Ťahyňa virus, a member of the California encephalitis group (9, 27, 64).

As yet clinical disease resulting from CEV infection in hares has not been reported. Experimental infection of European rabbits results in viremia and antibody formation but no clinical disease (63). Caged domestic rabbits have been shown to be excellent indicators of CEV activity since they develop CEV antibodies and serve as a source for virus isolation (102).

Wild rabbits of the genus *Sylvilagus* as well as hares (*L. californicus* and *L. americanus*) have been shown by anti-body surveys to be naturally susceptible to the western equine encephalomyelitis (WEE) virus (14, 174). It is not known whether the virus causes disease in these lago-morphs or whether they play an important role in the maintenance of the virus in nature. Antibodies to the eastern equine encephalomyelitis (EEE) virus and the St. Louis encephalitis (SLE) virus have also been detected in snowshoe hares (174).

The Silverwater virus (SWV), a togavirus of unknown pathogenicity, has been isolated from snowshoe hares and from *Haemaphysalis leporis-palustris* ticks taken from snow-shoe hares in Ontario and Alberta. The hare is believed to play a central role in the natural cycle of this virus (103, 174). No disease of hares resulting from infection with this virus has been reported.

Antibodies to the Čalovo virus, a member of the Bunyam-wera subgroup of togaviruses have been detected in Euro-pean hares, but no disease has been reported to result from infection with this virus.

While overt disease associated with togavirus infection in hares has not been reported, Youill *et al.* (174) have found that a significant decline in the snowshoe hare population occurs during periods of high togavirus activity. Further-more he presents evidence indicating that the mortality is in fact occurring in those hares which have been exposed to the viruses, CEV, WEE, and SWV being the predominant viruses in the populations which he studied.

IV. DISEASES POSSIBLY OF VIRAL ORIGIN

Infectious Vesicular Stomatitis

Sprehn in his 1956 edition of the German Language book on rabbit diseases "Kaninchen-Krankheiten" (154) des-cribes a highly fatal disease of rabbits which he refers to as "Stomatitis Vesiculosa Infectiosa." This disease, which may cause 50% mortality on infected premises, is character-ized by an acute vesicular stomatitis accompanied by exces-sive salivation which characteristically results in matting of the hair in the region of the mouth and in the area of the throat and the chest. Lesion in the oral mucosa range from "millet-seed" size to lentil-sized vesicles filled with clear fibrinous fluid. Lesions may also be observed on the lips, the tongue, and the hard palate, and may even occur on the external genitalia. Primary infection may be followed by secondary bacterial infections often resulting in necrosis of the lips and tongue which is accompanied by a fetid odor.

Infected animals show anorexia, depression, and in the later stages of the disease-marked debilitation often fol-lowed by death. This disease can be readily confused with stomatitis due to fungus-contaminated feed, chemical irritants, or poisonous plants.

Sprehn describes the etiological agent of this disease as a heat-sensitive virus measuring 100–150 nm in diameter. In the absence of further information on this virus the author considers this condition at this time as being a disease of uncertain etiology, "possibly of viral origin."

REFERENCES

1. Ahlström, C. G. (1938). The histology of the infectious fibroma in rabbits. *J. Pathol. Bacteriol.* **46**, 461–472.
2. Andrewes, C. H. (1928). A study of virus III with special reference to the response of immunized rabbits to reinoculation. *J. Pathol. Bacteriol.* **31**, 461–471.
3. Andrewes, C. H. (1936). A change in rabbit fibroma virus suggesting mutation. I. Experiments on domestic rabbits. *J. Exp. Med.* **63**, 157–172.
4. Andrewes, C. H., and Horstmann, D. M. (1949). The susceptibility of viruses to ethyl ether. *J. Gen. Microbiol.* **3**, 290–297.
5. Andrewes, C. H., and Miller, C. P., Jr. (1924). A filterable virus infec-tion of rabbits, II. Its occurrence in apparently normal rabbits. *J. Exp. Med.* **40**, 789–796.
6. Andrewes, C. H., and Pereira, H. G. (1967). "Viruses of Vertebrates." Williams & Wilkins, Baltimore, Maryland.
7. Andrewes, C. H., Thompson, H. V., and Mansi, W. (1959). Myxoma-tosis: Present position and future prospects in Great Britain. *Nature* (*London*) **184**, 1179–1180.
8. Aragão, H. de B. (1943). O virus do mixoma no coelho do mato (*Sylvilagus minensis*), sua transmissão pelso *Aedes scapularis* e aegypti. *Mem. Inst. Oswaldo Cruz* **38**, 93.
9. Bardos, V. (1965). On the ecological problems of Tahyňa and Čalovo virus. *Proc. Symp. Czech. Acad. Sci.* pp. 411–422.

10. Beard, J. W. and Rous, P. (1935). Effectiveness of the Shope papilloma virus in various American rabbits. *Proc. Soc. Exp. Biol. Med.* **33**, 191–193.

11. Bedson, H. S., and Duckworth, M. J. (1963). Rabbit pox: An experimental study of the pathways of infection in rabbits. *J. Pathol. Bacteriol.* **85**, 1–20.

12. Berry, G. P., and Dedrick, H. M. (1936). A method for changing the virus of rabbit fibroma (Shope) into that of infectious myxomatosis (Sanarelli). *J. Bacteriol.* **31**, 50–51.

13. Borg, K., and Bakos, K. (1963). Dissemination of myxomatosis by birds. *Nord. Veterinaer med.* **15**, 159–166.

14. Bowers, J. H., Hayes, R. O., and Hughes, T. B. (1969). Studies on the role of mammals in the natural history of western encephalitis in Hale County, Texas, *J. Med. Entomol.* **6**, 175–178.

15. Burgdorfer, W., Newhouse, V. F., and Thomas, L. A. (1961). Isolation of California encephalitis virus from the blood of a snowshoe hare (*Lepus americanus*) in western Montana. *Amer. J. Hyg.* **73**, 344–349.

16. Chambers, V. C., Hsia, S., and Ito, Y. (1966). Rabbit kidney vacuolating virus: Ultrastructural studies. *Virology* **29**, 32–43.

17. Chapple, P. J., and Bowen, E. T. W. (1963). A note on two attenuated strains of myxoma virus isolated in Great Britain. *J. Hyg.* **61**, 161–168.

18. Chapple, P. J., and Lewis, N. D. (1965). Myxomatosis and the rabbit flea. *Nature (London)* **207**, 388–389.

19. Chaproniere, D. M., and Andrewes, C. H. (1957). Cultivation of rabbit myxoma and fibroma viruses in tissues of non-susceptible hosts. *Virology* **4**, 351–360.

20. Christensen, L. R., Bond, E., and Matanic, B. (1967). Pockless rabbit pox. *Lab. Anim. Care* **17**, 281–296.

21. Coman, D. R. (1946). Induction of neoplasia *in vitro* with a virus. Experiments with rabbit skin grown in tissue culture and treated with Shope papilloma virus. *Cancer Res.* **6**, 602–607.

22. Constantin, T., Febvre, H., and Harel, J. (1956). Cycle de multiplication due virus fibrome de Shope *in vitro* (souche OA). *C. R. Acad. Sci.* **150**, 347–348.

23. Crawford, L. V., and Follett, E. A. C. (1967). A study of rabbit kidney vacuolating virus and its DNA. *J. Gen. Virol.* **1**, 19–24.

24. Dalmat, H. T. (1958). Arthropod transmission of rabbit papillomatosis. *J. Exp. Med.* **108**, 9–20.

25. Dalmat, H. T. (1959). Arthropod transmission of rabbit fibromatosis (Shope). *J. Hyg.* **57**, 1–29.

26. Dalmat, H. T., and Stanton, M. F. (1959). A comparative study of the Shope fibroma in rabbits in relation to transmissibility by mosquitoes. *J. Nat. Cancer. Inst.* **22**, 593–615.

27. Danielová, V., Kolman, J. M., Málková, D., Marhoul, Z., and Smetana, A. (1969). Natural focus of Ťahyňa virus in South Moravia. *Proc. Symp. Arbov. Calif. Bunyamwera Group, Slov. Acad. Sci. Bratislava* pp. 147–150.

28. De Maeyer, E. (1962). Organ cultures of newborn rabbit skin affected by rabbit papilloma virus. *Science* **136**, 985–986.

29. Dungern, von E., and Coca, A. F. (1909). Ueber hasensarkome, die in Kanincher wachsen und über das Wesen der Geschwulstimmunität. *Z. Immunitaetsforsch. Exp. Ther.* **2**, 391–398.

30. Duran-Reynals, F. (1940). Production of degenerative inflammatory or neoplastic effects in the newborn rabbit by the Shope fibroma virus. *Yale J. Biol. Med.* **13**, 99–110.

31. Duran-Reynals, F. (1945). Immunological factors that influence the neoplastic effects of the rabbit fibroma virus. *Cancer Res.* **5**, 25–39.

32. Dyce, A. L. (1961). Transmission of myxomatosis on the spines of thistles, *Cirsium vulgare* (savi) Ten. *CSIRO Wildl. Res.* **6**, 88–90.

33. Evans, C. A., Gorman, L. P., and Ito, Y. (1962). Antitumor immunity in the Shope papilloma-carcinoma complex of rabbits. II. Suppression of a transplanted carcinoma, VX7, by homologous papilloma vaccine. *J. Nat. Cancer Inst.* **29**, 287–292.

34. Evans, C. A., and Thomsen, J. J. (1969). Antitumor immunity in the Shope papilloma-carcinoma complex. IV. Search for a transmissable factor increasing the frequency of tumor regression. *J. Nat. Cancer Inst.* **42**, 477–484.

35. Fenner, F. (1953). Classification of myxoma and fibroma viruses. *Nature (London)* **171**, 562–563.

36. Fenner, F. (1958). The biological characters of several strains of vaccinia, cowpox and rabbit pox viruses. *Virology* **5**, 502–529.

37. Fenner, F. (1962). The reactivation of animal viruses. *Brit. Med. J.* **2**, 135–142.

38. Fenner, F. (1965). Viruses of the myxoma-fibroma subgroup of poxviruses. II. Comparison of soluble antigens by gel-diffusion tests, and a general discussion of the subgroup. *Aust. J. Exp. Biol. Med. Sci.* **43**, 143–156.

39. Fenner, F., and Chapple, P. J. (1965). Evolutionary changes in myxoma virus in Britain. An examination of 222 naturally occurring strains obtained from 80 counties during the period October–November, 1962. *J. Hyg.* **63**, 175–185.

40. Fenner, F., and McIntyre, G. A. (1956). Infectivity titrations of myxoma virus in the rabbit and the developing chick embryo. *J. Hyg.* **54**, 246–257.

41. Fenner, F., and Marshall, I. D. (1957). A comparison of the virulence for European rabbits (*Oryctolagus cuniculus*) of strains of myxoma virus recovered in the field in Australia, Europe and America. *J. Hyg.* **55**, 149–191.

42. Fenner, F., Poole, W. E., Marshall, I. D., and Dyce, A. L. (1957). Studies in the epidemiology of infectious myxomatosis of rabbits. VI. The experimental introduction of the European strain of virus into Australian wild rabbit populations. *J. Hyg.* **55**, 192–206.

43. Fenner, F., and Ratcliffe, F. N. (1965). "Myxomatosis." Cambridge Univ. Press, London and New York.

44. Fenner, F., and Sambrook, J. F. (1966). Conditional lethal mutants of rabbitpox virus. II. Mutants (*p*) which fail to multiply in PK-2a cells. *Virology* **28**, 600–609.

45. Fenner, F., and Woodroofe, T. M. (1953). The pathogenesis of infectious myxomatosis: The mechanism of infection and the immunological response in the European rabbit (*Oryctolagus cuniculus*). *Brit. J. Exp. Pathol.* **34**, 400–411.

46. Fenner, F., and Woodroofe, T. M. (1965). Changes in the virulence and antigenic structure of strains of myxoma virus recovered from Australian wild rabbits between 1950 and 1964. *Aust. J. Exp. Biol. Med. Sci.* **43**, 359–370.

47. Fisher, E. R. (1953). The nature and staining reactions of the fibroma-cell inclusions of the Shope fibroma of the rabbit. *J. Nat. Cancer Inst.* **14**, 355–364.

48. Francis, T., and Magill, T. P. (1935). Immunological studies with the virus of influenza. *J. Exp. Med.* **62**, 505–516.

49. Gordon, D. E., and Olson, C. (1968). Meningiomas and fibroblastic neoplasia in calves induced with the bovine papilloma virus. *Cancer Res.* **28**, 2423–2431.

50. Greene, H. S. N. (1933). A pandemic of rabbitpox. *Proc. Soc. Exp. Biol. Med.* **30**, 892–894.

51. Greene, H. S. N. (1934). Rabbit Pox. I. Clinical manifestations and course of the disease. *J. Exp. Med.* **60**, 427–440.

52. Greene, H. S. N. (1934). Rabbit Pox. II. Pathology of the epidemic disease *J. Exp. Med.* **60**, 441–457.

53. Greene, H. S. N. (1935). Rabbit Pox. III. Report of an epidemic with special reference to epidemiological factors. *J. Exp. Med.* **61**, 807–831.

54. Greene, H. S. N. (1953). The induction of Shope papilloma in transplants of embryonic rabbit skin. *Cancer Res.* **13**, 58–63.

55. Greene, H. S. N. (1953). The induction of the Shope papilloma in homologous transplants of embryonic rat skin. *Cancer Res.* **13**, 681–683.

56. Griffith, J. F., Kibrick, S., and Dodge, P. R. (1968). EEG studies of recurrent herpetic infection in rabbits. *Neurology* **18**, 308.

57. Grimes, G. M. (1968). Epidemiology of rabies in South Bavaria. *Mil. Bull. U.S. Army, Europe* **25**.

58. Grodhaus, G., Regnery, D. C., and Marshall, I. D. (1963). Studies on the epidemiology of myxomatosis in California. II. The experimental transmission of myxomatosis in brush rabbits (*Sylvilagus bachmani*) by several species of mosquitoes. *Amer. J. Hyg.* **77**, 205–212.

59. Gross, L. (1970). "Oncogenic Viruses." Pergamon, Oxford.

60. Hagen, K. W. (1966). Spontaneous papillomatosis in domestic rabbits. *Bull. Wildl. Dis. Ass.* **2**, 108–110.

61. Hahon, N. (1961). Smallpox and related pox-virus infections in the simian host. *Bacteriol. Rev.* **25**, 459–476.

62. Hammon, W. McD., and Reeves, W. C. (1952). California encephalitis virus, a newly described agent. I. Evidence of natural infection in man and other animals. *Calif. Med.* **77**, 303–309.

63. Hammon, W. McD., and Sather, G. E. (1966). History and recent reappearance of viruses in the California encephalitis group. *Amer. J. Trop. Med.* **15**, 199–204.

64. Hannoun, C., Panthier, R., and Corniou, R. (1969). Serological and virological evidence of the endemic activity of Tahyña virus in France. *Proc. Symp. Arbov. Calif. Bunyamwera Group, Slov. Acad. Sci., Bratislava* pp. 121–125.

65. Hartley, J. W., and Rowe, W. P. (1964). New *Papovavirus* contaminating Shope papillomata. *Science* **143**, 258–261.

66. Hellström, I., Evans, C. A., and Hellström, K. E. (1969). Cellular immunity and its serum mediated inhibition in Shope-virus-induced rabbit papillomas. *Int. J. Cancer* **4**, 601–607.

67. Herman, C. M., Kilham, L., and Warbach, O. (1956). Incidence of Shope's rabbit fibroma in cottontails at the Patuxent Research Refuge. *J. Wildl. Manage.* **20**, 85–89.

68. Hinze, H. C. (1968). Isolation of a new herpesvirus from cottontail rabbits. *Bacteriol. Proc.* p. 149.

69. Hinze, H. C. (1969). Rabbit lymphoma induced by a new herpesvirus. *Bacteriol. Proc.* p. 157.

70. Hinze, H. C. (1971). New member of the herpesvirus group isolated from wild cottontail rabbits. *Infec. Immunity* **3**, 350–354.

71. Hinze, H. C. (1971). Induction of lymphoid hyperplasia and lymphoma-like disease in rabbits by *Herpesvirus sylvilagus*. *Int. J. Cancer* **8**, 514–522.

72. Hinze, H. C., and Walker, D. L. (1964). Response of cultured rabbit cells to infection with the Shope fibroma virus. I. Proliferation and morphological alteration of the infected cells. *J. Bacteriol.* **88**, 1185–1194.

73. Hoff, G. L., Yuill, T. M., and Iversen, J. O. (1969). Snowshoe hares and the California encephalitis virus group in Alberta, 1961–1968. *Bull. Wildl. Dis. Ass.* **5**, 254–259.

74. Hu, C. K., Rosahn, P. D., and Pearce, L. (1936). Studies on the etiology of rabbit pox. III. Tests of the relation of rabbit pox to other viruses by crossed inoculations and exposure experiments. *J. Exp. Med.* **63**, 353–378.

75. Hurst, E. W. (1937). Myxoma and Shope fibroma. I. The histology of myxoma. *Brit. J. Exp. Pathol.* **18**, 1–15.

76. Hurst, E. W. (1937). Myxoma and Shope fibroma. II. The effect of intracerebral passage on the myxoma virus. *Brit. J. Exp. Pathol.* **18**, 15–23.

77. Hurst, E. W. (1937). Myxoma and the Shope fibroma. III. Miscellaneous observations bearing on the relationship between myxoma, neuromyxoma and fibroma viruses. *Brit. J. Exp. Pathol.* **18**, 23–30.

78. Hyde, R. R. (1942). Behavior of influenza A viruses in the rabbit, rat, and guinea pig. *Amer. J. Hyg.* **36**, 338–353.

79. Ishimoto, A., and Ito, Y. (1969). Specific surface antigen in Shope papilloma cells. *Virology* **39**, 595–597.

80. Ishimoto, A., Oota, S., Kimura, I., Miyake, T., and Ito, Y. (1970). *In vitro* cultivation and antigenicity of cottontail rabbit papilloma cells induced by the Shope papilloma virus. *Cancer Res.* **30**, 2598–2605.

81. Jansen, J. (1941). Tödliche Infektionen von Kaninchen durch ein Filtrierbares Virus. *Zentralbl. Bakteriol., Parasitenk., Infektionskr. Hyg.* **148**, 65–68.

82. Jansen, J. (1946). Immunity in rabbit plague. Immunological relationship with cowpox. *Antonie van Leeuwenhoek; J. Microbiol. Serol.* **11**, 139–167.

83. Jansen, J. (1947). Immuniteitsproblemen betreffende konijnen pest. *T; Diergeneesk.* **72**, 550–557.

83a. Jiran, E., Sladka, M., and Kunstýř, I. (1970). Myxomatose der Kaninchen. Beitrag zur virus-modifizierung. *Zentralbl. Veterinaer med.* **17**, 418–428.

84. Joiner, G. N., Jardine, J. H., and Gleiser, C. A. (1971). An epizootic of Shope fibromatosis in a commercial rabbitry. *J. Amer. Vet. Med. Ass.* **159**, 1583–1587.

85. Kass, S. J., and Knight, C. A. (1965). Purification and chemical analysis of Shope papilloma virus. *Virology* **27**, 273–281.

86. Kessel, J. F., Fisk, R. T., and Prouty, C. C. (1934). Studies with the California strain of the virus of infectious myxomatosis. *Proc. Pac. Sci. Congr., 5th, 1933* Vol. 4, pp. 2927–2939.

87. Kessel, J. F., Prouty, C. C., and Meyer, J. W. (1931). Occurrence of infectious myxomatosis in southern California. *Proc. Soc. Exp. Biol. Med.* **28**, 413–414.

88. Kidd, J. G., and Parsons, R. J. (1936). Tissue affinity of Shope papilloma virus. *Proc. Soc. Exp. Biol. Med.* **35**, 438–441.

89. Kidd, J. G., and Rous, P. (1940). Cancer deriving from virus papillomas of wild rabbits under natural conditions. *J. Exp. Med.* **71**, 469–493.

90. Kilham, L. (1956). Propagation of fibroma virus in tissue cultures of cottontail testes. *Proc. Soc. Exp. Biol. Med.* **92**, 739–742.

91. Kilham, L. (1957). Transformation of fibroma into myxoma virus in tissue culture. *Proc. Soc. Exp. Biol. Med.* **95**, 59–62.

92. Kilham, L. (1958). Fibroma-myxoma virus transformations in different types of tissue culture. *J. Nat. Cancer Inst.* **20**, 729–740.

93. Kilham, L., and Dalmat, H. T. (1955). Host-virus mosquito relations of Shope fibromas in cottontail rabbits. *Amer. J. Hyg.* **61**, 45–54.

94. Kilham, L., and Fisher, E. R. (1954). Pathogenesis of fibromas in cottontail rabbits. *Amer. J. Hyg.* **59**, 104–112.

95. Kilham, L., and Woke, P. A. (1953). Laboratory transmission of fibromas (Shope) in cottontail rabbits by means of fleas and mosquitoes. *Proc. Soc. Exp. Biol. Med.* **83**, 296–301.

96. Lafenètre, H., Cortez, A., Rioux, J. A., Pages, A. Vollhardt, Y., and Quatrefages, H. (1960). Enzootie de tumeurs cutanées chez le lièvre. *Bull. Acad. Vet. Fr.* **33**, 379.

97. Larson, C. L., Schillinger, J. E., and Green, R. C. (1936). Transmission of rabbit papillomatosis by the rabbit tick, *Haemaphysalis leporis palustris. Proc. Soc. Exp. Biol. Med.* **33**, 536–538.

98. Leinati, L., Mandelli, G., and Carrara, O. (1959). Lesioni cutanee nodulari nelle lepridella pianura padana. *Atti Soc. Ital. Sci. Vet.* **13**, 429.

99. Leinati, L., Mandelli, G., Carrara, O., Cilli, V., Castrucci, G., and Scatozza, F. (1961). Ricerche anatomo-istopatologiche e virologiche sulla malattia cutanea nodulare delle lepri padane. *Boll. Ist. Sieroter. Milan.* **40**, 295.

100. Lubke, H. (1968). Ten years of experiences in governmental control of myxomatosis of rabbits in Berlin. *Berlin Muenchen. Tieraerztl. Wochenschr.* **81**, 275–279.

101. McKercher, D. G., and Saito, J. (1964). An attenuated live-virus vaccine for myxomatosis. *Nature (London)* **202**, 933–934.

102. McKiel, J. A., Hall, R. R., and Newhouse, V. F. (1966). Viruses of the California encephalitis complex in indicator rabbits. *Amer. J. Trop. Med.* **15**, 98–102.

103. McLean, D. M., and Bryce, R. P. (1963). Powassan and Silverwater viruses: Ecology of two Ontario arboviruses. *J. Can. Med. Ass.* **88**, 182–185.

104. Marshall, I. D., and Regnery, D. C. (1960). Myxomatosis in a California brush rabbit (*Sylvilagus bachmani*). *Nature (London)* **188**, 73–74.

105. Marshall, I. D., and Regnery, D. C. (1963). Studies on the epidemiology of myxomatosis in California. III. The response of brush rabbits (*Sylvilagus bachmani*) to infection with exotic and enzootic strains of myxoma virus, and the relative infectivity of the tumors for mosquitoes. *Amer. J. Hyg.* **77**, 213–219.

106. Mead-Briggs, A. R. (1964). Observations on the rabbit flea, a vector of myxomatosis. *Ann. Appl. Biol.* **51**, 338–342.

107. Melnick, J. L. (1962). Papova virus group. *Science* **135**, 1128–1130.

108. Miller, C. P., Jr., Andrewes, C. H., and Swift, H. F. (1924). A filterable virus infection of rabbits. I. Its occurrence in animals inoculated with rheumatic fever material. *J. Exp. Med.* **40**, 773–787.

109. Morris, J. A., Saglam, M., and Bozeman, F. M. (1965). Recovery of a new syncytium virus from a cottontail rabbit. *J. Infec. Dis.* **115**, 495–499.

110. Moses, A. (1911). O virus do mixoma dos coelhos. *Mem. Inst. Oswaldo Cruz* **3**, 46.

111. Moulton, J. E., and Lau, D. (1967). Microspectrophotometry of deoxyribonucleic acid in Shope papilloma and derived carcinoma. *Amer. J. Vet. Res.* **28**, 219–223.

112. Mykytowycz, R. (1957). The transmission of myxomatosis by *Simulium melatum* Wharton. *CSIRO Wildl. Res.* **2**, 1–4.

113. Mykytowycz, R. (1958). Contact transmission of infectious myxomatosis of the rabbit *Oryctolagus cuniculus*. *CSIRO Wildl. Res.* **3**, 1–6.

114. Mykytowycz, R. (1961). Social behaviour of an experimental colony of wild rabbits, *Oryctolagus cuniculus*. IV. Conclusion: Outbreak of myxomatosis; third breeding season and starvation. *CSIRO Wildl. Res.* **6**, 142–155.

115. Nesburn, A. B. (1969). Isolation and characterization of a herpes-like virus from New Zealand albino rabbit kidney cell cultures: a probable reisolation of Virus III of Rivers. *J. Virol.* **3**, 59–69.

116. Newhouse, V. F., Burgdorfer, W., McKiel, J. A., and Gregson, J. D. (1963). California encephalitis virus. Serologic survey of small wild mammals in northern United States and Southern Canada and isolation of additional strains. *Amer. J. Hyg.* **78**, 123–129.

117. Noyes, W. F., and Mellors, R. C. (1957). Fluorescent antibody detection of the antigens of the Shope papilloma virus in papillomas of the wild and domestic rabbit. *J. Exp. Med.* **106**, 555–562.

118. Padgett, B. L., Moore, M. S., and Walker, D. L. (1962). Plaque assays for myxoma and fibroma viruses and differentiation of the viruses by plaque form. *Virology* **17**, 462–471.

119. Parsons, R. J., and Kidd, J. G. (1936). A virus causing oral papillomatosis in rabbits. *Proc. Soc. Exp. Biol. Med.* **35**, 441–443.

120. Parsons, R. J., and Kidd, J. G. (1943). Oral papillomatosis in rabbits. *J. Exp. Med.* **77**, 233–250.

121. Pearce, J. M. (1950). Cardiac lesions produced by viruses. *In* "The Pathogenesis and Pathology of Viral Diseases" (J. G. Kidd, ed.), pp. 107–133. Columbia Univ. Press, New York.

122. Pearce, J. M. (1960). Heart disease and filterable viruses. *Circulation* **21**, 448–455.

123. Pearce, L., Rosahn, P. D., and Hu, C. K. (1936). Studies on etiology of rabbit pox. I. Isolation of a filterable agent: Its pathogenic properties. *J. Exp. Med.* **63**, 241–258.

124. Pennington, T. H., and Follett, E. A. C. (1971). Inhibition of poxvirus maturation by rifamycin derivatives and related compounds. Rifamycin. *J. Virol.* **7**, 821–829.

125. Rao, A. R., McFadzean, J. A., and Squires, S. (1965). The laboratory and clinical assessment of an isothiazole thiosemicarbazone (M & B 7714) against pox viruses. *Ann. N.Y. Acad. Sci.* **130**, 118–127.

126. Rapp, W. R. (1969). Rabies in Kansas. *J. Kans. Med. Soc.* **70**, 483–486.

127. Rdzok, E. J., Shipkowitz, N. L., and Richter, W. R. (1966). Rabbit oral papillomatosis: Ultrastructure of experimental infection. *Cancer Res.* **26**, 160–165.

128. Reddick, R. A., and Lefkowitz, S. S. (1969). *In vitro* immune responses of rabbits with persistent adenovirus type 5 infection. *J. Immun.* **103**, 687–694.

129. Reisner, A. H., Sobey, W. R., and Conolly, D. (1963). Differences between the soluble antigens of myxoma viruses originating in Brazil and California. *Virology* **20**, 539–541.

130. Rendtorff, R. C., and Wilcox, A. (1957). The role of nematodes as an entry for viruses of Shope's fibromas and papillomas of rabbits. *J. Inf. Dis.* **100**, 119–123.

131. Renquist, D., and Soave, O. (1972). Preliminary report on the isolation and characterization of a virus from the respiratory tract of rabbits. *Lab. Anim. Sci.* **22**, 104–106.

132. Richter, W. R., Shipkowitz, N. L., and Rdzok, E. J. (1964). Oral papillomatosis of the rabbit: An electron microscopic study. *Lab. Invest.* **13**, 430–438.

133. Rivers, T. M. (1930). Infectious myxomatosis of rabbits. Observations on the pathological changes introduced by virus myxomatosum (Sanarelli). *J. Exp. Med.* **51**, 965–976.

134. Rivers, T. M., and Pearce, L. (1925). Growth and persistence of filterable viruses in a transplantable rabbit neoplasm. *J. Exp. Med.* **38**, 673–692.

135. Rivers, T. M., and Tillett, W. S. (1923). Studies on varicella. The susceptibility of rabbits to the virus of varicella. *J. Exp. Med.* **38**, 673–692.

136. Rivers, T. M., and Tillett, W. S. (1924). Further observations on the phenomena encountered in attempting to transmit varicella to rabbits. *J. Exp. Med.* **39**, 777–803.

137. Rivers, T. M., and Tillett, W. S. (1924). The lesions in rabbits experimentally infected by a virus encountered in the attempted transmission of varicella. *J. Exp. Med.* **40**, 281–287.

138. Rivers, T. M., and Ward, S. M. (1937). Infectious myxomatosis of rabbits. *J. Exp. Med.* **66**, 1–14.

139. Rosahn, P. D., Hu, C. K., and Pearce, L. (1936). Studies on the etiology of rabbit pox. IV. Test on the relation of rabbit pox to other viruses by serum neutralization experiments. *J. Exp. Med.* **63**, 379–396.

140. Rosen, L. (1963). Respiratory enterovirus and reovirus infections. *Arch. Gesmte Virusforsch.* **13**, 272–280.

141. Rous, P., and Beard, J. W. (1934). Carcinomatous changes in virus-induced papillomas of the skin of the rabbit. *Proc. Soc. Exp. Biol. Med.* **32**, 578–580.

142. Rous, P., and Beard, J. W. (1935). The progression to carcinoma of virus-induced rabbit papillomas (Shope). *J. Exp. Med.* **62**, 523–548.

143. Sanarelli, B. (1898). Das Myxomatogene Virus. Beitrag zum Studium der Krankheitserreger ausserhalb des Sichbaren. *Zentrallbl. Bakteriol., Parasitenk. Infektionskr., Abt. 1* **23**, 865–882.

144. Selbie, F. R., and Robinson, R. H. M. (1947). Serial transmission of infectious papillomatosis in the domestic rabbit. *Brit. J. Cancer* **1**, 371–379.

145. Shope, R. E. (1932). A transmissible tumor-like condition in rabbits. *J. Exp. Med.* **56**, 793–802.

146. Shope, R. E. (1932). A filtrable virus causing a tumor-like condition in rabbits and its relationship to virus myxomatosum. *J. Exp. Med.* **56**, 803–822.

147. Shope, R. E. (1935). Serial transmission of the virus of infectious papillomatosis in domestic rabbits. *Proc. Soc. Exp. Biol. Med.* **32**, 830–832.

148. Shope, R. E. (1936). Infectious fibroma of rabbits. III. The serial transmission of virus myxomatosum in cottontail rabbits, and cross-immunity tests with the fibroma virus. *J. Exp. Med.* **63**, 33–41.

149. Shope, R. E. (1937). Immunization of rabbits to infectious papillomatosis. *J. Exp. Med.* **65**, 607–624.

150. Shope, R. E., and Hurst, E. W. (1933). Infectious papillomatosis of rabbits. *J. Exp. Med.* **58**, 607–624.

151. Smith, M. H. D. (1952). The Berry-Dedrick transformation of fibroma into myxoma in the rabbit. *Ann. N.Y. Acad. Sci.* **54**, 1141–1152.

152. Sobey, W. R. (1969). Selection for resistance to myxomatosis in domestic rabbits (*Oryctolagus cuniculus*). *J. Hyg.* **67**, 743–754.

153. Sobey, W. R., Conolly, D., Haycock, P., and Edmonds, J. W. (1970). Myxomatosis. The effect of age upon survival of wild and domestic rabbits (*Oryctolagus cuniculus*) with a degree of genetic resistance and unselected domestic rabbits infected with myxoma virus. *J. Hyg.* **68**, 137–149.

154. Sprehn, C. (1956). "Kaninchenkrankheiten." Verlagshaus Reutlingen Oertel und Sporer, Reutlingen, Germany.

155. Stanley, N. F., Leak, P. J., Grieve, G. M., and Perret, D. (1964). The ecology and epidemiology of Reovirus. *Aust. J. Exp. Biol. Med. Sci.* **42**, 373–384.

156. Stone, R. S., Shope, R. E., and Moore, D. H. (1959). Electron microscope study of the development of the papilloma virus in the skin of the rabbit. *J. Exp. Med.* **110**, 543–546.

157. Subak-Sharpe, J. H., Timbury, M. C., and Williams, J. F. (1969). Rifampicin inhibits the growth of some mammalian viruses. *Nature (London)* **222**, 341–345.

158. Syverton, J. T. (1952). The pathogenesis of the rabbit papilloma-to-carcinoma sequence. *Ann. N.Y. Acad. Sci.* **54**, 1126–1140.

159. Syverton, J. T., and Berry, G. P. (1935). Carcinoma in the cottontail rabbit following spontaneous virus papilloma (Shope). *Proc. Soc. Exp. Biol. Med.* **33**, 399–400.

160. Syverton, J. T., Dascomb, H. E., Wells, E. B., Koomen, J., and Berry, G. P. (1950). The virus-induced papilloma-to-carcinoma sequence. II. Carcinomas in the natural host, the cottontail rabbit. *Cancer Res.* **10**, 440–444.

161. Takahashi, M., Kameyama, S., Kato, S., and Kamahora, J. (1958). A study of myxoma virus inclusions by fluorescein-labelled antibody. *Biken J.* **1**, 198–210.

162. Takahashi, M., Kameyama, S., Kato, S., and Kamahora, J. (1959). Immunological relationship of the poxvirus group. *Biken J.* **2**, 27–34.

163. Topacio, T., and Hyde, R. R. (1932). The behavior of rabbit virus III in tissue culture. *Amer. J. Hyg.* **15**, 99–126.

164. Vail, E. L., and McKenney, F. D. (1943). Diseases of domestic rabbits. *U.S., Fish Wildl. Serv. Conserv. Bull.* **31**.

165. Verlinde, J., and Wersinck, F. (1951). Manifestations of the laboratory epidemic of rabbit pox. *Antonie van Leeuwenhoek; J. Microbiol. Serol.* **17**, 232–236.

166. Weisbroth, S. H., and Scher, S. (1970). Spontaneous oral papillomatosis in rabbits. *J. Amer. Vet. Med. Ass.* **157**, 1940–1944.

167. Wildy, P. (1971). Classification and nomenclature of viruses. *Monogr. Virol.* **5**, 1–81.

168. Wilner, B. I. (1968). Characteristics of cell culture systems. Viral flora of rabbits. *Nat. Cancer Inst., Monogr.* **29**, 141–148.

169. Woodroofe, T. M., and Fenner, F. (1962). Serological relationships within the pox-virus group: An antigen common to all members of the group. *Virology* **16**, 334–341.

170. Woodroofe, T. M., and Fenner, F. (1965). Viruses of the myxoma-fibroma subgroup of the poxviruses. I. Plaque production in cultured cells, plaque-reduction tests, and cross protection tests in rabbits. *Aust. J. Exp. Biol. Med. Sci.* **43**, 123–142.

171. Yoshida, T. O., and Ito, Y. (1968). Immunofluorescent study on early virus-cell interaction in Shope papilloma *in vitro* system *Proc. Soc. Exp. Biol. Med.* **128**, 587–591.

172. Yuill, T. M. (1970). Myxomatosis and fibromatosis of rabbits, hares, and squirrels. *In* "Infectious Diseases of Wild Mammals" (J. W. Davis, L. H. Karstad, and D. O. Trainer, eds.), pp. 104–130. Iowa State Univ. Press, Ames.

173. Yuill, T. M., and Hanson, R. P. (1964). Infection of suckling cottontail rabbits with Shope's fibroma virus. *Proc. Soc. Exp. Biol. Med.* **117**, 376–380.

174. Yuill, T. M., Iversen, J. O., and Hanson, R. P. (1969). Evidence for arbovirus infections in a population of snowshoe hares. A possible mortality factor. *Bull. Wildl. Dis. Ass.* **5**, 248–253.

CHAPTER 11

Protozoal Diseases*

Steven P. Pakes

I. GENERAL INTRODUCTION

Although disease-free laboratory rats and mice have been developed in great numbers for biomedical research, there are few laboratory rabbits of comparable quality available. Rabbits harbor many real and potential patho-

gens, although few are more insidious than the protozoal organisms. Encephalitozoonosis and coccidiosis are the most important protozoal diseases of the laboratory rabbit, and the scientific literature contains numerous examples where they have caused great concern to investigators using this animal. In addition, coccidiosis is a significant economic factor in raising rabbits for food and fiber.

The purpose of this chapter is to review these and other protozoal diseases of rabbits. Emphasis will be placed on the morphology, life cycle, and transmission of the etiolog-

*Supported in part by Grants AI-09733 and RR-05463 from the National Institutes of Health.

ical agent and the clinical signs, pathology, diagnosis, treatment, prevention and control, and public health significance of the disease.

II. SPOROZOA

The most important sporozoa that infect rabbits are the coccidia. Coccidiosis is a major disease of rabbits and is caused by members of the genus *Eimeria*. Many rabbits are infected to some degree and there is little doubt that coccidiosis has been a complicating factor in many experiments using rabbits. Although the disease may be severe in rabbits, especially in the young, it is no longer the devastating disease that it once was. Chemotherapeutic procedures and more rigid sanitation methods are responsible for this.

Many species of *Eimeria* can infect members of the genera *Oryctolagus*, *Sylvilagus*, and *Lepus*, but only a few are important. One species, *E. stiedae*, parasitizes the liver, while there are several species that infect the different levels of the intestinal tract (Table I). Most infections are caused by 2 or more species, therefore, the precise role of the different species as pathogens is not clearly defined. The reviews of Becker (11), Levine (102), Pellérdy (133), and Rutherford (149) are recommended for detailed discussions.

Other sporozoa that have been reported in rabbits are *Cryptosporidium* sp. and *Hepatozoon cuniculi*.

A. *Eimeria stiedae* **(Lindemann, 1865; Kissalt and Hartmann, 1907)** (Synonyms: *Monocystis stiedae, Coccidium oviforme, Coccidium cuniculi*)

I. Introduction

As early as 1674, Leeuwenhoek noted and described bodies in the bile of rabbits (11), which Dobell (44) believed were oocysts of *Eimeria stiedae*. *Eimeria stiedae* infects the bile duct epithelium of members of the genera *Oryctolagus*, *Sylvilagus*, and *Lepus*. It is a common parasite of the domestic rabbit and is found throughout the world. In mild infections there may be no clinical signs, but the disease may be fatal, especially in young rabbits.

2. Morphology and Life Cycle

The oocysts are ovoid or ellipsoidal and measure 30 to 40 by 16 to 25 μm (133). One pole is flattened and contains a micropyle. The wall is smooth and yellow-orange. Internally there is a polar granule but no oocyst residuum. The sporulated oocysts contain 4 sporocysts, each containing 2 sporozoites, which is characteristic of the genus *Eimeria*. The ultrastructure of the various developmental forms of

E. stiedae has been described by Scholtyseck and others (64, 151–153).

The life cycle of *E. stiedae* is not definitely known, although several investigators have studied this aspect (69, 128, 146, 161–163). In 1933 Smetana (161–163) published an account of the life cycle and pathogenesis of the disease and, with few exceptions, it is generally accepted today. Sporulated oocysts are ingested by the rabbit and excyst in the duodenum. Sporozoites are thought to be liberated by enzymatic action. The sporozoites penetrate the intestinal mucosa, but how they are transported to the liver is not yet known. Smetana (162) reported that the sporozoites enter the liver via the portal blood system. Horton (69) and Rose (146) have described sporozoites in the mesenteric lymph nodes of rabbits experimentally infected with *E. stiedae*, suggesting lymphogenous transportation. Horton (69) suggested that the sporozoites are carried in lymphatic monocytes. Owen (128) reported viable sporozoites in the bone marrow, as well as in mesenteric lymph nodes of experimentally infected rabbits and favors the hematogenous route of migration. She showed that sporozoites reach the liver after 48 hours postinoculation. In any event, the sporozoites invade the epithelial cells of the bile ducts and begin schizogeny. Numerous merozoites are produced which infect contiguous epithelial cells. The exact number of asexual generations is not known for *E. stiedae*. In gametogony, microgametes and macrogametes are formed and after fertilization the oocyst develops. The unsporulated oocysts break out of the host cells, pass into the lumens of the bile ducts, and exit the body in the feces. Sporulation takes place outside the host's body. The prepatent period is approximately 15 to 18 days.

3. Transmission

Natural transmission of all *Eimeria* is by ingestion of sporulated oocysts. Pellérdy (134) induced hepatic coccidiosis in rabbits by intravenous, intraperitoneal, and intramuscular administration of *E. stiedae* oocysts, but the disease was always milder than that induced orally.

4. Clinical Signs

Many rabbits infected with *E. stiedae* show no clinical signs. In heavily infected rabbits signs noted are due to the interference of hepatic function and blockage of bile ducts. The rabbits may become anorectic and debilitated. Diarrhea, especially in the terminal stages of the disease, or constipation, may be noted. There may be hepatomegaly, resulting in an enlarged, pendulous abdomen, and icterus in advanced phases of the disease (7). Lund (105) noted that the enlarged liver often was 20% of the total body weight, as compared with about 3.7% in normal rabbits. Deaths are rare except in young rabbits with extremely heavy infections.

The clinical signs of the experimental disease have been reviewed by Rose (146) and Davies *et al*. (41). Other investigators have reported various changes in serum components of rabbits experimentally infected with *E. stiedae*. Dunlap and associates (45) observed an increase in serum β- and γ-globulin and β-lipoprotein and a decrease in the α-lipoprotein in experimentally infected rabbits. Smith and McShan (164) described an increase in the succinic dehydrogenase activity of the liver in early experimental infection, with a decrease in the enzyme near the twentieth day postinfection. Rose (146) indicated that there is a rise in the serum bilirubin, up to 3 to 5 mg/100 ml serum, as compared to a normal value of 0 mg/100 ml serum. The rise in serum bilirubin was first detected at 6 days postinoculation and increased until 20 to 24 days postinoculation, after which it fell.

Diehl (42) and Diehl and Kistler (43) observed vitamin E deficiency associated with experimental hepatic coccidiosis. They attributed the deficiency to malabsorption, impaired storage, and increased use or destruction of the vitamin due to hepatobiliary dysfunction.

5. Pathology

The liver is often grossly enlarged and bosselated with yellowish-white lesions of varying size. These lesions are dilated bileducts that contain a yellow exudate. The gallbladder may also be enlarged and contain exudate. One may note a fibrous capsule around some lesions and, due to this, the liver will be difficult to cut. Ascites may also be a feature.

Microscopically, there is destruction and regeneration of the bile duct epithelium, resulting in marked papillomatous hyperplasia, bile duct reduplication, and cystic enlargement of the bile ducts (Figs. 1 and 2). These changes are accompanied by infiltration of lymphocytes, plasma cells, and occasional epithelioid cells. Enlarged bile ducts may rupture releasing their contents which initiates a severe granulomatous response. Fibrosis may be a prominent feature of the lesion. The exudate is composed primarily

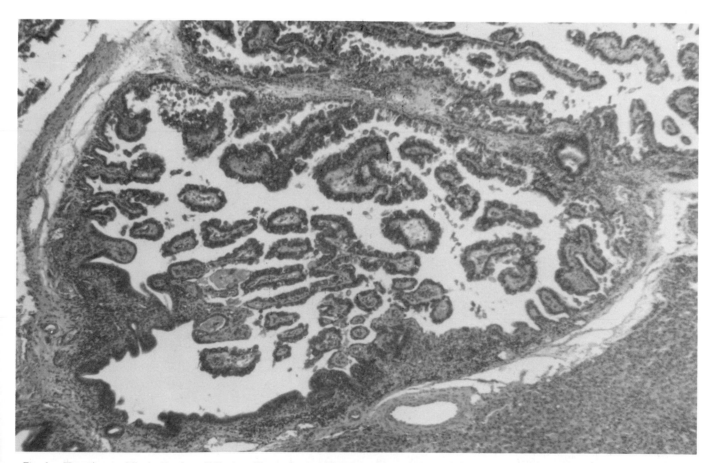

Fig. 1.   Hepatic coccidiosis. Section of bile duct illustrating papillomatous hyperplasia and inflammatory response. Hepatic cells are not involved. Hematoxylin and eosin. × 85.

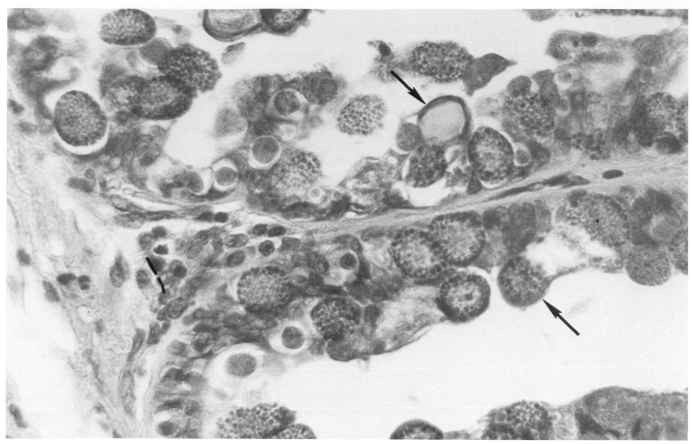

Fig. 2. Hepatic coccidiosis. Various developmental stages of *E. stiedae* in bile duct epithelium. Arrows indicate macrogametes and developing oocysts. A few mononuclear inflammatory cells are evident. Hematoxylin and eosin. × 900.

of oocysts, sometimes admixed with inflammatory cells. Biliary outflow may be obstructed by oocysts, resulting in distended bile ducts. There may be ischemic necrosis of the contiguous liver parenchyma due to compression of adjacent blood vessels by bile duct swelling, but the hepatic cells themselves are not parasitized.

6. Diagnosis

The oocysts of each *Eimeria* species have characteristic features (Table I) (Fig. 3) and are described in detail by Kessel and Jankiewicz (81) and Goodrich (58).

The feces are examined for oocysts using direct smear, flotation, or concentration-flotation methods. The number of oocysts found in the feces should be correlated with necropsy findings. In the acute disease, oocysts may not be shed in the feces, and necropsy examination is needed for the diagnosis. Exudate from the dilated bile ducts and gallbladder can also be examined for *E. stiedae* oocysts.

Several investigators have studied the immunological response of rabbits to *Eimeria* (5, 6, 16, 29, 31, 72, 147, 148, 165). The reviews of Horton-Smith *et al.* (72) and Kozar

(83) should be consulted for detailed information. Complement fixation (5, 31, 130, 147, 148) and a serum indirect fluorescent antibody method (29), using the gametocytes as antigens, appear to be promising serological tests, however, they currently are not considered practical in routine surveillance.

7. Treatment, Control, and Prevention

Numerous chemotherapeutic agents have been used to treat hepatic and intestinal coccidiosis (39, 47, 62, 136, 137) but only a few have been found to be effective. The most effective agents are the sulfonamides (61, 70, 78, 106, 170, 171). Sulfamerazine sodium administered continually as 0.02% in the drinking water will inhibit the development of *E. stiedae* (141). Sulfaquinoxaline sodium is also effective at a dose of 0.05% in the drinking water or 0.03% in the feed (32, 106). These drugs can be administered safely at these levels for prolonged periods. Sulfamethoxine fed in the diet for 7 consecutive days to ensure a dose of 75 mg/kg of body weight was found to be beneficial in treating hepatic and intestinal coccidiosis (170). The nitrofurans are unsatis-

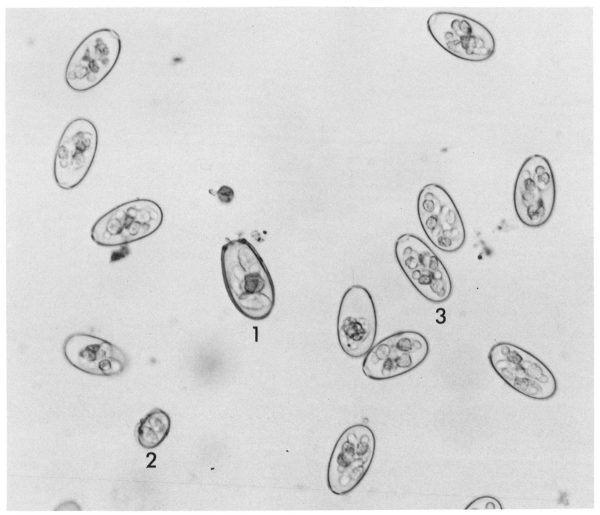

Fig. 3. Flotation preparation of rabbit feces; a typical mixed *Eimeria* infection. Species include *E. magna* (1), *E. perforans* (2), and *E. stiedae* (3). × 600. (Photograph courtesy of Dr. K. S. Todd, Jr.)

factory for hepatic coccidiosis, but Boch (18) has recommended nitrofurazone for intestinal coccidiosis at a rate of 0.5 to 1 gm/kg of body weight for prevention and 0.5 to 2 gm/kg for therapy.

The major role of chemotherapeutic agents may be to control the organisms until immunity develops. Davies and co-workers (41) found that a substantial degree of resistance follows a light infection of *E. stiedae*, and it is very rare for a previously infected rabbit to show clinical signs of the disease when challenged. The immunity resulting from mild infections is practically lifelong in rabbits (133). Niilo (125) was able to demonstrate acquired resistance in previously infected rabbits to challenge with large doses of sporulated *E. magna* oocysts.

Coccidiosis can be eliminated from rabbits, or, at least, kept at a low level by rearing and maintaining rabbits under scrupulously clean conditions. This may mean the use of expensive barrier systems for producing acceptable laboratory rabbits. Rabbits should be prevented from contacting infective feces or food and water contaminated with feces (71). The use of self-cleaning cages and external feeders and watering apparatus will alleviate this possibility. Cages and equipment should be thoroughly cleaned regularly, preferably with hot water and a suitable detergent. Some chemicals, such as 10% ammonia solution (73), are lethal to coccidial oocysts and can be used to advantage in rabbitries. Young animals should be separated from the dams as soon as possible, as suckling rabbits are more susceptible to coccidiosis than older animals (46). Flies and other vermin should be controlled in rabbitries since they may act as mechanical vectors (103). The recognition of the role the animal caretaker may play in spreading the disease from cage to cage is important and should be stressed to these individuals.

TABLE I
COMPARISON OF *EIMERIA* SP. INFECTING RABBITS

	Mean size of oocysts (μm)	Shape	Other distinguishing characteristics	Part of digestive tract affected	Prepatent period (days)	Pathogenicity
E. stiedae (Lindemann, 1865) Kisskalt & Hartmann, 1907	37 × 20	Ellipsoidal	Smooth, light-yellow wall; wide, thin micropyle; no residual body in oocyst; sporocyst with terminal knob (stiedae body)	Bile duct, epithelium	15–18	Variable
E. irresidua Kessel & Jankiewicz, 1931	38 × 26	Ovoid	Smooth, light-yellow wall; prominent micropyle; no residual body	Small intestine	7–8	Significant
E. magna Perard, 1925	35 × 24	Ovoid-ellipsoidal	Dark yellow-brown wall; prominent micropyle with lipping; large residual body	Jejunum, ileum	6–7	Significant
E. media Kessel, 1929	31 × 18	Ellipsoidal	Smooth, thick, light-pink wall; micropyle; large residual body	Small, large intestines	6–7	Moderate
E. perforans (Leuckart, 1879) Sluiter & Swellengrebel, 1912	21 × 15	Ellipsoidal	Smooth, colorless wall; indistinguishable micropyle; small residual body	Small intestine	5	Slight
E. exigua Yakimoff, 1934	15 × 13	Ovoid	Smooth wall; indistinguishable micropyle; no residual body	—	—	—
E. intestinalis Cheissen, 1948	27 × 18	Ellipsoidal	Smooth yellow wall; micropyle; large granular residual body	Ileum	10	Significant
E. matsubayashii Tsunoda, 1952	25 × 18	Ovoid	Smooth, light-colored wall no residual body	Small intestine, cecum	7	Slight
E. nagpurensis Gill & Ray, 1960	23 × 13	Barrel-shaped	Smooth, colorless wall; no micropyle; no residual body	—	—	—
E. neoleporis Carvalho, 1942 (*Eimeria coecicola* Cheissin, 1947)	39 × 20	Elongate ellipsoidal	Smooth, yellow wall; distinct micropyle; no residual body; sporocysts with terminal knob	Small intestine, cecum	12	Significant
E. piriformis Kotlan & Pospesch, 1934	29 × 18	Ellipsoidal	Smooth, light yellowish-brown wall; prominent micropyle; no residual body	Jejunum, ileum	9–10	Significant

8. Public Health Significance

The coccidia of the rabbit are not infectious to man.

B. Intestinal Coccidiosis

1. Introduction

The major characteristics of the intestinal coccidia that infect rabbits are listed in Table I. Many *Eimeria* occur in the intestinal tract of rabbits but the most important species are *E. irresidua*, *E. magna*, *E. media*, and *E. perforans* (104).

2. Morphology and Life Cycle

The morphological characteristics of the oocysts of the various *Eimeria* infecting rabbits are listed in Table I. The ultrastructure of rabbit coccidia has been studied by several investigators (64, 151–153, 155).

There are 3 phases in the life cycle of *Eimeria*: schizo-gony, gametogony, and sporogony. Generally, oocysts are ingested and sporozoites are liberated in the intestinal tract, where they enter the epithelial cells and multiply by schizogony. Usually 2 generations of 8 to 32 merozoites develop by schizogony. The next phase is gametogony which involves the formation of a large number of male microgametes, which are comma-shaped bodies, and the female macrogametes. After fertilization of the macro-gametes by the microgametes, sporogony begins. Oocysts develop and are extruded into the intestinal lumen and passed in the feces. Freshly voided unsporulated oocysts are not infective. They become infective after the formation of 4 sporocysts, each containing 2 sporozoites.

The works of Pellérdy (133), Davies *et al.* (41), and Rutherford (149) are suggested for specific information of the life cycles of the various *Eimeria*.

3. Transmission

Transmission is by ingestion of sporulated oocysts. Al-though rabbits are coprophagic, it is generally accepted that the soft feces eaten directly from the anus do not contain infectious oocysts.

4. Clinical Signs

The clinical signs exhibited by rabbits infected with intestinal coccidiosis vary considerably, depending on the age of the animal, relative susceptibility, degree of infection, and the causative organism or organisms (135). Often, infections are mild and no clinical signs are noted. In more severe infections, such as occurs in young rabbits, one may note that the animals lose weight or simply do not gain. There may be varying degrees of diarrhea, ranging from the intermittant type to profuse watery feces admixed with mucus and blood. The animals show intense thirst and deaths are attributed to dehydration and secondary bacterial infection. With very heavy primary infections, the rabbits may die before oocysts appear in the feces (52).

5. Pathology

In intestinal coccidiosis, lesions are seen in the small and large intestine, depending on the causative agent. The parasitized epithelial cells usually die, which may result in ulceration and a mixed mononuclear and polymorphonuclear exudate. This may be manifested grossly as multiple white areas in the wall of the intestinal tract. Figure 4 depicts the exudate and various stages of the developing organisms in cells.

6. Diagnosis

Examination of fecal preparations, as outlined in section A,6, and histological examination are used for diagnosing intestinal coccidiosis.

7. Treatment, Prevention, and Control

Methods of treatment, prevention, and control have been discussed in the section on hepatic coccidiosis. McPherson *et al.* (114) were able to eliminate intestinal coccidiosis from a rabbit breeding colony by sulfaquinoxaline treatment, strict sanitation methods, and elimination of infected animals.

8. Public Health Significance

As far as is known, none of the intestinal coccidia of rabbits infects man.

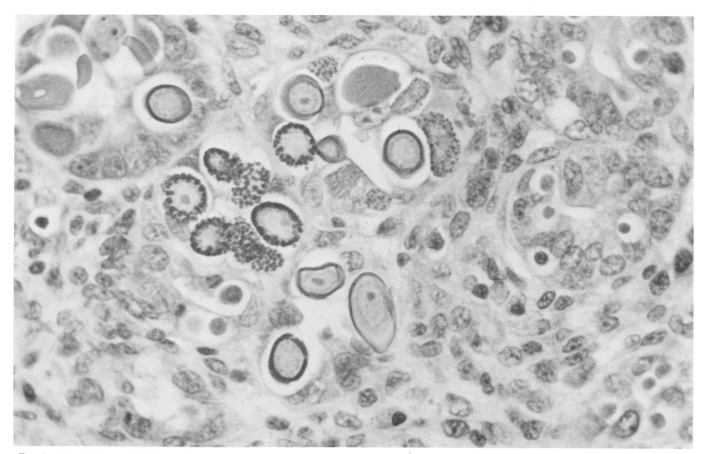

Fig. 4. Intestinal coccidiosis. An intense, mononuclear cell response to infection of epithelium with intestinal *Eimeria*. Developing oocysts predominate. Hematoxylin and eosin. × 800.

C. *Eimeria* of *Sylvilagus* and *Lepus*

Aside from *E. stiedae* and possibly *E. neoleporis* (26), there is little evidence that cross-transmission of *Eimeria* occurs among *Oryctolagus*, *Sylvilagus*, and *Lepus* (133). The host specificity of coccidia from *Lepus europaeus* and *O. cuniculus* was demonstrated by Pellérdy (132), who was not able to transmit the intestinal coccidia of domestic rabbits to hares, and vice versa. Detailed descriptions of the *Eimeria* of *Sylvilagus* and *Lepus* are contained in papers by Pellérdy (133) and Carvalho (27, 28).

D. *Cryptosporidium* sp.

In 1929 Tyzzer (174) found a *Cryptosporidium* in the intestinal tract of the rabbit, which was morphologically similar to *C. parvum*, a parasite of mice. He did not discuss the organism in detail and offered no comments of its pathogenicity in rabbits.

E. *Hepatozoon cuniculi* (**Sangiorgi, 1914**) (**Synonym:** *Leucocytogregarina cuniculi*)

Sangiorgi (150) described *Hepatozoon cuniculi* in domestic rabbits in Italy in 1914. He noted gametocytes in leukocytes and schizonts in the spleen. The intermediate host was not identified, although fleas have been incriminated in laboratory rodents. The leukocyte forms may resemble gametocytes of *Plasmodium*, but there is no pigment (52).

III. TOXOPLASMIDA

A. *Toxoplasma gondii* (**Nicolle and Manceaux, 1908**) (**Synonym:** *Toxoplasma cuniculi*, **Carini, 1911**)

1. Introduction

Splendore (167, 168) first reported toxoplasmosis in the domestic rabbit in 1908 and 1909, almost simultaneously with the discovery of the causative organism in the North African rodent, *Ctenodactylus gondi*, after which it was named (122–124). Since that time there have been several reports of spontaneous toxoplasmosis in the domestic rabbit from many parts of the world (17, 19, 24, 65, 80, 84, 88, 92, 100, 110, 117, 138, 140, 176, 178, 181, 184, 185), as well as in wild rabbits and hares (20, 33–35, 91). Other reports deal with the serological surveillance of toxoplasmosis in rabbits (14, 87). Antibody prevalence figures of up to 50% have been reported in clinically normal rabbits using the Sabin-Feldman dye test (14, 183).

It is interesting to note that there is a paucity of recent reports of toxoplasmosis in laboratory rabbits in the United States. This is no doubt partly due to the employment of stringent animal care practices.

Investigators must remain alert, however, since one of the forms of toxoplasmosis is latent. Although not definitely proved, the latent form can apparently be activated when rabbits are exposed to experimental procedures of various kinds (37, 85, 99).

2. Morphology and Life Cycle

Two forms of *T. gondii*, the proliferative phase and cysts, have been observed in the rabbit. Trophozoites, or proliferative organisms, may be intra- or extracellular and occur during acute phases of the disease. The extracellular forms are crescent- or banana-shaped, pointed on one end and blunted on the other, and measure 4 to 8 by 2 to 4 μm (102). Intracellular organisms are shorter and appear more blunt. The vesicular nucleus is located centrally or near the blunt end. Ultrastructurally, there is a conoid located at one pole and a nucleus that is surrounded by a double membrane (158). *Toxoplasma* sporozoites ultrastructurally resemble coccidia and *Plasmodium* (59).

Toxoplasma cysts develop intracellularly and are noted in the chronic or latent phase of the disease. The tough cyst wall contains the numerous merozoites in cells throughout the body.

With the discovery of infectious *Toxoplasma* oocysts in cats, the life cycle has been revised to include a sexual cycle with gametogony and sporogony (53a). However, in rabbits only the asexual cycle, consisting of the proliferative and cyst stages, is recognized. Trophozoites develop intracellularly by binary fission during the initial stages of the disease. As immunity develops or in latent disease the cyst phase becomes prominent (53).

3. Transmission

Toxoplasmosis is known to be transmitted transplacentally and by carnivorism. In addition, with the discovery of infectious oocysts in cats, fecal transmission to herbivorous animals, such as the rabbit, is likely.

4. Clinical Signs

The following is largely taken from the review of Siim and associates (160). Acute toxoplasmosis in rabbits is characterized by sudden anorexia and fever ($>$ 104°F) and increased respiratory rate. There is a serous or seropurulent ocular and nasal discharge. The rabbits become lethargic and, in a few days, central nervous signs of localized or generalized convulsions may occur. In some cases paralysis, especially in the hindquarters, develops. Death usually occurs after 2 to 8 days of illness.

In the chronic form of the disease, the course is pro-

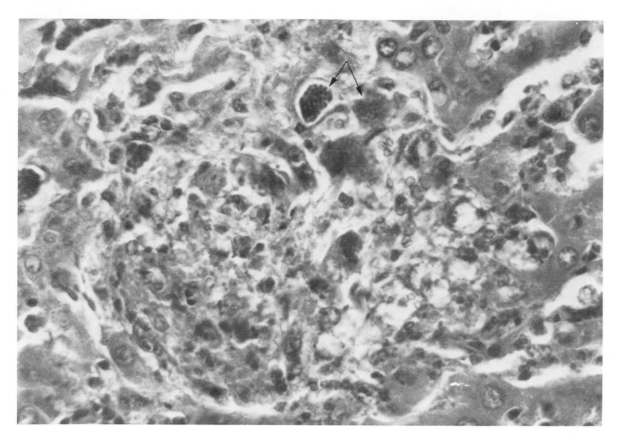

Fig. 5. Toxoplasmosis. Necrotic focus in liver of *Oryctolagus cuniculus.* Arrows indicate clusters of intracellular *T. gondii* organisms. Hematoxylin and eosin. × 1250.

tracted and the rabbits become anorectic and emaciated. Anemia is a common sequel. As the disease progresses, central nervous signs may occur, usually as paralysis of the hind parts. Death may be sudden, but many animals recover.

5. Pathology

In acute toxoplasmosis the lesions are generalized and are characterized by extensive necrosis of lymph nodes, spleen, liver, lungs, and heart (Fig. 5). Grossly, the organs may be swollen with multiple, necrotic foci (160). Histologically, the reticuloendothelial elements and vascular connective tissue are most often affected. There is marked cellular necrosis, with or without infiltration of inflammatory cells. *T. gondii* trophozoites are located intra- and extracellularly (Fig. 5). This form of the disease is found primarily in young animals.

In chronic toxoplasmosis, the gross lesions are more variable. There may be edematous enlargement of various organs and scattered necrotic foci. Microscopically, pronounced reticuloendothelial hyperplasia is present, particularly in lymph nodes, spleen, liver, and central nervous system (160). *Toxoplasma* is more difficult to demonstrate

than in the acute disease. Chronic toxoplasmosis is usually a disease of older rabbits.

In the latent form of the disease lesions are seen primarily in the form of cysts in the central nervous system with or without reaction. One may see gliosis and granulomatous encephalitis with nonsuppurative meningitis and perivascular cuffing. These lesions may be confused with encephalitozoonosis.

6. Diagnosis

The diagnosis of toxoplasmosis in rabbits may be made from a combination of histological examination of lesions, morphological identification of organisms, isolation of organisms, and serological methods. In acute toxoplasmosis Wright-Giemsa stained smears of peritoneal exudate, mesentery, or omentum can be examined for the characteristic organisms.

Toxoplasma can be isolated by intraperitoneal injection of mice or hamsters which are free of toxoplasmosis. Some animals should be pretreated with cortisone acetate subcutaneously at a dose of 2 mg once weekly in mice and 2.5 mg twice weekly in hamsters (52).

Various serological tests for toxoplasmosis are useful

in rabbits, including the Sabin-Feldman dye test, fluorescent antibody test (116), and complement fixation. The reader is referred to the reviews of Frenkel (52) and Siim et al. (160) for detailed information.

7. Treatment, Control, and Prevention

Several drugs have been used to inhibit trophozoites and cysts of toxoplasmosis in laboratory animals, including sulfadiazine, pyrimethamine, sulfones, tetracyclines, spiramycin, and 1, 2-dihydrotriazones. Frenkel (53) points out that because rabbits eat irregularly, the administration of proper levels of the drugs may be difficult, and parenteral administration may be necessary. In most cases, treatment of toxoplasmosis in rabbits is not practical.

Because toxoplasmosis is now known to be transmitted by ingestion of infectious oocysts from cats (53a), rabbits should be kept separated from cats. Precautions must be taken to insure that food and water for rabbits are not contaminated with cat excreta. Animal care personnel should be made aware of the role they may play in transmission of infectious oocysts. Most disinfectants will not inactivate the oocysts, but heating and drying will (53). If toxoplasmosis is known to be present in a rabbit colony, only sero-negative animals should be used for breeding (52).

8. Public Health Significance

Beverley and co-workers (15) used the Sabin-Feldman dye test to show that a high proportion of sera from people regularly handling rabbits contained significant titers of antibodies for *Toxoplasma*. A study of the epizootiology of *T. gondii* in rabbits in England suggests that rabbits may be an important host reservoir for the disease in man (88).

B. *Sarcocystis cuniculi* (**Brumpt, 1913**) (**Synonym:** *Sarcocystis leporum*)

1. Introduction

Sarcocystis was first described in rabbits in Germany in 1867 by Manz (109). In 1913 Brumpt (21) reported the protozoan in the European hare and named it *Sarcocystis cuniculi*. Crawley (38) described the organism in cottontail rabbits in 1914 in the United States and called it *Sarcocystis leporum*.

Sarcocystis has been described rarely in *Oryctolagus cuniculus* (91a), although it is common in *Sylvilagus* sp. (49, 102, 118, 169a). Levine (102) considers the organism infecting the 2 genera to be *S. cuniculi*.

2. Morphology and Life Cycle

Sarcocystis forms cysts in the muscles of the host. In the rabbit these cysts are up to 5 mm long and are limited by a wall that has radial spines called cytophaneres (Fig. 6). The cysts of *Sarcocystis* in rabbits are reported to be septate (51, 52, 102), however, well-defined septa were not identified in cysts found in a 3-month-old cottontail rabbit submitted to our laboratory, although they had the characteristic cytophaneres. It is possible that more than one species may infect the rabbit. Levine and Kocan (102a) have suggested that immature cysts may not be septate, the septa developing as the parasite matures. The cysts are filled with trophozoites, which are banana-shaped and slightly pointed on one end. The trophozoites usually measure 12 to 13 by 4 to 5 μm, but vary in length from 6 to 16 μm. In smears stained with Giemsa, the nucleus is visible near the blunter end. Glycogen granules surround the nucleus, while the rest of the cytoplasm is fairly clear (51).

Levine (102) discusses several differing accounts of the life cycle of *Sarcocystis*. Although the exact life cycle is not known there is general agreement that the life cycle is simple, without sexual stages. The trophozoites are motile when released from the cyst and are thought to reproduce by binary fission.

3. Transmission

The mode of transmission of *Sarcocystis* in rabbits is unknown although horizontal transmission via fleas and by ingestion of infective excreta has been suggested (49).

4. Clinical Signs

Light or moderate infections in rabbits are asymptomatic. In very heavy infections lameness may be evident.

5. Pathology

Lesions are found in cardiac and skeletal muscle, especially in the hindlegs, flanks, and loins (102). In heavy infections cysts may be seen grossly as multiple white streaks running in the direction of the muscle fibers. Microscopically, one often sees the intact cysts in the muscle without an inflammatory response (Fig. 6). However, degeneration of the cyst wall and invasion of released trophozoites may result in severe focal myocarditis and myositis, characterized by mineralization and infiltration of lymphocytes, plasma cells, eosinophils, and macrophages. Scarring may be prominent.

The cysts produce a strong endotoxin known as sarcocystin, which is lethal to rabbits when administered intravenously (48). The toxin has been associated with myositis in this disease (52) but its role in the pathogenesis has not been established.

6. Diagnosis

Heavy infections may be diagnosed grossly by the presence of the white streaks paralleling the direction of

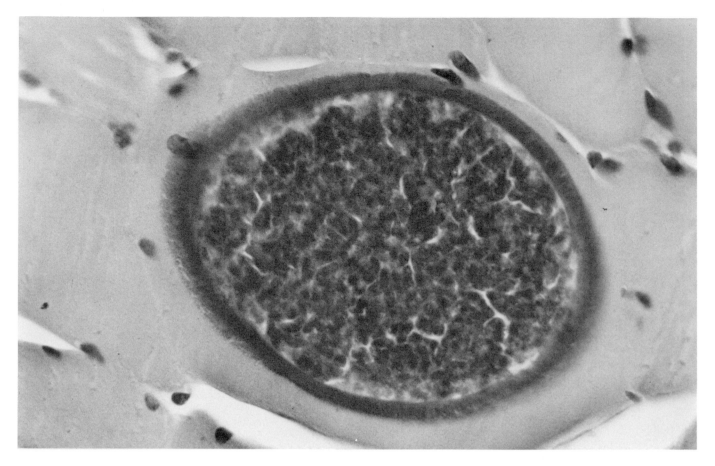

Fig. 6. Sarcocystis. Cyst in the skeletal muscle from a 3-month-old cottontail rabbit. There is an absence of septa and the radial spines (cytophaneres) of the cyst wall are prominent. Note the lack of cellular response to the parasite. Hematoxylin and eosin. × 800.

the muscle bundles. Smears made from excised pieces of infected muscle can be examined under the phase microscope for the motile banana-shaped organisms or can be stained with Giemsa (49).

Most diagnoses are made by microscopic examination of tissue sections. *Sarcocystis* in the rabbit has radial spines in the wall and usually is septate.

Arcay-de-Peraza (2) has attempted serological testing for *Sarcocystis* in ruminents using the dye test and complement fixation, precipitation, and agglutination procedures. The efficacy of these procedures for *Sarcocystis* in rabbits is not known.

7. Treatment, Control, and Prevention

An effective treatment for *Sarcocystis* has not been developed. The prevalence of *Sarcocystis* in laboratory rabbits is practically nil, due primarily to the methods now used to raise and maintain these animals. Although the mode of transmission is not known, isolation of laboratory rabbits from cottontails should be practiced.

8. Public Health Significance

The public health significance of this disease is not known. Human infection with *Sarcocystis* has been reported sporadically. Levine (102) feels that it is extremely doubtful that there is a species which infects man solely and suggests that human infections probably represent transmission from animals.

IV. MICROSPORIDA

Encephalitozoon cuniculi (**Synonym:** *Nosema cuniculi*)

1. Introduction

Encephalitozoonosis is a chronic, usually latent disease of rabbits caused by the microsporidan parasite, *Encephalitozoon cuniculi*. The disease is prevalent in many rabbit colonies, with reported ranges of 15 (172) to 76% (108). It has also been reported in a wild rabbit (79).

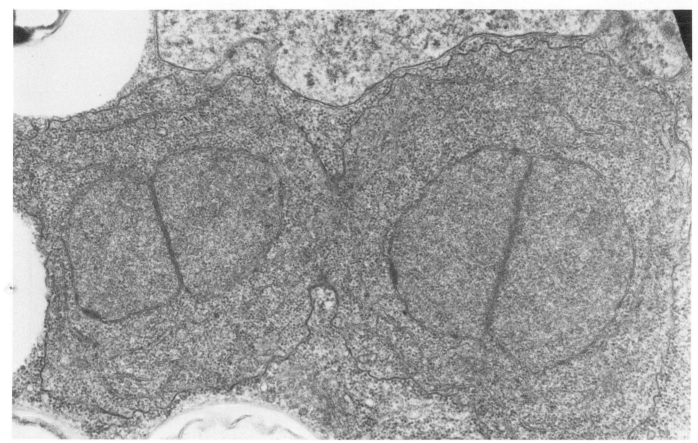

Fig. 7. Nosema bombycis, microsporidium of silkworm. Note 2 diplokarya in dividing form. × 25,000. [Photograph courtesy of Dr. Ann Cali (23).]

The disease is especially important as a complicating factor in the interpretation of experimental data. This aspect has been reviewed recently by Shadduck and Pakes (157).

The disease was first described in 1922 as an infectious encephalomyelitis causing motor paralysis in young rabbits (187), although other investigators had previously reported similar diseases in laboratory rabbits (22, 126, 172). Subsequently, additional workers reported the disease (13, 40, 57, 93, 98, 113, 127, 144, 154, 166, 173). In 1923 Levaditi *et al.* (94, 95) identified a protozoan organism in tissues of rabbits with granulomatous encephalitis. These workers suggested the organism was a microsporidium and named it *Encephalitozoon cuniculi*. Nelson (120), working with the organism in mice, demonstrated the presence of a polar filament, thereby confirming the microsporidan classification. In 1964, Weiser (182) and Lainson *et al.* (89) concluded independently that *Encephalitozoon* belongs in the family Nosematidae, and Lainson and co-workers (89) proposed that the organism infecting mammals should be named *Nosema cuniculi*. Generally, investigators have adhered to this conclusion. Recently, Cali (23) compared ultrastructurally the microsporidium isolated from the rabbit, which is by definition the type species for the mammalian organisms, with the type species for the genus *Nosema* (*N. bombycis*). A significant difference was noted in the nuclei of the 2 organisms. *Nosema bombycis* possesses diplokarya (Fig. 7) (2 nuclei in contact with one another) throughout the developmental cycle, while the rabbit organism does not contain diplokarya during any phase of development. The proliferative forms contain more than one nucleus but the nuclei are never attached to each other (Fig. 8). After spore formation has commenced a single nucleus is present in each sporoblast and spore. It has been proposed, therefore, to return to the original name *Encephalitozoon cuniculi*.

2. Morphology and Life Cycle

The mature spores of *E. cuniculi* are approximately 2.5 by 1.5 μm, oval, and possess a rather thick cell wall (52, 142). In stained preparations vacuoles may be seen at both poles or in the center of the spore (89). They stain poorly with hematoxylin and eosin and are best demonstrated with

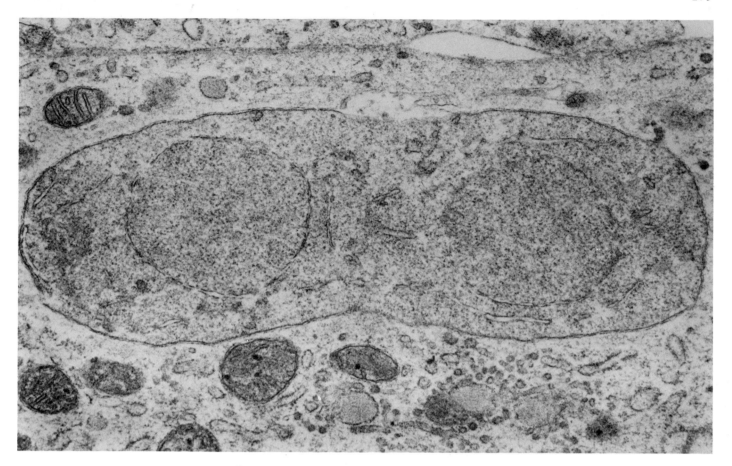

Fig. 8. Encephalitozoon cuniculi. Each segment of the dividing organism contains a single nucleus, as compared to the diplokarya of *N. bombycis.* × 29,700. [Photograph courtesy of Dr. Ann Cali (23).]

the Giemsa stain, Gram's stain, or Goodpasture's carbol fuchsin method. The staining characteristics of *Encephalitozoon* with different preparations is summarized and compared to *Toxoplasma* in Table II.

Ultrastructurally, the mature spores are characterized by a polaroplast at one pole, from which emanates the polar filament that coils around the inner wall (1, 4, 143, 169)

TABLE II

COMPARATIVE STAINING OF *ENCEPHALITOZOON* AND *TOXOPLASMA*[a]

Method	Encephalitozoon	Toxoplasma
Hematoxylin-eosin	Poorly stained	Moderately stained
Giemsa	Bright blue	Mauve
Periodic acid-Schiff	Small granules, positive	Larger granules positive
Gram	Positive	Negative
Weigert's method	Positive	Negative
Wilder's reticulin silver method	No cyst wall	Cyst wall present
Goodpasture (carbol fuchsin)	Deep magenta purple	Not stained

[a]According to Innes *et al.* (77)

(Fig. 9). The polar filament is characteristic of the order Microsporida. This structure may also be viewed by phase microscopy of wet preparations of spores. The filament often extrudes spontaneously, but application of various methods, including pressure, heat, ether, or hydrogen peroxide, stimulates this process (120, 142). The infectious unit, or sporoplasm, is often seen attached to the end of the polar filament.

The life cycle of *E. cuniculi* has not been studied extensively but the following generalizations were garnered from reports by Lainson *et al.* (89) and Petri (142). The sporoplasm is extruded from the spore, apparently on the end of the polar filament. The exact mechanism by which the sporoplasm enters the host's cells is not known but two possibilities have been suggested. The sporoplasm may be mechanically introduced into the cell by the force of the extruded polar filament, or the sporoplasm may enter the cell by some active process. Once the sporoplasm gains entrance into the cell, multiplication occurs. The proliferative forms multiply at the periphery of a vacuole formed in the cytoplasm of the host cell and mature into spores (Fig. 9). The vacuole becomes distended with dividing and

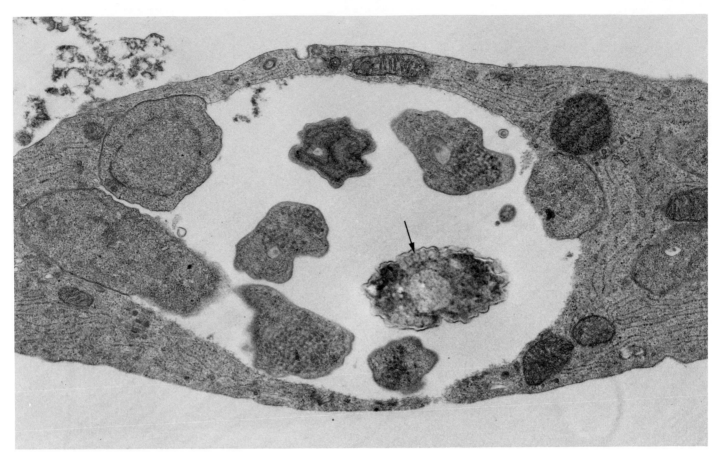

Fig. 9. Encephalitozoonosis. Electron micrograph of a rabbit choroid plexus cell with a cytoplasmic vacuole containing proliferative forms, sporoblasts and a spore of *E. cuniculi.* Arrow indicates cross sections of polar filament in early spore form. × 23,000.

maturating organisms and finally the cell ruptures releasing the spores to start the cycle again.

3. Transmission

Routes by which encephalitozoonosis is transmitted naturally are not definitely known. Most of the experimental work has been done in laboratory mice. The disease has been transmitted experimentally in mice and rabbits by oral administration of infectious material, by various parenteral routes, including intranasally, and by contact between normal and inoculated mice (76, 96, 121, 139, 142). *Encephalitozoon* organisms have been demonstrated in the urine of infected rabbits (188) and the disease has been transmitted to rabbits by oral administration of infectious urine (96). The results of these experiments strongly suggest that horizontal transmission, probably via infectious excreta, is important in the dissemination of the disease. Nelson (121) has also stated that ecto- and endoparasites may play a role in the natural transmission of encephalitozoonosis.

Vertical transmission of *Encephalitozoon* is strongly suspected (97, 121, 139), but has never been definitely proved. Typical lesions and organisms were noted in rabbits which were delivered by Caesarian section into a germfree environment and fed a sterile diet (75a). This observation greatly strengthens the possibility that transplacental infection can occur.

4. Clinical Signs

The disease is usually latent in rabbits. Occasionally infected rabbits may exhibit varied neurological signs, including convulsions, temors, torticollis, paresis, and coma (77, 130a).

5. Pathology

Lesions in the kidneys of infected rabbits are common. Grossly, they may be manifested as multiple, white pinpoint areas randomly scattered over the surface, or usually, as small (2 to 4 mm) indented gray areas on the cortical surface

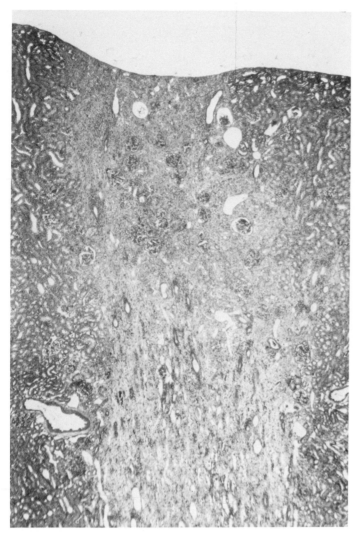

Fig. 10. Encephalitozoonosis. Intense fibrosis of kidney with interstitial infiltration of mononuclear inflammatory cells. Note depression of surface due to scarring. Hematoxylin and eosin. × 47.

(12, 86). The kidney will have a granular appearance if extensively involved. Microscopically, one may see granulomatous nephritis or degrees of interstitial infiltration of lymphocytes and plasma cells with fibrosis and tubular degeneration and dilatation (50). The scars often extend from the cortical surface to the medulla (Fig. 10).

Granulomatous encephalitis is the characteristic lesion of encephalitozoonosis of rabbits. Commonly, one observes randomly distributed, focal granulomas, characterized by a central area of necrosis, surrounded by lymphocytes, plasma cells, microglia, epithelioid cells, and sometimes giant cells (Fig. 11). In other instances, only dense accumulations of glial cells may be noted. Perivascular cuffing with lymphocytes and plasma cells is prominent and nonsuppurative meningitis, especially in areas closely associated with

brain lesions, may also be a feature of the disease. There appears to be no predilection sites and lesions occur in all areas of the brain, most often with a perivascular and periventricular distribution. Lesions of the spinal cord and cerebellum are rarely mentioned but may occur (82).

Although lesions in other tissues are rare, focal, nonsuppurative hepatitis and myocarditis have been reported in rabbits (82).

In many instances, organisms cannot be identified in the lesions. Robinson (145) did not observe organisms in 300 rabbits that had lesions compatible with encephalitozoonosis. When organisms are recognized they may be intra- or extracellular. In the kidneys they are located in tubular cells, often in the medulla, or they may be free in the lumens. Flatt and Jackson (50) found fewer organisms in kidneys with chronic interstitial nephritis and scarring than in those with acute lesions. In the brain organisms may be detected in masses in the necrotic centers of the granulomas (Fig. 12) or individually in areas of glial cell accumulations and in the meninges. One may also observe host cells greatly distended with organisms, often with no detectable inflammatory response (Fig. 13). Organisms from ruptured cells may initiate an intense cellular response (82).

6. Diagnosis

In the past, reliable methods for the diagnosis of encephalitozoonosis in the living host were not available. Therefore, initiation and maintenance of rabbit colonies which are known to be free of the disease are nearly impossible (74). Recently, Pakes and co-workers (129a) reported that a skin test, using antigen prepared by disrupting *E. cuniculi* organisms, appears to be very sensitive in detecting encephalitozoonosis in rabbits. Other investigators have reported that indirect fluorescent antibody techniques may be useful in detecting the disease in rabbits (29a, 77a). Attempts to develop other methods for the detection of humoral antibody in rabbits have not been successful (129).

Tissue suspensions and body fluids can be inoculated intraperitoneally into susceptible mice. Subcutaneous or intramuscular injection of 2.5 mg cortisone acetate once weekly in test mice facilitates diagnosis (77). The assay must be adequately controlled, since the test animals may be latently infected with encephalitozoonosis. Mice develop ascites in 2 to 3 weeks and smears of the ascitic fluid can be examined for the presence of *Encephalitozoon*. This is best done in methanol-fixed smears stained with Giemsa (52). Similarly, smears of tissue suspensions and body fluids from suspect animals can be examined directly.

Recently, Shadduck (156) reported that rabbit choroid plexus cell cultures support the growth of *E. cuniculi*. This system may be used to isolate the organism from suspect tissues, although the reliability of the method for detection

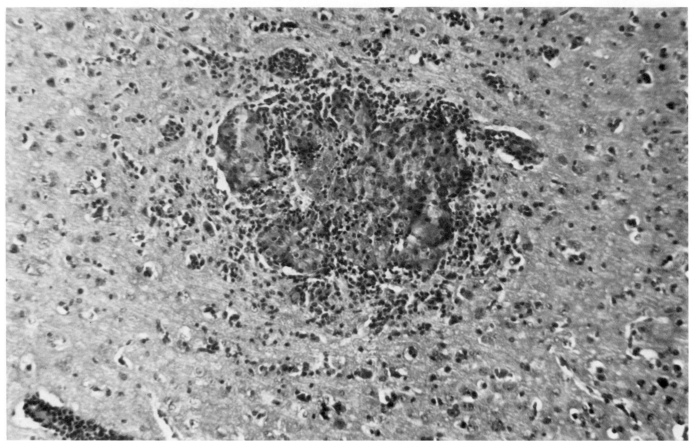

Fig. 11. Encephalitozoonosis. Typical necrotic granuloma in the grain of a rabbit. Perivascular cuffing of vessels with mononuclear inflammatory. cells is also evident. Hematoxylin and eosin. × 225.

of organisms in animal tissues is not yet known (157). Occasionally, the cultures may be spontaneously infected and confuse interpretation, therefore brains from rabbits used as donors of choroid plexus cells should be carefully examined histologically for lesions of encephalitozoonosis.

Encephalitozoonosis is usually diagnosed by histological identification of typical lesions and organisms in fixed tissue sections. The morphological characteristics of the lesions and organisms have been described previously.

7. Treatment, Control, and Prevention

There is no good chemotherapeutic agent for treating encephalitozoonosis. Controlling and preventing the disease is difficult because antemortem detection of infected animals is nearly impossible. Caesarean derivation of stock is apparently not reliable since transplacental transmission may occur.

Howell and Edington (74) have suggested initiating colonies from rabbits that are progeny of does found to be free of the disease by histological examination. The reliability of recognizing the disease by this method has never been

determined and these investigators found that 6% of the progeny from "uninfected" does had lesions of encephalitozoonosis.

8. Public Health Significance

Encephalitozoonosis has been diagnosed in a Japanese boy (112) and a Russian worker has described *Encephalitozoon*-like bodies in the brains of patients with lesions of multiple sclerosis (179, 180). Recently, the disseminated disease was diagnosed by light and electron microscopy in an immunologically compromised infant in the United States (109a). The true public health significance of this disease is not yet known.

V. FLAGELLATES AND AMOEBAE

A. *Chilomastix cuniculi* (**Fonseca, 1915**)

Chilomastix cuniculi is reported in the cecum of the rabbit and is apparently nonpathogenic. The trophozoite has 3

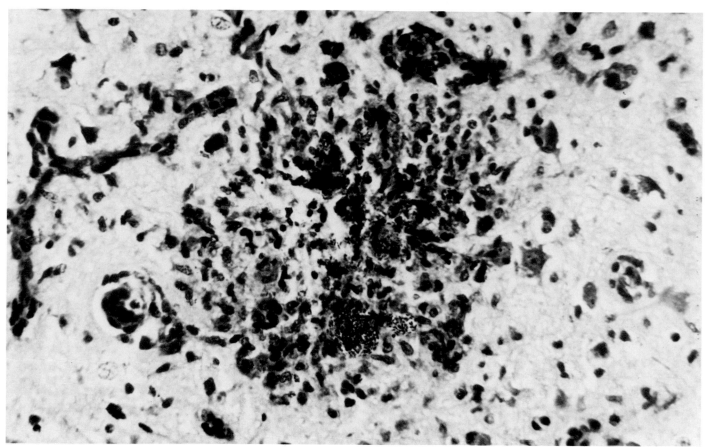

Fig. 12. Encephalitozoonosis. Granuloma with intra- and extracellular *E. cuniculi* organisms. Goodpasture stain × 475.

anterior flagella and measures 10 to 15 μm in length with ranges from 3 to 20 μm (119).

B. *Giardia duodenalis* (**Davaine, 1875**) (**Synonyms:**

Hexamita duodenalis, Lamblia cuniculi)

Giardia duodenalis occurs sporadically in the anterior small intestine of laboratory rabbits and apparently is not pathogenic (101, 102), although Willomitzer (186) strongly suspected that it contributed to the deaths of rabbits in Czechoslovakia. *Giardia* are bilaterally symmetrical flagellates with 2 anterior nuclei, 2 axostyles, and 4 pairs of flagella. The trophozoites of *G. duodenalis* measure approximately 16 × 9 μm and contain curved median bodies that lie transversely across the body (66). The cysts contain 2 to 4 nuclei and are passed in the feces.

Although some species of *Giardia* are pathogens in man, there is no evidence that the rabbit organism is of any public health significance.

Rabbits are used extensively to cultivate *Giardia* sp. for experimental purposes (115).

C. *Monocercomonas cuniculi* (**Tanabe, 1926**) (**Synonyms:**

Eutrichomastix cuniculi, Trichomastix cuniculi)

This apparently nonpathogenic flagellate is found in the cecum of the domestic rabbit. It is piriform and ranges from 5 to 14 μm long. The slender axostyle has a hyalin appearance and projects posteriorly from the body. There are 3 anterior and 1 posterior flagella (119).

D. *Retortamonas cuniculi* (**Collier and Boeck, 1926**)

(**Synonym:** *Embadomonas cuniculi*)

Retortamonas cuniculi occurs primarily in the cecum of rabbits. It is not common and is apparently nonpathogenic. The trophozoites measure approximately 7 to 13 by 5 to 10 μm and are usually ovoid, although occasionally they may have a tail-like process. The cysts are oval and measure 5 to 7 by 3 to 4 μm (36).

E. *Trypanosoma nabiasi* (**Railliet, 1895**)

This hemoflagellate has been found sporadically in wild *Oryctolagus cuniculi* in Europe. *Trypanosoma nabiasi* is

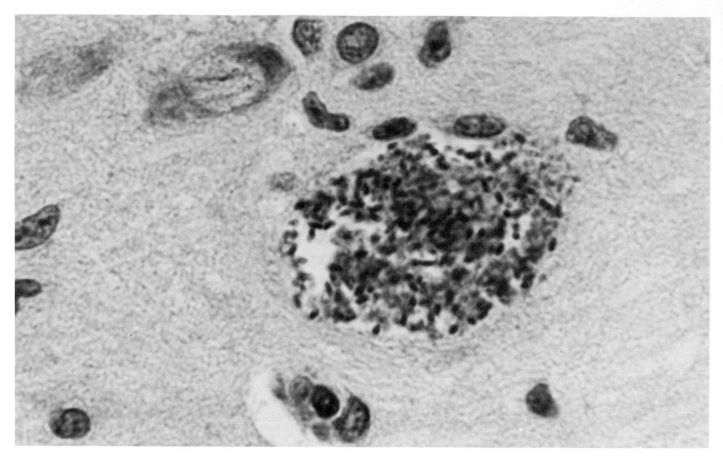

Fig. 13.   Encephalitozoonosis. Brain cell greatly distended with *E. cuniculi*. There is an absence of host response. Goodpasture stain × 1500.

24 to 28 μm long and develops in the gut of the intermediate host, *Spilopsyllus cuniculi*, the rabbit flea. Infection is probably by ingestion of the flea, with development in the blood sinuses of the spleen (60).

F. *Trypanosoma lewisi*-like Organism

Holliman (68) found a *T. lewisi*-like hemoflagellate in 6 of 30 wild rabbits (*Sylvilagus floridanus*). The parasites disappeared from the peripheral circulation after the hosts were confined for several weeks. Leishman bodies or other intermediate forms were not seen in tissue sections. The author felt that fleas were vectors.

G. *Entamoeba cuniculi* (Brug, 1918) (Synonyms: *Entamoeba coli* forma *cuniculi, Entamoeba muris*)

This nonpathogenic amoeba is commonly found in the cecum and colon of the rabbit (102). The trophozoites measure from 10 to 30 μm. The typical cysts are oval, have 8 nuclei and range from 7 to 21 μm in diameter, resembling

E. coli (67). Hoare (67) considers it to be identical to *E. muris*, a common nonpathogenic parasite of rodents.

VI. ORGANISMS OF UNCERTAIN CLASSIFICATION

Pneumocystis carinii

1. Introduction

Pneumocystis carinii is an organism of uncertain classification that is ubiquitous, infecting man and various laboratory animals, including rabbits. In man, pneumocystosis is associated with premature and early infancy, or with diseases and clinical therapy that depress the immunological mechanism. The prevalence of the disease in rabbits is unknown since it is usually latent (25). Mata (111) found the parasite in the lungs of 2 of 15 rabbits that had no gross lesions and showed no clinical signs. It has also been reported in hares (3). It may be recognized in rabbits after treatment with large doses of corticosteroids (159). Thus, pneumocystosis may play a complicating role in biomedical

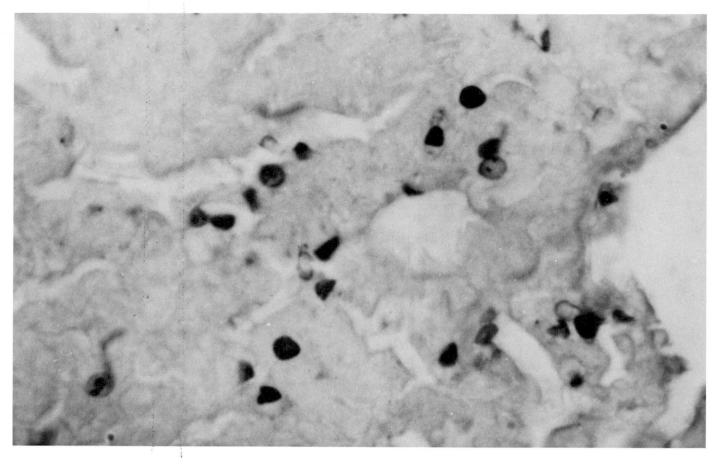

Fig. 14. Pneumocystosis. Dark-staining bodies are *Pneumocystis* in lungs of a rabbit. Grocott's methenamine silver stain. × 1600. (Photograph courtesy of Dr. J. K. Frenkel and The William and Wilkins Company.)

research. Also, rabbits so infected may serve as models for the human disease.

2. Morphology and Life Cycle

In smears or sections of lung tissues the organisms may occur singly as round, oval, or usually crescent-shaped trophozoites of about 1 μm or as cysts 8 to 12 μm in diameter, usually containing 8 trophozoites (107). The nuclei can be demonstrated with Giemsa's or hematoxylin stain. The cyst wall is demonstrated clearly by methanamine silver techniques and Gridley's fungus stain, which may interfere with staining of the nuclei (Fig. 14). The Gram stain and the Chalvardjian and Graw (30) modification of the Mowry fungus stain may also be used and they have the advantage of not interfering with nuclear stains (52). The free forms and cysts are found in alveoli filled with foamy material that stains intensely in the periodic acid-Schiff reaction (107).

Pneumocystis is generally regarded as a protozoan although its exact taxonomic position has never been established. A recent extensive ultrastructural study by Vavra and Kucera (177) suggests that the organism resembles more closely the fungi. Likewise, the life cycle of *Pneumocystis* has not been fully elucidated. Vanek and Jirovec (175) proposed that division is by multiple binary fission. Ultrastructural studies (9, 10, 63, 75) describe thin-walled trophozoite forms and thick-walled organisms which represent the true encysted stage of *Pneumocystis carinii*. It has been suggested that the thin-walled forms make up the foamy material recognized with the light microscope (9, 75).

3. Transmission

Direct contact appears to be necessary for transmission of the disease (52). Experiments by Sheldon (159), in which he infected rabbits by the intranasal route using infected human or rabbit lung, suggest the possibility of aerosol transmission. *In utero* transmission has been reported in man (131), and Frenkel (52) states that transplacental transmission must be considered in animals, giving rise to latent infections.

4. Clinical Signs

The disease is most often inapparent in rabbits. Frenkel *et al.* (54) describe dyspnea in cortisonized rats with the disease but Sheldon (159) observed no clinical signs in infected rabbits following cortisone treatment.

5. Pathology

Sheldon (159) studied the pulmonary lesions histologically in rabbits treated with cortisone and antibiotics and inoculated with material from infected human or rabbit lungs. Lesions ranged from a few foci of mild pneumonitis confined to subpleural areas to extensive interstitial pneumonia characterized by alveolar infiltration with macrophages, lymphocytes and occasionally plasma cells. Free or cyst-bound organisms were commonly seen enmeshed in a PAS positive foamy material in the alveoli. Occasionally cysts were seen to be phagocytized by macrophages. In animals and in man the lesions are usually confined to the lungs, although spleen and lymph node involvement has been described in man (8).

6. Diagnosis

Diagnosis depends on the demonstration of the organism in lung smears or tissue sections, using the tinctorial characteristics previously described. Direct and indirect fluorescein-labeled antibody methods for the specific demonstration of *Pneumocystis* in human lung sections have been used with excellent results (55, 56).

7. Treatment, Control, and Prevention

Frenkel *et al.* (54) describe the effectiveness of sulfadiazine in combination with pyrimethamine and other antiprotozoal drugs in the rat. The effectiveness of these drugs in rabbits is not known.

Caesarean derivation of animals and maintaining them behind suitable barrier systems has been suggested for preventing the disease (52), although the possibility of transplacental infection has not been ruled out.

8. Public Health Significance

The true public health significance of *Pneumocystis carinii* has not been established, although it has been suggested that naturally infected animals in man's environment may serve as a reservoir of the disease (90).

REFERENCES

1. Akao, S. (1969). Studies on the ultrastructure of *Nosema cuniculi*, a microsporidan parasite of rodents. *Jap. J. Parasitol.* **18**, 8–20.
2. Arcay-de-Peraza, L. (1966). The use of *Sarcocystis tenella* "spores" in a new agglutination test for sarcosporidiosis. *Trans. Roy. Soc. Trop. Med. Hyg.* **60**, 761–765.
3. Areán, V. M. (1971). Pulmonary pneumocystosis. *In* "Pathology of Protozoal and Helminthic Diseases" (R. A. Marcial-Rojas, ed.), pp. 291–317. Williams & Wilkins, Baltimore, Maryland.
4. Arison, R. N., Cassaro, J. A., and Pruss, M. P. (1966). Studies on a murine ascites producing agent and its effect on tumor development. *Cancer Res.* **26**, 1915–1920.
5. Bachman, G. W. (1930). Serological studies in coccidiosis of rabbits. *Amer. J. Hyg.* **12**, 624–640.
6. Bachman, G. W. (1930). Immunity in experimental coccidiosis of rabbits. *Amer. J. Hyg.* **12**, 641–649.
7. Bachman, G. W., and Menendez, P. E. (1930). Jaundice in experimental coccidiosis of rabbits. *Amer. J. Hyg.* **12**, 650–656.
8. Barnett, R. N., Hull, J. G., Vortel, V., Kralove, H., and Schwarz, J. (1969). *Pneumocystis carinii* in lymph nodes and spleen. *Arch. Pathol.* **38**, 175–180
9. Barton, E. G., Jr., and Campbell, W. G., Jr. (1967). Further observations on the ultrastructure of *Pneumocystis*. *Arch. Pathol.* **83**, 527–534.
10. Barton, E. G., Jr., and Campbell, W. G., Jr. (1969). *Pneumocystis carinii* in lungs of rats treated with cortisone acetate. *Amer. J. Pathol.* **54**, 209–236.
11. Becker, E. R. (1934). "Coccidia and Coccidiosis of Domesticated, Game and Laboratory Animals and of Man." Collegiate Press, Inc., Ames, Iowa.
12. Bell, E. T., and Hartzell, T. B. (1919). Spontaneous nephritis in rabbits and its relation to chronic nephritis in man. *J. Infec. Dis.* **24**, 628–635.
13. Bender, L. (1925). Spontaneous central nervous system lesions in the laboratory rabbit. *Amer. J. Pathol.* **1**, 653–656.
14. Berengo, A., DeLalla, F., Cavallini-Sampieri, L., Bechelli, G., and Cavallini, F. (1969). Prevalence of toxoplasmosis among domestic and wild animals in the area of Siena, Italy. *Amer. J. Trop. Med.* **18**, 391–394.
15. Beverley, J. K. A., Beattie, C. P., and Roseman, C. (1954). Human *Toxoplasma* infection. *J. Hyg.* **52**, 37–46.
16. Beyer, T. V. (1961). Immunity in experimental coccidiosis of the rabbit caused by heavy infective doses of *Eimeria intestinalis*. *Progr. Protozool., Czech. Acad. Sci. Publ.* p. 448.
17. Biering-Sørensen, U. (1953). Et tilfaelde af *Toxoplasmosis cuniculi* i Danmark. *Nord Veterinaermed.* **5**, 417–426.
18. Boch, J. (1957). Versuche zur Benandlung der Kaninchenkokzidiose mit Nitrofurazon (Furacin-W). *Berlin. Muenchen. Tieraerztl. Wochenschr.* **70**, 264–267.
19. Bourret, G. (1911). La toxoplasmose du lapin à St.-Louis du Sénégal. *Bull. Soc. Pathol. Exot.* **4**, 373–376.
20. Brug, S. L., Den Heyer, J. K., and Haga, J. (1925). Toxoplasmose du la in aux Indes Oriéntales Néerlandaises. *Ann. Trop. Med. Parasitol.* **3**, 232–238.
21. Brumpt, E. (1913). "Précis de Parasitologie." Masson, Paris.
22. Bull, C. G. B. (1917). The pathologic effects of steptococci from cases of poliomyelitis and other sources. *J. Exp. Med.* **25**, 557–580.
23. Cali, A. (1970). Morphogenesis in the genus *Nosema*. *Proc. Int. Colloq. Insect Pathol., 4th.* pp. 431–438.
24. Carini, A. (1911). Infection spontanée du pigeon et du chien due au "*Toxoplasma cuniculi*." *Bull. Soc. Pathol. Exot.* **4**, 518–519.
25. Carini, A., and Maceil, J. (1916). Ueber *Pneumocystis carinii*. *Zentralbl. Bakteriol., Parasitenk.*, **77**, 46.
26. Carvalho, J. C. M. (1942). *Eimeria neoleporis* n. sp., occurring naturally in the cottontail and transmissible to the tame rabbit. *Iowa State Coll. J. Sci.* **16**, 409–410.
27. Carvalho, J. C. M. (1943). The coccidia of wild rabbits of Iowa. I. Taxonomy and host-specificity. *Iowa State Coll. J. Sci.* **18**, 103–134.
28. Carvalho, J. C. M. (1943). The coccidia of wild rabbits of Iowa. II.

Experimental studies with *Eimeria noeleporis. Iowa State Coll. J. Sci.* **18**, 177–189.

29. Cerna, Z. (1966). Application of the indirect fluorescent antibody reaction for the demonstration of antibodies in coccidiosis of rabbits. *Zentralbl. Bakteriol., Parasitenk., Infektionskr. Hyg., Abt. I: Orig.* **199**, 264–267. (Engl. summ.).

29a. Chalupsky, J., Bedrnik, P., and Vavra, J. (1971). The indirect fluorescent antibody test for *Nosema cuniculi. J. Protozool.* **18**, Suppl., Pap. 177.

30. Chalvardjian, A. M., and Grawe, L. A. (1963). A new procedure for the identification of *Pneumocystis carinii* cysts in tissue sections and smears. *J. Clin. Pathol.* **16**, 383–384.

31. Chapman, J. (1929). A study of coccidiosis in an isolated rabbit colony. The clinical symptoms, pathology, immunology and therapy. *Amer. J. Hyg.* **9**, 389–429.

32. Chapman, M. P. (1948). The use of sulfaquinoxaline in the control of liver coccidiosis in domestic rabbits. *Vet. Med.* **43**, 375–379.

33. Christiansen, M. (1948). Toxoplasmose hos harer i Danmark. *Medlemsbl. Dan. Dyrlaegeforen.* **31**, 93–104.

34. Christiansen, M., and Siim, J. C. (1951). Toxoplasmosis in hares in Denmark. Serological identity of human and hare strains of *Toxoplasma. Lancet* **1**, 1201–1206.

35. Christiansen, M., and Siim, J. C. (1951). Toxoplasmosis in hares. *J. Amer. Med. Ass.* **147**, 93.

36. Collier, J., and Boeck, W. C. (1926). The morphology and cultivation of *Embadomonas cuniculi* n. sp. *J. Parasitol.* **12**, 131–140.

37. Cormio, A. (1933). Intorno alle toxoplasmosi spontanea dei conigli. *Pathologica* **25**, 87–91.

38. Crawley, H. (1914). Two new sarcosporidia. *Proc. Acad. Natur. Sci. Philadelphia* **66**, 214–218.

39. Cvetkovic, L., and Tomanovic, B. (1967). Study of amprolium in the treatment of coccidiosis of rabbits. *Vet. Glasn.* **21**, 607–612 (Engl. summ.).

40. Da Fano, C. (1924). Protozoan-like parasites in spontaneous encephalitis of rabbits. *J. Pathol. Bacteriol.* **27**, 333–336.

41. Davies, S. F. M., Joyner, L. P., and Kendall, S. B. (1963). "Coccidiosis." Oliver & Boyd, Edinburgh.

41a. Deschiens, P. R., Levaditi, J. C., and Lamy, L. (1957). Sur quelques aspects morphologiques de sarcosporidies de divers mammifères. *Bull. Soc. Pathol. Exot.* **50**, 225–228.

42. Diehl, J. F. (1960). Effect of hepatic coccidiosis infection in rabbits on tissue levels of vitamins A and E. *J. Nutr.* **71**, 322–326.

43. Diehl, J. F., and Kistler, B. G. (1961). Vitamin E saturation test in coccidiosis-infected rabbits. *J. Nutr.* **74**, 495–499.

44. Dobell, C. C. (1922). The discovery of the coccidia. *Parasitology* **14**, 342–348.

45. Dunlap, J. S., Dickson, W. M., and Johnson, V. L. (1959). Ionographic studies of rabbits infected with *Eimeria stiedae. Amer. J. Vet. Res.* **20**, 589–591.

46. Dürr, U., and Pellérdy, L. (1969). The susceptibility of suckling rabbits to infection with coccidia. *Acta Vet. (Budapest)* **19**, 453–462.

47. Dürr, U., and Schrecke, W. (1970). Chemotherapy of coccidiosis in rabbits. *Deut. Tieraerztl. Wochenschr.* **77**, 435–441.

48. El-Akkad, I. N., and Mandour, A. M. (1969). On some pharmacological and toxicological effects of a protozoan toxin "sarcocystin." *J. Egypt. Med. Ass.* **52**, 942–948.

49. Erickson, A. B. (1946). Incidence and transmission of *Sarcocystis* in cottontails. *J. Wildl. Manage.* **10**, 44–46.

50. Flatt, R. E., and Jackson, S. J. (1970). Renal nosematosis in young rabbits. *Pathol. Vet.* **7**, 492–497.

51. Frenkel, J. K. (1956). Pathogenesis of toxoplasmosis and of infections with organisms resembling *Toxoplasma. Ann. N.Y. Acad. Sci.* **64**, 215–251.

52. Frenkel, J. K. (1971). Protozoal diseases of laboratory animals. *In* "Pathology of Protozoal and Helminthic Diseases" (R. A.

Marcial-Rojas, ed.), pp. 318–369. Williams & Wilkins, Baltimore, Maryland.

53. Frenkel, J. K. (1971). Toxoplasmosis. *In* "Pathology of Protozoal and Helminthic Diseases" (R. A. Marcial-Rojas, ed.), pp. 254–290. Williams & Wilkins, Baltimore, Maryland.

53a. Frenkel, J. K., Dubey, J. P., and Miller, N. L. (1970). *Toxoplasma gondii* in cats: Fecal stages identified as coccidian oocysts. *Science* **167**, 893.

54. Frenkel, J. K., Good, J. T., and Shultz, J. A. (1966). Latent pneumocystis infection of rats, relapse, and chemotherapy. *Lab. Invest.* **15**, 1559–1577.

55. Gaedeke, R. (1960). Morphologische Antigen–Antikörper-Lokalisation mittels fluoreszierender Antikörper im Lungengewebe bei interstitieller Pneumonie. *Monatsschr. Kinderheilk.* **108**, 152.

56. Goetz, O. (1965). Fluorescenzmikroskopische Untersuchungen bei interstitieller plasmacellulärer Pneumonie. *Monatsschr. Kinderheilk.* **113**, 194.

57. Goodpasture, E. W. (1924). Spontaneous encephalitis in rabbits. *J. Infec. Dis.* **34**, 428–432.

58. Goodrich, H. P. (1944). Coccidian oocysts. *Parasitology* **36**, 72–79.

59. Graham, P. C. C., Baker, J. R., and Bird, R. G. (1962). Fine structure of cystic form of *Toxoplasma gondii. Brit. Med. J.* **1**, 83–84.

60. Grenwal, M. S. (1956). Life cycle of the rabbit trypanosome, *Trypanosoma nabiasi* Railliet, 1895. *Proc. Roy. Soc. Trop. Med. Hyg.* **50**, 2–3.

61. Hagen, K. W., Jr. (1958). The effects of continuous sulfaquinoxaline feeding on rabbit mortality. *Amer. J. Vet. Res.* **19**, 494–496.

62. Hagen, K. W., Jr. (1961). Hepatic coccidiosis in domestic rabbits treated with 2 nitrofuran compounds and sulfaquinoxaline. *J. Amer. Vet. Med. Ass.* **130**, 99–100.

63. Ham, E. K., Greenberg, S. S., Reynolds, R. C. and Singer, D. B. (1971). Ultrastructure of *Pneumocystis carinii. Exp. Mol. Pathol.* **14**, 362–372.

64. Hammond, D. M., Scholtyseck, E., and Miner, M. L. (1969). The fine structure of microgametes of *Eimeria stiedae, E. bovis,* and *E. auburnensis. J. Parasitol.* **53**, 235–247.

65. Harcourt, R. A. (1967). Toxoplasmosis in rabbits. *Vet. Rec.* **81**, 191–192.

66. Hegner, R. W. (1922). A comparative study of the Giardias living in man, rabbit, and dog. *Amer. J. Hyg.* **2**, 442–454.

67. Hoare, C. A. (1959). Amoebic infections in animals. *Vet. Rev. Annot.* **5**, 91–102.

68. Holliman, R. B. (1966). A *Trypanosoma lewisi*-like organism from the rabbit, *Sylvilagus floridanus,* in Virginia. *J. Parasitol.* **52**, 622.

69. Horton, R. J. (1967). The route of migration of *Eimeria stiedae* (Lindemann, 1865). sporozoites between the duodenum and bile ducts of the rabbit. *Parasitology* **57**, 9–17.

70. Horton-Smith, C. (1947). The treatment of hepatic coccidiosis in rabbits. *Vet. J.* **103**, 207–213.

71. Horton-Smith, C. (1958). Coccidiosis in domestic mammals. *Vet. Rec.* **70**, 256–262.

72. Horton-Smith, C., Long, P. L., Pierce, A. E., and Rose, M. (1963). Immunity to coccidia in domestic animals. *In* "Immunity to Protozoa" (P. C. C. Garnham, A. E. Pierce, and I. Roitt, eds.), pp. 273–295. Davis, Philadelphia, Pennsylvania.

73. Horton-Smith, C., Taylor, E. L., and Turtle, E. E. (1940). Ammonia fumigation for coccidial disinfection. *Vet. Rec.* **52**, 829–832.

74. Howell, J. McC., and Edington, N. (1968). The production of rabbits free from lesions associated with *Encephalitozoon cuniculi. Lab. Anim.* **2**, 143–146.

75. Huang, S-N., and Marshall, K. G. (1970). *Pneumocystis carinii* infection. A cytologic, histologic, and electron microscopic study of the organism. *Amer. Rev. Resp. Dis.* **102**, 623–635.

75a. Hunt, R. D., King, N. W., and Foster, H. L. (1972). Encephalitozoonosis: Evidence for vertical transmission. *J. Infec. Dis.* **126**, 212–214.

76. Iino, H. (1960). Studies on Encephalitozoon. III. Relationship between the natural infection in mice and environmental conditions. *Keio Igaku* **37**, 515–518 (Engl. summ.).

77. Innes, J. R. M., Zeman, W., Frenkel, J. K., and Borner, G. (1962). Occult, endemic encephalitozoonosis of the central nervous system of mice (Swiss-Bagg-O'Grady strain). *J. Neuropathol. Exp. Neurol.* **21**, 519–533.

77a. Jackson, S. J. (1972). Detection of antibodies in rabbit sera to *Nosema (Encephalitozoon) cuniculi* by the indirect fluorescent antibody technique. M.S. Thesis, University of Missouri, Columbia.

78. Jankiewicz, H. A. (1945). Liver coccidiosis prevented by sulfasuxidine. *J. Parasitol., Suppl.* **31**, 15.

79. Jungheer, E. (1955). Encephalitozoon encephalomyelitis in a rabbit. *J. Amer. Vet. Med. Ass.* **127**, 518.

80. Kardeván, A., and Kapp, P. (1957). A toxoplasmosis elöfordulásáva irányúló vizsgálatok háziállatainkban, Hazai toxoplasmatörzs izolálása. *Magy. Allatorv. Lapja* **12**, 17–22, Abstract in *Vet. Bull.* **27**, 569 (1957).

81. Kessel, J. F., and Jankiewicz, H. A. (1931). Species differentiation of the coccidia of the domestic rabbit based on a study of oocysts. *Amer. J. Hyg.* **14**, 304–324.

82. Koller, L. D. (1969). Spontaneous *Nosema cuniculi* infection in laboratory rabbits. *J. Amer. Vet. Med. Ass.* **155**, 1108–1114.

83. Kozar, Z. (1970). Toxoplasmosis and coccidiosis in mammalian hosts. *In* "Immunity to Parasitic Animals" (G. J. Jackson, R. Herman, and I. Singer, eds.), Vol. 2, pp. 871–912. Appleton, New York.

84. Kraus, R. (1926). Über Tierseuchen in Südamerika. *Seuchenbekaempfung* **3**, 150.

85. Krishnan, K. V. and Lal, J. C. (1933). A note on the finding of *Toxoplasma cuniculi* in two experimental rabbits. *Indian J. Med. Res.* **20**, 1049–1050.

86. Krishna Reddy, A. M. (1963). A case of *Encephalitozoon cuniculi* in a rabbit in Andhra Pradesh. *Indian Vet. J.* **40**, 400–401.

87. Kunstýr, I., Jíra, J., Princová, D., and Míka, J. (1970). Survey of toxoplasma antibodies in domestic rabbits. *Folia Parasitol. (Praha)* **17**, 277–280.

88. Lainson, R. (1955). Toxoplasmosis in England. I. The rabbit (*Oryctolagus cuniculi*) as a host of *Toxoplasma gondii. Ann. Trop. Med. Parasitol.* **49**, 384–396 and 413–416.

89. Lainson, R., Garnham, P. C. C., Killick-Kendrick, R., and Bird, R. G. (1964). Nosematosis, a microsporidial infection of rodents and other animals, including man. *Brit. Med. J.* **2**, 470–472.

90. Le Clair, R. A. (1967). *Pneumocystis carinii* and interstitial plasma cell pneumonia. A review. *Amer. Rev. Resp. Dis.* **96**, 1131–1136.

91. Le Pennec, J., and Vermeil, C. (1969). Hare toxoplasmosis in Vendee, *Bull. Soc. Pathol. Exot.* **62**, 522–528.

92. Le Pennec, J., Vermeil, C., and Le Pennec, J. J. (1966). On 2 further cases of cuniculine toxoplasmosis in Vendee. *Bull. Acad. Vet. Fr.* [N.S.] **39**, 109–114.

93. Levaditi, C., Nicolau, S., and Schoen, R. (1923). Encephalitis du lapin. *C. R. Soc. Biol.* **89**, 775–779.

94. Levaditi, C., Nicolau, S., and Schoen, R. (1923). L'agent étiologique de l'encéphalite epizootique du lapin (*Encephalitozoon cuniculi*). *C. R. Soc. Biol.* **89**, 984–986.

95. Levaditi, C., Nicolau, S., and Schoen, R. (1923). L'étiologie de l'encéphalite. *C. R. Acad. Sci.* **177**, 985–988.

96. Levaditi, C., Nicolau, S., and Schoen, R. (1924). L'étiologie de l'encéphalite epizootique du lapin, dans ses rapports avec l'étude expérimentale de l'encéphalite lethargique *Encephalitozoon cuniculi* (nov. spec.). *Ann. Inst. Pasteur, Paris* **38**, 651–712.

97. Levaditi, C., Nicolau, S., and Schoen, R. (1924). Virulence de l'*Encephalitozoon cuniculi* pour la souris. *C. R. Soc. Biol.* **90**, 194–196.

98. Levaditi, C., Nicolau, S., and Schoen, R. (1924). La microsporidiose du lapin, ses relations avec la rage. *C. R. Acad. Sci.* **178**, 256–258.

99. Levaditi, C., Schoen, R., and Sanchis-Bayarri, V. (1927). Encéphalité spontanée du lapin, provoquée par le *Toxoplasma cuniculi. C. R. Soc. Biol.* **97**, 1692–1693.

100. Levi della Vida, M. (1914). Alcune osservazioni sul Toxoplasma riscontrato in conigli nostrani. *Arch. Farmacol. Sper. Sci. Affini* **17**, 405–413.

101. Levine, N. D. (1957). Protozoan diseases of laboratory animals. *Proc. Anim. Care Panel* **7**, 98–126.

102. Levine, N. D. (1961). "Protozoan Parisites of Domestic Animals and of Man." Burgess, Minneapolis, Minnesota.

102a. Levine, N. D., and Kocan, R. Personal communication.

103. Lund, E. E. (1949). Considerations in the practical control of intestinal coccidiosis of domestic rabbits. *Ann. N.Y. Acad. Sci.* **52**, 611–620.

104. Lund, E. E. (1950). A survey of intestinal parasites in domestic rabbits in six counties in Southern California. *J. Parasitol.* **36**, 13–19.

105. Lund, E. E. (1954). Estimating relative pollution of the environment with oocysts of *Eimeria stiedae. J. Parasitol.* **40**, 663–667.

106. Lund, E. E. (1954). The effect of sulfaquinoxaline on the course of *Eimeria stiedae* infections in the domestic rabbit. *Exp. Parasitol.* **3**, 497–503.

107. Lyons, H. A., Vinijchaikul, K. and Hennigar, G. R. (1961). *Pneumocystis carinii* unassociated with any other disease. Clinical and pathological studies. *Arch. Intern. Med.* **108**, 929–936.

108. Malherbe, H., and Munday, V. (1958). *Encephalitozoon cuniculi* infection of laboratory rabbits and mice in South Africa. *J. S. Afr. Vet. Med. Ass.* **29**, 241–246.

109. Manz, W. (1867). Beitrag sur Kenntnis der Miescher'schen Schläuche. *Arch. Mikrosk. Anat. Entwicklungsmech.* **3**, 345–356.

109a. Margileth, A. M., Strano, A. J., Chandra, R., Neafie, R., Blum, M., and McCully, R. M. (1973). Disseminated nosematosis in an immunologically compromised infant. *Arch. Path.* **95**, 145–150.

110. Marotel, G., and Pierron, P. M. (1943). Deux notes de clinique parasitaire: Sur la coccidiose bovine et al toxoplasmose. *Rev. Med. Vet.* **94**, 112–116.

111. Mata, A. D. (1959). Latent infection by *Pneumocystis carinii* in rabbits in Mexico City. *Rev. Inst. Salubr. Enferm. Trop., Mex.* **19**, 357–360.

112. Matsubayashi, H., Koike, T., Mikata, I., Takei, H., and Hagiwara, S. (1959). A case of *Encephalitozoon*-like body infection in man. *Arch. Pathol.* **67**, 181–187.

113. McCartney, J. E. (1924). Brain lesions of the domestic rabbit. *J. Exp. Med.* **39**, 51–61.

114. McPherson, C. W., Habermann, R. T., Every, R. R., and Pierson, R. (1962). Eradication of coccidiosis from large breeding colony of rabbits. *Proc. Anim. Care Panel* **12**, 133–140.

115. Meyer, E. A. (1970). Isolation and axenic cultivation of Giardia trophozoites from the rabbit, chinchilla, and cat. *Exp. Parasitol.* **27**, 179–183.

116. Mocsari, E., and Szemeredi, G. (1969). Diagnostical value of the complement fixation test and the fluorescent antibody method in laboratory diagnosis of Toxoplasmosis. *Acta Vet. (Budapest)* **19**, 21–27.

117. Morel, P. (1954). Toxoplasmose du lapin. *Rec. Med. Vet.* **130**, 271–376.

118. Morgan, B. B., and Waller, E. F. (1940). A survey of the parasites of the Iowa cottontail (*Sylvilagus floridanus mearnse*). *J. Wildl. Manage.* **4**, 21–26.

119. Morgan, B. B., and Hawkins, P. A. (1952). "Veterinary Protozoology." Burgess, Minneapolis, Minnesota.

120. Nelson, J. B. (1962). An intracellular parasite resembling a microsporidian associated with ascites in Swiss mice. *Proc. Soc. Exp. Biol. Med.* **109**, 714–717.

121. Nelson, J. B. (1967). Experimental transmission of a murine microsporidian in Swiss mice. *J. Bacteriol.* **94**, 1340–1345.

122. Nicolle, C., and Manceaux, L. (1908). Sur une infection a corps de Leishman (ou organismes voisins) du gondi. *C. R. Acad. Sci.* **147**, 763–766.

123. Nicolle, C., and Manceaux, L. (1909). Sur un nouveau protozoaire du gondi. *C. R. Acad. Sci.* **148**, 369–372.

124. Nicolle, C., and Manceaux, L. (1909). Sur un protozoaire du gondi (Toxoplasma N. Gen.). *Arch. Inst. Pasteur Tunis* **2**, 97–103.

125. Niilo, L. (1967). Acquired resistance to reinfection of rabbits with *Eimeria magna*. *Can. Vet. J.* **8**, 201–208.

126. Oliver, J. (1922). Spontaneous chronic meningoencephalitis of rabbits. *J. Infec. Dis.* **30**, 91–94.

127. Oliver, J. (1924). Morphologic differentiation of meningoencephalitis of rabbits and epidemic (lethargic) encephalitis. *Arch. Neurol. Psychol. (Chicago)* **11**, 321–327.

128. Owen, D. (1970). Life cycle of *Eimeria stiedae*. *Nature (London)* **227**, 304.

129. Pakes, S. P., and Shadduck, J. A. Unpublished observations.

129a. Pakes, S. P., Shadduck, J. A., and Olsen, R. G. (1972). Diagnostic skin test for encephalitozoonosis in rabbits. *Lab. Anim. Sci.* **22**, 870–877.

130. Patterson, S. W. (1923). A complement-fixation test in coccidiosis of the rabbit. *Brit. J. Exp. Pathol.* **4**, 1–4.

130a. Pattison, M., Clegg, F. G., and Duncan, A. L. (1971). An outbreak of encephalomyelitis in broiler rabbits caused by *Nosema cuniculi*. *Vet. Rec.* **88**, 404–405.

131. Pavlica, F. (1962). The first observation of congenital pneumocystic pneumonia in a fully developed still-born child. *Ann. Paediat.* **198**, 177–184.

132. Pellérdy, L. (1956). On the status of the *Eimeria* species of *Lepus europaeus* and related species. *Acta Vet. (Budapest)* **6**, 451–467.

133. Pellérdy, L. (1965). "Coccidia and Coccidiosis." Hung. Acad. Sci., Budapest.

134. Pellérdy, L. (1969). Parenteral infection experiments with *Eimeria stiedae*. *Acta Vet. (Budapest)* **19**, 171–182.

135. Pellérdy, L. (1969). Zur Problematik der Kaninchenkokzidiosen. *Parasitol. Hung.* **2**, 176–186.

136. Pellérdy, L., and Babos, A. (1952). Investigations into the therapy of liver coccidiosis with atabrine, nitrofurazone and ultraseptl, together with attempts to measure the toxic doses of these compounds. *Acta Vet. (Budapest)* **2**, 281–287.

137. Pellérdy, L., and Szemeredi, G. (1965). Studies on the effectiveness of furazolidone against rabbit coccidiosis. *Magy. Allatorv. Lapja* **20**, 546–548.

138. Pelouro, J. T. (1943). Toxoplasmose espontanea em coelhos. *Repos. Lab. Patol. Vet., Lisboa* **5**, 241–246; Abstract in *Vet. Bull.* **15**, 387 (1945).

139. Perrin, T. L. (1943). Spontaneous and experimental encephalitozoon infection in laboratory animals. *Arch. Pathol.* **36**, 559–567.

140. Perrin, T. L. (1943). *Toxoplasma* and *Encephalitozoon* in spontaneous and in experimental infections of animals. A comparative study. *Arch. Pathol.* **36**, 568–578.

141. Peterson, E. H. (1950). The prophylaxis and therapy of hepatic coccidiosis in the rabbit by the administration of sulfonamides. *Vet. Med.* **45**, 170–172.

142. Petri, M. (1969). Studies on *Nosema cuniculi* found in transplantable ascites tumours with a survey of microsporidiosis in mammals. *Acta Pathol. Microbiol. Scand., Suppl.* **204**, 1–91.

143. Petri, M., and Schiødt, T. (1966). On the ultrastructure of *Nosema cuniculi* in the cells of the Yoshida rat ascites sarcoma. *Acta Pathol. Microbiol. Scand.* **66**, 437–446.

144. Pette, H. (1925). Über eine spontan beim Kaninchen auftretende encephalitische Erkrankung. *Berlin. Klin. Wochenschr.* **4**, 257.

145. Robinson, J. J. (1954). Common infectious disease of laboratory rabbits questionably attributed to *Encephalitozoon cuniculi*. *AMA Arch. Pathol.* **58**, 71–84.

146. Rose, M. E. (1959). A study of the life cycle of *Eimeria stiedae* (Lindemann, 1865) and the immunological response of the host. Ph.D. Thesis, Cambridge University.

147. Rose, M. E. (1959). Serological reactions in *E. stiedae* infection of the rabbit. *Immunology* **2**, 112–122.

148. Rose, M. E. (1961). The complement fixation test in hepatic coccidiosis of rabbits. *Immunology* **4**, 346–353.

149. Rutherford, R. L. (1943). The life cycle of four intestinal coccidia of the domestic rabbit. *J. Parasitol.* **29**, 10–32.

150. Sangiorgi, A. (1914). *Leucocytogregarina cuniculi* n. sp. *Pathologica* **6**, 49.

151. Scholtyseck, E. (1965). Elektronmikroskopische Untersuchungen über die schizogonie bei Coccidien (*Eimeria perforans* and *E. stiedae*). *Z. Parasitenk.* **26**, 50–62.

152. Scholtyseck, E., Hammond, D. M., and Ernst, J. V. (1966). Fine structure of the macrogametes of *E. stiedae*, *E. bovis*, and *E. auburnensis J. Parasitol.* **52**, 975–987.

153. Scholtyseck, E., and Piekarski, G. (1965). Elektronmikroskopische Untersuchungen an Merozoiten von Eimerian (*Eimeria perforans* and *E. stiedae*) und *Toxoplasma gondii*. Für systematischen Stellung von *T. gondii. Z. Parasitenk.* **26**, 91–115.

154. Seifreid, O. (1929). Zum Problem der spontanem und experimentellen Encephalitis beim Kaninchen. *Z. Infektionskr. Haustiere* **36**, 18–30.

155. Senaud, J., and Cerna, Z. (1969). Etude ultrastructurale des merozoites et de la schizogonie des coccidies (Eimeriina): *Eimeria magna* (Perard 1925) de l'intestin des lapins et *E. tenella* (Railliet et Lucet, 1891) des coecums des poulets, *J. Protozool.* **16**, 155–165.

156. Shadduck, J. A. (1969). *Nosema cuniculi: In vitro* isolation. *Science* **166**. 516–517.

157. Shadduck, J. A., and Pakes, S. P. (1971). Encephalitozoonosis (nosematosis) and toxoplasmosis. *Amer. J. Pathol.* **64**, 657–674.

158. Sheffield, H. G., and Melton, M. L. (1968). The fine structure and reproduction of *Toxoplasma gondii*. *J. Parasitol.* **54**, 209–226.

159. Sheldon, W. H. (1959). Experimental pulmonary *Pneumocystis carinii* infection in rabbits. *J. Exp. Med.* **110**, 147–160.

160. Siim, J. C., Biering-Sørensen, U., and Møller, T. (1963). Toxoplasmosis in domestic animals. *Advan. Vet. Sci.* **8**, 335–429.

161. Smetana, H. (1933). Coccidiosis of the liver in rabbits. I. Experimental study on the excystation of oocysts of *Eimeria stiedae*. *Arch. Pathol.* **15**, 175–192.

162. Smetana, H. (1933). Coccidiosis of the liver in rabbits. II. Experimental study on the mode of infection of the liver by sporozoites of *Eimeria stiedae*. *Arch. Pathol.* **15**, 330–339.

163. Smetana, H. (1933). Coccidiosis of the liver in rabbits. III. Experimental study of the histogenesis of coccidiosis of the liver. *Arch. Pathol.* **15**, 516–536.

164. Smith, B. F., and McShan, W. H. (1949). The effect of the protozoan parasite, *Eimeria stiedae*, on the succinic dehydrogenase activity of liver tissue of rabbits. *Ann. N.Y. Acad. Sci.* **52**, 496–500.

165. Smith, T. (1910). A protective reaction of the host in intestinal coccidiosis of the rabbit. *J. Med. Res.* **23**, 407–415.

166. Smith, T., and Florence, L. (1925). *Encephalitozoon cuniculi* as a kidney parasite in the rabbit. *J. Exp. Med.* **41**, 25–35.

167. Splendore, A. (1908). Un nuova protozoa parassita di conigli incontrato nelle lesioni anatomiche d'una malattia che ricorda in molti punti il kala-azar dell'uomo. *Rev. Soc. Sci. Sao Paulo* **3**, 109.

168. Splendore, A. (1909). Sur un nouveau protozoaire parasite du lapin. Deuxième note preliminaire. *Bull. Soc. Pathol. Exot.* **2**, 462–465.

169. Sprague, V., and Vernick, S. H. (1971). The ultrastructure of *Encephalitozoon cuniculi* (Microsporida, Nosematidae) and its taxonomic significance. *J. Protozool.* **18**, 560–569.

169a. Stringer, R. P., Harkema, R., and Miller, G. C. (1969). Parasites in rabbits in North Carolina. *J. Parasitol.* **55**, 328.

170. Tsunoda, K., Imai, S., Tsutsumi, Y., and Inouye, S. (1968). Clinical effectiveness of sulfamonomethoxine and sulfadimethoxine in spontaneous coccidial infections in rabbits. *Jap. J. Vet. Sci.* **30**, 109–117.

171. Tsunoda, K., Imai, S., Tsutsumi, Y., and Inouye, S. (1968). Intermittent medication of sulfadimethoxine and sulfamonomethoxine for the treatment of coccidiosis in domestic rabbits. *Tokyo Nat. Inst. Anim. Health Quart.* **8**, 74–80.

172. Twort, C. C., and Archer, H. E. (1922). Spontaneous encephalomyelitis of rabbits, and its relation to spontaneous nephritis, etc. *Vet. J.* **28**, 367–372.

173. Twort, C. C., and Archer, H. E. (1923). The experimental production of a fatal nephritis with a filter-passing virus of nervous origin. *Lancet* **204**, 1102–1106.

174. Tyzzer, E. E. (1929). Coccidiosis in gallinaceous birds. *Amer. J. Hyg.* **10**, 269–383.

175. Vanek, J., and Jirovec, O. (1952). Parasitäre pneumonie: "Interstitielle" Plasmazellenpneumonie der Fruhgeborenen, verursacht durch *Pneumocystis carinii. Zentralbl. Bakteriol., Parasitenk., Infektionskr. Hyg., Abt. 1: Orig.* **158**, 120–127.

176. van Sacegham, R. (1916). Observations sur des infections naturelles par *Toxoplasma cuniculi. Bull. Soc. Pathol. Exot.* **9**, 432–434.

177. Vavra, J., and Kucera, K. (1970). *Pneumocystis carinii* Delanoë, its ultrastructure and ultrastructural affinities. *J. Protozool.* **17**, 463–483.

178. Vermeil, C., LePennec, J., and Senelar, R. (1966). Toxoplasmosis in the Vendée: Contribution to the study of the reservoir of the organ-

ism: The rabbit. *Bull. Soc. Pathol. Exot.* **58**, 1040–1049 (Engl. summ.).

179. Viting, A. I. (1965). (The etiology of multiple sclerosis in light of findings of a morphologic study of the central nervous system (preliminary communication).) *Zh. Nervopatol. Psikhiat. S. S. Korsakova.* **65**, 1641–1645 (Russ.).

180. Viting, A. I. (1969). (The parasitic nature of diffuse or multiple sclerosis.) *Akad. Nauk., Leningr. Br. Parasitol.* **3**, 569–580 (Russ.).

181. Vivell, O., and Buhn, W. H. (1952). Über eine Toxoplasmose-epidemie in einen Tierstall mit kleinen Laboratoriumstieren. *Z. Hyg. Infektionskr.* **135**, 298–306.

182. Weiser, J. (1964). On the taxonomic position of the genus *Encephalitozoon.* Levaditi, Nicolou and Schoen, 1923 (Protozoa: Microsporidia). *Parasitology* **54**, 749–751.

183. Werner, H. (1966). Spontane Toxoplasma-infektion beim Hauskaninchen (*Oryctolagus cuniculus* L.) *Zentralbl. Bakteriol., Parasitenk., Infekionskr. Hyg., Abt. 1: Orig.* **199**, 259–263.

184. Wickham, M., and Carne, H. R. (1950). Toxoplasmosis in domestic animals in Australia. *Aust. Vet. J.* **26**, 1–3.

185. Wiktor, T. J. (1950). Toxoplasmose animale: Sur une épidémie des lapins et des pigeons à Stanleyville (Congo Belge). *Ann. Soc. Belg. Med. Trop.* **30**, 97–107.

186. Willomitzer, J. (1959). Notes on the incidence of the parasite *Lamblia cuniculi. Vet. Casopsis* **8**, 182–187 (Engl. summ.).

187. Wright, J. H., and Craighead, E. M. (1922). Infectious motor paralysis in young rabbits. *J. Exp. Med.* **36**, 135–140.

188. Yost, D. H. (1958). Encephalitozoon infection in laboratory animals. *J. Nat. Cancer Inst.* **20**, 957–960.

CHAPTER 12

Arthropod Parasites

Alan L. Kraus

I. INTRODUCTION

A. General

This chapter contains a review of the principal arthropod parasites described for the domestic laboratory rabbit (*Oryctolagus cuniculi*). Since wild species of rabbits and hares are sometimes used in research or are presented to diagnostic laboratories for evaluation, the common arthropod parasites of these species are also included. In addition, contact between wild or feral lagomorphs (or their ectoparasites) with those *Oryctolagus cuniculus* raised in an outdoor or semi-enclosed commercial breeding environment may invite transmission of arthropod parasites from wild to domestic species. In general, the arthropods affecting wild rabbits exhibit little host specificity and may be readily transmitted to the domestic species if the appropriate conditions exist.

Of the nine Classes of arthropods, only two appear to be significant as parasites of the rabbit: the Arachnida and the Hexapoda.

B. Arthropods as Agents of Disease

The arthropod parasites of any species may produce problems varying from merely annoyance to severe disease and even death. The Anoplura or sucking lice, for example, can produce merely an annoyance to the host, however, if the infestation is severe, blood loss and the resulting concomitants of anemia can be significant. Certain parasites may cause severe injury to the sensory organs. For example, the bot or warble flies may deposit their eggs on or about the head of the rabbit and the larvae which develop may produce severe damage to the ocular structures (ophthalmomyiasis). Myiasis at other sites is very common in the rabbit. Dermatoses may occur when the various species of mites, chiggers, or lice infest the rabbit. The host's allergic reaction to many of the ectoparasites also may play a role in the pathogenesis of clinical signs and pathological changes, and, not the least of one's concern, is the role that the various ectoparasites of the rabbit play as vectors of disease such as viral myxomatosis and tularemia.

The potential pathogenecity of these arthropods is enhanced by their ability to maintain and increase their populations due to their short generation time and because of the enormous numbers of offspring capable of being produced on or about the host. Their jointed appendages, and, in the case of many of the insects, wings allow these small animals considerable mobility both on and off the host. Their highly specialized mouth parts and means of eliminating body wastes and secretions allow for differing mechanisms for transferring microbial pathogens as well as producing primary or secondary (allergic or infectious) injury to the host.

Finally, it is most important to note that although there is a wealth of literature concerning the arthropod parasites that can occur on the rabbit, much less has been studied and reported concerning their potential for producing disease.

II. THE PRINCIPAL ARTHROPOD GROUPS

A. General

It is beyong the scope of this work to detail much of what is known about the anatomy, physiology, and life cycles of the parasites to be discussed. The reader is referred to general texts in medical entomology and zoology for such information (95, 108). General and anatomical features that will assist one in identifying the various species will be presented wherever possible. The physiology and life cycles of these parasites will be discussed only as they relate to disease production or transmission and control or eradication of the parasite.

B. The Arachnid Parasites of Rabbits (Mites and Ticks)

This class contains those commonly encountered ectoparasites of rabbits, the mites and ticks, as well as those of lesser importance, the pentastomes.

As opposed to other arachnids (spiders and scorpions), the mites and ticks do not have a segmented abdomen and it is imperceptibly fused to the cephalothorax. They neither have a distinct head nor do they have antennae. Adult acarines possess four pair of legs, whereas the larval and nymphal stages have only three pains. For a general consideration of mites of medical important the reader is referred to an article by Baker (9).

MITES

The mites parasitic on the rabbit are all found in one of five families: the Psoroptidae, Sarcoptidae, Cheyletidae, Listrophoridae, and Trombiculidae.

1. The Psoroptic Mange Mite of Rabbits (*Psoroptes cuniculi*)

a. HISTORICAL. Without question, more has been written on the psoroptic mange mite of rabbits than any other

single ectoparasite affecting this species. The disease caused by this mite is commonly known as psoroptic mange, ear mange, otoacariasis, ear canker, or psoroptic scabies.

The psoroptic mange mite of rabbits historically has been identified by a variety of names—*Dermatodectes cuniculi* (Delafond, 1859): *Psoroptes longirostris var. cuniculi* (Megnin, 1877); *Psoroptes communis var. cuniculi* (Raillet, 1893); and *Psoroptes equi var. cuniculi* (Neven-Lemaire, 1938). On the basis of Sweatman's review of the life history and validity of the species of mites in the genus *Psoroptes*, the correct designation of the rabbit ear mite is *Psoroptes cuniculi* (Delafond, 1859) Canestrini and Kramer, 1899 (224).

In addition to affecting rabbits, Sweatman also considers *P. cuniculi* to be the ear mite of the goat, sheep, horse, donkey, mule, and possibly *Gazella* (224). Roberts has reported acaritic otitis externa and dermatitis affecting other parts of the body caused by *P. cuniculi* in captive mule deer (185).

Most reviews of rabbit ectoparasites include *Chorioptes cuniculi* as a second ear mite of laboratory rabbits. The extensive review of the literature conducted prior to the writing of this chapter failed to locate one original article which described primary recovery of chorioptic mites from rabbits (2, 23, 58, 123, 205, 208, 209). In a review of mange in laboratory animals published in 1911, Low was at that time also unable to record *C. cuniculi* from rabbits (136). The entrance of *C. cuniculi* into the literature of rabbit diseases appears to have been the result of an English translation of a German textbook by Zürn (92). Zürn writing in 1874 about the chorioptic mange mite of the horse, goat, cow, sheep, and llama stated as an addendum to his paper.... "Ich nun nicht als *Dermatophagen* (*Chorioptes*) ausprechen konnte, sondern die zur Gattung *Dermatokoptes* (*Psoroptes*) gehorten...." This author may have written this addendum in haste and in so doing interchanged *Dermatophagen* (*Chorioptes*) and *Dermatokoptes* (*Psoroptes*) when attempting to report dermatophagid or chorioptic mange from the rabbit for the first time. The matter could have been further confused because *Otodectes cyanotus*, a rarely observed rabbit parasite (primarily found in dog, cat, ferret, and fox) which was at that time considered a chorioptic mite, may have been the parasite observed by Zürn. At a point in time, now nearly 100 years later, one can merely speculate, however, that the published literature of the succeeding 100 years has not clarified Zürn's observations. Interested readers are referred to the articles by Sweatman on the two genera—*Psoroptes* and *Chorioptes* (223, 224). Since present information on the occurrence of *Chorioptes* in the rabbit is, at best, inconclusive, it is suggested that this parasite not be considered as affecting rabbits until evidence to the contrary is published.

b. OCCURRENCE. i. GEOGRAPHICAL. *Psoroptes cuniculi*

has been reported from every continent and is probably present in all areas where rabbits are raised (137).

ii. INCIDENCE. Numerous surveys have been conducted of both commercially raised and wild lagomorphs to determine the incidence of *Psoroptes* infestation. Reports of surveys of wild (*Sylvilagus*) lagomorphs in the eastern United States (18), Wisconsin (90), North Carolina (221), California (94), Georgia (157), Virginia (140), and Illinois (217), did not contain mention of the presence of *Psoroptes cuniculi*, however. Reports from Great Britain (134) and New Zealand (28) also did not mention the presence of *Psoroptes cuniculi* in the wild (*Oryctolagus*) rabbits examined.

The absence of this parasite in wild lagomorphs is interesting in light of the relatively high incidence of psoroptes in many commercial rabbitries and in laboratory colonies (23, 53, 57, 58, 97, 98, 143).

There is no question, however, that psoroptic otoacariasis is the most common and costly ectoparasitic disease of laboratory rabbits.

c. DIAGNOSIS. i. CLINICAL SIGNS. A rabbit affected by psoroptic mites may be seen shaking its head or scratching at its head and ears with its rear feet due to the intense pruritis caused by *P. cuniculi*. Self-mutilation may lead to secondary infection with bacterial pathogens in or about the head and neck. These obligate, nonburrowing parasites have chelicerae (mouth parts) that both pierce and chew the epidermal layers of the skin of the host causing inflamation with exudation. In the early stages of the disease there is a dry whitish-gray to tan crusty exudate inside the ear at the bottom of the concha. As the disease progresses the crusty dry bran-like material thickens to as much as three-fourths of an inch or more and consists of desquamated epithelial cells. serum, inflamatory cells, mite feces, and the mites themselves. (See Fig. 1, top.) If the detritus and exudate are removed, the skin surface is found to be moist and red. There may be an offensive odor emanating from the ears. The ears are severely inflamed and may be very painful when touched.

Since the psoroptic mite is a nonburrowing mite, they may be readily seen in the detritus or on the raw oozing surface if the debris is removed. The mites are quite large and can be readily seen in many cases with the unaided eye. Otoscopic examination of lesions deep within the external ear canal may be necessary to visualize the mites, however.

It is a commonly held misconception that psoroptic mites are responsible for initiating otitis media with subsequent torticollis to the affected side following their penetration of the tympanum. Two reviews covering the literature over a span of 40 years provide ample evidence that otitis media in the rabbit is caused by an ascending infection via the eustachean tube and is most commonly caused by *Pasteurella multocida* (7,212). Evidence presented clearly absolves

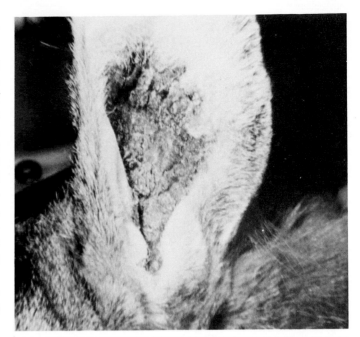

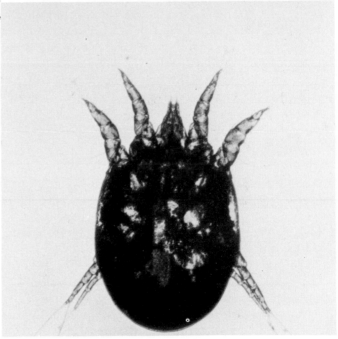

Fig. 1. Ear mange due to *Psoroptes cuniculi*, the common ear mite (top), *Psoroptes cuniculi* (bottom).

P. cuniculi of any important role in the pathogenesis of otitis media in the rabbit (75).

ii. LOCATION ON HOST. *Psoroptes cuniculi* is found almost exclusively on the inner epithelial surface of the pinna and concha of the ear. Low and others have described rabbits that also had lesions on the face, neck, and limbs in the absence of any other pathological agent but *Psoroptes cuniculi*. He has also reported dual infections of both psoroptic and sarcoptic mange in the same animal. Although sarcoptic mites were never found on the inner surface of the ears, both psoroptic and sarcoptic mites were identified on the face, outside of the ears, neck, limbs, abdomen, and genitals (136).

iii. IDENTIFICATION. Clinical samples of desquamated epithelium and exudate from affected ears usually reveal numerous mites. Many are found in attachment ("copulation") pairs. Upon closer examination under a disecting microscope, a positive identification of *P. cuniculi* can be readily made. (See Fig. 1, bottom.)

All stages in the life cycle of *P. cuniculi* may be found since this mite spends its entire life cycle on the affected host. The stages are the same for both male and female mites (egg, larvae, protonymph, and adult), however, in the female, the deutonymph and adult mites are referred to as the pubescent and ovigerous females. For a detailed discussion of the life history of psoroptic mites, the reader is referred to the article by Sweatman (224). This article contains detailed illustrations of the various life stages of *P. cuniculi*.

Psoroptes cuniculi adults are large (males: 431–547 μm long, by 322–462 μm wide; females: 403–749 μm long, by 351–499 μm wide) oval-shaped mites with well-developed legs bearing jointed pedicels. At the end of the pedicels are bell-shaped suckers or caruncles. (See Fig. 2.) The legs project beyong the margin of the body. The pedicel of the tarsal suckers too are long and are composed of three segments. The anus is terminal. *Psoroptes* have long and pointed chelicerae. Chorioptic mites, on the other hand, have short unjointed pedicels.

iv. CONTROL AND TREATMENT. Under favorable environmental conditions, a complete life cycle from egg to egg can be completed in a little less than three weeks and in untreated cases the numbers of *Psoroptes* increase astronomically. An intensely infected ear may contain as many as 10,000 mites (244).

Since psoroptic otoacariasis is readily transmitted between rabbits it is important to minimize spread by isolation of affected animals and by maintaining high standards of sanitation and personal hygiene to prevent fomites from serving as a source of spread.

Although many acaricidal preparations have been described in the literature (59, 137, 143, 227), mineral oil alone or in combination with an acaricide is efficacious. A commercially available acaricide—antibiotic combination—is also effective in treating psoroptic otoacariasis. Periodic prophylactic (therapeutic) treatment may be warranted in many laboratory or commercial colonies since early lesions may be undetected. Therapeutic treatment with the above agents should be preceded by a gentle cleansing of the ear and ear canal with a mild antiseptic-detergent mixture such as may be employed in preparing a skin site for surgery. Appropriate antibiotics may be employed when secondary bacterial invasion has caused a suppurative dermatitis in addition to the primary acaritic-

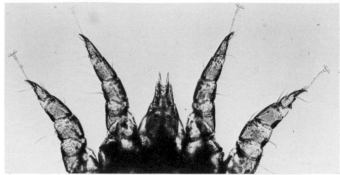

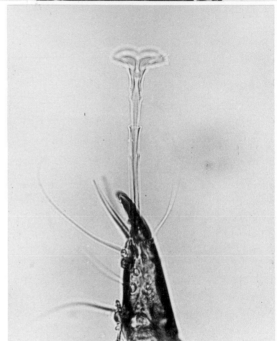

Fig. 2. *Psoroptes cuniculi*, mouth parts and anterior two pair of legs (top). *Psoroptes cuniculi*, jointed or segmented pedicel characteristic of the genus (bottom).

induced lesion. Antiinflamatory agents may also be desirable once the infection is under control.

The histological appearance of otoacariasis consists of a chronic inflamatory response with hypertrophy of the Malpighian layer, parakeratosis of the horny layer, and in places sloughing of the epithelium. The corneum is infiltrated with small round cells. It has been suggested that there is an allergic component in the pathogenesis of the lesions and, in fact, specific serum precipitins against mite antigens have been identified (46). Secondary bacterial infection can superimpose a chronic suppurative inflamatory response on the primary lesion.

v. ZOONOTIC AND VECTOR SIGNIFICANCE. None.

2. The Sarcoptid Mange Mites (*Notoedres cati* and *Sarcoptes scabiei*

a. HISTORICAL. Sarcoptic mites (of the genus *Sarcoptes*), commonly known as itch or scab mites, are parasites of a wide variety of mammalian hosts, including man, and are generally considered to be of one species, *Sarcoptes scabiei* (De Geer), although some authors subdivide the species into varieties, e.g., *Sarcoptes scabiei var. cuniculi.*

Notoedres cati, referred to in the older literature as *N. minor cuniculi,* affects dogs, cats, rats, and rabbits and produces lesions similar to those caused by *Sarcoptes scabiei,* and, therefore, the two related genera are considered together.

b. OCCURRENCE. i. GEOGRAPHICAL. *Sarcoptes scabiei* and *Notoedres cati* are cosmopolitan and ubiquitous (28, 90, 94, 140, 157).

ii. INCIDENCE. Although at one time these mange mites had been a common problem they are reported infrequently in both laboratory and wild lagomorphs (18, 24, 28, 58, 90, 94, 97, 102, 125, 136, 140, 157, 205, 208, 221).

c. DIAGNOSIS. i. CLINICAL SIGNS. These obligate burrowing mites tunnel through the skin where they are believed to ingest epithelial cells and suck lymph. Their presence causes an intense pruritis which incites the rabbit to rub and scratch at the infected areas. Initially a partial alopecia and a serous exudate develops which may be observed to be followed by the formation of a whitish-yellow crust of dried serum and epidermal debris. Continual self-mutilation contributes significantly to the appearance of skin lesions observed. Secondary bacterial infection is common.

Sarcoptic mange in rabbits may be a very severe disease and its spread on individual animals can lead to general debility, emaciation, and death within a matter of weeks. Notoedric mange usually resembles that caused by *S. scabiei* and only isolation and identification of the causative organism can differentiate the two agents.

Acaritic dermatitis in rabbits must be distinguished from dermatophytosis (see Chapter 18).

ii. LOCATION ON HOST. Sarcoptic and notoedric mange usually first appear on the nose and lips extending to the eyes, forehead, face, and occasionally the external genitalia. Occasionally it may occur on the lateral surface of the ears. These mange mites have not been reported from the inner aspect of pinna, however. Dual infections of *Sarcoptes* and *Psoroptes* in the rabbit have been reported with the sarcoptic mites infesting the outer surface and the psoroptic mites the inner surface of the pinna (136).

iii. IDENTIFICATION. Although these burrowing mites may be present on the surface of the lesion, a deep skin scraping is usually necessary to obtain individual specimens for examination and identification.

All stages of the life cycle of these mites may be observed in clinical specimens. Adult females are the most active tunnelers, while larvae, nymphs, and adult males tend to use existing tunnels or move freely over the skins surface where they are able to invade hair follicles.

Adult female mites actively burrow into the skin and may lay 40–50 ova. Larvae which have only three pairs of legs may continue to develop in these preexisting tunnels and pockets or migrate out to the skin surface where many die. Some may migrate and burrow into the stratum corneum where they feed and molt to form nymphs. The nymphs have four pairs of legs and may be distinguished from adults by the absence of genital apertures. Adult males and females develop approximately 17 days after the ova hatch. They mate and after 4–5 days the females begin to lay eggs. Infection is spread primarily by the actively motile first larval and first and third nymphal stages or fertilized adult females.

The forms of these mites which may be observed are ova, two larval stages, four nymphal stages, and adults. The mites are small, globose or round with short legs. The rear two pairs of legs do not project beyong the margins of the body. The female *Sarcoptes* is about twice as large as the male and measures from 303–450 μm in length, by 250–350 μm in width. *Notoedres cati* is slightly smaller than *Sarcoptes scabiei.* There are bell-shaped suckers or caruncles on pedicels of the tarsi of some or all of the legs. The pedicels of the tarsal suckers are not segmented. The anus of *Sarcoptes* is terminal, whereas in *Notoedres* it is dorsally located. (See Fig. 3.) *Sarcoptes* and *Notoedres* both have dorsal dentate spines, however, they are considerably larger in *Sarcoptes.*

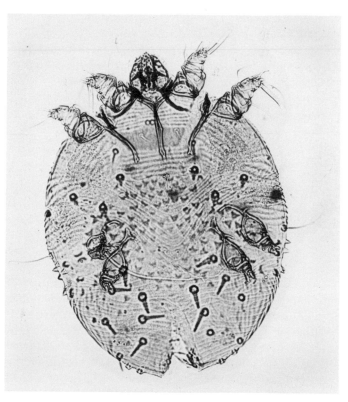

Fig. 3.   Sarcoptes scabiei, the sarcoptic mange mite.

iv. CONTROL AND TREATMENT. In naturally occurring cases sarcoptic and notoedric mange is spread most readily by direct contact. It is interesting to note, however, that Low was unable to produce disease in "healthy rabbits on the areas which are most usually "affected" after repeated transfer of "living *Sarcoptes* and ova in all stages of development and in considerable numbers" (136). This failure was not apparently due to any host resistance factors since when one of the rabbits was placed in a cage beside a diseased rabbit, it developed sarcoptic mange within a short period of time.

These mites can live only a few hours off their host as they are very susceptible to drying. In laboratory situations where standards of sanitation are high and identification and elimination of affected animals is practiced the disease can be readily eliminated. It is suggested that if a diagnosis is made of sarcoptic or notoedric acariasis the affected animals be removed from the colony. If it is absolutely necessary to keep the affected animals, they should be isolated from nonaffected animals and treated. It must be recognized that although many treatment regimes have been proposed and utilized for this parasite, the clinical course is invariably protracted and the results of treatment are not always good.

Treatments that have been suggested for sarcoptic and notoedric mange include: Dimethyl phthalate (152), 10% DDT in talc (87), the γ-isomer of benzene hexachloride, benzyl benzoate, and many others (79).

v. ZOONOTIC AND VECTOR SIGNIFICANCE. *Sarcoptes scabiei* is a parasite with little host specificity, and physiological variants can affect man (172, 255).

Notoedres cati do not become established in human skin but they may temporarily invade it. Repeated infection with these mites can prolong clinical symptoms in man.

3. The Cheyletid Fur Mite (*Cheyletiella* spp.)

a. HISTORICAL. It has only been within the last several years that the genus *Cheyletiella* has been clarified (76, 240). Prior to that time the genus was considered monospecific—containing only one species—*C. parasitovorax* (Megnin, 1878). Currently there are five species recognized in the genus: *C. parasitovorax* (Megnin, 1878), *C. takahasii* (Sasa and Kano, 1951) (201), *C. ochotonae* (Volgin, 1960) (240), *C. johnsoni* (Smiley, 1965) (211) affecting lagomorphs, and *C. yasguri* affecting dogs (76, 211).

The older veterinary literature contains descriptions of a mite named *Ewingella americana* (234). This designation is now considered synonymous with *C. parasitovorax* (45).

Although *C. parasitovorax* was originally thought not to be a parasite of the rabbit host itself but rather a predator on soft body mites of the genus *Listrophorus* (45), as originally suggested by Megnin in 1878, evidence has been produced by several investigators to prove that this species is indeed a primary parasite of the rabbit. Mykytowycz in 1948 gave the following evidence of the primary pathogenetic nature of *C. parasitovorax*. (a) He noted that *C. parasitovorax* is capable of transmitting viral myxomatosis, therefore, he concluded it must be feeding on the host. (b) He demonstrated the mite engorged with lymph. (c) He kept *C. parasitovorax* and *L. gibbus* together *in vitro* and did not observe *C. parasitovorax* preying on *L. gibbus*. (d) He also noted that *C. parasitovorax* and *L. gibbus* have a slightly different distribution on the rabbit itself, i.e., *C. parasitovorax* lives on the skin of the abdomen, back, and scapular regions, whereas *L. gibbus* is found primarily on the neck and ventral aspect of the tail. Additional investigators have demonstrated *C. parasitovorax* living on rabbits without concommitant parasitism with soft-bodied mites (17). Numerous other workers have confirmed the parasitic nature of *C. parasitovorax* (12, 76, 89). For an excellent general view of the cheyletids affecting species other than the rabbit, see the publication and bibliography of Foxx and Ewing (76).

It is interesting to note that *C. parasitovorax* was so named because of the then held notion that it was parasitic on other mites.

b. OCCURRENCE (GEOGRAPHICAL INCIDENCE). *Cheyletiella parasitovorax* has been reported in surveys taken in Canada, the eastern United States (New York, Maryland, Georgia, and North Carolina), the Midwest (Michigan and Wisconsin), and the West Coast (California) (45, 90, 142, 157, 221, 234).

This mite has also been reported as being widespread in Great Britain, and present in South Africa, New Zealand, and Australia (15, 28, 97, 150, 249).

With the exception of Great Britain, where out of 374 rabbits examined from a cross-section of England, Scotland, Wales, and from certain off shore islands in which 112 rabbits were positive for infection with *C. parasitovorax*, the literature contains very sketchy data on the natural incidence of this parasite.

Cheyletiella takahasii has been reported from Hokkaido, Japan (201), *C. ochotonae* from Canada (240), and *C. johnsoni* from Washington (211).

Cheyletiella parasitovorax, however, is probably the most commonly occurring member of the genus. *Cheyletiella parasitovorax* has been reported many times from the rabbit (12, 17, 28, 45, 53, 119, 150, 164, 169, 175, 201, 220, 234, 240), the dog (17, 89, 211, 241, 250), the cat (96, 119, 166, 175), and man (166).

Although the true incidence of *C. parasitovorax* in laboratory rabbits is not known, its small size may allow it to be overlooked in clinical and postmortem examinations and it may be, therefore, that the incidence is higher than is currently recognized.

c. DIAGNOSIS. i. CLINICAL SIGNS. It is likely that

many rabbits who harbor cheyletid mites do not have overt signs of disease (147). In heavy infestations, however, lesions are located primarily on the dorsal trunk in the scapular area although additional lesions may be found elsewhere in the body, especially the ventral abdomen.

These obligatory parasitic nonburrowing mites live in immediate association with the keratin layer of the epidermis. The affected area may be partially alopecic with the skin surface covered by a fine grayish-white slightly oily scale resembling an early stage of dermatomycosis. The underlying skin may be reddish and painful to the touch. The lesions do not appear to be intensely pruritic since the rabbits do not scratch themselves as they do in sarcoptic or notoedric mange (234).

Specimens of cheyletid mites may be obtained by brushing epidermal debris from the skin and hair with a clean stiff bristled brush (76). Skin scrapings may also be used when indicated.

Histopathological lesions are characterized by subacute nonsuppurative dermatitis with mild hyperkeratosis. Mites may be observed attached to the keratin layers of the epidermis. An inflammatory exudate consisting of many polymorphonuclear neutrophils and mononuclear phagocytes, lymphocytes, plasma cells, and a few eosinophils is observed (76).

Although *C. parasitovorax* may cause grossly and microscopically visible lesions according to a number of reports previously cited, other investigators have observed no lesions (45, 71, 244).

ii. LOCATION ON HOST.   Cheyletid fur mites are found primarily on the dorsum of the trunk in the area of the scapulae of the rabbit but in heavy infestations may be present elsewhere on the body.

iii. IDENTIFICATION.   This small (265–385 μm) ovoid to saddle-shaped mite is whitish to yellowish in color and is characterized by piercing chelicerae and large recurved palpal hooks or claws. (See Fig. 4.) The legs have comblike setae on their last segment (172). All stages of the life cycle of this mite can be found on the host. The eggs when layed are attached to the hairs or fur. The mites can be identified under a disecting microscope, however, it may be helpful and necessary at times to clear epidermal debris in the specimen with 5–10% potassium hydroxide solution.

Although the life cycle of *C. parasitovorax* has been described as follows, eggs hatch after a four-day incubation, larval stage ($7\frac{1}{2}$ days), nymphal stages ($9\frac{1}{2}$ days), and adult (14 days), the author has not presented experimental evidence of this life history (109). The life history of *C. yasguri*, however, has recently been elucidated and published by Foxx (76). Although it certainly has not been established, the life history of other cheyletids are more than likely similar to that described for *C. yasguri*. The stages of

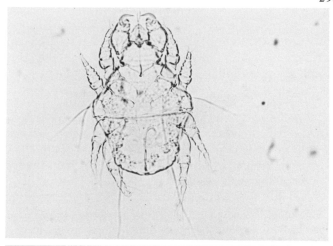

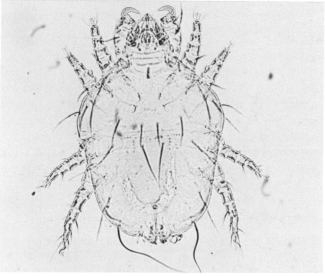

Fig. 4.   *Cheyletiella parasitovorax* fur mite, male (top). *Cheyletiella parasitovorax*, female (bottom).

development of *C. yasguri* are the same as originally reported by Humphrey's for *C. parasitovorax* (76, 109).

iv. CONTROL AND TREATMENT.   Although cheyletid fur mites may or may not produce clinically observable disease if present they probably can be eliminated with topical silica gel acaricides since they are nonburrowing and should be located where such agents can be effective. Although not reported, dichlorvos may be effective against this parasite.

v. ZOONOTIC AND VECTOR SIGNIFICANCE.   *Cheyletiella parasitovorax* is reported to be potentially parasitic for man (89, 172) (see Chapter 18).

Cheyletiella parasitovorax is a known vector of myxomatosis in Australia (see Chapter 10).

4. The Listrophorid Fur Mite (*Listrophorus gibbus*)

a. HISTORICAL.   *Listrophorus gibbus* was first described by Pagenstecher as fur mite of wild *Oryctolagus* in 1862

(171). Since that time, controversy as to its pathogenetic capabilities have occurred in the literature. Recent observations of heavy infestations in laboratory rabbits seem to indicate that this mite, although parasitic on the rabbit, is nonpathogenic (245).

b. OCCURRENCE. i. GEOGRAPHICAL. *Listrophorus gibbus* has been reported from many areas where wild or laboratory *Oryctolagus* rabbits are known to occur. They have been reported from Great Britain (97, 150), France (58), Germany (208), Eastern Europe, and the Soviet Union (242, 254). In addition there are reports documenting *L. gibbus* in New Zealand (28) and Australia (163). In the Western Hemisphere *L. gibbus* has been reported from a laboratory rabbit in Puerto Rico (225), wild *Lepus* in Texas (122), and most recently from domestic *Oryctolagus* in New York State (245).

ii. INCIDENCE. Although Weisbroth has recently reported on the incidence of *L. gibbus* in a few commercial rabbitries, he has speculated that because these mites are small and do not produce overt disease, they may be present in many colonies and are unnoticed (245). He also noted no hair pattern or color predilection of these mites as they were found on New Zealand White, Checker Hare Brown, New Zealand Red, Sable Champagne, and Dutch Belted rabbits (245). *Listrophorus gibbus*, therefore, has been reported from both wild and domestic *Orytolagus* and wild *Lepus* rabbits and hares in the United States.

c. DIAGNOSIS. i. CLINICAL SIGNS. Although early reports of clinical signs are present in the literature, the causal relationship to the lesions observed are uncertain. Currently, *L. gibbus* is considered to be a nonpathogenic obligate parasite of the rabbit (242, 245).

ii. LOCATION ON HOST. *Listrophorus gibbus* may be found primarily on the back and abdomen, however, they can possibly be found on other locations as well (242).

iii. IDENTIFICATION. Isolation of the mite in clinical specimens and identification under a disecting microscope is readily performed. Combing or brushing the hair of infested rabbits provides a good means of obtaining specimens of this mite for identification. All stages of the parasite may be encountered. *Listrophorus gibbus* is a small (males: 240 by 440 μm; females: 310 by 560 μm) ovoid mite with short legs and a dorsal projection extending out over its mouth parts. Males have powerful appearing clasping organs located at their posterior end. (See Figs. 5 and 6.)

iv. CONTROL AND TREATMENT. Since *L. gibbus* is an obligate parasite and spends its entire life cycle on the host, eradication or isolation and treatment of the affected animals should control this parasite.

Since the mite has not produced disease, the literature does not contain any specific suggestions as to the method

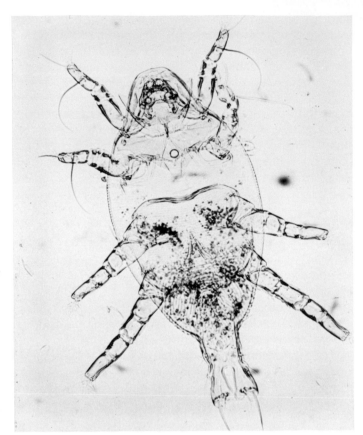

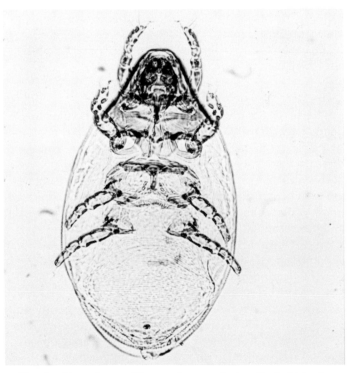

Fig. 5. *Listrophorus gibbus*, male, ventral view (top). *Listrophorus gibbus*, female, ventral view (bottom).

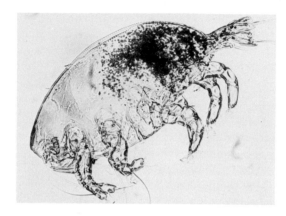

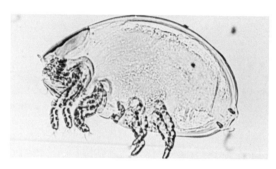

Fig. 6. *Listrophorus gibbus*, male, lateral view (top). *Listrophorus gibbus*, female, lateral view (bottom).

of treatment, however, acaricides used for other mites of rabbits or rodents may well be effective.

v. ZOONOTIC AND VECTOR SIGNIFICANCE. No known zoonotic diseases are transmitted by *L. gibbus*. It is not known to be a vector of any rabbit disease.

5. Trombiculid or Harvest Mites

a. HISTORICAL. Larvae of mites of the family *Trombiculidae* are known as chiggers, red bugs, or harvest bugs. Certain species of this genus are occasionally parasites of wild rabbits. Within the genus there is little host specificity. The most commonly encountered trombiculid mites in the rabbit are *T. autumnalis*, *T. cavicola*, *T. irritans*, and *T. microti*. *Tunga penetrans*, a tropicopolitan flea affecting man, swine, dogs, and wild burrowing mammals is also known commonly as a chigger or chigoe.

b. OCCURRENCE. Chiggers are notoriously erratic in their distribution over their ranges, however, they are known to occur throughout the world (116). They favor interfaces between strikingly different vegetative patterns. Little is known about the host ranges of the larval stages of the genus *Trombicula*. It is known, however, that *T. autumnalis* larvae feed on rabbits, voles, hedgehogs, and a variety of avian species. Species known to affect rabbits are reported in the following references (18, 24, 97, 116, 205, 215).

c. DIAGNOSIS. i. CLINICAL SIGNS. The trombiculid larvae do not burrow but attach to the surface of the skin after it is pierced by the cutting chelecerae. Saliva from the mite digests the epithelium and stimulates the formation of a stylostome, a tubular downgrowth of cells produced as a protective response by the host. The larva continues to ingest digested cells within the stylostome. This process induces an intense pruritis and the formation of discrete macules and pustules. The feeding process may last for weeks or even months, (239). Larval mites are found on their mammalian host primarily in the fall of the year as they prepare for overwintering on their host.

ii. LOCATION ON HOST. Sites of predilection for *T. autumnalis* include both the inner and outer surface of the ears, anus, canthi of the eyes, and feet. When the harvest mite is particularly numerous it can be found attached anywhere on the body particularly the ventral body surface (116).

iii. IDENTIFICATION. Since many of the trombiculids have been described by their larval state only and since biological variation among the chiggers is great, the reader is encouraged to consult Wharton and Fuller's "Manual of the Chiggers" (248).

Unfed larva are approximately 210 μm long and are deep red in color; fully fed larva reach a size of approximately 400 μm and are pale yellow in color. Larvae have three pairs of legs. There are plumose setae on each coxa and trochanter. The dorsal body setae are also almost always plumose. (See Fig. 7.)

iv. CONTROL AND TREATMENT. Since only the larval stage is parasitic in the rabbit, control measures aimed at killing or minimizing the other stages in the life cycle may be indicated.

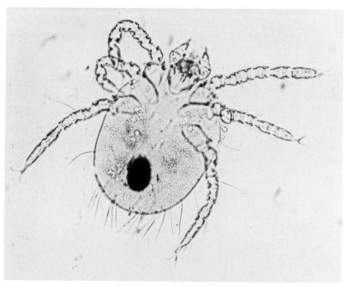

Fig. 7. *Trombicula cavicola* larva (6-legged).

Adults are free-living mites which feed on invertebrates (insects) and plants. Eggs are laid in the soil where they and insect eggs hatch in about 6 days to form deutovum. In another 6 days the actively parasitic hexacanth larva hatches and seeks out the rabbit host. Once on the host it feeds for one to several days (*T. autumnalis*), then drops to the ground to become the inactive nymphocrysalis. The nymph emerges and feeds on eggs of other arthropods until it is engorged. It than becomes inactive until it matures into the imago or adult mite.

Since maintenance of the trombiculid life cycle depends on both a supply of other arthropods as well as a suitable mammalian host, control of indigenous arthropods will in turn help control the trombiculid mites.

No treatment has been suggested in the literature for rabbits infested with these mites.

Trombiculid infestation occurs in wild populations and in commercial rabbitries with outdoor hutches but rarely gains entrance to or becomes a problem in the laboratory.

v. ZOONOTIC AND VECTOR SIGNIFICANCE. Trombiculid mites appear to have very low host specificity and may be transmitted to man as a secondary host (172).

6. Pentastomid Mites (*Linguatula serrata*)

a. HISTORICAL. The pentastomes or tongue worms as they are also called have been classified with the acarines (mites and ticks) because of certain similarities with the plant feeding family of mites known as the Eriophyidae (200). Some authors have described the pentastomes as "degenerate mites." Two families within the class Pentastomida that are of veterinary significance include the Porocephalidae containing two genera (*Porocephalus* and *Armillifer*) and the Linguatulidae containing the single genus, *Linguatula*. *Linguatula serrata* (Froelich, 1789) is the most common form affecting mammals, including the domestic and wild rabbit.

b. OCCURRENCE. i. GEOGRAPHICAL. Linguatulids are cosmopolitan in distribution and have been reported from wild *Oryctolagus* rabbits in New Zealand (28, 86), from North Carolina and Alabama cottontails (221), and from a laboratory rabbit in China (64). Although only 4 articles on linguatulid in rabbits were found and reviewed, it is important to note that wild rabbits are considered to be a natural intermediate host of the parasite (95). Other herbivorous invertebrates such as cattle and sheep may also serve as an intermediate host.

ii. INCIDENCE. Undoubtedly the few reports are not a true index of the incidence of *Linguatula serrata* in rabbits since as noted above the wild rabbit is considered as the natural intermediate host. One of the reasons that the incidence of linguatulids according to the rabbit literature

may be misleading is that in surveys of other acarines, the studies involved examinations designed to detect ectoparasites and not the developing forms of this tissue-migrating organism. Another factor may also be that the rabbit acts as only the intermediate host of the parasite. The definitive host is commonly the dog or other wild carnivorous predator and occasionally man.

c. DIAGNOSIS. i. CLINICAL SIGNS. Although affected animals are generally asymptomatic, heavy infestations with the larva or nymph of *L. serrata* may produce signs related to the malfunctioning of the organ(s) involved. They are usually diagnosed at necropsy and are generally considered to be incidental findings and not themselves the cause of death.

ii. LOCATION ON HOST. Faust reported that encysted linguatulid nymphs, averaging 4 mm in length by 0.8 mm in width, were found in the liver and lungs of a laboratory rabbit in Peking, China (64). Wild cottontail rabbits contained numerous nymphal forms in the lungs, mesenteric lymph nodes, and the liver. In one rabbit 75 nymphs were found in the lung, 65 in the mesenteric lymph nodes, and 11 in the liver (155).

iii. IDENTIFICATION. This parasite only spends a portion of its life cycle in the rabbit, however, an overall view of its life history is interesting and important. Adult *L. serrata* live in the respiratory tract of the dog, fox, wolf, and more rarely in man, goat, and sheep. The adult is a "tongue-shaped" pseudo-annulated legless parasite that is slightly convex dorsally and flattened ventrally. Males are approximately 2 cm long but females range from 8–13 cm. Ova are approximately $70 \times 90 \mu$m in size.

Eggs are expelled from the respiratory tract of the definitive host and when swallowed by the rabbit, hatch into larva which puncture the bowel wall and enter the mesenteric lymph nodes in which they develop into the infective nymphal stage.

The nymph resembles that adult described above but is approximately 5 mm long and has a whitish color. The nymphal form is usually found encysted and surrounded by a viscous opaque fluid.

The definitive host becomes infected when he ingests affected rabbits in whose body the infective nymphs of *L. serrata* have formed.

iv. CONTROL AND TREATMENT. Infection with this parasite can be prevented in controlled populations by eliminating the definitive host from the area, that is, the dog, fox, and wolf.

Treatment for the intermediate stages (larva and nymphal stages) has not been attempted.

v. ZOONOTIC AND VECTOR SIGNIFICANCE. Man can become infected presumably by eating improperly cooked rabbit meat in which infective nymphal stages of *L. serrata* have developed. Although most human cases are asymptomatic, a report of a human case of linguatulosis in which a partial obstruction of the colon occurred because of gross thickening of its parasite-infested wall has been published (32).

TICKS AFFECTING LAGOMORPHS

A number of both hard and soft ticks of the families Ixodidae and Argasidae are known to affect various wild leporids. Rarely these ticks may also affect domestic rabbits. While most of the species are of primary importance due to their capabilities as vectors of infectious disease, some of the hard ticks do produce primary and secondary insults to their hosts. Of all of the arthropods covered in this chapter, one of the ticks, *Haemaphysalis leporis-palustris*, is fairly commonly reported from wild leporids and occasionally from domesticated rabbits. This species will be the only one considered in any detail. The remaining ixodid and argasid ticks will be treated very briefly in this review.

1. The Continental Rabbit Tick (*Haemaphysalis leporis-palustris*)

a. HISTORICAL. Although a large number of different ixodid and argasid ticks are described in surveys of wild rabbit ectoparasites, only one, *H. leporis-palustris*, Packard, 1869, occurs with any frequency in domesticated rabbits (5, 6, 18, 41, 90, 113, 114, 140, 163, 165, 217). It is reported primarily, however, from wild *Lepus* and *Sylvilagus* rabbits.

b. OCCURRENCE. *H. leporis-palustris* is found throughout the United States and Canada. Those areas and states on or below a latitude with southern Missouri are favorable for year-round activity of this tick (176). This fact is important in that *H. leporis-palustris* is an important vector of both tularemia and Rocky Mountain Spotted Fever among rabbits. The farther one goes from this latitude the more markedly the population of this tick varies with the season, being noticably absent during the winter months in the northern states and reoccurring in the spring, with a peak occurring in the summer (43, 84, 113, 114, 176, 217).

No ticks were reported in a survey of wild rabbits in New Zealand (28). In other surveys reviewed, no reports of this tick were noted from Europe, Asia, or Africa.

In the United States, however, it is the principal ectoparasite of cottontail rabbits in Illinois (217) and in the eastern United States (18).

Because this ixodid tick is a three host-tick and spends relatively short periods of time on the rabbit host (only 3–5 days while feeding), it is infrequently encountered in domestic rabbits (84, 114).

Although *H. leporis-palustris* is primarily a parasite of the rabbit, hence its common name, it is not exclusively host-specific as it has been reported to occur on a number of species of ground-inhabiting birds, less frequently on domestic animals and dogs and cats. It rarely bites man (95). It is interesting to note that only larvae and nymphs have been reported on birds indicating that it probably does prefer the rabbit as its definitive host (22). It is very possible that migratory birds are responsible for dispersion of *H. leporis-palustris* northward in the spring and southward in the fall (217).

c. DIAGNOSIS. i. CLINICAL SIGNS. The primary danger of ticks to the host is their ability to act as a vector for a variety of infectious diseases, particularly tularemia and Rocky Mountain Spotted Fever. They are voracious blood suckers and repeatedly drop off the host, digest the blood meal, and then find new victims. Since but few ticks are found on rabbits at any one time, signs of chronic blood loss probably will not be evident. Cottontail rabbits affected with *Ixodes dentatus* have been known to develop *Staphyloccus* abscesses in areas where they have attached while feeding (18). It seems likely that a similar disease manifestation could possibly occur with *H. leporis-palustris*.

Investigations into the development of immunity to *H. leporis-palustris* have indicated that animals can become hypersensitive to this tick and subsequently become protected against the attachment of larval forms (184). This suggests that repeated infection in nature with *H. leporis-palustris* may confer a degree of immunity to subsequent reinfestation.

ii. LOCATION ON HOST. *Haemophysalis leporis-palustris* is found primarily on the head of the rabbit, especially on and in the ears, and back of the neck. Sometimes ticks of this species are located around the eyes, nose, and under the chin.

iii. IDENTIFICATION. This tick may be differentiated from other ixodid ticks by the presence of anal grooves surrounding the anus posterially, the long hypostome, and characteristic sharply pointed lateral angles marking the base of the mouth parts. These features only are present in all stages of the tick (larvae, nymph, and adult) and are diagnostic of this parasite (217). This tick also does not possess eyes. (See Fig. 8.)

Haemophysalis leporis-palustris is, as mentioned previously, a three-host tick in which each instar or stage requires a separate host for each blood meal. Although all stages can be found on wild rabbits, the larval and nymphal stages can infest a wide variety of birds (for example, brown thrasher, towhee, prairie chicken, olive backed thrush, tufted titmouse, quail, and swamp sparrow) which undoubtedly are responsible for moving these immature stages many hundreds of miles along their normal migration routes each year (217). The adult ticks, however, are apparently more host-specific and are not reported from nonlagomorph hosts.

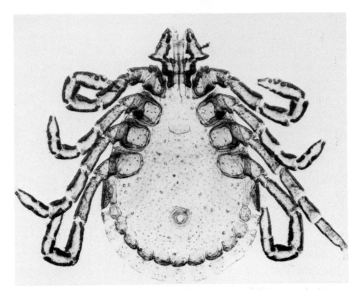

Fig. 8. Haemaphysalis leporis-palustris, the continental rabbit tick.

Fig. 9. Dermacentor parumaptertus, adult, the rabbit *Dermacentor*.

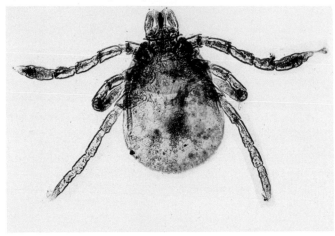

iv. CONTROL AND TREATMENT. Control measures in commercial, outdoor, or semienclosed rabbitries include attempts at exclusion of wild lagomorphs and birds from possible contact with domestic stock. Heavily infested pastures surrounding rabbitries can be burned and kept free of growth or kept cultivated.

Treatment of affected rabbits with any of the commercially available tick sprays or dips can be readily performed (79).

v. ZOONOTIC AND VECTOR SIGNIFICANCE. *Haemaphysalis leporis-palustris* rarely affects man and can, therefore, theoretically transmit to him the agents responsible for tularemia, Rocky Mountain Spotted Fever, and Q Fever (124, 253) (see Chapter 18). *Haemophysalis leporis-palustris* obviously also transmits tularemia, Rocky Mountain Spotted Fever, and Q Fever between wild lagomorphs.

2. The Eastern Rabbit Tick (*Ixodes dentatus*)

Although *I. dentatus* Marx, 1899, and other less frequently encountered *Ixodes* species occur in wild lagomorphs and have been known to incite secondary infections (18), they remain important only insofar as they may serve as biological vectors of tularemia, Rocky Mountain Spotted Fever, and Q Fever. *Ixodes dentatus* has been reported most frequently in the eastern United States (18, 140), however, this species has been identified from as far west as Illinois (191), Wisconsin (81), and Iowa (97). *Ixodes dentatus* also does not have eyes but may be distinguished from *H. leporis-palustris* in that it is in inornate and does not have the sharp angled basis capituli characteristic of *H. leporis-palustris*. Another distinguishing characteristic of the genus *Ixodes* is the anal groove which surrounds the anus anteriorly.

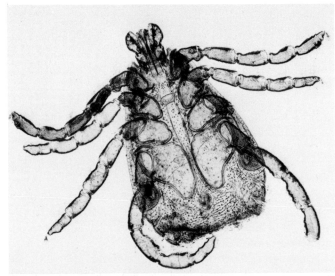

Fig. 10. Dermacentor andersoni, larva, (top). *Dermacentor andersoni*, adult, (bottom).

3. Other Ixodid (Hard Ticks) Affecting Lagomorphs

Although a variety of other hard ticks are reported from rabbit hosts, *Dermacentor parumapertus*, the rabbit *Dermacentor* (see Fig. 9), and *Ixodes dentatus*, the eastern rabbit tick, are perhaps the most common. Other species of *Dermacentor* found less commonly but yet important are *D. variabilis*, the American dog tick, *D. andersoni*, the Rocky Mountain Spotted Fever tick (see Fig. 10), *D. occidentalis*, the Pacific Coast tick. Other species of *Ixodes* reported from rabbits and hares include *I. neotomae* (217), *I. spinepalpis*, *I. sculptus*, *I. ricinus californicus*, and *I. muris* (22). *Amblyomma americana*, the Lone Star tick, *A. cayennense*, the cayenne tick, and *A. maculatum*, the Gulf Coast tick, are three members of the genus *Amblyomma* affecting wild rabbits and hares (22). In addition *Rhipicephalus sanguineus*, the brown dog tick, also can affect wild rabbits and occasionally man (22).

These parasites are far more important as vectors of various bacterial (tularemia and plaque), viral, (Russian summer–spring and encephalitis), and rickettsial (Rocky Mountain Spotted Fever, Q Fever, and Colorado tick fever) pathogens than as primary agents of disease. Although this is true, bites and blood sucking by these arthropods can lead to secondary infections with bacteria or the larvae of dipteran insects (myiasis) or debility and predisposition to other diseases as a result of the anemia produced when large numbers of ticks feed on the same host. Table I summarizes some important facts concerning these ticks.

4. Argasid (Soft) Ticks Affecting Lagomorphs

a. GENERAL. *Otobius lagophilus* (Cooley and Cohls, 1940) is the most commonly reported and studied argasid tick affecting lagomorphs (43). Other soft-bodied ticks including *Ornithodoros parkeri* (Cooley) and *O. turicata* are sometimes also reported (6, 217).

b. OCCURRENCE. The above species (*O. lagophilus*) has been reported primarily from the Pacific northwest from wild *Lepus* and *Sylvilagus* species in addition to other small animals, although Stannard (217) states that *O. lagophilus* occurs in northern areas west of the one-hundredth meridian (217). Only larval forms of *O. parkeri* have been removed from rabbits. Since no burrow or nest examinations were made and since these species are supposedly nest inhabitors they may be much more numerous than that indicated by studies of rabbit carcasses such as the above. In addition to lagomorphs, *O. lagophilus* has been reported to affect cats (4).

c. DIAGNOSIS. i. CLINICAL SIGNS. None. The importance of these ticks are related to their role as vectors of infectious diseases agents (173).

ii. LOCATION ON HOST. *Otobius lagophilus* is usually found on the face above the vibrissae.

iii. IDENTIFICATION. For a detailed consideration of the members of the family Argasidae, the reader is referred to the work of Cooley and Cohls (43a). The details of the life history and morphology of *O. lagophilus* is presented by Bacha (4) and by Cooley (43). *Otobius lagophilus* may be readily distinguished from *O. megnini*, the common spinose ear tick which affects primarily dogs, sheep, horses, and cattle.

In general, however, the argasid ticks have a leatherlike frequently mammillated integument. Sexual dimorphism is not marked. Only larval and nymphal stages are parasitic as the adult has only rudimentary mouth parts and does not feed.

iv. CONTROL AND TREATMENT. Control in wild populations has not been attempted. The presence of these ticks in domesticated rabbits has not been reported in other than experimental situations.

v. ZOONOTIC AND VECTOR SIGNIFICANCE. These ticks, as with any of the hemotophagous arthropods are potential carriers of zoonotic diseases and may act as vectors of rabbit diseases.

C. The Hexapod (Insect) Parasites of Rabbits (Lice, Flies, and Mosquitoes)

The class Hexapoda (Insecta) contains numerous members which are important parasites of rabbits. This class of arthropod is characterized by having a distinct head, thorax, and abdomen, three pairs of legs in adults, and typically two pairs of wings. The orders of this class which contain members of importance to students of lagomorph biology include the Anoplura (sucking lice), the Diptera (flies and mosquitoes), and the Siphonaptera (fleas). There are no representatives of the order Mallophaga (biting lice) known to infest the rabbit.

THE ANOPLURID PARASITES OF THE RABBIT (ORDER ANOPLURA, LEACH, 1815)

This large order contains over 225 species in five families, however, for the purpose of this review only one member of this order will be considered in any detail. For a comprehensive review of the sucking lice, the author suggests the classic work of Ferris (67) and the excellent illustrated key to the Anoplura by Stojanovich and Pratt (218). The single genus of sucking louse to be considered is a member of the family Hoplopleuridae, subfamily Polyplacinae (67). The genus is *Haemodipsus*, members of which are close relatives of *Polyplax* species, the common sucking lice of rats and mice.

Haemodipsus Species (The Sucking Louse of Rabbits)

a. HISTORICAL. *Haemodipsus ventricosus* (Denny, 1842) is by far the most commonly reported louse from *Orycto-*

TABLE I
TICKS AFFECTING LAGOMORPHS

Tick genus spp.	Common name	Lagomorph host	Geographical distribution	Identification
Amblyomma americanum	The Lone Star tick, 3-host tick	Wild marsh-rabbits and jackrabbits and swamp rabbits; rabbits rarely infected	States bordering Gulf of Mexico occassionally else-where in the southeast United States	Female easily recognized by conspicuous silver white spot at tip of scutum
A. cajennense	Cayenne tick	Cottontails	Extreme southern Texas on Gulf	Scutum ornate with pale mark-ings, coxa of female with internal spur about $\frac{1}{2}$ length of external spur
A. maculatum	Gulf Coast tick, 3-host tick	Cottontail and jack-rabbits	North, Central, and South America, Gulf Coast and coast south Atlantic States from South Carolina to Texas	Spurs on 2nd, 3rd, and 4th pairs of legs; pale markings more diffuse than on females of Lone Star tick
Dermacentor andersoni	RMSF tick	Wide variety of wild rabbits and hares	From western Nebraska and Black Hills of South Dakota to eastern slopes of the Cascades and from nor-thern Arizona and New Mexico up into British Co-lumbia and Manitoba, Canada	Similar to *D. variabilis* but adults in general have paler coloring and larger goblets on the spiracular plates
D. occidentalis	Pacific Coast tick	Wild rabbit	Western California and western Oregon	Basis capituli with conspicuous toothlike projections (cornus) on posterior margin
D. parumapertus	Rabbit dermacentor	Jackrabbits	Southwest United States west of 100th meridian and arid portions of northern Mexico, especially prevalent in western Texas, southern New Mexico, Arizona, and Cali-fornia	Spurs on coxa I are widely diver-gent and the scutum has deep large punctations
D. variabilis	American dog tick	Wild and domestic rab-bits	Widely distributed throughout United States with the excep-tion of Washington, Rocky Mountains, intermountain areas, also south Canada	Mouth parts and basis capituli subequal length, scutum with pale white or yellowish mark-ings, basis capituli parallelsided
Haemaphysalis leporis-palustris	The rabbit tick	Rabbits preferred by ground birds for larva and nymphs	One of most common and widely distributed ticks in North America—all over United States including southern Canada and Mexico	Sharply pointed lateral angle at base of mouth parts is unique; no eyes small inornate but with testorus
Ixodes angustus		Rarely on wild rabbits	Western Oregon and Washing-ton, northern California, southern British Columbia, especially along coast	See ref. 37
I. dentatus	Eastern rabbit tick	Cottontail rabbits preferred host	Along eastern coast of United States from Maryland to Massachusetts	
I. muris	The mouse tick	Rarely on cottontails	Southern Cape Cod	See ref. 37
I. ricinus californicus	California black-legged tick	Jackrabbits and cotton-tails	Pacific coast west of Cascades and southern United States	See ref. 37
I. sculptus	Ground squirrel tick	Rarely on wild rabbits	Illinios, Michigan, Louisiana, westward; Canada and northern Mexico	See ref. 37
I. spinepalpis	—	Wild rabbits	Northern States west of the 100th meridian	See refs. 37 and 191
I. neotomae		Wild rabbits	Southern states west of the 100th meridian	See refs. 37 and 191
Otobius megnini	Spinose ear tick	Cottontail rabbits and jackrabbits	Southern United States	Larva and nymph spinose; adult free-living and spinose, hypostome

Table I (*continued*)

Tick genus spp.	Common name	Lagomorph host	Geographical distribution	Identification
Otobius lagophilus	Rabbit spinose tick	Cottontail rabbits and jackrabbits	Pacific northwest and northern states west of 100th meridian	See ref. 37
Rhipicephalus sanguineus	Brown dog tick	Wild rabbits, rarely	Spotty distribution in eastern half of the United States but reported from Midwest, Southwest and California. Temperate regions of rest of world, on every continent	Reddish-brown hexagonal basis capituli with festoons, male has characteristic adanal and accessory plates

lagus rabbits in the United States. *Haemodipsus setoni* has been reported only from wild rabbits in this country (67).

b. OCCURRENCE. *Haemodipsus ventricosus* occurs in European rabbits (*Oryctolagus cuniculi*) in Europe, Africa, Australia, New Zealand, and the United States (28, 66, 67, 150, 163).

Haemodipsus setoni has been reported from the United States in states west of 100th meridian (94, 217). *Haemodipsus africanus* is a species reported only from *Lepus zuluensis* from Africa. *Haemodipsus lyriocephalus*, the type species of the genus, is a louse identified only in *Lepus timidis* and *Lepus eurapeus* in Europe. It is a little known species and only the female has been identified (66). Louse infestation of rabbits is generally uncommon and is usually associated with poor hygienic conditions.

c. DIAGNOSIS. i CLINICAL SIGNS. Rabbit lice are voracious blood suckers and may produce weakness, emaciation. and anemia in severe infestations. There may be a thinning and ruffled appearance to the fur with the skin showing signs of inflammation. In light or moderate infestations, close examination for the nits (ova) or lice themselves may be necessary in order to recognize that the rabbit is parasitized. Pruritis may be pronounced and the rabbit may be seen scratching and rubbing infected areas. Secondary infection of primary lesions may occur (86).

ii. LOCATION ON HOST. No specific anatomical predilection for these ectoparasites is presented in the literature.

iii. IDENTIFICATION. For a detailed and illustrated key to the Anoplura, the reader is again referred to the review by Stojanovich and Pratt (217). In general, if one finds an anoplurid louse on an *Oryctolagus* rabbit it is most likely to be *H. ventricosus*. All anoplurid lice possess a head that is narrower than the thorax and sucking mouth parts to contrast them with the mallophagid lice of birds which have a head wider than the thorax and chewing mouth parts. The genus *Haemodipsus* is further characterized by the absence of eyes, five segmented antennae, a small first pair of legs with slender claws but with a second and third pair of equal length, moderately stout, and with stout claws. In *H. ventricosus*, paratergal plates or pleurites are totally absent but

are present as minute vestiges and the third and sixth segments in others (67) (see Figs. 11 and 12).

All stages of *Haemodipsus* live in intimate association with the host. Although specific details of the life history of *Haemodipsus* are not known, their general development probably resembles those of other members of the order Anoplura. Nits or eggs are glued to the hair shaft at their base where they hatch in approximately 10–20 days becoming nymphs which resemble adults except that they are smaller and lack sexual dimorphism. Following several instars the nymphs mature and become adults. Both nymphs and adults are blood suckers.

iv. CONTROL AND TREATMENT. These obligate host-specific insects spend their entire life cycle on the rabbit and although they are agile and mobile, they are loathe to move from host to host unless there is prolonged and intimate contact, such as may occur with a doe and her litter. Organophosphates and naturally occurring insecticides (pyrethrins) are effective against these parasites. Silica aerogels or dichlorvos may also be efficacious. It is important that treatment be repeated at least twice at 10-day intervals in order to kill nymphs which have hatched and were previously unaffected by the initial treatment.

v. ZOONOTIC AND VECTOR SIGNIFICANCE. *Haemodipsus* species are extremely host-specific and no reports were reviewed in which these species were recovered from man. The blood sucking habits of *Haemodipsus* enables it to be a potential vector for any of the blood-borne rabbit diseases.

THE DIPTERAN PARASITES OF THE RABBIT (ORDER: DIPTERA)

As the name of the order indicates, all winged members have only one pair of wings with vestigal posterior pair being present only as minute knobbed organs known as halteres. The life cycle of these insects involves complete metamorphosis with the stages normally being, egg, larva, pupa, and adult.

The mouth parts of adult Diptera are exquisitely sophisticated and usually adapted to the feeding habit of the species. The larval forms of some flies are important because they invade tissues and organs (myiasis). Although there

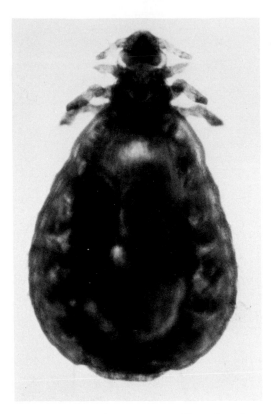

Fig. 11.   *Haemodipsus ventricosus* (Denny), male.

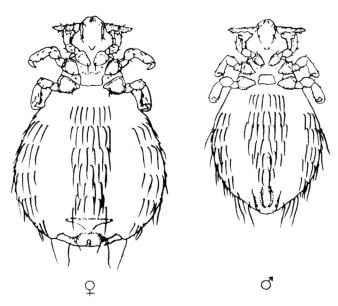

♀   ♂

Fig. 12.   *Haemodipsus ventricosus* (Denny), female and male, ventral surface. (Redrawn from Ferris.)

are approximately 75,000 known species of this order classified in some 140 families, relatively few families contain members which are of importance to lagomorphs (47). The various members of this order are important both because of the direct injury that they produce, e.g., the myiasis-producing flies, and because they act as vectors of disease, e.g., the mosquitoes. The families containing members of greatest significance for the rabbit include the following: Cuterebridae (*Cuterebra* species), Calliphoridae (*Wohlfahrtia* and *Cordylobia* species), Simulidae (*Simulium* species), and the Culicidae (mosquitoes in approximately 20 genera).

1.   The Cuterebrid Flies Affecting Rabbits

a. HISTORICAL AND GENERAL.   Members of the family Cuterebridae are known commonly as the rodent (and lagomorph) warble flies (48). They produce a striking form of parasitosis, known as myiasis. The most commonly reported genus is *Cuterebra*. Besides affecting lagomorphs and rodents, *Cuterebra* species also have been reported commonly from dogs, cats, mink kits, and foxes and infrequently from opossum, swine, asses, and occasionally man (13, 43, 54, 88, 108). The cuterebrid flies are obligatory myiasis-producing flies as opposed to the facultative forms of certain blow or flesh flies who have the ability to develop in either carrion or living animals.

Dermatobia hominus, another member of the family, is primarily a human parasite but has been reported from rabbits as well as a wide range of domestic and wild mammals and some birds (95).

b. OCCURRENCE.   i. GEOGRAPHICAL.   Members of the genus *Cuterebra* are restricted to the Western Hemisphere and have been reported from much of the United States and Canada (50). *Dermatobia* species, of much lesser significance to lagomorphs, are generally restricted to the neotropical regions including Mexico and Central and Northern South America (95).

Several *Cuterebra* species have been reported from rabbits including *C. cuniculi*, Clark, 1797 (205, 217, 236), *C. horripilum*, Clark, 1815 (48, 90, 91, 135, 217), *C. buccata*, Fabricius, 1776 (76, 90, 91, 157), and others (77, 180, 186, 221, 229).

ii. INCIDENCE.   Cuterebriasis may frequently occur in domestic rabbits reared out-of-doors or in otherwise unscreened quarters, however, no incidence data is available (235). A survey of wild cottontails in Virginia reported an incidence of *Cuterebra* larvae in 15% of 413 rabbits examined (140), whereas another in Georgia indicated only a 0.5% incidence in 215 cottontails examined (157). Other workers have described an incidence as high as 68% in a sample of black-tailed jackrabbits (110). Although as a general rule, there are usually few warbles on a given host, an individual jackrabbit in Arizona had 16 *Cuterebra* larvae (241). Geis reported on the incidence and affect of warble on southern Michigan cottontails (77).

In the north there is a marked seasonal incidence of myiasis which may explain any apparent discrepancy in the incidence reported in field surveys. The peak period occurs during the summer and early fall after which the larvae drop to the ground, pupate, and overwinter. In southern regions no periodicity in infestation is generally found and larvae may be recovered from wild rabbits the year round (229).

c. DIAGNOSIS. i. CLINICAL SIGNS. Although early lesions produced by these warbles may go undetected, the rabbit may be seen attempting to lick the affected area. Once the subcutaneous site becomes enlarged and swollen (2–3 cm), simple inspection and palpation of the affected area will reveal a swollen mass with a hole or fistula in the center of the swelling. This fistula is usually surrounded by moist matted fur. The lesion is usually very painful to the touch. Although infestations with a single warble may produce a spectacular lesion, e.g., ophthalmomyiasis, multiple warbles can cause considerable distress to the affected animal which may progress to debility and even death. Close examination of the encysted larvae may reveal movement of the larva and its spiracular plates (50). Secondary infection with bacteria may occur with a superimposed suppurative inflammatory response. Other myiasis-producing flies may become secondarily involved in a given animal, particularly the blowflies and flesh flies of the family *Calliphoridae*. When this occurs, death usually rapidly ensues.

Once the larvae is ready to pupate, it drops to the ground leaving a gaping hole in the host. (See Fig. 13.) The cavity rapidly closes over and heals in uncomplicated cases leaving a scar or area of hairless skin within about ten days.

Although not in a lagomorph species, Payne and Cosgrove have described the tissue changes seen following *Cuterebra* infestation (171a). Studies have also been conducted into the production of skin and ocular hypersensitivity and precipitating antibodies produced by botfly larvae in rabbits (102).

ii. LOCATION ON HOST. Although the site of initial entry of the larvae of *Cuterebra* is usually unknown, these larvae then migrate to preferred subcutaneous body sites. *Cuterebra horripilum*, for example, prefers to become established in the ventral cervical region of cottontail rabbits (91). *Cuterebra buccata* larvae may be found anywhere on the trunk, however, they prefer the interscapular, axillary, inguinal, and rump areas.

iii. IDENTIFICATION. Identification of the species of botflies or warble flies represented by the larval form is most difficult. In fact, over 50% of the larvae removed in

Fig. 13.   *Cuterebra* spp. lesion after removal of larva (top).

field studies are never specifically identified. Although some investigators have been successful in rearing the larvae to adulthood, very few have succeeded in duplicating a complete life cycle in the laboratory. A key to the genera of both adult and larval forms of myiasis-producing flies is given below.

KEY TO FAMILIES OF MYIASIS-PRODUCING FLIES IN NORTH AMERICA*

Mature Larvae

1. Larva robust and grublike, body only slightly tapering
   anteriorly 2
   Larva generally smooth, body conical, strongly tapering
   anteriorly 3
2. Body with finely developed spines located on ventral surface only or
   on the anterior margin of each segment dorsally—**Oestridae** (botflies
   and warble flies of wild and domestic ruminants)
   Body with stout, strong spines rather evenly distributed—**Cuterebridae** (rodent and lagomorph botflies)
3. Posterior spiracles flush with posterior face of anal segment—**Calliphoridae** (blowflies)
   Posterior spiracles deeply sunken in a rounded concavity—**Sarcophagidae** (flesh flies)

Adults

1. Large to moderate-sized flies, mouthparts reduced or inconspicuous.
   Body usually densely-haired, devoid of strong bristles, all vibrissae
   absent 2

* Adapted from Ref. 34.

Flies of moderate size, mouthparts well-developed and conspicuous. Body with strong bristles, all vibrissae present 3

2. Scutellum short, metascutellum strongly developed, mouthparts vestigal or reduced—**Oestridae** (botflies or warble flies of wild and domestic ruminants)

 Scutellum extending well beyond base of metanotum, metascutellum never developed, mouthparts present but inconspicuous—**Cuterebridae** (rodent and lagomorph botflies), e.g., *Cuterebra* species.

3. Two notopleural bristles present; body metallic blue or green—**Calliphoridae** (blowflies), e.g., *Cochliomyia* species.

 Three or more notopleural bristles present, body dark with diffuse spots or a grey, tesellated pattern on the abdomen—**Sarcophagidae** (flesh flies), e.g., *Wohlfahrtia vigil*

The life cycle of some of the *Cuterebra* have been investigated (25, 48, 80). Under natural conditions the mating of adult flies and egg laying takes place in the vicinity of the host's habitat and females proceed to actually deposit their eggs along runways, at entrances to burrows, or directly on the host. The ova hatch usually in response to rapid increases in environmental temperature and moisture possibly caused by the immediate presence of a host. The larvae gain entrance into the host by way of natural body openings (nares, mouth, anus) or unnatural openings (abrasions, puncture wounds). The larvae may stay at their entrance site for several days before migrating to their preferred body location (50) (see Fig. 14).

iv. CONTROL AND TREATMENT. Although myiasis is rather uncommon in laboratory rabbits it can occur when unscreened outdoor hutches are used. Myiasis in wild lagomorphs is generally an uncontrollable disease.

Fig. 14.   Cuterebra larva ecapsulated by host tissue (dissected free from host) (top). *Cuterebra* larva (bottom).

Treatment of affected rabbits involves physical removal of the larvae from the host under local anesthesia. It may be necessary to lance the skin to permit removal of the offending larvae, however, it has been stated that removal from intact cysts can be facilitated by injecting a few drops of chloroform into the fistulous sac followed by removal with a pair of fine-tooth forceps (235). Removal should be followed by cleansing of the would with a suitable detergent-antiseptic followed by the application of an appropriate topical antibiotic. In the absence of secondary infection, the disease is usually self-limiting unless multiple bots are present or they are critically located, e.g., on the eyes.

v. ZOONOTIC AND VECTOR SIGNIFICANCE.   There are reports of *Cuterebra* species causing myiasis in man including a case of ophthalomyiasis interna in which the *Cuterebra* larva was located in the anterior chamber of the eye (13, 14, 54). *Cuterebra* species are not known to be vectors of any disease.

2. Other Myiasis-Producing Flies

a. GENERAL.   Members of the family Calliphoridae, the blowflies or screwworms, and the family Sarcophagidae, known as flesh flies, are parasitic on domestic and wild rabbits.

The Calliphoridae is represented by *Cochliomyia hominovorax* (Coquerel, 1858), which is synonymous with *Cochliomyia americana* (Cushing and Patten, 1833), and *Callitroga americana*. It is commonly known as the screwworm fly.

Most members of the Sarcophagidae are saprophagous in their larval stages, however, some species are facultative myiasis-inducing parasites and one, *Wohlfahrtia vigil* (Walker, 1849), is an obligatory parasite of various wild mammals including cottontail rabbits and the jackrabbit (112, 186).

b. OCCURRENCE.   i. GEOGRAPHICAL.   Although *C. hominovorax* is a primarily neotropical species and is found most commonly in the southeastern and south central United States, it has been reported in states as far north as Wisconsin and Montana (34). Occasional reports come from the southwestern United States (131). It is capable of overwintering only in the southernmost areas, however (108).

Wohlfahrtia vigil is found in many areas of North America from as far north as Alaska to southern California in the west and New Jersey and New York in the east (34, 112, 186).

ii. INCIDENCE.   Screwworm infestations (*Cochliomyia* species) have been dramatically reduced as the result of a very effective eradication program of the United States Department of Agriculture in the southeastern United States.

Wohlfahrtia vigil primarily affects nestling rabbits and is most common in rabbits up to 14 days of age. Cottontail rabbits are thought to be natural reservoir of this parasite, which is then the cause of significant economic losses due to infestations in commercially raised mink kits and fox cubs. Since wild cottontails produce from 3 to 4 litters per year in many areas, an adequate supply of young hosts, are available from April–August of each year.

c. DIAGNOSIS. i. CLINICAL SIGNS. The clinical signs of infestation with these two parasitic larvae are not similar to that described for cuterebriasis. *Cochliomyia hominovorax* usually invades preexisting wounds and does not itself initiate the primary lesion. *Cochliomyia hominovorax* may, for example, lay its eggs adjacent to a cuterebrid lesion or more commonly near a gunshot or other wound (131, 186). *Cochliomyia hominovorax* infestations rapidly lead to emanciation and death of the host if untreated and are not self-limiting as are most cuterebrid infestations. Secondary bacterial infections with abscess formation are common and a foul-smelling reddish brown discharge commonly is present. The larva voraciously consume the rabbit's flesh with rapid debility and weakness leading to death many times at the hands of predatory carnivores. Newly laid batches of eggs may be observed adjacent to the lesions. The larva or maggots may be seen actively feeding in the wound. The characteristic circular pockets of feeding larva differentiate the screwworm lesion from other wounds merely invaded by bacterial pathogens.

Wohlfahrtia vigil also invades preexisting wounds but, in addition, larva may actually penetrate through the skin where it is sparsely haired and thin. *Wohlfahrtia vigil* has been reported from rabbits as young as one week of age and in young rabbits being reared out of doors (112). Eggs of *Wohlfahrtia* may be deposited on areas of intact skin but more frequently, like the screwworm, will be deposited near gunshot or other wounds. The many larvae that rapidly develop can produce generalized debility and death also. Over 40 larvae were found in a single rabbit which had lesions over one-third its body surface (251). Mortality in young animals is exceptionally high.

ii. LOCATION ON HOST. *Cochliomyia* species most commonly lay their ova adjacent to preexisting wounds so that the site of infestation may be anywhere on the host. They are found, however, most commonly on the face, neck, and rump regions of the body. *Wohlfahrtia vigil* females are larviporous and lay their eggs on or near nestling rabbits. The larva either infest preexisting wounds or actively penetrate the unbroken skin.

iii. IDENTIFICATION. While it is relatively easy to distinguish the cuterebrid from the calliphorid infestations on the basis of the appearance of the lesions, identification of the larvae within the calliphorid wounds is best handled by a specialist. A key to mature myiasis-producing larva on page 303. (Fig. 14) should enable one to speciate the larvae, however. Larval specimens should then be preserved in formalin or alcohol and submitted for identification. For those who wish to become familiar with the characteristics of dipteran larvae, they are referred to the works of Henning (93, 93a, 93b) and Knipling (117). It may be useful to attempt to rear some of the larvae to adulthood in order to aid in identification of the species involved (139).

The mature larva of *C. hominovorax* is cylindrical, strongly tapering, and encircled by bands of stout spines on each segment. The posterior spiracles are flush with the posterior face of the anal segments, however, the anterior spiracles have from 6 to 11 fingerlike branches. The mature larva or maggot is from 15 to 17 mm long (34, 93).

Cochliomyia hominovorax females lay from 200 to 400 ova adjacent to wounds. The eggs hatch under favorable conditions in from 12 to 21 hours and the larvae immediately penetrate the host and feed in characteristic oval pockets in the flesh. The larvae mature in 4–8 days after which they drop to the ground to pupate. The pupal stages last from one week in the summer to two months in the winter. Imagoes mate at approximately 3–4 days after emerging and begin to lay eggs when only six days old. The average complete life cycle takes approximately 3 weeks (34).

Mature *Wohlfahrtia vigil* larvae can best be identified by specialists, however, one can establish that a given larva is a member of the family Sarcophagidae by using the key in Fig. 14. Again it is useful to attempt to raise adult flies by allowing the larvae to mature and pupate in the laboratory (60). Specimens of both adult and larval forms can then be submitted for identification. Adult *W. vigil* flies mate, and gravid females deposit larva on or near nestling rabbits. The larvae attack the nestlings, usually killing them within a few days. Larvae mature rapidly (4–9 days) many times in the carcass of the now decreased rabbit, drop to the ground, and pupate. They emerge as adults approximately 10–12 days later. Cottontails have been found to be most susceptible between birth and two weeks of age (187).

iv. CONTROL AND TREATMENT. Control of myiasis in wildlife in general and rabbits in particular is a question of fly control and eradication. The massive efforts of the United States Department of Agriculture in screw worm eradication in the southeastern United States and especially Florida has undoubtedly brought about a tremendous decrease in myiasis in wild rabbits and other mammals. No fly control measures have been attempted for *Wohlfahrtia vigil*, however.

Control in domestic rabbits raised out of doors is probably simply a matter of providing hutches or out-buildings with protective screening, however, it is to be noted that in mink houses gravid flies have been known to deposit *W.*

vigil larvae through fine mesh screening that is thought to be effective against *C. hominovorax* (34).

Treatment of affected rabbits must be promptly administered. One must make daily observation of litters and treat affected animals by injection of small amounts of chloroform or hydrogen peroxide into the lesions. The maggots should be mechanically removed and the remaining wound treated with an antiseptic. Care must be taken to protect the wound from reinfestation by screening and/or application of a pine oil and larvacidal preparation to the wound. Workers in Russia have reported the use of various insecticidal agents which may be used in the treatment of rabbit myiasis (202).

V. ZOONOTIC AND VECTOR SIGNIFICANCE. *Cochliomyia hominovorax* and *Wohlfahrtia vigil* can both produce myiasis in man (34). Neither of these flies, however, are known to act as vectors of infectious diseases among rabbits.

3. Simulid Flies of Importance to the Rabbit

Members of the family Simulidae are commonly known as black flies. They will be treated only very briefly in this review. Their primary significance appears to be as vectors of myxomatosis in Australia. A report by Mykytowycz on the transmission of myxomatosis by *S. melatum*, Wharton, in which naturally occurring lesions due to the flies themselves are described and subsequent experimental work proving that *S. melatum* is capable of transmitting viral myxomatosis has been published (162). In addition to *S. melatum* workers in Scotland have seen and described *Simulium reptans* feeding on myxomatosis infected rabbits (27).

It is interesting to note that the primary lesions due to the simulid bite were substantial and were described as follows: "The ears of these rabbits had marked swellings which were, in some cases, as large as walnuts, causing the ear to hang down over the side of the head" (162). The experimentally induced simulid lesions in domestic rabbits were identical to the naturally occurring cases. The very short mouth parts of the simulid flies necessitate it getting very close to the skin of its host in order to feed. The most accessable parts of the body from this standpoint are the ears and nostrils.

4. Fleas (Order: Siphonaptera) Affecting the Rabbit

a. GENERAL. Although there are approximately 1500 known species of fleas, only about 25 have been reported as attacking wild or domestic lagomorphs and of these 25, only 2 are found with any degree of regularity in the United States. It is beyond the scope of this review to detail what is known about these many species of Siponaptera affecting the rabbit. For the purposes of this review, then, only the common species will be described, however, references to

these and other less common species are made in the following reviews (5, 28, 61, 70, 156, 163, 165, 178, 206, 217, 221).

The species which will be discussed in any detail are the following: *Cediopsylla simplex* (the common eastern rabbit flea), *Odontopsyllus mulpispinosus* (the giant eastern rabbit flea), *Hoplopsyllus glacialis affinis*, *Echidnophaga myremecobii* (the rabbit stick-tight flea), and *Spilopsylla cuniculi* (the common European rabbit flea). The other 20 or so species will be listed but not discussed.

These insects are hematophagous in the adult stage only. They are wingless insects with laterally compressed bodies which are covered with small spines. They vary greatly in size from 1 to 9 mm. Their legs are highly modified for jumping. On a weight–distance basis, a man would be able to jump $2\frac{1}{2}$ miles if he had the equivalent capabilities of the flea. The mouth parts of the flea are highly specialized for their host. Both sexes are hematophagous. Males are, as a rule smaller, often considerably smaller, than females. There is a wide variation in the degree of host specificity among the fleas (101).

b. OCCURRENCE. i. GEOGRAPHICAL. For a review of the distribution of the more common species of fleas affecting rabbits in the United States, the reader is referred to Table II. Of these species only two appear to occur with any frequency in the domestic rabbit. These two species are the common eastern rabbit flea, *C. simplex*, and the giant eastern rabbit flea, *O. multispinosus*. In Europe, the common European rabbit flea, *S. cuniculi*, is extremely important. It is not, however, found in the United States, Australia, or New Zealand (28, 134, 146, 149, 151, 163, 189). This finding is particularly interesting because *Oryctolagus* rabbits in the United States, Australia, and New Zealand came from European stock.

ii. INCIDENCE. The reader is referred to Table III on the siphonaptera affecting lagomorphs in which general statements concerning incidence in rabbits and key references are presented.

TABLE II
DISTRIBUTION OF COMMON FLEAS AFFECTING RABBITS IN THE UNITED STATES[a]

Species found east of the 100th meridian	*Cediopsylla simplex* (common eastern rabbit flea)
	Odontopsyllus multispinosus (giant eastern rabbit flea)
Species found west of the 100th meridian	
Northern States	*Cediopsylla inequalis* (rabbit flea)
	Odontopsyllus dentatus
	Hoplopsyllus glacialis affinis
Southern States	*Cediopsylla inequalis*
	Odontopsyllus dentatus
	Hoplopsyllus glacialis affinis

[a] Adopted from Stannard and Pietsch (217).

TABLE III
THE SIPHONAPTERA AFFECTING LAGOMORPHS

Species of flea	Host	Location	Comments	Ref.
Anomiopsyllus novomexicanensis	Cottontails	Southwestern United States		70
Cediopsylla simplex	Wild and domestic rabbits, common	See Table I	"Common Eastern Rabbit Flea"	153, 176, 191, 217
Ctenocephalides felis	Wild rabbits, very rare	Ubiquitous	Domestic cats are usual host	153, 176
Ctenophthalmus pseudagyrtes (Baker)	Cottontails, very rare	Southern Illinois	Moles are usual host	191, 217
Echidnophaga gallinacea	Domestic rabbits, occasionally	Southwestern United States	"Avian Sticktight Flea," birds are usual host	62, 70, 178, 205
Echidnophagea myremecobii	Wild and domestic Oryctolagus rabbits, common	Australia	"Australian Sticktight Flea," wild Oryctolagus rabbits are usual host	159, 179, 206
Epitedia wenmanni	Cottontails, very rare	Southern Illinois	Wild mice are the usual host	127
Hoplopsyllus glacialis affinis	Cottontails, common	United States west of 100th meridian		70, 176, 217
Megarthroglossus bisetis	Cottontails	Southwestern United States	—	70
Meringis bilsingi	Cottontails	Southwest United States		70
Meringis dipodomys	Cottontails	Southwestern United States		70
Meringis nidi	Cottontails	Southwest United States		70
Meringis rectus	Cottontails	Southwest United States		70
Nosopsyllus fasciatus (Basc)	Cottontails, very rare		"Northern Rat Flea"	176
Odontopsyllus multispinosus	Cottontails, rare	See Table I	"Giant Eastern Rabbit Flea"	176, 191, 217 217
Opistocrostis bruneri	"Wild Rabbit" Cottontails, very rare	Southern Illinois		
Orchopeus howardii	"Wild Rabbit," very rare	Southern Illinois	Tree squirrels are usual host, very rare	217
Pulex irritans	Cottontails	Southwest United States	"The Human Flea"	70
Polygenus gwyni	Cottontails	Southwest United States		70
Rhadinopsylla (Actenophthalmus) fraterna	Cottontails	Southwest United States		70
Spilopsyllus cuniculi	Wild and domestic rabbits, common	British Isles and European continent	"European Rabbit Flea"	1, 113, 125, 126, 128, 134, 146, 149, 151, 189
Thrassis fotus	Cottontails	Southwestern United States		70

The common eastern rabbit flea, *C. simplex*, is extremely common within its range. It is probably the second most common ectoparasite of the rabbit; second only to the continental rabbit tick (*Haemaphysalis leporis-palustris*). Twice as many female rabbits harbor *C. simplex* as male rabbits and, although only a few fleas are usually encountered on a given animal, up to 100 have been taken from individual rabbits. This flea may be found the year round even in the northern regions of its range (217). Peak incidence of *C. simplex* has been reported to coincide with the peak of the breeding season of the rabbit, i.e., the spring and fall of the year with the greatest numbers per animal being reported in the spring.

Odontopsyllus multispinosus primarily affects cottontail rabbits and one usually can detect only a few fleas on a given rabbit. It has a lower incidence than *Cediopsylla* within its range.

Echidnophaga myremecobii is primarily a parasite of wild rabbits although it has been reported from a variety of other wild rodents, a snake, and a fox in addition to an occasional isolation from a domestic dog or cat. *Echidnophaga gallinacea*, the avian stick tight flea, has been reported from wild lagomorphs but *E. myremecobii* has not been reported in birds. Heavy infestation of this flea in young rabbits has been reported to be responsible for some deaths (159).

c. DIAGNOSIS. i. CLINICAL SIGNS. The literature does not reveal much information concerning the clinical signs observed in rabbits infested with fleas. These ectoparasites usually produce only slight primary effects on their host. Signs attributable to the pruritis caused by development of hypersensitivity to flea antigens may be seen. Such signs as scratching and biting at affected areas with partial depilation and redness and rawness of the skin may be observed.

Secondary bacterial infection of the mutilated epidermis may occur in these cases. To my knowledge, no research has been reported which deals with the primary pathogenetic effects of fleas on lagomorph hosts. One can only draw inferences from studies conducted in other mammalian species and their fleas.

Most of the time there are very few individual fleas on the given host and they may be difficult to find, however, an attempt should be made to collect specimens and have them identified in the laboratory. In the absence of fleas on a given host, the diagnosis of flea dermatitis is almost impossible.

ii. LOCATION ON HOST. *Cediopsylla simplex* is found primarily on the ears around the face, top of the head, and back of the neck (217). Such distribution also appears to be similar for *S. cuniculi*. *Odontopsyllus multispinosus* is generally found on the back and hind end of the rabbit.

iii. IDENTIFICATION. For a detailed and thorough examination of the order Siphonaptera in which keys to the families, subfamilies, genera, and species of flea are presented with many excellent illustrations and photographs, the reader is referred to the classic work of Hopkins and Rothschild (103).

For the purposes of this review, however, certain general identifying features of the more common species are presented. If one is unable to readily identify given specimens and would like to clear and whole-mount fleas for critical study or referral, an excellent method has been published by Holland (101).

Cediopsylla simplex is within the same subfamily (subfamily: Spilopsyllinae) as are *Spilopsyllus* and *Hoplopsyllus*. *Cediopsylla* can be differentiated from *Hoplopsyllus* in that it has both genal and pronatal combs (or ctenidia) and from *Spilopsyllus*, which has not been reported from the United States, in that *Cediopsyllus* has a four-segmented labial palp as contrasted to a two-segmented one in *Spilopsyllus* (103) (see Figs. 15–19).

Odontopsyllus and *Cediopsylla* both have genal combs. In *Cediopsylla* the comb is contiguous with the eye and consists of 7 or 8 blunt black spines. *Odontopsyllus* has a heavier comb of toothlike or pointed spines located beneath the eyes. *Echidnophaga myremecobii* is identified by its angular frons and by the presence of an area of weak sclerotization above the pronotal angle.

d. LIFE CYCLE OF FLEAS. Fleas have evolved some rather remarkable adaptations to their hosts, which have recently been explored in the extensive work of Rothchild and Ford (188, 190–193, 195, 196) and Mead-Briggs. (147–149). Much of the data regarding seasonal incidence and the selective presence of more fleas on female rabbits (and on the young and in burrows) can be explained by studying the most interesting work of these authors. It has been

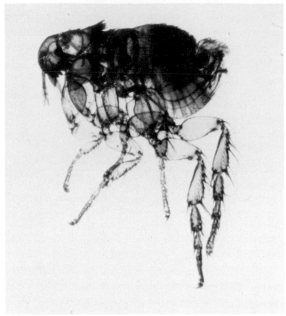

Fig. 15. Cediopsylla simplex, male, the common eastern rabbit flea (top). *Cediopsylla simplex,* female, (bottom).

observed also, that fleas occur in greater numbers on some species (e.g., jackrabbits). Such findings can be explained on the basis of the following work.

Fleas are insects which undergo complete metamorphosis. The various forms of the flea are ova, larva, pupae, and adult. In *Spilopsyllus cuniculi* (138, 188, 190) and *Cediopsylla simplex* (196) females do not undergo oogenesis with vitellogenesis and oviposition unless they are on or have fed on a pregnant doe in the later part of gestation or are on young nestling rabbits. Vitellogenesis in *S. cuniculi* begins in female fleas only. On males or immature females this pro-

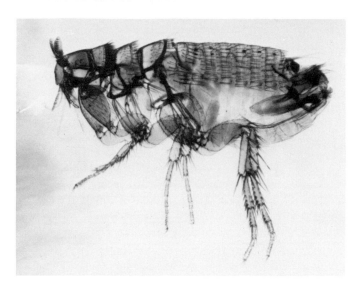

Fig. 16. Odontopsyllus multispinosus, male, the giant eastern rabbit flea.

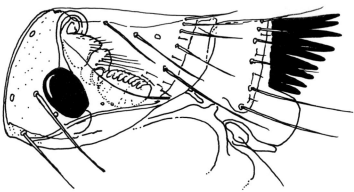

Fig. 18. Hoplopsyllus glacialis affinis, head and prothorax of male. (Redrawn from Hopkins and Rothchild.)

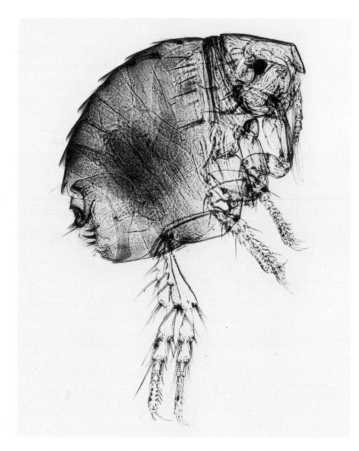

Fig. 17. Echidnophaga gallinacea, the avian sticktight flea.

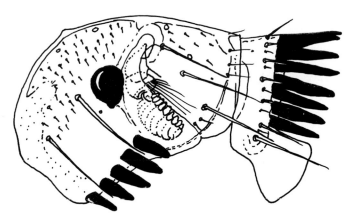

Fig. 19. Spilopsylla cuniculi, head and prothorax of male. (Redrawn from Hopkins and Rothchild.)

doe, but if they are removed to an estrous doe, the developing oocysts regress. If they then are transferred to newborn nestlings, the flea oocysts mature again and copulation between male and female fleas rapidly ensues (196). Under natural conditions, *S. cuniculi* moves from the ears of the doe to her face within a few hours after kindling, and then jumps onto the newborn rabbits where they feed.

After about 12 days, the fleas lay their eggs in the nest and the adults return to the doe. If she becomes pregnant, the fleas begin a new breeding cycle. This rather exquisitely developed life cycle in this specie of flea has been shown to be related to adrenocorticotropic hormone levels. It has been shown for example that immature fleas also produce oocysts if cortisone is sprayed on externally while the fleas are feeding on estrous (nonpregnant) does (196). Rothchild speculated in 1965 that such host parasite relationships are probably not unique and that the future will undoubtedly show many other such intimate interrelationships (188, 190). She did show that such was the case for *Cediopsylla simplex* in 1972 (196).

The flea eggs are layed in the burrow, and under favor-

cess does not take place. This insures that the fleas will have an ample supply of hosts (newborn or nestling rabbits) in the ensuing weeks. *Cediopsylla simplex* females develop chorionated eggs within three days if they are on a near term able conditions hatch within 10 days. The active larvae are negatively phototropic and seek out dark recesses in and about the burrow where they feed on the iron-rich feces expelled by adult fleas. The larvae are very susceptible to drying and thus do best in the relatively warm and humid conditions in the burrow. Following rapid development, the larvae pupate and emerge as adults in as short a time as 3 weeks (or as long as 3 years) if proper conditions exist.

i. CONTROL AND TREATMENT. No research has been reported in which control or treatment of flea infestations in wild lagomorphs has been described. Under most conditions in the wild, treatment is not practical. In domesticated rabbits treatment of both individuals and the adjacent environment with any one of a number of commercially available insecticides will control the problem (79). Prohibiting entry by wild lagomorphs into the common areas is essential to prevent reinfestation.

ii. ZOONOSES AND VECTOR SIGNIFICANCE. Fleas found on rabbits are both capable of transmitting various infectious diseases to man as well as serving as vectors of tularemia, Rocky Mountain Spotted Fever, myxomatosis, plague, and other diseases.

5. Mosquitoes of Importance in Lagomorph Biology

Numerous species of mosquitoes affect the lagomorphs. Their primary importance as ectoparasites of rabbits is related to their role as vectors for a wide variety of diseases and especially viral myxomatosis (3, 51, 65, 85, 129, 160, 228) and other viral diseases (99).

The mosquito is a mechanical vector only (for viruses) and has been described as merely a "flying pin." The mosquito need not feed on a viremic rabbit in order to transmit myxomatosis. High concentrations of virus are picked up on the mouth parts of mosquitoes feeding through myxomatosis skin lesions. In fact, a rabbit that has recently died from myxomatosis is often a better source of infectious material than the live rabbit since the mosquito repeatedly can bite the rabbit while trying to get a blood meal, therefore, contaminating his mouth parts with a higher concentration of virus.

Although the mosquito is only a mechanical carrier, it appears from the literature that there are differences in the ability of the various species of mosquitoes to transmit the virus. It has been suggested that structural differences in mouth part anatomy may account for this fact (51).

It is interesting to note that although mosquitoes and particularly *Aedes queenslandis, A. alboannulatus, Anopheles annulipes,* and *Culex annulirostris* are the principal vectors of myxomatosis in Australia, the rabbit flea, *Spilopsyllus cuniculi,* is the principal vector in Europe (128, 134).

Mosquitoes are distinguished by their characteristic wing veination, highly developed piercing and sucking mouth parts, and long and filamentous antennae with whorling or plumose hairs.

It is well beyond the scope of this work to review the biology of the numerous mosquitoes that are known to feed on lagomorphs. In one study alone in New South Wales, for example, over 9 different species of mosquitoes were collected which were shown to attack rabbits (128). Several of these were shown to contain myxomatosis virus.

Acknowledgments

The author wishes to thank Drs. Richard Stringer and Steven Weisbroth for supplying photographs of several of the parasites presented and to Dr. J. Ralph Lichtenfels and the National Animal Parasite Laboratory of the United States Department of Agriculture for allowing the author to use preserved specimens for photography. Thanks also are hereby given to the *Journal of Parasitology* for allowing republication of the photographs of *Listrophorus gibbus.*

REFERENCES

1. Allan, R. N. (1956). A study of the populations of the rabbit flea *Spilopsyllus cuniculi* (Dale) on the wild rabbit *Oryctolagus cuniculi* in Northeast Scotland. *Proc. Roy. Entomol. Soc. London, Ser. A* **31,** 145–152.
2. Anonymous. (1958). Care and management of laboratory animals. *Dep. Army Tech. Bull. Med.* **255.**
3. Aragao, H. B. (1943). The myxomatosis virus in wild rabbits (*Sylvilagus minensis*) and its transmission by *Aedes scapularis* and *Aedes aegypti. Mem. Inst. Oswaldo Cruz* **38,** 92–99.
4. Bacha, W. H., Jr. (1957). The life history of *Otobius lagophilus. J. Parasitol.* **43,** 560–561.
4a. Bacon, M. (1953). A study of arthropods of medical and veterinary importance in the Columbia Basin. *Wash., Agr. Exp. Sta., Bull.* **11,** 1–40.
5. Bacon, M., and Bacon, R. F. (1946). Notes on the distribution of Siphonaptera on wild rabbits of eastern and central Washington. *Northwest Sci.* **38,** 35–53.
6. Bacon, M., Drake, C. H., and Miller, N. G. (1959). Ticks (Acarina: Ixodoidea) on rabbits and rodents of eastern and central Washington. *J. Parasitol.* **45,** 281–286.
7. Baies, A., Suteu, I., and Klemm, W. (1968). *Notoedres* scabies of the golden hamster. *Z. Versuchstierk.* **1,** 251–257.
8. Baird, C. R., and Capell, K. J. (1969). Successful laboratory mating of two species of jackrabbit bot flies. *J. Med. Entomol.* **6,** 1966.
9. Baker, E. W. (1956). "A Manual of Parasitic Mites of Medical or Economic Importance." Natural Pest Control Association Publication.
10. Baker, K. P. (1969). Infestation of domestic animals with mite *Cheyletiella parasitovorax. Vet. Rec.* **84,** 561.
11. Banks, N. (1915). "The Acarina or Mites," Reprint No. 108. Bur. Entomol., Dep. Agr., U.S. Gov. Printing Office, Washington, D.C.
12. Barr, A. R. (1955). A case of "mange" of the domestic rabbit due to *Cheyletiella parasitovorax. J. Parasitol.* **41,** 323.
13. Beachley, R. G., and Bishopp, F. C. (1942). Report of a case of nasal myiasis due to a bot fly larva. *Vet. Med. Mon.* **69,** 41–42.
14. Beamer, R. H., and Penner, L. R. (1942). Observations on the life history of a rabbit cuterebrid, the larvae of which may penetrate the human skin. *J. Parasitol.* **28,** 25.
15. Bedford, G. A. H. (1932). A synoptic check-list and host-list of the ectoparasites found on South African mammalia, aves, and reptilia. *Rep. Vet. Res. Soc. S. Afr.* **18,** 223 and 523.

16. Bedford, G. A. H. (1943). *Haemodipsus africanus. Onderstepoort J. Vet. Sci. Anim. Ind.* **2**, 48.

17. Beesley, W. N. (1958). *Cheylietella parasitovorax* (Acarina: Trombidoidea) as a parasitic mite in Britain. *Parasitology* **19**, 14–16.

18. Bell, J. F., and Chalgren, W. S. (1944). Some wildlife diseases in the eastern United States. *J. Wildl. Manage.* **7**, 270–278.

19. Benjamini, E., and Feingold, B. F. (1970). Mammalian immunity to arthropods. *In* "Immunity to Parasitic Animals" (G. J. Jackson, R. Herman, and I. Singer, eds.), Vol. 2, pp. 1061–1134. Appleton, New York.

20. Bennet, G. F. (1955). Studies on *Cuterebra emasulator* Fitch 1859 (Diptera: Cuterebridae) and a discussion of the status of the Genus *Cephenemyia* ltr. 1818. *Can. J. Zool.* **33**, 75–98.

21. Bessalov, V. (1968). A focus of tularemia on Biriuchii Island in the Kherson Region. *Z. Mikrobiol.* **45**, 97–103.

22. Bishopp, F. C., and Trembley, H. L. (1945). Distribution and hosts of Certain North American ticks. *J. Parasitol,* **31**, 1–54.

23. Blaisdell, F. E. (1924). Note regarding the treatment of ear canker in rabbits. *Science* **69**, 429–430.

24. Blount, W. P. (1957). "Rabbits' Ailments." "Fur and Feathers," Idle, Bradford, England.

25. Boisvenue, R. J. (1958). Studies on the life history and ecology of *Cuterebra spp.* occurring in Michigan cottontails with systematic studies on cuterebrine larvae from other mammals. *Diss. Abstr.*, **19**, 14–16.

26. Bram, R. A., and Romanowski, R. D. (1970). Recognition of *Anaplasma marginale* (Theiler) in *Dermacentor andersoni* (Stiles) by the fluorescent antibody method. I. Smears of nymphal organs. *J. Parasitol.* **56**, 32–38.

27. Brown, P. W., Allan, R. M., and Shanks, R. L. (1956). Rabbits and myxomatosis in the northeast of Scotland. *Scot. Agr.* **35**, 4.

28. Bull, P. C. (1953). Parasites of the wild rabbit *Oryctolagus cuniculus* (L) in New Zealand. *N. Z. J. Sci. Technol.* **34**, 341–372.

29. Burbutis, P. P., and Mangold, R. E. (1956). A study of fleas of cottontail in New Jersey. *J. Wildl. Manage.* **20**, 217–218.

30. Butz, W. C., Stacy, L. D., and Heryford, N. N. (1971). Arachnidism in rabbits. *Arch. Pathol.* **91**, 97–100.

31. Caldwell, L. D. (1966). Marsh rabbit development and ectoparasites. *J. Mammal.* **47**, 527–528.

32. Cannon, D. A. (1942). Linguatulid infestation in man. *Ann. Trop. Med. Parasitol.* **36**, 160–166.

33. Capelle, K. J. (1970). Studies on the life history and development of *Cuterebra polita* (Diptera: Cuterebridae) in four species of rodents. *J. Med. Entomol.* **7**, 320–327.

34. Capelle, K. J. (1971). Myiasis, *In* "Parasitic Diseases of Wild Mammals" (J. W. Davis and R. C. Anderson, eds.), pp. 279–305. Iowa State Univ. Press, Ames.

35. Carpenter, S. J., and LaCasse, W. J. (1955). "Mosquitoes of North America (North of Mexico)." Univ. of California Press, Berkeley. pp. 279–305.

36. Chernesky, M. A. (1969). Powassan virus transmission by ixodid ticks infection after feeding on viremic rabbits injected intravenously. *Can. J. Microbiol.* **15**, 521–526.

37. Chernesky, M. A., and McLean, D. M. (1969). Localization of Powassan virus in *Dermacentor andersoni* ticks by immunofluorescence. *Can. J. Microbiol.* **15**, 1399–1408.

38. Chernysheva, T. F. (1967). A comparative study of multiplication of *Borrelia sogdianae* in the hemolymph of body lice. IV. *Med. Parazitol. Parazit. Bolez.* **36**, 478–482.

39. Chernysheva, T. F. (1968). Possibilities of transmission of tick relapsing fever by lice. V. The dynamics of *Borrelia sogdianae* multiplication in hemolymph of body lice. *Med. Parazitol. Parazit. Bolez.* **37**, 73–75.

40. Chubarian, K. H. A. (1966). Data of an experimental study of tickborne relapsing fever in laboratory and wild animals. *Zh. Eksp. Klin. Med.* **6**, 39–46.

41. Coles, E. H. (1967). "Veterinary Clinical Pathology." Saunders, Philadelphia, Pennsylvania.

42. Cooley, R. A. (1946). Notes on the tick, *Ixodes angustus* Neumann. *J. Parasitol.* **32**, 210.

43. Cooley, R. A., and Kohls, G. M. (1940). Two new species of Argasidae (Acarina: Ixodioidae). *Pub. Health Rep.* **55**, 925–933.

43a. Cooley, R. A., and Kohls, G. M. (1944). The Argasidae of North America, Central America and Cuba. *Amer. Midl. Natur., Monogr.* **1**, 1–152.

44. Cooley, R. A., and Kohls, G. M. (1945). "The Genus *Ixodes* in North America," N.I.H. Bull. No. 184. U.S. Pub. Health Serv., Washington, D.C.

45. Cooper, K. W. (1946). The occurrence of the mite, *Cheyletiella parasitovorax* (Megnin) in North America, with notes on its synonymy and parasitic habits. *J. Parasitol.* **32**, 480–482.

46. Culberston, J. T. (1935). Antibody production by the rabbit against an ectoparasite. *Proc. Soc. Exp. Biol. Med.* **32**, 1239–1240.

47. Curran, C. H. (1934). "The Families and Genera of North American Diptera." Ballou Press, New York.

48. Dalmat, H. T. (1943). A contribution to the knowledge of the rodent warble flies (Cuterebridae). *J. Parasitol.* **29**, 311–318.

49. Davis, G. E., Phillip, C. B., and Parker, R. R. (1934). The isolation from the Rocky Mountain wood tick (*D. andersoni*) of bacteria tolerance of low virulence for guinea pigs and domestic rabbits. *Amer. J. Hyg.* **19**, 449–456.

50. Davis, J. W., and Anderson, R. C., eds. (1971). "Parasitic Diseases of Wild Mammals." Iowa State Univ. Press, Ames.

51. Day, M. F. (1938). Factors influencing the transmissability of myxoma virus by mosquitoes. *J. Aust. Inst. Agr. Sci.* **21**, 145–151.

52. Delorme, P. M. (1927). Transmission Experimentale de *Sarcoptes scabiei var. cuniculi*, an Cynocephale (*Papio sphinx*, E. Geoff). *Bull. Soc. Pathol. Exot.* **19**, 899–900.

53. Deoras, P. J., and Patel, K. K. (1960). Collection of ectoparasites of laboratory animals. *Indian J. Entomol.* **22**, 7014.

54. Dixon, J. M., Winkler, C. H., and Nelson, J. H. (1971). Ophthalomyiasis interna caused by *Cuterebra* larva. *Amer. J. Ophthalmol* **71**, 415–516.

55. Dodd, K. (1970). *Cheyletiella yasguri*: Widespread infestation in a breeding kennel. *Vet. Rec.* **86**, 346–348.

56. Dorough, H. W., and Arthur, B. W. (1961). Systemic and contact insecticidal effectiveness of selected chemicals administered orally or dermally to rabbits. *J. Econ. Entomol.* **54**, 933–996.

57. Duckett, A. B. (1916). Little-known rabbit ear mite. *J. Amer. Vet. Med. Ass.* **48**, 726–730.

58. Dumas, J. (1953). "Les animaux de laboratoire." Editions Médicales Flammarion, Paris.

59. Elmes, B. G. T. (1945). DDT treatment of mange (scabies) in rabbits. *Lancet* **248**, 563.

60. Escile, J. L. (1965). Rearing and biology of *Wohlfahrtia vigil* (Diptera: Sarcophagidae). *Ann. Entomol. Soc. Amer.* **58**, 849.

61. Ewing, H. E., and Fox, I. (1943). The Fleas of North America. *U.S., Dep. Agr.*, Misc. Publ. **500**, 1–142.

62. Farlow, J. E., Burns, E. C., and Newsom, J. D. (1969). Seasonal distribution of some arthropod parasites of rabbits in Louisiana. *J. Med. Entomol.* **6**, 172–174.

63. Faust, E. C. (1927). Linguatulids (order Acarina) from man and other hosts in China. *Amer. J. Trop. Med.* **7**, 311–325.

64. Faust, E. C. (1928). Linguatulids in China. *China Med. J.* **42**, 107–108.

65. Fenner, F., Day, M. F., and Woodroofe, G. M. (1925). The mechanism of the transmission of myxomatosis in the European rabbit (*O. cuniculi*) by the mosquito *Aedes aegypti. Aust. J. Exp. Biol. Med. Sci.* **30**, 139–152.

66. Ferris, G. F. (1932). "Contributions toward a Monograph of the Sucking Lice," Part IV. Stanford Univ. Press, Stanford, California.

67. Ferris, G. F. (1951). The sucking lice. *Mem. Pac. Coast Entomol. Soc.* **1**, 320.

68. Field, G., Duplessis, R. J., and Breton, A. P. (1967). Progress report on laboratory rearing of black flies (Diptera: Simuliidae). *J. Med. Entomol.* **4**, 304–305.

69. Fiennes, R. N. (1968). Ecological concepts of stress in relation to medical conditions in captive wild animals. *Proc. Roy. Soc. Med.* **61**, 161–162.

70. Forcum, D. L., Rael, C. D., Wheeler, J. R., and Miller, B. E. (1969). Abundance of cottontails and their fleas at Red Bluff Ranch, New Mexico. *J. Wildl. Manage.* **33**, 422–424.

71. Ford, C. M. (1944). A non-parasitic mite in rabbits fur. *Vet. Rec.* **56**, 115.

72. Forstner, M. J. (1964). Course of the disease in Chamois goat and rabbit after infection with Chamois mange mite *Sarcoptes rupicaprae*. *Z. Parasitenk.* **25**, 16–17.

73. Fox, I., and Bayonna, I. G. (1968). Circulating precipitating antibodies in the rabbit from the bites of *Rhodnus prolixus* as shown by agar-gel tests. *J. Parasitol.* **54**, 1239–1240.

74. Fox, I., Bayonna, I. G., Umpierre, C. C., and Morris, J. M. (1967). Circulating precipitating antibodies in the rabbit from mite infection as shown by agar-gel test. *J. Parasitol.* **53**, 402–405.

75. Fox, R. R., Norberg, R. F., and Myers, D. D. (1971). The relationship of *Pasteurella multocida* to otitis media in the domestic rabbit (*Oryctolagus cuniculus*). *Lab. Anim. Care* **21**, 45–48.

76. Foxx, T. S., and Ewing, S. A. (1969). Morphologic features, behavior and life history of *Cheyletiella yasguri*. *Amer. J. Vet. Res.* **30**, 269–285.

77. Geis, A. D. (1957). Incidence and effect of warbles on southern Michigan cottontails. *J. Wildl. Mangae.* **21**, 94–95.

78. George, J. E. (1965). Drop-off rhythms of engorged rabbit ticks, *Haemaphysalis leporis-palustris* (Packard 1869). *Acari:Ixodidae. Diss. Abstr.* **26**, 553–554.

79. Georgi, J. R. (1969). "Parasitology for Veterinarians." Saunders, Philadelphia, Pennsylvania.

80. Gesztessy, T., and Nemesseri, L. (1970). Veterinary-hygienic aspects of mite contaminated feed. I. Experiments on chicken, guinea pigs, rabbits, and sheep. *Acta. Vet.* (*Budapest*) **20**, 29–33.

81. Gill, H. S., Rao, B. V., and Chhabra, R. C. (1968). Note on the occurrence of *Linguatula serata* (Frohlich, 1789) in domesticated animals. *Trans. Roy. Soc. Trop. Med. Hyg.* **62**, 506–508.

82. Gladenko, I. N., and Ostrenkii, E. S. (1967). On the toxicity of polychlorpinene in relation to animals. *Veterinariya* (*Moscow*) **44** (5), 87.

83. Gray, H. (1931). Some of the common diseases of the rabbit. *Vet. Rec.* **11**, 921–926.

84. Green, R. G., Evans, C. A., and Larson, C. L. (1943). A ten year population study of the rabbit tick, *Haemaphysalis leporis palustris*. *Amer. J. Hyg.* **38**, 260–281.

85. Grodhaus, G., Regnery, D. C., and Marshall, I. D. (1963). Studies in the epidemiology of myxomatosis in California. II. The experimental transmission in brush rabbits (*Sylvilagus bachmani*) by several species of mosquitoes. *Amer. J. Hyg.* **77**, 205–212.

86. Gurr, L. (1953). Some remarks on the possible insect vectors of myxomatosis in New Zealand. *N. Z. Sci. Rev.* **1**, 81–82.

87. Hagen, K. W. (1962). Disease of domestic rabbits. *U.S., Fish Wildl. Serv., Conserv. Bull.* **31**.

88. Hall, M. C. (1921). *Cuterebra* larvae from cats, with a list of those recorded from other hosts. *J. Amer. Vet. Med. Ass.* **12**, 480–484.

89. Hart, C. B., and Malone, J. C. (1958). The occurrence of the rabbit fur mite *Cheyletiella parasitovorax* (Megnin, 1878) on the dog. *Vet. Rec.* **70**, 991–993.

90. Hass, G. E. (1957). Ectoparasites of the Mearns cottontail in Wisconsin. *Diss. Abstr.* **17**, 2094.

91. Hass, G. E. and Dicke, R. J. (1958). On *Cuterebra horripilum*, Clark (Diptera: Cuterebridae) parasitizing cottontail rabbits in Wisconsin. *J. Parasitol.* **44**, 527–540.

92. Hayes, M. H. (1908). "Veterinary Pathology" [English translation from the text (in German) by Fredberger & Frohner, London].

93. Hennig, W. (1948). "Die Larvenformen der Dipteren," Part 1; pp. 1–186. Akademie-Verlag, Berlin.

93a. Hennig, W. (1950). "Die Larvenformen der Dipteren," Part 2; p. viii. Akademie-Verlag, Berlin.

93b. Hennig, W. (1953). "Die Larvenformen der Dipteren," Part 3; p. viii. Akademie-Verlag, Berlin.

94. Herman, C. M., and Jankiesicz, H. A. (1943). Parasites of cottontail rabbits on the San Joaquin Experimental Range, California. *J. Wildl. Manage.* **7**, 395–400.

95. Hermes, W. B., and James, M. T. (1961). "Medical Entomology," 5th ed. Macmillan, New York.

96. Hirst, S. (1917). On the occurrence of pseudo-parasitic mite (*Cheyletiella parasitovorax*, Megnin) on the domestic cat. *Ann. Natur. hist. Mus. Wien* **20**, 132–133.

97. Hirst, S. (1922). Mites injurious to domestic animals. *Bull. Mus.* (*Natur. Hist.*) *Econ. Ser.* No. 13.

98. Hite, J., Gladney, W., Lancaster, J. (1966). Biology of the Brown Recluse Spider. *Ark., Agr. Exp. Sta., Bull.* **117**.

99. Hitney, E., Jamnback, H., Means, R. G., Roz, A. P., and Rayner, G. A. (1969). California virus in New York state. Isolation and characterization of California encephalitis virus complex form *Aedes cinereus*. *Amer. J. Trop. Med.* **18**, 123–131.

100. Hoff, G. L., Yuill, T. M., Iversen, J. O., and Hanson, R. P. (1969). Snowshoe hares and the California encephalitis virus group in Alberta. *Wildl. Dis.* **5**, 254–259.

101. Holland, G. P. (1949). The Siphonaptera of Canada. *Can., Dep. Agr., Publ.* **817**, Tech. Bull. 70.

102. Holtman, D. F. (1953). Hypersensitivity in rabbits immune to the protein of bot-fly larvae. *J. Amer. Vet. Med. Ass.* **87**, 171–174.

103. Hopkins, G. H. E., and Rothschild, M. (1953). "An Illustrated Catalogue of the Rothschild Collection of Fleas (Siphonaptera) in the British Museum," Vol. 1. British Museum (Natur. Hist.), London.

104. Hopkins, G. H. E., and Rothschild, M. (1953). "An Illustrated Catalogue of the Rothschild Collection of Fleas (Siphonaptera) in the British Museum," Vol. 2. British Museum (Natur. Hist.), London.

105. Hopkins, G. H. E., and Rothschild, M. (1953). "An Illustrated Catalogue of the Rothschild Collection of Fleas (Siphonaptera) in the British Museum," Vol. 3. British Museum (Natur. Hist.), London.

106. Hopkins, G. H. E., and Rothschild, M. (1953). "An Illustrated Catalogue of the Rothschild Collection of Fleas (Siphonaptera) in the British Museum," Vol. 4. British Museum (Natur. Hist.), London.

107. Hopla, C. E. (1955). Observations on the life history of a rabbit tick. *J. Kans. Med. Soc.* **28**, 114–116.

108. Horsfall, W. R. (1962). "Medical Entomology." Ronald Press, New York.

109. Humphreys, M. (1958). *Cheyletiella parasitovorax* infestation in the dog. *Vet. Rec.* **70**, 442.

110. Ignoffo, C. M. (1961). Biology of *Cuterebra jellisoni* (Diptera: Cuterebridae) on *Lepus californicus deserticola* (Lagomorpha: Lepivoridae). *Ann. Entomol. Soc. Amer.* **54**, 509.

111. Jennings, W. L., Lewis, A. L., Sather, G. E., Hammon, W. M., and Bond, J. O. (1968). California encepthalitis-group viruses in Florida rabbits, report of experimental and sentinel studies. *Amer. J. Trop. Med.* **17**, 781–787.

112. Johansen, A. A. (1926). *Wohlfahrtia vigil* a parasite upon rabbits. *J. Parasitol.* **13**, 156.

113. Joyce, C. R., and Eddy, G. W. (1942). *Ixodes dentatus*. *Iowa J. Econ. Entomol.* **35**, 673.

114. Joyce, C. R., and Eddy, G. W. (1943). Host and seasonal notes on the rabbit tick, *Haemaphysalis leporis palustris*. *Iowa State Coll. J. Sci.* **17**, 205–212.

115. Kagramonov, A. I., Blagodarnyl, I. A. A., Madarebich, N. M.,

Blechman, I. M., and Iakunin, M. P. (1967). A possible carrier of tubercular infection. *Probl. Tuberk.* **45**, 60–64.

116. Keay, G. (1937). The ecology of the harvest mite (*Trombicula autumnalis*) in the British Isles. *J. Anim. Ecol.* **6**, 23–35.

117. Knipling, E. F. (1935). A key for blow fly larvae concerned in wound and cutaneous myiasis. *Ann. Entomol. Soc. Amer.* **32**, 376.

118. Knipling, E. F., and Rainwater, H. T., (1937). Species and incidence of dipterous larvae concerned in wound myiasis. *J. Parasitol.* **23**, 451.

119. Koffman, N., and Viriden, P. (1949). Nagra *Cheyletiella parasitovorax* fall hos katt och knin. *Nord. Veterinaermed.* **1**, 499–503.

120. Kohler, G., and Hoffman, G. (1967). Mites and lice as possible carriers of dematophytes. *Veterinarius* **34**, 107.

121. Kohls, G. M. (1939). Siphonaptera: Notes on synonymy of North American species of the genus *Hoplopsyllus*. *Baker Pub. Health Rep.* **54**, 2019–2023.

122. Kokernot, R. H., Calisher, C. H., Stannard, L. J., and Hayes, J. (1969). Arbovirus studies in the Ohio-Mississippi Basin, 1964–1967. VII. Lone Star Virus, a hitherto unknown agent isolated from the tick *Amblyomma americanum* (Linn). *Amer. J. Trop. Med.* **18**, 789–795.

123. Lane-Petter, W., Worden, A. N., Hill, B. F., Paterson, J. J., and Vevers, H. G., eds. (1967). "The U.F.A.W. Handbook On the Care and Management of Laboratory Animals," 3rd ed. Williams & Wilkins, Baltimore, Maryland.

124. LaPage, G. (1962). "Mönnings Veterinary Helminthology and Entomology." Williams & Wilkins, Baltimore, Maryland.

125. Lavoipierre, M. M. J. (1964). Mange mites of the genus *Notoedres* (Acari: Sarcoptidae) with descriptions of two new species and remarks on notoedric mange in the squirrel and the vole. *J. Med. Entomol.* **1**, 5–17.

126. Lawrence, R. F. (1954). Studies on the listrophorid mites (Sarcoptiformes) of Centetidae from Madagascar. *Mem. Inst. Sci. Madagascar, Ser. A* **9**, 130–149.

127. Lawrence, R. R. (1951). New fur mites of South African Mammals. *Ann. Natal. Mus.* **12**, 91–144.

128. Lee, D. J., Dyce, A. L., and O'Gower, A. K. (1957). Blood sucking flies (Diptera) and myxomatosis transmission of a mountain environment in New South Wales. *Aust. J. Zool.* **5**, 355 and 401.

129. LeGac, P. (1963). The role of the flea of the wild rabbit *Spilosyllus cuniculi* (Dale, 1878) in the preservation of the Mediterranean boutonneuse exanthematic fever virus. *C. R. Acad. Sci.* **257**.

130. Lesle, G. B., and Weetamn, D. F. (1969). The pharmacological actions of an experimental acaricide, NC 5016 (5,6-dichloro-1 phenoxy-carbonyl-2-trifluoromethyl benzimidazole). *Arch. Int. Pharmacodyn. Ther.* **177**, 105–21.

131. Lindquist, A. W. (1937). Myiasis in wild animals in southwestern Texas. *J. Econ. Entomol.* **30**, 735–40.

132. Litvishko, N. T., Kharchenko, O. N., and Tertyphnyi, A. A. (1965). *Haemodipsus ventricosus* infestation in rabbits. *Veterinariya (Moscow)* **42**,

133. Litvishko, N. T., Kharchenko, O. N., and Tertyphnyi, A. A. (1969). *Helmodipsus* lice on rabbits and their control. *Veterinarius* **36**, 115.

134. Lockley, R. M. (1954). The European rabbit flea, *Spilopsyllus cuniculi* as a vector of myxomatosis in Britain. *Vet. Rec.* **66**, 434–435.

135. Lopushinsky, T. (1970). Myiasis of nestling cottontail rabbit. *J. Wildl. Dis.* **6**, 98–100.

136. Low, C. R. (1911). An investigation into scabies in laboratory animals. *J. Pathol. Bacteriol.* **15**, 333–349.

137. Lund, E. E. (1951). Ear mange in domestic rabbits. *Amer. Rabbit J.* **21**, 67–69.

138. McDaniel, B. (1965). The subfamily Listrophorinae Gunther with a description of a new species of the genus *Listrophorus*, Pagestecher from Texas (Acarina: Listrophoridae). *Acarologia* **7**, 704–712.

139. McDonald, W. A. (1962). A method of rearing myiasis-producing fly larvae. *J. Entomol. Soc. S. Afr.* **25**, 149.

140. McGinnes, B. S. (1964). Parasites of cottontail rabbits in southwestern Virginia *Wildl. Dis.* **35**, 11.

141. McKenney, F. D. (1937). Infectious myxomatosis of domestic rabbits. *U.S., Bur. Biol. Serv., Wildl. Res. Leafl.* **BS-89**.

142. MacLulich, D. A. (1937). Fluctuations in the numbers of the varying hare (*Lepus americanus*). *Univ. Toronto Stud., Biol. Ser.* **43**, 1–136.

143. Marine, D. (1924). The cure and prevention of ear canker in rabbits. *Science* **60**, 158.

144. Markov, A. A. (1958). The effect of desiccant powders on ixodid ticks. *Veterinariya (Moscow)* **45**, 61.

145. Matumoto, M. (1969). Mechanism of perpetuation of animal viruses in nature. *Bacteriol. Rev.* **33**, 404–418.

146. Mead-Briggs, A. R. (1962). Observations on the rabbit flea: A vector of myxomatosis. *Ann. Appl. Biol.* **15**, 338–342.

147. Mead-Briggs, A. R. (1964). A correlation between development of the ovaries and of the midget epithelium in the rabbit flea *Spilospyllus cuniculi*. *Nature (London)* **201**, 1303–1304.

148. Mead-Briggs, A. R. (1964). Some experiments concerning the interchange of rabbit fleas *Spilosyllus cuniculi* (Dale) between living rabbit hosts. *J. Anim. Ecol.* **33**, 13–26.

149. Mead-Briggs, A. R. (1964). The reproductive biology of the rabbit flea *Spilosyllus cuniculi* (Dale) and the dependence of this species upon the breeding of its host. *J. Exp. Biol.* **41**, 371–402.

150. Mead-Briggs, A. R., and Hughes, A. M. (1965). Records of mites and lice from wild rabbits collected throughout Great Britain. *Annu. Mag. Natur. Hist.* **8**, 695–708.

151. Mead-Briggs, A. R., and Rudge, A. J. B. (1960). Breeding of the rabbit flea *Spilopsyllus cuniculi* (Dale): Requirement of a "factor" from a rabbit for ovarian maturation. *Nature (London)* **187**, 1136–1137.

152. Meek, M. W. (1944). Diseases and parasites of rabbits and their control. Reliable Fur Industries, Menteliello, California. *Vet. Bull. (London)* **14**, 2600.

153. Milalo, I. I. (1967). Precautions in the use of trichlormetapho-3. *Veterinariya (Moscow)* **44**, 95–97.

154. Mohr, C. O. (1961). The relation of rabbit tick populations to spacing in host populations. *J. Parasitol.* **47**, 605–607.

155. Moore, E. R., and Moore, G. C. (1947). The helminth parasites of cottontail rabbits in Alabama, with notes on the arthropod *Linguatula serrata*. *J. Mammal.* **28**, 279–284.

156. Morgan, B. B. (1940). A survey of the Iowa cottontail (*Sylvilagus floridanus mearnsi*). *J. Wildl. Manage.* **4**, 21–26.

157. Morlan, H. B. (1952). Host relationships and seasonal abundance of some southwest Georgia ectoparasites. *Amer. Midl. Natur.* **48**, 74–93.

158. Muers, J. G. (1924). The cattle tick (*Haemaphysalis bispinosa*): Investigations during 1923–1924. *N. Z. Dep. Agr. Bull.* **116**.

159. Mules, M. W. (1940). Notes on the life history and artifician breeding of the Australian "Stickfast" flea, *Echidonophaga myremecobii*, Rothschild. *Biol. Med. Sci.* **18**, 385–390.

160. Myers, K. (1955). The ecology of the mosquito vectors of myxomatosis (*Culex annulirostris* Skuse, and *Anopheles annulipes* Walk) in the Eastern Riverina. *J. Aust. Inst. Agr. Sci.* **21**, 250–253.

161. Myers, K. (1956). Methods of sampling winged insects feeding on the rabbit *Oryctolagus cuniculus* (L). *CSIRO Wildl. Res.* **1**, 45–48.

162. Mykytowycz, R. (1957). The transmission of myxomatosis by *Simulium melatum*, Wharton (Diptera: Simuliidae). *CSIRO Wildl. Res.* **2**, 1–4.

163. Mykytowycz, R. (1957). Ectoparasites of the wild rabbit, *Oryctolagus cuniculus* (L) in Australia. *CSIRO Wildl. Res.* **2**, 63–65.

164. Mykytowycz, R. (1957). Parasitic habit of the rabbit mite *Cheyletiella parasitovorax* (Megnin). *CSIRO Wildl. Res.* **2**, 164.

165. Mykytowycz, R. (1958). Ectoparasites of the wild rabbit *Oryctolagus cuniculus* (L) in Australia. *Vet. Bull. (London)* **28**, 475.

166. Olsen, J. J., and Roth, H. (1947). On the mite *Cheyletiella parasitovorax*, occurring on cats, as a facultative parasite of man. *J. Parasitol.* **33**, 444–445.

167. Osborne, H. G. (1947). Dimethyl phthalate for the treatment of mange in rabbits caused by *Notoedres cati var. cuniculi. Aust. Vet. J.* **24**, 52.

168. Ouchi, S., and Amada, M. (1961). Pathology of tularima in naturally, infected wild rabbits, especially relation between the primary affect and tick-bit, and state of the bacillus in tissues of the host. *Fukushima J. Med. Sci.* **11**, 203–227.

169. Oudemans, C. (1906). Revision des Chelelenes. *Mem. Soc. Zool. Fr.* **19**, 211–218.

170. Pabrai, P. E., Venkatachalam, K., and Ojha, K. N. (1950). The efficacy of tincture of tobacco in the treatment of mange in rabbits. *Indian J. Vet. Sci. Anim. Husb.* **20**, 13–15.

171. Pagenstecher, H. (1862). *Listrophorus gibbus*, Nebst Nachtraglichen Bemerkungen uber *Listrophorus leuckarti. Z. Wiss. Zool.* **11**, 156–161.

171a. Payne, J. A., and Cosgrove, G. E. (1966). Tissue changes following *Cuterebra* intestation in rodents. *Amer. Midl. Natur.* **75**, 205–213.

172. Pegg, E. J. (1970). Three ectoparasites of veterinary interest communicable to man. *Med. Biol. Illus.* **20**, 106–110.

173. Phillip, C. B., Bell, J. F., and Larson, C. L. (1955). Evidence of infectious diseases and parasites on a peak population of black-tailed jack rabbits in Nevada. *J. Wildl. Manage.* **19**, 225–233.

174. Pichen, E. G., Gerloff, R. K., and Burgdorfer, M. (1968). Spirochete from the rabbit tick, *Haemaphysalis leporis-palustris* (Packard). I. Isolation and preliminary characterization. *J. Bacteriol.* **95**, 291–299.

175. Pillers, A. W. N. (1925). *Cheyletiella parasitovorax*, Megnin causing lesions in domestic rabbits. *Vet. J.* **81**, 96–97.

176. Portman, R. W. (1944). Winter distribution of two ectoparasites of the cottontail rabbit in Missouri. *J. Econ. Entomol.* **37**, 541.

177. Pratt, H. D., and Littig, D. S. (1962). Ticks of public health importance and their control. *U.S., Pub. Health Serv.* **772**, Insect Contr. Ser., Part X.

178. Pratt, H. D., and Wiseman, J. S. (1962). Fleas of public health significance and their control. *U.S. Pub. Health Serv., Publ.* **772**, Insect Contr. Ser., Part VII.

179. Pridham, T. J. (1968). Common diseases of fur bearing animals. II. Diseases of chinchillas, nutria. *Can. Vet. J.* **7**, 84–87.

180. Pyckman, R. E., and Lindt, C. C. (1954). *Cuterebra lepivora* reared from *Sylvilagus audubonii sanctidiegi* in San Bernardino County, California. *J. Econ. Entomol.* **47**, 1146–1148.

181. Radford, C. D. (1935). New parasitic mites (Acarina) from rodents. *Parasitology* **35**, 161–166.

182. Radford, C. D. (1953). Four new species of "Harvest Mite" or "Chigger" and a new fur mite (Acarina: Trombiculidae and Listrophoridae). *Parasitology* **43**, 210–214.

183. Rasasco, M. E. (1957). Seasonal abundance of the tick *Dermacentor parumaptertus* on the blacktailed jack rabbit, with notes on other ectoparasites. *J. Mammal.* **38**, 485–490.

184. Rick, R. F. (1958). Studies on the reactions of laboratory animals with ticks. III. The Reactions of laboratory animals to sublethal doses of egg extracts of *Hemaphysalis leporis-palustris* Neumann. *Aust. J. Agr. Res.* **9**, 830–841.

185. Roberts, I. H., and Meleney, W. P. (1970). Ear-scab mites, *Psoroptes cuniculi* (Acarina: Psoroptidae) in captive mule deer. *J. Parasitol.* **56**, 1039–1040.

186. Roberts, R. A. (1953). Additional notes on myiasis in rabbits (Diptera: Calliphoridae, Sarcophagidae). *Entomol. Rev. (USSR)* **33**, 157–159.

187. Rogstad, O. J. (1966). Biology of penned cottontails. *J. Wildl. Manage.* **30**, 312.

188. Rothschild, M. (1965). Fleas. *Sci. Amer.* **213**, 44–53.

189. Rothschild, M. (1965). Myxomatosis and the rabbit flea. *Nature (London)* **207**, 1162–1163.

190. Rothschild, M. (1965). The rabbit flea and hormones. *Endeavour* **24**, 162–168.

191. Rothschild, M., and Ford, B. (1964). Breeding of the rabbit flea (*Spilospyllus cuniculi*, Dale) controlled by the reproductive hormones of the host. *Nature (London)* **201**, 103–105.

192. Rothschild, M., and Ford, B. (1964). Maturation and egg-laying of rabbit flea (*Spilopsyllus cuniculi*, Dale) induced by the external application of hydrocortisone. *Nature (London)* **203**, 210–211.

193. Rothschild, M., and Ford, B. (1965). Observations on gravid rabbit fleas (*Spilopsyllus cuniculi*, Dale) parasitising the hare (*Lepus europaeus*, Pallus), together with further speculations concerning the course of myxomatosis at Ashton, Northants. *Proc. Roy. Entomol. Soc. London* **40**, 117–190.

194. Rothschild, M. and Ford, B. (1966). Hormones of the vertebrate host controlling ovarian regression and copulation of the rabbit flea. *Nature (London)* **211**, 261–266.

195. Rothschild, M., and Ford, B. (1969). Does a pheromone-like factor from the nestling rabbit stimulate impregnation and maturation in the rabbit flea? *Nature (London)* **211**, 1169.

196. Rothschild, M., and Ford B. (1972). Breeding cycle of the flea *Cediopsylla simplex* is controlled by breeding cycle of host. *Science* **178**, 625–626.

197. Saggers, D. T., Clark, M. L., and Lichard, M. (1957). Trifluoromethylbenzimidazoles—a new family of acaricides. *Nature (London)* **215**, 275–276.

198. Saito, Y. (1962). Isolation of *Pasteurella tularensis* from ticks and chiggers parasitized on tularemia-wild hares, with some tularemia transmission experiment. *Acta Med. Biol. (Niigata)* **10**, 147–159.

199. Salomon, P. F., Catts, E. P., and Knox, W. G. (1970). Human dermal myiasis caused by rabbit bot fly in Connecticut. *J. Amer. Med. Ass.* **213**, 10035–10036.

200. Sambon, L. W. (1922). A synopsis of the family Linguatulidae. *J. Trop. Med. Hyg.* **25**, 188–206 and 391–428.

201. Sasa, M., and Kano, R. (1951). *Cheyletiella takahasii* n. sp., A new species of parasitic mite from Ochotuna of Hokkaido, Japan (Acarina: Cheylitidae). *Jap. J. Exp. Med.* **21**, 205–207.

202. Savelev, P. V., Voblikova, N. V., Mezenev, N. P., and Silkow, A. M. (1962). Trails with trichlorphon, fenchlorphos, dichloruos, and dimethoate against the reindeer warble fly. *Veterinariya (Moscow)* **39**, 74.

203. Savory, T. (1964). "Arachnida." Academic Press, New York.

204. Schaefer, R. E., and Steelman, C. D. (1969). Determination of mosquitoe hosts in salt march areas of Louisiana. *J. Med. Entomol.* **6**, 131–134.

205. Schwartz, B., and Shook, B. S. (1928). Rabbit parasites and diseases. *U.S., Dep. Agr., Farmers' Bulletin* **1568**.

206. Seddon, H. R. (1951). Disease of domestic animals in Australia. Part II. Fly, louse, and flea infestation. *Aust., Commonw., Dep. Health Serv., Publ.* No. 6.

207. Seidel, K. (1936). Die Krankheiten des Kaninchens (Diseases of rabbits). *Vet. Bull. (Berlin)* **6**, 480.

208. Seifried, O. (1937). "Die Krankheiten des Kaninchens," 2nd ed. Springer-Verlag, Berlin and New York.

209. Short, D. J., and Woodnott, D. P. (1969). "The I.A.T. Manual of Laboratory Animal Practice and Techniques," 2nd ed. Thomas, Springfield, Illinois.

210. Singh, K. R., and Anderson, C. R. (1968). Relation of *Haemaphysalis spinigera* larval infection rates and host viremia levels of Kyasanur Forest disease virus. *Indian J. Med. Res.* **56**, 137–141.

211. Smiley, R. L. (1965). Two new species of the genus *Cheyletiella* (Acarina: Cheyletidae). *Proc. Entomol. Soc. Wash.* **67**, 75–79.

212. Smith, D. T., and Webster, L. T. (1925). Epidemiological studies on respiratory infections of the rabbit. VI. Etiology of otitis media. *J. Exp. Med.* **41**, 275–282.

213. Smith, H. G., Goulding, R. L., and Priano, J. L. (1970). A tick cage, plastic collar, and adjustable restraining cage for use in rabbit ectoparasite-pesticide studies. *J. Econ. Entomol.* **63**, 330–331.

214. Smythe, R. H. (1962). "Animal Habit, the Things Animals Do." Thomas, Springfield, Illinois.

215. Soulsby, E. J. L. (1968). "Helminths, Arthropods, and Protozoa of Domesticated Animals." Williams & Wilkins, Baltimore, Maryland.

216. Specht, F. (1931). Inflammation of the middle ear, labryinth, and meninges in the rabbit resulting from mites in the auditory canal. *Arch. Ohren-, Nasen-Kehlkopfheik.* **128**, 103–114.

217. Stannard, L. J., and Pietsch, L. R. (1958). Ectoparasites of the cottontail rabbit in Lee County, northern Illinois. *Ill., Natur. Hist. Surv., Biol. Notes* **38**, 3–17.

218. Stojanovich, C. J., and Pratt, H. D. (1965). "Key to Anoplura of North America." USDHEW, PHS, Communicable Disease Center, Atlanta, Georgia.

219. Stone, A., Knight, K. L., and Starcke, H. (1959). "A Synoptic Catalogue of the Mosquitoes of the World (Diptera, Culicidae)." Entomol. Soc. Amer., Washington, D.C.

220. Stresser, H. (1963). Mange in rabbits due to *Cheyletiella*. Kleintier-Praxis **8**, 212–214.

221. Stringer, R. P., Harkema, R., and Miller, G. C. (1969). Parasites of rabbits in North Carolina. *J. Parasitol.* **55**, 328.

222. Sudia, W. D., Coleman, P. H., and Chamberlain, R. W. (1969). Experimental vector-host studies with tensaw virus, a newly recognized member of the Bunyamwera arbovirus group. *Amer. J. Trop. Med.* **18**, 98–102.

223. Sweatman, G. K. (1957). Life history, nonspecificity, and revision of the genus *Chorioptes*, a parasitic mite of herbivores, *Can. J. Zool.* **35**, 641–689.

224. Sweatman, G. K. (1958). On the life history and validity of the species in *Psoroptes*, a genus of mange mites. *Can. J. Zool.* **36**, 905–929.

225. Tamsitt, J. R., and Fox, I. (1970). Mites of the family Listrophoridae in Puerto Rico. *Can. J. Zool.* **48**, 398–399.

226. Tarshis, I. B. (1962). The use of silica aerogel compounds for the control of ectoparasites. *Proc. Anim. Care Panel* **12**, 217–258.

227. Templeton, G. S. (1943). Ear mange or ear canker. *Amer. Rabbit J.* **13**, 85.

228. Templis, C. H., Hayes, R. O., Hess, A. D., and Reeves, W. C. (1970). Blood-feeding habits of four species of mosquitoe found in Hawaii. *Amer. J. Trop. Med.* **19**, 335–341.

229. Test, F. H., and Test, A. R. (1943). Incidence of dipteran parasitosis in populations of small mammals. *J. Mammal.* **24**, 506–509.

230. Thompson, H. V., and Worden, A. N. (1956). "The Rabbit." Collins, London.

231. Thomsett, L. R. (1968). Zoonosis—*Cheyletiella parasitovorax. Brit. Med. J.* **3**, 93–95.

232. Turner, E. C. (1971). Fleas and lice. *In* "Parasitic Diseases of Wild Mammals" (J. W. Davis and R. C. Anderson, eds.) pp. 65–67. Iowa State Univ. Press, Ames.

233. Turner, E. C. (1971). Fleas and lice. *In* "Parasitic Diseases of Wild Birds" (J. W. Davis and R. C. Anderson, eds.), pp. 175–184. Iowa State Univ. Press, Ames.

234. Vail, E. L., and Auguston, G. F. (1943). A new ectoparasite (Acarina: Cheyletidae) from domestic rabbits. *J. Parasitol.* **29**, 419–421.

235. Vail, E. L., and McKenny, F. D. (1943). "Diseases of Domestic Rabbits." *U.S. Dept. Internal Fish Wildl. Serv. Conserv. Bull.* **31**, pp. 1–28. U.S. Gov. Printing Office, Washington, D.C.

236. Valadez, S. M. (1926). Pequena contribugion para la parasitologia Mexicana. *Mem. Rev. Soc. Cient. Antonio Alzate"* **45**, 1–12.

237. Vashkov, V. I., Testlin, V. M., Volkova, A. P., Brikman, L. I., Bessonova, I. V., Zhuk, E. B., Martynseva, M. N., and Yankovsky, E. Y. (1967). Development and study of filler composition of aerosol dispenser for controlling flying insects. *J. Hyg. Epidemiol., Microbiol., Immunol.* **11**, 301–308.

238. Vaughan, J. A., and Mead-Briggs, A. R. (1970). Host-finding behavior of the rabbit flea *Spilosyllus cuniculi* with special reference to the significance of urine as an attractant. *Parasitology* **61**, 397–410.

239. Vitzthum, H. G. (1929). Systematische Betrachtungen zur Frage der Trombidiose. *Z. Parasitenk.* **2**, 223–247.

240. Volgin, V. I. (1960). On the taxonomy of predatory mites of the family *Cheyletidae*. I. Genus *Cheletiella. Can. Parasitol.* **19**, 237–248.

241. Voorhies, C. T., and Taylor, W. P. (1933). The life histories and ecology of jack rabbits, *Lepus alleni* and *Lepus californicus* species, in relation to grazing in Arizona. *Ariz., Agr. Exp. Sta., Tech. Bull.* **49**, 470.

242. Wasylik, A. (1965). Studies on the European Hare. VII. Dynamics of *Listrophorus gibbus* Pagenstecher, 1862. *Acta Theriol.* **10**, 27–54.

243. Wegner, Z., and Eichler, W. (1968). Fuana of lice found on hares in the Poznan Provice (Poland). *Bull. Inst. Mar. Med. Gdansk.* **19**, 217–224.

244. Weisbroth, S. H. Personal communication.

245. Weisbroth, S. H., and Scher, S. (1971). *Listrophorus gibbus* (Acarina: Listrophoridae) an unusual parasitic mite from laboratory rabbits (*Oryctolagus cuniculus*) in the United States. *J. Parasitol.* **57**, 438–440.

246. Weisbroth, S. H., and Scher, S. (1971). *Microsporum gypseum* dermatophytosis in a rabbit. *J. Amer. Vet. Med. Ass.* **159**, 629–634.

247. Weitkamp, R. A. (1964). *Cheyletiella parasitovorax* parasitism in dogs. *J. Amer. Vet. Med. Ass.* **144**, 597–599.

248. Wharton, G. W., and Fuller, A. S. (1952). "A Manual of Chiggers," No. 4. Entomol. Soc. Wash. Washington, D.C.

249. Wormessley, H. (1941). Notes on the Cheyletidae (Acarina: Trombidoidea) of Australia and New Zealand with descriptions of new species. *Rec. S. Aust. Mus.* **7**, 51–64.

250. Yasgur, I. (1964). Parasitism of kennel puppies with the mite *Cheyletiella parasitovorax. Cornell Vet.* **54**, 406–407.

251. Yuill, T. M., and Eschle, J. L. (1963). Myiasis of penned nestling cottontails. *J. Wildl. Manage.* **27**, 477.

252. Yuill, T. M., Iverson, J. O., and Hanson, R. P. (1969). Evidence for arbovirus infection in a population of snowshoe hares—a possible mortality factor. *J. Wildl. Dis.* **5**, 248–253.

253. Yunker, C. E. (1964). Infections of laboratory animals potentially dangerous for man: Ectoparasites and other arthropods, with emphasis on mites. *Lab. Anim. Care* **14**, 455–465.

254. Zapletai, M. (1960). Kratke sdeleni o roztocich nadceledi Listrophoidae *Zool. Listy* **23**, 56–58.

255. Zurn, F. A. (1874). Raudmilben un Ohr derttunde and kie Kannichen. *Wochenschr. Tierheilk. Viehzucht* **18**, 277–283.

CHAPTER 13

Helminth Parasites

Richard B. Wescott

I. INTRODUCTION

Natural helminth infections do not present a serious problem in rabbits raised for biomedical research if proper husbandry practices are followed. Occasionally investigators find their experimental subjects infected with *Passaluris ambiguus*, the rabbit pinworm. From time to time other helminths, common in wild rabbits, are reported in domestic rabbits. In general, these natural infections are regarded as a nuisance and diagnosis is made because of curiosity. The actual importance of the infections is that several are excellent experimental models for parasitological research. In addition to such parasitism, many important helminth infections of man and domestic animals may be produced experimentally in domestic rabbits. A number of these also represent valuable experimental models. For this reason, both natural and experimental helminth infections of domestic and wild rabbits are considered in this chapter.

Consideration of natural helminth infections is limited to those parasites that are prevalent in wild rabbits or serve as experimental models. A short description of morphology is provided for diagnostic purposes and usefulness as research models is noted when appropriate.

The review of experimental helminth infections is limited to those that are useful research models or are important parasites of man and domestic animals. A number of these infections occur in wild rabbits under natural conditions but they are listed in the experimental infection category if the rabbit is not considered the normal host. The morphology of these parasites is not reviewed. Consideration is given only to their reported research value.

II. NATURAL INFECTIONS

The helminth parasites causing natural infections are divided into 3 major groups: nematodes, cestodes, and trematodes. The nematodes are most numerous and are further divided on the basis of family (96). When more than one member of a genus occurs, the most important infection is described in detail and other members of the genus are covered less extensively. The numbers of cestodes and trematodes are not of sufficient size to subdivide these groups on the basis of family to facilitate identification. The descriptions of the morphology of many of the parasites were obtained from Morgan and Hawkins (53). Should further information be required for identification, Appendix 1 should be consulted (31).

A. Nematodes

"Unsegmented animals, without appendages arranged on a regularly segmental plan; usually elongated, cylindrical, or filiform; with a body cavity in which the organs float; sexes usually separate" (96).

1. Trichostrongylidae

Males possess a cuticular bursa copulatrix, supported by rays; both sexes are more or less filiform worms with poorly developed buccal capsules (96).

a. *Obeliscoides cuniculi* (Graybill, 1923), rabbit stomach worm. i. MORPHOLOGY (Fig. 1; 42, 53, 96). Males are 10 to 14 mm in length, 230 μm in width, and have a typical bursa. Females are 15 to 18 mm long, 546 μm wide and have a pointed tail. The vulva is located in the posterior part of the body. No buccal capsule is present. Eggs are thin-shelled, oval, and measure approximately 80 × 45 μm.

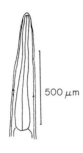

500 μm

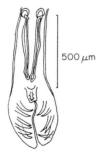

500 μm

Fig. 1. *Obeliscoides cuniculi*, anterior end (above), and ventral view of posterior end of male (below). After Morgan and Hawkins (53).

ii. LIFE CYCLE (3, 42, 53, 73). The life cycle is direct. Eggs hatch in feces from infected rabbits in about 30 hours and develop to the infective stage in 6 days. Rabbits are infected by infesting infective larvae. The parasite develops to maturity in the stomach in 16 to 20 days.

iii. PREVALENCE (4, 31, 54, 79). *Obeliscoides* is widely distributed in wild rabbits throughout the United States. Spontaneous infection does not appear to be a problem in laboratory rabbits.

iv. EXPERIMENTAL (33, 62, 63, 72, 73, 93, 95). This infection is similar in many ways to *Trichostrongylus* and *Ostertagia* infections of ruminants. Techniques have been described that allow the infection to be maintained in laboratory rabbits with relative ease. Although not extensively used to date, this parasite is a potentially valuable laboratory animal.

b. *Nematodirus leporis* (Chandler, 1924). i. MORPHOLOGY (53). Males are from 8 to 13 mm long and have a bursa with rounded lobes and parallel medio- and posterolateral rays. Spicules are from 650 μm to 1 mm in length. Females are from 16 to 20 mm long and produce large (160 to 180 μm × 80 to 90 μm), oval eggs.

ii. LIFE CYCLE (53). Adults are located in the small intestine. Life cycle is direct.

iii. PREVALENCE (4, 31, 53). This parasite appears to be widely distributed in wild rabbits in the United States. It has been reported in domestic rabbits, but does not present a problem of consequence.

iv. EXPERIMENTAL. Experimental *N. leporis* infections have not been reported.

v. REMARKS (4, 18, 24, 31, 53). There are several other species of *Nematodirus* that are found infrequently in wild rabbits. They include *N. neomexicanus*, *N. arizonensis*, and *N. triangularis*. All are apparently rather similar in appearance to *N. leporis*.

c. *Longistriata noviberiae* (Dikmans, 1935). i. MORPHOLOGY (22, 53). Males are from 2 to 5 mm long and 49 to 110 μm in diameter, have a bursa and 2 similar spicules measuring from 400 to 620 μm in length. Females are from 4 to 10 mm long, 60 to 100 μm wide, and have a vulva located about 200 μm from their posterior end. Females produce spherical eggs measuring 55 to 74 μm by 26 to 40 μm.

ii. LIFE CYCLE (53). The adult parasites are found in the small intestine. The life cycle is probably direct, but it has not been established.

iii. PREVALENCE (4, 31, 53). This nematode is widely distributed in wild rabbits throughout the United States. It has not been reported in domestic rabbits.

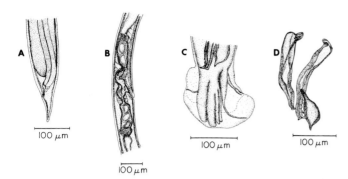

Fig. 2. *Trichostrongylus calcaratus*, (A) posterior extremity of female, (B) female in region of vulva, (C) bursa of male, and (D) spicules and gubernaculum. After Hall (44).

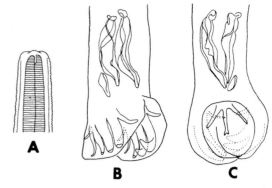

Fig. 3. *Trichostrongylus retortaeformis*, (A) anterior extremity ventral view, (B) lateral view of bursa, and (C) dorsal view of bursa. Approximate width of parasite is 115 μm. After Yorke and Maplestone (96).

iv. EXPERIMENTAL. No experimental use has been made of *Longistriata* spp.

d. *Trichostrongylus calcaratus* (Ransom, 1911). i. MORPHOLOGY (Fig. 2; 23, 53). Males are 4.7 to 6.6 mm long, 100 to 130 μm wide, bursate with asymmetrical dorsal rays, and have nearly equal spicules that measure 170 to 190 μm in length. Females are 5.8 to 7 mm long, 90 to 120 μm in diameter, and have a vulva located 850 μm to 1 mm from the tip of the tail. Females produce spherical, thin-shelled eggs 60 to 70 μm long by 30 to 36 μm wide.

ii. LIFE CYCLE (44, 53). Adult parasites are found in the small intestine. They produce eggs that are passed in the fecal material and hatch and develop to the infective stage in about 6 days. Infection takes place by ingestion of infective larvae.

iii. PREVALENCE (4, 31, 53, 79). This parasite occurs commonly in wild rabbits in the United States and has also been reported from domestic rabbits.

iv. EXPERIMENTAL (25, 67, 68, 78). Experimental infections have been produced and studied. The parasite apparently is not difficult to work with in the laboratory and represents a good model to study trichostrongyle infections, although it has not been used extensively for this purpose.

v. REMARKS (Fig. 3; 4, 18, 23, 25, 26, 31, 32, 41, 44, 53, 84, 96). *Trichostrongylus affinis*, *T. ransomi*, and *T. retortaeformis* also have been found in wild rabbits in the United States. *Trichostrongylus affinis* is the most prevalent of these and experimental infections with this parasite also have been produced in domestic rabbits.

e. *Graphidium strigosum* (Dujardin, 1845). i. MORPHOLOGY (Fig. 4; 44, 96). Body is blood red and filiform. Males are 8 to 16 mm long, 130 to 175 μm wide, bursate, and have filiform spicules 1.1 to 2.4 mm in length. Females are 11 to 20 mm long and 190 to 215 μm wide, have a conical tail

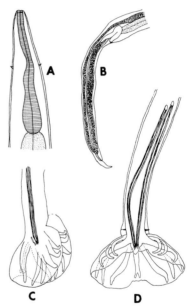

Fig. 4. *Graphidium strigosum*, (A) anterior extremity, ventral view, (B) posterior extremity of female, lateral view, (C) bursa, semilateral view, and (D) bursa, ventral view. Approximate width of parasite is 175 μm. After Yorke and Maplestone (96).

with a constriction before the tip. Eggs are ellipsoidal and measure 98 to 106 μm by 50 to 58 μm.

ii. LIFE CYCLE (44). Adults are usually found in the stomach but occasionally are observed in the small intestine. Life cycle is probably direct.

iii. PREVALENCE (4, 18, 28, 31, 32). This parasite has been reported infrequently from wild rabbits in the United States.

iv. EXPERIMENTAL (28, 84). *Graphidium strigosum* has been studied rather extensively in Australia in surveys that have correlated parasitism with age and sex of host and season of year.

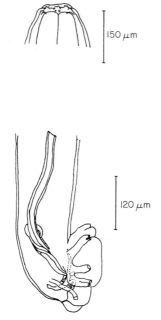

Fig. 5. *Protostrongylus boughtoni*, lateral view of anterior end (above) and lateral view of posterior end of male (below). After Morgan and Hawkins (53).

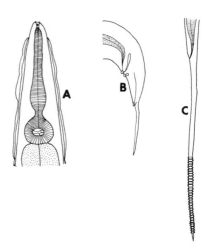

Fig. 6. *Passalurus ambiguus*, (A) anterior extremity, ventral view, (B) posterior extremity of male, lateral view, and (C) posterior extremity of female, lateral view. Approximate width of parasite is 500 μm. After Yorke and Maplestone (96).

2. Metastrongylidae

Males possess a cuticular bursa copulatrix, supported by rays; both sexes are more or less filiform worms with well-developed buccal capsules (96).

a. *Protostrongylus boughtoni* (Goble and Dougherty, 1943), rabbit lung worm. i. MORPHOLOGY (Fig. 5; 27, 39, 40, 44, 53). Males are 13 to 26 mm long and 160 to 250 μm in diameter, bursate and have spicules that measure 260 to 320 [m in length. Females are 21 to 36 mm long and 200 to 300 μm in diameter. The vulva is positioned about 200 μm from the tip of the tail. Eggs are elliptical, very thin-shelled, and measure 50 to 70 μm by 40 to 60 μm.

ii. LIFE CYCLE (53). Adult parasites are found in the bronchi of wild rabbits. First stage larvae are found in the trachea and feces. The remainder of the life cycle is unknown.

iii. PREVALENCE (53). Lung worms are found infrequently in wild rabbits in the United States and have not been reported in domestic rabbits.

iv. EXPERIMENTAL. Experimental infections have not been produced or studied.

v. REMARKS (18, 39, 40, 44, 53). Several other members of this genus apparently parasitize wild rabbits. *Protostrongylus pulmonalis* is perhaps the most prevalent of these.

3. Oxyuridae

Esophagus with a posterior bulbar enlargement; intestine without diverticula; caudal extremity of females usually prolonged into a finely pointed tail (96).

a. *Passaluris ambiguus* (Rudolphi, 1819), rabbit pin worm. i. MORPHOLOGY (Fig. 6; 19, 43, 44, 53, 96). Males are approximately 4.1 mm long and 300 μm in diameter, and have a single curved spicule 130 μm long. Females are 6.6 mm long and 500 μm wide with the vulva located at the anterior end. The posterior end of the body is distinctive. Posterior to the anus is a long tail that shows definite annular structures over its posterior portion. Both sexes have a simple mouth surrounded by 4 papillae and a posterior esophageal bulb. The females produce eggs measuring 103 × 43 μm that are slightly flattened on one side.

ii. LIFE CYCLE (17, 53). Adult worms are found in the cecum and large intestine. Females produce eggs that are oviposited in the morula stages and that initiate a direct life cycle.

iii. PREVALENCE (4, 31, 32, 79). This is a common parasite of wild and domestic rabbits.

iv. EXPERIMENTAL. *Passaluris* apparently has not been used as an experimental model.

v. REMARKS (35, 53). Although heavy infections have not been found to cause clinical disease, it may be desirable to eliminate infection. Piperazine adipate, given once daily for 2 days in food or water at the rate of 0.5 gm/kg for adult rabbits and 0.75 gm/kg for young rabbits, has been reported to be a highly effective treatment.

A similar parasite, *P. nonanulatus*, has been reported

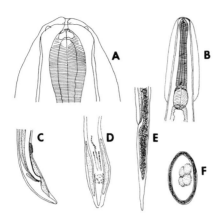

Fig. 7. *Dermatoxys veligera*, (A) head, ventral view, (B) anterior extremity, ventral view, (C) posterior extremity of male, ventral view, (D) posterior extremity of male, lateral view, (E) posterior extremity of female, lateral view, and (F) egg. Approximate width of parasite is 600 μm. After Yorke and Maplestone (96).

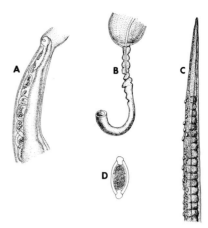

Fig. 8. *Trichuris leporis*, (A) region of female showing pocketing in vagina, (B) posterior extremity of male showing sheath and spicule, (C) anterior extremity of body showing cuticular plaques, and (D) egg. After Hall (44).

from wild rabbits on a few occasions. Its importance in laboratory rabbits is not known.

b. *Dermatoxys veligeria* (Rudolphi, 1819). i. MORPHOLOGY (Fig. 7; 21, 44, 53, 91, 96). Males are 8 to 11 mm long and 435 μm wide and have small spicules only 85 μm in length. Females are 16 to 17 mm long, 600 μm in diameter, and have a vulva located about 7 mm from the anterior end. Both sexes have a mouth with 3 well-developed lips and an esophagus that terminates in a bulb. Eggs are 110 × 50 μm and usually slightly flattened on one side.

ii. LIFE CYCLE (53). Adults are found in the cecum. The life cycle is probably direct.

iii. PREVALENCE (4, 31, 79). *Dermatoxys* infections occur fairly frequently in wild rabbits. The parasite is occasionally found in domestic rabbits but not as often as is *Passaluris*.

iv. EXPERIMENTAL. Experimental value of this parasite has not been studied.

v. REMARKS. *Dermatoxys* might be confused with *Passaluris* unless care is taken in examination of specimens.

4. Trichuridae

Anterior portion of body filiform, the esophagus consisting of a delicate tube running, in part of its length at least, through the center of a chain of single cells (96).

a. *Trichuris leporis* (Froelich, 1789), rabbit whip worm. i. MORPHOLOGY (Fig. 8; 44, 53, 87). The anterior part of the body of both adult males and female parasites is very slender and longer than the posterior part, which is much thicker and contains the reproductive and digestive structures. Males are 29 to 32 mm long and 430 μm in diameter, and have a single spicule approximately 3 mm long. Females

are about 32 mm long and 1 mm in diameter. They produce characteristic eggs about 56 μm long with polar plugs.

ii. LIFE CYCLE (53). The adult parasites are found in the cecum and large intestine. Females produce eggs which are eliminated in the feces of the host. The remainder of the life cycle has not been studied but is probably direct.

iii. PREVALENCE (4, 31, 53, 87). Infections have been reported in both wild and domestic rabbits in the United States. Although prevalence is high in wild rabbits, infections are uncommon in domestic rabbits.

iv. EXPERIMENTAL. This infection has not been produced and studied experimentally.

v. REMARKS. (4). Several other whip worms of lesser importance have been reported from wild rabbits. *Trichuris sylvilagi* is perhaps the most common of these.

5. Filariidae

Mouth usually simple; cuticle usually smooth or finely striated transversely; oviparous or viviparous (96).

a. *Dirofilaria scapiceps* (Leidy, 1886). i. MORPHOLOGY (Fig. 9; 44, 53). Males are from 12 to 15 mm long, 310 to 375 μm wide, and have 2 spicules of unequal length. Females are 25 to 30 mm long, 750 μm wide, and produce slender, filiform embryos.

ii. LIFE CYCLE (46, 53). Adult parasites are found under the skin of the lumbar region, on the medial side of the tarsus, and in the subcutaneous tissues of the forelegs and hindlegs. They produce microfilaria that have been described and used to experimentally infect mosquitoes. Apparently after a developmental period, the cycle is completed when infected mosquitoes feed on a susceptible host.

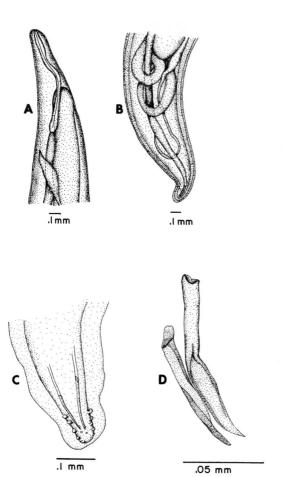

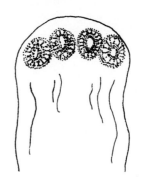

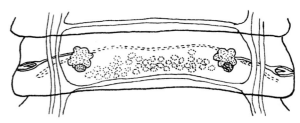

Fig. 10. *Cittotaenia variabilis*, scolex (above), mature proglottid (below). Width of proglottid is approximately 10 mm. After Morgan and Hawkins (53).

Fig. 9. *Dirofilaria scapiceps*, (A) anterior extremity of female, (B) posterior extremity of female, (C) posterior extremity of male, and (D) spicules. After Hall (44).

iii. PREVALENCE (2, 4, 31, 79). Infections are common in wild rabbits in the United States but have not been reported in domestic rabbits.

iv. REMARKS (4, 12, 29, 64). A second filarial worm, *D. uniformis*, is also found in wild rabbits in the United States. This parasite has been shown to infect mosquitoes experimentally and has been used to study influence of filarial worms on the immune response of the host. These parasites may eventually be developed into a useful experimental model.

B. Cestodes

Wild rabbits are frequently definitive and intermediate hosts for a number of tapeworms. The life cycles of these parasites practically precludes accidental infection of domestic or laboratory rabbits. Only modest coverage of these parasites is justified in this section. Further information can be obtained in references 5, 7, 47, 80 and 90.

1. Definitive Hosts

a. *Cittotaenia variabilis* (Stiles, 1895). i. MORPHOLOGY (Fig. 10; 5, 47, 53). The adult parasite may reach 450 mm in length and 10.5 mm in width, containing as many as 750 proglottids. The scolex is unarmed.

ii. LIFE CYCLE (17, 53, 80). The adult parasites are found in the small intestine. The eggs that they produce may use oribatid mites as intermediate hosts.

iii. PREVALENCE (4, 31, 53, 79). This infection is widely distributed throughout wild rabbits in the United States and on rare occasions has been reported in domestic rabbits.

iv. REMARKS (4, 31, 53). Several other similar parasites, including *C. ctenoides*, *C. pectinata*, *C. perplexa*, and *Schizotaenia americana*, are found occasionally in wild rabbits in the United States. They can be differentiated from *C. variabilis* on the basis of morphological differences. None of these parasites has been used as a laboratory model.

b. *Raillietina salmoni* (Stiles, 1895). i. MORPHOLOGY (53). The adult parasites are approximately 86 mm long and 3 mm wide. The proglottids are detectable 800 μm behind the scolex and have genital pores that are irregularly alternate. The scolex has a rostellum with a double row of small hooks.

ii. LIFE CYCLE (53). The adult parasites are found in the small intestine and an ant intermediate host is probably involved in the life cycle.

iii. PREVALENCE (4, 31, 53, 79). This parasite is found infrequently in wild rabbits and has not been reported in domestic rabbits.

iv. REMARKS (4, 7, 53, 79). *Raillietina retractilis* and *R. loeweni* are similar parasites that are occasionally found in wild rabbits. *Raillietina loeweni* infections have been studied and are a useful experimental model.

2. Intermediate Hosts

a. *Taenia pisiformis* (Bloch, 1780). i. MORPHOLOGY (31, 53). The stage found in rabbits is a cysticercus measuring up to 18 mm in diameter. Most cysticerci are found in the liver, although a few may be found attached to the mesentery. Usually they cause little damage.

ii. PREVALENCE (4, 31, 53, 79). Infections are very common in wild rabbits and found occasionally in domestic rabbits.

iii. REMARKS. Experimental use of *T. pisiformis* is included in the section on experimental infections where parasites of its definitive host, the dog, are considered.

b. *Taenia serialis* (Gervais, 1847). i. MORPHOLOGY (31, 53). The stage found in rabbits is a coenurus. These develop in the connective tissue of muscle and may reach 40 to 50 mm in diameter.

ii. PREVALENCE (4, 31, 45, 53). Infections are less frequent in wild rabbits than *T. pisiformis* and extremely rare in domestic rabbits.

iii. REMARKS. The dog is the definitive host for *M. serialis* and this parasite also will be mentioned in the section on small animals.

C. Trematodes

The rabbit appears to be the natural host for few trematodes, but may serve as a reservoir host for several flukes that are important infections of domestic ruminants. No spontaneous trematode infections have been reported in laboratory rabbits. For discussion, these parasites are grouped into natural host and reservoir host categories.

1. Natural Host

a *Hasstilesia tricolor* (Stiles and Hassal, 1894). i. MORPHOLOGY (Fig. 11; 13, 53, 61). The adult flukes are from 710 to 745 μm long and approximately 565 μm wide, have a cuticle covered with small spines and an oral and ventral

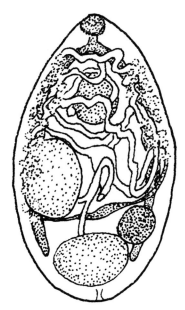

Fig. 11. *Hasstilesia tricolor*, adult fluke is approximately 725 μm in length. After Morgan and Hawkins (53).

sucker. The parasite, when fresh, shows 3 distinct shades of color. Typical operculated eggs measuring 22 × 14 μm are present in the uterus.

ii. LIFE CYCLE (61). The adult parasites are found in the anterior part of the small intestine. They produce eggs that infect a land snail, *Vertigo ventricosa*. Snails liberate cercaria 12 weeks after infection which are thought to encyst on vegetation and become metacercaria. Rabbits are infected by ingesting the metacercaria which develop to maturity in about 3 weeks in the host.

iii. PREVALENCE (4, 53, 61, 79). *Hassitilesia* is a common infection of wild rabbits in the United States, but has not been reported in domestic rabbits.

iv. EXPERIMENTAL. Specimens of this parasite are often used in teaching morphology of trematodes in basic parasitology courses.

v. REMARKS. A second member of this genus, *H. texensis*, has been reported in wild rabbits. It is not as prevalent as *H. tricolor*.

2. Reservoir Host

a. *Fasciola hepatica* (Linnaeus, 1758). i. REMARKS (4, 6, 18, 58). This parasite apparently can complete its life cycle in rabbits and they may constitute a reservoir for these infections in ruminants. *Fasciola hepatica* will be discussed more fully under the section on experimental infections.

b. *Dicrocoelium dendriticum* (Rudolphi, 1819). i. RE-MARKS (4, 6). This parasite has been found from time to time in wild rabbits but the importance of rabbits as reservoir hosts is not known.

III. EXPERIMENTAL INFECTIONS

The helminth parasites covered in this section are considered natural infections of hosts other than rabbits. Some appear to use wild rabbits as reservoir hosts, but the importance of the rabbit in this role is not great. That the rabbit may serve as a laboratory model for the study of these parasitisms is of more interest. The experimental helminth infections are grouped by host type for ease of discussion. Most emphasis is placed on experimental applications. If information regarding morphology, treatment, or life cycle of these parasites is desired, it can be readily obtained in current veterinary or medical parasitology texts (16, 76).

A. Ruminants

1. Nematodes

a. *Strongyloides papillosus* (Wedl, 1856). This parasite has been found in wild rabbits under natural conditions (53) and experimental infections of domestic rabbits have been produced and studied by a number of investigators (1, 38, 77, 86, 94). The percentage of infective larvae that develop to adult parasites in experimentally infected rabbits is lower than in the natural host (94), indicating that rabbits are somewhat refractory to infection. However, patent infections can be produced with little difficulty. Studies to date, in rabbits, have revealed differences in infectivity of *S. papillosus* and *S. ransomi* (77) and that the rabbit model is useful to study the pathogenesis and host response to these infections (38).

b. *Trichostrongylus colubriformis* (Giles, 1892) and *Trichostrongylus axei* (Cobbold, 1879). Both of these infections have been experimentally produced and studied in laboratory rabbits (15, 60, 71, 74, 92). *Trichostrongylus colubriformis* appears to infect rabbits more readily than *T. axei* (15, 60). One investigator (71) has used the *T. axei*-rabbit model to study the immune response of the host to this parasite and another (74) has studied the distribution of *T. colubriformis* in the intestine of experimentally infected rabbits. Available evidence suggests that the rabbit may be used to further advantage in the study of both of these important infections of domestic ruminants.

c. *Cooperia punctata* (V. Linstow, 1907) and *Cooperia curticei* (Railliet, 1895). Experimental infections of *C. punctata* have been produced and studied in some detail in domestic rabbits (10, 11, 51). Up to 10% of infective larvae, given orally, may develop to adult worms in this host. The prepatent period is from 11 to 16 days and the patent period may range from 3 to 38 weeks. Young male rabbits appear to be the most suitable hosts.

Cooperia curticei has not been studied as extensively. Experimental infections with this parasite have been produced in laboratory rabbits (92) but their value as an experimental model remains to be established.

d. *Ostertagia circumcincta* (Stadelmann, 1894). One investigator (92) has successfully produced *O. circumcincta* infections in domestic rabbits. If further study proves this to be a usable system, it might have considerable application because of the economic importance of this parasite.

e. *Nematodirus spathiger* (Railliet, 1896). Domestic rabbits appear to be fairly resistant to infection with *N. spathiger*. One report (74) indicates only 1 of 6 rabbits experimentally exposed developed patent infections. More work is needed to evaluate the rabbit as an experimental host.

f. *Dictyocaulus viviparus* (Bloch, 1782). Domestic rabbits have been used to study plasma enzyme changes associated with this infection (69). The investigators using the system felt that the rabbit was a good host for their purposes. The parasite probably does not develop to patency in rabbits.

2. Trematodes

a. *Fasciola hepatica* (Linnaeus, 1758) and *Fasciola gigantica* (Cobbold, 1885). The laboratory rabbit has been used by many investigators (20, 48, 59, 89) to study *F. hepatica* infection and at least 2 reports (6, 58) suggest that wild rabbits serve as reservoirs for this infection. The pathogenesis of *F. hepatica* in rabbits has been described in detail (89) and experimentally infected rabbits have been used to test anthelmintics (59). The value of this model system has been established and its use probably will be expanded in the future.

Fasciola gigantica has not been studied nearly as extensively in rabbits as *F. hepatica*. One report (57) indicates that this fluke does develop in rabbits and that the rabbit is a suitable host for some types of experimentation.

b. *Dicrocoelium dendriticum* (Rudolphi, 1819). *Dicrocoelium dendriticum* is found occasionally in wild rabbits and one report (6) suggests the rabbit serves as a reservoir host for this infection. To date, experimental infections have not been studied in domestic rabbits.

B. Swine

1. Nematodes

a. *Ascaris suum* (Goeze, 1782).   The rabbit often has been used to study aspects of the migratory phase of the life cycle of *A. suum*. These studies include differentiation of *A. suum* from *Ascaris lumbricoides* (36), behavior of *A. suum* in hosts upon reexposure to the parasite (49), and examination of the histopathology produced by migrating larval stages (83). The rabbit appears to be a good host for these types of studies and this model will undoubtedly be used more extensively in the future (8).

b. *Trichinella spiralis* (Owen, 1835).   Domestic rabbits can be infected readily with *T. spiralis*. The rabbit host for this infection appears well-suited for immunological work (66) and the pathogenesis of *T. spiralis* has also been studied successfully in rabbits (30). However, many host animals can be infected with *T. spiralis* and investigators have a choice of many hosts for their purposes.

c. *Strongyloides ransomi* (Schwartz and Alicata, 1930). Experimental infections of *S. ransomi* in domestic rabbits have been produced and studied on at least 2 occasions (77, 86). One report (77) indicates that this parasite is less successful infecting rabbits than *S. papillosus*. However, patent infections were produced and the rabbit appears to be a suitable host in which to study the infection.

d. *Oesophagostomum dentatum* (Rudolphi, 1830).   Berger and Ribelin (9) have described the lesions produced by *O. dentatum* in experimentally infected rabbits. The parasite apparently does not reach patency in the rabbit but the model had some value for study of the histotropic stages of the infection.

e. *Hyostrongylus rubidis* (Hassall and Stiles, 1892), *Ascarops strongylina* (Rudolphi, 1819), and *Physocephalus sexalatus* (Molin, 1860).   These stomach worms of swine have been studied in a preliminary way in laboratory rabbits (14, 88). Several of them reach patency in rabbits which encourages further exploration of the use of this host.

C. Small Animals

1. Nematodes

a. *Ancylostoma caninum* (Ercolani, 1895) and *Toxocara canis* (Werner, 1782).   Laboratory rabbits have been used to study both of these nematode infections of the canine (25, 34). The rabbit appears to be of value in certain immunological studies of these infections, but neither infection completes its life cycle in the rabbit, which limits the usefulness of this host.

b. *Nippostrongylus brasiliensis* (Travassos, 1914). *Nippostrongylus brasiliensis* is a natural infection of rats. This parasite, in the rat host, has been used extensively as a laboratory model. On several occasions (81, 85) rabbits have been used to study immunological aspects of this parasitism. However, few *N. brasiliensis* larvae develop to maturity in rabbits and the use of this system in the future probably will not be great.

c. *Angiostrongylus cantonensis* (Chen, 1935).   During the past few years *A. cantonensis*, normally a lung worm of rats, has been found to produce serious disease in people. When rabbits are infected experimentally, the parasite produces lesions in the central nervous system somewhat similar to those produced in man (70). Although little work has been done to date using the rabbit as an experimental host for this infection, the human health implications may encourage further study.

2. Cestodes

a. *Taenia pisiformis* (Bloch, 1780).   Wild rabbits serve as intermediate hosts for this relatively common parasitism of dogs. Laboratory rabbits are also readily infected and have been used rather extensively to investigate the host response to the infective stage, a cysticercus, of this cestode (37, 52, 55, 56).

b. *Taenia serialis* (Gervais, 1847).   Wild rabbits also serve as intermediate hosts for this parasite of dogs. The prevalence of *T. serialis* is much lower than *T. pisiformis*. The importance of this infection from an experimental point of view is that the infective stage is a coenurus. This stage develops in laboratory rabbits (45) and probably could be studied with little difficulty, should this be desired.

D. Man

1. Nematodes

a. *Necator americanus* (Stiles, 1902) and *Ancylostoma duodenale* (Dubini, 1843).   Laboratory rabbits have been experimentally infected with human hookworms (97). *Necator* develops to maturity in this host and appears to be especially suitable for further development as a model system. The world-wide importance of hookworm infections in man encourages further investigation of this condition.

b. *Ascaris lumbricoides* (Linnaeus, 1758).   The laboratory rabbit has proven to be a valuable experimental host for both *A. lumbricoides* and *A. suum*. In fact, rabbits have been used to help differentiate these 2 species (36). Although morphologically identical, these parasites behave differently when experimentally introduced into rabbits. The rabbit also appears to be a good host in which to study the immune response to *A. lumbricoides* (75). Since this parasite remains an important human infection, further investigation utilizing the rabbit as an experimental host can be anticipated.

2. Cestodes

Taenia solium (Linnaeus, 1758). The rabbit has been used successfully to study cerebral cysticercosis (50). Apparently, cysticerci of *T. solium* develop readily in rabbits, and the model has experimental application.

3. Trematodes

a. *Schistosoma mansoni* (Sambon, 1907), *S. hematobium* (Sambon, 1907), and *S. japonicum* (Katsurada, 1904). Laboratory rabbits are a useful host to study certain immunological problems of human schistosomiasis (65). More extensive use of rabbits for this purpose can be expected in the future because of the magnitude of importance of human infections.

b. *Clanorchis sinensis* (Cobbold, 1875). Experimental infections of *C. sinensis* have been produced and studied in rabbits (82). The rabbit appears to be a good host and the parasite develops to adult with little difficulty.

IV. CONCLUSIONS

Wild rabbits are natural hosts for a wide variety of helminths. Domestic rabbits are seldom infected with these parasites and when such infections occur, they usually do not represent a serious problem to the producer.

Domestic rabbits may be experimentally infected with a number of helminths of wild rabbits and other animals. Several of these parasitisms represent valuable experimental models.

The most promising experimental models for which rabbits are considered a natural host are *Obeliscoides cuniculi*, *Trichostrongylus calcaratus*, *T. affinis*, *Dirofilaria uniformis*, and *Taenia pisiformis*.

The most promising experimental models for which rabbits are considered "abnormal" hosts are *Trichostrongylus colubriformis*, *T. axei*, *Cooperia punctata*, *Strongyloides papillosus*, *S. ransomi*, *Ostertagia circumcincta*, *Ascaris suum*, *A. lumbricoides*, *Trichinella spiralis*, *Necator americanus*, *Schistosoma japonicum*, *Fasciola hepatica*, and *Clonorchis sinensis*.

REFERENCES

1. Abdel-Gawad, A. R. (1968). On the experimental infection of domestic rabbits with *Strongyloides papillosus* (Wedl, 1850) from sheep. *J. Egypt. Vet. Med. Ass.* **28**, 61–68.

2. Alicata, J. E. (1929). The occurrence of *Dirofilaria scapiceps* in rabbits. *J. Parasitol.* **15**, 287.

3. Alicata, J. E. (1932). Life history of the rabbit stomach worm *Obeliscoides cuniculi. J. Agr. Res.* **44**, 401–419.

4. Andrewes, C. L. (1969). Parasitism and other disease entities among selected populations of cottontail rabbits (*Sylvilagus floridanus*). Ph.D. Thesis, University of Georgia, Athens.

5. Arnold, J. G. (1938). A study of the anoplocephaline cestodes of North American rabbits. *Zoologica (New York)* **23**, 31–53.

6. Bailenger, J., Tribouley, J., Amyot, B., and Duret, J. (1967). Importance des léporidés comme réservoirs sauvages cans l'épidémiologie des distomatoses á *Fasciola hepatica* et *Dicrocoelium dendriticum. Helm. Abstr.* **36**, 34 (abstr.).

7. Bartel, M. H., and Hansen, M. F. (1967). Biological investigations on the cestode *Raillietina (R.) loweni* Bartel and Hansen, 1964. *Trans. Amer. Microsc. Soc.* **86**, 9–15.

8. Berger, H. (1971). Experimentally induced patent infections of *Ascaris suum* in rabbits. *J. Parasitol.* **57**, 344–347.

9. Berger, H., and Ribelin, W. E. (1969). Pathology of the swine nodular worm, *Oesophagostomum dentatum*, in rabbits. *J. Parasitol.* **55**, 1099–1101.

10. Besch, E. D. (1964). The effects of time and temperature on the infectivity of third-stage larvae of *Cooperia punctata* (Trichostrongylidae) in the domestic rabbit, *Oryctolagus cuniculus* L. *Amer. J. Vet. Res.* **25**, 535–537.

11. Besch, E. D. (1965). Biology of *Cooperia punctata* (Nematoda Trichostrongylidae) in the domestic rabbit. *J. Parasitol.* **51**, 139–144.

12. Bray, R. L., and Walton, B. C. (1961). The life cycle of *Dirofilaria uniformis* Price and transmission to wild and laboratory rabbits. *J. Parasitol.* **47**, 13–22.

13. Chandler, A. C. (1929). A new species of trematode worms belonging to the genus *Hasstilesia* from rabbits in Texas. *Proc. U. S. Nat. Mus.* **75**, 1–5.

14. Chowdhury, N., and Pande, B. P. (1969). Intermediate hosts of *Ascarops strongylina* in India and development of the parasite in rabbits and guinea pig. *Indian J. Anim. Sci.* **39**, 139–148.

15. Ciordia, H, Bizzell, W. E., Porter, D. A., and Dixon, C. F. (1966). The effect of culture temperature and age on the infectivity of the larvae of *Trichostrongylus axei* and *T. colubriformis* in rabbits and guinea pigs. *J. Parasitol.* **52**, 866–870.

16. Craig, E. D., Faust, P. F., and Jung, R. C. (1970). "Clinical Parasitology." Lea & Febiger, Philadelphia, Pennsylvania.

17. Cushnie, G. H. (1954). The life cycle of some helminth parasites of the rat, mouse, and rabbit. *J. Anim. Tech. Ass.* **5**, 22–25.

18. Czaplinska, D., Czaplinski, B., Rutkowska, M., and Zebrowska, D. (1965). Studies on the European hare. IX. Helminth fauna in the annual cycle. *Acta Theriol.* **10**, 55–78.

19. Danheim, B. L., and Ackert, J. E. (1929). On the anatomy of the nematode *Passalurus ambiguus* (Rudolphi). *Trans. Amer. Microsc. Soc.* **48**, 80–85.

20. Dargie, J. D., Holms, P. H., McLean, J. M., and Mulligan, W. (1968). Pathophysiology of fascioliasis in the rabbit. Studies on albumin turnover. *J. Comp. Pathol.* **78**, 101–105.

21. Dikmans, G. (1931). An interesting larval stage of *Dermatoxys veligera. Trans. Amer. Microsc. Soc.* **50**, 364–365.

22. Dikmans, G. (1935). New nematodes of the genus *Longistriata* in rodents. *J. Wash. Acad. Sci.* **25**, 72–81.

23. Dikmans, G. (1937). A note on the members of the nematode genus *Trichostrongylus* occurring in rodents and lagomorphs with descriptions of two new species. *J. Wash. Acad. Sci.* **27**, 203–209.

24. Dickmans, G. (1937). Two new species of the nematode genus *Nematodirus* (Trichostrongylidae) from rabbits. *Proc. Helminthol. Soc. Wash.* **4**, 65–67.

25. Dixon, C. F. (1965). Failure of infective trichostrongyloid larvae to

establish patent infection in rabbits by skin penetration. *J. Parasitol.* **51**, 145–148.

26. Dixon, C. F. (1965). Infection of domestic rabbits and gerbils with *Trichostrongylus affinis* (Graybill, 1924). *J. Parasitol.* **51**, 299.

27. Dougherty, E. C., and Goble, F. C. (1946). The genus *Protostrongylus*, Kamenskii, 1905. (Nematode: Metastrongylidae), and its relatives: Preliminary note. *J. Parasitol.* **32**, 7–16.

28. Dunsmore, J. D., Dudzinski, M. L. (1968). Relationship of numbers of nematode parasites in wild rabbits, *Oryctolagus cuniculus* (L.), to host, sex, age, and season. *J. Parasitol.* **54**, 462–474.

29. Duxbury, R. E., Moon, A. P., and Sadun, E. H. (1961). Susceptibility and resistance of *Anopheles quadrimaculatus* to *Dirofilaria uniformis*. *J. Parasitol.* **47**, 687–691.

30. Edwards, J. L., and Hood, C. I. (1962). Studies on the pathogenesis of cardiac and cerebral lesions of experimental trichinosis in rabbits. *Amer. J. Pathol.* **40**, 711–720.

31. Erickson, A. B. (1947). Helminth parasites of rabbits of the genus *Sylvilagus. J. Wildl. Manage* **11**, 255–263.

32. Evans, W. M. (1940). Observations on the incidence of some nematode parasites of the common rabbit, *Oryctolagus cuniculus. J. Parasitol.* **32**, 67–77.

33. Fernando, M. A. (1969). Hemoglobins of parasitic nematodes. II. Electro-phoretic analysis of the multiple hemoglobins of adults and developmental stages of the rabbit stomach worm, *Obeliscoides cuniculi. J. Parasitol.* **55**, 493–497.

34. Fernando, S. T. (1968). Immunological response of rabbits to *Toxocara canis* infection. *Parasitology* **58**, 91–103.

35. Festisov, V. I. (1968). (Treatment and chemopropyeanis of Passalarus infection of rabbits.) *Helminthol. Abstr.* **37**, 278 (abstr.).

36. Galvin, T. J. (1968). Development of human and pig *Ascaris* in the pig and rabbit. *J. Parasitol.* **54**, 1085–1091.

37. Gemmell, M. A. (1965). Immunological responses of the mammalian host against tape worm infections. II. Species specificity of hexacanth embryos in protecting rabbits against *Taenia pisiformis. Immunology* **8**, 270–280.

38. Geyer, E. (1965). Pathologie und immunität der strongyloidiasis des kaninchens. *Helminthol. Abstr.* **34**, 377. (abstr.).

39. Goble, F. C., and Cheatum, E. L. (1944). Notes on the lungworms of North American leporidae. *J. Parasitol.* **30**, 119–120.

40. Goble, F. C., and Dougherty, E. C. (1943). Notes on the lungworms (genus *Protostrongylus*) of varying hares (*Lepus americanus*) in eastern North America. *J. Parasitol.* **29**, 397–404.

41. Graybill, H. W. (1924). A new species of roundworm of the genus *Trichostrongylus* from the rabbit. *Proc. U.S. Nat. Mus.* **66**, 1–3.

42. Graybill, H. W. (1924). *Obeliscoides*, a new name for the nematode genus *Obeliscus. Parasitology* **16**, 317.

43. Habermann, R. T., and Williams, F. P. (1968). The identification and control of helminths in laboratory animals. *J. Nat. Cancer Inst.* **20**, 979–1008.

44. Hall, M. C. (1916). Nematode parasites of mammals of the orders Rodentia, Lagomorpha, and Hyracoidea. *Proc. U.S. Nat. Mus.* **50**, 1–258.

45. Hamilton, A. G. (1950). The occurrence and morphology of *Coenurus serialis* in rabbits. *Parasitology* **40**, 46–49.

46. Highby, P. R. (1943). Vectors, transmission, development, and incidence of *Dirofilaria scapiceps* (Leidy, 1886) (Nematoda) from the snowshoe hare in Minnesota. *J. Parasitol.* **29**, 253–259.

47. Honess, R. F. (1963). Unarmed cestodes of Wyoming rabbits. *Univ. Wyo., Publ.* **28**, 7–21.

48. Jennings, F. W., Mulligan, W., and Urquhart, G. M. (1965). Some isotopic studies on the blood loss associated with *Fasciola hepatica* infections in rabbits. *Trans. Roy. Soc. Trop. Med. Hyg.* **49**, 305.

49. Jeska, E. L., Williams, J. F., and Cox, D. F. (1969). *Ascaris suum*: Larval returns in rabbits, guinea pigs, and mice after low-dose exposure to eggs. *Exp. Parasitol.* **26**, 187–192.

50. Kepski, A., Szlaminski, Z., and Zapart, W. (1963). Serological tests in experimental cerebral cysticercosis in rabbits. *Acta Parasitol. Pol.* **11**, 133–143.

51. Leland, S. E., and Wallace, L. J. (1966). Development to viable egg production in the rabbit duodenum of parasitic stages of *Cooperia punctata* grown *in vitro. J. Parasitol.* **52**, 280–284.

52. Leonard, A. B., and Leonard, A. E. (1941). The intestinal phase of the resistance of rabbits to the larval of *Taenia pisiformis. J. Parasitol.* **27**, 375–378.

53. Morgan, B. B., and Hawkins, P. A. (1949). "Veterinary Helminthology." Burgess, Minneapolis, Minnesota.

54. Morgan, B. B., and Waller, E. F. (1940). A survey of parasites of the Iowa cottontail (*Sylvilagus floridanus mearnsi*). *J. Wildl. Manage.* **4**, 21–26.

55. Nemeth, J. (1965). Immunological study of rabbit cysticercosis. I. The suitability of agar gel diffusion precipitation and indirect haemagglutination tests. *Z. Immunitaets-Allergieforsch.* **128**, 468–482.

56. Nemeth, J. (1970). Immunological study of rabbit cysticercosis. II. Transfer of immunity to *Cysticercus pisiformis* (Bloch, 1780) with parenterally administered immune serum or lymphoid cells. *Acta Vet.* (*Budapest*) **20**, 69–79.

57. Obara, J., Sonoda, A., and Watanabe, S. (1964). *In vivo* uptake of racobalimin-60 by *Fasciola gigantica* and its effect on experimentally infected rabbits. *Exp. Parasitol.* **15**, 471–478.

58. Olsen, O. W. (1948). Wild rabbits as reservoir hosts of the common liver fluke, *Fasciola hepatica*, in southern Texas. *J. Parasitol.* **34**, 119–123.

59. Ray, R. J. (1969). The effect of simultaneous treatment with hexachlorophene and nitroxynil upon immature *Fasciola hepatica* in rats and rabbits. *Res. Vet. Sci.* **10**, 405–408.

60. Rohrbacker, G. H. (1960). The effect of green feed and ascorbic acid on single experimental infections of *Trichostrongylus axei* (Cobbold, 1879) in the laboratory rabbit. *Amer. J. Vet. Res.* **21**, 138–143.

61. Rowan, W. B. (1955). The life cycle and epizootiology of the rabbit trematode, *Hasstilesia tricolor* (Stiles and Hassall, 1894), Hall, 1916 (Trematoda: Brachylaemidae). *Trans. Amer. Microsc. Soc.* **74**, 1–21.

62. Russell, S. W., Baker, N. F., and Raizes, G. S. (1966). Experimental *Obeliscoides cuniculi* infections in rabbits: Comparison with *Trichostrongylus* and *Ostertagia* infections in cattle and sheep. *Exp. Parasitol.* **19**, 163–167.

63. Russell, S. W., Ward, B. C., and Baker, N. F. (1970). *Obeliscoides cuniculi*: Comparison of gastric lesions in rabbits with those of bovine ostertagiosis. *Exp. Parasitol.* **28**, 217–225.

64. Sadun, E. H., Duxbury, R. E., Gore, R. W., and Stechschulte, D. J. (1964). Homologous passive cutaneous anaphylactic and fluorescent antibody reactions in rabbits infected with *Dirofilaria uniformis. J. Infec. Dis.* **117**, 317–326.

65. Sadun, E. H., and Gore, R. W. (1967). Relative sensitivity and specificity of soluble antigens (metabolic and sematic) and whole cercarial in fluorescent antibody tests for schistosomiasis in humans and rabbits. *Exp. Parasitol.* **20**, 131–137.

66. Sadun, E. H., Mota, I., and Gore, R. W. (1968). Demonstration of homocytotropic reagin-like antibodies in mice and rabbits infected with *Trichinella spiralis. J. Parasitol.* **54**, 814–821.

67. Sarles, M. P. (1932). Development of an acquired resistance in rabbits by repeated infection with an intestinal nematode, *Trichostrongylus calcaratus* Ransom. *J. Parasitol.* **19**, 61–82.

68. Sarles, M. P. (1934). Production of fatal infestations in rabbits with *Trichostrongylus calcaratus* (nematode). *Amer. J. Hyg.* **19**, 86–102.

69. Shelty, S. N., Himes, J. A., and Edds, G. T. (1970). Plasma enzyme changes associated with experimental *Dictyocaulus viviparus* infection in rabbits. *Amer. J. Vet. Res.* **31**, 2251–2260.

70. Shoho, C. (1966). Observations on rats and rabbits infected with *Angiostrongylus cantonensis* (Chen). *Brit. Vet. J.* **122**, 251–258.

71. Sinha, P. K. (1967). Active immunity against *Trichostrongylus axei* in sheep and rabbits. *Amer. J. Vet. Res.* **28**, 793–798.

72. Sollod, A. E., and Allen, J. R. (1971). Immunity to *Obeliscoides cuniculi*: Chemical suppression and passive transfer. *J. Parasitol.* **57**, 339–343.

73. Sollod, A. E., Hayes, T. J., and Soulsby, E. J. L. (1968). Parasitic development of *Obeliscoides cuniculi* in rabbits. *J. Parasitol.* **54**, 129–132.

74. Sommerville, R. J. (1963). Distribution of some parasitic nematodes in the alimentary tract of sheep, cattle, and rabbits. *J. Parasitol.* **49**, 593–599.

75. Soulsby, E. J. L. (1958). Studies on the heterophile antibodies associated with helminth infections. I. Heterophile antibodies in *Ascaris lumbricoides* infection in rabbits. *J. Comp. Pathol. Ther.* **68**, 71–81.

76. Soulsby, E. J. L. (1965). "Textbook of Veterinary Clinical Parasitology." Davis, Philadelphia, Pennsylvania.

77. Stewart, T. B. (1963). Environmental factors affecting the survival and development of *Strongyloides ransomi* with special reference to its free-living stages. *Diss. Abstr.* **24**, 907.

78. Stoll, N. R. (1932). Note on re-infection under "natural" conditions with a gut nematode of the rabbit. *J. Parasitol.* **19**, 54–60.

79. Stringer, R. P., Harkema, R., and Miller, G. C. (1969). Parasites of rabbits in North Carolina. *J. Parasitol.* **55**, 328.

80. Stunkard, H. W. (1941). Studies on the life history of the anoplocephaline cestodes of hares and rabbits. *J. Parasitol.* **27**, 299–325.

81. Sulzer, A., and Goodchild, C. (1963). Serology of rats and rabbits exposed to *Nippostrongylus brasiliensis* (Travassos, 1914). *J. Parasitol. Suppl.* **49**, 33.

82. Sun, T., Chou, S. T., and Gibson, J. B. (1968). Route of entry of *Clonorchis sinensis* to the mammalian liver. *Exp. Parasitol.* **22**, 346–351.

83. Taffs, L. F. (1965). Immunological studies on experimental infections of guineapigs and rabbits with *Ascaris suum*, Goeze, 1782. IV. The histopathology of the liver and lung. *J. Helminthol.* **39**, 297–302.

84. Taylor, E. L. (1935). Do nematodes assist bacterial invasion of the host by wounding the wall of the intestinal tract? *Parasitology* **27**, 145–151.

85. Thorsen, R. E. (1953). Infection of rabbits with a rat nematode *Nippostrongylus muris*. *J. Parasitol.* **39**, 575.

86. Timm, W. (1955). Untersuchungen mit *Strongyloides ransomi* (Schwartz and Alicata, 1930) und *S. papillosus* (Wedl, 1856) beim kaninchen. *Helminthol. Abstr.* **24**, 440 (abstr.).

87. Tiner, J. D. (1950). Two new species of *Trichuris* from North America, with redescriptions of *Trichuris opaca* and *Trichuris leporis* (Nematoda: Aphasmidia). *J. Parasitol.* **36**, 350–355.

88. Tromba, F. B., and Douvres, F. W. (1958). Cross transmission of nematodes of domestic animals. III. Preliminary observations on the infection of goats and rabbits with *Hyostrongylus rubidus*. *J. Parasitol.* **44**, 209.

89. Urquhart, G. M. (1956). The pathology of experimental fascioliasis in the rabbit. *J. Pathol. Bacteriol.* **71**, 301–310.

90. Wardle, R. A., and McLeod, J. A. (1952). "The Zoology of Tapeworms." Univ. of Minnesota Press, Minneapolis.

91. Wetzel, R. (1931). On the biology of the fourth-stage larva of *Dematoxys veligera* (Rudolphi, 1819), Schneider, 1866, an oxyurid parasitic in the hare. *J. Parasitol.* **18**, 40–43.

92. Wood, J. B., and Hansen, M. F. (1960). Experimental transmission of ruminant nematodes of the genera *Cooperia*, *Ostertagia* and *Haemonchus* to laboratory rabbits. *J. Parasitol.* **46**, 775–776.

93. Worley, D. E. (1963). Experimental studies on *Obeliscoides cuniculi*, a trichostrongylid stomach worm of rabbits. I. Host-parasite relationships and maintenance in laboratory rabbits. *J. Parasitol.* **49**, 46–50.

94. Worley, D. E., and Barrett, R. E. (1964). Infectivity and migratory behavior of an ovine strain of *Strongyloides papillosus* in laboratory animals. *J. Parasitol., Suppl.* **50**, 34–35.

95. Worley, D. E., and Thompson, P. E. (1963). Experimental studies on *Obeliscoides cuniculi*, a trichostrongylid stomach worm of rabbits. II. Anthelmintic studies in the Dutch rabbit. *J. Parasitol.* **49**, 51–54.

96. Yorke, W., and Maplestone, P. A. (1962). "The Nematode Parasites of Vertebrates." Hafner, New York.

97. Yoshida, Y., and Fukutome, S. (1967). Experimental infection of rabbits with human hook worm, *Necator americanus*. *J. Parasitol.* **53**, 1067–1073.

APPENDIX 1

KEYS TO THE HELMINTHS OF COTTONTAIL AND MARSH RABBITS[a]

TREMATODA

1. In the liver or bile ducts .... *Fasciola hepatica*
2. In the intestine ....(3)
3. Oral sucker 69 to 74 μm long by 87 to 93 μm wide; uterine coils loosely arranged .... *Hasstilesia tricolor*
   Oral sucker 86 to 115 μm long by 120 to 128 μm wide; uterine coils compactly arranged (solid massing of eggs) .... *Hasstilesia texensis*

CESTODA

1. Larva encysted in muscles or internal organs, or free in body cavity ....(2)
   Adults in intestine ....(3)
2. Larva a fluid-filled vesicle (coenurus) containing many heads. In the muscles ....*Multiceps* spp.
   (Larvae of *M. serialis* and *M. packii* not distinguishable.)
   Larva a bladder worm (Cysticercus) with little fluid and only one head. In internal organs or free in body cavity .... *Taenia pisiformis*
3. Rostellum with two circles of small hammer-shaped hooks. One set of reproductive organs in each segment ....(4)
   Rostellum without hooks. Two sets of reproductive organs in each segment ....(5)
4. Genital pores unilateral; one egg in each egg capsule. Hooks 40 to 120 in number 12 μm long .... *Raillietina retractilis*
   Genital pores generally alternate; 3 to 15 eggs in each egg capsule. Hooks 120 in number 20 μm long .... *Raillietina stilesiella*
5. Cirrus pouch extending mediad of longitudinal excretory canal ...(6)
   Cirrus pouch lying laterad of longitudinal excretory canal ....(7)
6. Cirrus pouch just barely crossing longitudinal excretory canal and with a maximal length of 0.64 mm; maximal body length, 100 mm .... *Cittotaenia perplexa*
   Cirrus pouch extending mediad of excretory canal and with a maximal length of 1.8 mm; maximal body length, 220 mm .... *Cittotaenia pectinata americana*
7. Maximal number of segments, 750; first appearance of female genital ducts between 45–50 segments; first indication of male genital primordia and ducts between 75–100 and 76–105 segments, respectively .... *Cittotaenia ctenoides*
   Maximal number of segments, 750; first appearance of female ducts between the 95–105 segments; male genital primordia and ducts originating at 125 and 175 segments, respectively. *Cittotaenia variabilis*

NEMATODA

1. Esophagus a fine tube imbedded in a column of glandular cells. Eggs with opercular plugs ....(2)
   Esophagus not imbedded in a column of glandular cells. Eggs without opercular plugs ....(3)
2. Anterior part of body longer than posterior, which is much thicker. In the intestine .... *Trichuris leporis*
   Anterior part of body shorter than posterior, which is only slightly thicker. In the liver .... *Capillaria hepatica*

[a] From Erickson (31).

3. Bursa, if present, with rays ....(4)
   Bursa, if present, without rays ....(17)
4. Bursa not conspicuous; parasites of respiratory tract ....(5)
   Bursa conspicuous; parasites of gastro-intestinal tract ....(6)
5. Spicules of male 156–180 μm long. Vagina of female 597–1110μm long. Provagina not prominent ....*Protostrongulus pulmonalis*
   Spicules of male 260–320 μm long. Vagina of femal 1900–2400 μm long. Provagina prominent ....*Protostrongylus boughtoni*
6. Genitalia of the female double ....(7)
   Genitalia of the female single ....(16)
7. Spicules short (less than 0.55 mm long) and stout with crests and protuberances ....(8)
   Spicules long and filiform 0.65 to 1.7 mm ....(12)
8. Ventroventral and lateroventral rays approximating at their tips; spicules cleft distally ending in two barbed processes; gubernaculum absent ....*Obeliscoides cuniculi*
   Ventroventral and lateroventral rays not approximating at tips; spicules not cleft distally; gubernaculum present ....(9)
9. Spicules more than 175 μm long, asymmetrical; distal end of right spicule smooth, of left spicule serrated .... *Trichostrongylus calcaratus*
   Spicules less than 175 μm ....(10)
10. Spicules 130 to 155 μm distal ends with two blunt recurved hooks; anus to tip of tail of female 140 to 165 μm

   ....*Trichostrongylus affinis*
   Distal ends of spicules without blunt, recurved hooks; anus to tip of tail of female less than 140 μm ....(11)
11. Spicules 130 to 140 μm distal ends recurved ending in sharp points and with three projections on inner side of each spicule

   ....*Trichostrongylus ransomi*
   Spicules equal, 135 to 145 μm; distal end of spicules not provided with projections as above .... *Trichostrongylus colubriformis*
12. Spicules with multiple tips; gubernaculum present ... *Graphidium strigosum* Spicules with simple tips; gubernaculum absent ....(13)
13. Lobes of bursa rounded; medio- and posterolateral rays parallel. (14)

Lobes of bursa triangular; medio- and postero-lateral rays divergent ....(15)
14. Spicules ending in a fingerlike process ....*Nematodirus leporis*
    Spicules ending in a sharp point ....*Nematodirus neomexicanus*
15. Combined length of muscular portions of ovejectors including sphincters 800 to 850 μm; vulva 3.6 to 4 mm. from tip to tail ....*Nematodirus triangularis*
    Combined length of muscular portions of ovejectors including sphincters 500 to 520 μ; vulva 2.0 2.2 mm. from tip to tail ....
    ....*Nematodirus arizonensis*
16. Genitalia of the female single ....*Longistriata novibariae*
17. Esophagus not dilated posteriorly into a bulb ....(18)
    Esophagus dilated posteriorly into a bulb ....(19)
18. Head without lips or a cephalic collarette; parasites usually of leg joint ....*Dirofilaria scapices*
    Head with two lips and a cephalic collarette; parasites of the stomach ....*Physaloptera* sp.
19. Males with a circular preanal sucker, with a chitinous rim; caudal alae supported by 12 pairs of papillae; eggs thick-shelled, ellipsoidal ...*Heterakis gallinae* Males without prenal sucker; eggs thinshelled, ellipsoidal ....(20)
20. Posterior end of male short and blunt, with row of transverse comb-like crests on ventral surface; bulb not separated from rest of esophagus by constriction ....*Dermatoxys veligira*
    Posterior end of male long and pointed; no comb-like crests; bulb separated from rest of esophagus by constriction ....(21)
21. Gubernaculum present; vagina usually extruded through a cuticular tube which projects from body of worm and vulva occurs on end of the tube ....*Wellcomia evoluta*
    Gubernaculum absent; vagina not extruded through a tube ....(22)
22. Tail of old females with circular cuticular thickenings; males with more than two pairs of circum-anal papillae ...*Passalurus ambiguus*
    Tail of old females without cuticular thickcnings; males with two pairs of circum-anal papillae ....*Passalurus nonanulatus*

CHAPTER 14

Neoplastic Diseases

Steven H. Weisbroth

INTRODUCTION

A. Preliminary Considerations

Several earlier reviews have summarized the literature dealing with spontaneous neoplasia in the domestic rabbit, *Oryctolagus cuniculus* (33, 63, 65, 75, 78, 84, 197, 204, 252, 285). Since that time, particularly in the last 30 years, a considerable amount of new data has accumulated on this subject. The data include a large number of case reports, descriptions of biological aspects of certain single tumor types, and papers dealing with experimentation involving certain recurrent tumor types. There has been no recent comprehensive effort to gather this data into a single compilation so as to give an overview of the neoplastic process in the various rabbit species. That is the primary goal of this chapter: to review and analyze this data.

Sufficient information is available now to enable the description of an oncological profile as a function of age and sex for *Oryctolagus*. Whereas a few years ago naturally occurring neoplasms in the rabbit were generally thought to be sporadic and unusual, the case can be made for a regular pattern in the incidence of the various tumor types now known to be in a spectrum naturally associated with the rabbit. More information is available with regard to this process in *Oryctolagus* than is the case with species of *Lepus* (hares) or *Sylvilagus* (cottontail rabbits), but this perhaps is fortunate since *Oryctolagus* is the major research species of the lagomorphs. This should not be surprising since most of the data has been provided from rabbit colonies associated with major research institutions.

The main emphasis of this chapter is to review the "naturally occurring" tumors of *Oryctolagus, Sylvilagus,* and *Lepus,* i.e., those neoplasia without known antecedent triggering mechanisms. Carcinogenic triggers are known to include certain chemicals, ionizing radiation, oncogenic viruses, and in at least one important case, chronic inflammation subsequent to syphlitic orchitis. How do such tumors relate in terms of tissue predilection, tissue response, and incidence to those that occur naturally? In pondering the same issue, Brues [cited by Upton (315)] was prompted to ask, "How and why do species differ in the spontaneous incidence of a neoplasm and are such differences correlated with variations in susceptibility to induction of the *same* neoplasm by radiation?" This is one of the fundamental questions of experimental oncology and it remains largely unanswered to this day. Nonetheless, evidence is available that bears on this question. We will try to show that when the various tumors are analyzed with regard to type of antecedent (i.e., naturally occurring or spontaneous versus chemical versus physical versus viral, etc.), it will be seen that there is little correlation with regard to tissue response (i.e., responding cell types, pathological architecture), tissue predilection, or biological behavior. This is to say, there appears to be a cluster of tumor types associated with each antecedent class, differing markedly in biological characteristics, and seemingly joined in fact only by the common thread of the neoplastic state. This, in my opinion, is important in the selection of oncological models in the rabbit. As an example, the following question can be posed; which model for human endometrial carcinoma is the more relevant, that induced by local application in the rabbit uterus of methylcholanthrene which induces first squamous metaplasia of the endometrium and then squamous carcinoma (18, 221) or that which occurs spontaneously and procedes from endometrial dysplasia to adenocarcinoma (18)? In this case the answer is forced: the naturally occuring model. It should be obvious that the selection of a model follows from the question asked, rather than in the reverse order, although this rule is not always followed (221). Excellent naturally occurring models for human cancer do exist in the rabbit and these will be pointed out as they occur in the text.

B. Influence of Genetic Predisposition

Only fragmentary information is available with regard to the influence of genetic constitution on the neoplastic process in the rabbit. It is clear, however, the further the genome is probed, characterized, and understood in a species, the awareness of this influence becomes of greater significance. This is especially so when line-breeding and inbreeding are practiced for the purpose of promoting the occurrence of tumors. This has been the pattern of development in domestic poultry (147) and in the mouse, not only with regard to spontaneous tumors but also with reference to neoplasia induced by oncogenic viruses (123) and ionizing radiation (316). It is likely that this pattern will be followed as the genetic characterization of the rabbit progresses and several lines of evidence lead to this conclusion.

In the large rabbit colony (average daily census of about 800 rabbits), maintained by Harry Greene between 1933 and 1955 for the purpose of studying the genetic influence on many disease processes, particularly cancer, records were kept throughout this period on many stocks (14 "pure" breeds), stock crosses, and familial lines. This data permitted conclusive evidence to be drawn with regard to the inheritance of susceptibility to adenocarcinoma of the uterus (115); to the familial distribution of mammary carcinoma of two distinct (papillary versus adenoid) types (106); and to the inheritance of embryonal nephroma (112) in *Oryctolagus*. In the Phipps colony maintained for some 30 years by Max Lurie in which inbred lines were established for studying the inherited resistance or susceptibility to tuberculosis (208), the familial relationship to the occurrence of adenocarcinoma of the uterus has been tracked with observations similar to those of Greene, especially in regard to incidence as a function of age and genetic constitution (2, 149).

The Brown-Pearce carcinoma is also known to be heritably influenced both in terms of transplantability and with regard to those antigens carried by it that are recognized by host rabbits. As examples, Casey (52) found that Blue Cross and Lilac Cross breeds of *Oryctolagus* possess natural resistance to growth of Brown-Pearce carcinoma, whereas Chocolate Dutch rabbits were shown to be more susceptible, and in fact still remain as the favored transplantation breed for this tumor. Kidd *et al.* demonstrated that the carcinoma contained normal tissue antigens and antigens specific to the tumor (100, 159, 175) and showed further that certain host breeds recognized only the tumor antigen(s) as foreign (e.g., Blue Cross and Chocolate Dutch), whereas others recognized in terms of antibodies produced in response, both types or neither.

From the large rabbit colony maintained for the study of genetics at the Jackson laboratory several recent reports have indicated that susceptibility to certain malignancies of the lymphoreticular system are inherited (within strains) according to a pattern consistent with that of an autosomal recessive gene (or alternatively, that of a vertically transmitted C-type RNA virus). This pattern was shown to be true for lymphosarcoma in the WH strain (97), and for an NZB-like disease with features of autoimmune hemolytic anemia and thymoma in the X strain (98).

All of these findings appear to indicate that spontaneous neoplasia encountered only sporadically in random-bred rabbits are also found with more predictable certainty in certain inbred strains, usually with increased frequency. The pattern of certain neoplasms associated with certain inbred lines (and of their inherited susceptibility and resistance) is well established in the mouse and on its way to being a demonstrated characteristic of the rabbit as well. It should be pointed out that the differentials in time between generations and the high cost of colony maintenance account in substantial measure for the differences in pace with which characterization of the neoplastic process has proceded in the mouse as compared to the rabbit.

C. Incidence of Neoplastic Disease

Forearmed with the knowledge that both age and genetic constitution are important determinants in assessing the incidence of cancer in any species, the question of incidence becomes a difficult one to deal with in the rabbit. This is a familiar problem in veterinary oncology; most animals in the sample group are not left to die a "natural" death in old age, but rather are killed for economic reasons during adolescence or young adulthood. For the purposes of grappling with this issue we would like to describe the tumor series collected in the laboratory of the author over a period of five years and drawn from a necropsy series of approximately 600 consecutive rabbit accessions. It illustrates in microcosm most of the problems encountered in collecting data on the incidence of cancer in this species. One may well question the relevance of a calculated incidence based on accessions drawn substantially from a population of random-source university research animals used for a variety of experimental purposes. It is first necessary to put such a population into perspective.

The life span of normal healthy *Orytolagus* rabbits under domestication is approximately 84–96 months (7–8 years). Most of those to be found in a university research colony are between 4 and 24 months of age; essentially juveniles and young adults. Following the period of experimentation, animals without further direct purpose are disposed of. The same factors are operative in the management of commercial colonies as well because the reproductive efficiency ordinarily declines rather quickly after the second year of life. This fact alone imposes great constraints on the variety and frequency with which spontaneous neoplasms are encountered because in no species are juveniles a particularly cancer-prone age group. Neoplasia induced by known antecedents whether viral, chemical, or physical have purposely been excluded from this series. We are considering here only "spontaneous" neoplasms.

We have recorded a total of 16 rabbits with naturally occurring neoplasms from a total (in this series) of 599 rabbit (*Oryctolagus*) accessions. This gives an overall incidence of about 2.6% with neoplasms. When the data are examined more closely, however, it becomes clear that the overall incidence is reduced and the mean age skewed to the left because of the preponderance of young animals in the sample group. It should be noted here, however, that the incidence of 2.6% compares in magnitude with similar compilations of the incidence of rabbit tumors from other laboratories. Rive *et al.* (260) reported an incidence of 0.8–2% in their series. Boycott found four tumors in 150 rabbits or an incidence of 2.7% (34), Bell and Henrici described two from 400 or 0.50% (28), seven tumors were found by Polson from a total of 560 rabbits, or 1.3% (252), two tumors were reported by Sprehn (300) from an autopsy series of 147 to give 1.3%, and finally Loliger has reported 19 tumors from a series of 1500 rabbit accessions to give an incidence of 1.3%. Some authors have suggested that most of the rabbits in these sample groups (with a range of incidences from 0.5% to 2.7%), were necropsied at too young an age to be in a cancer-prone age group and that this observation has tended to give an artificially low and misleading picture of the true incidence of neoplasms in this species (75, 140, 201), and in this I concur.

In 1958, Greene (115) presented data to show that 16.7% of a total of 849 female rabbits (dying of various causes) in his colony were found to have adenocarcinoma of the uterus. It is important to note here that the rabbits in this colony were permitted to live until death from natural causes, and that the average age of these rabbits was substantially greater (> 4 years of age) than those in the surveys mentioned previously. Greene believed his stocks (for this series) to be generally typical of other rabbit stocks and not particularly cancer susceptible. The importance of the age factor in this group is further stressed by noting his observation that the incidence (for females) of this tumor in rabbits 2–3 years of age was 4.2%, whereas in rabbits 5–6 years of age the incidence rose to 79.1%. Others, notably Ingalls (149), have observed the age-associated increase in frequency of this tumor and it will be explored in greater detail in Section II, A. The factors of age and sex are further analyzed in Table I using data from our laboratory. You will note that two age classes have been delineated for this analysis; those younger and those older than two years of age. In comparing these two age groups it is clear that a dramatic increase (7 times) in the incidence of neoplastic disease is initiated after the second year of life. The incidence was approximately the same (about 8%) for both males and females over two years of age, but the incidence was determined by different types of tumors in the two sexes (see Table II). In the younger age group, however, the incidence for tumors was approximately six times greater in females than males (1.00% compared to 0.16%). This, as we will see below, is almost entirely due to adenocarcinoma of the uterus. In fact, if the latter entity were subtracted out of the series for females less than two years of age, the incidence for them would not be much greater than that for young males.

Table II summarizes the various tumor types in my series as a function of age and sex. The tumors are ranked in terms of decreasing frequency. In assessing the significance of these relative frequencies and the relevance of the sample size from which they are drawn, a striking correlation can be drawn from the rank of the tumors in Table II and the frequency with which they occur as subjects for reports in the literature. (See Table III.)

Unquestionably, adenocarcinoma of the uterus is the most frequent and important of the spontaneous neoplasms of *Oryctolagus*. It has been the subject of intensive investigation and is well understood biologically. The next three tumors, lymphosarcoma, embryonal nephroma, and bile duct adenoma, have been reported in the literature with frequency as well (in that order) but are less well understood. These first four tumor types account for the bulk of those observed as spontaneous neoplasms, and this preeminence is reflected in the literature as well.

There are two further points that might be drawn from data in Table II. The first is that to my knowledge, osteochondroma and rectal papilloma have not previously been reported and are recorded here as new tumors from *Oryctolagus*. Secondly, it may be observed that in Table I 16 rabbits with tumors are listed, whereas in Table II there

TABLE I
AGE-SEX FACTORS IN NEOPLASTIC INCIDENCE[a]

Category	N	Neoplasms	% of total N with neoplasm	% of age-sex category with neoplasm	% of age category with neoplasm	
M(−)	217	1	0.16	0.46	(+)	(−)
M(+)	54	5	0.83	9.26		
Total M	271	6	0.99			
F(−)	275	6	1.00	2.18		
F(+)	53	4	0.66	7.54		
Total F	328	10	1.66			
Grand total	599	16	2.65		8.4	1.4

[a]Where N = Number of acessions, M = male, F = female, (−) = less than 2 years of age, (+) = more than 2 years of age.

TABLE II
INCIDENCE OF VARIOUS TUMOR TYPES[a]

Tumor type	No.	M(−)	M(+)	F(−)	F(+)
Uterine adenocarcinoma	5			4	1
Lymphosarcoma	4		1	2	1
Embryonal nephroma	2	1		1	
Bile duct adenoma	2		1		1
Osteosarcoma	1		1		
Osteochondroma	1				1
Leiomyoma	1		1		
Basal cell adenoma	1				1
Rectoanal papilloma	1		1		
	18	1	5	7	5

[a]Where M = male, F = female, (−) = less than 2 years of age, (+) = more than 2 years of age.

are 18 primary tumors. The discrepancy lies in the fact that two animals had multiple primary tumors. Drawing on a much higher sample size, Greene and Strauss (121), some 25 years ago, observed that the incidence of multiple primary tumors among those rabbits with at least one primary tumor was no greater than the incidence of single primary tumors in the larger population from which the tumor group was drawn. He statistically inferred from this that the presence of one neoplastic focus did not predispose toward the development of another. Within the limits of an admittedly smaller group, the incidence of multiple primary tumors (2 out of 16, or 12%) was approximately equal to that of the same age group (8.4%) in my series and confirms Greene's observations. Both of the individuals in the present series were over two years of age, one of each sex. In both cases a benign epithelial tumor (basal cell adenoma and rectoanal papilloma) was associated with a malignant epithelial tumor, uterine adenocarcinoma and bile duct carcinoma, respectively. Multiple primary tumors have been observed in the rabbit by others as well (252).

TABLE III

NATURALLY OCCURRING TUMORS OF *Oryctolagus cuniculus*:
REPORTS TO THE LITERATURE

Tumor site and type	No. of cases	Ref.
I. Reproductive		
A. Female		
1. Uterus		
a. Adenocarcinoma	1	Lack 1900 (192)
	1	Shattock 1900 (287)
	1	Wagner 1905 (320)
	1	Selinow 1907 (286)
	4	Boycott 1911 (34)
	1	Marie and Aubertin 1911 (212)
	1	Leitch 1911 (196)
	1	Katase 1912 (162)
	13	Stilling and Beitzke 1913 (302)
	2	Paine and Peyron 1918 (242)
	1	Dible 1921 (73)
	5	Polson 1927 (252)
	1	Rusk and Epstein 1927 (271)
	1	Koyama 1927 (190)
	1	Usawa 1930 (317)
	1	Watrin and Floretin 1930 (323)
	1	Cutler 1934 (67)
	2	Medlar and Sasano 1937 (216)
	1	Twort 1937 (313)
	5	Orr and Polson 1938 (241)
	1	Witherspoon 1938 (330)
	142	Greene 1938–1958 (105, 110, 115, 116, 118, 120, 121)
	15	Burrows 1940 (46)
	1	Biesele 1945 (31)
	1	Dobberstein and Tamaschke 1958 (75)
	3	Lombard 1959 (202)
	2	Merriam *et al.* 1960 (221)
	86	Ingalls *et al.* 1964 (149)
	2	Cotchin 1964 (64a)
	8	Baba and von Haam 1967 (18)
	2	Loliger 1968 (201)
	1	Flatt 1969 (94)
	5	Weisbroth 1972 (324)
b. Leiomyoma, leiomyosarcoma	4	Stilling and Bietzke 1913 (302)
	1	Polson 1927 (252)
	12	Greene 1958 (115)
	2	Ingalls *et al.* 1964 (149)
	1	Irving *et al.* 1967 (150)
2. Vaginal squamous–columnar junction		
a. Squamous cell carcinoma	3	Greene *et al.* 1947 (119)
3. Ovary		
a. Hemangioma	1	Greene and Strauss 1949 (121)
B. Male		
1. Testicle		
a. Seminoma	1	Paine and Peyron 1918 (242)
b. Interstitial cell carcinoma	1	DeFaria 1961 (72)
c. Adenocarcinoma	1	Hoffman 1954 (140)
d. Teratoma	1	Meier *et al.* 1970 (218)
II. Digestive tract and its organs		

Table III (*continued*)

Tumor site and type	No. of cases	Ref.
A. Stomach		
1. Adenocarcinoma	1	Schmorl 1903 (279)
	1	Schultze 1913, 1914 (280, 281)
	1	Greene and Strauss 1949 (121)
2. Leiomyosarcoma	1	Colella 1925 (64)
B. Intestine		
1. Leiomyosarcoma	1	Greene and Strauss 1949 (121)
C. Sacculus rotundus		
1. Papilloma	1	Polson 1927 (252)
D. Rectal squamo–columnar junction		
1. Papilloma	1	Weisbroth 1972 (324)
E. Liver		
1. Bile duct system		
a. Carcinoma	1	Schweizer 1888 (283)
	1	Niessen 1913, 1927 (233, 234)
	1	Greene and Strauss 1949 (121)
	1	Kaufmann 1973 (165)
	2	Weisbroth 1972 (324)
III. Respiratory		
A. Lung		
1. "Carcinoma simplex"	1	Schmorl 1903 (279)
2. "Epitheliome primitif"	1	Petit 1909, 1910 (249, 250)
3. "Cancer primitif"	1	Barile 1920 (21)
VI. Central nervous system		
A. Medulla oblongata		
1. Teratoma	1	Shima 1908 (288)
B. Cervical sympathetic trunk		
1. Neurinoma	1	Salaskin 1929 (273)
V. Urinary tract		
A. Kidney		
1. Embryonal nephroma	1	Lubarsch 1905 (207)
	1	Nurnberger 1912 (237)
	2	Bell and Henrici 1916 (28)
	1	Scott 1917 (284)
	1	Colella 1925 (64)
	1	Oberling 1927 (238)
	1	Polson 1927 (252)
	2	Eisler 1938 (81)
	1	Miyadi 1940 (222)
	4	Greene 1943 (112)
	1	Flir 1952 (95)
	2	Loliger 1968 (201)
	2	Weisbroth 1972 (324)
2. Renal carcinoma	1	Kaufman and Quist 1970 (166)
B. Urinary bladder		
1. Leiomyoma	1	Weisbroth 1972 (324)
VI. Mammary glands		
A. Female		
1. Papilloma	19	Greene 1939 (107)
	2	Burrows 1940 (46)
	17	Greene and Strauss 1949 (121)
2. Adenocarcinoma	1	Bashford 1911 (24)
	1	Marie and Aubertin 1911 (212)
	1	Polson 1927 (252)
	1	Ball and Douville 1929 (20)

Table III (*continued*)

Tumor site and type	No. of cases	Ref.
	1	Heiman 1937 (131)
	19	Greene 1939 (106)
	10	Pearce and Greene 1938 (247)
	1	Burrows 1940 (46)
	1	Lombard 1962 (203); Lombard and Goulard, 1960 (205)
VII. Skin		
A. Squamous epithelium		
1. Papilloma (nonviral)	1	Dobberstein and Tamaschke 1958 (75)
2. Squamous cell carcinoma	1	Greene and Brown 1943 (117)
	1	Pearce (cited by Greene 1943) (112)
	3	Greene and Strauss 1949 (120)
	1	Carini 1941 (47)
B. Adnexa		
1. Basal cell cacinoma	1	Groth 1955 (124)
	1	Weisbroth 1972 (324)
VIII. Bone, cartilage, connective tissue		
A. Skull		
1. Mandible and maxilla		
a. "Round cell sarcoma"	2	Katase 1912 (162)
	1	Schultz 1913, 1914, (280, 281)
b. Adamantinoma	1	Orr 1936 (239)
c. Osteosarcoma	1	Weisbroth and Hurvitz 1969 (325)
B. Appendicular skelaton		
1. Osteochondroma	1	Weisbroth 1972 (324)
2. Osteosarcoma	1	Salm and Field 1965 (274)
C. Subcutaneous tissues		
1. "Sarcoma"	1	Bashford 1911 (23)
2. "Spindle cell sarcoma"	1	Kato 1925 (164)
IX. Lymphoreticular system		
A. Lymph tissue		
1. Generalized lymphosarcoma	1	Aberastury and Dessy 1903 (1)
	1	Schultze 1913, 1914 (280, 281)
	1	Zschocke 1914 (337)
	1	Feldman 1926, 1927, 1932 (87, 88, 89)
	1	Greene and Strauss 1949 (121)
	1	Llambes and Mendez 1954 (198)
	1	Zbindin and Studer 1957 (335)
	12	Loliger 1966, 1968 (200, 201)
	1	Van Kampen 1968 (318)
	29	Fox *et al*. 1970 (97)
	1	Hayden 1970 (130)
	1	Kraack 1971 (191)
	4	Weisbroth 1972 (324)
	1	Flatt 1972 (93)
	1	Ubertini 1972 (313a)
2. Hodgkins-like lymphosarcoma	1	Medlar and Sasano 1937 (216)
B. Thymus		
1. Thymoma, thymosarcoma	1	Drieux and Poisson 1938 (77)
	1	Blanchard *et al*. 1939 (32)
	1	Orr 1939 (240)
	4	Greene and Strauss 1949 (121)
	3	Fox *et al*. 1971 (98)

Table III (*continued*)

Tumor site and type	No. of cases	Ref.
C. Brain		
1. Myeloma, plasma cell	1	Pascal 1961 (245)
D. Spleen		
1. Epithelioma	1	Ball 1926 (19)
X. Endocrine glands		
A. Adrenal		
1. Adenoma	1	Boycott and Pembrey 1912 (35)
2. Carcinoma	1	Hueper and Ichnowski 1944 (144)
B. Pituitary		
1. Eosinophilic adenoma	1	Polson 1927 (252)
2. Teratoma	1	Margulies 1901 (211)
C. Thyroid		
1. Carcinoma	1	Dinges and Kovac 1972 (74)
XI. Miscellaneous		
A. Malignant melanoma	1	Sustmann 1922 (304)
	1	Brown and Pearce 1926 (44)
	1	Holz and Heutgens 1955 (143)
B. Disphragmatic endothelioma	1	Polson 1927 (252)
C. Metastatic "abdominal sarcoma"	1	Wallner 1921 (321)
	1	Baumgarten 1906 (25)
D. Retroperitoneal fibroleiomyoma	1	Raso 1936 (253)
E. Unknown site		
1. Myxosarcoma	2	Polson 1927 (252)
2. "Spindle cell sarcoma"	1	Polson 1927 (252)
3. Encephaloid carcinoma	1	Polson 1927 (252)

II. NEOPLASMS OF *ORYCTOLAGUS* SP.

This section includes a tabular summarization of the reported naturally occurring neoplasms in *Oryctolagus*, description and illustration of the gross and microscopic morphology of these tumors, an assessment of their biological behavior, and an estimate of the value they might have as model systems for cancer research. The summarization of reports to the literature involving naturally occurring tumors of *Oryctolagus* is detailed in Table III. The descriptions (to follow) of these tumors are in turn drawn from the reports in Table III and from my personal experience and that of others.

A. Tumors of the Female Reproductive Tract

1. Adenocarcinoma of the Uterus

It is worth emphasizing again that this tumor is unquestionably the most frequently encountered neoplasm of *Oryctolagus*. It has received the greatest attention in the literature and has been the subject of several well-organized and intensively investigated longitudinal studies (18, 115, 149). These studies all emphasize that in order to understand the behavior of this tumor it is necessary to view the histological events in a time-lapse context as part of a continuum that begins with an initially normal (appearing) endometrium and culminates in metastasizing adenocarcinoma. The concept of cancer arising *sui generis* in the epithelium of the rabbit uterine fundus as a sudden transformation of normal cells is not supported by the evidence. On the contrary, most investigators have emphasized that metastatic fundic carcinoma is the final stage of an evolutionary process; and that the identifying biological and morphological attributes are not present initially, but develop gradually over a period of time. Two main lines of evidence lead to this conclusion. The first is circumstantial and inferential. Female breeder rabbits with adenocarcinoma of the uterus, when studied retrospectively invariably have a case history that includes a period of reproductive disturbance prior to detection of the tumor (115, 116, 120). Fertility is diminished, litter size is reduced, there are more stillborn births, and desertion by the mother is more common. Dystocia, litter retention *in utero*, abdominal pregnancy, and fetal resorptions are also more likely. In general, the period of altered reproductive behavior precedes clinical detection (palpation) by some 6–10 months. Early tumors are palpated as persistent uterine nodules, about the size of a 10-day fetal cyst when first noted. They are usually multiple and present in both horns. In 25% of Greene's cases cystic changes were detectable in the mammary glands as coincident findings (115, 118, 120). The growth rate of the tumor varies considerably between individuals, but in most cases progresses from approximately 1 cm in size initially to that of a mass 5 cm or larger in 6 months. In some cases the tumors do not enlarge prior to metastasis. In Greene's work, the duration in time between clinical detection and death from metastasis averaged 12–24 months (118). In the 12 females studied by Ingalls *et al.* (149), although 11 rabbits developed adenocarcinoma of the uterus, metastasis did not occur during the three-year period between initial detection and death in the case of the majority (8/11) of tumors. The tumor appears to bear no relationship to degree of parity. Adams reported no differences in the incidence of the tumor in breeder rabbits as opposed to aged virgin rabbits (2). My cases all occurred in nulliparous females. This evidence, taken with the known age-related factors in the incidence of this tumor suggest that it develops slowly in most cases.

The other line of evidence is more direct and has been accumulated by employing the technique of periodical laparotomy and uterine biopsy (pioneered by Greene) over the life span of individual rabbits (2, 18, 115, 149). This work has not only supported the clinical observations mentioned above but has permitted sequential study of the changes encountered in the evolution of adenocarcinoma

of the uterus. The normal endometrium of the rabbit is one that may be characterized as follows (16, 18): (a) There is good differentiation between surface and glandular epithelium. (b) Glands are regularly distributed and without evidence of hypersecretion. (c) There are no cellular or architectural abnormalities. (d) The stroma is moderate in amount and well developed (e) There is no substantial amount of inflammation. (f) Papillary and polypoid changes in the superficial epithelium are not present. (g) Both ciliated and nonciliated cells can be distinguished. (h) With the use of specific histochemical stains, the presence of alkaline phosphatase localized to the apical (luminal) aspect of epithelial cells can be demonstrated. Acid phosphatase is demonstrable in the cytoplasmic lysosomes of these cells. Both enzymes are present in the fundic stroma.

Against this normal background, a number of endometrial abnormalities have been characterized. They include endometriosis, papillary, cystic, and adenomatous hyperplasia, endometritis, pyometra, and senile atrophy (18, 219, 221). Each, of course, is characterized by a typical histological appearance. There is general agreement that adenocarcinoma develops from the glandular epithelium and preserves an identifiable adenomatous appearance of irregular glandular elements embedded in a richly vascular myxoid stroma throughout its course. The tumors are usually initially situated in the mucosal folds adjacent to the mesometrial insertion, and most frequently arise in multicentric foci irregularly spaced in both uteri.

One key question is whether or not any of the aforementioned endometrial abnormalities can be identified as precancerous lesions. There is no uniformity of opinion on this. In Greene's studies (105, 110, 115, 116, 118, 120) based on 140 cases of fundic carcinoma in which cancer could be anticipated by monitored changes detected during sequential biopsy, every case, without exception, was said to have been preceded by cystic endometrial hyperplasia. A more recent, but technically similar study has been done by Baba and von Haam (18). They, too, describe antecedent endometrial lesions, dysplasia, and carcinoma *in-situ*, but their findings could not support cystic hyperplasia as a preneoplastic change in spontaneous adenocarcinoma. In the case of the Phipps colony (similarly studied by sequential biopsy) preneoplastic changes included hyperplasia and cystic degeneration (2, 149). These lesions were said to progress to focal anaplasia and then carcinoma.

Greene believed and marshalled considerable data to substantiate his thesis that both carcinoma of the breast and adenocarcinoma of the uterus were reflective of profound and preexisting endocrinological disturbance in the rabbit (116). The evidence included an incidence of uterine tumors of 48 and 34% in transmitters (genetic) of pituitary dwarfism and cretinism, respectively, compared to an incidence of 14% in the remainder of the sample group. No tumors were found in animals of the Belgian or Rex breeds, and in the other breeds tumors occurred in the following order (of decreasing frequency): Tan, French Silver, Havana, Dutch, Marten, English, Chinchilla, Beveran, Sable, Himalayan, Polish. Greene found a parallelism between the incidence of toxemia of pregnancy and the incidence of fundic carcinoma in these various breeds. Forearmed with the knowledge that disorders of pregnancy characterized the history of animals that subsequently developed uterine tumors, and the relationship between the incidence of tumors and of toxemia of pregnancy, Greene concluded that a relationship existed between the milder disorders and toxemia of pregnancy. He reasoned that if the liver were damaged, as it would be, by repeated sublethal bouts of toxemia, it would be less able to detoxify endogenously produced estrogenic hormones, and the receptor tissues therefore (uterus and breast) rendered more vulnerable to the carcinogenic activity of estrogenic hormones. Additional evidence included the finding of endocrine gland changes; e.g., small fetal adenoma-like thyroids, pronounced accumulations of lipoid material in the zona fasciculata of the adrenal, hyperplasia of the pars intermedia of the pituitary in adenocarcinoma-bearing rabbits, and the statistically significant association of mammary carcinoma in greatest frequency with adenocarcinoma of the uterus (when multiple primary tumors were present). Finally we may cite his observation that breast changes other than neoplastic ones, e.g., epithelial hyperplasia, adenosis, and cyst formation almost invariably accompanied adenocarcinoma of the uterus (121).

These observations have prompted experimentation relating to the carcinogenicity of estrogen in the rabbit and the issue has recently been reviewed, along with other carcinogens active in the rabbit uterus (18). There is little experimental evidence to support the role of estrogen as a factor in spontaneous endometrial adenocarcinoma in the rabbit. Moreover, the model of estrogen-induced glandular hyperplasia that develops in the uterus is known not to be a precancerous lesion (16, 18) and to have histochemical changes not characteristic of adenocarcinoma (17). The evidence is conflicting, however, because others have reported induction of uterine carcinoma in the rabbit with estrogenic hormones (219). The data is extremely difficult to interpret because of the naturally high incidence of adenocarcinoma of the uterus in untreated rabbits. It is known that estrogen enhances the carcinogenic effect of certain chemicals (18) but in these cases (as in carcinogen alone), the sequence of squamous metaplasia leading to squamous carcinoma bears little morphological relationship to the tumors observed to occur naturally.

All investigators have emphasized that the evolutionary course of fundic carcinoma is relatively slow, but inexorable. Histologically the events of progression are characterized by increasing degrees of dedifferentiation and anaplasia with increase of the vascular, myxoid stroma

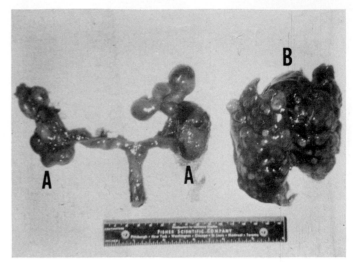

Fig. 1. Uterine adenocarcinoma. Note multiple neoplastic foci in uterus (A) and lungs (B).

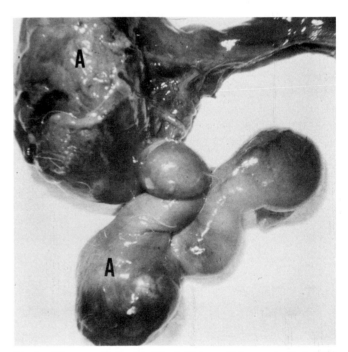

Fig. 2. Close-up of uterus in Fig. 1. Note multiple neoplastic foci (A).

often at the expense of the glandular elements. Carcinomatous endometrium also is distinguished from nonneoplastic changes by loss of such cellular elements as cilia and secretory vesicles, and histochemically demonstrated absence of the enzymes (alkaline and acid phosphatases) normally found both in the normal and hyperplastic uterus. Areas of necrosis are not uncommon in mature tumors and Greene correlated the onset of necrotic areas with attainment of the ability to metastasize (118). The tumor is illus-

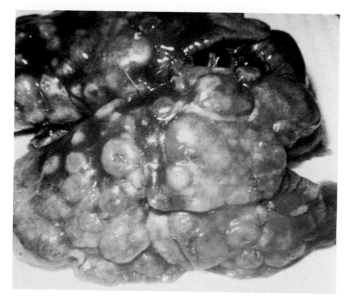

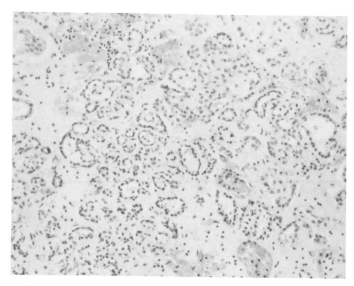

Fig. 5. Photomicrograph of postmetastatic uterine adenocarcinoma. Compare loss of tubular organization and stromal increase with Fig. 4.

Fig. 3. Close-up of lungs in Fig. 1. Almost total replacement of pulmonary tissues by metastatic nodules of uterine adenocarcinoma.

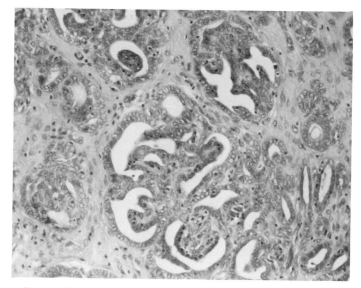

Fig. 4. Photomicrograph of premetastatic uterine adenocarcinoma with only local invasion. Note well-organized tubule formations.

for serial rabbit passage, for tissue culture, and for autologous transplantation have been established several times in the past (31, 105, 111, 148a, 204), Autotransplants are illustrated in Figs. 6–9. Ability to metastasize appears to be a requisite stage of development before passage to homologous or heterologous hosts is successful (111, 115). Greene pioneered the technique of using the anterior chamber of the eye as a shielded site for tumor growth with the use of this cancer (120).

2. Leiomyoma and Leiomyosarcoma

Both leiomyomas and leiomyosarcomas have been reported from rabbits as sporadic and usually incidental findings. Although the benign form is more common,

trated in Figs. 1–5. Excellent photographs may be studied in the papers by Greene (118, 121).

A definite hereditary component has been referred to above as a determinant in the incidence of this tumor both between and among breeds. Notwithstanding this, however, the incidence has been noted to exceed 50% in certain colonies (of random breed females) kept past the fifth and sixth year of life. This is a useful feature for those seeking model systems for adenocarcinoma of the uterus because it means that material for study can be obtained with certainty, and almost at a predictable rate (18a). Tumor lines

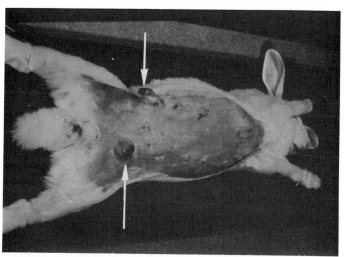

Fig. 6. Rabbit with autotransplants (arrows) to skin of naturally occurring uterine adenocarcinoma. Photograph courtesy of T. H. Ingalls.

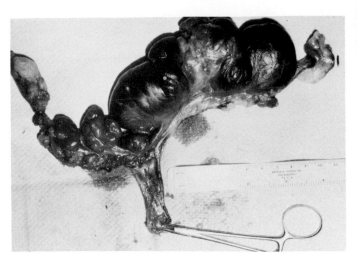

Fig. 7. Carcinomatous uterus of rabbit in Fig. 6. Autotransplants made from this tissue. Photograph courtesy of T. H. Ingalls.

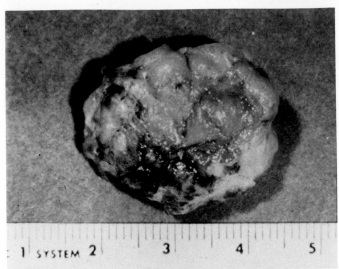

Fig. 9. Close-up of excised autotransplant of uterine adenocarcinoma. Note hemorrhagic areas. Photograph courtesy of T. H. Ingalls.

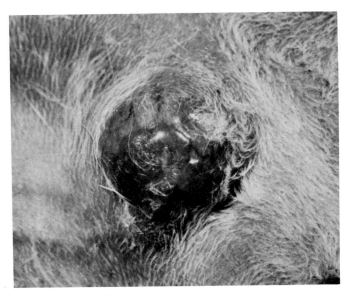

Fig. 8. Close-up of autotransplant seen in Fig. 6. Photograph courtesy of T. H. Ingalls.

metastasizing leiomyosarcomas have been infrequently reported (252). As in other anomen*, these smooth muscle tumors are found most frequently in association with the digestive and female reproductive tracts (see Table III). The histology is typical for the same tumor in other species; whorls and bundles of fusiform cells arranged in a criss-crossing pattern. Malignant forms generally appear less regular, with more abundant nuclei and more mitotic figures. They are generally reported from older rabbits (over 2 years of age).

_____

* An acronym for: animals-other-than-man, coined by Dr. R. W. Leader (193a).

3. Squamous Cell Carcinoma

Squamous cell carcinoma of the vaginal squamous—columnar junction is a comparatively rare tumor. It has been described only by Greene (119) who reported 3 cases from a necropsy series of 1100 females over 2 years of age. It is an important tumor, however, because in the rabbit, the columnar epithelium of the uterus is continued out over the cervix into the vagina. The transition to squamous epithelium occurs at the level of the urethral meatus. This location would, in the rabbit, appear to be the homolog of the external cervical os in the human female; as that is where the epithelial transition occurs in woman. The difference between these two species in the location of the transitional zone emphasizes the importance of the junction as a predisposing factor in carcinogenesis. The 3 animals were not related genetically and all were over 4 years of age. Interestingly enough, all 3 were the bearers of multiple primary tumors including fundic adenocarcinoma in all three cases and mammary neoplasia in the case of two. In all three cases the fundic tumor was in a nonmetastatic stage of development but the vaginal squamous carcinomas had metastasized via lymph nodes and/or peritoneal implants to other organs. Two of the tumors were used for transplantation studies and at least 14 anterior chamber-passaged generations were obtained from them.

B. Papillomatosis

Papillomas in the rabbit, not known to be induced by oncogenic viruses have an interest, out of proportion to their incidence. The prominence of the Shope papilloma (of *Sylvilagus*) as a research model on the one hand, and the oral papilloma (of *Oryctolagus*) on the other, have tended to produce in the popular mind an assumption that all papillomas in the rabbit are associated with papova group viruses. Dobberstein and Tamaschke briefly described papillomatosis of the skin in *Oryctolagus* (75). They were not able to pass the papilloma in transplantaion experiments. Details of the histology were not given. Greene (107) described papillomatosis as a transitory phase in the development of one of two types of mammary carcinoma in the rabbit. In the papillomatous form, similar in pathogenesis to Schimmelbusch's disease in women, the stage is set for the onset of tumor formation by preexisting cystic disease in the breast tissues. Tumors in cystic breasts arose as papillomata from the epithelial lining of dilated acinar and ductal walls which, after continued growth, formed multiple anastomosing radicles with the production of acinus-like structures. Greene described these structures as anastomosing papillomata rather than cyst adenomas in order to distinguish them from tumors of the second type. The latter originated in previously normal mammary tissue, appeared to arise as proliferations of true acini, and were classified as adenomas (107). It should be observed that the transition of papilloma to carcinoma observed spontaneously here, in the absence of known oncogenic DNA viruses, casts some suspicion on the role of DNA viruses in the same (appearing) process in which Shope papillomas in certain cases progress to carcinoma (262, 265). The same histological progression, i.e., from ductal papilloma to papillary adenocarcinoma is known in other species, e.g., the dog (226).

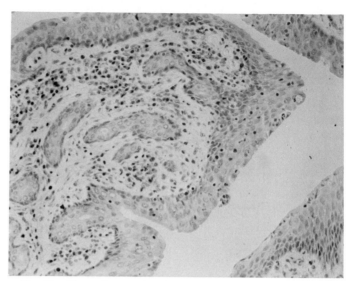

Fig. 11. Photomicrograph of papillomatous frond from tissue in Fig. 10. Inclusion bodies are absent.

Nonviral papillomas of the digestive tract also have been infrequently observed in the rabbit. The single previously reported case is that of a polyp of the sacculus rotundus (252). In my series one fungating cauliflowerlike papilloma was observed at the rectoanal junction (Figs. 10 and 11) of a 5-year-old male rabbit. It was well differentiated and benign. Inclusion bodies could not be demonstrated and transmission experiments were unsuccessful.

C. Tumors of the Liver

Bile duct adenoma (or cyst adenoma) and its malignant derivative, the bile duct adenocarcinoma (or cholangiocarcinoma) are relatively common as spontaneous tumors of *Oryctolagus* (see Table III for reported cases). In the gross, the bile duct adenoma may appear as a massive solitary growth or as multiple foci of variable size. The lesions are sharply circumscribed from normal areas of the liver and are not capsulated. The tumor is frequently described as a cystadenoma because the gross picture of multiple interloculating cysts filled with fluid of a honeylike consistency is matched by the histological patern of variably differentiated ductal forms (see Fig. 12–16). The stroma may appear myxoid in certain areas and fibrotic in others.

The tumor is of some interest because it may be reflective of the hyperirritability known to be associated with rabbit bile duct epithelium. As long ago as Schweizer's observation (of coccidia in a liver tumor) in 1888 (283), it has been known that any of a variety of noxious stimuli may induce the proliferation of bile duct epithelium in the rabbit (Fig. 17). Particularly, bile duct hyperplasia has been associated with *Eimeria steidae* infection (see Chapter 11). One is

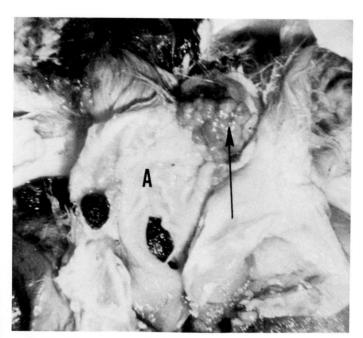

Fig. 10. Gross photograph of rectoanal papilloma. Note opened colon and rectum (A) and papilloma (arrow).

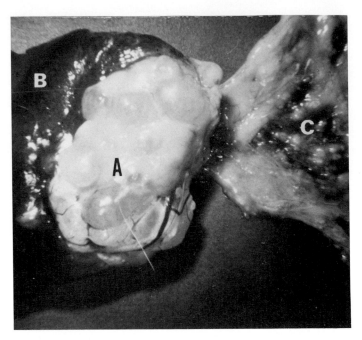

Fig. 12. Bile duct adenocarcinoma (A) sharply circumscribed from liver (B). Note metastatic nodules on mesentery to the right (C).

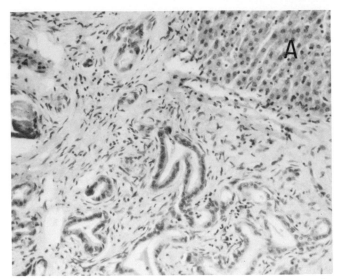

Fig. 14. Bile duct adenocarcinoma. Note fibrous stroma surrounding tubular formations. Tumor adjoins normal liver (A) without capsule.

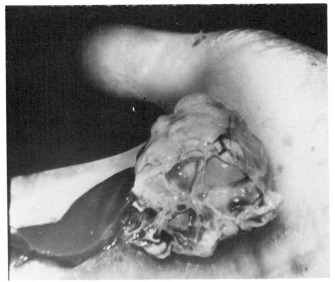

Fig. 13. Bile duct adenocarcinoma with cut surface. Note interloculating fluid-filled cysts in tumor. Normal liver to the left.

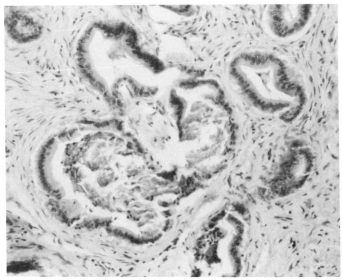

Fig. 15. High power photomicrograph of bile duct adenocarcinoma. Note irregularity of ductal epithelium.

tempted to speculate that the genesis of bile duct carcinoma may originate in initially hyperplastic epitherium. This view is consistent with the observation (of the author) that early developing and small bile duct adenomas are well differentiated compared to those that are large, presumably more developed and of longer duration. The bile ducts in non-adenomatous areas of the liver are normal. The histological pattern of the tumor may vary from that of a hamartoma in the most differentiated forms, to that of a cystic honeycombed carcinoma with less obviously differentiated tubular forms. Metastases in the form of peritoneal, diaphragmatic, and mesenteric implants are not uncommon.

All known cases have been studied retrospectively from postmortem tissues and have been encountered only sporadically as incidental findings at necropsy. The tumor appears to have little potential as a research model primarily because of the difficulty of case finding. The tumor has not, to my knowledge, been induced experimentally.

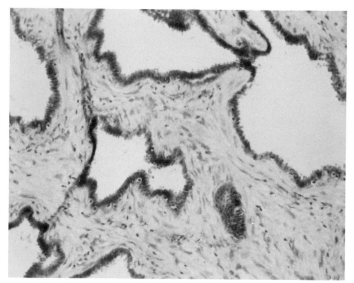

Fig. 16. Photomicrograph of cystic formations in bile duct adenocarcinoma.

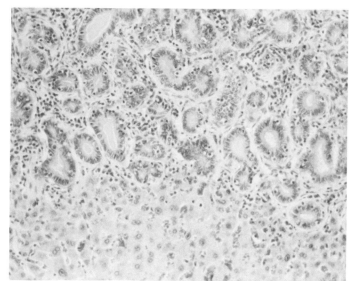

Fig. 17. Photomicrograph of bile duct hyperplasia of unknown causation. Compare regularity of tubular formations with those of Figs. 15 and 16.

D. Tumors of the Kidney

1. Embryonal Nephroma

Embryonal nephromas are comparatively common tumors in *Oryctolagus* and have been described many times (see Table III). In the review by Dobberstein and Tamaschke (75) the incidence of this tumor was second only to uterine malignancies. They are most often encountered as incidental findings at necropsy; indeed, all described cases have been benign and not obviously a cause of interference with normal renal function. Cases have been described from both young (18–24 months)

and old (60–72 months) rabbits. In my experience the nephromas do tend to increase in size very slowly, since the tumors from aged rabbits (2 cm diameter) are larger in general than those from younger trabbits (1 cm or less). The tumors appear grossly as whitish, sharply circumscribed nodules of tissue projecting above the cortical surface (Figs. 18 and 19). Certain authors (75) have suggested the tumors exhibit polarity, favoring the cranial pole

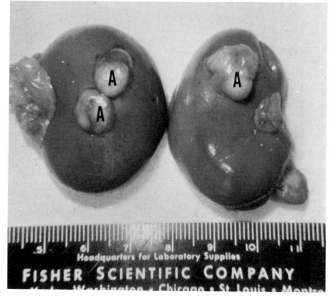

Fig. 18. Bilateral nonpolar renal embryonal nephromas seen as raised subcapsular nodules (A).

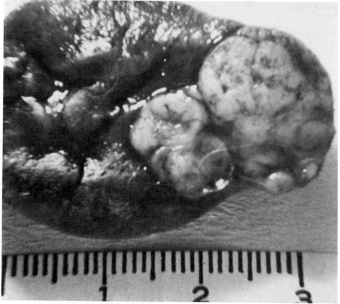

Fig. 19. Nodules of polar renal embryonal nephroma seen in cut section. Note sharp circumscription from adjacent renal tissue.

of the kidney, but this is not invariable as the cases reported here were not polar. The tumors may be single or multiple and appear to occur uni-or bilaterally with equal frequency in either kidney.

Histologically, the embryonal nephroma of *Oryctolagus* is remarkably similar to that of man (112) and swine (226). The basic histological pattern is that of cylindrical cells arranged in the tubular formations of an adenoma with spindle-shaped or round stromal elements in indifferent arrangement. The tubular formations are not uniform and many appear cystic. One distinctive feature is the pseudo-glomerulus structures seen within tubular formations in the

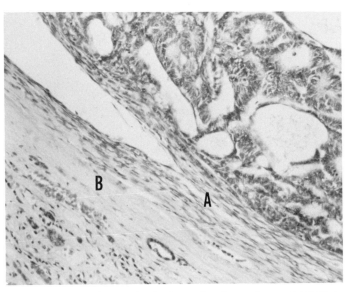

Fig. 20. Photomicrograph of embryonal nephroma (upper right). Note fibrosis (A) and compression (B) of adjacent renal tissue.

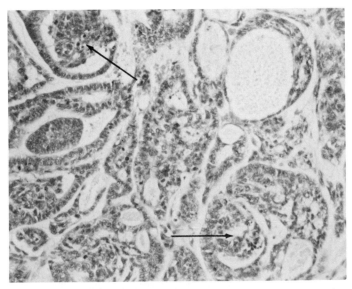

Fig. 21. Photomicrograph of embryonal nephroma. Note pseudo-glomeruli (arrows).

more differentiated areas. Muscle, cartilage, and other tissues common to human Wilm's tumors have not been seen in the rabbit analog. The tumors arise from the cortex and as they expand radially they project above the cortex distally, and compress the surrounding tissue, proximally. These relationships are illustrated in Figs. 20 and 21.

The incidence of embryonal nephroma in the human is limited almost exclusively to childhood, and the tumor is clinically characterized by a rapidly fatal course. In spite of the remarkable similar histological appearance and youthful incidence, the rabbit nephroma, on the other hand, is routinely benign and not known to metastasize. This point deserves some emphasis becasue of the great stress placed on the inference of biological behavior, e.g., malignancy, from morphological characters. It is one thing to point out similar patterns of histopathological architecture between animal and human tumors, but yet another to predict biological behavior in one species based on information gained from another.

Because the tumor occurs in other species with biological properties more similar to those of the human Wilm's tumor (226), the embryonal nephroma of *Oryctolagus* offers little of interest as a nephroma model. In the four cases encountered incidentally by Greene (112), successful transplants were established in about 20% of experimental rabbits heterografted either in the testicle or anterior chamber of the eye. It is interesting that the very slow rate of growth characteristic of *in situ* spontaneous growths was carried forth as a property of transplanted tumors as well, even into the fourth transplant generation.

2. Renal Carcinoma

Renal carcinoma (clear cell carcinoma or hypernephroma) while rare in most animal species has been reported in the rabbit only once (166). In that case the tumor appeared to arise from the cortex but was histologically dissimilar to the embryonal nephroma. Grossly the tumor presented as a large smooth-surfaced mass arising from the lateral-posterior aspect of the right kidney. There was a large (10-cm) cyst on one side, but in cut section the remainder of the tumor was solid although there were necrotic areas. No metastases were observed. The histological appearance varied from that of cells ranging in shape from low cuboidal to spindle-shaped and arranged in irregular tubular formations, to that of solid sheets and nests of the same cell type. The cells were characterized by eosinophilic cytoplasm, often with extensive vacuolations (clear cells). This pattern, easily distinguished from that of the nephroma, makes it unlikely that the renal carcinoma is preceded by or related in any other direct way to the embryonal nephroma. The renal carcinoma is illustrated in Figs. 22 and 23.

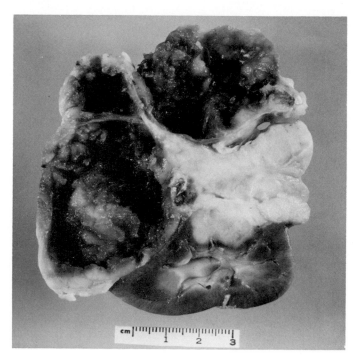

Fig. 22. Renal carcinoma. Note normal renal tissue in lower aspect. Photograph courtesy of A. F. Kaufmann.

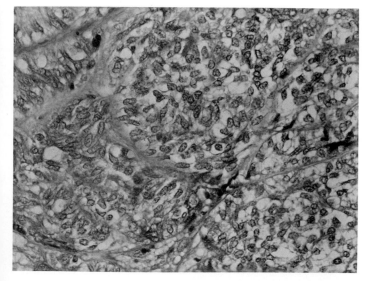

Fig. 23. Renal carcinoma. Note clear cells. Photomicrograph courtesy of A. F. Kaufmann.

E. Tumors of the Mammary Gland

Mammary carcinomatosis has been reported with fair frequency in laboratory rabbits (see Table III). The tumor appears to arise in multiparous females in their third and fourth years of life most commonly. Greene described 23 cases of mammary carcinomatosis that occurred exclusively in two families (of the Belgian and English breeds) from his colony (106–109). Of them, 19 were of a type associated with antecedent abnormalities of breast tissue, and four appeared to arise, *de novo*, without preceding mammary lesions. The other cases reported in the literature had not been followed clinically, and their presentation did not include a description of antecedents (references in Table III).

Greene studied the progressive changes leading to invasive adenocarcinoma of the breast by sequential biopsy over the life span of individual rabbits (106–109). In one family of the Belgian breed, 19 rabbits were observed to have breast tissue with the occurrence of cystic disease, benign neoplasia, and invasive adenocarcinoma as succeeding events. Considerable individual variation was observed in the time required to pass through these successive stages; varying from 10 to 29 months. The age at which the onset of cystic mastitis was first detected varied from 10 to 44 months of age but was most common in females 2–3 years of age. All cases were observed in multiparous females. Numerous animals of unrelated lines were foster-nursed by tumor-bearing rabbits and held under observation for long periods of time; without the occurrence of mammary tumors. This, together with the low incidence of tumors in daughters of tumor mothers was evidence against the passage in the milk (analogous to the Bittner virus of mice) of a virus influencing tumor development.

Cystic disease was preceded by a variable period of intense mammary congestion which waxed and waned, apparently according to irregular cyclical events in the estrous cycle. Cysts were observed to arise from dilated acini and ducts. Cysts arose in many areas simultaneously and could be palpated *in situ* as shotlike structures. Cyst epithelium was flattened in the majority of cases. As the cystic disease progressed, more and more of the acini in a given field became involved, and coalesence of cysts became more common. Fibrous connective tissue accumulated around dilated acini and frank cysts as this stage progressed.

After the variable period of cystic disease, early neoplastic changes in the form of small, intracystic sessile growths of cystic epithelium, or as uniradicular papillomas with stalks of attachment, would supervene. The purely epithelial growths were not found in later stages and it was assumed that they were capable of inducing a connective tissue response and appear in subsequent sections as papillomata. Papillomata were found in all breasts and all stages of development were frequently seen in different sections of the same breast. The papillomas rapidly became multiradicular, their branches uniting with each other to form epithelial-lined spaces, resembling acini. The walls of these structures in turn gave rise to papillary forms, resulting in an extremely complex structure comparable to that observed in human breasts and referred to as intracystic adenomata (107). The tumor was not at this stage invasive and growth occurred by expansion along the cyst or duct of origin into other radicles of the gland. Eventually, cellular changes occurred that signaled the onset of malignancy. These changes occurred in widely separated areas of mammary tissue, and included disorder in the layers of lining papillomatous epithelium, irregularity in size and shape of cells, increasing nuclear hyperchromaticity, and more commonly observed mitotic figures. These late characteristics of papillomata in some cases persisted for more than six months before the invasive stages began.

Invasive mammary adenocarcinoma was the inexorable outcome of the papillomatous stage. Released cells would form neoplastic structures in their new locations identical to those of the parent tumor. In some instances, further dedifferentiation occurred and the cells proliferated in solid, medullary sheets. It was characteristic that all areas did not undergo invasive progression and in different areas all stages from cystic disease to metastatic adenocarcinoma could be observed. Invasion of the carcinomatous tumor to muscle and connective tissue surrounding the mammary gland occurred commonly in the invasive stage. Invasion (or metastasis) was not accompanied by an inflammatory response. Metastasis occured in 3 cases; and in all of these the lungs and regional lymph nodes were involved. Other organs observed as metastatic sites (in different animals) included thoracic organs, kidney, liver, adrenal, pancreas, ovary, and bone marrow.

In contrast to the picture just described of preceding cystic disease, papillomata, and carcinoma, 4 animals in a single family of the English breed were observed to develop mammary adenomas as atypical proliferations of previously normal acini. The 4 tumors appeared to originate similarly and consisted of numerous small (but otherwise normal-looking) acini divided by fibrous bands into large lobules. As growth progressed the greater part of the adenoma was made up of small, closely packed acini in an atypical arrangement. Frequently there were areas of disorganization in which there was no acinar formation and the structure resembled that of a medullary carcinoma simplex. In other areas the acini were distended with eosinophilic material and occasionally coalesence to form macroscopic cysts was observed. Two of the four cases were metastatic to thoracic and abdominal organs and to the regional lymph nodes.

While the dominating lesion in these animals was the mammary carcinoma (of either type), other pathological changes found in the body at necropsy reflected the physiological disturbances observed clinically and were of such a nature as to suggest that the prescence of neoplasia in the breast formed only a local manifestation of more profound constitutional disorder. Three organs of known endocrine activity appeared to be consistently involved in mammary carcinoma preceded by cystic disease; the adrenals, the pituitary, and the uterus. In the case of carcinoma arising in the absence of cystic disease, the adrenals and uterus were affected by the same changes in animals with antecedent cystic disease but the pituitary was normal. Adrenal changes included the presence of numerous small areas, especially in the zona reticularis, of large, pale-staining, highly vacuolated cells with pyknotic nuclei. The pituitary was enlarged in all cases with preexisting cystic disease. The enlargement, up to 3–5 times normal size, was proportional to the degree of breast anaplasia, and appeared to be due to hyperplasia of chromophobe cells of the anterior pituitary. The uterus of rabbits with mammary carcinoma was consistently involved with cystic hyperplasia of the endometrium. Greene believed that the observed changes, e.g., cystic hyperplasia of the endometrium, pituitary hyperplasia, adrenal changes, and breast engorgement, were all consistent with a basic syndrome of hyperestrinism. In support of this view he pointed out the frequent involvement of mammary tissues (50%) in the case of uterine adenocarcinoma and the unusual frequency with which tumors of these glands were associated. Of the 23 cases of mammary carcinoma, six also had uterine cancer. Furthermore, he argued, the similar endocrine gland changes (in the adrenal and pituitary especially) supported his inference of a common etiological factor. In his view the factor was hormonal distrubance manifesting primarily as hyperestrinism (108, 116). These findings, and the conclusions as well, were shared by Burrows (46) who followed 25 female rabbits up through the 900th day of life. Of them, 15 developed uterine adenocarcinomas and of the 15, 3 developed mammary carcinomas. The rabbits were frequently observed to be pseudopregnant due to the constant presence of males in the room. The mammary carcinomas were of the type described by Greene as being preceded by cystic mastopathy. In different areas all changes described by Greene, including intracystic papillomata and adenocarcinoma, were observed. More importantly, however, Greene's observations of (a) cystic mastopathy, (b) cystic hyperplasia of the endometrium, (c) ovarian changes in which the bulk of the organ was changed to luteal tissue (ovaries were cystic in Greene's studies), lipoid degeneration of the adrenal similar to that induced in experimental mice by injection with estrogen, and (e) the unusually frequent association of uterine and mammary tumors as multiple primary neoplasia were all confirmed by Burrows (46). They led him to the same conclusion, namely, that the tumors of the uterus and mammae were predetermined by more fundamental endocrine changes, perhaps induced by pseudopregnancy but in any case simulating hyperestrinism. Notwithstanding these impressive investigations, the fact remains however that neither uterine nor mammary adenocarcinomas of the type that occur spontaneously have been experimentally induced by hyperestrinism and the role of estrogen in their induction remains a moot question.

The cases of Greene and Burrows aside, most cases of mammary carcinomatosis reported in the literature have been encountered as sporadic findings with unknown history. The histology of these, however (131, 247, 252) was consistent with those described by Burrows and by Greene as the type preceded by cystic mastopathy, including in progression (or in different areas of the same or other breasts) intracystic papillomata or adenocarcinoma with all gradating forms. The case I observed (and submitted by Arnold F. Kaufmann) was of this type. It is illustrated in Fig. 24.

F. Tumors of the Skin

1. Squamous Cell Carcinoma

Despite the frequency with which squamous cell tumors have been induced experimentally in the skin of *Oryctolagus*, spontaneous tumors of this tissue are correspondingly rare. More will be said of the role of oncogenic viruses in this regard further ahead. Similarly, the Brown-Pearce tumor, which originated in the scar of an experimentally induced *Treponema pallidum* chancre in the skin of the scrotum, will be described in detail later. The Brown-Pearce tumor, originally described as a squamous cell carcinoma, became one of the most extensively investigated neoplasms in cancer research.

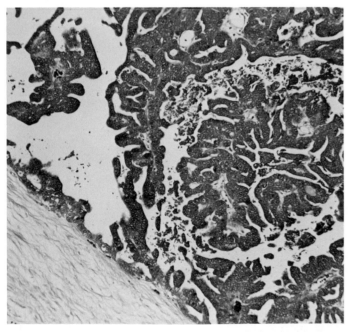

Fig. 24. Mammary adenocarcinoma. Photomicrograph courtesy of A. F. Kaufmann.

The literature has revealed only several reported cases of naturally occurring squamous cell carcinoma (47, 117, 121). There appears to be little to relate those described as having a common denominator of significance in their origin. All known cases have arisen in different locations, literally from the head (117) to the tail [Pearce, cited by Greene (117)]. Most were histologically typical of squamous carcinomas in other species. Of the six cases described, only one (117) was metastatic. This particular tumor was investigated by Greene through a series of experiments involving sequential biopsy and transplantation via anterior (eye) chamber preparations and testicular injection (117). It is of interest that "takes" did not occur in homologous hosts until the tumor (in the propositus) had achieved sufficient autonomy to metastasize. This observation is a frequently reported theme in Greene's extensive investigations and deserves to be reiterated here.

Basal cell adenomas (carcinomas) similarly are comparatively rare in *Oryctolagus*, having seemingly been observed only once previously (124). The tumors were in both cases benign and seen in relatively old (more than four years) rabbits. The case seen by the present author consisted of 2 small (2 cm × 1 cm) pedunculated adenomas of multicentric origin, both on the right abdominal dorsum, projecting above the skin (Fig. 25). The basal cell tumors were coincident with adenocarcinoma of the uterus. Histologically the tumors were similar to those described from other species, especially canine (226). The surface was covered with hyperkeratotic stratified squamous epithelium. The adenoma was composed of clumps or nests of epithelial (basilar) cells in a coarse, collagenous, often acellular stroma. The tumor is further illustrated in Fig. 26. It appears to have little potential as a model because of its low and sporadic incidence, although they have been produced experimentally (145).

G. Tumors of Bone and Cartilage

As mentioned earlier as a characteristic of squamous cell carcinoma, spontaneous tumors of bone are extremely rare in *Oryctolagus*, although they may be induced with comparative ease experimentally. Particularly, X-irradiation, bone-seeking radioactive isotopes, and beryllium salts have been successfully employed in the experimental induction of bone tumors in *Oryctolagus* (22, 66, 145, 157, 272, 282, 309–311, 314). A transplantable bone sarcoma has been reported in the Russian literature (155, 167, 188, 189) and will be described further in Section VI,C. The "round cell tumors" of bone reported by Katase (162) are difficult to evaluate in terms of present-day descriptive terminology for osteosarcomas because of the insufficient description provided by that author. Orr (239) described a multicystic tumor of the mandible that he classified as an

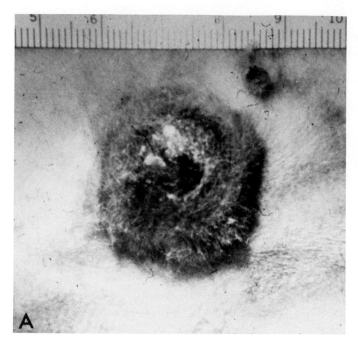

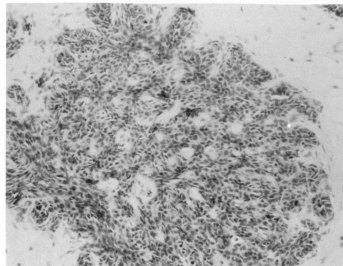

Fig. 26.   Basal cell adenoma. Note clump of neoplastic basilar cells surrounded by sparsely cellular collagenous stroma.

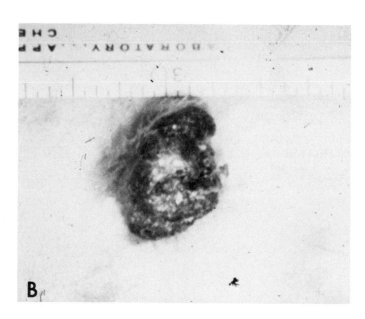

Fig. 25.   Basal cell adenoma. (A), (B), composite illustration of multicentric origin in same rabbit.

adamantinoma. The cysts were lined with stratified squamous epithelium and numerous desquamated cells were seen in the cyst cavities. The cysts were situated within a loose matrix of fibrous tissue. Other areas of the tumor consisted of solid alveolar masses of cells of squamous (non-keratinized) type. The description and illustrations provided by Orr were not typical of adamantinomas in other species (226).

In all, there appear to be only two previously reported genuine cases of naturally occurring osteosarcoma in *Oryctolagus* (274, 325), and in this publication I have included one additional case. These cases all occurred in aged rabbits, over five years old. Those previously reported arose in the body of the mandible and the twelfth and thirteenth ribs, respectively (274, 325). The new case reported here arose as a slowly growing mass that appeared to originate in the proximal tibia, but bridged the joint with the distal femur, causing its ankylosis. No metastases were observed. When these three tumors are considered as a group, it appears that the range of structural variation observed histologically in rabbit osteosarcomas is typical of that observed in other species (226). The range includes an eburnetic, sclerotic tumor of bone only (325), a tumor predominately of bone and calcifying osteoid but with some cartilaginous differentiation (274), and one with both bone and substantial cartilaginous elements (324). Of the 3 tumors, 2 were metastatic to thoracic organs (274, 325). The histological appearance of the metastatic foci were identical to those of the primary growths. It is noteworthy that the variation in bony versus cartilaginous elements and the histopathological architecture of these tumors were seemingly indistinguishable from those reported above as induced by beryllium salts, including the pattern of thoracic metastasis. The beryllium-induced tumor appears to act as an excellent model for those observed naturally. In one case (325) the naturally occurring osteosarcoma was associated with an elevated serum alkaline phosphatase value. The osteosarcomas are illustrated in Figs. 27–31.

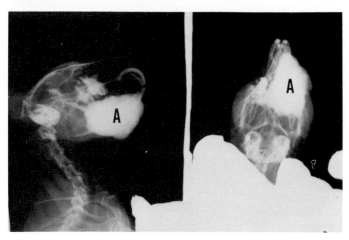

Fig. 27. Radiographs of mandibular osteosarcoma (A) in lateral (left) and ventrodorsal (right) perspective.

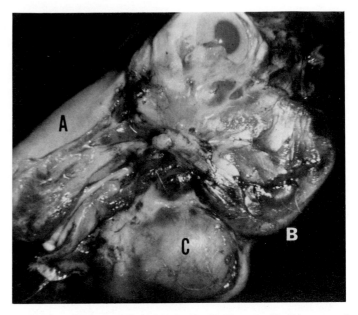

Fig. 29. Skinned skull of rabbit in Fig. 28. Note maxilla (A), ramus of mandible (B), and osteosarcoma (C).

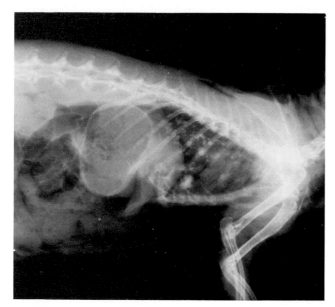

Fig. 28. Radiograph of thoracic "cotton ball" osteosarcoma metastases.

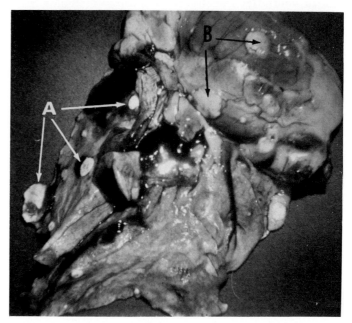

Fig. 30. Thoracic organs with metastatic osteosarcoma. Compare with Fig. 28, noting metastases in lung (A) and pericardium (B).

H. Tumors of Lymphorecticular Tissues

1. Lymphosarcoma

Lymphosarcoma appears to be the second most commonly encountered neoplastic entity in *Oryctolagus*. Most cases seem to have been encountered and studied retrospectively from postmortem material. This has impeded progress in characterization of the lymphoid tumors of *Oryctolagus* because until recently the known cases have occurred sporadically and with unknown histories. All case reports have consensus in establishing the lymphosarcoma as a tumor of juvenile and young adult rabbits, most cases in the age group of 8–18 months. Fox and his co-workers have emphasized that in their cases the hematological picture is uniformly aleukemic, as have been the two cases from the laboratory of this author where diagnosis was established prior to death. The former group has emphasized that clinical laboratory findings include anemia, depressed hematocrit, low hemoglobin (attribut-

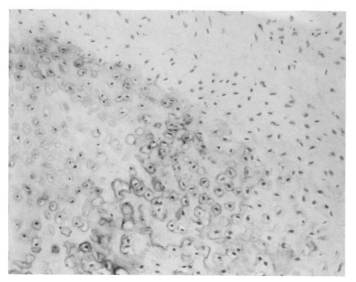

Fig. 31. Photomicrograph of ankylosing osteochondroma of femo-rotibial joint. Note gradual loss of lacunization toward right.

able to infiltrative obliteration of marrow by tumor cells) and, late in the course, elevated blood urea nitrogen (attributable to infiltrative destruction of renal cortex). The 4 cases seen in our laboratory have had remarkably similar lesions to those described in the literature, almost to the point of monotony. From these cases and those described in the literature, we have established a tetrad of gross lesions the presence of which at necropsy we consider pathognomonic for lymphosarcoma in *Oryctolagus*. The tetrad consists of: (a) greatly enlarged kidneys, light tan in color with irregular lumpy surface, and thickened whitish cortex but normal medulla on cut surface; (b) hepatomegaly with diffuse pattern of small (0.5 mm) pale foci; (c) splenomegaly; and (d) lymphadenopathy.

Similarly with the histological picture, although practically any organ of the body may be susceptible to lymphomatous infiltrates, a regular pattern has been observed in the way certain organs are almost routinely involved.

Infiltration of the structures of the eye, particularly the choroid, ciliary body, iris, and anterior chamber has been observed in two of the four cases observed by this author and in one case submitted by Dr. Ronald Flatt. This lesion has not been described by other investigators.

The lymphoid depots of the gastrointestinal tract, including the pharyngeal tonsils, Peyer's patches, and mesenteric lymph nodes, appear to be routinely involved in the sarcomatous process. Particularly in the stomach, raised and circumscribed nodules 0.5–1 cm in diameter may be observed, frequently in a state of ulceration.

In the lung the lymphosarcoma lesion appears to take the form of exaggerated peribronchial lymphoblastic infiltrates, obliterating the normal aggregates of lym-

phoid tissue associated with that location. The interalveolar septa may be infiltrated as well.

Two patterns of hepatic infiltration have been seen in the cases I have observed. Most commonly infiltration appears to proceed by periportal accumulations which extend radially by sinusoidal infiltration. Such livers may be extensively involved with over 50% of hepatic tissues obliterated by solid lymphosarcoma masses. This pattern is the one most frequently described in the literature as well. In one case, however, the infiltration appeared to involve a diffuse pattern not obviously periportal in distribution.

Renal lesions are distinctive and those observed by the author have been typical of those reported in the literature. The cortex appears to be extensively involved by an interstitial infiltration of neoplastic cells. It is worth emphasizing that the glomeruli and cortical tubules are not engulfed and obliterated by the infiltrative process to any significant degree and stand out in relief as widely separated structures. The renal medulla ordinarily appears to be spread.

The ovary and adrenals are organs usually involved in the sarcomatous process, frequently to the point where only scattered remnants of organ-distinctive tissue may persist.

The spleen has been uniformly involved in all the cases I have seen and most of those described in the literature, but was essentially normal in the case reviewed by Van Kampen (318). The lymphosarcomatous spleen appears to enlarge by radial extension of infiltrated white pulp which imparts an intensely multinodular appearance to a stained section. The red pulp frequently demonstrates hematopoietic activity, an apparent reflection of decreased functional ability of infiltrated bone marrow.

Virtually all of the lymph nodes in the body may be involved in the neoplastic process, often bringing to prominence lymph nodes difficult to discern in normal rabbits. The normal architecture of the node usually is entirely obliterated by masses of infiltrating neoplastic lymphoblasts. The lymphoblasts are typically larger than normal lymphocytes, ranging in size from 5 to 35 μm in diameter. The nuclei are large and basophilic. The cytoplasm is scanty. Two or more nucleoli may be observed. Regardless of the organ, mitotic figures are common and indicative of the rapidity with which lymphosarcomas progress in *Oryctolagus*. Lymphosarcoma is illustrated in Figs. 32–41.

Medlar and Sasano (216) described a single case of retroperitoneal sarcomatosis grossly involving the appendix, adrenals, mesentery, pancreas, and left ureter in a solitary large neoplastic mass. Microscopically, the tumor mass itself and focal infiltrates in such organs as spleen, liver, heart, and adrenals appeared to vary in different fields from the typical appearance of Hodgkins disease (in man) to that of lymphosarcoma. The kidneys were not involved, and the description as a whole was not typical of lymphosarcoma in the rabbit. It is possible that the

Fig. 32. Characteristic gross lesions of lymphosarcoma: splenomegaly, hepatomegaly, and renal infiltration.

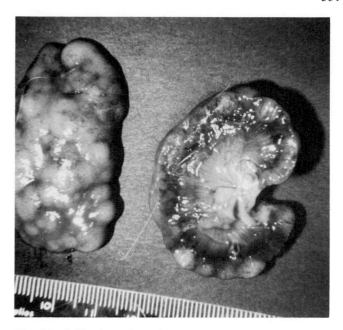

Fig. 34. Infiltration of renal cortex in lymphosarcoma. Note irregular "lumpy" subcapsular contour.

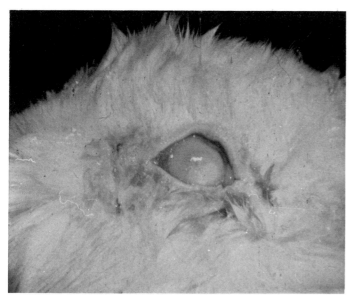

Fig. 33. Infiltration of orbital structures in lymphosarcoma.

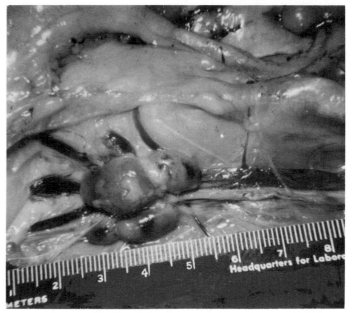

Fig. 35. Infiltration of iliac lymph nodes in lymphosarcoma.

"abdominal sarcomas" of Wallner (321) and of Baumgarten (25) were of this type.

Although most previously reported cases of lymphosarcoma have been sporadic, there is increasing evidence that it does not occur randomly. Loliger (200, 201) reported 10 cases (from a necropsy series of 602) from a single rabbitry; too great an incidence for random distributions. Richard Fox and his co-workers at the Jackson Laboratory have recently described 29 cases of lymphosarcoma occurring in the WH strain of rabbit maintained at the Laboratory (97, 217). The pathological descriptions of these tumors

appear to be typical for lymphosarcomas described by other investigators, as well as those observed here. Analysis of their data revealed a pattern of incidence that led them to the conclusion that an autosomal recessive gene (designated *ls*) confers, in homozygosity, susceptibility to vertically transmitted lymphosarcoma. This must be considered as an extremely important finding because the means are now at hand to mount a meaningful attack on the characterization of yet another mammalian lympho-

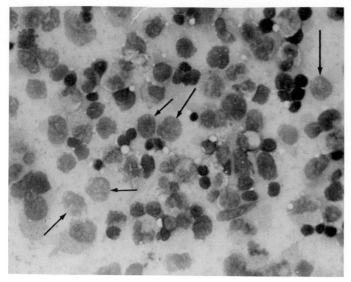

Fig. 36. Smear of bone marrow (femur) in lymphosarcoma. Note frequency of sarcomatous lymphoblasts (arrows).

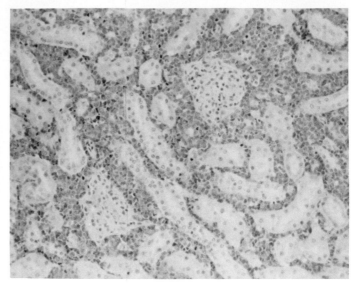

Fig. 38. Infiltration of renal cortex in lymphosarcoma. Note interstitial distribution with sparing of tubules and glomeruli.

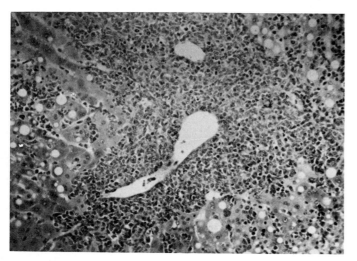

Fig. 37. Periportal infiltration of the liver in lymphosarcoma.

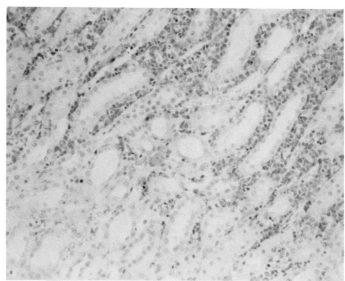

Fig. 39. Renal corticomedullary junction with interstitial infiltration of cortex (upper right) and sparing of medulla (lower left).

sarcoma. Attempts by Fox and his colleagues and others (313a) to demonstrate C-type oncoviruses have so far been unsuccessful.

2. Thymoma

Tumors of the thymus have been observed infrequently in the rabbit on a sporadic and usually postmortem basis. This was the case in those observed by Blanchard *et al.* (32), Drieux and Poisson (77), Orr (240), and Greene and Strauss (121). These cases all occurred in mature rabbits between 1 and 4 years of age. The case described by Orr was metastatic to thoracic organs and abdominal lymph nodes. It was classified (by Orr) as a neoplasm of thymic reticulum cells. Complicating the diagnosis of thymic sarcoma in rabbits is my observation that hyperplasia of the

thymus in adult rabbits is not uncommon; and enlargement of the organ up to 3–4 times normal size without acquisition of neoplastic characters can lead to a gross diagnosis of tumor that may not be supported by careful microscopic observation (see Figs. 42 and 43). Three cases of thymic hyperplasia were observed in our necropsy series of approximately 600. Richard Fox and his colleagues have recently described 3 cases of thymoma in association with systemic immunopathy (98, 217). The disease occurred in the X strain and was characterized in homozygotes (for a newly designated gene, *ha*) by hemolytic anemia. Disturbances in the peripheral blood included polychromatophilia, anisocytosis, decreased numbers of erythrocytes, in-

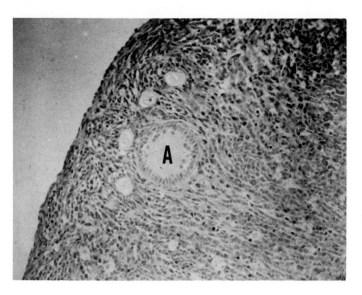

Fig. 40. Infiltration of ovary in lymphosarcoma with obliteration of most structures. Note follicle (A).

creased numbers of lymphocytes, and the presence of numerous nucleated erythrocytes. The erythrocytes of *haha* individuals were Coomb's test positive. Lesions and signs included petechiation of most tissues, darkened feces, icterus, hemoglobinemia, pale liver, hemorrhagic spleen and kidney, hemoglobinuria, and edema of most tissues. The age at death in homozygotes ranged from 1.5 to 12 months with an average of 4–5 months. Thymoma occurred only in protracted cases. The authors submitted evidence to show that the gene designated *ls* in strain WH rabbits (associated with lymphosarcoma) and the gene designated *ha* in strain X rabbits (associated with hemolytic anemia and thymoma) may in fact be the same gene (at the same locus) expressing itself differently in the two substrate strains. Supporting evidence included not only common ancestry for the 2 strains, but also the fact that blood cells from *lsls* individuals were Coomb's test positive and 21% of all individuals in both strains were Coomb's test positive. The data presented were compatible with a concept of gene-

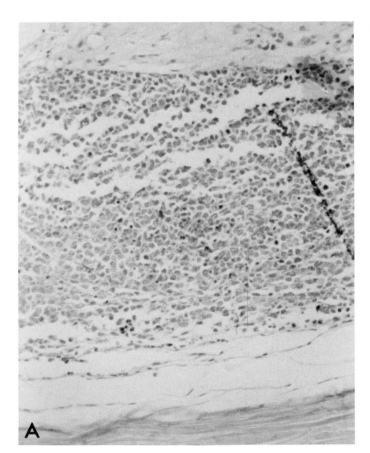

Fig. 41. Composite photomicrograph at the same magnification with normal retina (B) and infiltration of the retina in lymphosarcoma (A).

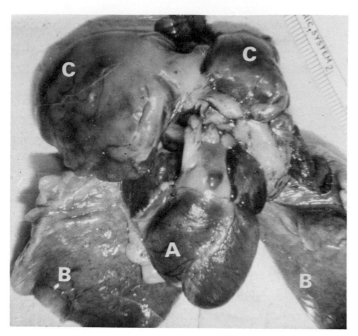

Fig. 42. Thymic hyperplasia. Note heart (A), lungs (B), and thymic lobes (C).

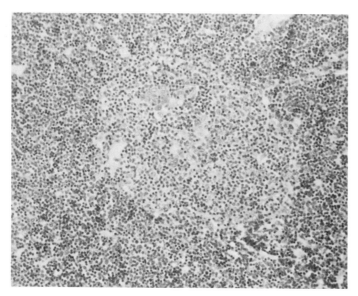

Fig. 43. Photomicrograph of hyperplastic thymus in Fig. 42. Note normal appearance of gland and thymic nodule.

tic susceptibility to vertically transmitted oncogenic virus analogous to the pattern now known for the murine C-type RNA genome. The authors suggest that the presence of this viral genome in the WH strain is expressed as lymphosarcoma, while in the X strain it is manifested as an NZB-like immunopathy culminating in thymoma when sublethal and chronic. This is an extremely important finding because the variables of age, sex, and genetic background can be purposefully manipulated in order to gain insight into the mechanism of this disease. Although key kidney lesions

were lacking in *haha* rabbits, the authors believed the condition to be an excellent model for systemic lupus erythematosus and to have most of the pathological features of the canine and murine models. Enhancing the value of the rabbit model is the genetic association with lymphosarcoma.

3. Plasma Cell Myeloma

A single case of plasma cell myeloma has been described from the rabbit (245) where it occurred as a large solitary mass lying on the left ventral side of the brain, extending from the medullopontine groove to the pyramidal decussation and projecting laterally to the edge of the cerebellum. It is not known if the tumor had the functional characteristics (gammopathy) of plasma cells myelomas in other species because diagnosis was in this case established retrospectively on solely histopathological criteria. The tumor as presented was not typical for plasma cell myelomas of other species (226).

4. Spleen

Ball (19) described a single case of tumor of the spleen that appeared to have the characteristics of a fetal rest of intestinal epithelium complete with goblet cells. The tumor was small (5 mm), well organized, and benign. The author did not believe the tumor tissue originated in the spleen but could not find a primary site elsewhere. It should be clear that this was a tumor in, but not a tumor of, the spleen and should be classified as an entodermal rest or hamartoma.

I. Tumors of the Endocrine Glands

Tumors of the endocrine glands have been rarely reported in *Oryctolagus*. In the pituitary gland one teratoma has been described (211) as a space-occupying tumor not associated with symptomatology. The tumor was approximately 1 cm in diameter, was located at the pituitary infundibulum, and had infiltrated the brain. Histologically the tumor was cystic and contained elements of mucus-secreting columnar epithelium, hairs, smooth muscle cells, connective tissue, and glands resembling those found in gastric epithelium and cartilage. Polson (252) reported an adenoma of the anterior pituitary. The tumor was small (2 mm), clearly demarcated from the normal tissue, and not associated with functional symptomatology. Histologically it was classified as an adenoma of pituitary eosinophilic cells.

Carcinoma of the adrenal cortex has been reported twice in the literature. The case described by Boycott and Pembrey occurred in a feral *Oryctolagus* rabbit, was bilateral, and appeared to involve the anterior pole of each kidney and the corresponding adrenal cortex (35). The normal tissue in the case of each adrenal was separated from the tumor by a well-developed capsule. The cell type of the tumor was described by the authors as "round cell." It was not characteristic of adrenal cortical tumors described later from the rabbit or from other species (226). Carcinoma of the adrenal cortex (also occurring bilaterally) was described as an incidental finding (not associated with symptomatology) in a male rabbit of unknown age by Hueper and Ichniowski (144). These tumors were located primarily in the adrenal glands which were in fact largely replaced by tumor tissue consisting of sheets of very large foamy cells with small oval nuclei. The tumors had metastasized radially to the subserosa of the duodenum and periadrenal fat and connective tissue. Cholesterol crystal clefts were scattered throughout the tumor tissue. This carcinoma was typical of adrenal cortical tumors reported in other species (226). The authors emphasize that the tumor tissue was not separated from the normal adrenal cortex by a capsule. This observation is perhaps worth emphasizing because I have observed that nodular hyperplasia (Fig. 44) of the adrenal cortex is a very common lesion in *Oryctolagus*, but in all cases the nodules appear to be well demarcated from the remainder of the organ by a capsule.

J. Melanomas

Malignant melanomas have been recorded at least 3 times in the literature. The case described by Brown and Pearce arose in the eye of a 10-month-old male rabbit that had been infected with *T. pallidum* some two months previously (44). The melanoma had infiltrated structures of the eye including the choroid, iris, and ciliary body.

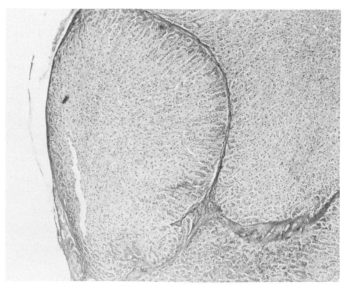

Fig. 44. Photomicrograph of nodular hyperplasia of the adrenal.

The authors believe the treponemal infection to have been of significance in predetermining the neoplasm. The cases described by Sustmann (304) and Holz and Heutgens (143) were more typical of malignant melanomas in other species. The case of Sustmann occurred in a 2.5-year-old male rabbit, while that of Holz and Heutgens occurred in a rabbit one year of age. The latter case is of interest because typical black tumor nodules were present in most abdominal and thoracic organs and the proximate cause of symptomatology was paralysis related to neoplastic infiltration of the vertebral column.

III. NEOPLASMS OF *SYLVILAGUS* SP.

Although neoplasms induced by oncogenic DNA viruses have been frequently reported as spontaneous findings in in *Sylvilagus* rabbits, reports describing other tumors in cottontail rabbits have appeared only rarely. Tumors of the former type (Shope fibromas and Shope papillomas and their malignant derivatives) are described in further detail below and in Chapter 10.

It is generally believed that neoplasms of all types have a very low incidence in wildlife. Numerous factors contribute to such an observation, but some we may cite include the facts that the rigors of life in the wild are such that rabbits rarely survive into old age; that most animals that die naturally are not available for prosection; that ill and tumor-bearing rabbits are rarely encountered by hunters; and also that *Sylvilagus* rabbits have not been successfully domesticated so as to enable the longitudinal lifetime observation of colonized cottontails. Accordingly, the infrequent reports of neoplasms (of nonviral origin) in *Sylvilagus* may be regarded less as indicative of a low tumor rate, but more as an almost total state of ignorance with regard to tumor epidemiology in wild cottontail populations and the infrequent survival of such populations into tumor-prone age groups. Several reports describing nonviral tumors in *Sylvilagus* rabbits have been summarized in Table IV. The reader will note that the bulk of these tumors (8/9) are lymphosarcomas and an embryonal nephroma; tumors associated with adolescence in *Oryctolagus* rabbits.

The published description of the embryonal nephroma by Lopushinsky and Fay (206) was not typical of the same tumor type in *Oryctolagus* rabbits. Although it had not metastasized, it was an extremely large tumor (55 gm) that had almost completely replaced all normal renal structures in the affected kidney. The descriptions of lymphosarcomas by Lopushinsky and Fay, however, appeared to be similar to those of *Oryctolagus*, involving diffuse infiltrative destruction of the liver, spleen, and stomach. The extent of renal involvement was not reported. The lymphosarcoma reported by Bell and Chalgren was more

TABLE IV
NATURALLY OCCURRING TUMORS OF *Sylvilagus* sp.:
REPORTS TO THE LITERATURE

Tumor site and type	No. of cases	Ref.
A. Urinary tract		
1. Kidney		
a. Embryonal nephroma	1	Lopushinsky and Fay 1967 (206)
B. Lymphoreticular system		
1. Lymph tissue		
a. Generalized lympho-sarcoma	6	Lopushinsky and Fay 1967 (206)
b. Hodgkins-like lymphosarcoma	1	Bell and Chalgren 1943 (29)
C. Connective tissue		
1. Subcutaneous tissues		
a. Lipoma	1	Lopushinsky and Fay 1967 (206)

of a Hodgkins type with sarcomatous hyperplasia of cervical, mesenteric, axillary, and cervical lymph nodes, with sparing of the liver, spleen, and kidney (29). Histological descriptions and illustrations were not provided.

IV. NEOPLASMS OF *LEPUS* SP.

The comments made previously with regard to the difficulty of case finding and assessing the true incidence of neoplasms in *Sylvilagus* rabbits apply equally to *Lepus* rabbits; most observations are obtained in a casual random way from wild populations. Although an insufficient number of cases have been described and reported to enable a confident profile of tumors in this genus to be outlined, preliminary evidence (Table V) indicates the tumor distribution in *Lepus* to be similar to that of *Oryctolagus*. One important exception should be noted, however, and that is the finding of myeloid leukemia in six separate accessions of *Lepus* rabbits reported from France and Switzerland (see Table V). The gross pattern of splenomegally, hepatomegally, and generalized lymphadenopathy was typical for all six cases. Although none were observed prior to death the infiltration of most major organs and tissues with neoplastic cells typical of the myeloid series suggests a leukemic state.

Another exception that should be mentioned includes the ovarian teratomas found in Europe (104) and Australia (96) by investigators surveying wild *Lepus* populations. This tumor has not been observed in *Oryctolagus* (or *Sylvilagus*) rabbits and is rare in most rodent populations. Flux has suggested that the high incidence of ovarian cysts and teratomas in wild hares may be attributed to the proximity of these animals to agricultural insecticides, e.g., DDT (96).

TABLE V
NATURALLY OCCURRING TUMORS OF *Lepus* sp.:
REPORTS TO THE LITERATURE

Tumor site and type	No. of cases	Ref.
A. Female reproductive tract		
1. Uterus		
a. Fibroma	12	Flux 1965 (96)
2. Vaginal squamo–columnar junction		
a. Squamous cell carcinoma	1	Cheatum and Bunting 1951 (57)
3. Ovary		
a. Teratoma	1	Goyon 1959 (104)
	6	Flux 1965 (96)
B. Urinary tract		
1. Kidney		
a. Renal carcinoma	1	Cheatum and Bunting 1951 (57)
C. Lymphoreticular system		
1. Lymph tissue		
a. Generalized lympho-sarcoma	1	Piening and Wermerssen 1965 (251)
b. Hodgkins-like lymphosarcoma	1	Salomon 1933 (275)
	1	Nicod and Burgisser 1961, 1964 (231, 232)
c. Myeloid leucosis	5	Burgisser 1957 (45)
	1	Guillon *et al.* 1963 (125)
D. Mammary gland		
1. Adenocarcinoma	1	Flux 1965 (96)
E. Connective tissue		
1. Subcutaneous tissue		
a. Fibroma	1	Flux 1965 (96)
F. Liver		
1. Hemangioma	4	Haberman, cited by Cheatum and Bunting 1951 (57)

V. NEOPLASMS ASSOCIATED WITH ONCOGENIC VIRUSES

Proliferative and truly neoplastic changes associated with certain pox and papova group viruses in rabbits have played an important role in furthering the understanding of viral oncogenesis as a basic biological process. The natural history, host range, epidemiology, virology, pathology, and immunology of these viruses has been reviewed in Chapter 10. Additional reviews with important bibliographical resources include the monographs by Fenner (90), Gross (123), Moulton (226), and the texts edited by Dalton and Haguenau (71) and Stewart *et al.* (301). All of these viruses (with the exceptions of hare fibroma and oral papilloma viruses), namely, myxomatosis, Shope fibroma viruses, Shope papilloma virus, the herpesvirus lymphoma of Hinze, and the Dos Santos hepatoma, have been discovered and limited in natural distribution to the American continents.

These viruses and the lesions they induce have been among the most intensively investigated agents in the entire field of viral oncogenesis. They are of great historical interest because early concepts of viral oncogenesis in mammalian species were formed on the basis of experimentation with the Shope fibroma, Shope papilloma, and the Shope papilloma-derived carcimomas principally. They remain as important models of viral tumorigenesis and the concept of coevolution between host and virus was generated largely from investigations employing these viruses.

A. Rabbit Oral Papilloma

The rabbit oral papilloma virus (ROPV) is the only virus of the group having the domestic rabbit (*Oryctolagus cuniculus*) as the natural host. The virus causes warty growths (papillomas) on the ventral aspect of the tongue anterior to the frenulum, principally but occasionally also on the near lingual epithelium of the oral cavity and gingiva. The lesions were originally described in the 1930's by Parsons and Kidd from domestic rabbits obtained in the vicinity of New York City (178, 243, 244). Interest in the virus lay dormant until it was rediscovered in the Midwest by Rdzok, Shipkowitz, and Richter in the mid-1960's (254, 259), and later by Weisbroth and Scher in domestic rabbits from Long Island, New York, and Massachusetts (326). The virus exhibits an unusual degree of fastidiousness in the choice of tissue and host being found naturally only in the oral epithelium of *Oryctolagus* rabbits. ROPV can be transferred by cell free filtrates to oral epithelium of *Oryctolagus* and *Sylvilagus* rabbits (244, 326). Hamster cheek pouch epithelium and chicken and duck egg chorioallantoic membrane (CAM) are refractory to ROPV as is also the case with epithelium of conjunctiva, nares, vagina, and rectum of *Oryctolagus* rabbits (324a). The virus is transmitted in oral secretions containing sloughed epithelial cells from the warts. These virus-containing discharges are ingested by susceptible rabbits, especially those still in the nest box, and infection occurs in the abraded epithelium of the tongue. The role of abrasion is underscored as important because of our observation that rabbits with maloccluded teeth commonly have greater numbers of warts.

Following a latent period of 2–4 weeks (in experimental infections and longer under natural conditions), the virus induces the formation of structurally typical nonpigmented papillomas. They grow slowly over a period of 3–9 months maturing from the initial broadly based domelike shape to pedunculated and roughly folded cauliflowerlike masses of 4–6 mm when mature. Rejection of the papilloma occurs when the rabbit becomes sufficiently immune and proceeds by a chronic inflammatory attack at the base of the papil-

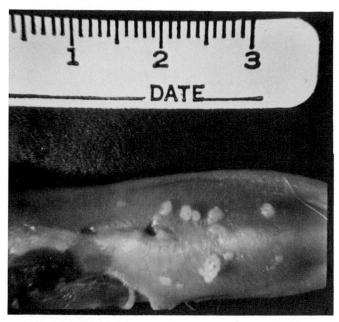

Fig. 45. Oral papillomatosis. Young, mostly sessile papillomas on ventrum of tongue.

loma with cell types characteristic of delayed hypersensitivity reactions. Oral papillomas of the rabbit are not known to undergo the carcinomatous transformation characteristic of Shope papillomas, although this sequence does occur in oral papillomas of the dog (322).

Oral papillomas in the gross are illustrated in Figs. 45–47 and microscopically in Figs. 48–50. Excellent illustrations of ROPV ultrastructure can be found in the publications of

Fig. 46. Oral papillomatosis. Mature califlowerlike pedunculated papillomas on ventrum of tongue near frenulum and on lingual epithelium at apex of mandible.

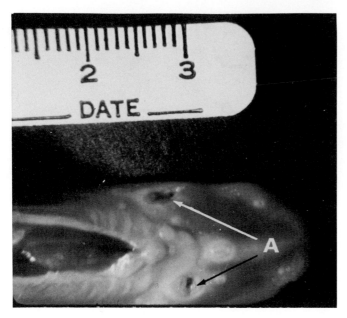

Fig. 47. Oral papillomatosis. Older regressing papillomas. Note ulcers (A) where rejected papillomas have sloughed.

Fig. 49. Photomicrograph of ulcer following rejection of papilloma. Note normal epithelium (A) and chronic inflammatory cells at base of papilloma site.

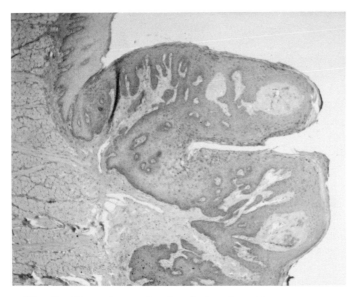

Fig. 48. Photomicrograph of typical oral papilloma. Note sharp transition to normal squamous epithelium at upper left.

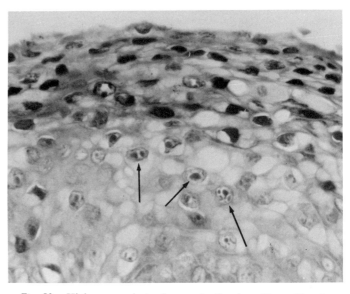

Fig. 50. High power photomicrograph of oral papilloma. Note basophilic intranuclear inclusion bodies (arrows) in cells approximately 6–8 rows down from surface.

Rdzok *et al.* (254) and Richter *et al.* (259). The architecture of rabbit oral papillomas is typical for that of papillomas in other species. Basophilic intranuclear inclusion bodies (demonstrable with hematoxylin and eosin or Giemsa stains) are found in epithelial cells below the cornfied strata (Fig. 50). The virus is not known to be immunologically related to Shope papilloma or fibroma viruses or to antigens of Brown-Pearce carcinoma (326).

B. Shope Papilloma and Papilloma-Derived Carcinomas

The Shope papilloma is induced by infection with the Shope papilloma virus (SPV). The natural infection was originally limited to *Sylvilagus floridanus* rabbits in states bordering the Mississippi River, Texas, and Oklahoma. Dr. Richard Shope who first characterized papillomatosis of cottontail rabbits did so with naturally infected rabbits from Iowa (291, 293, 294). The condition is not natural to eastern cottontails (*S. floridanus*) but is now known to be endemic in California cottontails, probably *S. bachmani*

(126). The disease may occasionally occur naturally in domestic (*Oryctolagus*) rabbits (126). The natural disease is characterized by the formation of cutaneous pigmented papillomas 0.5–1.0 cm wide at the base and which project 0.5–1.0 cm or higher above the skin line. Spontaneously infected cottontails may carry from 1 to 10 warts with some predilection for the skin of the inner thighs, ventral abdomen, neck, and shoulders. Infections in domestic rabbits were reported to occur on the ears and eyelids (126). Experimental infections will induce papillomas in jackrabbits, snowshoe rabbits, and hares (27, 178).

The papilloma can be experimentally transmitted with suspensions of papilloma cells or by cell-free extracts of them which are scarified onto the skin of susceptible rabbits. Following a latent period of approximately 10–12 days papillomas begin to develop visibly. Pretreatment of the skin (at the same or other sites) with tar, mixtures of turpentine and acetone, or 20-methylcholanthrene is known to enhance susceptibility of the skin to SPV (99, 266–269). The papillomas persist for varying periods of time depending on host factors. In *Sylvilagus* rabbits spontaneous regression occurs in approximately 36% of cases naturally acquired, usually by 12 months after infection (336). According to Syverton (305–308), malignant transformation of the papilloma to squamous cell carcinoma occurs in at least 25% of those cases acquired naturally and observed over 12 months from infection.

The virus is recoverable from naturally acquired or experimental papillomas of *Sylvilagus* rabbits up till the time they undergo regression or malignant transformation. Generally less virus is recoverable from experimental infections of *Sylvilagus* rabbits than from those naturally acquired. *Oryctolagus* rabbits are susceptible to induction of papillomas; however, infective SPV is generally absent or present in too low a titer for extracts of them to be infectious for either *Sylvilagus* or *Oryctolagus* rabbits. That the SPV is present in a noninfectious (masked or latent) form in *Oryctolagus* papillomas can be demonstrated, however, by the induction of SPV-specific neutralizing and complement-fixing antibodies in animals immunized with extracts of papillomas from domestic rabbits and was the key observation leading Shope to postulate viral masking (168–171, 173, 292). In a similar pattern, infectious SPV is generally absent from the carcinomas that develop from persistent papillomas in either naturally or experimentally infected rabbits of either species (305). This has led to the concept of masked or "immature" noninfective (but antigenic) SPV, localized with the use of fluorescent antibody technique and electronmicroscopy, respectively, by Noyes (235, 236) and Stone (303) as granular material in the nucleoli of cells in the basal germinal layers of the papilloma. Mature, spherical, viral particles with infective and antigenic properties are found in the upper layers of the papilloma in the nucleus of cells beginning to undergo keratinization. Inclusion bodies visible with the light microscope and typical of other wart viruses have not been observed (264, 291).

The Shope papilloma appears histologically similar in both naturally acquired and experimental infections (301) and is typical of papillomas in other species as well. Good descriptions of the histopathology may be found in original paper by Shope and Hurst (291) and also in the monographs by Moulton (226) and Stewart *et al.* (301). The papilloma consists of elongated retelike pegs of squamous epithelium surrounding central cores of connective tissue. The exterior surface is keratinized and often builds up to form cutaneous horns. The papilloma is sharply delineated where it meets the normal squamous epithelium. The well-defined basal layer is formed of enlarged cells often in palisade arrangement. A mild inflammatory infiltrate is ordinarily found in the dermal layers underlying the papilloma. The epithelium is thickened in the area of the papilloma and the granular layer merges gradually into the cornified layers which are often incompletely keratinized (parakeratosis). Mitotic figures are common. Melanin pigment granules are commonly found in tumor cells of pigmented rabbits but are absent in albino rabbits (26). The Shope papilloma is illustrated in Figs. 51 and 52.

The malignant potential of the seemingly benign cutaneous papillomas was recognized early by Rous and Beard (263) who found that papillomatous fragments implanted to muscle or certain internal organs would develop into squamous cells carcinomas and acquire in-

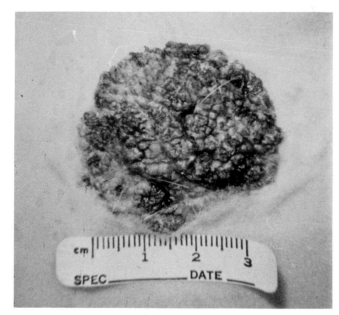

Fig. 51. Experimental Shope papilloma of approximately 3 months duration.

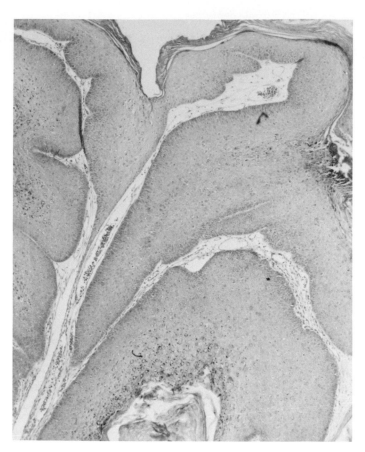

Fig. 52. Photomicrograph of Shope papilloma. Note hyperkeratinization of surface at top of photograph. Same papilloma as Fig. 51.

vasive properties. These same authors (262, 265) observed spontaneous change from papilloma to squamous cell carcinoma in domestic rabbits with tumors maintained 200 days or longer after infection. It was thought originally that this change occurred only in *Oryctolagus* rabbits, but the observations of Syverton and his coworkers (305–308) demonstrated that carcinomatous change occurred in 25% of either naturally acquired or experimental papillomas in *Sylvilagus* rabbits maintained over 200 days; and it is now recognized that this change occurs in 75% of Shope papillomas in *Oryctolagus* rabbits persisting beyond 6 months.

Infectious SPV was absent in the carcinomas derived from either *Sylvilagus* or *Oryctolagus* rabbits although incomplete or masked virus could be demonstrated by the antibody response to carcinoma (173, 177). In this regard the development of SPV appeared to be similar to that within papillomas of *Oryctolagus* rabbits.

Masking or incomplete and noninfectious SPV formation is not the final stage of the papilloma-to-carcinoma sequence, however. The $V \times 2$ (V2, $V-2$, or V_2) carcinoma is an illustration of this principle. It originated in the carcinomatous transformation of Shope papillomas carried for

9–11 months by a Dutch Belted male rabbit (179). The axillary lymph node of this rabbit contained metastatic carcinoma and it was minced to provide transplantation fragments. The success of transplantation averaged only about 5%, even in genetically related Dutch Belted rabbits until after the fifth transplant generation. The tumor, described originally as a squamous cell carcinoma became increasingly anaplastic in the course of serial transfer, but at least through the twenty-second transplant generation (1940), it unfailingly induced neutralizing antibodies against SPV in homografted rabbits. Serological testing in host rabbits was abandoned during World War II, and when it was resumed (subsequent to the forty-sixth transplant generation) it was found that antibody to SPV was no longer present in the sera of rabbits bearing the carcinoma (270). Ginder and Friedewald further clarified this observation, since repeated, by their report that although tumor-specific antibodies not reacting with normal host tissues were present in such sera, SPV-specific antibodies were not (103). More recently, Ito and his colleagues (and others as well) have shown that although phenolic nucleic acid extracts of papillomas and other papilloma-derived carcinomas (e.g., $V \times 7$) retain ability to induce papillomas in susceptible hosts, such extracts of $V \times 2$ carcinomas do not (55, 139, 151–154).

The $V \times 2$ tumor was in its 203rd transplant generation in 1970. Metastasis to lung and lymph node is common but rare to other sites. The apparent barrier to unrestricted metastasis has been the subject of several investigations (83, 92, 331). Greene (114) found the host range of transplantability to include brain of guinea pig, mouse, rat, and hamster, anterior chamber of guinea pig eye and subcutaneous tissues and testis of hamster and mouse.

The $V \times 2$ tumor invades surrounding skeletal muscle and areolar connective tissue both in the primary site and from its lymph node metastases. It is not ordinarily encapsulated but does induce chronic inflammation in invaded tissues. It is organized histologically into sheets and nodules surrounded by reticular fibers. The tumor is separated by thin cords of stroma with capillaries and inflammatory cells. There may be some necrosis but little or no hemorrhage. Nuclei are hyperchromatic, uniform in size and shape, and giant cells are infrequent. The tumor is known to secrete a parathormone-like substance that depresses the renal tubular reabsorbtion of phosphorous and causes hypophosphatemia, hypercalcemia, and dystrophic calcification of the tumor and soft tissues (101, 295, 319, 327–329). The tumor is illustrated in Fig. 53, and the excellent photomicrographs in the monograph by Stewart *et al.* can be used as reference (301). Other carcinomas (e.g., $V \times 7$) have been developed in like manner from Shope papillomas by the same and other laboratories. Some have used newborn rabbits to develop strains of greater malignancy (91, 298).

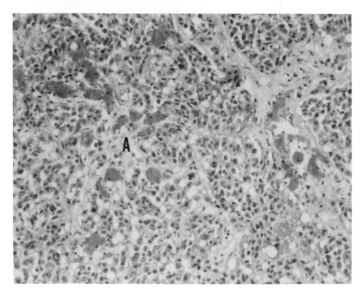

Fig. 53. Photomicrograph of Shope papilloma-derived Vx2 carcinoma. Note fractured muscle bundles (A) and sheetlike architecture.

C. Shope Fibroma

Few host-virus antagonisms have been characterized within the context of coevolution to the extent that they have between the various species of rabbits on the one hand and the poxviruses of the myxomafibroma subgroup on the other. Myxomatosis is reviewed in detail in Chapter 10 because of its taxonomically closer position in terms of clinicopatholigical consequences to the classical cytopathogenic poxviruses. It is impossible, however, to discuss the rabbit fibroma viruses without also including the myxomatosis viruses because they are related genetically and immunologically. This relationship has been well reviewed in the monographs by Fenner and Ratcliffe (90), Gross (123), and also the chapter by Febvre (86).

Fibroma-like proliferative lesions of the skin are found, under natural conditions in the wild, enzootically distributed among the three principal species of *Sylvilagus* rabbits on the American continents. The histology of these lesions in the natural host is essentially identical and the discussion below is based mainly on the morphology of the Shope fibroma because it has been studied in greatest detail. The biological characteristics of these viruses (causing fibromas in the natural host) is known to be different, however, when injected into the domestic (European) *Orytolagus* rabbit. Thus the fibroma of the South American tropical forest rabbit, or tapeti (*S. brasiliensis*), when inoculated into *Oryctolagus* rabbits causes classic myxomatosis. The Brazilian strain of myxomatosis virus responsible for these effects in domestic rabbits was recognized presumptively as a virus by Sanarelli (276) and confirmed some 15 years later by Moses (225); almost 50 years before

it was realized that the enzootic reservoir for the virus was the fibroma of the tapeti (15). Aragao was the first to realize the role of mosquito vectors in transmitting the disease from wild *Sylvilagus* rabbits to the domestic rabbit (14). In a similar pattern it is now recognized that the Californian strain of myxomatosis is carried enzootically as a fibroma-like disease of the brush rabbit, *Sylvilagus bachmani* (213, 215). Outbreaks of myxomatosis have occurred in California since 1930 (167) and this is now understood to occur when mosquitoes fed on brush rabbit fibromas transmit the disease to colonies of domestic rabbits (122, 214). The Brazilian strain of myxomatosis is infectious for at least 3 species of North American cottontails (255). The California strain, however, is sufficiently virulent only for *S. bachmani* (of the 6 native Californian leporids) to act as a reservoir host (256).

The fibroma of the eastern cottontail rabbit, *Sylvilagus floridanus*, is known by observation and serological survey to be enzootically distributed in wild rabbit populations of the eastern and midwestern states (132, 206, 334). This fibroma is not known to cause a rapidly fatal myxomatosis-like disease when injected into *Oryctolagus* rabbits, but determines instead a fibroma with many of the characteristics common to that of the natural host. Richard Shope was the first to investigate the nature of the rabbit fibromas that now bear his name. They were found originally on the feet of a cottontail rabbit shot in the vicinity of Princeton, New Jersey, when the Rockefeller Institute (in part) was located there (189). Transmission experiments with cell free extracts revealed that the lesion, with histological characteristics of a fibroma, was infectious, probably of viral origin, and would cause at the point of inoculation fibromatous tumors in both *Oryctolagus* and *Sylvilagus* rabbits similar to those found naturally in wild *S. floridanus* populations in eastern states. He recognized early (290) that the cottontail fibroma virus was immunologically related to the myxomatosis virus of Sanarelli.

It has been reliably established that a variety of hematophagous arthropods including fleas and mosquitoes act as transmitting agents for the disease under natural conditions (14, 70, 122, 148, 183, 185, 214). Shope in personal communications to Ludwik Gross (123) and Fenner and Ratcliffe (90) reported that *Oryctolagus* rabbits penned on the ground (or on platforms) in a fibroma-enzootic area in southern New Jersey also developed typical fibromas when exposed (presumably) to arthropod vectors. It has been shown that mosquitoe vectors remain infective up to 5 or 6 weeks after feeding on virus-bearing tumors and that the virus is localized to the head parts of the vector (185). It remains uncertain whether the vector is simply a "flying pin" mechanically transferring virus from one host to another or if the virus undergoes multiplication in the vector (86). Skin-piercing vectorism of some type is apparently necessary since infected rabbits can be maintained in the laboratory without risk of contagion to nearby cages (86). Acting on a suggestion by Shope, it has been shown recently that nematodes appear under experimental conditions incapable of transmitting fibroma virus (257, 258).

Shope fibromatosis is limited in natural distribution to *Sylvilagus* and *Oryctolagus* rabbits. Guinea pigs, mice, rats, chickens, and a variety of wild animals (cohabiting the natural distribution of *S. floridarus* are known to be resistant (132, 290). Experimental host systems are known to include the CAM of embryonating chicken and duck eggs (141, 142, 296) and inclusion body-producing tissue culture cells of several types (10, 56, 85, 180–182). The fibroma virus-infected tissue cultures of Hinze and Walker were shown to induce tumors following their inoculation into hamster cheek pouch (138). The brain of intracerebrally inoculated suckling mice will also support fibromatosis (69).

Newborn rabbits of both *Sylvilagus* and *Orytolagus* species are considerably more susceptible than those older than 2–3 weeks of age (80, 158, 334). This effect appears to be related to the weak ability of newborn rabbits to immunologically react to the presence of virus before the twelfth to thirteenth day after birth (80), whereas in the adult, immunity to reinfection begins 1–2 days after inoculation of the virus (86). Duran-Reynals observed with the use of neonatal rabbits that one of several consequences [including: (a) generalized fatal inflammatory disease; (b) local or generalized fibromatosis; or (c) generalized sarcomatosis] could supervene depending on the dose of virus and age of host (79). These results have since been confirmed (128, 186) in part and form the basis of a sensitive assay and virus titration (127).

Generalized metastatic fibromatosis or sarcomatosis may be induced in experimental hosts by factors that modify the immunological response. These factors are known to include mixture of the virus inoculum with hyaluronidase (127, 246), mucin (62), cortisone (128), X-irradiation (60, 61, 68), azoproteins (59), and immunosuppressants (5). Several modifications including infection with *Herpesvirus cuniculi* (7) or Semliki Forest virus (102) and treatment with heparin (133) are known to enhance tumor rejection or depress tumor development.

The concurrent treatment of experimental *Oryctolagus* hosts with carcinogens, most notably tar, along with fibroma virus injections has been found to enhance the expression of fibromatosis. In a series of experiments, tar was injected subcutaneously or intramuscularly either prior to inoculation with Shope fibroma virus or at the same time (4). If the virus was injected intradermally, tumors developed which resembled those of cottontail rabbits and rejection was delayed until 2–6 months. Subcutaneous injections of virus resulted in the production of enormous fibrosarcomas that were locally invasive. Intravenous inoculation of virus induced generalized fibromatosis as a counterpart to progressively enlarging fibromas at the site of tar injec-

tion. Somewhat similar results (to tar) were obtained with 3,4-benzopyrene and methylcholanthrene.

The clinical picture in the natural host, *S. floridanus*, is somewhat different than that obtaining in *Oryctolagus* rabbits; the course is prolonged by comparison and the lesion histopathology slightly different. Inoculation of Shope fibroma virus into the skin of cottontail rabbits causes the development of a proliferant fibroma with the first detectable skin thickening observed on the seventh to eighth day after inoculation (184). The fibromas reach their maximal size (of about 4–6 cm) in about 5 weeks. They then remain static for a variable period of time, but thereafter undergo necrotic involution and slough. The tumors usually persist for a period of 4–5 months (184), but may occasionally remain on the host up to one year before rejection. In this regard the pattern of development is compressed in experimental *Oryctolagus* infection; the tumors often reaching full development within 8–12 days and usually not persisting beyond 30 days (86, 163).

The histology of the lesions has been well described (3, 146, 184, 289). Lesion development begins from dermal centers composed of capillary elements from which endothelial and adventitial cells migrate out and in concert with undifferentiated mesenchymal cells form the proliferation that develops into the fibroma (86). The fully developed fibroma (at 10 days in *Oryctolagus* and 35 days in *Sylvilagus*) is composed largely of spindle- or rounder polygonal-shaped fibroblasts with abundant cytoplasm. Inclusion bodies in the cytoplasm of the fibroma cells are a constant feature of *Sylvilagus* tumors, but variable in *Oryctolagus* tumors (86, 158). They are apparent as coccoid-sized particles in the cytoplasm of fibroma cells after the first week in cottontail fibromas and gradually increase in size and coalesce to occupy much of the cytoplasmic volume as the fibroma develops (184). The inclusions are best demonstrated by the Giemsa stain, Mann technique, or PAS stain (86, 184). The inclusions may appear eosinophilic or basophilic depending on the stain (86). Involvement of the epidermis overlying the fibroma depends to some degree on host factors. Epidermal changes develop to a greater extent in the natural host, *Sylvilagus* (184). In *Oryctolagus* fibromas the epidermis typically becomes thickened and edematous (or necrotic); in *Sylvilagus* fibromas the epidermis begins to elongate and project downward rete pegs of epidermis into the dermal fibroma after the third week of development. The epidermal involvement reaches maximal development by the fifth week and the cytoplasm of these epidermal cells as well contains eosinophilic inclusion bodies (184). The histological picture in *Sylvilagus* rabbits has been likened to that of human molluscum contagiosum lesions (86, 184). An interesting recent investigation, apparently the first to report naturally occurring Shope fibromatosis in *Oryctolagus* rabbits (158), described lesions intermediate in character between myxomatosis and classic

Shope fibromas, underscoring again the spectrum of viruses in this group.

An inflammatory infiltrate directed toward the base of the fibroma is a constant feature in both *Sylvilagus* and *Oryctolagus* hosts. The infiltrate is composed largely of lymphocytes, plasma cells, macrophages, and neutrophils. It is known that the degree of inflammation is in part determined by genetic characteristics of the virus. The original strain of the virus isolated by Shope is called the OA strain. A strain he isolated later (297) determining larger fibromas in *Oryctolagus* rabbits with lesser inflammation has been termed the "Boerlage" strain. A sample of OA strain fibroma virus sent to C. H. Andrewes in England was found after transport and storage in glycerine to determine fibromas in *Oryctolagus* smaller than those typical for OA virus and with a more intense inflammatory reaction (6). This relatively stable change in the virus, now termed the IA strain of Shope fibroma virus, was investigated by both Shope and Andrewes (6, 11, 294). It was found that one lot of glycerinated OA virus had characteristics intermediate in character between those of OA and IA virus; and, further, could be duplicated by inoculation of a mixture of OA and IA virus. It is likely that the strains detected are specific mutants (12), and it has been observed that they arise repeatedly (86). Smith found that fibroma tissues stored in glycerine lost both infectivity titer and tumor-producing characteristics. Rapid passaging in experimental hosts would restore infectivity titers but not the tumor-producing properties (297). This is believed to be due to a specific modifying (or selecting) property of glycerine since it has been commonly observed that storage in buffered saline at $-70°C$ maintains the pathogenicity as a stable character (86).

The morphology of experimental Shope fibroma is illustrated in Figs. 54–56. Excellent photographs illustrating the gross and microscopic features of Shope fibromas are to be found in the original articles by Shope (290), Ahlstrom (3), Hurst (146), and Kilham (184). The ultrastructure of inclusion bodies and the virus particles within them can be found in the following references (30, 86, 199).

D. Hare Fibroma

The European hare (*Lepus europaeus*) is the natural host for the only member of the myxoma-fibroma subgroup of poxviruses observed to have arisen outside of the Americas. The fibroma has been recognized as a benign disease of hares in Europe since the original description in 1909 by von Dungern and Coca from Germany (319a). Since that time it has been observed on several occasions in the Po Valley of Italy (194, 220) and from the Mediterranean coast of France (193).

The hare fibroma virus causes the formation of small fibromas of the skin when injected into *Oryctolagus* rabbits

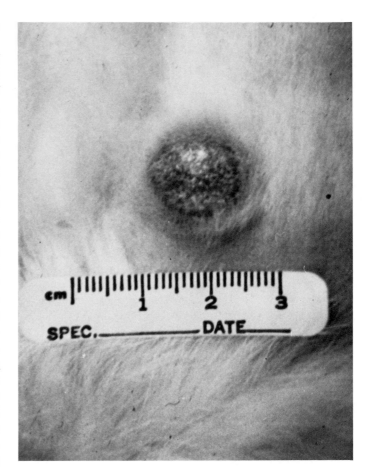

Fig. 54. Experimental Shope fibroma of approximately 15 days duration.

which do not appear until the twelfth day after inoculation (90, 194). When injected into newborn domestic rabbits, larger fibromas are produced which are histologically indistinguishable for Shope fibromas (194). *Sylvilagus* rabbits, in preliminary results, appear to be refractory to hare fibroma virus (90). The hare fibroma virus is related to the myxoma viruses, however. Myxoma-immune *Oryctolagus* rabbits are protected against hare fibroma induction, and conversely, a mild, generalized form of myxomatosis is produced in hare fibroma-immune rabbits (332). Hares appear to be refractory to Shope fibroma virus (90).

Although the epizootics cited above indicate a seasonal incidence (late summer and fall), which would be consistent with arthropod vectorism, the method of transmission is unknown. Fibromas have not been observed in hares examined in the winter thus the reservoir for the virus between epizootics is similarly not known (193, 194, 220).

The fibromas are approximately 1–3 cm in size in the natural host and most frequently found on the ears, legs, and eyelids. The tumors are reddish-gray in color during the growth phase and become dry and whitish as regression ensues (333). The histopathological details are almost identical to those of the Shope fibroma (193, 194).

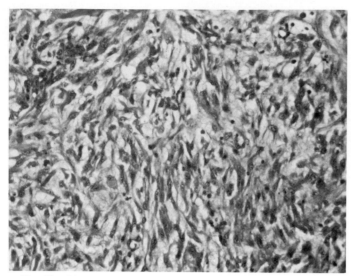

Fig. 55. Photomicrograph of Shope fibroma composed of sheets of spindle-shaped fibroma cells.

E. Hinze Herpesvirus Lymphoma

Hinze reported in 1968 the isolation of a new herpesvirus from *S. floridanus* in Wisconsin (134). In a series of subsequent investigations it was shown that the original isolate (CHV-1), recovered from the pooled kidney tissue cultures of three apparently healthy weanling cottontails, would induce cytopathogenic effects in rabbit tissue cultures and a lymphoproliferative disease in cottontail rabbits (135–137). The virus, subsequently called *Herpesvirus sylvilagus*, was shown to be immunologically distinct from other herpesviruses (136), including a newly isolated strain of *Herpesvirus cuniculi* (herpesvirus III) (230). Tissue cultures of kidney cells (from either *Sylvilagus* or *Oryctolagus* rabbits) within 1–2 days after inoculation develop syncytial masses of cells with characteristic eosinophilic intranuclear Type A inclusion bodies. The inclusion bodies stain well in hematoxylin and eosin preparations (136).

Cottontail rabbits inoculated by the subcutaneous or intraperitoneal routes uniformly develop a chronic viremia that appears to persist indefinitely. It was shown that viremic virus was confined to the formed elements of the blood (137). These rabbits also developed a lymphocytosis that rose from the normal value of $6000/m^3$ to a maximum peak value of $12,000–15,000/mm^3$ by the fourth to eighth week after infection. The differential count on these bloods showed an absolute lymphocytosis as well, with a percentage of lymphocytes increasing from the normal value of 55 to 97% in some cases. Large, immature, and abnormal lymphocytes (but not blast forms) made their appearance in the periperal blood during the second week and increased in proportion to make up over 50% of the lymphocyte population by the fourth to sixth week. The hematogram of

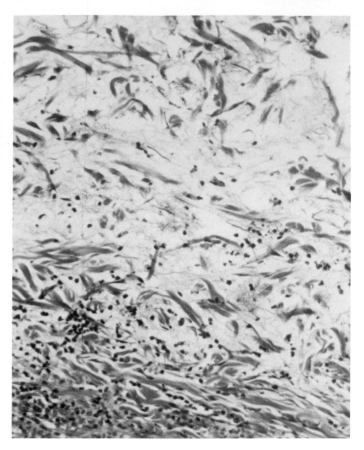

Fig. 56. Photomicrograph of Shope fibroma at base to show inflammatory infiltrate and edema.

animals surviving the 18-week observation period had gradually returned to normal cellular patterns by the fourteenth week although the white cell count remained elevated ($7000–8000/mm^3$) and the presence of abnormal lymphoid cells was persistent.

All of the infected rabbits, both juveniles and adults, appeared to develop a triad of generalized hyperplastic lymphadenopathy, splenomegaly, and renal enlargement. These gross findings were correlated on the microscopic level with hyperplasia of the lymphoid elements in all tissues in which they are normally resident. The dominant cell type was the immature lymphocyte in varying stages of development. These cells were actively dividing and had large vesicular nuclei with prominent nucleoli and abundant cytoplasm. The intensity of hyperplasia increased through the eight week after infection. Infiltration of other tissues with these cells included liver (65%) and myocardium (32%). In 10% of the adults and 27% of the juveniles the lymphoproliferative aspects were increased in degree so as to fulfill most of the criteria for malignant lymphoma. The normal architecture of lymph node and spleen was obliterated by proliferating blast cells and masses of immature lymphocytes. In these animals the range of infiltrated tissues was extended to include lung, pancreas, submaxillary (salivary?) gland, and intestinal wall. Hinze correlated the loss of normal architecture in lymph nodes (rather than the degree of hyperplasia) as the chief determinant in massive infiltration of other tissues. Inclusion bodies were not observed in tissues from infected cottontails. *Oryctolagus* rabbits were refractory to experimental infection. Infected tissue cultures are illustrated in Fig. 57, and representative infected organs in Figs. 58–59.

Hinze analogized the pattern in rabbits of benign lymphocytosis in some individuals and malignant lymphomatosis-like disease in others to that obtaining with the Epstein-Barr herpesvirus in infectious mononucleosis and Burkitt's lymphoma of man (137). Of additional interest is the pathological relationship (if any) between the cottontail lymphoma and lymphomatosis of *Oryctolagus* rabbits. They differ in several respects; most notably in the observation that lymphomatosis appears to be uniformly aleukemic in *Oryctolagus* (97). The pattern of lymphocytosis appears to be that of a clone arrested at the same, definite stage of immaturity in *Oryctolagus* in contrast to the varying stages of development in the abnormal lymphocytes of *Sylvilagus* lymphomatosis. It would also be of interest to know whether the *Sylvilagus* lymphomas found as spontaneous cases in wild rabbits by Lopushinsky and Fay (206) were also induced by *Herpesvirus sylvilagus* infection. The published description was histologically consistent and the rabbits derived from a nearby geographical area (Michigan).

F. Dos Santos Hepatoma

In a single publication, Dos Santos reported 35 fatal cases of hepatoma in an autopsy series of 120 rabbits (*Ory-*

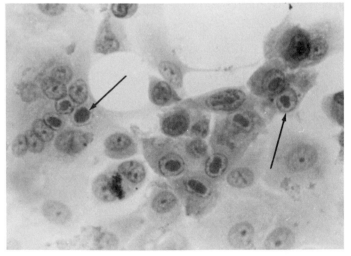

Fig. 57. Photomicrograph of *Herpesvirus sylvilagus*–infected rabbit kidney (DRK) tissue culture. Note basophilic (hematoxylin and eosin) intranuclear inclusion bodies (arrows). Photograph courtesy of H. C. Hinze.

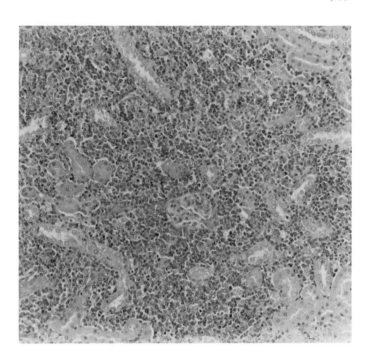

Fig. 58. Photomicrograph of kidney of *Sylvilagus* rabbit 4 weeks after *Herpesvirus sylvilagus* infection (10^5 PFU given subcutaneously). Note sparing of tubules and glomeruli. Compare for similarity with Fig. 38. Photograph courtesy of H. C. Hinze.

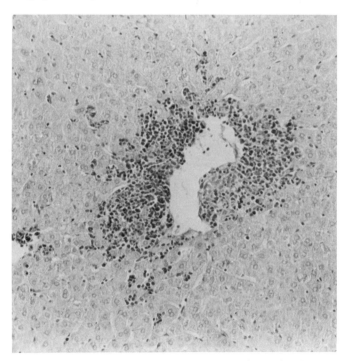

Fig. 59. Photomicrograph of liver of same rabbit as Fig. 58. Note periportal infiltration and compare with Fig. 37. Photograph courtesy of H. C. Hinze.

ctolagus) from the same breeding colony in Brazil (76). The hepatoma was limited to the liver in 10 cases but metastases were observed in 25, most commonly to the lung. The tumors were more common in females by a ratio of 1:6.

Thirty-two of the tumors were specifically hepatic cell hepatomas, 2 were mixtures of hepatic cell and bile duct elements, and 1 appeared to be a bile duct carcinoma. In 10 of the tumors intranuclear inclusion bodies were observed in hepatoma cells. The latter finding led the author conclude that the tumors were induced by a virus.

VI. TRANSPLANTABLE NEOPLASMS USED AS MODEL SYSTEMS

A. Kato Sarcoma

The older literature contains occasional references to a transmissible tumor discovered as a naturally occurring sarcoma by T. Kato in 1916 (164). The tumor, which bears his name, was originally described as a spindle cell sarcoma with pronounced metastatic properties. It was used in a series of experiments designed to characterize the nature of metastasis (228, 229, 312), interaction of the sarcoma with certain endocrine glands (129, 210, 227), and several biochemical properties (223, 224) between 1918 and 1938. It does not appear in the literature after the reports by Mori (223, 224), and it is doubtful if the sarcoma has been maintained and still available.

B. Andrewes Rabbit Sarcoma I (RSI)

In the course of work previously described in which the object of study was the generalization of fibromatosis in tar-treated rabbits (4), Andrewes and Ahlstrom happened upon an unusual occurrence. In one rabbit given two intramuscular injections of tar $4\frac{1}{2}$ months apart and 15 topical treatments of tar during the intervening period, a tumor was observed to occur at the intramuscular tar site 7 days following the intravenous injection of OA strain Shope fibroma virus. In a preliminary communication, the tumor was described as a transplantable fibroma (9), but subsequent changes led to its reclassification as a sarcoma (8). The tumor was passed for a series of 12 transplantation generations in rabbits during the "fibroma phase." During this phase the tumor had the histological characteristics of a typical fibroma: elongated, spindle-shaped cells arranged in bundles and whorls with abundant fibrils. As a rule the tumor grew only to an average size of 20 × 10 × 10 mm and central necrosis would begin by the fourteenth day.

The tumor, which had begun to routinely fail by the twelfth transplant generation, was restored by injection of OA fibroma virus into rabbits with regressing fibromas during the thirteenth passage. During the next 5 transplant generations (the sarcoma phase), the tumor became less differentiated, more cellular, fibrils were less frequent,

and mitoses more numerous. The tranplanted sarcoma underwent progressive growth in only about one-third of inoculated rabbit hosts, but in them metastasis was not infrequent; most commonly to the inguinal and iliac lymph nodes.

All attempts to demonstrate either the presence of fibroma virus (in RSI) or anti-fibroma antibodies in rabbits bearing RSI fibrosarcomas met with failure. Inclusion bodies typical for fibroma virus were never observed in RSI. Interestingly it was found several years later (7) that the sarcoma itself had become infected with *Herpesvirus cuniculi* [virus III of Rivers and Tillett (261)]. Virus III was found not to influence the course of RSI transplant growth; although it did abort or ameliorate the course of OA fibroma virus infection, apparently by the process of the interference phenomenon. This finding further strengthened the concept that fibroma virus was no longer detectable in RSI fibrosarcomas. After the 1940 report by Andrewes the tumor does not appear in the literature and it is assumed that it is now extinct.

C. Kondrateva Osteogenic Sarcoma

In previous discussion on naturally occurring osteogenic sarcomas in *Oryctolagus* (Sections II, G), it was pointed out that transplantable bone sarcomas have been reported from the Russian literature. The first of these was described by A. F. Kondrateva in the course of several experiments (188, 189). The tumor arose 890 days after a paraffin pellet containing 9,10-dimethyl-1,2-benzanthracene had been inserted into a bone. It was found that the tumor was transplantable; by 1958 was in its fortieth transplantation generation, and had an overall success rate in transplantation of at least 75%. The tumor typically caused death 30–95 days after transplantation by both extensive local invasion and metastasis. The incidence of pulmonary metastasis was 100% in early transplant generations but became less frequent in later generations. Transplantation of the tumor to bony tissues was more successful (virtually 100%) than it was to soft tissues. Similarly, the rate of tumor growth was faster in bony tissues although tumors generally obtained a larger size in soft tissues. Tumor growth (and success of transplantation) was higher in muscle than in liver, and transplantation to testicle or subcutaneous sites was not generally successful. The morphology of the tumor was similar irrespective of the host tissue in which it was transplanted. The morphology of the tumor was described as that of an osteolytic osteosarcoma (188).

The Kondrateva osteosarcoma has been used to some degree in experimental oncology. It has been exploited as a model for the study of tumor vascularization by making use of the technique of intraosseous phlebography (187). The intraarterial injection of a contrast medium (Diodon) has been reported to exert a marked inhibitory effect on this tumor culminating in regression and cure (155). Several similar tumors including the LOI osteoblastic osteosarcoma and the UzIKEM osteolytic osteosarcoma have been reported in Russian journals and used for a variety of purpose (161). The relation of the LOI and UzIKEM tumors to the Kondrateva osteosarcoma is not certain.

D. Brown-Pearce Carcinoma

This tumor, which has been widely accepted as one of the most useful of the transplantable rabbit tumors, was observed to arise from the scrotal chancre of an experimental rabbit. Despite the exhaustive description provided by Brown and Pearce (36–43) an understanding of the antecedent tissue from which the carcinoma originated is destined to remain somewhat ambiguous. For that reason it is worth describing the circumstances of the rabbit in which the tumor arose in some detail. Wade Brown and Louise Pearce of The Rockefeller Institute were intensively engaged during the years prior to 1920 in the study of syphilis using as a model for this disease rabbits injected with *Treponema pallidum*. In June, 1916, one particular rabbit was injected intrascrotally with testicular tissues from another rabbit infected previously with *T. pallidum*. Typical syphilis chancres were observed to develop at each of the scrotal sites. In the intervening period between 1916 and 1919, the testicles of this rabbit became fibrotic and atrophic. In April 1919 the rabbit developed late secondary generalized cutaneous lesions and, by 1920, an eczematoid dermatitis in the skin of the left scrotum. This process became chronic but was marked by exacerbations and remissions. In October of 1920, the appearance of a nodule was noted at the site of the primary chancre in the left scrotum. The nodule, which rapidly increased in size, proved to be a malignant tumor. Treponemes, while demonstrable in a popliteal lymph node as late as a month prior to the appearance of the nodule, could not be demonstrated in dark-field preparations from nodule biopsies. The nodule was removed twice by surgical excision but recurred.

The rabbit was killed in March 1921. Postmortem dissection revealed tumor metastases in the following locations; liver, bone, bone marrow, lung, spleen, kidneys, left testicle, inguinal adipose tissue, and the iliac, inguinal, and retroperitoneal lymph nodes. The eczematoid dermatitis of the scrotum was observed to have extended carnially to involve the skin of the abdominal ventrum and chronic degenerative and inflammatory changes were found as well in the epithelium of the tongue oral cavity, esophagous, and respiratory tract. Elsewhere, the skin was found to have fibrotic changes with atypical pilofollicular proliferations. The endocrine glands were abnormal and the rabbit had chronic bilateral orchitis. At necropsy then, this rabbit from which the Brown-Pearce carcinoma was derived,

presented a variety of abnormalities some of which were due to the effects of malignancy, while others were clearly attributable to chronic treponematosis.

Brown and Pearce believed that the tumor originated in the skin of the scrotum, but their conclusion has been challenged by others on several grounds (301). The original descriptors found that the tumor varied histologically in its various metastatic locations; being sarcomatous in some, while appearing more like a carcinoma in others. For this reason they believed it possible that two primary neoplasms arose simultaneously, one in the scrotum and the other in the testicle. In analyzing the findings by Brown and Pearce, Stewart *et al.* (301) agreed with the conclusion that a primary neoplastic process in the scrotum or testicle (or both) would likely metastasize to the inguinal, iliac, and retroperitoneal lymph nodes and to the inguinal adipose tissue as well. The latter group believed it highly improbable, however, that a primary neoplasm of the scrotal skin would metastasize to the long bones of the extremities as well as to the vertebrae and pelvic bones. In further support of their view that a cutaneous origin was improbable, Stewart *et al.* cited the varied histological appearance which in some tumors even included mucus-producing cells (301).

Neoplastic tissue from the left inguinal lymph node was minced and intratesticular transplants established in host rabbits. Brown and Pearce characterized the biological behavior of the transplant tumor in a series of publications (37–43) which still remain as fundamental descriptions. From the first it was noted that variable results were the expectation. Spontaneous regression was common, even after the onset of metastasis. The course might be fulminating with death 6–7 weeks after inoculation or prolonged to more than six months. Most commonly death occurred 8–12 weeks after transplantation. The tumor was remarkably stable and after some 15 years of continuous transplantation, Casey (52, 54) was to write that the tumor had undergone no change with regard to metastatic distribution. It was noted that animal hosts marked by tumor regression inhibited the growth of subsequent transplants.

Brown and Pearce reported close to 100% success when transplanting the tumor to testicle, 80–90% when inoculated by the intramuscular or intradermal routes, and 20–25% when transplanted to subcutaneous, intravenous or intraperitoneal locations. Nearly 100% of inoculations by the intracerebral or intraocular routes were also successful. Neither dessicated tumor preparations nor cell-free extracts were successful in establishing tumors (248). Green and his colleagues, who pioneered the technique of anterior chamber preparations as a shielded location for tumor transplants, reported 100% success with the Brown-Pearce carcinoma (82, 277). Growth in that location was rapid, often filling the chamber within 7–10 days and producing metastases within 4–6 weeks. Metastasis was more common in cold weather, regression more likely in warm weather

(82). Greene (113) explored the use of heterologous hosts for this tumor and reported initial growth following transplantation to testicle, eye, or subcutaneous sites in rats, mice, and hamsters, but regression was the invariable rule in these hosts. No growth was observed in guinea pig hosts. Growth of the tumor is supported on embryonating chicken CAM (278). Even today variability in the behavior of this tumor is experienced by investigators working in different laboratories and even in the same laboratory host rabbits may die as early as 10 days after transplantation or persist for over 100 days (301).

Casey observed in 1932 that rabbits injected with killed and preserved Brown-Pearce carcinoma tissues were rendered more susceptible to malignant consequences when later inoculated with the tumor (48,49). He believed there were "enhancing meterials" or an "XYZ factor" in the preserved tumors which were responsible for the specificity of this effect since it occurred only after pretreatment of the host with killed homologous tumor preparations (53). At the same time he believed the XYZ factor induced constitutional changes in the host since the effects lasted as long as seven months after injection (50, 51). This series of observations provided important support for Nathan Kaliss who marshalled the evidence for a general theory of the immunological enhancement of tumor growth. Kaliss, in a review of this evidence (166), emphasized the conditions required for maximal expression of immunological enchancement and the dependency of the phenomenon on late appearing IgG antibody.

Even before it was realized that the XYZ factor(s) were immunogens of the tumor, it was generally recognized that there were other immunological consequences of transplantation. The carcinoma has been used extensively as an important model for the study of tumor immunity (299). Kidd, and others, have shown that saline extracts of the tumor contained antigens that fixed compliment in the presence of sera from rabbits bearing the tumor (12,58, 82,100,156,172). Over a succession of experiments it became clear that the tumor possessed both tissue antigens shared with normal tissues and also tumor-specific antigens. In turn, rabbits implanted with the tumor developed antibodies with specificity against either tumor-specific or both classes of antigens. The genetic constitution of the host was important in determining the antibody response; Blue Cross and Chocolate Dutch rabbits forming antibodies only against the tumor-specific antibodies, other breeds if they produced antibody at all, producing both types (209). It has been generally concluded that the Brown-Pearce carcinoma contains antigens different from those of normal tissues. It has been demonstrated that the antigen is probably a ribonucleoprotein (209) and that it is located on tumor cell microsomes (176). Zilber has reported that certain methods of antigenic extraction, e.g., methanol precipitation, preferentially favor release of tumor-specific

antigens (336). Tumor-specific antibodies are known to inhibit culture of the tumor in vitro (174,175).

The tumor itself is soft, friable, has areas with hemorrhage, and may be necrotic, particularly toward the center. Microscopically the tumor has a medullary pattern and may be defined as an anaplastic carcinoma. The architecture is monotonous and consists of cords of carcinoma cells supported in a scanty stroma of connective tissue with a few reticular fibers and blood vessels. The tumor cells are typically round or oval with hyperchromatic centrally located nuclei in a slightly basophilic but pale cytoplasm.

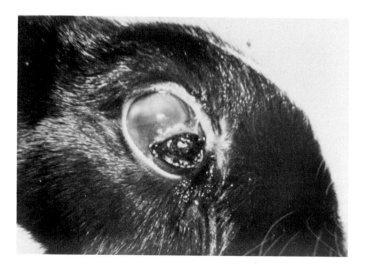

Fig. 60. Anterior chamber preparation of Brown-Pearce carcinoma of approximately 30 days duration. Note corneal perforation by tumor Preparation courtesy of A. M. Jonas.

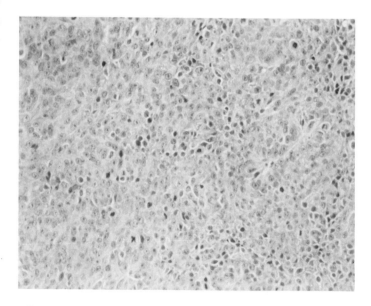

Fig. 61. Photomicrograph of Brown-Pearce carcinoma. Note frequent mitotic figures.

The prominent nuclear membrane encloses 2–4 acidophilic nucleoli, dense chromatin bars, and a lacelike meshwork. Mitotic figures are common. Necrosis is more common centrally than at the periphery. The tumor appears to provoke little of an inflammatory reaction unless regression supervenes. The tumor is illustrated in Figs. 60–61. Excellent illustrations of the tumor are to be found in the original reports by Brown and Pearce, the reviews by Snell (299) and Stewart *et al.* (301), and the investigation by Appel *et al.* (13).

REFERENCES

1. Aberastury, M. A., and Dessy, S. (1903). Un caso de sarcomatosis en el conejo. *Rev. Sud. Amer. Cienc. Med.* **1**; Review in *Z. Krebsforsch.* **1**, 257 (1904).

2. Adams, W. M., Jr. (1962). The Natural History of Adenocarcinoma of the Uterus in the Phipps Rabbit Colony. M. Med. Sci. Thesis, Henry Phipps Institute, University Pennsilvania.

3. Ahlstrom, C. G. (1938). The histology of the infectious fibroma in rabbits. *J. Pathol. Bacteriol.* **46**, 461–472.

4. Ahlstrom, C. G., and Andrewes, C. H. (1938). Fibroma virus infection in tarred rabbits, *J. Pathol. Bacteriol.* **47**, 65–86.

5. Allison, A. C., and Friedman, R. M. (1966). Effects of immunosuppressants on Shope rabbit fibroma. *J. Nat. Cancer Inst.* **36**, 859–868.

6. Andrewes, C. H., and Shope, R. E. (1936). A change in rabbit fibroma virus suggesting mutation. I. Experiments on domestic rabbits. *J. Exp. Med.* **63**, 157–172.

7. Andrewes, C. H. (1940). The occurrence of virus III in rabbits in the lesions of infectious fibroma and of transplantable sarcoma. *J. Pathol. Bacteriol.* **50**, 227–240.

8. Andrewes, C. H., and Ahlstrom, C. G. (1938). A transplantable sarcoma occurring in a rabbit inoculated with tar and infectious fibroma virus. *J. Pathol. Bacteriol.* **47**, 87–99.

9. Andrewes, C. H., Ahlstrom, C. G., Foulds, L. and Gye, W. E. (1937). Reaction of tarred rabbits to infectious fibroma virus (Shope). *Lancet* **2**, 893–895.

10. Andrewes, C. H., and Chaproniere, D. M. (1957). Propagation of rabbit myxoma and fibroma viruses in a guinea pig sarcoma. *Virology* **4**, 346–350.

11. Andrewes, C. H., and Shope, R. E. (1936). A change in rabbit fibroma virus suggesting mutation. III. Interpretation of findings. *J. Exp. Med.* **63**, 179–184.

12. Appel, M., Saphir, O., Janota, M., and Strauss, A. A. (1942) Complement-fixing antibodies (Brown-Pearce Carcinoma) in the blood serum and in the aqueous of the chamber of the eye. *Cancer Res.* **2**, 576–578.

13. Appel, M., Saphir, O., and Strauss, A. A. (1941). Morphologic alterations in the regressing Brown-Pearce tumor and their relation to changes due to irradiation. *Arch. Pathol.* **31**, 317–325.

14. Aragao, H. B. (1942). Sensibilidade do coelho do mato ao do mixoma; transmissao pelo *Aedes scapularis* e pelo *Steomyia. Brasil-Med.* **56**, 207–209.

15. Aragao, H. B. (1942). O virus do mixoma no coelho-domato (*Sylvilagus* minenses) sua transmissao pelos aedes scapularis et aegypti. *Mem Inst. Oswaldo Cruz* **38**, 93–99.

16. Baba, N., Vidyarthi, S., and von Haam, E. (1970). Nonspecific phosphatases of rabbit endometrial carcinoma. *Arch. Pathol.* **90**, 65–71.

17. Baba, N., and von Haam, E. (1967). Ultramicroscopic changes in the endometrial cells of spontaneous adenocarcinoma of rabbits. *J. Nat. Cancer Inst.* **38**, 657–672.

18. Baba, N., and von Haam, E. (1967). Experimental carcinoma of the endometrium. *Progr. Exp. Tumor Res.* **9**, 192–260.

18a. Baba, N., and von Haam, E. (1972). Animal model: Spontaneous adenocarcinoma in aged rabbits. *Amer. J. Pathol.* **68**, 653–656.

19. Ball, N. D. (1926). An epithelial tumour in the spleen of a rabbit. *J. Pathol.* **29**, 239–240.

20. Ball, V., and Douville, M. (1929). Cancer de la mamelle chez une lapine angora. *J. Med. Vet.*

21. Barile (1920). Cancer primitif du poumon chez le lapine. "Revue Pathol Comparée" [cited by Courteau (65)].

22. Barnes, J. M., Denz, F. A., and Sissons, H. A. (1950). Beryllium bone sarcomata in rabbits. *Brit. J. Cancer.* **4**, 212–222.

23. Bashford, E. (1911). Sarcoma of the subcutaneous tissues in a rabbit. *Annu. Rep. Imp. Cancer Res. Found., 9th* p. 6.

24. Bashford, E. (1911). A cancer of the mammary gland. *Lancet* **2**, 315.

25. Baumgarten, A. (1906). Uber einen malignen tumor mit ausgebreiteter metastasen bildung bei einem kaninchen. *Zentrabl. Pathol.* **17**, 769.

26. Beard, J. W. (1935). Conditions determining melanosis of a virus-induced rabbit papilloma (Shope). *Proc. Soc. Exp. Biol. Med.* **32**, 1334–1336.

27. Beard, J. W., and Rous, P. (1935). Effectiveness of the Shope papilloma in various American rabbits. *Proc. Soc. Exp. Biol. Med.* **33**, 191–193.

28. Bell, E., and Henrici, A. T. (1916). Renal tumors in the rabbit. *J. Cancer Res.* **1**, 157–167.

29. Bell, J. F., and Chalgren, W. S. (1943). Some wildlife diseases in the eastern United States. *J. Wildl. Manage.* **7**, 270–278.

30. Bernhard, W., Bauer, A., Harel, J., and Oberling, C. (1954). Les formes intracytoplasmiques du virus fibromateus de Shope. Etude di coupes ultrafines au microscope électronique. *Bull. Cancer* **61**, 423–444.

31. Biesele, J. J. (1945). Chromosomal enlargement in neoplastic rabbit tissues. *Cancer Res.* **5**, 179–182.

32. Blanchard, L., Poisson, J., and Drieux, H. (1939). Pathologie comparée des tumeurs du thymus son intérêt pour l'histogénèse. *Rec. Med. Vet.* **115**, 392–413.

33. Blount, W. P. (1957). "Rabbits' Ailments." "Fur and Feathers," Idle, Bradford, London.

34. Boycott, A. E. (1911). Uterine tumors in rabbits. *Proc. Roy. Soc. Med.* **4**, 225–232.

35. Boycott, A. E., and Pembrey, M. S. (1912). A bilateral sarcoma in a wild rabbit. *J. Pathol. Bacteriol.* **17**, 130.

36. Brown, W. H., and Pearce, L. (1923). Studies based on a malignant tumor of the rabbit. I. The spontaneous tumor and associated abnormalities. *J. Exp. Med.* **37**, 601–630.

37. Brown, W. H., and Pearce, L. (1923). Studies based on a malignant tumor of the rabbit. II. Primary transplantation and elimination of a coexisting syphilitic infection. *J. Exp. Med.* **37**, 631–646.

38. Brown, W. H., and Pearce, L. (1923). Studies based on a malignant tumor of the rabbit. III. Intratesticular transplantation and clinical course of the disease. *J. Exp. Med.* **37**, 799–810.

39. Brown, W. H., and Pearce, L. (1923). Studies based on a malignant tumor of the rabbit. IV. The results of miscellaneous methods of transplantation with a discussion of factors influencing transplantation in general. *J. Exp. Med.* **37**, 811–828.

40. Brown, W. H., and Pearce, L. (1923). Studies based on a malignant tumor of the rabbit. V. Metastases. Part 1. Description of the lesions with especial reference to their occurrence and distribution. *J. Exp. Med.* **38**, 347–366.

41. Brown, W. H., and Pearce, L. (1923). Studies based on a malignant tumor of the rabbit. V. Metastases. Part 2. Description of the lesions with especial reference to their occurrence and distribution. *J. Exp. Med.* **38**, 367–384.

42. Brown, W. H., and Pearce, L. (1923). Studies based on a malignant

tumor of the rabbit. V. Metastases. Part 3. Factors that influence occurrence and distribution. *J. Exp. Med.* **38**, 385–406.

43. Brown, W. H., and Pearce, L. (1924). Studies based on a malignant tumor of the rabbit. VI. Variations in growth and malignancy of transplanted tumors. *J. Exp. Med.* **40**, 583–618.

44. Brown, W. H., and Pearce, L. (1926). Melanoma (sarcoma) of the eye in a syphilitic rabbit. *J. Exp. Med.* **43**, 807–814.

45. Burgisser, H. (1957). La leucose du lièvre. *Schweiz. Arch. Tierheilk.* **99**, 141–149.

46. Burrows, H. (1940). Spontaneous uterine and mammary tumors in the rabbit. *J. Pathol. Bacteriol.* **51**, 385–390.

47. Carini, A. (1941). Carcinoma espinocelular espontaneo de la piel de un conejo, con metastasis en los ganglios y en el pulmon. *Bol. Inst. Med. Exp. Estud. Trat. Cancer, Buenos Aires* **18**, 689–702.

48. Casey, A. E. (1932). Experimental enhancement of malignancy in the Brown-Pearce rabbit tumor. *Proc. Soc. Exp. Biol. Med.* **29**, 816–819.

49. Casey, A. E. (1933). A species limitation of an enhancing material derived from a mammalian tumor. *Proc. Soc. Exp. Biol. Med.* **30**, 674–677.

50. Casey, A. E. (1934). The experimental alteration of malignancy with an homologous mammalian tumor material. I. Results with intratesticular inoculation. *Amer. J. Cancer* **21**, 760–775.

51. Casey, A. E. (1934). The experimental alteration of malignancy with an homologous mammalian tumor material. II. Intracutaneous inoculation of preserved material. *Amer. J. Cancer* **21**, 776–780.

52. Casey, A. E. (1939). Distribution of metastases in the Brown-Pearce tumor. I. In standard breeds of rabbits. II. Effects of 15 years of transplantation. III. Effect of homologous material. *Proc. Soc. Exp. Biol. Med.* **40**, 223–234.

53. Casey, A. E. (1941). Experiments with a material from the Brown-Pearce tumor. *Cancer Res.* **1**, 134–135.

54. Casey, A. E., and Pearson, B. (1939). Distribution of metastases in the Brown-Pearce tumor. IV. Effects of site of inoculation. *Proc. Soc. Exp. Biol. Med.* **40**, 234–237.

55. Chambers, V. C., and Ito, Y. (1964). Morphology of Shope papilloma virus associated with nucleic acid-induced tumors of cottontail rabbits. *Virology* **23**, 434–436.

56. Chaproniere, D. M., and Andrewes, C. H. (1957). Cultivation of rabbit myxoma and fibroma viruses in tissues of nonsusceptible hosts. *Virology* **4**, 351–365.

57. Cheatum, E. L., and Bunting, H. (1951). Malignant neoplasms in varying hares. *Cornell Vet.* **41**, 136–140.

58. Cheever, F. S. (1940). A complement-fixing antibody in sera of rabbits bearing Brown-Pearce carcinoma. *Proc. Soc. Exp. Biol. Med.* **45**, 517–522.

59. Claude, A. (1939). The enhancing effect of azoproteins on the lesions produced by vaccine virus, Shope fibroma virus and the agent transmitting chicken tumor. *J. Exp. Med.* **69**, 641–648.

60. Clemmesen, J. (1938). Shope fibroma in X-rayed rabbits. *Acta Pathol. Microbiol. Scand.* **38**, 47–48.

61. Clemmesen, J. (1939). The influence of roentgen radiation on immunity to Shope fibroma virus. *Amer. J. Cancer* **35**, 378–385.

62. Clemmesen, J., and Anderson, E. K. (1942). The influence of "mucin 1701 W" on infection with Shope fibroma and vaccinia viruses. *Acta Pathol. Microbiol. Scand.* **19**, 173–182.

63. Cohrs, P., Jaffe, R., and Meesen, H., eds. (1958). "Pathologie der Laboratoriumstiere," pp. 477–493. Springer-Verlag, Berlin and New York.

64. Collela, C. (1925). Di un leimioma dello stomatico in un coniglio. Di um cistadenoma papilifero del rone in un coniglio. *Clin. Vet.* 549.

64a. Cotchin, E. (1964). Spontaneous uterine cancer in animals. *Brit. J. Cancer* **18**, 209–210.

65. Courteau, R. (1935). Pathologie Comparée des Tumeurs chez les Mammifères Domestiques, D. V. M. Thesis, pp. 152–168. Le François, Paris.

66. Coventry, M. B., Maher, F. T., Janes, J. M., and Dahlin, D. C. (1959). The effect of high cholesterol diet (egg yolk) on beryllium-induced osteogenic sarcoma in the rabbit. *Proc. Staff Meet. Mayo Clin.* **34**, 543–547.

67. Cutler, O. I. (1934). Adenocarcinoma of the uterus in a rabbit. *Amer. J. Cancer* **21**, 600–603.

68. Dalmat, H. T. (1958). Effects of X-rays and chemical carcinogens on infectivity of domestic rabbit fibromas for arthropods. *J. Infec. Dis.* **102**, 153–157.

69. Dalmat, H. T. (1958). Passage of Shope's rabbit fibroma virus through one-day-old mice. *Proc. Soc. Exp. Biol. Med.* **97**, 219–220.

70. Dalmat, H. T., and Stanton, M. F. (1959). A comparative study of the Shope fibroma in rabbits in relation to transmissibility by mosquitoes. *J. Nat. Cancer Inst.* **22**, 593–615.

71. Dalton, A. J., and Haguenau, G., eds. (1962). "Ultrastructure of Tumors Induced by Viruses." Academic Press, New York.

72. DeFaria, J. F. (1961). Tumor de celulas intersticiais do testiculo em coelho. *Arq. Inst. Biol. Anim. (Brazil)* **4**, 127–131.

73. Dible, J. (1921). Un cas de tumeur de l'uterus chez la lapine. *J. Pathol. Bacteriol.* **24**, 3.

74. Dinges, H. P., and Kovac, W. (1972). Ein metastasierendes schildrusen carcinom beim kaninchen. *Z. Versuchstierk.* **14**, 197–204.

75. Dobberstein, J., and Tamaschke, C. (1958). Die spontantumoren beim kaninchen. *In* "Pathologie der Laboratoriumstiere" (P. Cohrs, R. Jaffe, and H. Meessen, eds.), pp. 477–493. Springer-Verlag, Berlin and New York.

76. Dos Santos, J. A. (1961). Sobre a occorencia de um hepatoma enzootico entre coelhos. Nota Previa. *Arq. Inst. Biol. Anim. (Brazil)* **4**, 133–167.

77. Drieux, H., and Poisson, J. (1938). Tumeur de thymus chez le lapin. *Bull. Acad. Vet. Fr.* [N.S.] **11**, 000

78. Dumas, J. (1953). "Les animaux de laboratoire." Editions Médicales, Flammarion, Paris.

79. Duran-Reynals, F. (1940). Production of degenerative inflammatory or neoplastic effects in the newborn rabbit by the Shope fibroma virus *Yale J. Biol. Med.* **13**, 99–110.

80. Duran-Reynals, F. (1945). Immunological factors that influences the neoplastic effects of the rabbit fibroma virus. *Cancer Res.* **5**, 25–39.

81. Eisler, B. (1938). Uber zwei embryonale nephrome beim Kaninchen. *Z. Krebsforsch.* **48**, 235–240.

82. Ellerbrook, L. D., Rhees, M. C., Thornton, H., and Lipincott, S. W. (1952). Complement fixation in animal neoplasia. II. Development and duration of the reaction in New Zealand White rabbits carrying Brown-Pearce carcinomas. *J. Nat. Cancer Inst.* **12**, 937–946.

83. Engzell, U., Rubio, C., Tjernberg, B., and Symeonidis, A. (1968). The lymph node barrier against V × 2 cancer cells before, during and after lymphography. A preliminary report of experiments on rabbits. *Eur. J. Cancer.* **4**, 305–313.

84. Fardeau, F. (1931). Les tumeurs spontanées chez le lapin. Dissertation, Paris.

85. Faulkner, G. H., and Andrewes, C. H. (1935). Propagation of a strain of rabbit fibroma virus in tissue-culture. *Brit. J. Exp. Pathol.* **16**, 271–276.

86. Febvre, H. (1962). The Shope fibroma of rabbits. *In* "Ultrastructure of Tumors Induced by Viruses." (A. J. Dalton and F. Haguenau, eds.), pp. 1–59. Academic Press, New York.

87. Feldman, W. H. (1926). A study of the tumor incidence in the lower animals. *Amer. J. Pathol.* **2**, 545–556.

88. Feldman, W. H. (1927). The primary situation of 133 spontaneous tumours in the lower animals. *J. Cancer Res.* **2**, 436.

89. Feldman, W. H. (1932). "Neoplasms of Domesticated Animals," Mayo Clini. Monogr. Saunders, Philadelphia, Pennsylvania.

90. Fenner, F., and Ratcliffe, F. N. (1965). "Myxomatosis." Cambridge Univ. Press, London and New York.

91. Fischer, R. G., and Syverton, J. T. (1951). The virus-induced papil-

loma-to-carcinoma sequence. IV. Carcinomas in domestic rabbits infected while *in-utero*. *Cancer Res.* **11**, 737–740.

92. Fisher, B., and Fisher, E. (1966). Transmigration of lymph nodes by tumor cells. *Science* **152**, 1397–1398.

93. Flatt, R. E. (1972). Lymphosarcoma not previously reported. Personal communication.

94. Flatt, R. E. (1969). Pyometra and uterine adenocarcinoma in a rabbit. *Lab. Anim. Care* **19**, 398–401.

95. Flir, K. (1952). Die primeren Nierengeschwulste der Haussaugetiere. *Wiss. Z. Humboldt-Univ. Berlin, Math.-Naturwiss. Reihe* **2**, 93.

96. Flux, J. E. C. (1965). Incidence of ovarian tumors in hares in New Zealand. *J. Wildl. Manage.* **29**, 622–624.

97. Fox, R. R., Meier, H., Crary, D. D., Myers, D. D., Norberg, R. F., and Laird, C. W. (1970). Lymphosarcoma in the rabbit: Genetics and pathology. *J. Nat. Cancer Inst.* **45**, 719–730.

98. Fox, R. R., Meier, H., Crary, D. D., Norberg, R. F., and Myers, D. D. (1971). Hemolytic anemia associated with thymoma in the rabbit: Genetic studies and pathologic findings. *Oncology* **25**, 372–382.

99. Friedewald, W. F. (1944). Certain conditions determining enhanced infection with the rabbit papilloma virus. *J. Exp. Med.* **80**, 65–76.

100. Friedewald, W. F., and Kidd, J. G. (1945). Induced antibodies that react *in vitro* with sedimentable constituents of normal and neoplastic tissue cells. Presence of the antibodies in the blood of rabbits carrying various transplanted cancers. *J. Exp. Med.* **82**, 21–40.

101. Gertner, H. R., Jr., Wilson, J. R., and Woodward, E. R. (1964). Parathormone bioassay in hypercalcemic tumor rabbits. *Proc. Soc. Exp. Biol. Med.* **116**, 177–178.

102. Ginder, D. T., and Friedewald, W. F. (1951). Effect of Semliki Forest virus on rabbit fibroma. *Proc. Soc. Exp. Biol. Med.* **77**, 272–276.

103. Ginder, D. T., and Friedewald, W. F. (1952). V_2 carcinoma in the rabbit eye. *Cancer Res.* **12**, 429–432.

104. Goyon, M. (1959). Teratome ovarian avec metastases pulmonaires chez la hase. *Rec. Med. Vet.* **135**, 651–655.

105. Greene, H. S. N. (1939). Uterine adenomata in the rabbit. II. Homologus transplantation experiments. *J. Exp. Med.* **69**, 447–466.

106. Greene, H. S. N. (1939). Familial mammary tumors in the rabbit. I. Clinical history. *J. Exp. Med.* **70**, 147–158.

107. Greene, H. S. N. (1939). Familial mammary tumors in the rabbit. II. Gross and microscopic pathology. *J. Exp. Med.* **70**, 159–166.

108. Greene, H. S. N. (1939). Familial mammary tumors in the rabbit. III. Factors concerned in their genesis and development. *J. Exp. Med.* **70**, 167–184.

109. Greene, H. S. N. (1940). Familial mammary tumors in the rabbit. IV. The evolution of autonomy in the course of tumor development as indicated by transplantation experiments. *J. Exp. Med.* **71**, 305–324.

110. Greene, H. S. N. (1942). Uterine adenomata in the rabbit. III. Susceptibility as a function of constitutional factors. *J. Exp. Med.* **73**, 273–292.

111. Greene, H. S. N. (1942). Transplantation of mammalian tumors. I. The transfer of rabbit tumors of alien species. *J. Exp. Med.* **73**, 461–474.

112. Greene, H. S. N. (1943). The occurrence and transplantation of embryonal nephromas in the rabbit. *Cancer Res.* **3**, 434–440.

113. Greene, H. S. N. (1949). Heterologous transplantation of the Brown-Pearce tumor. *Cancer Res.* **9**, 728–735.

114. Greene, H. S. N. (1953). The heterologous transplantation of the V-2 carcinoma. *Cancer Res.* **13**, 610–612.

115. Greene, H. S. N. (1958). Adenocarcinoma of the uterine fundus in the rabbit. *Ann. N. Y. Acad. Sci.* **75**, 535–542.

116. Greene, H. S. N. (1965). Diseases of the rabbit. *In* "The Pathology of Laboratory Animals" (W. W. Ribelin and J. R, McCoy, eds.), pp. 330–348. Thomas, Springfield, Illinois.

117. Greene, H. S. N., and Brown, W. H. (1943). A transplantable squamous cell carcinoma in the rabbit. *Cancer Res.* **3**, 53–64.

118. Greene, H. S. N., and Newton, R. I. (1948). Evolution of cancer of the uterine fundus in the rabbit. *Cancer* **1**, 88–99.

119. Greene, H. S. N., Newton, H. L., and Fisk, A. A. (1947). Carcinoma of the vaginal wall in the rabbit. *Cancer Res.* **7**, 502–510.

120. Greene, H. S. N., and Saxton, J. A., Jr. (1938). Uterine adenomata in the rabbit. I Clinical history, pathology and preliminary transplantation experiments. *J. Exp. Med.* **67**, 691–708.

121. Greene, H. S. N., and Strauss, J. S. (1949). Multiple primary tumors in the rabbit. *Cancer* **2**, 673–691.

122. Grodhaus, G., Regnery, D. C., and Marshall, I. D. (1963). Studies in the epidemiology of myxomatosis in California. II. The experimental transmission of myxomatosis in brush rabbits. (*Sylvilagus bachmani*) by several species of mosquitoes. *Amer. J. Hyg.* **77**, 195–204.

123. Gross, L. (1970). "Oncogenic Viruses," 2nd ed. Pergamon, Oxford.

124. Groth, W. (1955). Basilome der haut beim hund und kaninchen. *Z. Krebsforsch.* **60**, 361–372.

125. Guillon, J. C., Groulade, P., and Vallee, A. (1963). Leucose myéloide chez un lièvre. *Bull. Acad. Vet. Fr.* [N.S.] **38**, 157.

126. Hagen, K. W. (1966). Spontaneous papillomatosis in domestic rabbits. *Bull. Wildl. Dis. Ass.* **2**, 108–110.

127. Harel, J. (1956). Titration et étude du cycle de multiplication in vivo du virus fibromateux de Shope dans les tumeurs malignes du lapereau. *C. R. Soc. Biol.* **150**, 139–142.

128. Harel, J., and Constantin, T. (1954). Sur la malignité des tumeurs provoquées par le virus fibromateux de Shope chez le lapin nouveau-né et le lapin adulte traité par des doses massives de cortisone. *Bull. Cancer* **41**, 482–497.

129. Hayashi, T. (1930). The relationship of the hypophysis and the growth of the tumor (rabbit sarcoma). *Trans. Jap. Pathol. Soc.* **20**, 661–663.

130. Hayden, D. W. (1970). Generalized lymphosarcoma in a juvenile rabbit. A case report. *Cornell Vet.* **60**, 73–82.

131. Heiman, J. (1937). Spontaneous mammary carcinoma in a rabbit. *Amer. J. Cancer* **29**, 93–101.

132. Herman, C. M., Kilham, L., and Warbach, O. (1956). Incidence of Shope's rabbit fibroma in cottontail at the Patuxent Research Refuge. *J. Wildl. Manage.* **20**, 85–89.

132a. Herr, A. (1947). Über das vorkommen von Myomen bei unseren Haussaugtieren. Dissortatism, Hanover.

133. Higginbotham, R. D., and Murillo, G. J. (1965). Influence of heparin on resistance of rabbits to infection with fibroma virus. *J. Immunol.* **94**, 228–233.

134. Hinze, H. C. (1968). Isolation of a new herpesvirus from cottontail rabbits. *Bacteriol. Proc.* p. 147.

135. Hinze, H. C. (1969). Rabbit lymphoma induced by a new herpesvirus. *Bacteriol. Proc.* p. 157.

136. Hinze, H. C. (1970). New member of the herpesvirus group isolated from wild cottontail rabbits. *Infec. Immunity* **3**, 350–354.

137. Hinze, H. C. (1971). Induction of lymphoid hyperplasia and lymphoma-like disease in rabbits by *Herpesvirus sylvilagus*. *Int. J. Cancer* **8**, 514–522.

138. Hinze, H. C., and Walker, D. L. (1964). Response of cultural rabbit cells to infection with the Shope fibroma virus. I. Proliferation and morphological alteration of the infected cells. *J. Bacteriol.* **88**, 1185–1194.

139. Hodes, M. E., Palmer, C. G., Beatty, L. E., Swenson, M. K., and Hubbard, J. D. (1963). Infectivity experiments with nucleic acids of the Shope rabbit papilloma and derived carcinomas. *J. Nat. Cancer Inst.* **30**, 1–15.

140. Hoffman, J. A. (1954). Hodenkrebs bei einem kaninchen. *Tieraertzl. Wochenschr.* **67**, 350–352.

141. Hoffstadt, R. E., Omundsun, D. V., and Donaldson, P. (1941). Cultivation of the virus of infections myxoma on the chorioallantoic membrane of the developing duck embryo. *J. Infec. Dis.* **68**, 213–219.

142. Hoffstadt, R. E., Pilcher, K. S. (1941). A study of IA and OA strains

of Shope's fibroma virus with special reference to the Berry transformation. *J. Infec. Dis.* **68**, 67–72.

143. Holz, K., and Heutgens, W. (1955). Multiple melanombildungen bei einem kaninchen. *Deut. Tieraerztl. Wochenschr.* **62**, 146–148.

144. Hueper, W. C., and Ichniowski, C. T. (1944). Carcinoma of the adrenal cortex in a rabbit. *Cancer Res.* **4**, 176–178.

145. Hulse, E. V. (1969). Osteosarcomas, fibrosarcomas and basal-cell carcimonas in rabbits after irradiation with gamma-rays or fission neutrons: An interim report on incidence, site of tumours and RBE. *Int. J. Radiat. Biol.* **16**, 27–30.

146. Hurst, E. W. (1938). Myxoma and the Shope fibroma. IV. The histology of Shope fibroma. *Aust. J. Exp. Biol. Med. Sci.* **16**, 53–64.

147. Hutt, F. B. (1958). "Genetic Resistance to Disease in Domestic Animals." Cornell Univ. Press (Comstock), Ithaca, New York.

148. Hyde, R. R. (1936). Concerning the transmission of the fibroma virus (Shope) of rabbits. *Amer. J. Hyg.* **24**, 217–226.

148a. Ingalls, T. H. Personal communication.

149. Ingalls, T. H., Adams, W., Lurie, M. B., and Ipsen, J. (1964). Natural history of adenocarcinoma of the uterus in the Phipps rabbit colony. *J. Nat. Cancer Inst.* **33**, 799–806.

150. Irving, C. C., Wiseman, R., and Young, J. M. (1967). Carcinogenicity of 2-acetylaminofluorene and *N*-hydroxy-2-acetylaminofluorene in the rabbit. *Cancer Res.* **27**, 838–848.

151. Ito, Y. (1960). A tumor-producing factor extracted by phenol from papillomatous tissue (Shope) of cottontail rabbits. *Virology* **12**, 596–601.

152. Ito, Y. (1970). Induction of papillomas in rabbits with nucleic acid extracts from Vx7 carcinomas. *Brit. J. Cancer* **24**, 535–541.

153. Ito, Y., and Evans, C. A. (1965). Induction of tumors in domestic rabbits with nucleic acid preparations from partially purified Shope papilloma virus and from extracts of the papillomas of domestic and cottontail rabbits. *J. Exp. Med.* **114**, 485–500.

154. Ito, Y., and Evans, C. A. (1965). Tumorigenic nucleic-acid extracts from tissues of a transplantable carcinoma, Vx7. *J. Nat. Cancer Inst.* **34**, 431–437.

155. Ivanov, V. I., and Pavlov, K. A. (1963). (The effect of intra-arterial injection of diodon in experimental rabbit sarcoma). *Vop. Onkol.* **9**, 67–70.

156. Jacobs, J. L., and Houghton, J. D. (1941). Complement fixation tests on rabbits with Brown-Pearce carcinoma. *Proc. Soc. Exp. Biol. Med.* **47**, 88–90.

157. Janes, J. M., Higgins, G. M., and Herrick, J. G. (1954). Beryllium induced osteogenic sarcoma in rabbits. *J. Bone Joint Surg.* **36**, 543–552.

158. Joiner, G. N., Jardine, G. H., and Gleiser, C. A. (1971). An epizootic of Shope fibromatosis in a commercial rabbitry. *J. Amer. Med. Vet. Ass.* **159**, 1583–1587.

159. Kalfayan, B., and Kidd, J. G. (1953). Structural changes produced in Brown-Pearce carcinoma cells by means of a specific antibody and complement. *J. Exp. Med.* **97**, 145–162.

160. Kaliss, N. (1970). Dynamics of immunologic enhancement. *Transplant. Proc.* **2**, 59–67.

161. Karimov, Z. N., and Subkhankulova, F. B. (1963). (The content of serum protein fractions in transplantable osteogenic sarcoma of rabbits.) *Eksp. Med. Akad. Nauk SSR* **4**, 188–191.

162. Katase, T. (1912). Demonstration verschiedener geschwulste bei tieren (zitierte). *Verh. Jap. Pathol. Ges.* **2**, 89.

163. Kato, S., Mityamoto, H., Takahshi, M., and Kamahora, J. (1963). Shope fibroma and rabbit myxoma viruses. II. Pathogenesis of fibromas in domestic rabbits. *Biken J.* **6**, 135–143.

164. Kato, T. (1925). On a transplantable sarcoma of the rabbit. *Far East. Ass. Trop. Med. Trans. Congr., 6th, 19* pp. 933–942.

165. Kaufman, A. F. (1973). Cholangiocarcinoma not previously reported. Personal communication.

166. Kaufman, A. F., and Quist, K. D. (1970). Spontaneous renal carcinoma in a New Zealand White rabbit. *Lab. Anim. Care.* **20**, 530–532.

167. Kessel, J. F., Prouty, C. C., and Meyer, J. W. (1931). Occurrence of infectious myxomatosis in southern California. *Proc. Soc. Exp. Biol. Med.* **28**, 413–414.

168. Kidd, J. G. (1938). Antigenicity and infectivity of extracts of virus-induced rabbit papillomas. *Proc. Soc. Exp. Biol. Med.* **37**, 657–658.

169. Kidd, J. G. (1938). Immunological reactions with a virus causing papillomas in rabbits. I. Demonstration of a complement-fixation reaction: Relation of virus-neutralizing and complement-binding antibodies. *J. Exp. Med.* **68**, 703–724.

170. Kidd, J. G. (1938). Immunological reactions with a virus causing papillomas in rabbits. II. Properties of the complement-binding antigen present in extracts of the growths: Its relation to the virus. *J. Exp. Med.* **68**, 725–736.

171. Kidd, J. G. (1938). Immunological reactions with a virus causing papillomas in rabbits. III. Antigenicity and pathogenicity of extracts of the growths of wild and domestic species: General discussion. *J. Exp. Med.* **68**, 737–760.

172. Kidd, J. G. (1938). A complement-binding antigen in extracts of the Brown-Pearce carcinoma of rabbits. *Proc. Soc. Exp. Biol. Med.* **38**, 292–295.

173. Kidd, J. G. (1941). The detection of a "masked" virus (the Shope papilloma virus) by means of immunization. Results of immunization with mixtures containing virus and antibody. *J. Exp. Med.* **74**, 231–244.

174. Kidd, J. G. (1944). Suppression of the growth of the Brown-Pearce tumor by a specific antibody. *Science* **99**, 348–350.

175. Kidd, J. G. (1946). Suppression of growth of Brown-Pearce tumor cells by a specific antibody. With a consideration of the nature of the reacting cell constituent. *J. Exp. Med.* **83**, 227–250.

176. Kidd, J. G. (1946). Distinctive constituents of tumor cells and their possible relation to phenomena of autonomy, anaplasia and cancer causation. *Cold Spring Harbor Symp. Quant. Biol.* **11**, 94–112.

177. Kidd, J. G., Beard, J. W., and Rous, P. (1936). Serological reactions with a virus causing rabbit papillomas which become cancerous. I. Tests of the blood of animals carrying the papilloma. *J. Exp. Med.* **64**, 63–78.

178. Kidd, J. G., and Parsons, R. J. (1936). Tissue affinity of Shope papilloma virus. *Proc. Soc. Exp. Biol. Med.* **35**, 438–441.

179. Kidd, J. G., and Rous, P. (1940). A transplantable rabbit carcinoma originating in a virus-induced papilloma and containing the virus in masked or altered form. *J. Exp. Med.* **71**, 813–838.

180. Kilham, L. (1956). Propagation of fibroma virus in tissue cultures of cottontail tests. *Proc. Soc. Exp. Biol. Med.* **92**, 739–742.

181. Kilham, L. (1957). Transformation of fibroma into myxoma virus in tissue culture. *Proc. Soc. Exp. Biol. Med.* **95**, 59–62.

182. Kilham, L. (1958). Fibroma-myxoma virus transformation in different types of tissue cultures. *J. Nat. Cancer Inst.* **20**, 729–738.

183. Kilham, L., and Dalmat, H. T. (1955). Host–virus–mosquitoe relations of Shope fibromas in cottontail rabbits. *Amer. J. Hyg.* **61**, 45–54.

184. Kilham, L., and Fisher, F. R. (1954). Pathogenesis of fibromas in cottontail rabbits. *Amer. J. Hyg.* **59**, 104–112.

185. Kilham, L., and Woke, P. A. (1953). Laboratory transmission of fibromas (Shope) in cottontail rabbits by means of fleas and mosquitoes. *Proc. Soc. Exp. Biol. Med.* **83**, 296–301.

186. Kirchstein, R. L., Rabson, A. S., and Kilham, L. (1958). Pulmonary lesions produced by fibroma viruses in squirrels and rabbits. *Cancer Res.* **18**, 1340–1344.

187. Kleinman, D. L., and Chernomordikova, M. F. (1964). (Technique for using intraosseous venography for studing venous outflow into transplantable tumors.) *Vop. Onkol.* **10**, 81–87.

188. Kondrateva, A. F. (1958). (A transplantable strain of an osteogenic sarcoma in rabbits.) *Tr. Inst. Onkol., Akad. Med. Nauk SSSR* **2**, 78–91.

189. Kondrateva, A. F., and Melnikov, R. A. (1965). (Transplants of

osteogenic rabbit sarcoma in various tissues and organs.) *Vop. Onkol.* **11**, 63–69.

190. Koyama, M. (1927). Ein Fall von uterusadenom beim Kaninchen, Nebst einem Befund uber Hypertrophie und Fettsekretion der Milchdruse dieses Tieres. *Gann* **21**, 7.

191. Kraack, J. (1971). Beitrag zum vorkommen der leukose bei kaninchen. *Kleintierpraxis* **16**, 181–183.

192. Lack, H. I. (1900). A preliminary note on the experimental production of cancer. *J. Pathol. Bacteriol.* **6**, 154–157.

193. Lafenetre, H., Cortez, A., Rioux, J. A., Pages, A., Vollhardt, Y., and Quartrepages, H. (1960). Enzootie de tumeurs cutanées chez le lièvre. *Bull. Acad. Vet. Fr.* [N.S.] **33**, 379–389.

193a. Leader, R. W., and Leader, I. (1971). "Dictionary of Comparative Pathology and Experimental Biology." Saunders, Philadelphia, Pennsylvania.

194. Leinati, L., Mandelli, G., and Carrara, O. (1959). Lesioni cutanee nodulari nelle lepri della pianura padana. *Atti Soc. Ital. Sci. Vet.* **13**, 429–435.

195. Leinati, L., Mandelli, G., Carrara, O., Cilli, V., Castrucci, G. and Scatozza, F. (1961). Ricerche anatomo-istopathologiche e virologiche sula mattia cutanea nodulare delle lepri padane. *Boll. Ist. Sieroter. Mina.* **40**, 295–328.

196. Leitch, A. (1911). Carcinoma of the uterus in a rabbit. *Proc. Roy. Soc. Med.* **5**, 1–17.

197. Lesbouyries, G. (1963). "Pathologie du lapin." Librairie Malonine S. A., Paris.

198. Llambes, J. J., and Mendez, J. G. (1954). Linfosarcoma gastrointestinal del conejo. *Arch. Cubanos Cancerol.* **13**, 51–58.

199. Lloyd, B. J., Jr., and Kahler, H. (1955). Electron microscopy of the virus of rabbit fibroma. *J. Nat. Cancer. Inst.* **15**, 991–994.

200. Loliger, H. C. (1966). Uber das Vorkommen on leukosen beim kaninchen. *Berlin. Muenchen. Tieraertzl. Wochenschr.* **79**, 192–194.

201. Loliger, H. C. (1968). Spontane Geschwulste und leukosen beim kaninchen. *Z. Versuchstierk.* **10**, 55–61.

202. Lombard, C. (1959). Nouvelle observation de cancer uterin chez la lapine. *Bull. Acad. Vet. Fr.* [N.S.] **32**, 447–451.

203. Lombard, C. (1962). Deux cas de tumeurs du gibier. *Bull. Acad. Vet. Fr.* [N.S.] **35**, 39–43.

204. Lombard, C. (1962). "Cancerologie comparée. Cancer spontane. Cancer expérimental." Doin, Paris.

205. Lombard, C., and Goulard, G. (1960). Nouvelle observation de cancer mammaire chez la lapine avec tentative de breffe. *Bull. Acad. Vet. Fr.* [N.S.] **33**, 335–339.

206. Lopushinsky, T., and Fay, L. D. (1967). Some benign and malignant neoplasms of Michigan cottontail rabbits. *Bull. Wildl. Dis. Ass.* **3**, 148–151.

207. Lubarsch, O. (1905). Uber einem grossen Nierentumor beim Kaninchen. *Zentralbl. Pathol.* **16**, 342–345.

208. Lurie, M. B. (1964). "Resistance to Tuberculosis: Experimental Studies in Native and Acquired Defensive Mechanisms." Harvard Univ. Press, Cambridge, Massachusetts.

209. MacKenzie, I., and Kidd, J. G. (1945). Incidence and specificity of the antibody for a distinctive constituent of the Brown-Pearce tumor. *J. Exp. Med.* **82**, 41–64.

210. Maeda, K. (1930). Relation between thymus gland and growth of tumor (rabbit sarcoma). *Trans. Jap. Pathol. Soc.* **20**, 659–661.

211. Margulies, H. (1901). Teratom der hypophyse beim Kaninchen. *Neurol. Zentralbl.* **00**, 1027.

212. Marie, P., and Aubertin, C. (1911). Cancer de l'utérus chez une lapine de neuf ans. *Bull. Ass. Fr. Etude Cancer* **4**, 253–255.

213. Marshall, I. D., and Regnery, D. C. (1960). Myxomatosis in a California brush rabbit (*Sylvilagus bachmani*). *Nature (London)* **118**, 73–74.

214. Marshall, I. D., and Regnery, D. C. (1963). Studies in the epidemiology of myxomatosis in California. III. The response of brush rabbits (*Sylvilagus bachmani*) to infection with exotic and enzootic strain myxoma virus and the relative infectivity of the tumors for mosquitoes. *Amer. J. Hyg.* **77**, 213–219.

215. Marshall, I. D., Regnery, D. C., and Grodhaus, G. (1963). Studies in the epidemiology of myxomatosis in California. I. Observation on two outbreaks of myxomatosis in coastal California and the recovery of myxoma virus from brush rabbit (*Sylvilagus bachmani*). *Amer. J. Hyg.* **77**, 195–204.

216. Medlar, E. M., and Sasano, K. T. (1937). An interpretation of the nature of Hodgkin's disease. III. Report of a neoplasm in the rabbit which corresponds closely to Hodgkin's disease in man. *Amer. J. Cancer* **29**, 102–110.

217. Meier, H., and Fox, R. R. (1973). Hereditary lymphosarcoma in WH rabbits and hemolytic anemia associated with thymoma in strain X rabbits. *Bibl. Haematol. (Basel)* **39**, 72–92.

218. Meier, H., Myers, D. D., Fox, R. R., and Laird, C. W. (1970). Occurrence, pathological features, and propagation of gonadal teratomas in inbred mice and in rabbits. *Cancer Res.* **30**, 30–34.

219. Meissner, W. A., Sommers, S. C., and Sherman, G. (1957). Endometrial hyperplasia, endometrial carcinoma and endometriosis produced experimentally by estrogen. *Cancer* **10**, 500–509.

220. Mello, U. (1929). Di una affezione neoplasica ad andamenio epizootico nelle lepri. *Ann. Sta. Sper. Lotta Mal. Inform. Bestiame Piem Lig.* **2**, 47.

221. Merrian, J. C. Easterday, C. L., McKay, D. G., and Hertig, A. T. (1960). Experimental production of endometrial carcinoma in the rabbit. *Obstet. Gynecol.* **16**, 253–262.

222. Miyadi, T. (1940). Uber einen fall von grosser Nierengeschwulst beim Kaninchen. *Gann* **34**, 374–375.

223. Mori, G. (1939). The oxidase in the various transplantable tumors. *Acta Dermatol.-Kyoto* **33**, 2–8.

224. Mori, G. (1939). The peroxidase in the various transplantable tumors. *Acta Dermatol.-Kyoto* **33**, 25–31.

225. Moses, A. (1911). O virus do mixoma dos coelhos. *Mem. Inst. Oswaldo. Cruz* **3**, 46–53.

226. Moulton, J. E. (1961). "Tumors in Domestic Animals." Univ. of California Press, Los Angeles, California.

227. Murohara, N. (1930). Relation between the thyroid gland and the growth of the tumor (rabbit sarcoma). *Trans. Jap. Pathol. Soc.* **20**, 655–659.

228. Nakano, M. (1929). Experimentelle studien über die transplantation des kaninchensarkomas mittel bluttransfusion. *Gann* **23**, 114–124.

229. Nakano, M. (1930). Transplantation of a rabbit sarcoma by blood transfusion. *Trans. Jap. Pathol. Soc.* **20**, 721–731.

230. Nesburn, A. B. (1969). Isolation and characterization of a herpeslike virus from New Zealand albino rabbit kidney cell cultures; a probable reisolation of virus III of Rivers. *J. Virol.* **3**, 59–69.

231. Nicod, J. L., and Burgisser, H. (1961). Lymphadenopathie gigantofulliculaire (Brill-Symmers) chez le lievre. *Pathol. Microbiol.* **24**, 409–414.

232. Nicod, J. L., and Burgisser, H. (1964). Lymphomes malin chez le lièvre et le chat. *Pathol. Microbiol.* **27**, 1–7.

233. Niessen, V. (1913). Ein Fall von Krebs beim Kaninchen. *Deut. Tieraertzl. Wochenschr.* **21**, 637.

234. Niessen, V. (1927). Ein Fall von Leberkrebs beim Kaninchen auf experimenteller Grundlege. *Z. Krebsforsch.* **24**, 272.

235. Noyes, W. F. (1959). Studies on the Shope rabbit papilloma virus. II. The location of infective virus in papillomas of the cottontail rabbit. *J. Exp. Med.* **109**, 423–428.

236. Noyes, W. F., and Mellors, R. C. (1958). Fluorescent antibody detection of Shope papilloma virus in papillomas of the wild and domestic rabbit. *J. Exp. Med.* **106**, 555–562.

237. Nurnberger, L. (1912). Uber einen tumoren in der kaninchenniere vom Typus der embryonalen Drusengeschwulste des Menschen. *Beitr. Pathol.* **52**, 523–539.

238. Oberling, C. (1927). Sarcome embryonnaire (adeno-sarcome) du rein chez un lapin. *Bull. Ass. Fr. Etude Cancer* **16**, 708–710.

239. Orr, J. W. (1936). Adamantinoma of the jaw in a rabbit. *J. Pathol. Bacteriol.* **42**, 703–704.

240. Orr, J. W. (1939). A malignant tumor of the thymus in a rabbit. *Amer. J. Cancer* **35**, 269–274.

241. Orr, J. W., and Polson, C. F. (1938). Uterine cancer in the rabbit. *Amer. J. Cancer* **32**, 114–125.

242. Paine, and Peyron (1918). Seminome du lapine. Bull. Ass. Fr. Etude *Cancer* **11**, 547.

243. Parsons, R. J., and Kidd, J. G. (1936). A virus causing oral papillomatosis in rabbits. *Proc. Soc. Exp. Biol. Med.* **35**, 441–443.

244. Parsons, R. J., and Kidd, J. G. (1943). Oral papillomatosis of rabbits: a virus disease. *J. Exp. Med.* **77**, 233–250.

245. Pascal, R. R. (1961). Plasma cell myeloma in the brain of a rabbit. *Cornell Vet.* **51**, 528–535.

246. Pearce, J. M., and LaSorte, A. F. (1954). Effects of an anti-hyaluronidase substance and of hyaluronidase on growth of virus-induced fibroma. *Proc. Soc. Exp. Biol. Med.* **86**, 573–577.

247. Pearce, L., and Greene, H. S. N. (1938). Clinical and genetic observations on spontaneous mammary carcinoma of the rabbit. *Amer. J. Pathol.* **14**, 655–657.

248. Pearce, L., and Murphy, J. B. (1927). Further observation on the inability to transmit a rabbit neoplasm by cell-free methods. *J Exp. Med.* **46**, 205–211.

249. Petit, G. (1909). Cancer du poumon. *Bull. Ass. Fr. Etude Cancer* **2**, 25.

250. Petit, G. (1910). Tumeur epiploigue provenant d'un pancréas accessoire. *Trans. Conf. Int. Etude Cancer, 2nd,* p. 209.

251. Piening, C., and Wermerssen, H. (1965). Leukose bei einem Hase. *Deut. Tieraertzl. Wochenschr.* **72**, 488–489.

252. Polson, C. (1927). Tumors of the rabbit. *J. Pathol. Bacteriol.* **30**, 603–614.

253. Raso (1936). Su di uno caso di fibroleimioma retroperiotoneale del coniglio. *Nuova Vet.* **14**, 305 [cited by Herr (132a)].

254. Razok, E. J., Shipkowitz, N. L., and Richter, W. R. (1966). Rabbit oral papillomatosis: Ultrastructure of experimental infection. *Cancer Res.* **26**, 160–165.

255. Regnery, D. C. (1971). The epidemic potential of Brazilian myxoma virus (Lausanne strain) for three species of north American cottontails. *Amer. J. Epidemiol.* **94**, 514–519.

256. Regnery, D. C., and Marshall, I. D. (1971). Studies in the epidemiology of myxomatosis in California. IV. The susceptibility of six leporid species to California myxoma virus and the relative infectivity of their tumors for mosquitoes. *Amer. J. Epidemiol.* **94**, 508–513.

257. Rendtorff, R. C. (1961). The transmission of rabbit fibromas by nematodes. *J. Parasitol.* **47**, 185.

258. Rendtorff, R. C., and Wilcox, A. (1957). The role of nematodes as an entry for viruses of Shope's fibromas and papillomas of rabbits. *J. Infec. Dis.* **100**, 119–123.

259. Richter, W. R., Shipkowitz, N. L., and Rdzok, E. J. (1964). Oral papillomatosis of the rabbit: An electron microscopic study. *Lab. Invest.* **13**, 430–438.

260. Rive, M., Levaditi, J. C., and Varenne, H. (1959). Neuf cas de cancer chez le lapin angora. *Rec. Vet. Med.* **135**, 31–43.

261. Rivers, T. M., and Tillett, W. S. (1923). Studies on varicella. The susceptibility of rabbits to the virus of varicella. *J. Exp. Med.* **38**, 673–691.

262. Rous, P., and Beard, J. W. (1934). Carcinomatous changes in virus-induced papillomas of the skin of the rabbit. *Proc. Soc. Exp. Biol. Med.* **32**, 578–580.

263. Rous, P., and Beard, J. W. (1934). A virus-induced mammalian growth with the characters of a tumor (the Shope rabbit papilloma) I. Growth on implantation within favorable host. *J. Exp. Med.* **60**, 701–722.

264. Rous, P., and Beard, J. W. (1934). A virus-induced mammalian

growth with the characters of a tumor (the Shope rabbit papilloma). III. Further characters of the growth: General discussion. *J. Exp. Med.* **60**, 741–766.

265. Rous, P., and Beard, J. W. (1935). The progression to carcinoma of virus-induced rabbit papillomas (Shope). *J. Exp. Med.* **62**, 523–548.

266. Rous, P., and Friedewald, W. F. (1941). The carcinogenic effect of methylcholanthrene and of tar on rabbit papillomas due to a virus. *Science* **94**, 495–496.

267. Rous, P., and Friedewald, W. F. (1944). The effect of chemical carcinogens on virus-induced rabbit papillomas. *J. Exp. Med.* **79**, 511–538.

268. Rous, P., and Kidd, J. G. (1936). The carcinogenic effect of a virus upon tarred skin. *Science* **83**, 468–469.

269. Rous, P., and Kidd, J. G. (1940). The activating, transforming, and carcinogenic effects of the rabbit papilloma virus (Shope) upon implanted tar tumors. *J. Exp. Med.* **71**, 787–812.

270. Rous, P., Kidd, J. G., and Smith, W. E. (1952). Experiments on the cause of the rabbit caecinomas derived from virus-induced papillomas. II. Loss by the Vx2 carcinoma of the power to immunize hosts against the papilloma virus. *J. Exp. Med.* **96**, 159–174.

271. Rusk, G. Y., and Epstein, N. (1927). Adenocarcinoma of the uterus in a rabbit. *Amer. J. Pathol.* **3**, 235–240.

272. Sabin, F. R., Doan, C. A., and Forkner, C. E. (1932). The production of osteogenic sarcomata and the effects on lymph nodes and bone marrow of intravenous injectious of radium chloride and mesothorium in rabbits. *J. Exp. Med.* **55**, 267–290.

273. Salaskin, D. A. (1929). Ein seltener fall von bosartiger geschwulst beim kaninchen. *Z. Krebsforsch.* **30**, 371–379.

274. Salm, R., and Field, J. (1965). Osteosarcoma in a rabbit. *J. Pathol. Bacteriol.* **89**, 400–402.

275. Salomon, S. (1933). Geschwulste beim Hasen. *Berlin Tieraertzl. Wochenschr.* **49**, 37–38.

276. Sanarelli, G. (1898). Das myxomatogene Virus. Beitrag zum Studium Krankheitserreger ausserhalb des Sichtbaren. *Zentrabl. Bakteriol., Parasitenk. Infektienskr., Abt. 1* **23**, 865–873.

277. Saphir, O., Appel, M., and Strauss, A. A. (1941). Growth of the Brown-Pearce carcinoma in the anterior chamber of the eyes of tumor-immune rabbits. *Cancer Res.* **1**, 545–547.

278. Schectman, A. M., Cohen, M. J., and Berkowitz, E. C. (1950). Culture of Brown-Pearce carcinoma in the embryonated egg. *Proc. Soc. Exp. Biol. Med.* **74**, 784–789.

279. Schmorl, G. (1903). Carcinome des poumon chez le lapine. Tumeur primitive de l'estomac. *Verh. Deut. Ges. Pathol.* **6**, 136.

280. Schultze, W. H. (1913). Beobachtungen an einem transplantablen Kaninchensarkom. *Verh. Deut. Ges. Pathol.* **16**, 358–362.

281. Schultze, W. H. (1914). Transplantables Kaninchensarkom und leukamie. *Verh. Deut. Ges. Pathol.* **17**, 382–386.

282. Schurch, O., and Uehlinger, E. (1931). Experimentelles knochensarkom nach radium-bestrahlung bei einem kaninchen. *Z. Krebsforsch.* **33**, 476–484.

283. Schweizer, F. (1888). Uber ein Cystadenoma papilliferum in einer kaninchenleber. *Arch. Pathol. Anat. Physiol. Klin. Med.* **113**, 209.

284. Scott, A. (1917). Tumors of the kidney in rabbits. *J. Cancer Res.* **2**, 367–369.

285. Seifried, O. (1937). "Die Krankheiten des Kaninchens." Springer-Verlag, Berlin and New York.

286. Selinow (1907). Uterus Krebs bei einem weiblichen Kaninchen. *Charkowsi Med. J.* **4**, 23; Abstract in *Zentrabl. Pathol.* **19**, 122 (1908).

287. Shattock, S. G. (1900). A specimen of "spontaneous" carcinoma of the uterus in the rabbit, with a criticism of the mechanical hypothesis of the origin of caecinoma. *Trans. Pathol. Soc. London* **51**, 56–65.

288. Shima, (1908). Teratom im Kaninchenhirn. *Berl. Vienna Neurol. Inst.* **14** [cited by Dobberstein and Tamaschke (75)].

289. Shope, R. E. (1932). A transmissible tumor-like condition in rabbits. *J. Exp. Med.* **56**, 796–802.

290. Shope, R. E. (1932). A filterable virus causing a tumor-like con-

dition in rabbits and its relationship to Virus myxomatosum. *J. Exp. Med.* **56**, 803–822.

291. Shope, R. E. (1933). Infectious papillomatosis of rabbits. With a note on the histopathology by E. Weston Hurst. *J. Exp. Med.* **58**, 607–624.

292. Shope, R. E. (1935). Serial transmission of virus of infectious papillomatosis in domestic rabbits. *Proc. Soc. Exp. Biol. Med.* **32**, 830–832.

293. Shope, R. E. (1936). A change in rabbit fibroma virus suggesting mutation. II, Behavior of the variant virus in cottontail rabbits. *J. Exp. Med.* **63**, 173–178.

294. Shope, R. E. (1937). Immunization of rabbits to infectious papillomatosis. *J. Exp. Med.* **65**, 219–231.

295. Shtacher, G. (1969). Selective renal involvement in the early development of hypercalcenia and hypophosphatemia in Vx2 carcinoma-bearing rabbits: Studies on serum and tissues alkaline phosphatase and renal handling of phosphorus. *Cancer Res.* **29**, 1512–1518.

296. Smith, M. H. D. (1948). Propagation of rabbit fibroma virus in the embryonated egg. *Proc. Soc. Exp. Biol. Med.* **69**, 136–140.

297. Smith, M. H. D. (1952). The Berry-Dedrick transformation of fibroma into myxoma in the rabbit. *Ann. N. Y. Acad. Sci.* **54**, 1141–1152.

298. Smith, W. E., Kidd, J. G., and Rous, P. (1952). Experiments on the cause of the rabbit carcinomas derived from virus-induced papillomas. I. Propagation of several of the cancers in sucklings, with etiological tests. *J. Exp. Med.* **95**, 299–318.

299. Snell, G. D. (1953). Transplantable tumors. *In* "The Physiopathology of Cancer" (F. Homburger and W. H. Fishman, eds.), pp. 338–391. Harper (Hoeber), New York.

300. Sprehn, C. (1953). Die myxomatose und andere wichtige krankheiten der kaninchen. *Kleintierzuechter, Thuring* **7**, 419.

301. Stewart, H. L., Snell, K. C., Dunham, L. C., and Schlyen, S. M., eds. (1959). "Transplantable and Transmissible Tumors of Animals," Atlas Tumor Pathol., Sect. XII, No. 40. Armed Forces Inst. Pathol., Washington, D. C.

302. Stilling, H., and Beitzke, H. (1913). Uber uterustumoren bei kaninchen. *Virchows Arch. Pathol. Anat. Physiol.* **214**, 358–380.

303. Stone, R. S., Shope, R. E., and Moore, D. H. (1959). Electron microscope study of the development of the papilloma virus in the skin of the rabbit. *J. Exp. Med.* **110**, 543–546.

304. Sustmann (1922). Multiple melanombildungen beim kaninchen. *Deut. Tieraerztl. Wochenschr.* 402.

305. Syverton, J. T., and Berry, G. P. (1935). Carcinoma in the cottontail rabbit, following spontaneous virus papilloma (Shope). *Proc. Soc. Exp. Biol. Med.* **33**, 399–400.

306. Syverton, J. T., Dascomb, H. E., Koomen, J., Jr., Wells, E. B., and Berry, G. P. (1950). The virus induced papilloma-to-carcinoma sequence. I. The growth pattern in natural and experimental infections. *Cancer Res.* **10**, 379–384.

307. Syverton, J. T., Dascomb, H. E., Wells, E. B., Koomen, J., Jr., and Berry, G. P. (1950). The virus induced papilloma-to-carcinoma sequence. II. Carcinomas in the natural host the cottontail rabbit. *Cancer Res.* **10**, 440–444.

308. Syverton, J. T., Wells, E. B., Koomen, J., Jr., Dascomb, H. E., and Berry, G. P. (1950). The virus induced rabbit papilloma-to-carcinoma sequence. III. Immunological tests for papilloma virus in cottontail carcinomas. *Cancer Res.* **10**, 474–482.

309. Tapp, E. (1966). Beryllium induced sarcomas of the rabbit tibia. *Brit. J. Cancer* **20**, 778–783.

310. Tapp, E. (1969). Osteogenic sarcoma in rabbits following subperiosteal implantation of beryllium. *Arch. Pathol.* **88**, 89–94.

311. Tapp, E. (1969). Changes in rabbit tibia due to direct implantation of beryllium salts. *Arch. Pathol.* **88**, 521–529.

312. Tomozawa, S. (1931). Contribution of the knowledge of the metastasis of rabbit sarcoma. *Trans. Jap. Pathol. Soc.* **21**, 748–754.

313. Twort, C. C. (1937). Carcinoma of the uterus of the rabbit with splenic metastasis. *J. Pathol. Bacteriol.* **44**, 492.

313a. Ubertini, T. R. (1972). Etiological study of a lymphosarcoma in a domestic rabbit. *J. Nat. Cancer Inst.* **48**, 1507–1511.

314. Uehlinger, E. (1943). Ueber experimentelle geschwulsterzeugung durch radioaktive substanzen. *Helv. Med. Acta* **10**, 694–695.

315. Upton, A. C. (1968). Radiation carcinogenesis. *Methods Cancer Res.* **4**, 53–82.

316. Upton, A. C., and Furth, J. (1958). Host factors in the pathogenesis of leukemia in animals and in man. *Proc. Nat. Cancer Conf., 3rd, 1957* pp. 312–324.

317. Usawa, T. (1930). (Spontaneous multiple adenocarcinoma of the uterus in a rabbit.) *Sei-I-Kai Med. J.* **49**, 5.

318. Van Kampen, K. R. (1968). Lymphosarcoma in the rabbit. A case report and general review. *Cornell Vet.* **58**, 121–129.

319. Vogel, S. D., Enneking, W. F., and Thomas, W. C. (1967). Effect of thyroparathyroidectomy on hypercalcemia associated with malignancy. *Endocrinology* **80**, 404–408.

319a. von Dungern, E., and Coca, A. F. (1909). Uber Hasensarkome, die in Kaninchen wachsen und ueber dans Wesen der Geschwulstimmunitat. *Z. Immunitaetsforsch. Exp. Ther.* **2**, 391–398.

320. Wagner, G. A. (1905). Uber multiple tumoren in uterus des kaninchens. *Zentrabl. Pathol.* **16**, 131–135.

321. Wallner, A. (1921). Uber einen fall von transplantablem kaninchensarkom. *Z. Krebsforsch.* **18**, 215–225.

322. Watrach, A. M., Small, E., and Case, M. T. (1970). Canine papilloma: Progression of oral papilloma to carcinoma. *J. Nat. Cancer Inst.* **45**, 915–920.

323. Watrin, J., and Floretin, P. (1930). Tumor uterine et secretion lactee chez la lapine. *C. R. Soc. Biol.* **104**, 1286–1288.

324. Weisbroth, S. H. (1972). Tumors of *Oryctolagus* newly reported in this publication.

324a. Weisbroth, S. H. Unpublished observations.

325. Weisbroth, S. H., and Hurvitz, A. (1969). Spontaneous ostoegenic sarcoma in *Oryctolagus cuniculus* with elevated serum alkaline phosphatase. *Lab. Anim. Care* **19**, 263–266.

326. Weisbroth, S. H., and Scher, S. (1970). Spontaneous oral papilloma in rabbits. *J. Amer. Vet. Med. Ass.* **157**, 1940–1944.

327. Wilson, J. R., Gertner, H. R., Jr., Edward, B. S., and Woodward, E. R. (1964). Calcification in the rabbit with Vx2 carcinoma. *Exp. Med. Surg.* **22**, 338–348.

328. Wilson, J. R., Merrick, H., Vogel, S. D., and Woodward, E. R. (1965). Hyperparathyroid-like state in rabbits with the Vx2 carcinoma. Further studies. *Amer. Surg.* **31**, 145–152.

329. Wilson, J. R., Merrick, H., and Woodward, E. R. (1961). Hypercalcemia simulating hyperparathyroidism induced by Vx2 carcinoma of rabbit. *Ann. Surg.* **154**, 485–490.

330. Witherspoon, J. T. (1938). Two spontaneous uterine tumors in a rabbit: An hormonal investigation. *Amer. J. Cancer* **33**, 389–393.

331. Wood, S., Jr., Baker, R. R., and Johnson, J. H. (1967). Failure of low molecular weight dextrans to alter frequency of lung metastasis. Report of the V2 carcinoma of the rabbit. *Cancer* **20**, 281–285.

332. Woodroofe, G. M., and Fenner, F. (1965). Viruses of the myxomafibroma subgroup of the poxviruses. I. Plaque production in cultures cells, plaque reduction tests, and cross protection tests in rabbits. *Aust. J. Exp. Biol. Med. Sci.* **43**, 123–142.

333. Yuill, T. M. (1970). Myxomatosis and fibromatosis of rabbits, hares, and squirrels. *In* "Infectious Diseases of Wild Animals" (J. W. Davis, L. H. Karstad, and D. O. Trainer, eds.), pp. 104–130. Iowa State Univ. Press, Ames, Iowa.

334. Yuill, T. M., and Hanson, R. P. (1964). Infection of suckling cottontail rabbits with Shope's fibroma virus. *Proc. Soc. Exp. Biol. Med.* **117**, 376–380.

335. Zbindin, G., and Studer, A. (1957). Generalisierte Veranderungen des lymphatischen systems beim kaninchen. Demonstration eines falles. *Schweiz. Z. Pathol. Bakteriol.* **20**, 710.

336. Zilber, L. A. (1958). Specific tumor antigens. *Advan. Cancer Res.* **5**, 291–329

337. Zschocke (1914). Pleuropneumonie und Leukamie bei Kaninchen. *Berl. Veterinarw. Sachsen.* **58**, 93.

CHAPTER 15

Inherited Diseases and Variations

J. Russell Lindsey and Richard R. Fox

I. INTRODUCTION

The role of genetics in disease states of *Oryctolagus* rabbits has been seriously neglected by producers and users of these animals. Commercial rabbit breeders, like other livestock producers, have been most intent on merely eliminating the unusual or abnormal animal as quietly as possible because of economic considerations. For the most part, biomedical investigators have simply accepted the rabbit as *the* animal for a fairly restricted group of experimental purposes, such as immunology and reproductive physiology, without giving serious thought to other potential uses. Thus, it is largely because of the remarkable contributions of a handful of immortals in the field of rabbit genetics (such as W. E. Castle, H. Nachtsheim, H. S. N. Greene, and P. B. Sawin) that sufficient knowledge has been accrued for this chapter.

The rabbit is of special value to biomedical research because of the rapidly increasing number of genetically defined model systems now available in some carefully maintained stocks. These models are all the more valuable because the species already has been proved in the laboratory and is so well characterized in a few areas, e.g., skeletal morphogenesis (165, 168), pelage and coat color variants (Chapter 1), and serological traits (Chapter 7). The animals, the breeders, and ultimately, all mankind stand to benefit from studies utilizing these unique experimental models and others to be discovered in the future.

At the present time there are approximately 70 known genetic loci in the rabbit (see Chapter 1 for complete listing). About one-third of these have to do with hair coat and another one-third control blood groups and antibody production. In this chapter attention will be focused on most of the remaining loci and genetic traits which have been recognized, namely, a large number of disease conditions and physiological or anatomical variations. They are divided into two broad groups: (a) those controlled by approximately 38 mutant genes and (b) some 10 additional conditions recognized as most likely familial or polygenic, but presently lacking fully documented modes of inheritance. Unfortunately, information on these diverse subjects varies from fragmentary to marginally complete. Only rarely does the information available permit meaningful statements about prevalence in present day commercial stocks of rabbits.

II. CONDITIONS CONTROLLED BY SINGLE (MUTANT) GENES

Those diseases and variations (excluding coat color, pelage, and serological variants) controlled by single or mutant genes fall naturally into eight groups according to organ system affected or physiological characteristics of the condition, as will be seen in the paragraphs below. The name(s) of individual traits will be given as subheadings, followed by the possible genotype(s) of affected animals.

A. Behavioral Mutants

1. Acrobat *(ak/ak)*

This unusual behavioral pattern was first reported in a French strain of rabbit by Letard (83, 84). He considered it the result of a simple recessive gene which he designated *ak*. Similar behavior has been observed in Dutch rabbits in England (152). Conversations with breeders indicate that it also has been observed rarely in the United States.

These animals have the ability to walk on all four feet in a perfectly normal manner, but frequently go into a handstand position and proceed on the forefeet while maintaining the hindquarters in a vertical position (Fig. 1). No physiological explanation is known. The animals observed by Letard apparently suffered from cataracts and lens dislocations when they became several months old, considerably later than onset of the behavior ascribed to the acrobat gene.

2. Epilepsy (Audiogenic Seizures) *(ep/ep)*

The spontaneous occurrence of epileptiform seizures in the Viennese White rabbit *(v/v)* has been the subject of several reports by Nachtsheim (114, 118). Seizures occurred most often at 6 to 8 weeks of age. The typical attack was characterized by four phases. First, there was a brief preconvulsive period of restlessness or excitement terminated by falling unconscious. The second, or tonic phase was characterized by complete body rigidity and nystagmus due to generalized tonic contraction of musculature including external muscles of the eyes. The tonic phase

Fig. 1. Dutch rabbit (at left) demonstrating behavior ascribed to acrobat *(ak/ak)* mutation. (From *Bibliographia Genetica* **17**, 229, 1958, by permission of Dr. Roy Robinson and Martinus Nijhoff Publishers.)

lasted only a few seconds and was replaced by a third and longer phase of moaning associated with the decrease of body rigidity, opening of the eyes, and spasmodic chewing movements with salivation. Phase two and three were sometimes combined to form a convulsive stage. The first three phases usually lasted about 15 seconds, or seldom up to 1 minute. In the final or recovery phase the animal righted itself again, appearing groggy at first. Animals usually appeared entirely normal within 3 to 4 minutes after the beginning of each attack.

Nachtsheim (118) observed seizures in association with a variety of normal events such as handling or feeding of animals, cleaning of cages, or unusual noises such as whistling. His conclusion, based on observations without controlled studies, was that visual or auditory stimuli precipitated the attacks. He did, however, attribute seizure susceptibility to a single recessive gene. He further postulated linkage between the *ep* and *v* loci in his strain of rabbits. This was not observed, however, by Ross *et al.* (158) at The Jackson Laboratory.

Subsequent investigations have confirmed many of Nachtsheim's preliminary observations as others have administered controlled sound stimuli to animals of several strains. Such studies have included AC and ACEP strains (2), the Small Silvery strain (147), and crossbred animals (66). In addition, Hohenboken and Nellhaus (65) have bred a number of susceptible lines using animals of the Beveren breed, a larger white rabbit with blue eyes. The latter authors were unable to disprove that susceptibility was due to a single recessive gene with incomplete penetrance. However, their data favored the idea of susceptibility being a threshold character under the influence of several conditioning genes. Ross *et al.* (158) have shown that seizure incidence is related inversely to age.

B. Neuromuscular Mutants

1. Ataxia *(ax/ax)*

In 1936 Sawin *et al.* (160) observed a few animals in a family of Chinchilla rabbits that showed incoordination at about 2 months of age. They became progressively worse and died 2 weeks after the onset. Selected matings revealed an autosomal recessive mode of inheritance for this unusual disease (160) generally known as "hereditary ataxia of the rabbit" or "Sawin-Anders ataxia." Subsequently, the mutant gene *(ax)* has been maintained at the Jackson Laboratory in the AX strain. The mutation has not been reported in any other stock. Affected animals have a normal karyotype. Heterozygotes are essential to perpetuating the gene and can be recognized only by test matings with known heterozygotes.

Homozygous recessive animals are clinically normal up until adolescence. On the average, first signs of the disease appear at about $2\frac{1}{2}$ months of age and include reduced inclination to move, alternate stepping of the hindfeed instead of hopping, stiffness or hyperextension of one hindleg with most of the weight being carried by the other leg, and a tendency for resting animals to brace themselves against the side of the cage. As the disease progresses, there is increasing incoordination and fatiguability, appearance of a coarse tremor of the head and neck during attempts to eat or drink, lateral nystagmus, increased tendon reflexes, bilateral flexion of the thighs in response to patellar tendon stimulation, prostration, purposeless paddling movements of the legs, opisthotonus, hypothermia, and death (1b, 160). In early generations after discovery of the mutation the disease usually followed a progressive course ending in death 12 to 15 days after onset, but in much later generations, as those available at present, approximately half the affected animals live 20 days beyond the date of first symptoms, and a few live as long as 143 days.

The anatomical changes in hereditary ataxia have been studied extensively by methods of light and electron microscopy (1b, 122–124). Lesions are bilaterally symmetrical and occur predominantly in parts of the brain derived embryologically from the rhombic lip of the pontine flexure. Changes appear first and reach greatest severity in the cochlear, vestibular, and central cerebellar nuclei. Lesions of lesser intensity occur in the trapezoid body, lateral lemniscus, reticular substance, motor V and VII nuclei, and surrounding reticular substance of the brain. Coalescence of swollen glial processes accounts for the spongy appearance which characterizes affected nuclei in the light microscope (Fig. 2). The major alterations in

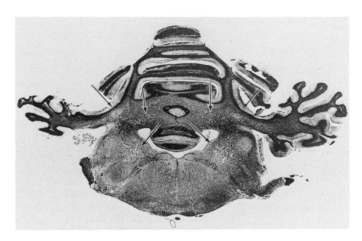

Fig. 2. Histological section through brainstem and cerebellar nuclei of rabbit *(ax/ax)* killed after 16 days of clinical ataxia. Note spongy appearance of the medial and lateral cerebellar and Deiters' nuclei (arrows) due to edema and accumulation of glycogen. Hematoxylin and eosin, × 4. (From O'Leary *et al.*, *J. Comp. Neurol.* **145**, 310, 1972, used by permission of the authors and publisher.)

these nuclei are edema and accumulation of glycogen in cells of the neuropil, particularly in astrocytes, and secondary degeneration of axon terminals (Fig. 3).

The neurons proper are generally well preserved but a few may show degenerative changes. There may be splitting of the myelin lamellae or complete loss of myelin around nerve fibers in affected areas. Secondary degenerative changes also have been observed in mossy and climbing terminals and in the cerebellar cortex. In the average case severity of anatomical changes correlates poorly with symptoms during the first 10 days after onset and only moderately well thereafter.

There can be no doubt that accumulation of glycogen in selected nuclei of the brain is of major significance in pathogenesis of hereditary ataxia. No specific biochemical defect primary to this event has been demonstrated, however, although various aspects of protein, carbohydrate, and inositide metabolism in brains of affected animals have been investigated (30, 150, 151, 184). Of considerable comparative interest is the fact that recently a case of cerebral glycogenosis with principal lesions in the neuropil was reported in a child (149).

2. Tremor (Shaking Palsy) *(tr/tr)*

Nachtsheim (107, 118) discovered a type of shaking palsy in rabbits of a German breed. It was transmitted as a simple recessive trait and was designated by the gene symbol *tr* for tremor.

The disease made its appearance at 10–14 days of age and, at first, was characterized by a fine tremor involving the entire body and head. The tremors became coarse with passage of time. They fully abated during complete rest

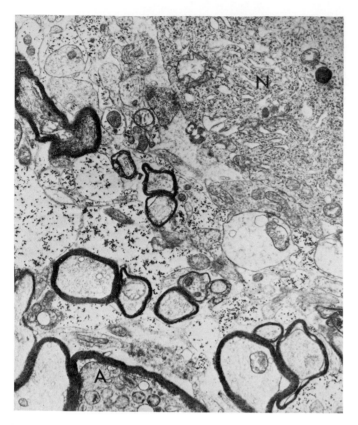

Fig. 3. Electron micrograph from vestibular nucleus of *ax/ax* rabbit after 89 days of clinical ataxia. A neuron (N) and its axosomatic synapses are essentially normal. Multiple clear glial processes heavily laden with glycogen comprise the neuropil. Many normal axons and one degenerating axon (A) are present. Lead hydroxide stain, × 7800. (From O'Leary *et al., Arch. Neurol.* **13**, 253, 1965, by permission.)

and were exaggerated in the presence of sudden noises. Swallowing was said to be affected but weight gain was similar to that of normal littermates. Flaccid paralysis, beginning in the hindlegs and later affecting the front legs, appeared in the second month of life. This was followed in the third month of life by complete paralysis and death due to debilitation and infection of decubital ulcers. Some bucks and does had a mild form of the disease and were able to copulate successfully after sexual maturity but the males proved to be sterile because of poor spermatogenesis.

According to Nachtsheim (118), a serious reduction in number of neurons in the neostriatum and globus pallidus has been observed by Ostertag. A complete description of the anatomical lesions in rabbits affected by shaking palsy appears to be lacking (70).

3. Paralytic Tremor *(pt/* —male; *pt/pt* female)

Another neurological disorder of genetic origin has been discovered in a European strain of chinchilla rabbits in recent years (126, 128). This disease of X-linked recessive

inheritance was recognized in a stock maintained by the Experimental Laboratory of Comparative Neurology of the Polish Academy of Sciences at Minsk-Mazowiecki. After consultation with Dr. E. Osetowska (127) who discovered this mutation, the authors have here designated the trait as "paralytic tremor" (*pt*) to distinguish it from tremor (*tr*) described above.

Affected individuals are recognized during the first week of life by the presence of a coarse tremor which is said to resemble the involuntary movements seen in Parkinson's disease of man. In addition, there is increased general muscle tone and exaggeration of tendon reflexes. The clinical course is variable. In some cases there is rapid increase in severity of the tremor and progressive spastic paralysis of all limbs by 4 to 6 weeks of age accompanied by urinary incontinence and development of decubital ulcers. About 40% of affected animals have this more progressive form of the disease and die between 6 and 7 months of age. In contrast, some affected animals reach an early stationary stage in their disease and an occasional one shows improvement after signs of rapid progression. Survivors are capable of normal reproductive functions.

The morphological changes in this disease are rather distinctive, consisting mainly of neuroaxonal degeneration starting in basal ganglia and progressing outward to involve the medulla, cerebellum, and cerebral cortex. Demyelination of cerebro-cerebellar and cerebro-tectospinal tracts are considered secondary. Neuronal changes are of two types. The first includes swelling, chromatolysis, and removal of both large and small neurons (Figs. 4 and 5). In early stages there is often an associated edema of oligodendroglia (Fig. 5). The second neuronal lesion, found in

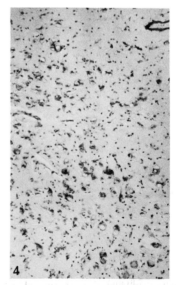

Fig. 4. Thalamus showing acute degenerative changes of paralytic tremor. Note swelling and chromatolysis of neurons and diffuse gliosis. Cresyl violet, × 72.

55% of 77 brains studied histologically, was termed pseudo-calcification (Figs. 6 and 7). Neurons thus affected stain positively by silver, periodic acid-Schiff, and von Kossa's methods. It is not clear how the two types of neuronal

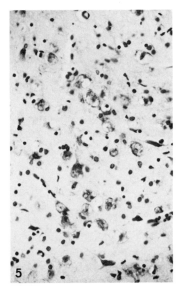

Fig. 5. Higher magnification of field similar to Fig. 4. Swelling of neurons and glial elements in attributed to edema. Cresyl violet, × 180.

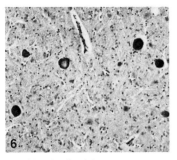

Fig. 6. "Pseudocalcification" of large neurons, medulla of 6-week-old rabbit with paralytic tremor (*pt/*-male). There is reduction in number of neurons and diffuse gliosis. Hematoxylin and eosin, × 86.

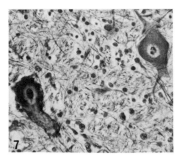

Fig. 7. Medulla of 1-month-old rabbit with paralytic tremor showing early "pseudocalcification" of a large neuron (at left). Holmes silver stain, × 216.

degeneration are related, but the two together are considered specific for the disease (126).

The paralytic tremor (*pt*) described by Osetowska and Wisniewski (126, 128) is distinguished from tremor (*tr*) described by Nachtsheim (107) by its earlier appearance, sex-linked inheritance, and the occurrence of spastic, instead of flaccid, paralysis.

4. Hydrocephalus *(hy/hy)*

Hydrocephalus is listed here as being due to a single mutant gene although this is not a fully satisfactory explanation for many instances of the natural disease, or even for both of the reported examples of inherited hydrocephalus. Nachtsheim (115) studied a family of rabbits in which hydrocephalus segregated as an autosomal recessive (*hy*) but was frequently associated with dwarfing (*nan*) and brachygnathia. A second (and different?) inherited hydrocephalus was observed by da Rosa (157) to behave as an autosomal trait with incomplete dominance. In his animals hydrocephalus was commonly associated with several anomalies including anophthalmia, microphthalmia, ectopia of the eyeball, coloboma of the iris and choroid, and cataracts. Unfortunately, the studies of Nachtsheim (115) and da Rosa (157) were based mainly on detection of cases of hydrocephalus by gross examinations. Undoubtedly, this method failed to recognize mild cases which survived and always appeared normal. Cases of hydrocephalus without a clear pattern of inheritance are known to occur sporadically in many contemporary stocks (159).

Whether abnormal vitamin A metabolism has been a factor in any of the above examples of hydrocephalus remains a moot question, but it is well established (61) that experimental vitamin A deficiency in pregnant does regularly leads to birth of hydrocephalic young (see Chapter 16, page 407 for further details). Also, Greene (55) has claimed to have seen a form of hydrocephalus in rabbits based on an inherited defect in vitamin A metabolism.

5. Lethal Muscle Contracture *(mc/mc)*

Lethal muscle contracture was described by Sawin (159). Animals are stillborn or die shortly after birth because or inability to ambulate or nurse normally. Both the fore- and hindlimbs may be held in rigid extension, but most cases actually show only mild stiffening of limbs. Grossly there appears to be atrophy of musculature. The majority of the affected animals also have hydrocephaly and cleft palate (23a). The microscopic and biochemical pathology is unknown.

6. Syringomyelia *(sy/sy)*

The German literature contains a number of studies of rabbits with hereditary syringomyelia. The genetic analyses of Nachtsheim (106, 118) established the trait as a simple recessive and the pathological changes were described by Ostertag (129–131). All affected animals were descendants of one buck of the Castorrex breed which had been imported from France in 1925.

The clinical character of the disease was extremely varied. Onset was usually after 1 month of age but varied from birth to 1½ years. The first symptom was usually a stiffening of the hindlegs in an asymmetrical manner so that one leg showed spastic paralysis while the other was only slightly lame. In more advanced cases the front legs also became affected. Affected animals often assumed a position in sternal recumbency with some or all of the legs splayed laterally. Problems related to incapacitation, loss of normal bladder and bowel functions, and decubital ulcers were the major causes of death.

The typical finding (130, 131) in affected animals was tubular cavitation of the spinal cord extending over several body segments (the definition of syringomyelia). But, because of the extreme variability in time of clinical onset and pathological severity of the disease, detection sometimes required histological examination of cords from apparently normal animals after they had reached 1 year of age, a procedure not regularly practiced by either of the authors who have reported on this disease. Such variation in methodology has raised questions as to whether this disease is indeed one of simple autosomal recessive inheritance (118).

C. Ocular Mutants

1. Buphthalmia (Syn. Hydrophthalmia, Congenital or Infantile Glaucoma) *(bu/bu)*

This mutant is recognized as one of the most common inherited diseases of contemporary domestic rabbits, although information on frequency of its occurrence in various breeds is generally lacking. It is regularly seen in some New Zealand White stocks bred for laboratory purposes (60).

According to Dorland's Medical Dictionary, the term buphthalmia is of Greek origin and literally means "ox eye." Its use as a descriptive term indicative of enlargement of the eye was in vogue late in the nineteenth century and in the earlier part of the present century. In recent times, buphthalmia seems to have become *the* accepted term for this particular disease of rabbits, perhaps because terms of such distinctive flavor have special appeal to geneticists!

Buphthalmia has been observed in rabbits of many stocks (4, 50, 60, 146, 153, 173, 180, 183) the world over. Grossly visible abnormalities of the eye may appear as early as two or three weeks of age but most appear much later. Initially, one may see either an increased anterior chamber size with a clear cornea or a slight cloudiness and very

8

Fig. 8. Normal and buphthalmic rabbits. Compare the clear eye of the normal animal on the left with the dull, ground-glass appearance of the cornea of the affected animal at right.

delicate bluish tint of the cornea (Fig. 8). Subsequent changes include progressive opacity and flattening of the cornea, increased prominence of the eyeball, and conjunctivitis. There may be ulceration and, possibly, traumatic rupture of the cornea with scarring and vascularization. General health, appetite, and libido are said to be reduced (47, 111).

Most of the eye lesions in buphthalmia seem to be related to abnormalities in production and removal of aqueous from the anterior chamber of the eye as in congenital glaucoma of man. For this reason buphthalmic rabbits serve as a useful experimental model of the human disease. A major difference is that the scleral coat of the rabbit is less mature at onset of increasing pressures, thus permitting enlargement (i.e., buphthalmia).

Buphthalmia is inherited as an autosomal recessive trait, but attempts to analyze the effect of the *(bu)* gene have proved rather baffling. First of all, there is incomplete penetrance so that clinical disease appears in some but not all homozygotes *(bu/bu)*. The eye lesions may be unilateral or bilateral. Age of onset of recognizable disease is extremely variable as is also true of various physiological parameters concerned with production and removal of fluid from the anterior chamber of the eye. This variable age of onset may account for some of the incomplete penetrance. In general, however, there is demonstrable decrease in outflow of aqueous from the anterior chamber of the eye by three months of age (Fig. 9) and an increase in intraocular pressure by five months of age (38, 76, 99, 100). Even at birth there is said to be an absence or underdevelopment of outflow channels in the ciliary body and sclera (60). As the disease progresses, there is widening and even complete obliteration of the angle, thickening of Descemet's membrane, increased corneal diameter, increased vascularity and opacity of the cornea, atrophy of the ciliary processes,

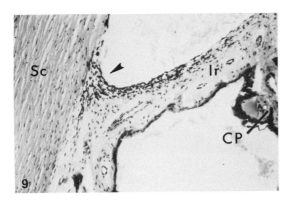

Fig. 9. Section of eye from an adult Chinchilla rabbit with buphthalmia (*bu/bu*). The anterior chamber is above and the posterior chamber below. The sclera (Sc), ciliary process (CP), and iris (Ir) are identified. Trabecular canals for drainage of the anterior chamber of the eye are completely lacking (arrow). Hematoxylin and eosin, ×86.

glaucomatous excavation of the optic disc, and, possibly, as a secondary effect, atrophy of the optic nerve. Striking differences in cytological features and turnover rate of corneal epithelial cells have been observed in buphthalmics as compared to normal rabbits (5, 33, 178, 179).

Recently, Fox *et al.* (38) obtained suggestive evidence that buphthalmic males have reduced spermatogenesis. Also, the latter authors compared the lesions of buphthalmia with those of vitamin A deficiency and suggested that the pleiotropic effects of the (*bu*) gene might involve an abnormality of vitamin A metabolism. Administration of ascorbic acid has been reported to have no effect on buphthalmics although it apparently facilitates outflow of anterior chamber fluid in normal rabbits (121). Glucose metabolism in buphthalmic animals is probably normal, however, buphthalmic rabbits have been shown to be significantly more resistant to insulin shock than normal controls (40).

As already mentioned, incomplete penetrance is an important factor in the transmission of buphthalmia. There apparently is another mechanism whereby something less than the expected number of buphthalmics are produced. Hanna *et al.* (60) found a deficiency of affected offspring in large litters and an excess of large litters without affected offspring.

2. Cataracts (*cat*-1/*cat*-1) or (*Cat*-2/*Cat*-2; *Cat*-2/*cat*-2)

At least two types of inherited cataracts have been reported in rabbits. Nachtsheim and Gürich (119) described one type (gene symbol *cat*-1) which was transmitted as a simple recessive. Slight bilateral dullness was present in the posterior wall of the lens at birth and subsequently progressed to complete opacity of the lens by 5 to 9 weeks of age. Rate of progression was slower if animals were fed dry diets as opposed to foods with high water content. A second

hereditary cataract (gene symbol *Cat*-2) has been described by Ehling (29). It was considered an incomplete dominant with 40 to 60% penetrance. In contrast to the other form of cataract, the one described by Ehling was often unilateral in occurrence.

3. Cyclopia *(cy/cy)*

Menschow (103) observed a total of 3 animals in which the two normal eyes were replaced by a single large eye in the usual position of the mouth. On the basis of suggestive data from analysis of the pedigree, he considered these cyclopian monsters to be caused by a recessive gene, *cy*. Subsequent cases were seen by Nachtsheim (118), but were not considered hereditary in origin.

4. Red Eye *(re/re)*

Magnussen (93, 94) discovered this mutation in a family of Chinchilla rabbits. Animals with "red eyes" had prominent red pupils associated with reduced pigmentation of the iris. Iris colors varied from light blue to gray-brown. Animals with both chinchilla and agouti coat colors were found to have affected eyes and an associated dilution of coat color. These defects were transmitted by a single autosomal recessive gene. Difficulty was encountered in maintaining the mutant gene as affected animals were more susceptible to infections.

The findings of increased red reflex of eyes ("red eye"), reduced pigmentation of the iris and hair coat, and increased susceptibility to infections constitute the classic triad of partial albinism which in recent years has proved so useful in recognizing the Chediak-Higashi syndrome in human patients and several species of animals (132). Thus, the mutation described by Magnussen (93, 94) could prove to be of much interest to comparative medicine. Unfortunately, it is not known whether his original stock is still in existence.

5. Keratitis Associated with the French Rex Mutation (r_1/r_1)

Animals homozygous for the pelage mutation known as French Rex also have deformed eyelashes. The constant irritation of the cornea by these deformed hairs has been reported to cause keratitis (82, 108).

D. Oral Cavity Mutants

1. Mandibular Prognathism *(mp/mp)*

Mandibular Prognathism (Brachygnathia, Hypognathia, Malocclusion, Walrus Teeth, Buck Teeth) (*mp/mp*) is probably the most common of the known inherited diseases in the rabbit, as few laboratory or commercial breeding stocks are free of the problem. The term mandibular prognathism refers simply to an abnormally long mandible relative to the length of the maxilla. The condition is known to occur in many other species, the most notable being man, dog, and sheep.

Mandibular prognathism in rabbits is rather unique because of peculiarities of the dentition in this species. The dental formula for the normal rabbit (22) is I_1^2, C_0^0, PM_2^3, M_3^3. There is a single pair of large chisel-shaped lower incisors. The upper incisors include a similar pair of large or primary incisors and a pair of much smaller secondary incisors or "peg teeth" positioned directly behind the primaries. Normally, the lower incisors occlude behind the primary incisors and against the secondary incisors which protrude only slightly beyond the gingiva. The lower incisors wear at approximately right angles to the secondary incisors and the primaries wear along the arc followed by the lower incisors. This is due in part to the fact that the enamel layer in the rabbit incisor tooth is not distributed evenly around the tooth. The anterior edge has the heaviest layer with the sides next and little if any enamel on the posterior surface (184a). Thus, a sharp cutting edge is present at the labial surface of upper primary and lower incisors (Figs. 10 and 11). The upper primaries curve posteriorly and laterally whereas the lower incisors grow in an anteromesial direction. Between the incisors and first premolars there is an edentulous space known as the diastema. The premolars and molars meet evenly to form a flat occlussal surface (34, 185).

The large incisors of the rabbit grow throughout life, at average rates on the order of 2.0 mm and 2.4 mm per week for uppers and lowers, respectively. On an annual basis this is approximately four inches for each upper incisor and five inches for each lower incisor (175). Optimal occlusion of incisors is critical for counterbalancing of this rapid extrusive growth against normal attrition. Absence of good occlusion from any cause in the rabbit quickly leads to overgrowth of incisors and/or cheek teeth. Malocclusion may result from a variety of hereditary and environmental causes.

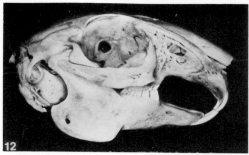

Figs. 12 and 13. Skull (Fig. 12) and mouth (Fig. 13) of a 10-week-old *mp/mp* rabbit with edge-to-edge occlusion of the primary incisors. As such affected animals become a few weeks older, the lower incisors move past the uppers resulting in full expression of mandibular prognathism.

In the rabbit with hereditary mandibular prognathism (*mp/mp*) malocclusion of incisors usually makes its appearance after the third week of life. At first there often is an edge-to-edge bite with blunting of incisural cutting edges (Figs. 12 and 13), this being replaced shortly by positioning of the lower incisors anterior to the uppers. In accordance with their usual planes of growth, the primary incisors tend to curl within the mouth while the lower incisors grow anteriorly to protrude from the mouth (Figs. 14 and 15). The upper incisors may pierce the gingival or buccal mucosa leading to ulceration and/or abscess formation. Impaired closure of the mouth also interferes with normal attrition of molars and premolars. If the incisors are not clipped, the animal has difficulty eating and may die of inanition.

Fox and Crary (34) recently reported evidence that mandibular prognathism is inherited as an autosomal recessive trait with incomplete (81%) penetrance. Affected animals were found to have reduced length measurements for the skull and maxillary diastema, without significant deviation from normal length of mandibles. These authors pos-

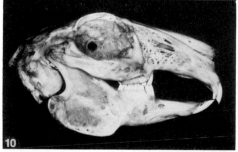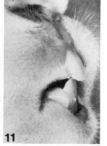

Figs. 10 and 11. Lateral view of skull (Fig. 10) and mouth (Fig. 11) of a young adult rabbit showing normal apposition of incisor teeth. Normal occlusion brings the lower incisors to rest against the upper second incisors or "peg teeth."

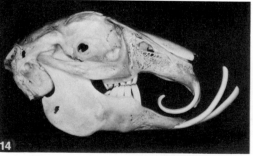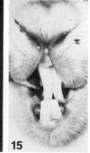

Figs. 14 and 15. Rabbits approximately 3 months of age with extreme mandibular prognathism (*mp/mp*). Impaired closure of the mouth due to overgrown incisors has resulted in reduced wear of jaw teeth. The upper large incisor on the distal side of the skull has been broken off. Also, this skull lacks "peg teeth," possibly representing an example of the dominant trait known as "absence of 2nd incisors" (I^2/I^2 or I^2/i^2). Note the normal grooves of the anterior surface of the upper first incisors of rabbits in Figs. 13 and 15.

tulated, as the mechanism for action of the abnormal gene (*mp*), differential growth of dorsal and basal skull bones with a resulting anterior displacement of the mandible. The condition in rabbits was compared to class III malocclusion, a similar abnormality in man (120).

Mandibular prognathism with its attendant malocclusion of incisors is to be distinguished from the condition involving malocclusion of premolars and molars which has been reported in a few older rabbits (71, 148, 189). However, it is tempting to speculate that the latter disease too has a genetic basis.

2. Absence of Second Incisors $(I^2/I^2; I^2/i^2)$

The absence of second incisors (i.e., peg teeth) was reported by Nachtsheim (113, 118) to occur in all races of domestic rabbits and occasionally in wild specimens. It was considered dominant. Instead of normal peg teeth, affected animals developed small rudimentary structures which resembled teeth but, in most instances, disappeared very shortly. There were no ill effects of not having the peg teeth. A possible example of this trait is shown in Fig. 14.

3. Supernumerary Second Incisors *(isup/isup)*

A mutation considered much less common than the absence of second incisors was that of supernumerary or increased numbers of peg teeth beyond the usual single pair. Such teeth appeared between members of the first pair of peg teeth and were always a bit larger than the first pair. Inheritance was considered to be by a single recessive gene (*isup*) with low penetrance (110, 111, 118, 154).

E. Skeletal Mutants

1. Achondroplasia *(ac/ac)*

Achondroplasia in the rabbit was first observed in pure-bred Havana rabbits, and described by Brown and Pearce (11, 138, 139). It is similar to the disease in man, cattle, and dogs.

Affected rabbits (*ac/ac*) are stillborn or die very shortly after birth. They have disproportionately short limbs, a broad short squarish head with prominent calvarium, protruding tongue, a comparatively short and flaring thorax, and a large bulging abdomen. Skin, muscle, and internal organs are essentially normal in size, thus being too large for the supportive system. Developmental retardation is manifest in many areas of the skeleton by absence of ossification in some bones and reduced rate of development of those bones which are ossified. In general, development of such newborn animals in strain AC is that of normal 21- to 23-day fetuses.

The pleiotropic effects of the *ac* gene are traceable to a disturbance of mesenchyme or its precursor, probably chemical in nature, which results in retardation of growth and disorganization of cartilage cells during periods of active growth. Death may be attributable to the inability of the supportive layers of the lung, mesenchyme in origin, to support respiration (26). Microscopic study of *ac/ac* rabbit cartilage shows an increased number of dead cartilage cells and a nonrandom distribution of these cells when compared to normal littermates (177). Recent organ culture studies of rabbit cartilage provide evidence for a defect in glucose utilization of achondroplastic rabbits. Glucose incorporation is elevated in *ac/ac* tissue, whereas incorporation of isotopes from sulfate and thymidine is not (176). A preliminary report suggests that achondroplasia in the rabbit may be a result of defective oxidative-energy formation and may represent the first biochemical description of a genetically determined phosphorylation deficiency in a mammalian system (7, 91).

2. Chondrodystrophy *(cd/cd)*

Another lethal mutation, chondrodystrophy was first observed in 1970 in strain III rabbits (36). Both sexes are affected. It appears to be similar to both the human metatropic and thanatophoric dwarfs (6). Grossly, these animals resemble the achondroplastics *(ac/ac)*, but they are more muscular, fatter, and the tongues do not protrude. They differ from the dachs *(Da/Da)* chondrodystrophy in many respects. The skeleton as a whole appears well-matured, but with irregularities of the centra. The long bones are grossly malformed. The ends of the diaphyses are very broad, the shafts bowed and very short. Epiphyseal, distal rib, and thyroid cartilages are all very much overgrown.

The progeny ratio of 51 normal to 18 affected from transmitter by transmitter matings suggests autosomal recessive inheritance (37a).

3. Dachs *(Da/Da)*

The dachs mutation in the rabbit occurred in a rabbitry in California and was first described by Crary and Sawin (25). It is a less severe chondrodystrophy than the *ac/ac*. Affected animals are perfectly viable but the disease is often accompanied by severe crippling due to malformation of the acetabulum and head of the femur. At birth affected animals are readily distinguishable by an extra papilla at the base of the ear, a diagnostic feature that is present from as early as 12 to 13 days gestation when the ear is just forming (23). In older animals, the shortness of the limbs becomes more pronounced and the animals have a peculiar ear carriage due to the downward rotation of the back of the skull associated with premature fusion of the spheno-occipital synchondrosis (167). These skeletal changes, in association with absent or poorly developed cartilages at the base of the ear, disturb the normal muscle attachments and thus the mobility of the ear (78). Peculiar and distinctive features of these animals are the misshapen scapula and the development of the dens in relation to either the basioccipital or the atlas (163). Development in the occipito-atlantal area involves extra ossification sites suggestive of a more primitive form of development involving a half somite which in normal animals disappears except for the very tip of the dens (168).

Because of their similarity in appearance to the Pelger (Pg/Pg) rabbit, the hematology of dachs animals has been investigated for changes in the white blood cells similar to those observed in the Pelger anomaly. None was found (44a).

4. Dwarf (Dw/Dw) or (nan/nan)

Pituitary dwarfism, the oldest known dwarfism in the rabbit, has been reported both in this country (54, 56) and in Europe (77, 111, 174). The European (gene symbol *nan*) and the American (gene symbol *Dw*) mutations appear to be similar and may well be the same mutation occurring in different stocks. Both were originally reported as autosomal recessive mutations symbolized *nan* and *dw*, respectively. However, since the heterozygous *Dw/dw* could be recognized when compared with the normal because of its reduced ear length and body size (24a) with a disproportionally greater brain and cord (78a), the mutant symbol was changed from *dw* to *Dw* (79, 159). Thus the mutation is now considered a semidominant lethal. The affected animals rarely live more than a few days, although in certain cases some have survived for as long as 5 weeks. The affected animals are perfectly proportioned miniatures, approximately one-third the size of normal littermates (Fig. 16), with one exception. The central nervous system (CNS) is disproportionately large (79). Thus, the animals appear to have rather bulging foreheads and protuberant eyes, and the skulls of those surviving more than 2 or 3 days have numerous large foramina (165). Cause of death probably is attributable either to the disproportionately large CNS or to hydronephrosis which develops in most of these animals.

Another dwarf mutation occurred in 1957 at The Institute for Human Genetics, The University of Münster. It was described by Degenhardt (27) as a single autosomal recessive and designated *zw*. Hydrocephaly was observed as a common occurrence with the *zw* gene. These animals are similar to the *Dw* and *nan* mutations and may well be a further mutation of the same gene.

5. Hereditary Distal Foreleg Curvature (fc/fc)

Pearce (136, 137) described this deformity of the forelegs in rabbits of the Beveren, Belgian, French Silver, and Dutch breeds (see Chapter 1 for explanation of gene symbol *fc*). Beginning at 2 to 3 weeks of age the forelegs were observed to bow inwardly while the paws deviated laterally in a "seal-

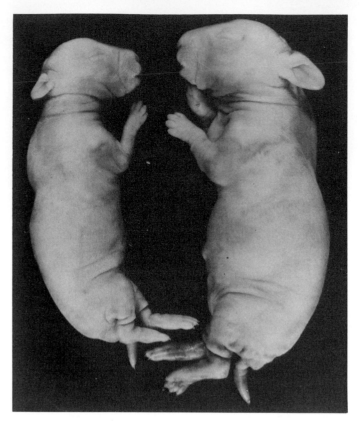

Fig. 16. Dwarf (Dw/Dw) and normal (dw/?) newborn male sibs of strain III_{Dw}. Weights were 34 and 60 gm, respectively. Gestation was 32 days. Note striking contrasts in ear and body sizes. Picture is × 1.6 normal size.

like" manner. The condition usually reached full expression by 2 to 3 months of age. There was marked widening of the distal epiphysis in each ulna. Histologically, this epiphysis showed retarded calcification and an abundance of osteoid tissue. The abnormally wide plate of cartilage usually diminished in thickness by about 2½ months of age and became normal histologically by about 4 months of age. The forelegs subsequently remained deformed permanently by the inward bowing of radii and ulnae, and lateral deviation of the paws. Bones of the carpal joints and paws were considered normal.

Pearce (136, 137) appears to have identified the effects of a single autosomal recessive gene which must at least strongly predispose to the foreleg deformity. However, among rabbit breeders a similar condition is widely recognized to occur in young rabbits raised on smooth surfaces as when the occasional doe persistently removed the bedding from her hutch. Affected rabbits a few weeks of age and very similar to those described by Pearce have been observed by one of us (J.R.L.) to return to normal when removed and kept in cages with wire floors. The relationship of such cases to the mutation described by Pearce is not clear.

6. Brachydactylia *(br/br)*

This anomaly in rabbits is transmitted by an autosomal recessive gene and is characterized mainly by shortening of the digits (53, 57). Severity of the disease was extremely variable, ranging from minor brachydactylia to complete acheiropodia (Figs. 17–19), with involvement of one or all feet. The forefeet were more frequently and severely affected. In the more extreme cases the metacarpals and metatarsals were also distorted and shortened. Some cases also had loss of the distal third of the ears, slight blunting of the ear tip or slight scalloping of the lateral border of the ear.

The more important events leading to brachydactylia occur during embryogenesis. Greene and Saxton (57) observed that the limb buds and ears developed normally until the eighteenth day of gestation when dilatation of blood vessels, hemorrhage, and necrosis occurred. The resulting stub was subsequently covered by scar tissue. These studies were extended by Inman (68) who found endothelial swelling and other degenerative changes in blood vessels of limb buds and ears as early as the twelfth day of gestation.

7. Spina Bifida *(sb/sb)*

Spina bifida is an unusually complex mutation discovered in the AC strain of Dutch rabbits (24). It is due to a lethal, autosomal recessive gene *(sb)*. Affected animals are stillborn. The bifid spine extends from the posterior calvarium into the tail. The open spine usually is covered by a thin layer of skin. The majority of animals also have harelip, cleft palate, severe kyphosis, and ventral deviation of the tail. Other defects involving bone also are observed. In addition, 80% or more of affected animals have one or more abnormalities of internal soft tissues, particularly in the cardiovascular, respiratory, and urogenital systems (24).

Spina bifida also has been reported in a variety of forms which may or may not be inherited (118, 159). Examples of some of these variations are seen in Fig. 20.

8. Hypoplasia Pelvis *(hyp/hyp)*

This mutation was discovered in English Checkered rabbits by Nachtsheim (109, 112, 116). It was originally called "spastic spinal paralysis" (gene symbol *sp*), but the name was changed later when Scherer (170) demonstrated that the lesions were in the skeleton instead of the nervous system (118). It was considered to have typical autosomal recessive inheritance.

The disease was congenital in appearance. The main clinical manifestation was inability to move the hindlegs so that animals characteristically dragged their hindlimbs. As animals became older they apparently developed some ability to elevate the body by contraction of the back

Fig. 17. An example of brachydactylia involving both forefeet, although not proved to be due to the *br* gene.

Fig. 18. Radiograph of foot of rabbit shown in Fig. 17. There is almost complete acheiropodia.

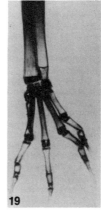

Fig. 19. Radiograph of normal rabbit's paw for comparison with Fig. 18.

[Figs. 17, 18, and 19 are from Nachtsheim, Erbpathologie der Nagetiere, *In* "Pathologie der Laboratoriumstiere" (P. Cohrs, R. Jaffe, and H. Meesen, eds.). Springer-Verlag, Berlin. Used by permission of the publisher.]

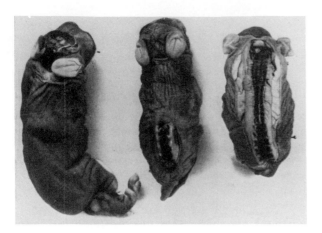

Fig. 20. Three examples of spina bifida in newborn rabbits. The entire length of the spine is affected in the animal at the right while in the other two only portions of the spine appear to be involved. [From Nachtsheim, Erbpathologie der Nagetiere. *In* "Pathologie der Laboratoriumstiere" (P. Cohrs, R. Jaffe, and H. Meesen, eds.). Springer-Verlag, Berlin. Used by permission of the publisher.]

muscles and the forelimbs. Diseased animals were usually symmetrical in appearance. The typical position was sternal recumbency with the thighs held against the body and the paws pointing laterally at right angles to the spine.

The described lesions involved the pelvis and femurs. The entire pelvis was markedly reduced in size. The inferior and superior rami of the pubis were shortened and the ischium was almost entirely lacking, resulting in poor development of the acetabulum. The femoral heads no longer functioned as articular surfaces, but were either poorly developed or missing completely. The greater trochanters were reduced also. The femurs were often much shorter than normal (118).

9. Femoral Luxation *(lu/lu)*

This condition was described in Chinchilla rabbits by da Rosa (156). Animals were normal until 2 to 4 months of age when they usually began carrying one leg in a laterally extended position supporting the weight on the other three legs. Some cases were bilateral. The basic lesion was subluxation of hip joints which was attributed to development of flattened and shallow acetabula. The heads of the femurs were considerably smaller than normal. The pattern of inheritance was consistent with an autosomal recessive gene except in a few instances where it appeared to behave as a partial dominant.

10. Osteopetrosis *(os/os)*

This autosomal recessive mutation was discovered in a Dutch stock of rabbits and extensively studied by Pearce and Brown (140) and Pearce (133–135). This same line, now designated as the OS strain, is maintained at The Jackson Laboratory.

At birth affected animals are normal in appearance except for slightly reduced body size and abnormalities of the teeth. The incisors, particularly the uppers, are retarded in development, abnormally shaped, and usually rudimentary in appearance. One or all incisors may be completely absent. At necropsy changes similar to those in incisors are observed in molars. Even at birth radiographs reveal a marked increase in density of the skeleton with practically no internal detail in bones throughout the body. For the first 2 to 3 weeks of life affected rabbits grow at a rate only slightly less than normal. Thereafter, growth ceases, animals become cachectic, have diarrheal stools, and usually die about 4 or 5 weeks of age.

The outstanding anatomical findings are persistence of spongy bone and failure to form normal marrow cavities. Those marrow cavities which do form are filled with undifferentiated mesenchymal tissue (183a). The spongy bone is composed of bony trabeculae having an irregular arrangement and variable staining. Although little compact bone is present, the bones are harder and more brittle than normal. Only sparse hematopoietic tissue ever appears in the bones and, despite persistence of centers of extramedullary hematopoiesis in other tissues such as liver and spleen, there is progressive macrocytic anemia with thrombocytopenia.

Pearce (133, 135) also observed that osteopetrotic rabbits had enlarged parathyroids and suggested that this finding was in some way related to the cause of the disease. Affected animals initially had normal serum calcium levels which declined with age. Serum phosphorus levels were low initially but slightly exceeded normal values by the fourth week of age. These and other findings in osteopetrotic rabbits bear a close similarity to so-called marble bone disease (syn., Albers-Schönberg disease) of man and the gray lethal mutation of mice (8). Familial osteopetrosis also has been reported in calves (81). Although definitive evidence is not available except for the mouse (96), it seems extremely probable that all of these diseases eventually will be explained by a defect(s) in parathormone and/or thyrocalcitonin regulation of bone metabolism.

F. Hematological Mutants

1. Atropinesterase *(As/As; As/as)*

Some rabbits have in their serum an enzyme known as atropinesterase which is capable of hydrolyzing atropine and other tropine esters (48, 49, 87), thus explaining the astonishing reports going back to about 1852 that rabbits can thrive on a diet of belladonna leaves (73).

Presence of serum atropinesterase is controlled by a semidominant gene *(As)* so that three genotypes are recognizable: *As/As*, high level of enzyme; *As/as*, reduced level of enzyme; and *as/as*, absence of enzyme (86, 166).

Actual absence of the enzyme in homozygous recessive animals, rather than presence of the enzyme in inactive form, has been supported by failure to demonstrate an immunologically related protein in serum from such rabbits (95). In animals homozygous or heterozygous for the *As* gene the enzyme first appears in the serum at about 1 month of age; it occurs at higher levels in females than in males. Atropinesterase is linked with the gene *E* for extension of black pigment in the coat (166).

There are no known harmful effects of an absence of serum atropinesterase on the health of rabbits. However, presence or absence of the enzyme makes a very significant difference in the response to atropine, regardless of whether it is given by injection or applied topically to the eye (186).

2. Red Cell Esterases

Starch gel electophoresis has made it possible to identify in the blood of rabbits three systems of esterase isozymes. Each of these systems has three phenotypes (*A*, *B*, and *AB*) controlled by a pair of codominant autosomal alleles designated by the following gene symbols: *Es-1*^{A}, *Es-1*^{B}, *Es-2*^{A}, *Es-2*^{B}; and *Es-3*^{A}, *Es-3*^{B} (59, 172). Originally all three systems of esterases were assumed to be in red blood cells, but recent studies indicate that the *Es-2* locus controls isozymes actually derived from platelets (172). The *Es-1* and *Es-2* loci are closely linked (171).

There is no conclusive evidence that any particular genetic constitution with regard to these esterases has a detrimental effect on health of rabbits. However, it is of some interest that animals with the genotype *Es-1*^{A}/*Es-1*^{B} have been reported to be more resistant to mucoid enteritis (58).

3. Pelger *(Pg/pg)*; Super-Pelger *(Pg/Pg)*

In 1928, a Dutch physician named Pelger noted that two of his patients had leukocytes with much reduced nuclear segmentation. Later, Huët disproved Pelger's mistaken idea that this was associated with tuberculosis and showed the trait to be inherited in man as an autosomal dominant. Subsequently, the condition has been referred to as either the "Pelger" or "Pelger-Huët" anomaly and has been recognized as an inherited trait in many animals including rabbits, dogs, and mice (118). In the rabbit the condition is now recognized as a partial dominant (152). In addition, the transient appearance of "pseudo-Pelger" leukocytes in peripheral blood has been seen as an acquired condition associated with inflammatory processes in several species (125).

The Pelger anomaly is fairly well known in rabbits because of studies over many years concerned with a single stock of animals established by the Swiss hematologist Undritz in 1939 (117). Two types of animals are known.

Those designated as "Pelgers" are heterozygotes (*Pg/pg*) and are normal except that nuclei of their neutrophils have a reduced number of lobes, usually one or two lobes or, rarely, as many as three. In homozygotes (*Pg/Pg*), called "super-Pelgers," all leukocytes have round, hyperchromatic nuclei which actually appear pyknotic with usual stains. Apparently because of fetal deaths, only about 18% of theoretically expected super-Pelgers are born. Most of these die during or shortly after birth. Super-Pelgers are smaller than normal at birth due to marked reduction in length of long bones. The animal rests on the chest with the forelimbs extending laterally when in the upright position. Only "paddling" movements are made with the deformed limbs. Excessive salivation and emaciation usually appear. Very rarely super-Pelgers live to maturity. One such buck has been produced and was used successfully in breeding trials which helped to confirm the pattern of incomplete dominant inheritance in rabbits.

The skeletal abnormality of the super-Pelger is considered a chondrodystrophy. Long bones are extremely shortened (including phocomelia). This is particularly noticeable in the ribs which are said to be thickened "like mushrooms" at the costochondral end. This rib deformity is held to be responsible for death from asphyxiation in most live born super-Pelgers. In the older animals the epiphyses remain thickened and often there is fusion of the radius and ulna.

4. Lymphosarcoma *(ls/ls)*

In view of the fact that lymphosarcoma generally has been considered rare in rabbits (182), the recent discovery of its common occurrence in strain WH rabbits (42–44, 102) seems highly significant. Furthermore, its reported pattern of transmission as an autosomal recessive gene (*ls*), compatible with concepts both of genetic susceptibility to the neoplasm and vertical transmission of a virus, heralds the finding of a new model of potential importance in experimental oncogenesis.

The disease manifests itself at an average age of eight months (range: 2 to 27½ months) with appearance of anorexia, lethargy, and weight loss. Most affected animals die between 5 and 13 months of age. Terminally there is severe anemia with total RBC's reaching as low as 2×10^6 and hematocrits 10%. Hematological examination of rabbits with lymphosarcoma reveals essentially an aleukemic picture. Even though the total WBC's are only 2000–10,000 mm^3, a shift in cell types is observed with 80% or more being lymphoblasts. The lymph nodes are increased in size and have a preponderance of lymphoblasts (Figs. 21 and 22). Advanced cases often have elevated blood urea nitrogen levels coinciding with occurrence of massive infiltration of malignant lymphoid cells into the kidneys (Fig. 23). Other anatomical findings include lymphoma-

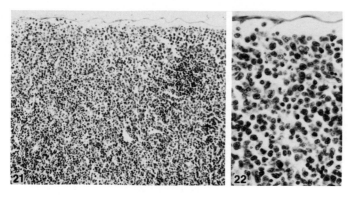

Figs. 21 and 22. Mesenteric lymph node from a female *ls/ls* rabbit killed at 8½ months of age. Entire node is almost completely overrun by cancer cells considered to be lymphoblasts. Numerous mitotic figures are seen in Fig. 22. Hematoxylin and eosin, ×86 and 216, respectively.

tous infiltrates in Peyer's patches, liver, spleen and, sometimes, in adrenals (Fig. 24), gonads, and thymus. The bone marrow from such animals is similar to that in the Hodgkins-type sarcoma observed in man (102). Rabbit lymphosarcoma is described in further detail in Chapter 14.

5. Hemolytic Anemia *(ha/ha)*

Fox *et al.* (41, 42, 44) and Meier and Fox (102) have very recently reported an hereditary hemolytic anemia occurring in the X strain of rabbits at The Jackson Laboratory. Approximately 60 cases had been observed in this stock over a period of years. Pedigree analysis suggested autosomal recessive inheritance and the gene symbol *(ha)* was assigned.

Clinically, the disease was characterized by anemia with Coombs positive RBC, hemoglobinemia, petechial

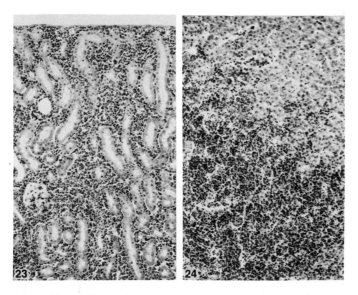

Figs. 23 and 24. Kidney (Fig. 23) and adrenal (Fig. 24) from same rabbit as Figs. 21 and 22 showing infiltrates of cancerous lymphoblasts. Hematoxylin and eosin, ×86.

hemorrhages on mucous membranes, and weakness, with death occurring on the average between 4 and 5 months of age. Severe icterus was seen in some. The peripheral blood was characterized by reduced hematocrit, polychromatophilia, anisocytosis, increased nucleated RBC's, and lymphocytosis. Autopsy findings included jaundice, visceral hemorrhages, hemoglobinuria, swelling of kidneys, and subcutaneous edema. Histologically there was widespread erythrophagocytosis, edema, and accumulated iron pigment in visceral organs. Some animals also had visceral lymphadenopathy and thymoma. No lesions suggestive of glomerulonephritis or arteritis were observed.

It is particularly interesting that the X strain of rabbits with hemolytic anemia share a common ancestry with the WH strain known to develop lymphosarcoma *(ls)*. Indeed, *(ha)* and *(ls)* might be one gene with the ascribed lesions merely representing different phenotypic expressions (44). Both lesions are of much interest because of analogies to such diseases as systemic lupus erythematosus of man and dogs, various lesions occurring in NZB mice and hemolytic anemias of other species.

G. Genitourinary Mutants

1. Hypogonadia *(hg/hg)*

A type of inherited sterility in the ACEP strain of rabbits at The Jackson Laboratory has been attributed (35, 164) to an autosomal recessive gene *(hg)*. Normal *(Hg/Hg)* and carrier *(Hg/hg)* animals are distinguishable histologically and by breeding tests from affected animals *(hg/hg)* which fall into two anatomical classes: "mosaic" and completely "abnormal." The "mosaic" in this instance is defined as an animal having a proportion (1 to almost 100%) of abnormal gonadal tissue. Less affected "mosaics" mate and produce young.

Gonad size of completely "abnormal" animals is markedly reduced as testes weigh only approximately 225 mg compared to 1700 mg for normal males, and ovaries weigh about 12 mg in contrast to 150 mg for normal females. Gonad size of "mosaic" males fluctuates in direct relationship to percent of abnormal tissue present.

Microscopically, the testes of completely "abnormal" males have small seminiferous tubules lined only by a single row of Sertoli cells (Fig. 25c). "Mosaics" have varying proportions of tubules with active spermatogenesis (Fig. 25b). "Abnormal" and "mosaic" testes have relatively more interstitial tissue and thicker tunica albuginea. Spermatogenesis tends to be retarded in "mosaics" compared to normal animals. Completely "abnormal" ovaries are small and devoid of follicles and are composed of tubules thought to be of mesonephric origin, connective tissue, "epithelioid cells," and scanty luteal type tissue (Fig. 25f). Variability in normal ovarian architecture makes it difficult to classify

TESTES   OVARIES

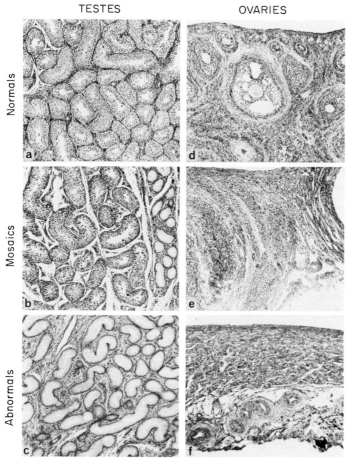

Fig. 25 a–f. Histological sections of gonads from young adult rabbits demonstrating effects of the hypogonadia gene (*hg*). Figs. 25 a–c are normal, "mosaic," and "abnormal" testes, respectively. Figs. 25 d–f show the ovarian counterpart of normal, "mosaic," and "abnormal," respectively. Note that Figs. 25 e and f are full thickness sections of ovaries. Hematoxylin and eosin, × 32.

low grade "mosaics" (Fig. 25e). Affected females also have reduced development of mammary glands and external genitalia.

The actual mechanism of action of the *hg* gene is unknown. Presumably there is a chemical defect which interferes with embryogenesis or germ cell differentiation. Regardless of the actual mechanism, this mutant provides an interesting model bearing a degree of similarity to certain types of gonadal dysgenesis in man (35, 164).

2. Renal Agenesis (*na/na*)

Da Rosa (155) reported a new autosomal recessive mutation, symbolized *na*, suppressing the development of either kidney. Four cases had the right kidney missing and five the left. He postulated that there might have been prenatal deaths where both kidneys were absent. Homozygous mutants had reduced vitality.

In addition, Rosa gave a minimum of data suggestive of abberent mesonephric structures attached to the gonad on the side of the absent kidney. This occurred in three of the five rabbits living for three weeks or more. He concluded that only one gene was responsible for the effects on both the metanephros and the mesonephros.

Greene (55) observed a similar or perhaps identical anomaly in rabbits of the Havana breed. In males the testicle also was frequently absent on the side missing the kidney. In affected females there was shortening or absence of the ipsilateral uterine horn while the ovaries both remained intact.

3. Renal Cysts (*rc/rc*)

Fox *et al.* (39) have reported the occurrence of multiple small cysts in kidneys of strain III$_{vo}$ rabbits. Cysts were detectable only by necropsy as there was no adverse affect on health or preliminary renal function tests. Significant numbers of cysts were never found in animals less than one month of age but, thereafter, they were found in increasing numbers up to approximately six months of age. Pedigree analysis gave evidence consistent with autosomal recessive inheritance with incomplete (76%) penetrance.

Affected kidneys have from one to hundreds of cysts up to 1 mm in diameter limited in location to the subcapsular region of the cortex (Figs. 26 and 27). Microdissection studies showed the cysts to be of tubular origin as can be

Fig. 26. Gross kidneys from animal with renal cysts (*rc/rc*). × 1.4 normal size.

Fig. 27. Cross section through kidney shown in Fig. 26. H & E, × 1.4.

demonstrated also in fortuitous histological sections (Fig. 29). The larger cysts usually are lined by a single layer of flat to low cuboidal epithelium surrounded by little or no fibrous connective tissue. Elsewhere cysts have the appearance of primitive ductules and tubules in abundant cellular mesenchymatous stroma (Fig. 28). The latter areas resemble closely the histological findings of renal cortical dysplasia in man (9, 75). The histology, the subcapsular location of lesions, and the age at which cysts first appear all tend to point toward an interruption of development in the late stages of renal organogenesis. It is of at least passing interest that similar subcapsular renal cysts have been produced experimentally in the rabbit by a single injection of long-acting adrenal corticosteroids given at birth (145).

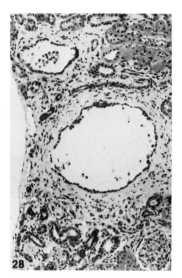

Fig. 28. Outer renal cortex of rabbit with renal cysts showing two cysts surrounded by collagenous connective tissue containing many dysplastic tubules. Parts of more normal nephrons are seen above and below. H & E, ×86.

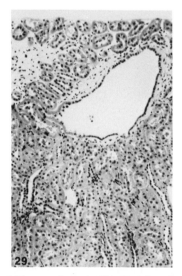

Fig. 29. Same kidney as shown in Fig. 28. A fortuitous section shows a subcortical renal cyst continuous with a normal portion of a renal tubule. H & E, ×86.

H. Other Mutants

1. Yellow Fat (*y/y*)

The fat of most rabbits is white, but in certain breeds an occasional individual is found to have yellow fat. This generally is considered objectionable in animals used for meat purposes (12, 143). The yellow color occurs in animals homozygous recessive for the gene (*y*). Inheritance of yellow fat is linked with the *B* and *C* loci for coat color (12, 13, 143, 144).

Animals homozygous recessive for yellow fat (*y/y*) reportedly lack a specific enzyme in the liver necessary to metabolize xanthophylls, a group of carotenoid pigments in plants which are incapable of serving as provitamins of vitamin A (187, 188). As a result, these rabbits are the only known mammals to selectively store xanthophylls (105). Of course, for yellow fat to be expressed in *y/y* animals it is necessary for plant materials containing xanthophyll pigments to be included in their diet. Depending on the level of such pigments in the diet, color of the fat may vary from light yellow to orange (64). Storage of xanthophylls is of no known significance to general health of rabbits.

2. Adrenal Hyperplasia (*ah/ah*)

An autosomal recessive gene has been reported causing extreme proliferation of the adrenal cortex and apparent elimination of the adrenal medulla (37). The condition is lethal perinatally and the progeny ratio of 100 normal to 38 affected from transmitter by transmitter matings (37a) suggests autosomal recessive inheritance. This condition is neither sex linked nor sex limited. A high percentage of the affected animals have clubbed forefeet. Internally, the testes are normal sized but undescended with no caput epididymus; the ovaries are thin. The adrenals are 10 to 20 times normal size. Histological examination of serial sections reveals remnants of the zona glomerulosa, extensive proliferation of the zona fasciculata, but complete absence of medullary cells. Preliminary histological examination of other organs has not revealed any abnormality.

III. FAMILIAL OR POLYGENIC CONDITIONS

In this section we will describe those conditions which apparently are inherited but not explained as single gene effects. The entities included are generally accepted as conforming to familial or polygenic patterns of inheritance

Fig. 30. Two littermate rabbits with the "droopy ear" trait. (Courtesy Dr. C. K. Chai.)

but, in a few instances, the evidence favoring a genetic basis is not fully convincing. No chromosomal abberations have been described in rabbits.

1. Abnormal Ear Carriage

Rabbits normally carry their ears erect pointing dorsolaterally, or in a relaxed position against the dorsum of the trunk and pointing posterolaterally. Chai and Clark (16), while studying a number of inbred lines of rabbits, observed a condition designated as "droopy ear" in which the ears were allowed to droop downward and anteroventrally from their bases (Fig. 30). It is thought to be due to more than one gene.

Droopy ear is distinguished from the "lop ear" or "flop ear" characteristic of certain breeds in which the ears, perhaps due to their excessive weight, simply hang downward from their attachments on the head. This trait too is thought to be controlled by multiple factors (14).

2. Abnormalities in Shape of the Calvarium

The studies of Greene (51, 55) focused attention on deviant shapes of the calvarium in rabbits. The more important examples are explained by craniosynostosis, the premature closure of one or more cranial sutures with compensatory growth at other sutures during embryonic development and the period of active growth to adulthood. Four distinct types of deformity are recognized in the rabbit depending on the suture or sutures affected:

a. *Oxycephaly*—the deformity explained by fusion of both coronal sutures and the sagittal suture. The resulting skull has a raised peak at the bregma and sharp ridges occupy the normal positions of the sagittal and coronal sutures. The frontal bones are shortened in the anteroposterior dimension. The parietal bones are markedly flattened.

b. *Trigonocephaly*—deformity due to fusion of both coronal sutures. The bregmatic peak is absent; a transverse ridge marks the site of the fused coronal sutures (Figs. 31 and 32). The sagittal suture is normal in appearance. The parietal bones may or may not be flattened.

c. *Plagiocephaly*—deformity due to fusion of either the left or right coronal suture. The result is disproportionate growth on the fused and normal sides. The fused side has the characteristics of trigonocephaly. Parietal and frontal bones are shorter and nasal bones longer in the anterior-posterior dimension on the affected side. The sagittal and frontal sutures form a curve convex to the normal side.

d. *Scaphocephaly*—deformity due to fusion of the sagittal suture. The normal location of the sagittal suture is

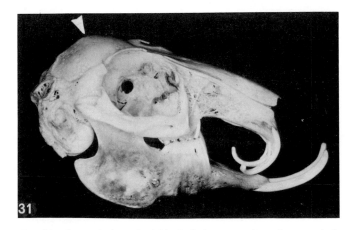

Fig. 31. Lateral view of rabbit skull demonstrating trigonocephaly, the cranial deformity resulting from bilateral fusion of the coronal sutures early in development. Note the resulting transverse ridge (arrow) which marks the normal site of the right coronal suture. This specimen also demonstrates extreme mandibular prognathism (*mp/mp*). Compare with Figs. 10, 12, and 14 which have normal calvaria, with or without mandibular prognathism.

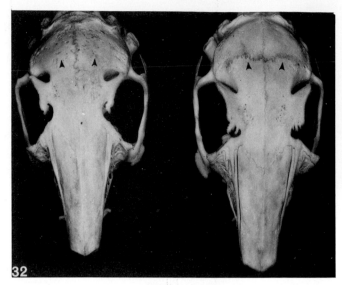

Fig. 32. Dorsal view of skull with trigonocephaly (at left, same as Fig. 31) and normal skull (at right). Coronal sutures are present in skull at right, absent from skull at left (arrows).

marked by a longitudinal ridge bounded on either side by flattened parietal bones.

None of the deformities of the calvarium are known to affect the brain except for slight differences in shape. Premature fusion of sutures may alter the morphogenesis of other bones in the head including those in the base of the skull. The prevalance of these cranial deformities in modern rabbit stocks is unknown.

The inheritance of craniosynostosis is not clear. Greene (51) considered the deformities definitely recessive to normal but was unable to specify the number of genes responsible.

3. Acromegaly

A condition thought to have the essential features of acromegaly has been observed by Hu and Greene (67) in animals derived from an inbred stock of Dutch rabbits. At one or two weeks of age the skin of the neck, shoulders, and chin of affected animals became thickened and slightly reddened. This rapidly spread so that the skin of the entire body became thickened and thrown into loose, transverse folds. This was followed by stiffening and induration of the skin, and appearance of white scales and crusts on the skin surface. The hair was at first normal but became coarse, sparse, and stubby. Affected animals initially were exceptionally large, but rapidly lost weight and died in a week or 10 days from onset. Skeletal overgrowth apparently did occur in some animals. The mode of inheritance was uncertain.

4. Aortic Arteriosclerosis

Spontaneous arteriosclerosis of the aorta in rabbits was first described by Israel (72) in 1881. Since that time many additional reports have appeared (10, 74, 85, 92, 104).

Although the exact cause of this arterial lesion remains obscure, there is ample evidence that genetic factors play an important role in its expression. Lesions have been observed in practically all breeds and even in wild rabbits. However, among the breeds there are distinct differences in incidence. The Dutch breed in the United States (45) and the "Danish Country strain" of albino rabbit in Europe (46) have low (10%) incidences of aortic arteriosclerosis, but incidences of 40% or greater are common in many contemporary breeds including laboratory strains of New Zealand White (45). Diseased aortas have been observed as early as six weeks of age. It has been suggested that these early lesions also may have a genetic basis, as their induction has been attributed to one or more factors in milk from some does rather than to the genotype of the young (31).

In the typical case, lesions occur in all major arteries but predominate in the ascending aorta and arch. Grossly, the appearance of the intimal surface is quite variable: longitudinal striations or furrows, raised plaques with or without central depressions, intimal nodules, or an admixture of these. In advanced cases the aortic wall may have an egg shell consistency due to mineralization of the wall. Histologically, the lesions are located in the inner one-third to one-half of the media.

Initially there is fragmentation and disruption of the elastic lamina with the intercellular accumulation of acid mucopolysaccharide. Subsequently, there is mineralization of elastic fibers, collagen fibrils, and intercellular spaces resulting in what appears to be the equivalent of Mönckeberg's medial sclerosis in man. Degenerative and proliferative changes may occur, ultimately going to cartilagenous or osseous metaplasia. Proliferative changes in the overlying intima are sometimes observed. The lesions are free of lipid substances such as those occurring in the ubiquitous atherosclerotic vascular disease of man (46, 62, 63, 169).

Many questions remain unanswered about the specificity of the anatomic lesions seen in so-called "aortic arteriosclerosis" of rabbits. For example, the lesions in this disease and vitamin D toxicity are remarkably similar (see Chapter 16), raising the question of dietary factors in aortic arteriosclerosis. The need for clear delineation of etiologies and types of vascular lesions in the rabbit is further supported by the sizeable number of instances in which naturally occurring arteriosclerotic lesions of rabbits reportedly have been erroneously interpreted to be the result of experimental procedures (45).

5. Familial Neoplasias

Lymphosarcoma occurs in rabbits of the WH strain and has been attributed to an autosomal recessive gene, *ls* (see above). A large number of additional neoplastic diseases are known to occur in certain families of rabbits and have been discussed elsewhere in this volume (see Chapter 14).

6. Hereditary Premature Senescence

This intriguing syndrome was observed in a colony of Belgian hares by Pearce and Brown (141, 142) whose studies included 185 affected animals from 20 generations and covered 15 years. The disease was considered familial as other breeds of rabbits were maintained under the same conditions, yet remained free of similar symptoms.

Beginning as early as 1 month of age affected animals developed a senile appearance characterized by loss of sheen, dryness, and thinning of the coat. In subsequent months and years other changes appeared, including irregular coat length and variable alopecia of the trunk and legs. The skin became dry, thin, and inelastic with an abundant branlike desquamation most noticeable on the ears. Decubital ulcers, granulomas, and thick calluses were commonly observed on the plantar surfaces of the feet and to a lesser extent on the palmar surfaces. Ophthalmia with conjunctivitis, keratitis, and excess lacrimation were common. There were numerous reproductive abnormalities. Infertility and sterility were commonly seen by 1 year of age. In addition, does frequently had fetal resorptions, abortions, delayed parturition, stillborn litters, desertion of litters, and abnormal mammary function. Ultimately, there was severe, progressive muscle atrophy, emaciation, and weakness ending in death. A few spontaneous remissions were noted just prior to the stage of muscle atrophy and emaciation.

The authors divided the disease into two forms, acute and chronic, depending on whether the animals died before or after $2\frac{1}{2}$ years of age. They considered the disease an accelerated form of senility, possibly useful as a model in studies of aging. Impressive pathological changes were mentioned as occurring in the cardiovascular, renal, and endocrine systems but, unfortunately, these findings, the genetic data, and the stock of affected rabbits have been lost through demise of both authors.

7. Resistance/Susceptibility to Infection

Much has been learned about hereditary influences on ability of rabbits to withstand experimental infections due to *Mycobacterium tuberculosis*, largely through the monumental studies of Max B. Lurie and his colleagues at the Henry Phipps Institute in Philadelphia (88, 89). In general, their studies compared the relative abilities of several families of rabbits (bred brother × sister for approximately

10 generations and subsequently line bred for many generations) to eliminate or suppress standard doses of human or bovine tubercle bacilli. The different families of rabbits varied greatly in response and were classified as either resistant, susceptible, or intermediate between the two extremes. Resistance was found to be most directly dependent on ability of host macrophages to engulf and inhibit the growth of bacilli in their cytoplasm. The family differences in macrophage efficiency were attributable to differences in endocrine balances, involving particularly thyroid and adrenal hormones. The genetic determinants of resistance were considered multiple genes (90).

Genetic factors also are known to influence very dramatically the susceptibility of rabbits to certain virus infections. Greene (52) observed striking breed differences in mortality during an epizootic of rabbit pox. The natural selection of resistance to myxomatosis has been observed in the wild rabbit population since release of myxoma virus in Australia in 1950 (32, 97, 98), and this has been largely duplicated by selective breeding of domestic rabbits (181). The inheritance of resistance to virus infections clearly is controlled by multiple genes, but the mechanisms by which resistance is accomplished remain obscure.

8. Scoliosis (Including Torticollis or Wry Neck)

Scoliosis is a term meaning lateral curvature of the spine. Serious investigations of this problem are very few. In the limited races of rabbits studied to date, it is considered familial, perhaps polygenic in inheritance. Anatomically, the findings in affected rabbits have included missing, reduced, or extra half vertebral units and bifurcations or fusions of ribs. Depending on the race of rabbit in which the anomaly occurs, vertebrae in either the cervicothoracic, thoracic, or lumbar regions of the spine may be deformed (162, 165).

Torticollis (or "wry neck") is a term which refers to twisting of the neck with associated abnormal positioning or carriage of the head. It generally implies an etiology involving muscular, neurological, or inner ear disease. Although the term has been applied frequently to diseased rabbits in the past, very little is known about its underlying cause in this species. An infectious cause for torticollis has been generally recognized. It is associated with *Pasteurella multocida* infection of the middle ear (44b). One should be careful to distinguish torticollis from the more specific scoliosis due to vertebral defects as described above.

9. Splay Leg

The term "splay leg" is the descriptive name commonly applied to rabbits which lack the ability to adduct one or all legs and come to a standing position. Affected animals assume the typical "splay leg" appearance (Fig. 33) in ventral

Fig. 33. Rabbit with "splay leg." Although mentally alert, affected animals are unable to adduct the legs and stand.

recumbency and, at best, are able to move horizontally for only short distances by making weak, clumsy movements of the legs. In more severe cases all legs may be completely paralyzed. Affected animals usually appear mentally alert.

Although common usage of the term presently connotes a specific disease, it seems more likely that "splay leg" is simply the leading clinical manifestation of several disease entities of rabbits, possibly including the mutants syrinomyelia (*sy/sy*), hypoplasia pelvis (*hyp/hyp*), femoral luxation (*lu/lu*), and hereditary distal foreleg curvature (*fc/fc*) described above. Recent observations have added further possibilities.

Innes and O'Steen (69) studied 10 cases in rabbits of the New Zealand White and Dutch breeds and advanced the idea that "splay leg" is a recessively inherited achondroplasia restricted in its pathological effects to the hip and shoulder joints. Subluxation and dislocation were regarded as late sequelae. In contrast, Arendar and Milch (3) found that the primary pathological defect in 30 cases from a strain of inbred Lop rabbits was anteversion (forward tipping) of the femoral neck. This was sometimes secondarily associated with subtrochanteric torsion of the femoral shaft. As a consequence of both processes, the femoral heads projected from anterior rather than medial surfaces of the femoral shafts. The result in every instance was subluxation of the hip joint. Complete luxation was observed only once. Arendar and Milch also considered the disease in their rabbits a recessive trait.

Thus, "splay leg" remains a descriptive clinical term without clearly established pathological meaning(s). Systematic investigation of the problem is badly needed.

10. Waltzing (Circling)

Cogan (19) reported on a litter of 8 rabbits among which 5 exhibited circling behavior, presumably of genetic origin

like similar phenomena known in several strains of mice. But, unfortunately genetic transmission could not be tested as Cogan's rabbits would never breed. The affected litter had been produced by a mating between a New Zealand White buck and a White Beveren doe. Cole and Steele (20) had previously reported on a single waltzing male of uncertain ancestral stock.

When at rest, the animals described by Cogan usually held the head tilted 45° to either side and ears flat against the back. After being aroused they circled repeatedly to one side or the other for up to 7 rotations, all the while nodding the head up and down through an arc of about 30°. Animals suspended by the rear quarters permitted the head to hang freely in vertical alignment with the back rather than in normal dorsoflexion. Affected animals were further characterized by generalized decrease in muscle tone and inability to make many postural adjustments like normal animals. No neurological or inner ear lesions were found by histological methods. It is of interest that 4 of the waltzers had coloboma of the iris and a fifth had corneal opacities, but these findings were presumably unrelated to the waltzing behavior.

11. Hypertension

Inherited strain differences in blood pressure have been observed in rabbits (44c). A population of hypertensive rabbits has been developed by selection (1) with a high percentage of the progeny in subsequent generations also having hypertension (1a). Direct arterial pressure measurements in this selected population of hypertensive rabbits showed the mean systolic blood pressure to be 30 mm Hg higher than normotensive stock rabbits. These hypertensive rabbits also had larger Traube-Hering waves than did normotensive rabbits (1a). A review of the literature (172a) suggests that the heritability of this condition in various laboratory species is of the order of 20%.

12. Miscellaneous Abnormalities

A large number of congenital malformations have been observed in instances where inbreeding of rabbits has been practiced (15, 18, 165). These have included conjoined twinning in Y strain rabbits (17), thoracogastroshisis in Lurie's C line (22a, 159), hypospadias in the AX strain (R.R.F.), and many other conditions which almost certainly have a genetic basis.

There still remain a number of entities which logically deserve mention in the interest of completeness of this chapter, although some of them are not necessarily pertinent to a consideration of disease states. Investigators interested in anatomical variants in rabbits should be aware of the wealth of information on genetic aspects of morphogenesis compiled by Sawin and many collaborators over the past 40 years. Their studies are particularly extensive in the area of normal and abnormal skeletal development (80, 165), but also include descriptions of anatomical variations involving branching of the aorta (28), anatomy of the vena cava (101), size and shape or complete absence of the gallbladder (161), and many other observations of interest to geneticists and pathologists.

IV. SOURCES OF RABBITS WITH INHERITED DISEASES AND VARIATIONS

In the past, means of communication about rabbits bearing genetically determined abnormalities were, at the very best, haphazard and inefficient. Perhaps this explains why several extremely interesting mutant genes and heritable diseases of the rabbit seem to have disappeared entirely shortly after their original description. Because of the importance of preserving such valuable stocks and making them available for biomedical research, the Institute of Laboratory Animal Resources of the National Academy of Sciences–National Research Council (2101 Constitution Avenue, Washington D.C. 20418) maintains up-to-date listings of special stocks of rabbits, as well as other animals, available nationally and internationally.

In the United States there presently are only two major resources of genetically defined stocks of rabbits. By far the larger assortment of special genetic stocks is maintained at The Jackson Laboratory in Bar Harbor, Maine. In addition, Dr. Carl Cohen (Center for Genetics, University of Illinois Medical Center, Chicago, Illinois 60612) maintains a large number of serological mutants.

Acknowledgments

This work was supported in part by National Institutes of Health research grants RR-00251 and RR-00463 from the Division of Research Resources, HD-01496 from the National Institutes of Child Health and Human Development, research funds from the Veterans Administration, a grant from The Council for Tobacco Research, U.S.A., and in part by income from the Endowment Funds of The Jackson Laboratory.

The authors gratefully acknowledge the assistance of Mrs. Dorcas D. Crary and Dr. Henry J. Baker in preparation of the manuscript and and photographs, respectively.

REFERENCES

1. Alexander, N., Hinshaw, L. B., and Drery, D. R. (1954). Development of a strain of spontaneously hypertensive rabbits. *Proc. Soc. Exp. Biol. Med.* **86**, 855–858.
1a. Alexander, N., Hinshaw, L. B., and Drery, D. R. (1956). Further observations on development of a colony of spontaneously hypertensive rabbits. *Proc. Soc. Exp. Biol. Med.* **92**, 249–253.
1b. Anders, M. V. (1945). The histopathology of a new type of hereditary loss of coordination in the domestic rabbit. *Amer. J. Anat.* **76**, 183–199.

2. Antonitis, J. J., Crary, D. D., Sawin, P. B., and Cohen, C. (1954). Sound-induced seizures in rabbits. *J. Hered.* **45**, 279–284.
3. Arendar, G. M., and Milch, R. A. (1966). Splay-leg—a recessively inherited form of femoral neck anteversion, femoral shaft torsion and subluxation of the hip in the laboratory Lop rabbit: Its possible relationship to factors involved in so-called "congenital dislocation" of the hip. *Clin. Orthopaed. Relat. Res.* **44**, 221–229.
4. Aurrichio, G., and Wistrand, P. (1959). The osmotic pressure in aqueous humor of rabbits with congenital glaucoma. *Acta Ophthalmol.* **37**, 340–343.
5. Babino, E. J., Jr., and Fox, R. R. (1967). Buphthalmia in the rabbit: Effect of age and repeated testing on the cornified cell count diagnostic technique. *Proc. Soc. Exp. Biol. Med.* **126**, 216–217.
6. Bailey, J. A., II. (1971). Forms of dwarfism recognizable at birth. *Gen. Orthopaed.* **76**, 150–159.
7. Bargman, G. J., Mackler, B., and Shepard, T. H. (1972). Studies of oxidative energy deficiency. I. Achondroplasia in the rabbit. *Arch. Biochem. Biophys.* **150**, 137–146.
8. Bateman, N. (1954). Bone growth: A study of the grey-lethal and microphthalmic mutants of the mouse. *J. Anat.* **88**, 212–262.
9. Bernstein, J. (1968). Developmental abnormalities of the renal parenchyma—renal hypoplasia and dysplasia. *In* "Pathology Annual," Vol. 3, pp. 213–247. Appleton, New York.
10. Bragdon, J. H. (1952). Spontaneous atherosclerosis in the rabbit. *Circulation* **5**, 641–648.
11. Brown, W. H., and Pearce, L. (1945). Hereditary achondroplasia in the rabbit. I. Physical appearance and general features. *J. Exp. Med.* **82**, 241–260.
12. Castle, W. E. (1933). The linkage relations of yellow fat in rabbits. *Proc. Nat. Acad. Sci. U.S.* **19**, 947–950.
13. Castle, W. E. (1940). "Mammalian Genetics." Harvard Univ. Press, Cambridge, Massachusetts.
14. Castle, W. E., and Reed, S. C. (1936). Studies of inheritance in lop-eared rabbits. *Genetics* **21**, 297–309.
15. Chai, C. K. (1970). Effect of inbreeding in rabbits—skeletal variations and malformations. *J. Hered.* **61**, 2–8.
16. Chai, C. K., and Clark, E. M. (1967). Droopy-ear, a genetic character in rabbits. *J. Hered.* **58**, 149–152.
17. Chai, C. K., and Crary, D. D. (1971). Conjoined twinning in rabbits. *Teratology* **4**, 433–444.
18. Chai, C. K., and Degenhardt, K.-H. (1962). Developmental anomalies in inbred rabbits. *J. Hered.* **53**, 174–182.
19. Cogan, D. G. (1943). The waltzing (circling) phenomenon in rabbits. *J. Comp. Psychol.* **35**, 111–117.
20. Cole, L. J., and Steele, D. G. (1922). A waltzing rabbit. *J. Hered.* **13**, 290–294.
21. Colyer, F. (1936). "Variations and Diseases of the Teeth of Animals," 1st ed. John Bale and Sons, Ltd., London.
22. Craigie, E. H. (1960). "Bensley's Practical Anatomy of the Rabbit," 8th ed. Univ. of Toronto Press, Toronto.
22a. Crary, D. D. (1951). A thoraco-gastroschisis in the rabbit. *Anat. Rec.* **109**, 368 (abstr.).
23. Crary, D. D. (1964). Development of the external ear in the dachs rabbit. *Anat. Rec.* **150**, 441–448.
23a. Crary, D. D. Personal communication.
24. Crary, D. D., Fox, R. R., and Sawin, P. B. (1966). Spina bifida in the rabbit. *J. Hered.* **57**, 236–243.
24a. Crary, D. D., and Sawin, P. B. (1949). Morphogenetic studies of the rabbit. VI. Genetic factors influencing the ossification pattern of the limbs. *Genetics* **34**, 508–523.
25. Crary, D. D., and Sawin, P. B. (1952). A second recessive achondroplasia in the domestic rabbit. *J. Hered.* **43**, 254–259.
26. Crary, D. D., and Sawin, P. B. (1963). Morphogenetic studies of the rabbit. XXXII. Qualitative skeletal variations induced by the *ac* gene (achondroplasia). *Amer. J. Anat.* **113**, 9–23.

27. Degenhardt, K.-H. (1960). Die genetische und morphologische Analyse spezieller Entwicklungsstörungen in einem Stamm inge-züchteter Hermelin-Kaninchen. *Akad. Wiss. Lit., Mainz, Abh. Math.-Naturwiss. K.* No. 12, 919–988.

28. Edmonds, H. W., and Sawin, P. B. (1936). Variations in the branches of the aortic arch in rabbits. *Amer. Natur.* **70**, 48–49.

29. Ehling, U. (1957). Untersuchungen zur kausalen Genese erblicher Katarakte beim Kaninchen. *Z. Konstitutionslehre* **34**, 77–104.

30. Eliasson, S. G., Scarpellini, J. D., and Fox, R. R. (1967). Inositide metabolism in rabbit hereditary ataxia. *Arch. Neurol.* **17**, 661–665.

31. Feigenbaum, A. S., and Gaman, E. M. (1967). Influence of mother's milk on incidence of spontaneous aortic lesions in weanling rabbits. *Proc. Soc. Exp. Biol. Med.* **124**, 1020–1022.

32. Fenner, F. (1953). Changes in the mortality rate due to myxomatosis in the Australian wild rabbit. *Nature (London)* **172**, 228–230.

33. Fox, R. R., and Babino, E. J., Jr. (1965). Buphthalmia in the rabbit: A test for early diagnosis. *Proc. Soc. Exp. Biol. Med.* **119**, 229–233.

34. Fox, R. R., and Crary, D. D. (1971). Mandibular prognathism in the rabbit: Genetic studies. *J. Hered.* **62**, 23–27.

35. Fox, R. R., and Crary, D. D. (1971). Hypogonadia in the rabbit: Genetic studies and morphology. *J. Hered.* **62**, 163–169.

36. Fox, R. R., and Crary, D. D. (1971). A new recessive chondrodys-trophy in the rabbit. *Teratology* **4**, 245–246 (abstr.),

37. Fox, R. R., and Crary, D. D. (1972). A lethal recessive gene for adrenal hyperplasia in the rabbit. *Teratology* **5**, 255.

37a. Fox, R. R., and Crary, D. D. Unpublished data.

38. Fox, R. R., Crary, D. D., Babino, E. J., Jr., and Sheppard, L. B. (1969). Buphthalmia in the rabbit. Pleiotropic effects of the (*bu*) gene and a possible explanation of mode of gene action. *J. Hered.* **60**, 206–212.

39. Fox, R. R., Krinsky, W. L., and Crary, D. D. (1971). Hereditary cor-tical renal cysts in the rabbit. *J. Hered.* **62**, 105–109.

40. Fox, R. R., Laird, C. W., Evans, R. W., and Sheppard, L. B. (1971). Buphthalmia in the rabbit: Glucose metabolism. *J. Hered.* **62**, 294–296.

41. Fox, R. R., Meier, H., and Crary, D. D. (1971). Genetic predisposi-tion to tumors in the rabbit. *Naturwissenschaften* **58**, 457.

42. Fox, R. R., Meier, H., Crary, D. D., Myers, D. D., Norberg, R. F., and Laird, C. W. (1970). Hereditary lymphosarcoma and anemia in rabbits. *Teratology* **3**, 200 (abstr.).

43. Fox, R. R., Meier, H., Crary, D. D., Myers, D. D., Norberg, R. F., and Laird, C. W. (1970). Lymphosarcoma in the rabbit: Genetics and pathology. *J. Nat. Cancer Inst.* **45**, 719–729.

44. Fox, R. R., Meier, H., Crary, D. D., Norberg, R. F., and Myers, D. D. (1971). Hemolytic anemia associated with thymoma in the rabbit: Genetic studies and pathological findings. *Oncology* **25**, 372–382.

44a. Fox, R. R., and Norberg, R. F. Unpublished observations.

44b. Fox, R. R., Norberg, R. F., and Myers, D. D. (1971). The relation-ship of *Pasteurella multocida* to otitis media in the domestic rabbit (*Oryctolagus cuniculus*). *Lab. Anim. Care* **21**, 45–48.

44c. Fox, R. R., Schlager, G., and Laird, C. W. (1969). Blood pressure in 13 strains of rabbits. *J. Hered.* **60**, 312–314.

45. Gaman, E. M., Feigenbaum, A. S., and Schenk, E. A. (1967). Spon-taneous aortic lesions in rabbits. III. Incidence and genetic factors. *J. Atheroscler. Res.* **7**, 131–141.

46. Garbarsch, C., Matthiessen, M. E., Helin, P., and Lorenzen, I. (1970). Spontaneous aortic arteriosclerosis in rabbits of the Danish Country strain. *Atherosclerosis* **12**, 291–300.

47. Geri, G. (1954). Considerazioni e ricerche sull'eredita dell' idrof-talmia nel coniglio. *Rich. Sci.* **24**, 2299–2315.

48. Glick, D. (1940). Properties of tropine esterase. *J. Biol. Chem.* **134**, 617–625.

49. Glick, D., and Glaubach, S. (1941). The occurrence and distribution of atropinesterase, and the specificity of tropinesterases. *J. Gen. Physiol.* **25**, 197–205.

50. Greaves, D. P., and Perkins, E. S. (1951). Buphthalmos in the rabbit. *Brit. J. Ophthalmol.* **35**, 232–233.

51. Greene, H. S. N. (1933). Oxycephaly and allied conditions in man and in the rabbit. *J. Exp. Med.* **57**, 967–976.

52. Greene, H. S. N. (1935). Rabbit pox. IV. Susceptibility as a function of constitutional factors. *J. Exp. Med.* **62**, 305–329.

53. Greene, H. S. N. (1935). Hereditary brachydactylia and associated abnormalities in the rabbit. *Science* **81**, 405–407.

54. Greene, H. S. N. (1940). A dwarf mutation in the rabbit. *J. Exp. Med.* **71**, 839–856.

55. Greene, H. S. N. (1965). Diseases of the rabbit. *In* "The Pathology of Laboratory Animals (W. E. Ribelin and J. R. McCoy, eds.), pp. 330–350. Thomas, Springfield, Illinois.

56. Greene, H. S. N., Hu, C. K., and Brown, W. H. (1934). A lethal dwarf mutation in the rabbit with stigmata of endocrine abnormality. *Science* **79**, 487–488.

57. Greene, H. S. N., and Saxton, J. A., Jr. (1939). Hereditary brachy-dactylia and allied abnormalities in the rabbit. *J. Exp. Med.* **69**, 301–314.

58. Grunder, A. A., Rollins, W. C., Stormont, C., and Casady, R. B. (1968). A note on differential mortality rates in young rabbits of esterase phenotypes A, AB, and B. *Anim. Prod.* **10**, 221–222.

59. Grunder, A. A., Sartore, G., and Stormont, C. (1965). Genetic variation in red cells esterases of rabbits. *Genetics* **52**, 1345–1353.

60. Hanna, B. L., Sawin, P. B. and Sheppard, L. B. (1962). Recessive buphthalmos in the rabbit. *Genetics* **47**, 519–529.

61. Harrington, D. D., and Newberne, P. M. (1970). Correlation of maternal blood levels of vitamin A at conception and the incidence of hydrocephalus in newborn rabbits. An experimental animal model. *Lab Anim. Care* **20**, 675–680.

62. Haust, M. D., and Geer, J. C. (1970). Mechanism of calcification in spontaneous aortic arteriosclerotic lesions of the rabbit. *Amer. J. Pathol.* **60**, 329–346.

63. Haust, M. D., and More, R. H. (1965). Spontaneous lesions of the aorta in the rabbit. *In* "Comparative Atherosclerosis" (J. C. Roberts and R. Straus, eds.), pp. 255–275. Harper, New York.

64. Hirzel, R. (1935). Note on the effect of condition on the colour of body fat. *J. Agr. Sci.* **25**, 541–544.

65. Hohenboken, W. D., and Nellhaus, G. (1970). Inheritance of audio-genic seizures in the rabbit. *J. Hered.* **61**, 107–112.

66. Horak, F. (1965). Selection of strains of rabbits sensitive to an epileptogenic sound stimulus. *Physiol. Bohemoslov.* **14**, 495–501.

67. Hu, C. K., and Greene, H. S. N. (1935). A lethal acromegalic muta-tion in the rabbit. *Science* **81**, 25–26.

68. Inman, O. R. (1941). Embryology of hereditary brachydactyly in the rabbit. *Anat. Rec.* **79**, 483–505.

69. Innes, J. R. M., and O'Steen, W. K. (1957). Splayleg in rabbits—an inherited disease analogous to joint dysplasia in children and dogs. *Lab. Invest.* **6**, 171–186.

70. Innes, J. R. M., and Saunders, L. Z. (1962). Inherited diseases and cogenital anomalies. "Comparative Neuropathology," p. 327. Academic Press, New York.

71. Ireson, H. (1968). A preliminary report on an abnormal dental con-dition in rabbits. *J. Inst. Anim. Tech.* **19**, 36–39.

72. Israel, O. (1881). Experimentelle Untersuchung über den Zusam-menhang zwischen Nierenkrankheiten und Secundären Verände-rungen des zirkulations-systems. *Arch. Pathol. Anat. Physiol. Klin. Med.* **86**, 299–321.

73. Kalow, W. (1968). Pharmacogenetics in animals and man. *Ann. N.Y. Acad. Sci.* **151**, 694–698.

74. Kesten, H. D. (1935). Early incidence of spontaneous medial de-generation ("arteriosclerosis") in the aorta of the rabbit. *Arch. Pathol.* **20**, 1–8.

75. Kissane, J. M. (1966). Renal dysplasia. *In* "Pathology of the Kidney"

(by R. H. Heptinstall), pp. 76–83. Little, Brown, Boston, Massachusetts.

76. Kolker, A. E., Moses, R. A., Constant, M. A., and Becker, B. (1963). The development of glaucoma in rabbits. *Invest. Ophthalmol.* **2**, 316–321.

77. Kröning, F. (1939). Ein neuer Fall von erblichem Zwergwuchs beim Kaninchen. *Biol. Zentralbl.* **59**, 148–160.

78. Lamb, N. P., and Sawin, P. B. (1963). Morphogenetic studies of the rabbit. XXXIII. Cartilages and muscles of the external ear as affected by the dachs gene (*Da*). *Amer. J. Anat.* **113**, 365–388.

78a. Latimer, H. B., and Sawin, P. B. (1955). The weight of the brain, of its parts and the weight and length of the spinal cord in the rabbit (Race X). *J. Comp. Neurol.* **103**, 513–540.

79. Latimer, H. B., and Sawin, P. B. (1955). Morphogenetic studies of the rabbit. XIII. The influence of the dwarf gene upon organ size and variability in race X. *Anat. Rec.* **123**, 447–466.

80. Latimer, H. B., and Sawin, P. B. (1967). Morphogenetic studies of the rabbit. XXXIX. Ponderal correlation coefficients of the bones from two races of rabbits. *Anat. Rec.* **159**, 29–32.

81. Leipold, H. W., Doige, C. E., Kaye, M. M., and Crebb, P. H. (1970). Congenital osteopetrosis in Aberdeen Angus calves. *Can. Vet. J.* **11**, 181–185.

82. Letard, E. (1929). La Kératite des lapins Castorrex et Rex de couleur. *Rev. Vet., Toulouse* **81**, 419–425.

83. Letard, E. (1935). Une mutation nouvelle chez le lapin. *Bull. Acad. Vet. Fr.* [N.S.] **8**, 608–610.

84. Letard, E. (1943). Troubles de la locomotion et troubles de la vision chez le lapin. Liasion héréditaire. *Bull. Acad. Vet. Fr.* [N.S.] **16**, 184–192.

85. Levin, I., and Larkin, J. H. (1910). Spontaneous lesions in the rabbit. The early stages of the spontaneous arterial lesions in the rabbit. *Proc. Soc. Exp. Biol. Med.* **7**, 109–110.

86. Lévy, J. (1946). Transmission héréditaire de la tropanolestérase. *C.R. Soc. Biol.* **140**, 823–825.

87. Lévy, J., and Michel E. (1938). [Enzymatic hydrolysis of atropine.] *C.R. Soc. Biol.* **129**, 820–822.

88. Lurie, M. B. (1964). "Resistance to Tuberculosis: Experimental studies in Native and Acquired Defensive Mechanisms." Harvard Univ. Press, Cambridge, Massachusetts.

89. Lurie, M. B., and Dannenberg, A. M., Jr. (1965). Macrophage function in infectious disease with inbred rabbits. *Bacteriol. Rev.* **29**, 466–476.

90. Lurie, M. B., Zappasodi, P., Dannenberg, A. M., Jr., and Weiss, G. H. (1952). On the mechanism of genetic resistance to tuberculosis and its mode of inheritance. *Amer. J. Hum. Genet.* **4**, 302–314.

91. Mackler, B., Bargman, G. J., and Shepard, T. H. (1972). Etiology of achondroplasia in the rabbit: A defect in oxidative energy metabolism. *Teratology* **5**, 261.

92. Maegraith, B. O., and Carleton, H. M. (1939). Aortic arteriosclerosis in rabbits. *J. Pathol. Bacteriol.* **48**, 33–40.

93. Magnussen, K. (1952). Beitrag zur Genetik und Histologie eines isolirten Augen-Albinismus beim Kaninchen. *Z. Morphol. Anthropol.* **44**, 127–135.

94. Magnussen, K. (1954). Beitrag zur Genetik und Histologie eines isolirten Augen-Albinismus beim Kaninchen. II. *Z. Morphol. Anthropol.* **46**, 24–29.

95. Margolis, F., and Feigelson, P. (1964). Genetic expression and developmental studies with rabbit serum atropinesterase. *Biochim. Biophys. Acta* **90**, 117–125.

96. Marks, S. C., Jr., and Walker, D. G. (1969). The role of the parafollicular cell of the thyroid gland in the pathogenesis of congenital osteopetrosis in mice. *Amer. J. Anat.* **126**, 299–314.

97. Marshall, I. D., and Douglas, G. W. (1961). Studies in the epidemiology of infectious myxomatosis of rabbits. VIII. Further observations on changes in the innate resistance of Australian wild rabbits exposed to myxomatosis. *J. Hyg.* **59**, 117–122.

98. Marshall, I. D., and Fenner, F. (1958). Studies in the epidemiology of infectious myxomatosis of rabbits. V. Changes in the innate resistance of Australian wild rabbits exposed to myxomatosis. *J. Hyg.* **56**, 288–302.

99. McMaster, P. R. B. (1960). Decreased aqueous outflow in rabbits with hereditary buphthalmia. *Arch. Ophthalmol.* **64**, 388–391.

100. McMaster, P. R. B., and Macri, F. J. (1967). The rate of aqueous humor formation in buphthalmic rabbit eyes. *Invest. Ophthalmol.* **6**, 84–87.

101. McNutt, C. W., and Sawin, P. B. (1943). Hereditary variations in the vena cava inferior of the rabbit. *Amer. J. Anat.* **72**, 255–289.

102. Meier, H., and Fox, R. R. (1973). Hereditary lymphosarcoma in WH rabbits and hereditary hemolytic anemia associated with thymoma in strain X rabbits. *Bibl. Haematol. (Basel)* **39**, 72–92.

103. Menschow, G. B. (1934). Fattori letali nel coniglio cincilla. *Riv. Coniglicolt.* **6**, 8–9.

104. Miles, A. B. (1907). Spontaneous arterial degeneration in rabbits. *J. Amer. Med. Ass.* **49**, 1173–1176.

105. Moore, T. (1957). "Vitamin A," p. 143. Amer. Elsevier, New York.

106. Nachtsheim, H. (1931). Über eine erbliche Nervenkrankheit (Syringomyelie) beim Kaninchen. *Z. Petztier-Rauchwarenkd.* **3**, 254–259.

107. Nachtsheim, H. (1934). Schüttellähmun-gein Beispiel fur ein einfach mendelndes rezessives Nervenleiden beim Kaninchen. *Erbarzt* **1**, 36–38.

108. Nachtsheim, H. (1934). Kurzhaarkaninchen—drei genotypisch verschiedene Mutanten mit dem gleichen Phänotypus. *Erbarzt* **1**, 97–102.

109. Nachtsheim, H. (1936). Die Genetik einiger Erbleiden des Kaninchens, verglichen mit ähnlichen Krankheiten des Meschen. *Deut. Tieraerztl. Wochenschr.* **44**, 742–746.

110. Nachtsheim, H. (1936). Erbliche Zahnanomalien beim Kaninchen. *Zuechtungskunde* **11**, 273–287.

111. Nachtscheim, H. (1937). Erbpathologie des Kaninchens. *Erbarzt* **4**, 25–30 and 50–55.

112. Nachtsheim, H. (1937). Erbpathologische Untersuchungen am Kaninchen. *Z. Vererbungslehre* **73**, 463–466.

113. Nachtsheim, H. (1938). Erbpathologie der Haustiere. I. Organe des äusseren Keimblattes. *Z. Erbpathol., Rassenhyg. Grenzgeb.* **2**, 58–104.

114. Nachtsheim, H. (1939). Krampfbereitschaft und Genotypus. I. Die Epilepsie der Weissen Wiener Kaninchen. *Z. Konstitutionslehre* **22**, 791–810.

115. Nachtsheim, H. (1939). Erbleiden des Nervensystems bei Säugetieren. *In* "Handbuch der Erbbiologie des Menschen" (K. H. Bauer, E. Hanhart, and J. Lange, eds.), Vol. 3, pp. 36–99. Springer-Verlag, Berlin and New York.

116. Nachtsheim, H. (1940). Erbpathologie des Stützgewebes der Säugetiere. *In* "Handbuch der Erbbiologie des Menschen" (K. H. Bauer, E. Hanhart, and J. Lange, eds.), Vol. 5, pp. 1–55. Springer-Verlag, Berlin and New York.

117. Nachtscheim, H. (1950). The Pelger-anomaly in man and rabbit. A Mendelian character of the nuclei of leukocytes. *J. Hered.* **41**, 131–137.

118. Nachtsheim, H. (1958). Erbpathologie der Nagetiere. *In* "Pathologie der Laboratoriumstiere" (P. Cohrs, R. Jaffe, and H. Meesen, eds.), pp. 310–452. Springer-Verlag, Berlin and New York.

119. Nachtsheim, H., and Gürich, H. (1939). Erbleiden des Kaninchenauges. I. Erbliche Nahtbändchentrübung der Linse mit nachfolgendem Kernstar. *Z. menschl. Vereb. Konstitutionslehre* **23**, 463–483.

120. Nepola, S. R. (1969). The intrinsic and extrinsic factors influencing the growth and development of the jaws: Heredity and functional matrix. *Amer. J. Orthodont.* **55**, 499–505.

121. Noah, V. B., and Geeraets, W. J. (1971). The effect of ascorbic acid on the facility of outflow in normal and buphthalmic rabbits. *Acta Ophthalmol.* **49**, 410–417.

122. O'Leary, J. L., Fox, R. R., Smith, J. M., and Inukai, J. (1972). Ultrastructural alterations in vestibular and cerebellar nuclei of the ataxic rabbit. *J. Comp. Neurol.* **145**, 307–352.

123. O'Leary, J. L., Smith, J. M., Fox, R. R., Harris, A. B., and O'Leary, M. B. (1968). Hereditary ataxia of animals. *Arch. Neurol.* **19**, 34–46.

124. O'Leary, J. L., Smith, J. M., Harris, A. B., and Fox, R. R. (1968). Animal prototypes in hereditary ataxia. *In* "The Central Nervous System" (O. T. Bailey and D. E. Smith, eds.), Int. Acad. Pathol. Monog. No. 9, Chapter 7, pp. 124–156. Williams & Wilkins, Baltimore, Maryland.

125. Osburn, B. I., and Glenn, B. L. (1968). Acquired Pelger-Huët anomaly in cattle. *J. Amer. Vet. Med. Ass.* **152**, 11–16.

126. Osetowska, E. (1967). Nouvelle maladie héréditáire du lapin de laboratoire. *Acta Neuropathol.* **8**, 331–344.

127. Osetowska, E. (1972). Personal communication.

128. Osetowska, E., and Wiśniewski, H. (1966). Ataxie familiale du lapin, différente de la maladie heréditaire de Sawin-Anders. *Acta Neuropathol.* **6**, 243–250.

129. Ostertag, B. (1930). Weitere Untersuchungen über vererbbare Syringomyelie des Kaninchens. *Deut. Z. Nervenheilk.* **116**, 147–150.

130. Ostertag, B. (1930). Die Syringomyelie als erbbiologisches Problem. *Verh. Deut. Pathol. Ges.* **25**, 166–174.

131. Ostertag, B. (1934). Neuere Ergebnisse bei der vererbbaren Syringomyelie des Kaninchens. *Atti. Congr. Mond. Pollicolt.*, Vol. 3, pp. 526–532.

132. Padgett, G. A., Holland, J. M., Prieur, D. J., Davis, W. C., and Gorham, J. R. (1970). The Chediak-Higashi syndrome: A review of the disease in man, mink, cattle and mice. *In* "Animal Models for Biomedical Research," Vol. III, pp. 1–12. Inst. Lab. Anim. Sci., Nat. Acad. Sci., Washington, D. C.

133. Pearce, L. (1948). Hereditary osteopetrosis of the rabbit. II. X-ray, hematologic and chemical observations. *J. Exp. Med.* **88**, 597–620.

134. Pearce, L. (1950). Hereditary osteopetrosis of the rabbit. III. Pathologic observations; skeletal abnormalities. *J. Exp. Med.* **92**, 591–600.

135. Pearce, L. (1950). Hereditary osteopetrosis of the rabbit. IV. Pathologic observations; general features. *J. Exp. Med.* **92**, 601–624.

136. Pearce, L. (1960). Hereditary distal foreleg curvature in the rabbit. I. Manifestations and course of the bowing deformity; genetic studies. *J. Exp. Med.* **111**, 801–822.

137. Pearce, L. (1960). Hereditary distal foreleg curvature in the rabbit. II. Genetic and pathological aspects. *J. Exp. Med.* **111**, 823–830.

138. Pearce, L., and Brown, W. H. (1945). Hereditary achondroplasia in the rabbit. II. Pathological aspects. *J. Exp. Med.* **82**, 261–280.

139. Pearce, L., and Brown, W. H. (1945). Hereditary achondroplasia in the rabbit. III. Genetic aspects, general considerations. *J. Exp. Med.* **82**, 281–295.

140. Pearce, L., and Brown, W. H. (1948). Hereditary osteopetrosis of the rabbit. I. General features and course of disease; genetic aspects. *J. Exp. Med.* **88**, 579–596.

141. Pearce, L., and Brown, W. H. (1960). Hereditary premature senescence. I. Chronic form; general features, *J. Exp. Med.* **111**, 485–504.

142. Pearce, L., and Brown, W. H. (1960). Hereditary premature senescence. II. Acute form; general features. *J. Exp. Med.* **111**, 505–516.

143. Pease, M. (1928). Yellow fat in rabbits, a linked character? *Z. Indukt. Abstamm.- Vererbungsl.* **2**, Suppl., 1153–1156.

144. Pease, M. (1930). The inheritance of yellow fat in rabbits. *Verh. Int. Kaninchenzuechter-Kongr., 1st, 1900* pp. 91–95.

145. Perey, D. Y. E., Herdman, R. C., and Good, R. A. (1967). Polycystic renal disease: A new experimental model, *Science* **158**, 494–496.

146. Pichler, A. (1910). Spontanes Glaukom (Hydrophthalmus) beim Kaninchen. *Arch. Verg. Ophthalmol.* **1**, 175—177.

147. Pobisch, R. (1957). Familiäres Auftreten von epileptiformen Anfällen bei einer nicht-leuzistischen Kaninchenrasse. *Wien. Tieraerzt. Monatsschr.* **44**, 193–199.

148. Pollock, S. (1951). Slobbers in the rabbit. *J. Amer. Vet. Med. Ass.* **119**, 443–444.

149. Résibois-Grégoire, A., and Dourov, N. (1966). Electron microscopic study of a case of cerebral glycogenosis. *Acta Neuropathol.* **6**, 70–79.

150. Robinson, N. (1970). Enzyme changes in the hereditary ataxic rabbit. *Acta Neuropathol.* **14**, 326–337.

151. Robinson, N. (1970). Glucose ^{14}C metabolism in rabbit hereditary ataxia. *Arch. Neurol.* (*Chicago*) **22**, 445–449.

152. Robinson, N. (1958). Genetic studies of the rabbit. *Bibliogr. Genet.* **17**, 229–558.

153. Rochon-Duvigneaud, A. (1921). Un cas de buphthalmie chez le lapin: Etude anatomique et physiologique. *Ann. Ocul.* **158**, 401–414.

154. Rohloff, R. (1945). Entwicklungsgeschichtliche Untersuchungen über erbliche Anomalien der Incisiven bei *Oryctolagus cuniculus.* L.; zugleich Mitteilung von Beobachtungen an den Rudimentärzähnchen. Inaugural-Dissertation, University of Berlin.

155. Rosa, F. M. Da (1943). Agenesia de um rim, uma nova mutação no coelho. *Rev. Med. Vet.* (*Lisboa*) **38**, 349–363.

156. Rosa, F. M. Da (1945). Uma nova mutação, luxação congénita da anca, no coelho. *Rev. Med. Vet.* (*Lisboa*) **40**, 1–23.

157. Rosa, F. M. Da (1946). Hidrocefalia, uma nova mutação no coelho. *Rev. Med. Vet.* (*Lisboa*) **41**, 1–55.

158. Ross, S., Sawin, P. B., Denenberg, V. H., and Volow, M. (1963). Effects of previous experience and age on sound-induced seizures in rabbits. *Int. J. Neuropharmacol.* **2**, 255–258.

159. Sawin, P. B. (1955). Recent genetics of the domestic rabbit. *Advan. Genet.* **7**, 183–226.

160. Sawin, P. B., Anders, M. V., and Johnson, R. B. (1942). "Ataxia," a hereditary nervous disorder of the rabbit. *Proc. Nat. Acad. Sci. U.S.* **28**, 123–127.

161. Sawin, P. B., and Crary, D. D. (1951). Morphogenetic studies of the rabbit. X. Racial variations in the gallbladder. *Anat. Rec.* **110**, 573–590.

162. Sawin, P. B., and Crary, D. D. (1955). Congenital scoliosis in the rabbit. *Anat. Rec.* **121**, 2 (abstr.).

163. Sawin, P. B., and Crary, D. D. (1957). Morphogenetic studies of the rabbit. XVII. Disproportionate adult size induced by the *Da* gene. *Genetics* **42**, 72–91.

164. Sawin, P. B., and Crary, D. D. (1962). Inherited hypogonadia in the rabbit. *Anat. Rec.* **142**, 325 (abstr.).

165. Sawin, P. B., and Crary, D. D. (1964). Genetics of skeletal deformities in the domestic rabbit (Oryctolagus cuniculus). *Clin. Orthop. Relat. Res.* **33**, 71–90.

166. Sawin, P. B., and Glick, D. (1943). Atropinesterase, a genetically determined enzyme in the rabbit. *Proc. Nat. Acad. Sci. U.S.* **29**, 55–59.

167. Sawin, P. B., Ranlett, M., and Crary, D. D. (1959). Morphogenetic studies of the rabbit. XXV. The spheno-occipital synchondrosis of the dachs (chondrodystrophy) rabbit. *Amer. J. Anat.* **105**, 257–280.

168. Sawin, P. B., Ranlett, M., and Crary, D. D. (1962). Morphogenetic studies of the rabbit. XXIX. Accessory ossification centers at the occipitovertebral articulation of the dachs (chondrodystrophy) rabbit. *Amer. J. Anat.* **111**, 239–258.

169. Schenk, E. A., Gaman, E., and Feigenbaum, A. S. (1966). Spontaneous aortic lesions in rabbits. I. Morphologic characteristics. *Circ. Res.* **19**, 80–88.

170. Scherer, H. J. (1944). "Vergleichende Pathologie des Nervensystems der Säugtiere," Thieme Leipzig.

171. Schiff, R. (1970). The biochemical genetics of rabbit erythrocyte esterases. Histochemical classification. *H. Histochem. Cytochem.* **18**, 709–721.

172. Schiff, R., and Stormont, C. (1970). The biochemical genetics of rabbit erythrocyte esterases: Two new esterase loci. *Biochem. Genet.* **4**, 11–23.

172a. Schlager, G. (1972). Spontaneous hypertension in laboratory animals. *J. Hered.* **63**, 35–38.

173. Schloesser, C. V. (1886). Acutes secundarglaucom beim Kaninchen. *Z. Vergl. Augenheilk.* **4**, 79–83.

174. Schnecke, C. (1941). Zwergwuchs beim Kaninchen und seine Vererbung *Z. Konstitutionslehre* **25**, 427–457.

175. Shadle, A. R. (1936). The attrition and extrusive growth of the four major incisor teeth of domestic rabbits. *J. Mammal.* **17**, 15–21.

176. Shepard, T. H., and Bass, G. L. (1971). Organ-culture studies of achondroplastic rabbit cartilage: Evidence for a metabolic defect in glucose utilization. *J. Embryol. Exp. Morphol.* **25**, 347–363.

177. Shepard, T. H., Fry, L. R., and Moffett, B. C., Jr. (1969). Microscopic studies of achondroplastic rabbit cartilage. *Teratology* **2**, 13–22.

178. Sheppard, L. B., and Shanklin, W. M. (1968). Corneal epithelium changes in rabbit congenital glaucoma. *Amer. J. Ophthalmol.* **65**, 406–413.

179. Sheppard, L. B., Shanklin, W. M., Harris, T. H., and Fox, R. R. (1971). A histologic study of regenerating epithelium of normal and buphthalmic rabbit cornea. *Ophthalmol. Res.* **2**, 116–125.

180. Smith, J. A. (1944). The blood-aqueous barrier in hydrophthalmic rabbits. *Ophthalmologica* **108**, 293–297.

181. Sobey, W. R. (1969). Selection for resistance to myxomatosis in domestic rabbits (*Oryctolagus cuniculus*). *J. Hyg.* **67**, 743–754.

182. Van Kampen, K. R. (1969). Lymphosarcoma in the rabbit. A case report and general review. *Cornell Vet.* **58**, 121–128.

183. Vogt, A. (1919). Hydrophthalmos héréditáire chez le lapin. *Clin. Ophthalmol.* **23**, 667–670.

183a. Walker, D. D., and Fox, R. R. Unpublished data.

184. Weinstein, W. J., Sawin, P. B., Fox, R. R., and O'Leary, J. L. (1964). Hereditary ataxia in the rabbit: Amino acid analyses of blood and brain. *J. Nerv. Ment. Dis.* **139**, 120–125.

184a. Weisbroth, S. H. Personal communication.

185. Weisbroth, S. H., and Ehrman, L. (1967). Malocclusion in the rabbit: A model for the study of the development, pathology and inheritance of malocclusion. *J. Hered.* **58**, 245–246.

186. Werner, G. (1965). Fermentdefekte als Ursache Unterschiedlicher Mydriatischer Wirksamkeit von Atropin und Cocain bei Kaninchen. *Naumyn-Schmeidebergs Arch. Exp. Pathol. Pharmakol.* **251**, 320–334.

187. Willimott, S. G. (1928). On the pigment of the fat of certain rabbits. *Biochem. J.* **22**, 1057–1059.

188. Wilson, W. K., and Dudley, F. J. (1946). Fat colour and fur colour in different varieties of rabbit. *J. Genet.* **47**, 290–294.

189. Zeman, W. V., and Fielder, F. G. (1969). Dental malocclusion and overgrowth in rabbits. *J. Amer. Vet. Med. Ass.* **155**, 1115–1119.

CHAPTER 16

Nutrition and Nutritional Diseases of the Rabbit

Charles E. Hunt and Daniel D. Harrington

I. COMPARATIVE NUTRITION

A. Comparative Physiology of Digestion

The rabbit is a nonruminant herbivore, i.e., it has a simple monogastric digestive system with well-developed functional cecum and large intestine. Rabbits are habitual avid ingesters of their own feces, a practice referred to by various descriptive terms such as coprophagy, reingestion, refection, and pseudo-rumination. This normal physiological phenomenon has been studied in the rabbit by numerous investigators (59, 60, 98, 112, 129, 146, 164, 170, 208, 219, 222) and occurs also in the rat, guinea pig, and other rodents. Several conclusions of nutritional impor-

tance may be made from these investigations. Rabbits produce two kinds of feces, a soft mucous form at night and a firm pellet during the day. The soft night feces are virtually completely reingested directly from the anus and represent 30–80% of the total daily excreta (59, 129). Soft night feces have protein, ash, and fiber composition similar to normal cecal contents (60, 112) and originate from the cecum as result of rhythmic contractions peculiar to the rabbit and not yet fully understood. Prevention of reingestion in controlled experiments has demonstrated that utilization of dietary protein and dry matter is improved by normal coprophagy (112, 219). Furthermore, removal of the cecum interrupts cyclical reingestion and results in poor utilization of sodium and potassium (98). Finally, there is ample evidence that the dietary requirement for many of

the vitamins is reduced or eliminated by the functional cecum and reingestion. For example, rabbits fed diets low in
pantothenic acid or riboflavin developed no signs of
vitamin deficiency and excreted many times the quantity
of these vitamins ingested (170). Also, the soft feces contain several times the quantity of niacin, riboflavin, pantothenic acid, and vitamin B$_{12}$ present in hard feces (129).
Consequently, the rabbit has a greatly enhanced potential
vitamin B supply by practicing reingestion.

An ecological advantage also may be realized through
reingestion. When wild rabbits are frightened into remaining in their burrow they can survive a week or more without
food or water by recycling the intestinal contents (208, 222).
Experimental evidence substantiates these observations.
In normal and fasting rabbits the cecum functions as a
reservoir for food materials (30).

All of these findings are of particular importance to investigators interested in nutritional studies in the rabbit.
Attempts to produce deficiency of nutrients required in
minute amounts, such as trace minerals, could be greatly
prolonged by normal recycling of intestinal contents. Also,
experiments designed to determine protein or B vitamin
requirements must take into account intestinal synthesis
and reingestion of these substances in soft feces. The phenomenon of reingestion is also of obvious interest in experiments where drugs and substances such as cholesterol are
fed and in studies of parasite infection.

B. Discussion of Nutrient Requirements

Although the rabbit has been used extensively in metabolic and various other studies, knowledge of nutrient
requirements of this species is incomplete. In fact, there
seem to have been no studies of the rabbit's requirement for
vitamins D and K, essential fatty acids, folic acid, iron,
zinc, or iodine. Other vitamins and minerals could be listed
for which only limited attempts have been made to produce
deficiency status and none to determine requirements.

Data in Table I summarize results of experiments which
were designed to study the specific nutrients listed. In some
instances quantitative data were obtained, in others estimates were derived from response of animals in deficiency
experiments. It is pertinent to note that essentially all these
estimates (Table I) were obtained under totally different
experimental conditions, especially with respect to type of
diet fed. Futhermore, the animals studied were usually in
the stage of rapid growth or young adults. It is apparent
(Table I) that the rabbit has a considerably higher requirement for potassium than the rat (236). The guinea pig,
another herbivore, also demands a high level of dietary
potassium (185). The rabbit also has an unusually high
requirement for niacin (Table I). This is peculiar in view
of the extent of intestinal synthesis and consequent sparing

TABLE I
ESTIMATED NUTRIENT REQUIREMENTS OF GROWING RABBITS[a]

Nutrient[b]	Percent of diet	Ref.
Protein[c]	16–20	74, 100, 111, 156
Fat[c]	5–10	74, 218
Potassium	0.6–0.9	109
Calcium	0.22–0.40	33
Phosphorus	0.22	148
Magnesium	0.03–0.04	130
Manganese	0.002–0.004	204
Copper	0.0003–0.0006	114
Niacin	0.01–0.02	255
Choline	0.13	107

Nutrient	μg/100 gm Diet	Ref.
Vitamin E (dl, α-tocopherol)	2400–2800	108
Vitamin A (11-cis-retinol)	33–44	175
Pyridoxine	100	110
Thiamine	Little or none	179, 185
Riboflavin	Little or none	129, 170
Pantothenic acid	Little or none	129, 170
Ascorbic acid	None	91

[a] Requirements listed are extrapolated from data provided in references
based on an intake of 50–60 gm dry food/kg body wt/day. Balance of other
nutrients and normal coprophagy is assumed.

[b] For the following nutrients no estimate of the requirement has been
made: essential fatty acids, vitamins D, K, and B$_{12}$, biotin, folic acid, iron,
zinc, cobalt, iodine, molybdenum, and selenium.

[c] Absolute requirements for these substances are not known. The
references provide evidence that levels listed are adequate for growth
and maintenance.

of dietary source of many B vitamins, including niacin (129).
Known or estimated requirements of the rabbit for other
minerals and vitamins (Table I) are similar to those agreed
on for the rat (236).

Determination of protein or amino acid requirements of
the rabbit is complicated by bacterial synthesis of amino
acids and protein in the gut, especially in the cecum, and
reingestion of the modified dietary nitrogen in the soft night
feces. It has been shown that the soft feces contain a substantially greater amount of protein than the hard fecal
pellets which are not normally reingested (60). The effect
on nitrogen metabolism of recycling intestinal contents is
somewhat analogous to ruminant digestion, i.e., through
reingestion the rabbit has opportunity to utilize protein
synthesized by intestinal flora and this has been confirmed
by evidence that coprophagy does enhance protein digestibility and nitrogen retention (112, 219). There is also
evidence that, in contrast to other mammals, the rabbit
absorbs considerable protein from the large intestine (112).

Specific studies of the suitability of various proteins for
rabbit diets are not easy to interpret because of other variables, e.g., form of diet, minerals and vitamins supplied,
type and amount of fat, carbohydrate, etc. A variety of protein sources have been utilized in semipurified and purified

diets for rabbits including dried whole or skim milk, casein, soybean meal or isolated soy protein, and peanut meal. Hove and Herndon (111) reported better growth with soybean meal than casein in young New Zealand White and California White rabbits. Addition of gum arabic or agar resulted in significant improvement in the growth rate of rabbits fed the casein diet. Gaman *et al.* (74) reported remarkable success using a pelleted purified diet containing 20% isolated soy protein plus 0.2% DL-methionine. Peanut meal is limited in methionine and therefore has been a useful protein source in studies of choline deficiency in the rabbit (106, 107).

Qualitative dietary requirements of the rabbit for certain amino acids have been studied in recent years. There is evidence that arginine (73, 175), lysine, and methionine (73) are essential for the growing rabbit. Further studies by Adamson and Fisher (1), using a chemically defined amino acid diet, have confirmed the essentiality of arginine, lysine, and methionine, and, in addition, have demonstrated a need for tryptophan, leucine, isoleucine, threonine, valine, phenylalanine, and histidine. Addition of glycine to the diet was necessary for rapid growth. These studies have provided a basis for assessment of quantitative amino acid requirements in the rabbit and, therefore, an opportunity to clarify knowledge of the protein needs of this species.

C. Use of Natural and Purified Diets

One of the weakest links in experimental design of animal studies is failure to take into account the importance of nutritional methodology. This is particularly true when experiments are performed that require use of special diets (metabolic studies, nutritional deficiencies or excesses) or addition of drugs or other substances to the diet. The inconsistencies introduced into an experiment by using inadequate or imbalanced diets or feeding unpurified commercial rations to control groups and purified diet to experimental groups have been extensively reviewed and discussed by Greenfield and Briggs (86). These criticisms are valid but the underlying principles generally have not been applied in research utilizing rabbits because of incomplete knowledge of nutrient requirements of this species and, until recently, unavailability of a satisfactory purified diet.

Three basic types of diets are utilized by experimentalists, namely, natural or practical, purified, and chemically defined. The natural diet includes ingredients such as cereals, alfalfa, yeast, or dried liver. Natural diets may contain unknown substances and quality and quantity of nutrients are difficult to control from one mix to another. This type of diet has the advantage of being easily formulated and is readily accepted by the animal. A laboratory mixed natural diet has proved useful in studies of vitamin A deficiency in the rabbit (88, 90).

Purified diets incorporate ingredients of known composition such as isolated soy protein or casein, refined sucrose, vegetable oil, and mixtures of pure vitamins and minerals. Purified diets are the type most commonly used in experimental nutrition. Chemically defined diets are homogeneous mixtures of chemically pure substances, e.g., amino acids, reagent grade minerals, pure vitamins, etc.

Many attempts have been made, since the early 1930's, to raise rabbits on "simplified" or purified rations. Cow's milk is low in copper, iron, and manganese, and, consequently, diets of whole or skim milk supplemented with certain minerals or vitamins were used to induce deficiencies of these elements in the rabbit (61, 203–206). However, this type of diet, even though supplemented with vitamins and minerals, did not produce normal gains and often resulted in other evidence of deficiencies such as fatty and/or cirrhotic liver (204, 205, 209, 220). A gel form of skim milk-based diet supplemented with methionine, glycine, and mineral and vitamin mixes has been used successfully in rabbit studies without pathological changes in liver or other organs (114, 115). Other workers have used semipurified rations that were based on casein but required addition of natural foods such as alfalfa, kale, liver, or fruit or vegetable juice to obtain maximum growth rate or weight gain (100–102, 253).

Recently, Gaman and associates (74) have published an adequate purified diet for rabbits based on isolated soy protein (Table II). Rabbits fed this ration for periods of 4 months to 2 years maintained excellent condition and some were exhibited and scored well in competition (74). A complete purified diet that has proved useful in our laboratory (113) is presented in Table III. This agar gel diet is a modification of a purified diet used successfully for guinea pigs by Navia and Lopez (167). Results of unpublished observations referred to throughout this chapter were obtained by feeding this diet to weanling Dutch rabbits. The gel diet was made available to young rabbits at 3 weeks of age and they grew to maturity (6 months or older) with weight gains equivalent to those simultaneously reared on commercial pellets. When rabbits have reached 3 months of age on the diet the supplemental methionine, arginine, and glycine may be deleted without affecting growth or condition of the animals. No rabbits have developed diarrhea or died while being fed the complete gel diet. All animals have been autopsied and no gross or histopathological lesions attributable to nutritional disease have been observed (113). Neither of these purified diets (Tables II and III) differ radically from several others previously tested (100, 111), except that they are in pelleted or gel form. This factor may contribute importantly to their success since many experiments attest to the fact that rabbits do not adapt readily to powdered diets.

TABLE II
A PELLETED PURIFIED DIET SATISFACTORY FOR GROWTH, MAINTENANCE, AND REPRODUCTION OF THE RABBIT[a]

Ingredient	Amount (%)
Soy protein (assay protein C-1)[b]	20.0
DL-Methionine	0.2
Glucose monohydrate	26.0
Dextrin	5.0
Corn starch	20.9
Cellulose	16.0
Corn oil	5.0
Mineral mix[c]	6.6
Vitamin mix[d]	0.2
Choline chloride (70% solution)	0.1
Antioxidant[e]	0.025
Water (for pelleting)	5.0

[a]Gaman et al. (74, 74a).

[b]Skidmore Sales and Distributing Co., Inc., Cincinnati, Ohio.

[c]Provides the following in mg/100 gm of diet: $CoCl_2 \cdot 6H_2O$, 0.35; $CuSO_4 \cdot 5H_2O$, 3.46; $MnSO_4 \cdot H_2O$, 8.11; $ZnSO_4$, 16.9; $FeC_6H_5O_4 \cdot 14H_2O$, 70.6; $(NH_4)_6Mo_7O_{24} \cdot 4H_2O$, 2.27; KI, 1.0; in gm/100 gm; K_2HPO_4, 1.0; $KHCO_3$, 1.0; $NaHCO_3$, 0.8; NaCl, 0.5; $CaCO_3$, 1.25; $CaHPO_4$, 1.0; $MgSO_4$, 1.0.

[d]Provides the following in mg or IU/100 gm of diet: thiamine HCl, 2.5; riboflavin, 1.6; calcium pantothenate, 2.0; pyridoxine HCl, 0.6; biotin 0.06; folic acid, 0.4; menadione, 0.5; vitamin B_{12}, 0.002; ascorbic acid, 25; Niacin, 15; vitamin A, 1000 IU; vitamin D_3 (Cholecalciferol), 60 IU; dl-α-tocopherolacetate, 5 IU.

[e]Santoquin, 50% dry concentrate, Monsanto Company, St. Louis, Missouri.

TABLE III
A PURIFIED AGAR GEL DIET FOR RABBITS

Ingredient	Amount (%)
Casein (vitamin-free)	20.00
L-Arginine	0.20
DL-Methionine	0.20
Glycine	0.20
Sucrose (powdered, confectionery)	48.00
Cellulose[a]	11.56
Vegetable oil [b]	10.00
Mineral mix[c]	4.04
Magnesium oxide (powder)	0.40
Potassium acetate	2.50
Vitamin mix[d]	1.00
Choline chloride	0.10
Agar[e]	1.80

[a]Alphacel, Nutritional Biochemicals Corp., Cleveland, Ohio.

[b]Soybean and cottonseed oil (Wesson), Hunt-Wesson Foods, Inc., Fullerton, California.

[c]Provides the following in gm/100 gm of diet: $CaCO_3$, 2.05; KH_2PO_4, 0.83; $MgSO_4$, 0.075; $MgCO_3$, 0.10; NaCl, 0.28; $FePO_4$, 0.16; KIO_3, 0.0038; $MnSO_4 \cdot H_2O$, 0.08; $ZnSO_4 \cdot 7H_2O$, 0.0025; $CuSO_4$, 0.0036; $CoCl_2 \cdot 6H_2O$, 0.0030; $AlK(SO_4)_2 \cdot 12H_2O$, 0.0007; NaF, 0.0004; KCl, 0.45.

[d]Provides the following in mg or IU/100 gm of diet (mixed in corn starch): inositol, 100; thiamine, 1; riboflavin, 1; pyridoxine hydrochloride, 1; Ca pantothenate, 3; niacin, 10; folic acid, 1; menadione, 1; biotin, 0.00002; vitamin B_{12} (0.1% in mannitol), 3; dl-α-tocopherol, 8 IU; vitamin A acetate, 1500 IU; vitamin D_2 (Calciferol), 50 IU.

[e]Bacty grade, powdered, General Biochemicals, Chagrin Falls, Ohio. The agar is dissolved in 100 ml boiling distilled water and mixed with the ingredients minus the vitamins until cool (60°C). The vitamins are then mixed in, the diet is allowed to gel and is refrigerated in plastic containers.

II. VITAMINS

A. Deficiency and Excess of Fat-Soluble Vitamins

1. Vitamin A

As postulated by Wald in the mid-1930's, vitamin A (11-cis-retinol) is required by man and lower vertebrates as a prosthetic group for visual pigments, and this is presently the only biological function of the vitamin which has been clearly defined (67, 233, 234). However, numerous investigators have described clinical signs and pathological and biochemical changes in deficient animals and their tissue isolates which makes it apparent that vitamin A has other metabolic roles. Research in the past two decades has demonstrated that this vitamin may influence the metabolism, biosynthesis, or tissue content of simple and complex carbohydrates (249), proteins (51, 248), nucleic acids (37, 257), steroids, and phospholipids (123, 126, 160, 245). While attempts to explain these observations have led a number of laboratories to undertake a search for specific systems in which vitamin A or one of its metabolites may participate as a cofactor or effector, results to date have been inconclusive (4, 172, 173).

Among the large number of compounds reported to posses pro-vitamin A activity, β-carotene has the highest biopotency of naturally occurring plant carotenoid precursors of vitamin A (13, 175). Cleavage and conversion of carotenoids to vitamin A is accomplished principally in the mucosa of the intestine by the enzyme carotene 15,15'-dioxygenase (79, 80, 174), while absorbed carotenoids are thought to be converted in the liver where an oxygenase with similar capabilities has been found (174). Activity of the intestinal enzyme is particularly high in the rabbit and it is estimated that an adult is probably capable of converting 750 to 2500 μg of β-carotene to vitamin A/kg body weight/day (175).

Fifty μg of β-carotene/kg body weight/day is reported to meet the rabbit's requirement for vitamin A for growth and reproduction (182). Experimentally determined figures for vitamin A itself are not available, but Olson and Lakshmanan (175) suggest that the rabbit's requirement is somewhere between that of the rat (30 μg/kg) and man (10 μg/kg), or about 20 μg/kg body weight/day. In the absence of experimental data, this latter figure seems reasonable and may serve as a useful guide. Accordingly, rations for growing rabbits consuming 50 to 60 gm/kg body weight/day (4) should contain the minimal equivalent of 333 to 440 μg of retinol (1100 to 1300 IU of vitamin A) per kg of diet. In the experience of the authors, this level of supplementation seems adequate.

Factors which may modify the dietary vitamin A requirement include length of time the diet is stored and its content of peroxidizable fatty acids or antioxidants such as vitamin E (46, 47, 150, 162); both carotene and vitamin A are particularly susceptible to oxidative destruction. There is also evidence to suggest that additional vitamin A may be necessary in rabbit colonies where hepatic coccidiosis is a concern. Diehl (52) reported that extensive liver damage caused by this protozoan disease was accompanied by a decrease in liver vitamin A reserves.

a. VITAMIN A DEFICIENCY.   The clinical signs of vitamin A deficiency in rabbits have been well documented. Retarded growth and, in severe cases, loss of weight occurs (182). Mellanby (154) reported that in advanced cases spontaneous movements decrease and rabbits may finally refuse to move. Deficient rabbits also may develop signs of neurological disturbance similar to those seen in parasitic otitis media, viz., circling with their heads turned to one side, wobbling from side to side, and in the extreme instance, falling sideways or backward with an inability to right themselves. Head retraction, limb paralysis, and occasionally convulsions may be seen (88, 135, 182).

Eye lesions are reported in both adult and young animals a few weeks of age (88, 147, 154, 182). In the adult this may be the first clinical sign of deficiency. A dull, whitish patch or band appears on the surface of the cornea, usually at or near the center and running parallel between the eyelids. The cornea becomes cloudy, rough, and dry in appearance (Fig. 1) and a dry, crusty exudate begins to accumulate around the eyes. Pigmentation of the bulbar conjunctiva may be seen near the limbus. Unless vitamin A therapy is instituted at this stage, the condition advances to a full-blown keratitis with possibly iridocyclitis, hypopyon, and permanent blindness (89). Retinal lesions, which appear as white dots on ophthalmoscopic examination of the fundus, have also been reported in rabbits fed deficient diets alone or deficient diets supplemented with retinoic acid (207).

As in other species, infertility and reproductive failures are to be expected in vitamin A-deficient does. The studies of Lamming et al. (133) indicate that ovum abnormalities, cleavage failure, and degeneration of the ova prior to implantation may result in decreased breeding efficiency in deficient colonies. Even when fertilization and subsequent uterine nidation are successful, early fetal death and resorption, abortion, and the presentation of stillborn and/or congenitally malformed young may occur (90, 134, 135). Young born to asymptomatic marginally deficient females may on the other hand appear normal at birth but

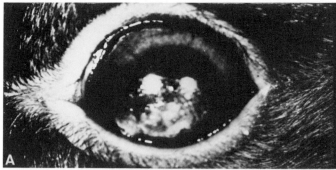
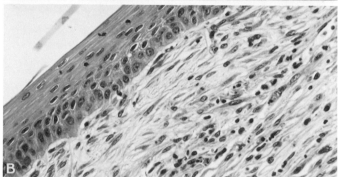

Fig. 1. Experimental vitamin A deficiency. A. Rough, cloudy, opaque cornea of adult rabbit. B. Section of cornea showing keratinization of epithelium and infiltration of stroma by inflammatory cells. Hematoxylin and eosin, ×176.

develop hydrocephalus and other clinical signs of deficiency within a few weeks postpartum. Although the occurrence of hydrocephalus is related to the period of time does are maintained on deficient diets prior to breeding (158), the incidence of hydrocephalus in different colonies and in litters within a single colony varies. This is thought to be due possibly to differences in the maternal vitamin A stores and thus, the rate of depletion of individual animals (90). (See Chapter 15 for other causes of hydrocephalus.)

Clinical studies of deficient rabbits have revealed lowered serum and liver vitamin A concentration. Lamming et al. (134) observed that low maternal plasma (4 μg/100 ml or less) and liver (less than 0.5 μg/gm) concentrations of vitamin A were associated with a significant reduction in the number of live fetuses and an increase in uterine fetal resorption sites. Liver levels, as might be expected, are also reduced in young born to deficient females (135). Harrington and Newberne (90) reported further that a correlation exists between a doe's blood level of vitamin A at conception and the frequency of hydrocephalus in offspring at birth (Fig. 2). The critical serum concentrations below which a significant incidence of hydrocephalus was seen in their studies were found to be between 20 and 30 μg/100 ml. When maternal blood levels were below 20 μg/100 ml of serum at conception, all of the young in 70% of the litters were hydrocephalic at birth. Young born to females with

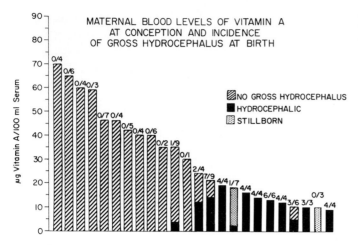

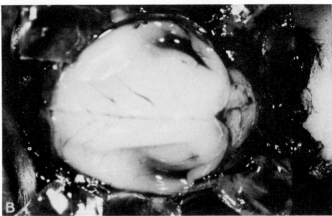

Fig. 2. Rabbit maternal serum vitamin A levels at conception and litter incidence of gross hydrocephalus at birth. Ratios give litter incidence of hydrocephalus in neonates. [From Harrington and Newberne (90, p. 679). Used by permission of authors and publisher.]

Fig. 3. Experimental vitamin A deficiency. A. Dome-shaped cranium in six-week-old rabbit with hydrocephalus. Note also the severe corneal involvement. B. Collapse of cerebral hemispheres in newborn hydrocephalic rabbit after removal of cranial bones.

blood levels above 40 μg/100 ml generally exhibit no gross hydrocephalus at birth (89), but cerebrospinal fluid pressure may be elevated (120 to 110 mm water) and dilatation of the cerebral ventricles can be demonstrated in radiographs or at autopsy a few weeks after birth (31,157, 158). Administration of vitamin A to such rabbits is reported to result in a rapid decline in cerebrospinal fluid pressure which as a rule continues until the pressure reaches normal limits (less than 100 mm water) (157). Mann *et al.* (147) noted that the onset of eye lesions also correlated with serum vitamin A levels. Corneal lesions generally made their appearance when blood levels of vitamin A fell to about 10% of normal, i.e., 10 to 15 IU/100 ml of plasma in their experiments.

Gross lesions seen in rabbits at autopsy are commonly restricted to the eyes and brain. However, pneumonia and nephritis, due presumably to reduced resistance and secondary infection, are not uncommon in cases of chronic vitamin A deficiency. In newborn and animals less than a few weeks of age born to deficient females, internal hydrocephalus may be the only gross lesion present. The anterior fontanel of the cranial vault in these rabbits is wide and soft and consists of poorly ossified connective tissue. The dorsal portion of the head may show a prominent bulge (89, 90). Portions of the cerebellum have been found herniated through the foramen magnum (135) and the cerebral hemispheres may be so thin and friable as to appear nearly transparent and collapse when the cranial bones are reflected (88, 90). In growing animals that survive more than a few weeks, enlargement of the head may be inapparent or completely absent but hydrocephalus (Fig. 3) is readily diagnosed after removal of the ossified cranial bones (88, 158). Hydrocephalus has not been reported in adult rabbits but the eye lesions described above may be found at autopsy in

both mature animals and deficient young which have lived for a few months after birth. Phillips and Bohstedt (182) also reported that the molar teeth of deficient rabbits may be unevenly worn.

Microscopically, eye lesions have been found to consist of keratinization of the epithelium of the cornea (Fig. 1) and bulbar conjunctiva with occasional cystic degeneration and erosion of the corneal epithelial cells. Secondary lesions include edema, vascularization, and inflammatory cell infiltration of the corneal stroma and iridocyclitis with an accumulation of polymorphonuclear cells in the anterior chamber of the eye (hypopyon). Masses of desquamated epithelial cells and necrotic debris may be found accumulating in the fornix conjunctivae (88, 147). The keratinized epithelium here may be quite thick and contain keratohyaline granules (88). In severe cases, the normally occurring mucus-secreting cells of the conjunctiva are absent (88, 147). Retinal lesions attributable to vitamin A deficiency were described by Sorsby *et al.* (207). Most of their rabbits had been maintained on a deficient ration supplemented with vitamin A acid (retinoic acid). The lesions observed consisted of a thinning of the outer nuclear layer and an accumulation of an eosinophilic debris between the outer rod segments and the pigmented epithelium. Macrophage-like cells were occasionally present, some of which appeared to be phagocytizing this material.

Microscopic studies of the nervous system have disclosed that demyelination may be found in the brain and spinal cord as well as in the peripheral and cranial nerves (88, 154, 182). Purkinje cell degeneration in the cerebellum, chromatolysis and other degenerative changes of the neurons in the reticular formation, medulla oblongata, spinal ganglia, and horns of the spinal cord are also described (Fig. 4) (88, 154, 182). In hydrocephalics, stretching and sloughing of the ependymal cells lining the cerebral ventricles with subependymal gliosis are sometimes seen (88). Phillips and

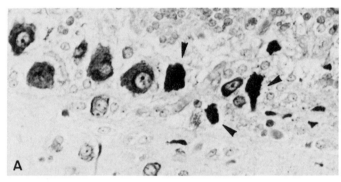

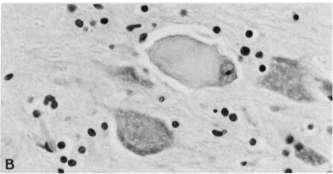

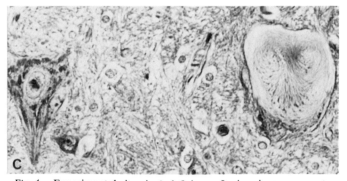

Fig. 4. Experimental vitamin A deficiency. Lesions in nervous system of young hydrocephalic rabbits. A. Section of cerebellum showing line of Purkinje cells. Degenerated cells are heavily stained and irregularly shaped (arrows). Luxol fast blue-cresyl Echt violet, × 176. B. Spinal cord; large neuron shows classic Nissl's chromatolysis and peripheral displacement of nucleus. Hematoxylin and eosin × 176. C. Reticular formation of brain with large neuron showing central chromatolysis. Luxol fast blue-cresyl Echt violet, × 280.

Bohstedt (182) reported vacuolization with a slight hypertrophy of these same cells which cover the choroid plexus.

Ultrastructural studies of the choroid plexus from vitamin A-deficient hydrocephalic rabbits have been conducted by Tennyson and Pappas (217) and Witzel and Hunt (246). Engorgement of choroid vessels (217), subepithelial hemorrhage, persistence of whorled myelinlike figures in the endoplasmic reticulum (246) and a lack of orderly arrangement of the usual parallel arrays of endoplasmic reticulum (217) were noted. The endothelial lining of choroid capillaries was also stretched with an apparent weakening of endothelial cell junctions (246). Cytoplasmic inclusions associated with the whorled endoplasmic reticulum also were observed by Witzel and Hunt (246) and reported to be more common in hydrocephalics. According to Tennyson and Pappas (217), the nuclei of the choroidal epithelial cells of hydrocephalics show more pronounced irregularities in shape and the mitochondria of these cells are large and frequently pleomorphic. These latter investigators also examined the absorptive (pinocytotic) function of the choroidal epithelium and found a reduction of the uptake of thorium dioxide particles injected into the cerebral ventricles.

A number of investigators have observed microscopic changes in cranial bone of deficient rabbits similar to those seen in vitamin A-deficient dogs and cattle. Harrington (88) and Mellanby (154) both noted a thickening of the basilar portion of the occipital bone. Carton *et al.* (31) reported an over-abundance of cartilage in the bones of the skull but no histological difference in the process of osteogenesis. Millen and co-workers (159) also examined the bones of the vault and base of the skull and stated that ossification was incomplete. They noted further that no evidence of bone overgrowth and thus possibly bony compression of the nervous system could be found.

Microscopic lesions of the kidney have been described as consisting of mild parenchymatous degeneration of the proximal convoluted tubules; interstitial inflammation with fibrosis, congestion, and edema; and squamous metaplasia of the transitional epithelium of the renal pelvis (182). Except for a possible slight congestion of sinusoids and swelling of Kupfer cells (182), liver lesions were not usually seen.

In view of our present lack of more specific information as to the roles vitamin A plays in metabolism, the genesis of the diverse pathology and clinical signs manifested in the deficient rabbit remains uncertain and speculative. The neurological disturbances and central nervous system (CNS) pathology reported in many species have, for example, been attributed to increased intracranial pressure brought about by (a) alterations in the position and activity of osteoblasts and osteoclasts which lead to abnormal bone growth and compression of the CNS (72, 154); (b) a cessation of bone growth with a continued expansive growth of the CNS within more limited bony confines (247); (c) over-production of cerebrospinal fluid (CSF) with a relative insufficiency of the cerebral aqueduct (159); and (d) under-absorption of CSF (18, 26, 168) due to alteration in the mucopolysaccharide content of the dura mater (43) and the components of the archnoid villi (95). Fell and Mellanby (65) proposed that the observed effects of vitamin A on epithelium and possibly cells of mesenchymal origin, e.g., osteoblasts and osteoclasts, could be related to a controlling influence of the vitamin on cell differentiation. The meta-plastic and keratinizing changes seen in epithelial covered structures of deficient rabbits would then, according to this hypothesis, represent expressions of an alternative pathway of epithelial cells differentiation—keratin versus mucus production—in the presence of low tissue concentrations of vitamin A. Evidence in support of the concept has been presented by Fell (65, 122, 181) and others (96).

b. VITAMIN A TOXICITY. Except for the work of Thomas *et al.* (221), who used the rabbit as a model in their *in vivo* investigations of the effects of hypervitaminosis A on cartilage, little is known about vitamin A toxicity in rabbits. These workers reported that the administraton of a single dose of 400,000 units orally or a daily intraperitoneal injection of 1,000,000 units to rabbits for one to seven days resulted in a loss of ear rigidity with partial collapse and curling of the distal portion. Most of their animals receiving the high dose also either failed to gain or decreased in body weight. Loss of hair was usual, especially around the mouth and paws in those given daily injections for five or more days. Because of variability in the staining characteristics of ear cartilage, microscopic changes here were equivocal. These investigators did note, however, a decrease in the size of chondrocytes of the femur and a thinning of the epiphyseal plates of this bone along with a reduction or loss of the basophilic, metachromatic, and alcian blue staining properties of the cartilage matrix of both the articular surface and epiphyseal plates. Except for slight fatty change in the liver and some evidence of calcification in the kidneys of a few animals, no other lesions were reported. Further studies of cartilage in this same report, demonstrated that the administration of excess vitamin A to rabbits gave rise not only to an increase in serum cobalt-precipitable materials (presumably chondroitin sulfate) and in crude extract-ed cartilage chrondromucoprotein nitrogen/hexosamine ratios, but also a loss of previously incorporated sulfur-35 from cartilage with an accompanying elevation of sulfur isotope in serum and urine. These findings led Thomas and co-workers (221) to suggest that massive amounts of vita-min A administered to rabbits appear to effect a loss of chondroitin sulfate from cartilage possibly by the activation of a proteolytic enzyme or enzymes with properties similar to those of papain which they had also studied.

2. Vitamin D

It is generally accepted that vitamin D or, as now seems certain, an *in vivo* formed hydroxylated metabolite of vitamin D (50) plays a significant role in the absorption of calcium (Ca) from the intestinal tract. While phosphorus (P) absorption and retention may be adversely affected in deficient animals, these effects are believed to be secondary and indirect responses to alterations in the metabolism of Ca (48, 92, 121). The exact mechanism by which vitamin D exerts its influence on Ca absorption is uncertain. Evidence for an increase in the permeability of the intestinal mucosa to Ca as well as a stimulation of the active transport of Ca across the intestinal wall in response to vitamin D have both been reported in studies with isolated segments of intestine from deficient animals (239). Other studies also have demonstrated the appearance of Ca-binding proteins (237) and a Ca-dependent adenosine triphosphatase in the mucosa of the intestine of depleted animals subsequent to treatment with vitamin D (151). The relationship between the latter compounds and the above effects of vitamin D on intestinal mucosa permeability and the active transport of Ca remains to be determined.

Aside from its effects on Ca absorption from the intestine, vitamin D is also thought to play a major role in bone mineral mobilization. The inability of exogenous para-thyroid hormone (PTH) to effectively elevate serum Ca in deficient rats has led some to suggest that vitamin D is necessary for the action of PTH on bone mineral mobiliza-tion (48, 152). An absolute dependence of PTH on vitamin D has not been conclusively demonstrated. Both PTH and certain metabolites of vitamin D, e.g., 25-hydroxycholecal-ciferol (25-HCC) and its 1,25-dihydroxy derivative, have been shown to cause resorption of bone in tissue culture (50, 184). The concentrations of 25-HCC and PTH required to accomplish this are, however, greatly reduced when both are present together; a finding that suggests a synergistic relationship (184). Further studies of the interactions of these agents as well as their relationship to calcitonin, which has been shown to decrease bone mobilization in-duced by vitamin D and by PTH (152), are obviously needed. A further action of vitamin D on bone has been suggested by Canas and his associates (27). These investi-gators observed an increase in the incorporation of tritiated proline into bone hydroxyproline in rachitic chicks after treatment with vitamin D. This occurred before serum concentrations of Ca had returned to normal. These in-vestigators postulated that the increased synthesis of bone collagen matrix observed was in response to vitamin D it-self rather than an effect of elevated blood levels of Ca.

A number of substances are known to possess vitamin D activity, but only two, vitamins D_2 (calciferol) and D_3 (cholecalciferol), are considered of any practical significance in nutrition. Differences in the activities of these two forms of vitamin D are well known in some species but neither their relative potencies nor their minimal requirements for the rabbit have been established. Although no naturally occurring cases of rickets or vitamin D deficiency-associated osteomalacia have been reported in this species, the development of rickets in rabbits fed experimental diets deficient in vitamin D and housed under conditions which minimize exposure to sources of ultraviolet irradiation indicates that, in the growing laboratory rabbit at least, some requirement for exogenous vitamin D exists. For this reason, vitamin D supplementation of experimental and commercial rations is universally practiced. Considering

the known toxicity of vitamin D in conjunction with the absence of information relevant to the rabbit's need for it, a cautious approach to the use of high concentrations of vitamin D in diet formulations is warranted. Indeed, one cannot help but wonder how much of the so-called naturally occurring cardiovascular disease reported in laboratory rabbits (75, 194) is actually due to the routine incorporation of substantial quantities of vitamin D into experimental and stock rations (See Figs. 5 and 6). Obviously, comprehensive investigations aimed at answering this question are justified particularly in view of the fact that this species enjoys widespread use not only as a common laboratory animal for diverse studies, but also as an animal model for the investigation of cardiovascular disease itself. For further discussion of etiology of cardiovascular disease in rabbits, see Chapter 15.

a. VITAMIN D DEFICIENCY. Experimental rickets has been produced in growing rabbits by a number of investigators (78, 127, 155, 163). Clinical signs reported include the development of potbellies and muscular weakness as well as deformities of the chest and legs with enlargement of the costochondral junctions of the ribs and of the epiphyses of the leg bones. Goldblatt and Moritz (78) obtained radiographs of deficient rabbits fed a purified ration over a five-

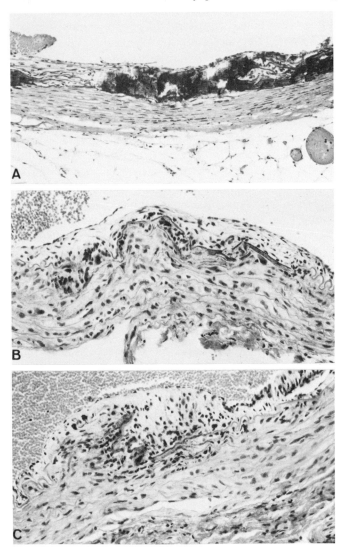

Fig. 5. Experimental vitamin D toxicity. Young (four-month-old) Dutch rabbits were given 1×10^5 IU vitamin D_3 intramuscularly every other day for a total of six doses (600,000 IU). A. Aorta with extensive mineralization of inner half of media. Hematoxylin and eosin, $\times$ 100. B. Section of aorta showing intimal thickening, calcification of internal elastic membrane, and inflammatory cell infiltration. Hematoxylin and eosin, $\times$ 160. C. Another section of aorta with degenerative changes similar to those in B. Hematoxylin and eosin, $\times$ 160.

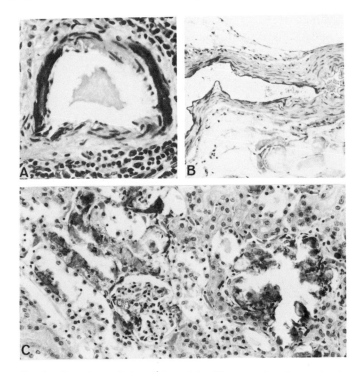

Fig. 6. Experimental vitamin D toxicity. Tissue sections from animals described in Fig. 5. A. Muscular artery in splenic follicle showing mineralization of smooth muscle wall. Hematoxylin and eosin, $\times$ 160. B. Muscular visceral artery (pancreas) with extensive mineralization of internal elastic membrane. Hematoxylin and eosin, $\times$ 120. C. Kidney section showing deposits of mineral within the glomerular tuft, tubular epithelial cells, and tubular lumens. Hematoxylin and eosin, $\times$ 170.

week period. They observed a progressive widening of the epiphyseal plates of the leg bones (distal radius, ulna and femur, and proximal tibia) which was attributable to hypertrophy of cartilage cells and defective calcification of osteoid being laid down. Evidence of these changes was apparent in radiographs of some animals as early as one week after exposure to deficient diets. Other investigators (127, 155) also reported radiographic evidence of poor calcification of teeth in rachitic rabbits.

Studies have shown that serum alkaline phosphatase may be elevated in vitamin D-deficient rabbits (127), whereas serum concentrations of Ca may be reduced depending on the Ca content of diets fed (78, 127, 163). These workers also observed a reduction in serum P in rachitic rabbits but their diets were low or deficient in this element. Kato (127) failed to indicate the levels of Ca and P present in his rachitogenic diet but did observe positive responses in both serum Ca and P after rabbits were allowed access to diets containing 4% Ca lactate. The effect of this treatment on circulating alkaline phosphatase was, however, variable. In contrast, subcutaneous administration of vitamin D_2 or D_3 was found to elevate both serum Ca and P and reduce serum alkaline phosphatase. Kato (127) also examined the influence of parathyroid extract on these blood components and reported a gradual increase in serum inorganic phosphorus. However, Ca was unaffected and, while there was a slight decrease in serum alkaline phosphatase levels in some animals, this effect was transient in nature.

Histological findings in the bones of vitamin D-deficient rabbits are reported to be typical of rickets as seen in other species (78). The metaphyses of long bones have an excessive proliferation and disarray of hypertrophied chondrocytes. The zone of provisional calcification may be absent and an excessive amount of osteoid found surrounding the trabeculae of the primary spongiosa and on the endosteal and subperiosteal surfaces of cortical bone in the region of the epiphyses. These bone lesions resemble those of molybdenum toxicity and of copper deficiency in ascorbic acid-fed rabbits. Mellanby and Killick (155) and Kato (127) reported that poor calcification of teeth is evident also in sections from vitamin D-deficient rabbits.

b. VITAMIN D TOXICITY. Both vitamins D_2 and D_3 are toxic to rabbits but their relative potencies have not been established. Ringler and Abrams (188) reported the occurrence of accidental vitamin D toxicity in a colony of rabbits fed a pelleted ration which on analysis was found to contain vitamin D at levels 23 units/gm. This ration was also deficient in vitamin E. Vitamin D toxicity has been produced experimentally in rabbits by intramuscular administration of massive amounts of irradiated ergosterol (71, 94). Hass et al. (94) found that the severity of gross and microscopic lesions seen under experimental conditions appeared to be a function of dosage received, the interval

between doses, and the duration of treatment. The occurrence of simultaneous kidney infection, i.e., chronic purulent pyelonephritis, was also believed to be a factor. In rabbits with such renal disorders, the dose required to produce significant generalized calcinosis was reduced from between 500 and 600 thousand units over a six-day period to 300 thousand units. Clinical signs of toxicity seen were anorexia and loss of weight followed by death. These investigators stated that no consistent correlation was found between the pathological changes observed and blood serum levels of Ca, P, or cholesterol. Friedman and Roberts (71) observed a significant increase in the antirachitic activity (rat bioassay) of serum obtained from does and their newborn when the does were given large intramuscular doses of irradiated ergosterol during pregnancy. A total of 1.5 million units was administered in divided doses over the 30-day gestation period. Serum calcium levels of offspring of these females also was significantly elevated. No similar increase in the serum calcium level of the does was detected, nor did the administration of vitamin D appear to have any significant influence on serum concentrations of phosphorus, cholesterol, total protein and albumin, or serum alkaline phosphatase activity. Ringler and Abrams (188), on the other hand, reported hypercalcemia, hyperphosphatemia, glycosuria with normal blood sugar, and increased blood urea nitrogen in rabbits receiving diets containing toxic levels of vitamin D and deficient in vitamin E.

The occurrence of extensive and generalized calcification of soft tissues at necropsy is strong evidence for vitamin D toxicity and/or deficiency or excess of certain minerals. In rabbits which have received large amounts of vitamin D for any significant period of time, mineralization of small, medium, and large arteries is common (188). The proximal aorta seems to be particularly susceptible and gross lesions here may be conspicuous (71, 94). Varying degrees of grossly discernible calcification may be observed in the kidneys, skeletal muscle, heart, stomach, and lower intestinal tract, as well as in the mucosa and cartilaginous ring of the tracheobronchial tree. In advanced cases of vitamin D toxicity, bones may be brittle and deformed due to resorption with a resulting increase in fragility.

Microscopic studies of vitamin D toxicity in rabbits have revealed that mineralization of elastic and fibroelastic tissues of the cardiovascular system is extensive (Fig. 5). Vessels within the central nervous system and liver were not, however, found to be affected by Hass and co-workers (94). In cardiac and skeletal muscle, mineral deposits were found both intra- and extracellulary in muscle fibers although intracellular deposits are more usual in the heart. The above changes may or may not be preceded by inflammatory cell reaction and muscle fiber degeneration. Hass et al. (94) stated that mineralization of smooth muscle was widespread. Lesions are found frequently in the media of the systemic arterial

(Fig. 6) and pulmonary venous systems, in the renal pelvis, and in the muscularis of the alimentary tract and the tracheobronchial tree.

In the spleen, fibroelastic tissue and smooth muscle, as well as collagen, were found to have undergone mineralization. Similar changes in the fibroelastic tissue of the dermis were observed. In kidneys, individual and fused spherical, concentric laminated mineral deposits may be present as may renal tubular dilation with microcyst formation (Fig. 6). Resulting kidney damage gives rise to mononuclear cell infiltration and interstitial fibrosis. Other organs and tissues in which mineral deposition is seen are the thyroids, thymus, pancreas, submaxillary salivary gland, alveolar septa of the lungs, and the sclera of the eyes. Microscopic bone changes seen are those of replacement of previously ossified bone by massive amounts of osteoid and proliferating fibroblasts. These elements may fill in and obliterate portions of the marrow cavity.

Friedman and Roberts (71) examined the effects of vitamin D on the aorta of offspring born to female rabbits which received massive amounts of activated ergosterol throughout gestation. The does received the vitamin intramuscularly in divided doses every other day during pregnancy for a total 1.5, 2.5, 3.5, or 4.5 million units. All females receiving the 2.5, 3.5, or 4.5 million units treatment died within 65 days after their initial injection and, among this group, those that conceived either aborted during the first 12 days of pregnancy or delivered macerated fetuses. Aortas from these does had focal areas of calcification, degeneration, and necrosis in the vessel wall. Similar but less severe lesions were found in aortas of females given the lowest dose, i.e., 1.5 million units.

Four of 18 young born to females in the 1.5 million unit treatment group were autopsied shortly after birth and had abnormal aortas (71). An annular protrusion was present at the superior margin of the aortic sinuses which resulted in a narrowing (stenosis) of the lumen of the aorta at this level. Ten additional young from this treatment group survived 2 to 20 days after birth. Three of these had gross stenosis of the aorta at the same site. One animal also had a nonobstructing fibrous band stretched across the lumen of the vessel at the superior margin of the sinuses. Microscopically, the narrowing of the lumen of the vessel was found to have resulted from a localized thickening of the aortic media. In addition to an apparent increase in the number of medial elastic fibers and smooth muscle cells, foci of degeneration and round cell infiltraton were seen. Six rabbits from the 1.5 million-unit treatment group were autopsied at three months of age. These animals had been nursed by control foster does from birth and were thereafter fed 250 units of vitamin D_2 per day. Supravalvar aortic stenosis was not seen in these animals, but the proximal aorta did show generalized irregularities of the wall with degeneration and calcification in the media. These inves-

tigators did not report any prenatal effect of treatment with vitamin D on the development of other tissues and organs.

Because of the similarity in vascular, renal, and/or cardiac and skeletal muscle micropathology, lesions of vitamin D toxicity in rabbits must be differentiated from those of deficiency of vitamin E, magnesium, or copper.

3. Vitamin E

Over the past few decades a body of evidence has accumulated which has led to the suggestion by some investigators (258) that vitamin E, or more correctly, tocopherol compounds with vitamin E activity, function primarily, if not exclusively, as biological antioxidants. Lines of evidence frequently cited in support of this hypothesis include (a) the observed influence of elevated dietary levels of highly unsaturated fatty acids as opposed to less unsaturated fats on the vitamin E requirements and the development of lesions in various species including the rabbit (20, 21, 32, 45, 104), (b) the ability of other nonspecific antioxidants to prevent or at least delay the onset of some clinical signs and pathology in different species (54–56, 200), and (c) the noted increased liability of lipid-containing subcellular organelles from deficient rabbits to oxidative destruction together with the demonstration of significant concentrations of thiobarbituric acid reactive ("peroxidation") products in their tissues (216, 258). Tappel (215) has summarized and discussed these and other findings as an advocate for the antioxidant theory, while Green and Bunyan (85) have recently reviewed and taken issue with its legitimacy. Alternative roles for vitamin E, such as its participation in oxidation-reduction reactions in the respiratory chain or as a donor or a component of nonenzymatic hydrogen-transferring systems have also been considered (19, 230).

The vitamin E (α-tocopherol) requirement of the rabbit has been variously estimated as being from 0.32 to 1.4 mg/kg body weight/day (62, 108, 144). Hove and Harris (108), who examined the relative activity of the tocopherol optical isomers dl-α- and dl-γ-tocopherol versus d-α- and d-γ-tocopherol in rabbits, concluded that the d-enantiomorphs were more active than racemic mixtures when the alleviation of creatinuria due to vitamin E deficiency-induced muscular dystrophy was used as the test for biopotency. Using this same criterion, they also compared and found that among the d forms, α-tocopherol was 3 to 5 times as active as β- and γ-tocopherol. Based on Hove and Harris' (108) creatinuria curative doses of 1.1 and 1.4 mg/kg body weight/day for d-α- and dl-α-tocopherol, respectively, diets for growing rabbits consuming 50 to 60 gm/kg body weight/day should contain a minimum of 19 to 22 mg of d-α- or 24 to 28 mg of dl-α-tocopherol/kg of ration.

Factors which may be expected to modify the vitamin E requirement of the rabbit include parasitism and the composition and concentration of fatty acids present in their diet. Diehl (52) and Diehl and Kistler (53) reported that, as with vitamin A, infection of rabbits with coccidia (*E. stiedae*) adversely affects their liver, skeletal muscle, and serum concentrations of tocopherol. Impaired uptake and/or storage, or increased utilization or destruction of the vitamin were presented as possible mechanisms which might account for the lowered liver levels found in infected animals.

A number of investigators have pointed out the influence of unsaturated dietary fatty acids on the acceleration of the onset of vitamin E deficiency and on vitamin E requirements in various species (45, 104, 149). Borgman (20) compared the effects of relatively saturated (lard) and unsaturated (cod-liver oil) dietary sources of lipid on the development of skeletal muscle lesions and the fatty acid composition of various tissues in rabbits fed vitamin E-deficient semipurified diets. The influence of dietary fats on fatty acid percentages in various tissues was marked, particularly in adipose tissue and skeletal muscle, and while skeletal muscle lesions were found in rabbits fed either lard or cod-liver oil, muscle degeneration was comparatively more severe in the latter group. These findings were confirmed in expanded studies by Borgman (21, 22) using diets containing oleic and linoleic acid and low in vitamin E. Rabbits fed an oleic acid-containing diet and receiving no supplemental vitamin E had no skeletal muscle degeneration at 12 weeks compared to moderate to severe degeneration in animals fed the same diet with linoleic acid for only 8 weeks (21). Borgman (20, 22) and others (104) took special note of the seemingly obvious relationship between tissue and dietary fatty acid composition and their effect on vitamin E requirements and the development of E-deficiency associated lesions.

a. VITAMIN E DEFICIENCY. Signs of vitamin E deficiency commonly reported in rabbits are those of stiffness, progressive muscular weakness, reduced food consumption, and loss of weight, followed by prostration and death. MacKenzie and McCollum (144), who carefully followed the clinical course of the disease in young rabbits, described three stages. The first was characterized by creatinuria, plateauing of weight gain, and a decline in food intake. The appearance of physical signs marked onset of the second stage. In some rabbits, the front legs were held stiff with the head slightly retracted, sometimes for hours, while in others the forelegs were placed well under the body between the hindlegs. Some rabbits, however, had none of these signs but were easily laid on their side and were slow in righting themselves. Weight loss continued during this stage at an accelerated rate and food consumption became drastically reduced with complete anorexia by the end of this period or in the early part of the last stage. MacKenzie and McCollum (144) described the third and final stage as

follows, "This, the stage of acute dystrophy, lasted 1 to 4 days and terminated in death. The animals were now readily pushed off of their feet, and regained an upright position only after a violent struggle. Some animals died while exhibiting these symptoms, while others were completely prostrated for several days before death. Such animals when picked up seemed devoid of all body tonus." A similar set of clinical signs was observed by Eppstein and Morgulis (62). In addition to general deterioraton in condition, Borgman (22) in his studies of E-deficient rabbits also reported signs of central nervous system involvement: circling, disturbances in equilibrium, holding the head to one side, and prostration with legs in a sprawled position. These latter signs are essentially the same as those sometimes seen in either vitamin A deficiency or otitis media.

The experimental effects of vitamin E deficiency on reproduction in the rabbit apparently have not been examined. Ringler and Abrams (188), however, recently reported a high incidence of neonatal mortality and infertility in does in a colony receiving a ration which was marginally deficient in vitamin E (16.7 mg of α-tocopherol/ kg of ration). Confounding the picture in this particular case, however, was the presence of toxic levels of vitamin D in the diet, 23,000 IU/kg of ration.

Clinical chemistry studies of deficient rabbits have revealed a vitamin E-reversible increase in erythrocyte susceptibility to *in vitro* hemolysis test (188, 189). Serum creatine phosphokinase levels (188, 189), frequently used as an indicator of cardiac and skeletal muscle degeneration (both of which occur in deficient rabbits), as well as urinary creatine/creatinine ratios are reported to be increased (108, 144). Myocardial damage also may be detected by alterations in the electrocardiogram tracing of rabbits made deficient experimentally (24, 76).

Gross morphological changes in the vitamin E-deficient rabbit are commonly restricted to the skeletal musculature, but cardiac muscle is not immune. The paravertebral group, diaphragm, masseter, and voluntary muscles of the rear legs are frequently affected (77, 120, 144). Goettsch and Pappenheimer (77), who reported on the pathology of vitamin E deficiency muscular dystrophy in rabbits and guinea pigs, noted the atrophy and extreme pale appearance of the musculature and made special mention of the tendency toward calcification of necrotic fibers. Innes and Yevich (120), in their report on spontaneous nutritional muscular dystrophy in rabbits, described lesions in the psoas muscle group as consisting of minute pallid spots intermingled with hemorrhagic streaks and large patches of friable yellowish necrotic tissue. Gross lesions in the heart apparently were not seen in their specimens, although microscopic lesions were occasionally found (see Fig. 8). Bragdon and Levine (24) stated, however, that in severe cases myocardial lesions may be seen at autopsy as circumscribed gray areas in the walls of the ventricles, papillary muscles, and occasionally in the auricles. It should be mentioned at this point that, from a diagnostic point of view, the above findings are not pathognomonic for vitamin E deficiency. Similar gross cardiac and skeletal muscle lesions also may be observed in choline- or potassium-deficient rabbits.

Histopathological studies of brains from vitamin E-deficient rabbits having neurological signs have disclosed no significant lesions (22). Although Borgman (22) reported microscopic evidence of testicular degeneration, specific morphological changes were not described. According to Bragdon and Levine (24), myocardial lesions were represented by both myodegeneration and inflammatory reactions, viz., loss of striations and coagulation necrosis of muscle fibers; nuclear pyknosis and karyorrhexis; and edema, hemorrhage, and mononuclear cell infiltration. Granules of dystrophic calcification were sometimes found. Gatz and Houchin (76) also described the occurrence of droplets within muscle fibers which were neither acid-fast nor osmic acid positive.

The microscopic alterations in skeletal muscle have been detailed by several investigators (77, 120, 188, 228, 229). Lesions here are in general characterized by early myocyte swelling, hyaline degeneration with loss of striations, clumping of sarcoplasm, and fragmentation or atrophy of muscle fibers (Fig. 7). Vacuolar degeneration and the accumulation of lipid deposits and calcified granules (Von Kossa positive) occur in some fibers (Fig. 7). The inflammatory cell infiltrate accompanying the above changes is composed of both heterophiles and macrophages with the latter predominating. Macrophages containing pigment and phagocytized debris may also be found within sarcolemma tubes. These various lesions may be observed also in heart (Fig. 8). Regeneration of muscle is reported to occur by amitotic division of undifferentiated cells with syncytial cell formation, while healing by replacement fibrosis may also be seen in cases of long duration. According to Van Vleet et al. (229), the earliest alterations seen in electron microscopic studies are those associated with mitochondria. These organelles undergo swelling, fragmentation of cristae, and formation of free intramitochondrial membrane profiles and show an accumulation of dense intramitochondrial granules.

Zalkin et al. (259) and Van Vleet et al. (229) have suggested that myodegeneration in vitamin E deficiency is due to a decrease in tissue antioxidant levels. The prime initiator of muscle degeneration is, in the opinion of Zalkin et al. (259), the release of hydrolytic enzymes from lysosomes by peroxidative disruption of their membranes. Van Vleet et al. (229) have questioned this on the basis of their electron microscopic studies of muscle from deficient rabbits. They suggested that the increase in lysosome marker enzymes reported by the above workers was due primarily to a secondary inflammatory influx of lysosome-laden macro-

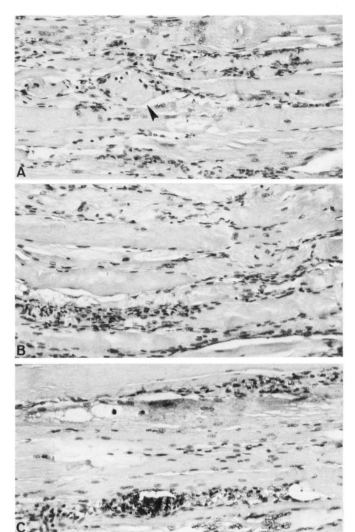

Fig. 7. Experimental vitamin E deficiency (nutritional muscular dystrophy). Weanling Dutch rabbits were fed a purified diet (Table III) marginal in vitamin E for 10 months. A, B, C. Sections of skeletal muscle demonstrating: A. Swelling, loss of striations, hyalinization, and necrosis (arrow) of fibers. B. Similar changes including proliferation of myocyte nuclei and inflammatory cell (heterophil) infiltration. C. Myofibril degeneration with accumulation of lipid and inflammatory cells. All photos hematoxylin and eosin, × 160.

phages rather than the peroxidation of constituent muscle fiber lysosomes. The earliest changes seen by them were in the mitochondria which they postulated occurred in response to uncontrolled lipid peroxidation, free radical formation, and a resulting increase in mitochondrial lipoprotein membrane permeability with a subsequent loss of structural and functional membrane integrity. Disturbances in membrane integrated enzyme reactions, such as those demonstrated in isolated liver mitochondria preparations from deficient rabbits (258), are then thought to lead to further metabolic disturbances and degenerative changes in other organelles dependent on the normal function of affected mitochondria.

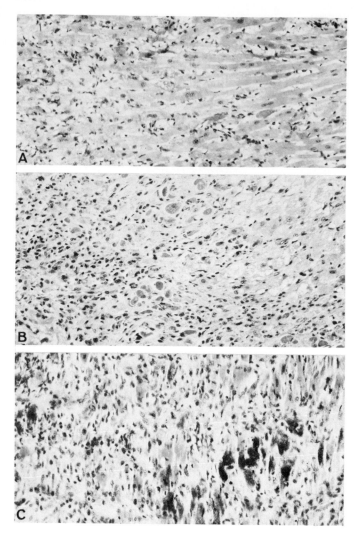

Fig. 8. Experimental vitamin E deficiency. Tissue sections from rabbits described in Fig. 7. A, B, and C are sections of myocardium demonstrating: A. Necrosis and associated inflammation. B. Lipid deposition in myofibrils and fibrosis. C. Mineralization of muscle fibers, infiltration by inflammatory cells, and fibrosis. All photos hematoxylin and eosin, × 160.

Another condition, "yellow fat disease" or "pansteatitis," which has not been described in vitamin E deficiency in experimental rabbits has been reported by Jones *et al.* (124) in wild European rabbits indigenous to islands off the English coast. No muscle lesions were apparently seen, but the sublumbar and perirenal fat of affected animals was described as being mottled, varying in color from orange-yellow to dark brown, and, unlike normal fat, firm in consistency. Jones and co-workers (124) indicated that microscopically the fat was composed of variable size adipose tissue cells, separated into lobules by connective tissue strands containing nonacid-fast brown pigment granules. Neutral fat, free fatty acids, and phospholipids were demonstrated by histochemical means. Much of the material present was reported to be insoluble in chloroform-methanol. Gas-liquid chromatography separation

of fatty acids extracted from the adipose tissue revealed significant concentrations of unsaturated fatty acids with 20 or more carbon atoms (125). These investigators also were able to produce essentially the same histological picture in fat of New Zealand White rabbits in the laboratory by feeding a ration containing 20% herring oil and no supplemental vitamin E. Small brown granules of pigment were present in the heart, liver, and kidneys of these rabbits. Supplementing the same diet with high levels of vitamin E, 250 mg of *dl*-α-tocopherol/kg of diet, prevented the adipose tissue lesions. It was therefore concluded that "yellow fat disease" in rabbits is due to a combined effect of vitamin E deficiency and a high intake of unsaturated fats. This was believed to occur in the wild rabbits from the island studied (Skokholm) during periods when the forage was insufficient to support a large population and the animals were forced to rely on unusual sources of food containing poly-unsaturated fatty acids, e.g., fish, earthworms, young sea birds, and sea bird eggs. This same condition has also been recently described in wild hares in New Zealand (140a).

b. VITAMIN E TOXICITY. The authors are unaware of any reports of vitamin E toxicity in the rabbit.

4. Vitamin K

Vitamin K compounds have been implicated in the biosynthesis of four blood coagulation factors: factors II (prothrombin), VII, IX, and X (176, 212). Consequently, deficiency signs and pathological changes seen in animals are characterized by hemorrhagic disorders traceable to failures in coagulation mechanisms (196). While a need for vitamin K by the rabbit has been demonstrated under experimental conditions (161), its specific dietary requirement has not been determined. A source of vitamin K, usually synthetic menadione (2-methyl-1,4-naphthoquinone), is therefore normally incorporated into the vitamin supplement of experimental diets used in nutritional studies of the rabbit. The authors have fed both practical and purified rations containing menadione at a level of 4 mg/kg of diet to rabbits for extended periods of time (up to 2 years) without adverse effects on reproduction (see below) or the development of lesions attributable to a deficiency of vitamin K.

No naturally occurring field or spontaneous laboratory cases of vitamin K deficiency have been reported in rabbits. The probability of deficiency in colonies fed commercially available rations and maintained under usual laboratory conditions is, in fact, considered negligible. Vitamin K is widely distributed in plant products, particularly green leafy materials such as alfalfa, commonly used in the formulation of practical type stock rabbit rations. Moreover, results of studies in the rat (10), another species that practices coprophagy, suggest that the rabbit may be able to meet its requirement by ingestion of soft feces which contain vitamin K synthesized by intestinal flora.

a. VITAMIN K DEFICIENCY. Female rabbits fed purified diets deficient in vitamin K for as short a period as 40 days had increased prothrombin time and aborted (161). Vaginal bleeding, which persisted for 2 to 4 days, was observed to occur 10 to 14 days after breeding. Histological studies of plancentas revealed areas of recent hemorrhage as well as older clots. Administration of vitamin K resulted in a return of prothrombin time to normal within 3 days and uneventful pregnancies when the treated does were subsequently rebred (161).

b. VITAMIN K TOXICITY. There are no reports of studies of vitamin K toxicity in rabbits.

B. Deficiency of Water-Soluble Vitamins, Choline, and Essential Fatty Acids

1. Niacin (Nicotinic Acid)

Nicotinate (niacin) and its amide, nicotinamide, occur in all organisms. The vitamin functions, in the form of nicotinamide nucleotides, as coenzyme (electron carrier) in numerous oxidation-reduction reactions.

Nicotinate can be synthesized from tryptophan in the liver, and therefore, like choline and vitamin D, it is not a vitamin in the strict sense of the word. However, efficiency of conversion of tryptophan to nicotinate is low, and, as a practical matter, dietary protein usually does not provide sufficient tryptophan to meet tissue needs for both protein and nicotinate synthesis. Consequently, many nonruminant species require a quantity of preformed nicotinate in the diet.

Studies in the rabbit indicate that this animal does require exogenous nicotinate in addition to that produced by tissue and intestinal synthesis. Rabbits fed purified diets which contained 20% casein and negligible nicotinate developed anorexia and lost weight (171, 255). No other signs or lesions were observed which could be attributed specifically to nicotinate deficiency. Urinary and fecal excretion of the vitamin far exceeded intake in deficient rabbits indicating that synthesis was taking place in the tissues and digestive tract (171). Furthermore, addition of free tryptophan to the niacin-deficient diet stimulated growth far more than feeding the amino acid as supplemental protein.

These studies indicate that: (a) a relative deficiency of nicotinate can be produced in rabbits fed purified diets deficient in the preformed vitamin but adequate in protein, (b) rabbit tissues and intestinal flora are capable of converting tryptophan to nicotinate, and (c) free tryptophan in the feed is more readily utilized for nicotinate synthesis in the rabbit than is tryptophan in peptide linkage (casein).

2. Pyridoxine

The substituted pyridine ring of pyridoxal phosphate is an example of a structure that cannot be synthesized by animal tissues. Pyridoxine, as pyridoxal phosphate, is the coenzyme for transaminases, certain dehydratases, amino acid decarboxylases, serine transhydroxymethylase, the hydroxymethyl transfer during glycine oxidation, kynureninase, 5-aminolevulinate synthase, and glycogen phosphorylase (143).

Although recycling of intestinal contents by coprophagy may provide sufficient quantities of some B vitamins for normal growth in the rabbit, signs of pyridoxine deficiency were seen in rabbits fed a purified diet devoid of the vitamin (110). Weanling rabbits fed the deficient diet gained less than 5 gm/day and more than half of them died at an average of 100 days. Optimum growth on this diet required addition of 1 μg pyridoxine per gram of food (about 40 μg/rabbit/day). Gross changes observed were skin lesions and neurological disturbances. Scaling and thickening of the skin around the ears and conjunctivitis and incrustations of the nose and paws occurred also. Neurological signs included seizures and paralysis. Rabbits generally recovered completely from seizures but succumbed after the onset of paralysis. A mild anemia, prolonged clotting time and abundant urinary excretion of xanthurenic acid were associated with pryidoxine deficiency in the rabbit (110).

The effects of pyridoxine deficiency in other animals (69) involve the skin, erythropoietic tissues, and nervous system as observed in the rabbit. Also, increased excretion of xanthurenic acid has been observed in rats and swine (69) fed pyridoxine-deficient diets. This response would be anticipated since pyridoxine participates (kynureninase) in the degradative pathway of kynurenine to nicotinic acid (44).

Although the studies of Hove and Herndon (110) have not been repeated it appears that the rabbit has a dietary requirement for pyridoxine.

3. Choline

Choline is usually listed as one of the B vitamins although it is alleged that choline deficiency cannot be produced in animals receiving diets adequate in protein (243), since choline is readily synthesized in animal cells as phosphatidylcholine from serine and methionine. Choline functions in nerve impulse conduction at synaptic junctions as acetylcholine and as a constituent of phosphatidylcholine, it is part of lipoproteins used in triglyceride transport. Phosphatidylcholine is also quantitatively the most important phospholipid structural component of membranes (143).

A minimal dietary requirement for choline in the rabbit has been established (0.13%, Ref. 107) but there have been few studies of uncomplicated choline deficiency in this species (106, 211). Lesions often associated with choline deficiency, including fatty liver and cirrhosis, have been reported in rabbits fed various diets. Unfortunately the diets used by some investigators were ill-defined with respect to vitamin and mineral or protein content (187, 209, 210), and it is therefore questionable whether a deficiency of choline per se was responsible for the liver changes observed. A further complication was introduced by the effects of hepatic coccidiosis in at least one of the studies referred to (187). This infection is not uncommon in rabbits and induces cystic biliary hyperplasia which may be accompanied by increased fibrous connective tissue.

Clinical signs of choline deficiency, including poor growth, moderate anemia, and death, were reported by Hove and co-workers (107) in rabbits fed a purified diet containing 0.006% choline. In chronically deficient rabbits, progressive muscular dystrophy (MD) developed accompanied by creatinuria (> 40 mg/kg/day), reduced urinary creatinine, and muscular weakness (106). Further studies of MD in choline-deficient rabbits by other investigators (211) confirmed these findings and provided additional biochemical data.

Gross lesions reported in choline-deficient rabbits by Hove's group were fatty and cirrhotic livers and pale atrophic leg muscles (106, 107). Microscopically the liver lesions were characterized by fatty change in hepatocytes, fibrosis with pseudolobulation, and bile duct proliferation. Similar liver lesions have been reported in rats fed choline-deficient diets for long periods (93, 214). Lesions in skeletal muscle included loss of striation, hyaline degeneration, and loss of fibers with increase in connective tissue. These latter changes in skeletal muscle are similar to those seen in MD induced by vitamin E or potassium deficiency in the rabbit (106). The authors excluded vitamin E deficiency as the cause of MD by feeding it in excess of normal requirements and analyzing plasma and tissues for the vitamin.

Acute hemorrhagic necrosis of the renal cortex, which occurs in weanling rats fed a choline-deficient diet, was not observed in young choline-deficient rabbits (107). However, about 50% of rabbits that died from choline deficiency had renal lesions consisting of tubular necrosis and attempted regeneration as evidenced by intratubular proliferation of epithelial cells.

It is well known that the choline requirement in laboratory animals is related to certain other dietary constituents, namely, methionine, cystine, vitamin B_{12}, and folic acid. Methionine contributes to synthesis of choline via transmethylation (57) and cystine promotes fatty liver in animals fed diets marginal in choline or methionine (87, 224). The rabbit experiments described above utilized this knowledge to advantage, i.e., the choline-deficient diet was not only limiting in methionine (30% peanut meal + 6% casein) but contained additional cystine (0.1%) as well (107, 211). Indeed, addition of methionine to the choline-deficient diet stimulated growth considerably (107). Casein contains

four times as much methionine as peanut meal and the authors (113) have observed no clinical signs or lesions of choline deficiency in young rabbits fed a purified gel diet for five months which contained 20% casein and no added choline or methionine. These observations are in agreement with our opening statement; that animals fed diets adequate (quantity and quality) in protein will not develop choline deficiency.

The pathogenetic mechanisms leading to fatty liver and subsequent cirrhosis in choline-deficient animals are still not fully understood. It is known that synthesis and elaboration of lipoproteins by the liver is a physiological function which is essential for fat transport and that various dietary insults, including choline deficiency, will block this process. The mechanism of this block has not been elucidated but it probably is related to a failure in the assembly of triglyceride-transporting lipoproteins which results in an accumulation of triglyceride in the hepatic cell (141, 177).

Muscular dystrophy resulting from choline deficiency has not been reported in the rat although myocardial lesions do occur (193). Skeletal and cardiac muscle derive energy from fatty acids when glycogen supplies are depleted and it has been shown that the rate of long-chain fatty acid oxidation is decreased in heart muscle of choline-deficient rats (42). Furthermore, the adenosine triphosphate (ATP) and creatine content of dystrophic muscle was decreased markedly in choline-deficient rabbits (Table IV). These biochemical effects suggest a defect in energy metabolism of muscle in choline deficiency which could explain the weakness and lesions observed in rabbits. Deficiency of the neuromuscular transmitter, acetylcholine, may also be involved.

4. Essential Fatty Acids

Mammals synthesize fatty acids and other lipids from carbon-containing molecules derived from metabolism of carbohydrates and amino acids. They also digest and assimilate fats present in the diet. There are, in fact, certain unsaturated fatty acids that must be supplied in the diet because in mammals there is no enzyme system capable of forming a double bond beyond carbon atom nine in the

TABLE IV

CHANGES IN CHEMICAL CONSTITUENTS OF SKELETAL MUSCLE FROM RABBITS FED A CHOLINE-DEFICIENT DIET[a]

Group	Constituent (mg/gm wet tissue)			
	ATP	Creatine	DNA	RNA
Normal	0.38 ± 0.08[b]	3.33 ± 0.33	0.30 ± 0.06	0.38 ± 0.08
Dystrophic	0.21 ± 0.06	2.22 ± 0.38	1.00 ± 0.15	1.28 ± 0.12

[a] Data adapted from Srivastava et al. (211).
[b] Standard deviation.

fatty acid chain (143). For example, linoleic (*cis*-$\Delta^{9,12}$-octadecadienoic) and linolenic (*cis*-$\Delta^{9,12,15}$-octadecatrienoic) acids are essential fatty acids required as building blocks for a variety of other unsaturated fatty acids which are components of structural lipids.

The dietary requirement for essential fatty acids in the rabbit has not been determined. However, signs of fatty acid deficiency in rabbits have been induced inadvertently by feeding rations containing only saturated fats (132, 244) or deliberately by feeding diets containing cholesterol only (180) or no fat (2). Ahluwalia and associates (2) fed a purified fat-free ration to young male New Zealand rabbits for 14 weeks. The animals gained an average of about 19 gm/day and began losing hair after 10 weeks on experiment. Feed efficiency was compared with control animals fed a commercial ration and, consequently, is meaningless since the diets differed not only in fat content but in many other respects, probably including palatability and digestibility. The authors also reported testicular changes consisting of degeneration and apparently atrophy of seminiferous tubules in fatty acid-deficient rabbits.

Thacker (218) has studied the effect of varying dietary fat levels of a purified diet fed *ad libitum* on growth in Dutch rabbits. Rabbits fed 10% or more fat as vegetable oil gained significantly more weight than those fed 5% fat. Palatability was apparently better for diets containing 10% or more fat since the increased weight gains were the result of greater caloric intake (218). In contrast to these results, Gaman and associates (74) have found 5% corn oil to be adequate in a purified ration fed to rabbits for periods up to two years.

C. Vitamins for Which There Is No Evidence of Simple Dietary Deficiency

1. Vitamin B$_{12}$ (Cyanocobalamin)

Vitamin B$_{12}$ functions as cobamide coenzyme in conversion of methylmalonyl coenzyme A to succinyl coenzyme A and in transfer of methyl groups from 5-methyltetrahydrofolate to homocysteine (143). Cobalt (Co) is an integral part of the B$_{12}$ molecule and this is the only known function for the element in animals. As is the case with other animals, rabbits are unable to synthesize vitamin B$_{12}$ and depend on microorganisms to supply the requirements of their tissues. However, under normal circumstances of synthesis by intestinal flora, coprophagy (129, 192), and adequate Co intake, a deficiency of the vitamin is unlikely. In fact, deficiency of Co or vitamin B$_{12}$ in rabbits has not been reported. The requirement for Co is apparently very small (223) although the level of vitamin B$_{12}$ in serum, urine, and feces of rabbits increased markedly when supplemental Co was fed (201).

2. Folic Acid

Only limited information is available concerning folic acid metabolism in the rabbit. Olcese and co-workers (170) found that rabbits fed diets low in folic acid nevertheless excreted the vitamin in the urine in amounts which greatly exceeded intake. Urinary excretion was markedly reduced when sulfasuxidine was fed. This seems to be the only report of folic acid studies in the rabbit and indicates that this vitamin is probably synthesized by intestinal flora under normal circumstances.

3. Riboflavin and Pantothenic Acid

Results of experiments designed to study intestinal synthesis of riboflavin and pantothenic acid in the rabbit indicate that this animal has little or no need for an exogenous supply of these B vitamins (129, 170). Growing rabbits fed diets low in riboflavin (33 μg/100 gm diet) or pantothenic acid (31 μg/100 gm diet) excreted several times the quantity of these vitamins ingested per day and grew as well as controls over a period of 112 days (170). Furthermore, it has been shown that the soft feces contain 3–4 times as much riboflavin and 6 times as much pantothenic acid as hard feces (129). These studies imply that under normal conditions of intestinal synthesis and reingestion of soft feces the rabbit has no dietary requirement for riboflavin or pantothenic acid.

4. Biotin

Little is known about biotin metabolism in the rabbit. Apparently no naturally occurring deficiency of the vitamin has been reported. Metabolic balance studies demonstrated that considerably more biotin was excreted per day in the urine and feces than was ingested (170). More studies need to be done but these results indicate intestinal synthesis of biotin may modify the dietary requirement for this vitamin in the rabbit.

Raw egg white contains a protein anti-biotin substance called avidin (58). Lease and associates (139) fed a diet which contained 40% egg white to rabbits and induced clinical signs now attributed to biotin deficiency including scaliness and flaking of skin, loss of hair on the back, lips, eyelids, and tail, conjunctivitis, and redness of skin. These skin lesions are similar to those observed in rats fed egg white in the diet (69).

5. Thiamine

Thiamine pyrophosphate is the coenzyme for several enzymes that catalyze either oxidative or nonoxidative decarboxylation of α-keto acids such as pyruvate. Consequently, this vitamin is essential in carbohydrate metabolism and deficiency signs develop in man and nonruminant animals when it is not supplied in the diet.

Attempts to produce thiamine deficiency in the rabbit have met with limited success (9, 179). The reasons for this probably are related to the type and form of purified diet used and, more importantly, reingestion of soft feces which contain considerable amounts of thiamine (185). The authors (113) have fed a thiamine-deficient purified diet (in gel form, based on 20% vitamin free casein and 65% sucrose) to young Dutch rabbits for four months with no attempt to prevent coprophagy. No clinical signs of deficiency were evident, no apparent effect on growth was observed, and no lesions were seen.

Other workers observed evidence of thiamine deficiency in three of seven weanling rabbits fed a purified ration (185). These animals developed mild locomotor ataxia involving the hindquarters after 66–146 days on the diet. No comment was made concerning lesions in tissues of these animals. In a subsequent experiment weanling rabbits were fed the same diet with and without added thiamine or the thiamine antagonist neopyrithiamine (185). No neurological signs were observed in rabbits fed the thiamine-free or thiamine-supplemented diets during the 78-day study. However, six of seven rabbits fed the thiamine-deficient diet supplemented with neopyrithiamine developed gross signs including ataxia, paralysis, convulsions, coma, and death. The neurological signs could be reversed by parenteral administration of thiamine or prevented by supplementing the diet with the vitamin. Other evidence that neopyrithiamine induced thiamine deficiency in the rabbits included a marked decrease in thiamine level of brains and about a 15-fold increase in urinary excretion of thiamine.

In conclusion, the studies referred to above indicate that: (a) in the rabbit, dietary deficiency of thiamine does not result in anorexia and weight loss as is observed in other species, (b) the rabbit attains normal growth even when the only source of thiamine is the normally ingested soft feces, and (c) neurological manifestations of thiamine deficiency in the rabbit can be induced by feeding neopyrithiamine, a thiamine antagonist, in the diet.

6. Vitamin C (Ascorbic Acid)

The literature concerning ascorbic acid and the rabbit has been reviewed by Harris et al. (91) who concluded that rabbits fed a vitamin C-free diet for periods up to six months gained as well and remained as healthy as controls that received the vitamin. During the test period the animals fed the deficient diet excreted vitamin C continuously in the urine and total amounts of the vitamin in various organs at autopsy were greater than in organs of control animals killed at the beginning of the experiment. These observations were interpreted as evidence that the rabbit can synthesize vitamin C and has no dietary requirement for the vitamin. Other workers have successfully used purified diets which contained no ascorbic acid (98, 111) and the

authors have observed no signs of scurvy in groups of young rabbits fed a purified agar gel diet for periods up to 12 months (113).

III. MINERALS

A. Deficiency

1. Calcium and Phosphorus

Calcium (Ca) is present in all cells and is especially concentrated as crystalline phosphates (hydroxyapatite) in calcified tissues. Calcium apparently is required for normal function of most membranes and has defined roles in the blood clotting mechanism and muscular contraction.

Phosphorus (P) also is ubiquitous in living cells. It is a constituent of numerous molecules including nucleic acids, proteins, lipids, and carbohydrates. Phosphorus is an important constituent of high energy compounds and, as noted above, it combines with calcium to form hydroxyapatite which gives hardness to bone and teeth.

Minimum Ca and P requirements of the rabbit have been determined in recent years and found to be 0.22% of the diet for Ca (33) and the same for P (148). Chapin and Smith also have studied tolerance of the rabbit to high levels of these elements in the diet (34, 35). Diets containing 4.5% Ca (0.3% P) did not adversely affect weight gains in growing rabbits or significantly alter reproductive performance of does (34). However, the data indicate a dietary Ca level of 0.45–1.5% resulted in better overall reproductive performance than a level of 2.5–4.5% (Table V).

Rabbits fed a diet high in P and relatively low in Ca (Ca:P = 0.5) developed hypertrophy and hyperplasia of parathyroids and maintained markedly increased serum levels of parathyroid hormone for periods up to two and a half years (12). In other studies it has been shown that the rabbit will tolerate 1.5% P in the diet if Ca is at this level or higher (35). These investigators concluded that the limit to which the Ca:P ratio can be reduced below unity depends on concentration of P; as the level of P is

TABLE V
DIETARY CALCIUM AND REPRODUCTION IN RABBIT DOES[a]

Criteria of performance	Dietary calcium (%)	
	0.45–1.5	2.5–4.5
Number of does per group	15	15
Average number of offspring per litter	6.5	7.7
Number of does weaning litters	13	8
Litters not accepted by doe	0	6
Average number weaned per litter	5.1	3.7
Average weight of young at 9 weeks (gm)	1909	1612

[a]Adapted from Chapin and Smith (34).

increased it becomes essential to raise Ca to at least the same level.

Calcium-deficient rabbits become progressively lethargic, pot-bellied, and weak with loss of appetite (33). The authors (113) have observed these signs in young rabbits fed a purified diet which contained approximately 0.05% Ca. In addition, the animals pulled out and ingested their fur. Plasma Ca characteristically falls from a normal level of about 15 to 7 mg/100 ml before death (33). These investigators did not observe tetany although tetany accompanied by hypocalcemia has been reported in rabbits fed a diet containing less than 0.005% Ca (Table VI). It is remarkable that serum Ca in the rabbit readily reflects dietary Ca level. This phenomenon also was observed in studies of the effect of high dietary Ca (4.5%) in rabbits. In some animals on this regimen serum Ca exceeded 20 mg/100 ml in less than 20 days although the vitamin D level of the diet was not excessive (660 IU D_2/kg of diet; Ref. 34). Apparently serum Ca is not strictly controlled by parathyroid-calcitonin-vitamin D interrelationships as it is in other animals. This deserves further study.

Lesions seen in Ca-deficient rabbits include lens opacities (213) and rib fractures. Although Chapin and Smith (33) and the authors (113) have observed evidence of fractures, there were no overt signs of rickets such as enlarged costochondral junctions or wide epiphyses of long bones. In fact, the epiphyseal plate of femur and tibia was of normal width with a distinct line between the zone of calcifying cartilage and metaphysis. However, there were few bone trabeculae in the metaphysis and those present were short compared to normal. In our experiments there was no gross or microscopic evidence of parathyroid hyperplasia in calcium-deficient rabbits (113).

2. Magnesium

In the years that have passed since Leroy (140) first demonstrated that magnesium (Mg) was required for the growth of mice, an untallied number of studies have been undertaken in an effort to elucidate the functions of Mg and the role of this element as an activator of biological

catalysts is now widely recognized (231, 232, 235). Most enzymes catalyzing the transfer of phosphorus from adenosine triphosphate to an acceptor molecule or from a phosphorylated compound to adenosine diphosphate are, for example, known to be activated by Mg. Magnesium is also required in enzyme systems which utilize thiamine pyrophosphate as a coenzyme. Other enzymes, such as enolase and some proteolytic enzymes, e.g., leucine amino peptidase, have likewise been found to be activated by Mg. There is some evidence to suggest further that Mg, in addition to other divalent cations, may have a function in protein synthesis, perhaps being necessary for the maintenance of the operational configuration of amino acid-transferring proteins and the structure of ribosomes on which amino acids are aligned during protein synthesis (231). Studies of the physiological and pharmacological actions of Mg in *in vivo* and *in vitro* systems have led to the conclusion that Mg also has a function in nerve impulse transmission and muscle contraction and relaxation. Rook and Storry (191) and Walser (235) have reviewed the evidence for the latter function and discussed its possible relationships to the origin of the neuromuscular signs seen in Mg deprivation in various species.

Kunkel and Pearson (130), using clinical manifestations, growth, and blood levels of Mg as criteria, estimated the dietary Mg requirement of the growing rabbit at 30 to 40 mg/100 gm of diet. Values were determined using supplemental Mg in the form of the sulfate. This form had previously been shown in the rat (131) to be more available than Mg supplied as carbonate, oxide, or as it occurs in wheat plant. No studies of factors which might modify the Mg requirement of the rabbit have come to the author's attention. High dietary levels of protein, cholesterol, lactose, and calcium and phosphorus, are, however, known to increase requirements or to aggravate the effects of Mg deficiency in other species (39, 165, 235). Colby and Frye (38) also reported that elevated dietary potassium hastened the onset and severity of Mg deficiency in rats but Grace and O'Dell (81) recently found that in guinea pigs excess potassium has an ameliorative effect on deficiency. Hegsted *et al.* (97), who studied the influence of environmental temperature on Mg requirements of rats, reported that animals housed in a cold room had twice the requirement of animals maintained at a more usual room temperature.

a. MAGNESIUM DEFICIENCY. Clinical signs of experimentally induced Mg deficiency in rabbits have been described by several groups of investigators. Reports suggest that the type of manifestations seen and their time of occurrence after exposure of animals to deficient diets is dependent on such factors as age of rabbits used and level of Mg of diets fed. Aikawa and David (3), for example, placed adult domestic rabbits on a ration sypplying 8 mg of Mg/100 gm of diet and tap water containing 0.6 mEq of

TABLE VI
EFFECTS OF ACUTE CALCIUM DEFICIENCY IN GROWING RABBITS[a]

No. of animals	Average Ca intake mg/day	Average serum Ca mg/100 ml	Lens opacities	Tetany
Controls				
3	30–60	8.33	No	No
3	600–700	10.98	No	No
Deficient				
3	0.08	4.08	Yes	Severe

[a]Adapted from Swan and Salit (213).

Mg/liter. The appearance and behavior of these animals remained normal until the fourth week when some began to lose hair over the back, hindlegs, and tail. The hair coat lost its luster and became ragged in appearance. When, on the other hand, young rabbits weighing approximately 1.8 kg were placed on a diet containing less than 1 mg Mg/100 gm of diet and water with 0.1 mEq Mg/liter, hyperirritability was observed in 4 to 5 days and tachycardia, vasodilation, and alopecia between the second and third week. Continuation on this regime resulted in an increase in the severity of these signs and the animals lost weight and stopped eating after 4 to 6 weeks.

Woodward and Reed (250) used New Zealand White rabbits weighing between 1.7 and 3.0 kg and a ration containing 5.6 mg Mg/kg of diet. The most consistent clinical sign noted by these workers was a blanching of ears which occurred 4 to 5 days after animals were placed on experiment. A few also showed alopecia and changes in the texture and luster of their fur. Anorexia occurred after 10 to 12 days and weight loss by the third week was slightly in excess of 1% of initial body weight. Their animals became extremely lethargic after 20 days on experiment and 5 of 6 fed the deficient diet died between the twenty-third and twenty-fifth day.

Kunkel and Pearson (130) using weanling New Zealand White rabbits were able to produce clinical signs of Mg deficiency which included hyperexcitability, convulsions, and retarded growth within a 3- to 6-week period by feeding partially purified rations containing 6 mg or less Mg/100 gm of ration. These animals became extremely emaciated and most were dead by the ninth week. The addition of Mg to the diet of the remaining animals resulted in an immediate resumption of growth. In another experiment, five weanling female rabbits fed a partially purified ration supplying 6 mg Mg/100 gm of diet survived for 20 to 25 weeks. A decrease in growth rate was noted by the tenth week and convulsive seizures were seen in one animal as early as the sixteenth week of deficiency. The others ultimately developed similar seizures which could be induced by unusual noise. One animal was observed to survive 18 such audiogenic convulsions. After the eighteenth week attempts were made to breed 3 of the deficient females to normal males. Several tries were required in each case before the male was accepted. Death occurred in all females 8 to 10 days after mating. Necropsy of one disclosed that although fertilization had apparently taken place, fetal death and resorption had occurred.

The most characteristic change in serum chemistry values seen in uncomplicated experimental Mg deficiency in animals is a lowering of Mg concentration with little or no change in ciculating calcium. The same is true of Mg deficiency in rabbits (3, 23, 130). Aikawa and David (3) found an increase in serum creatinine and blood urea nitrogen which they considered supportive evidence of renal damage. The elevation in creatinine could also have been due to muscle degeneration, and this possibility, while not explored, was not excluded by these workers. Other findings compatible with liver damage were reported, viz., elevated blood levels of total and conjugated bilirubin, lactic dehydrogenase, and serum glutamic oxalacetic transaminase. Unexplained increases in mean total leukocytes and serum cholesterol also were found.

Studies of the effects of Mg deficiency in rabbits on tissue distributions of Mg, Ca, Na, K, and Cl have been reported by various investigators (3, 23, 250). Notable decreases in Mg have been found in bone, lung, and segments of the intestinal tract (3), but not in the brain, cerebrospinal fluid, or skeletal muscle (3, 250). Calcium concentrations have been reported to increase in liver (3) and skeletal muscle (3, 250). Woodward and Reed (250) also found that while total brain Ca was unaffected in Mg-deficient rabbits, the ultrafiltrable (dialyzable) fraction of the total Ca in this tissue did increase. These changes in conjunction with a significant reduction in brain extracellular (inulin) space indicated marked accumulation of intracellular Ca had occurred. A similar increase in intracellular Na and Cl accompanied by a decrease in intracellular K was detected. As with Ca, however, the total brain content of these elements was not significantly altered. Aikawa and David (3), on the other hand, found significant increases in the Na content of brain as well as of bone (cortical), liver, lung, skin, spleen, stomach, and testes of deficient rabbits. Potassium levels were decreased in the bone and kidneys of their rabbits but elevated in the large intestine.

Studies of the gross and microscopic pathology of Mg deficiency in rabbits have been limited. Kunkel and Pearson (130) reported that the kidneys in one group of weanling rabbits maintained on a deficient ration for several weeks were mottled and bloody in appearance at autopsy. Microscopic changes were not reported. Barron et al. (11) examined tissue sections of the cerebellum, heart, and kidneys of weanlings fed purified rations containing graded levels of Mg (5 to 85 mg/100 gm of ration) for a 10-week period. No cardiac pathology was reported. Rabbits receiving diets providing 5 to 10 mg of Mg/100 gm of diet, however, had cerebellar lesions which consisted of large numbers of fusiform, dense and basophilic-staining Purkinje cells without visible nuclei. Other Purkinje cells had eccentric positioned nuclei and showed varying degrees of chromatolysis. In animals receiving diets containing the lowest concentrations of Mg, degeneration of renal tubular epithelial cells and fibrosis of the corticomedullary region were seen. The renal glomeruli were, in addition, often enlarged to twice the diameter of those of control animals and frequently found to contain an amorphous acidophilic-staining material which displaced the glomerular tufts peripherally.

The authors have examined tissues from a number of young rabbits fed a purified Mg-deficient ration (113).

While no significant gross lesions were observed at necropsy, histopathological studies revealed degenerative changes in the kidneys, heart, and skeletal muscle similar to those seen in vitamin E deficiency and vitamin D toxicity. In the kidneys, individual and fused clusters of small spherical mineral deposits which stained blue-black in hematoxylin-eosin sections and positive for Ca with alizarin red were found both within the lumens of the cortical and medullary tubules and within the tubular epithelial cells (Fig. 9). The deposits acted as a nidus from which larger irregularly shaped deposits developed by accretion of additional mineral. The expansive growth and encroachment of these deposits on surrounding tissue elements gave rise to focal areas of necrosis and fibrosis.

Microscopic cardiac and skeletal muscle lesions seen in Mg-deficient rabbits in these studies consisted of areas of muscle fiber mineralization, necrosis, and granulomatous type inflammatory cell response. In the heart, individual and small groups of muscle fibers containing numerous Ca-containing particles (alizarin red positive) were encountered surrounded by normal appearing muscle. The deposition of mineral in these fibers did not appear in most instances to have been preceded by either degenerative changes or inflammatory cell reaction (Fig. 10). In other areas, more advanced lesions with groups of necrotic muscle fibers undergoing calcification were found sur-

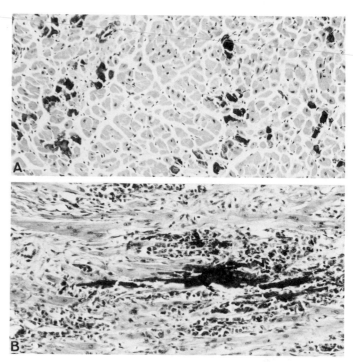

Fig. 10. Experimental magnesium deficiency. A. Transverse section of myocardial fibers, many of which are mineralized. Hematoxylin and eosin, × 160. B. Section of myocardium showing mineralization of myofibrils and inflammatory cell infiltration. Hematoxylin and eosin, × 160.

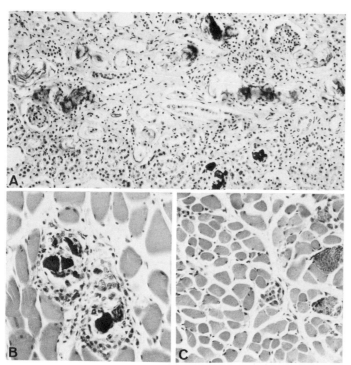

Fig. 9. Experimental magnesium deficiency in young Dutch rabbits. A. Section of kidney with clusters of mineral in lumens of renal tubules. Hematoxylin and eosin, × 120. B. Skeletal muscle with granulomas which contain a central core of mineral. Hematoxylin and eosin, × 170. C. Section of skeletal muscle showing several myofibrils in various degenerative states. Hematoxylin and eosin, × 120.

rounded by actively proliferating mononuclear cells (Fig. 10). Anitschkow type myocytes were present in some. Skeletal muscle lesions were similar to those seen in the heart. Here, however, evidence of muscle degeneration and necrosis (fiber swelling and coagulation, clumping, and fragmentation of sarcoplasm) was commonly seen to precede the actual deposition of mineral (Fig. 9). Each lesion usually involved only one to three fibers. Fiber degeneration and mineralization was accompanied by a proliferation of histiocytes which in some lesions formed multinucleated syncytial type giant cells. A few polymorphonuclear cells were occasionally found. Older lesions in skeletal muscle appeared as small granulomas which sometimes contained a central core of calcified material (Fig. 9).

3. Manganese

Manganese (Mn) and other divalent cations are known to activate many enzymes *in vitro*. In recent years specific biochemical roles have been demonstrated for Mn in carbohydrate metabolism. For example, there is evidence that Mn catalyzes glucosamine-serine linkages in mucopolysaccharides of cartilage (84, 190). Furthermore, pyruvate carboxylase was found to be a Mn metalloenzyme (197) and there is evidence that the element plays a role in glucose utilization (63).

Smith and associates have published observations on Mn as a nutrient in the rabbit which indicate that 1–2 mg/day (0.002–0.004% of diet) is adequate for maximum growth (204). Milk is low in Mn and these studies utilized a milk diet supplemented with iron and copper. Weanling rabbits fed the diet did not grow as well as controls and developed crooked front legs (61, 204, 205). The weight, density, length, breaking strength, and ash of bones were less in Mn-deficient animals than in controls fed 5 mg Mn/day (Table VII). Similar results were obtained using the paired-feeding technique (61). Histological changes in proximal humerus included thinning of the epiphyseal plate with fewer spicules of calcified cartilage and less trabecular bone than controls (205). All bone changes were prevented by feeding 0.3 mg Mn/day although more was required for maximum growth (204).

Bone alterations resulting from Mn deficiency have been observed in several species of animals. Congenital deficiency of the element in rats and guinea pigs results in skeletal abnormalities accompanied by decreased levels of mucopolysaccharide precursors in epiphyseal cartilage (117). Studies in the chick have shown that Mn deficiency results in a reduction in mucopolysaccharide content of several tissues with an especially marked decrease in chondroitin sulfate of epiphyseal cartilage (137, 138). These observations may explain the bone lesions seen in Mn-deficient rabbits.

4. Potassium

Potassium (K) is the major intracellular base ion and as such plays an important role in regulation of pH within the cell. In addition, the K ion is an activator of pyruvic kinase and several other enzymes involved in cellular syntheses (243). Potassium also is essential to muscle contraction and neuronal conduction and affects muscle irritability (103, 243).

Potassium deficiency in the rabbit was referred to by Wooley (252) and Wooley and Mickelsen (254) but severe deficiency with resultant muscular dystrophy (MD) has been described only by Hove and Herndon (109). These latter authors found the K requirement of the rabbit to be between 0.6 and 0.9% of the diet. This is considerably higher than the established requirement of the rat (0.18% of the diet; Ref. 128, 198).

The average survival time of adult rabbits fed a diet which contained 0.009% K was 37 days (109). During this period the animals developed various signs of nervous derangement including trembling and apparent fright. Many (66%) of these rabbits developed terminal creatinuria and MD that progressed to generalized flaccid paralysis. Young rabbits fed diets containing 0.3% or less K also became paralyzed with concomitant creatinuria. Gross lesions observed at autopsy included atrophy and white streaks in skeletal muscle, pale hearts with evidence of necrosis and scarring, focal hemorrhages in gastric mucosa, thin translucent intestines, and swollen and pale kidneys. The authors concluded that there are many similarities in the MD induced in rabbits by deficiency of potassium, choline, or vitamin E including extensive damage to myocardium and skeletal muscle and progressive creatinuria.

Effects of K deficiency have been studied also in rats, mice, dogs, and calves. Tissues most commonly affected were myocardium, skeletal muscle, and kidney (69). Rats fed a K-deficient diet developed necrosis and scarring of the myocardium and necrosis of the renal tubular epithelium (68, 195). Dogs became paralytic and rats and dogs (69) developed degenerative lesions in skeletal muscle fibers associated with cellular infiltration and proliferation of sarcolemma nuclei. These observations provide evidence for common effects of K deficiency in several species, including the rabbit. The biochemical lesions which result in these tissue changes have not been clarified entirely but undoubtedly relate to the cellular functions of K referred to above and perhaps others not yet discovered.

5. Zinc

Zinc (Zn), a divalent cation, is known to be a constituent of several metalloenzymes. The element has a role in the metabolism and/or synthesis of deoxyribonucleic acid (DNA), ribonucleic acid (RNA), and protein, which also may result from its involvement as a prosthetic group or cofactor for enzymes essential to these processes. High concentrations of Zn are found in bone (15) although these stores apparently are not immediately available to tissues of fetuses carried by female rats fed diets deficient in Zn during pregnancy (119).

Information concerning Zn deficiency or the Zn requirement of rabbits is sparse. Possibly this is because of difficulties encountered in devolping purified diets that are deficient in Zn; most protein sources contain this element. Recently a study was conducted in which breeding does were deprived of Zn during gestation by feeding a diet containing 0.2–0.3 mg Zn/100 gm of feed (5). This resulted

TABLE VII
EFFECTS OF MANGANESE DEFICIENCY ON BONES OF RABBITS[a]

Parameters	Control[b]	Deficient[c]
Weight of dry, fat-free humeri (gm)	1.21	0.92
Density of humeri (gm/ml)	1.04	0.82
Length of humeri (mm)	59.7	52.6
Breaking strength of ulnae (lbs)	13.5	9.0
Ash in dry, fat-free humeri (%)	61.4	55.7

[a] Adapted from Smith *et al.* (205).
[b] Mean values for six animals.
[c] Mean values for seven animals.

in failure to deliver or extended parturition with retained placentae and small offspring which usually failed to survive. Studies of Zn deficiency in pregnant rats have produced teratogenic effects with high incidence of malformations affecting every organ system of offspring (118). No malformations were observed in offspring born by Zn-deficient does in Apgar's studies (5).

Young rabbits fed a diet low in Zn ceased to grow after two weeks and developed partial alopecia, scaley skin, soreness around the mouth, and wet matted hair on the lower jaw and ruff (5). In another report (82), weanling rabbits were fed a diet which was not adequately described but contained an "extremely high" concentration of calcium and 2.5 mg Zn/100 gm supplied by soybean meal. The authors found that rabbits fed this diet retained a much greater percentage of orally administered ^{65}Zn than those fed the same diet which contained supplemental zinc oxide. Apparently Zn in soybean meal was not assimilated by the rabbits.

Growth retardation and various changes in skin and testes have been assoicated with Zn deficiency in animals and man (183, 225, 226), yet it still is not possible to correlate biochemical functions with pathological lesions of the deficiency. As is apparent from our brief discussion above, much remains to be done with regard to Zn requirement and metabolism in the rabbit.

6. Iron

Practically nothing is known about the requirement for iron (Fe) or metabolism of this important element in the rabbit. Smith and co-workers (206) fed a milk diet to rabbits and induced development of microcytic and hypochromic anemia which was not corrected by feeding supplemental iron or copper alone. The authors are not aware of any studies in which purified diets were used to determine the actual requirement of Fe in the rabbit.

7. Copper

Copper (Cu) functions as a cofactor for several mammalian enzymes including cytochrome oxidase, tyrosinase, and amine oxidase. In animals, dietary deficiency of Cu results in reduced cytochrome oxidase activity of many tissues and loss of pigmentation of feathers and hair, thought to be due to decreased tyrosinase activity. An amine oxidase is apparently involved in oxidation of lysine residues which condense to form elastin cross-links (178) and reduced activity of this enzyme probably explains the occurrence of aortic rupture associated with elastin defects seen in Cu-deficient chicks (99).

The minimal dietary requirement for Cu in the rabbit has not been established although a level of 3–6 parts per million (ppm) maintained growth and blood hemoglobin levels for up to five months (114).

Signs of Cu deficiency in rabbits include achromotrichia, alopecia, dermatosis, and hypochromic, microcytic anemia (203, 206). Tissue changes associated with Cu deficiency in growing rabbits include reduced cytochrome oxidase activity in heart and liver and increased concentration of iron in liver (115). Gross and microscopic evidence of myocardial mineralization and apparent failure of formation or maintenance of elastin in the media of aorta (Fig. 11) have also been reported (114). Calcification of the internal elastica and media of muscular arteries also was seen in Cu-deficient animals. These changes are claimed to be spontaneous or age-related phenomena in rabbit aorta (75, 194). These authors have made a valid point with respect to the variety of lesions found in aortas of "normal" rabbits fed commercial rations. However, such diets may not provide optimum quantities of nutrients and may result in toxicities or deficiencies that adversely affect the cardiovascular system (188). Furthermore, as is obvious in the present review, very little is known about the nutrient requirements of the rabbit and deficiency of vitamin E, Mg, or Cu or excess of vitamin D or Mo may result in mineralization of soft tissue including heart and blood vessels. These observations emphasize the need for further research on the requirement for these nutrients and their

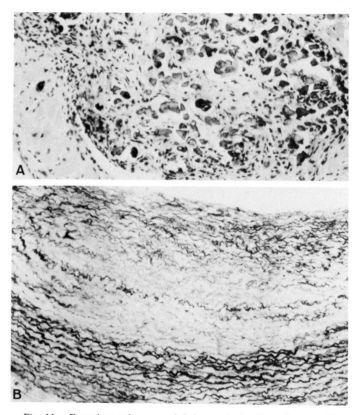

Fig. 11. Experimental copper deficiency. A. Section of myocardium which shows extensive mineralization and fibrosis. Hematoxylin and eosin, × 160. B. Aorta with fragmentation and loss of elastic lamellae in media. Hematoxylin and eosin, × 160. [From Hunt and Carlton, (114, p. 388). Used by permission of publisher.]

interrelationships with respect to production of cardiovascular lesions in the laboratory rabbit.

Bone lesions attributed to uncomplicated Cu deficiency have been observed in several species of animals. Disrupted bone growth in Cu-deficient rabbits has, however, been observed only when ascorbic acid was added to the diet (115). In these animals the radii and ulnae became bowed in a few weeks and ossification centers of long bones had thickened uneven epiphyseal plates and few bone trabeculae in the metaphyses. Molybdenosis results in similar bone lesions and possible relationships between these substances are discussed under that heading.

8. Iodine

It seems that iodine (I) deficiency per se has not been studied in the rabbit nor has the dietary requirement been established for this species. However, goiter was produced inadvertently in a large rabbit colony maintained for studying experimental syphilis (36). These animals, mostly adult males, were fed oats, hay, and fresh cabbage daily as the sole source of water. The investigators were not aware that cabbage contains goitrogenic substances, a fact which was established some years later (8). Thyroids of rabbits autopsied after 200 days on this diet averaged more than 2 gm (Table VIII) compared to an average weight of 0.233 gm in a series of 644 normal rabbits (25). However, animals with large, easily palpated thyroids showed no change in behavior or condition. The microscopic appearance of the enlarged glands was that of diffuse parenchymatous goiter. There was extensive hyperplasia of epithelial elements without colloid formation and marked vascularity with occasional focal accumulation of lymphoid cells. In other studies on this colony of rabbits it was determined that heat production was about 16% lower in goitrous than in normal rabbits (242). The greatest reduction in metabolic rate was observed in animals with the largest thyroid glands.

Daily oral administration of I to these goitrous rabbits resulted in hyperactivity, rapid weight loss, and death, usually within a week (241). In every case the metabolic rate increased steadily until death occurred. The thyroid gland of animals that survived a week or longer decreased in size. Autopsy revealed extreme emaciation with depleted body fat depots. Histological examination of thyroids from animals that received I revealed parenchymatous goiters in various stages of involution. The amount of colloid was greatly increased compared to sections of enlarged thyroids from untreated rabbits.

IODINE TOXICITY. The effect of excess I on pregnancy in the rabbit has been studied by Arrington et al. (7). Their data are summarized in Table IX. Two hundred fifty ppm or more of I fed for two days prior to parturition resulted in significantly more deaths of young than those of controls. Young were of normal size but died within a few hours after birth with no evidence of nursing, although mammary glands of females fed 500 ppm of I appeared to contain a normal quantity of milk at autopsy. These studies did not provide an explanation for the toxic effect of I on the newborn. Similar dietary levels of the element fed to pregnant hamsters and swine did not adversely affect the young (7). However, coprophagy was not prevented in the rabbits and the authors suggested that reingestion of the feces may have resulted in greater net absorption of dietary I in these animals.

B. Toxicity

1. Molybdenum

The only known molybdenum-containing metalloproteins in animal tissues are xanthine oxidase and aldehyde oxidase. A dietary need for the element has been established in the rat and possibly in the chick and lamb (225). There is no evidence at present to suggest that molybdenum (Mo) is an essential nutrient for the rabbit. Molybdenum toxicity has been produced by feeding 0.1% or more

TABLE VIII

RELATION OF THYROID WEIGHT TO TIME IN RABBITS
FED FRESH CABBAGE AS SOURCE OF WATER[a]

Duration of observation (days)	No. of animals	Thyroid weights (gm)		
		Minimum	Maximum	Mean
1–20	14	0.1	0.5	0.18
101–120	37	0.4	3.0	1.47
201–220	21	0.7	9.8	3.56
301–340	4	1.1	5.0	2.42
421–460	10	1.5	36.0	12.00
501–540	12	1.1	22.1	6.77
601–700	6	1.2	18.5	6.20
701–800	7	4.2	43.0	19.50
801–1000	7	2.8	35.7	11.70

[a]Data adapted from Chesney et al. (36).

TABLE IX

SURVIVAL OF RABBITS FROM DOES FED IODINE DURING PREGNANCY[a]

Dietary I (ppm)	I feeding before parturition (days)	No. of litters	Average no./litter	% Surviving	
				3 Days	4 Weeks
0.55 (Control)	—	10	4.3	91	91
250	2	9	6.2	30	30
500	2	15	4.5	29	29
500	5	19	4.2	3	3
500	10	5	4.8	0	0
1000	5	23	3.9	4	4

[a]Data adapted from Arrington, et al. (7).

TABLE X
MOLYBDENUM TOXICITY IN WEANLING RABBITS[a]

Treatment (%Mo)	No. of rabbits	Anemia	Alopecia, dermatosis	Front leg abnormality	Average survival (days)
Control (0.0003)	5	0	0	0	—[c]
0.05	5	0	0	0	—
0.1	5	4	4	2	—
0.2	5	5	4	4	44
0.4	2	2	0[b]	1	30

[a] Adapted from Arrington and Davis (6).

[b] Deaths occurred before alopecia developed.

[c] No deaths in control, 0.05 or 0.1% Mo groups during a 12-week growth period.

of the element to rabbits in a commercial ration (6). This resulted in anorexia, weight loss, anemia, alopecia, dermatosis, bone abnormalities, and death in 30–50 days (Table X). These observations were confirmed and radiographic evidence of epiphyseal fractures was found in rabbits fed 0.08–0.10% Mo (142).

The anemia associated with molybenosis has been characterized in rabbits fed 0.19% Mo (227). Hemoglobin concentration and packed cell volume were decreased significantly by 18 days and reticulocytes disappeared from peripheral blood of affected animals. Reduced numbers

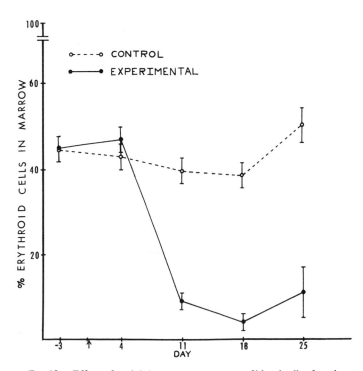

Fig. 12. Effect of molybdenum on percentage of blood cells of erythrocytic series in bone marrow. Molybdate feeding commenced at day zero (arrow). Vertical bars represent standard error of the mean. [Adapted from Valli *et al.* (227, p. 440). Used by permission of authors and publisher.]

of normoblasts and reticulocytes in marrow (Fig. 12) were associated with a general decrease in cellularity of this tissue and concomitant increase in M:E ratio (control = 1–2; affected = 10).

Valli *et al.* (227) described histological changes in muscle and bone of rabbits fed Mo. Degeneration with necrosis occurred in skeletal muscle of hindlimbs and severe focal myocardial degeneration was observed also. The heart lesions appear similar to those seen in Cu-deficient rabbits (114) and may be a manifestation of induced Cu deficiency. In long bones from rabbits fed Mo there was longitudinal widening of epiphyses with apparently normal chondrocytes but disorderly proliferation of cartilage. In addition, there was marked reduction in amount of trabecular bone in the metaphyses. These changes are very similar to those seen in bones of rabbits fed ascorbic acid in a Cu-deficient diet. Signs of Cu deficiency are accentuated by feeding ascorbic acid (28, 29, 116), and an analogous adverse effect on Cu metabolism may be induced by Mo. Evidence pointing to a relationship between dietary levels and metabolism of Mo and Cu is summarized by Underwood (225). For example, poor growth and deaths occurred in young rats fed 80 ppm Mo in a low Cu diet but these effects were prevented by increasing the Cu level to 35 ppm (40). The same relationship has been observed in cattle (66). Also, in rabbits fed 2000 ppm Mo and 200 ppm Cu no signs of molybdenosis were seen during a period of four months (6). These observations indicate that the toxicity of Mo may be mediated by effects on Cu metabolism.

2. Fluorine

We are not aware of any experimental evidence that demonstrates a dietary requirement for fluorine (F) in the rabbit. However, a study has been reported on skeletal changes associated with chronic fluorosis in this species (240). Adult rabbits were provided drinking water which contained 500 ppm F. Lesions developed in various bones and were most pronounced after 22 days on supplemental fluorine.

Diffuse exostoses resulted in gross thickening of the mandible and long bones of affected rabbits. Histological examination revealed: (a) an increase in number of fibroblasts in periosteum and osteoblasts on newly formed trabeculae, (b) exostoses consisting of coarsely bundled woven bone and, (c) resorption beginning with enlargement of Haversian canals in the endosteal half of the bone and becoming confluent cavities. However, resorption was not an invariable feature of fluorosis in the rabbit. Other workers have found that overproduction of osteoid is a feature of skeletal fluorosis (41) and mucopolysaccharide production in bones is affected as well (14). These findings and bone lesions observed in rabbits indicate a direct effect of excess F on osteogenesis per se.

IV. SUMMARY

The rabbit, a herbivore, has some peculiarities of its digestive tract and processes that distinguish it from other animals species. It has a large functional cecum, and bacterial syntheses within this organ and the large intestine modify the rabbit's need for dietary nutrients, especially with regard to B vitamins. Habitual reingestion of soft night feces, which originate in the cecum, reduces the exogenous requirement for B vitamins and probably increases efficiency of utilization of other nutrients, particularly protein. Recycling of intestinal contents is of great importance to experimentalists performing metabolic studies involving nutrients or other compounds administered orally.

Present knowledge of nutrient requirements in the rabbit is at best fragmentary. To expand this knowledge will require utilization of complete, satisfactory purified diets. Two such diets are presented in this chapter. They do not differ markedly in composition from those tested in the past but are fed in forms (pellet or gel) readily accepted by the rabbit and result in growth and gains equivalent to those attained by feeding natural commercial rations.

Deficiences of certain nutrients have shown the rabbit to be a useful model in studies of metabolic disease, e.g., nutritional muscular dystrophy and congenital hydrocephalus, resulting from vitamin E and A deficiency, respectively. Also, the almost traditional use of rabbits in cardiovascular research demands an evaluation of vitamin D requirements, differential effects of vitamin D_2 and D_3 and the importance of mineral balance (especially Ca, P, Mg, K, and Cu) in relation to vascular disease and vitamin D status in this species.

Acknowledgment

This work was supported in part by Research Grant RR00463 from the Animal Resources Branch of the National Institutes of Health, Bethesda, Maryland.

REFERENCES

1. Adamson, I., and Fisher, H. (1971). The amino acid requirement of the growing rabbit: Qualitative needs. *Nutr. Rep. Int.* **4**, 59–64.
2. Ahluwalia, B., Pincus, G., and Holman, R. T. (1967). Essential fatty acid deficiency and its effects upon reproductive organs of male rabbits. *J. Nutr.* **92**, 205–214.
3. Aikawa, J. K., and David, A. P. (1969). ^{28}Mg studies in magnesium-deficient animals. *Ann. N.Y. Acad. Sci.* **162**, 744–757.
4. Anonymous, (1970). Vitamin A and glucocorticoid biosynthesis. *Nutr. Rev.* **28**, 190–192.
5. Apgar, J. (1971). Effect of a low zinc diet during gestation on reproduction in the rabbit. *J. Anim. Sci.* **33**, 1255–1258.
6. Arrington, L. R., and Davis, G. K. (1953). Molybdenum toxicity in the rabbit. *J. Nutr.* **51**, 295–304.
7. Arrington, L. R., Taylor, R. N., Jr., Ammerman, C. B., and Shirley, R. L. (1965). Effects of excess dietary iodine upon rabbits, hamsters, rats, and swine. *J. Nutr.* **87**, 394–398.
8. Astwood, E. B., Greer, M. A., and Ettlinger, M. G. (1949). L-Vinyl-2-thiooxazolidone, and antithyroid compound from yellow turnip and from *Brassica* seeds. *J. Biol. Chem.* **181**, 121–130.
9. Baglioni, A. (1936). Sulla particolari resistenza del coniglio a disturbi dell'avitaminosi B_1 e sulla possibilità di una elaborazione endogena di vitamine. *Arch. Fisiol.* **35**, 362–376; cited in *Nutr. Abstr. Rev.* **6**, 637 (1937).
10. Barnes, R. H., and Fiala, G. (1959). Effects of the prevention of coprophagy in the rat. VI. Vitamin K. *J. Nutr.* **68**, 603–614.
11. Barron, G. P., Brown, S. O., and Pearson, P. B. (1949). Histological manifestations of a magnesium deficiency in the rat and rabbit. *Proc. Soc. Exp. Biol. Med.* **70**, 220–223.
12. Baumann, E. J., and Sprinson, D. B. (1939). Hyperparathyroidism produced by diet. *Amer. J. Physiol.* **125**, 741–746.
13. Beeson, W. M. (1965). Relative potencies of vitamin A and carotene for animals. *Fed. Proc., Fed. Amer. Soc. Exp. Biol.* **24**, Part 2, 924–926.
14. Bélanger, L. F., Visek, W. J., Lotz, W. E., and Comar, C. L. (1957). The effects of fluoride feeding on the organic matrix of bones and teeth of the pigs as observed by autoradiography after *in vitro* uptake of CA^{45} and S^{35}. *J. Biophys. Biochem. Cytol.* **3**, 559–566.
15. Bergman, B. (1970). The zinc concentration in hard and soft tissues of the rat. *Acta Ondontol. Scand.* **28**, 425–440.
16. Best, C. H., and Huntsmann, M. E. (1932). The effects of the components of lecithine upon deposition of fat in the liver. *J. Physiol. (London)* **75**, 405–412.
17. Best, C. H., and Huntsmann, M. E. (1935). The effect of choline on the liver fat of rats in various states of nutrition. *J. Physiol. (London)* **83**, 255–274.
18. Bitman, J., Cecil, H. C., Connolly, M. R., Miller, R. W., Okamoto, M., and Wrenn, T. R. (1962). Effect of vitamin A deficiency on efflux of sodium thiocyanate from cerebrospinal fluid. *J. Dairy Sci.* **45**, 879–881.
19. Boguth, W. (1969). Aspects of the action of vitamin E. *Vitam. Horm. (New York)* **27**, 1–15.
20. Borgman, R. F. (1964). Fat metabolism as influenced by dietary fats and avitaminosis E in rabbits. *Amer. J. Vet. Res.* **25**, 543–546.
21. Borgman, R. F. (1964). Fatty acid composition as influenced by dietary fatty acids and vitamin E status in the rabbit. *J. Food Sci.* **29**, 20–24.
22. Borgman, R. F. (1966). The effect of feeding rabbits a vitamin E-low diet containing oleic acid. *Amer. J. Vet. Res.* **27**, 809–813.
23. Bradbury, M. W. B., Kleeman, C. R., Bagdoyan, H., and Berberian, A. (1968). The calcium and magnesium content of skeletal muscle, brain, and cerebrospinal fluid as determined by atomic absorption flame photometry. *J. Lab. Clin. Med.* **71**, 884–892.
24. Bragdon, J. H., and Levine, H. D. (1949). Myocarditis in vitamin E-deficient rabbits. *Amer. J. Pathol.* **25**, Part 1, 265–271.
25. Brown, W. H., Pearce, L., and Van Allen, C. M. (1925). Organ weights of normal rabbits. *J. Exp. Med.* **43**, 733–741.
26. Calhoun, M. C., Hurt, H. D., Eaton, H. D., Rousseau, J. E., Jr., and Hall, R. C., Jr. (1962). Rates of formation and absorption of cerebrospinal fluid in bovine hypovitaminosis A. *A. Dairy Sci.* **50**, 1486–1494.
27. Canas, F., Brand, J. S., Neuman, W. F., and Terepka, A. R. (1969). Some effects of vitamin D_3 on collagen synthesis in rachitic chick cortical bone. *Amer. J. Physiol.* **216**, 1092–1096.
28. Carlton, W. W., and Henderson, W. (1964). Skeletal lesions in experimental copper-deficiency in chickens. *Avian Dis.* **8**, 48–55.
29. Carlton, W. W., and Henderson, W. (1965). Studies in chickens fed

a copper-deficient diet supplemented with ascorbic acid, reserpine and diethylstilbestrol. *J. Nutr.* **85**, 67–72.

30. Carmichael, E. B. Strickland, J. T., and Driver, R. L. (1945). The contents of the stomach, small intestine, cecum and colon of normal and fasting rabbits. *Amer. J. Physiol.* **143**, 562–566.

31. Carton, C. A., Pascal, R. R., and Tennyson, V. M. (1961). Hydrocephalus and vitamin-A-deficiency in the rabbit: General considerations. *In* "Disorders of the Developing Nervous System" (W. S. Fields and M. M. Desmond, eds.), pp. 214–266. Thomas, Springfield, Illinois.

32. Century, B., and Horwitt, M. K. (1960). Role of diet lipids in the appearance of dystrophy and creatinuria in the vitamin E-deficient rat. *J. Nutr.* **72**, 357–367.

33. Chapin, R. E., and Smith, S. E. (1967). Calcium requirement of growing rabbits. *J. Anim. Sci.* **26**, 67–71.

34. Chapin, R. E., and Smith, S. E. (1967). The calcium tolerance of growing and reproducing rabbits. *Cornell Vet.* **57**, 480–491.

35. Chapin, R. E., and Smith, S. E. (1967). High phosphorus diets fed to growing rabbits. *Cornell Vet.* **57**, 492–500.

36. Chesney, A. M., Clawson, T. A., and Webster, B. (1928). Endemic goitre in rabbits. I. Incidence and characteristics. *Bull. Johns Hopkins Hops.* **43**, 261–277.

37. Christopher, E., and Braun-Falco, O. (1968). Stimulation der epidermalen DNA-synthese durch vitamin A-säure. *Arch. Klin. Exp. Dermatol.* **232**, 427–433.

38. Colby, R. W., and Frye, C. M. (1951). Effect of feeding various levels of calcium, potassium and magnesium to rats. *Amer. J. Physiol.* **166**, 209–212.

39. Colby, R. W., and Frye, C. M. (1951). Effect of feeding high levels of protein and calcium in rat rations on magnesium deficiency syndrome. *Amer. J. Physiol.* **166**, 408–412.

40. Comar, C. L., Singer, L., and Davis, G. K. (1949). Molybdenum metabolism and interrelationships with copper and phosphorus. *J. Biol. Chem.* **180**, 913–922.

41. Comar, C. L., Visek, W. J., Lotz, W. E., and Rust, J. H. (1953). Effects of fluorine on calcium metabolism and bone growth in pigs. *Amer. J. Anat.* **92**, 361–390.

42. Corredor, C., Mansbach, C., and Bressler, R. (1967). Carnitine depletion in the choline-deficient state. *Biochim. Biophys. Acta* **144**, 366–374.

43. Cousins, R. J., Eaton, H. D., Rousseau, J. E., Jr., and Hall, R. C., Jr. (1969). Biochemical constituents of the dura mater in vitamin A deficiency. *J. Nutr.* **97**, 409–418.

44. Dalgliesh, C. E. (1956). Interrelationships of tryptophan, nicotinic acid and other B vitamins. *Brit. Med. Bull.* **12**, 49–51.

45. Dam, H. (1962). Interrelations between vitamin E and polyunsaturated fatty acids in animals. *Vitam. Horm.* (*New York*) **20**, 527–540.

46. Dam, H., Prange, I., and Søndergaard, E. (1952). The effect of certain substances on vitamin A storage in the liver of the rat. *Acta Pharmacol. Toxicol.* **8**, 23–29.

47. Davies, A. W., and Moore, T. (1941). Interaction of vitamins A and E. *Nature* (*London*) **147**, 794–796.

48. De Luca, H. F. (1967). Mechanism of action and metabolic fate of vitamin D. *Vitam. Horm.* (*New York*) **25**, 315–367.

49. De Luca, H. F. (1969). Recent advances in the metabolism and function of vitamin D. *Fed. Proc., Fed. Amer. Soc. Exp. Biol.* **28**, 1678–1689.

50. De Luca, H. F. (1971). Vitamin D: A new look at an old vitamin. *Nutr. Rev.* **29**, 179–181.

51. De Luca, L., Little, E. P., and Wolf, G. (1969). Vitamin A and protein synthesis by rat intestinal mucosa. *J. Biol. Chem.* **244**, 701–708.

52. Diehl, J. F. (1960). Effect of hepatic coccidiosis infection in rabbits on tissue levels of vitamins A and E. *J. Nutr.* **71**, 322–326.

53. Diehl, J. F., and Kistler, B. G. (1961). Vitamin E saturation test in coccidiosis-infected rabbits. *J. Nutr.* **74**, 495–499.

54. Draper, H. H., and Csallany, A. S. (1958). Action of *N,N'*-diphenyl-*p*-phenylenediamine in tocopherol deficiency disease. *Proc. Soc. Exp. Biol. Med.* **99**, 739–742.

55. Draper, H. H., Goodyear, S., Barbee, K. D., and Johnson, B. C. (1958). A study of the nutritional role of antioxidants in the diet of the rat. *Brit. J. Nutr.* **12**, 89–97.

56. Draper, H. H., and Johnson, B. C. (1956). *N,N'*-diphenyl-*p*-phenylenediamine in the prevention of vitamin E deficiency in the lamb. *J. Anim. Sci.* **15**, 1154–1157.

57. Du Vigneaud, V., Cohn. M., Chandler, J. P., Schenck, J. R., and Simmonds, S. (1941). The utilization of the methyl group of methionine in the biological synthesis of choline and creatine. *J. Biol. Chem.* **140**, 625–641.

58. Eakin, R. E., Snell, E. E., and Williams, R. J. (1941). The concentration and assay of avidin, the injury producing protein in raw egg white. *J. Biol. Chem.* **140**, 535–543.

59. Eden, A. (1940). Coprophagy in the rabbit. *Nature* (*London*) **145**, 36–37.

60. Eden, A. (1940). Coprophagy in the rabbit: Origin of "night" faeces. *Nature* (*London*) **145**, 628–629.

61. Ellis, G. H., Smith, S. E., and Gates, E. M. (1947). Further studies of manganese deficiency in the rabbit. *J. Nutr.* **34**, 21–31.

62. Eppstein, S. H., and Morgulis, S. (1941). The minimum requirement of rabbits for *dl*-α-tocopherol. *J. Nutr.* **22**, 415–424.

63. Everson, G. J., and Shrader, R. E. (1968). Abnormal glucose tolerance in manganese-deficient guinea pigs. *J. Nutr.* **94**, 89–94.

64. Feaster, J. P., and Davis, G. K. (1959). Sulfate metabolism in rabbits on high molybdenum intake. *J. Nutr.* **67**, 319–323.

65. Fell, H. B., and Mellanby, E. (1953). Metaplasia produced in cultures of chick ectoderm by high vitamin A. *J. Physiol.* (*London*) **119**, 470–488.

66. Ferguson, W. S., Lewis, A. H., and Watson, S. J. (1943). The teart pastures of Somerset. I. The cause and cure of teartness. *J. Agr. Sci.* **33**, 44–52.

67. Fisher, K. D., Carr, C. J., Huff, J. E., and Huber, T. E. (1970). Dark adaptation and night vision. *Fed. Proc., Fed. Amer. Soc. Exp. Biol.* **29**, Part 3, 1605–1638.

68. Follis, R. H., Jr., Orent-Keiles, E., and McCollum, E. V. (1942). The production of cardiac and renal lesions in rats by a diet extremely deficient in potassium. *Amer. J. Pathol.* **18**, 29–40.

69. Follis, R. H., Jr. (1958). "Deficiency Disease." Thomas, Springfield, Illinois.

70. Frank, I., Hadeler, U., and Harder, W. (1951). Zur ernahrungsphsiologie der nagetiere, über die bedeutung der coecotrophie und die zusammensetzung der coecotrophe. *Pfluegers Arch. Gesamte Physiol. Menschem Tiere* **253**, 173–180.

71. Friedman, W. F., and Roberts, W. C. (1966). Vitamin D and the supravalvar aortic stenosis syndrome. The transplacental effects of vitamin D on the aorta of the rabbit. *Circulation* **34**, 77–86.

72. Gallina, A. M., Helmboldt, C. F., Frier, H. I., Nielsen, S. W., and Eaton, H. D. (1969). Bone growth in the hypovitaminotic A calf. *J. Nutr.* **100**, 129–142.

73. Gaman, E., and Fisher, H. (1970). The essentiality of arginine, lysine and methionine for the growing rabbit. *Nutr. Rep. Int.* **1**, 57–64.

74. Gaman, E., Fisher, H., and Feigenbaum, A. S. (1970). An adequate purified diet for rabbits of all ages. *Nutr. Rep. Int.* **1**, 35–48.

74a. Gaman, E., Fisher, H., and Feigenbaum, A. S. Personal communication.

75. Garbarsch, C., Matthiessen, M. E., Helin, P., and Lorenzen, I. (1970). Spontaneous aortic arteriosclerosis in rabbits of the Danish country strain. *Atherosclerosis* **12**, 291–300.

76. Gatz, A. J., and Houchin, O. B. (1947). Histological observations on the vitamin E-deficient rabbit heart. *Anat. Rec.* **97**, 337.

77. Goettsch, M., and Pappenheimer, A. M. (1931). Nutritional muscular dystrophy in the guinea pig and rabbit. *J. Exp. Med.* **54**, 145–166.

78. Goldblatt, H., and Moritz, A. R. (1925). Experimental rickets in rabbits. *J. Exp. Med.* **42**, 499–506.

79. Goodman, D. S., and Huang, H. S. (1965). Biosynthesis of vitamin A with rat intestinal enzymes. *Science* **149**, 879–880.

80. Goodman, D. S., Huang, H. S., Kanai, M., and Shiratori, T. (1967). The enzymatic conversion of all-*trans* β-carotene into retinal. *J. Biol. Chem.* **242**, 3543–3554.

81. Grace, N. D., and O'Dell, B. L. (1970). Interrelationship of dietary magnesium and potassium in the guinea pig. *J. Nutr.* **100**, 37–44.

82. Graham, E. R., and Telle, P. (1967). Zinc retention in rabbits: Effects of previous diet. *Science* **155**, 691–692.

83. Grebner, E. E., Hall, C. W., and Neufeld, E. P. (1966). Glycosylation of serine residues by a uridine diphosphate-xylose: Protein xylosyltransferase from mouse mastocytoma, *Arch. Biochem. Biophys.* **116**, 391–398.

84. Grebner, E. E., Hall, C. W., and Neufeld, E. F. (1966). Incorporation of D-xylose-C¹⁴ into glycoprotein by particles from hen oviduct. *Biochem. Biophys. Res. Commun.* **22**, 672–677.

85. Green, J., and Bunyan, J. (1969). Vitamin E and the biological antioxidant theory. *Nutr. Abstr. Rev.* **39**, 321–345.

86. Greenfield, H., and Briggs, G. M. (1971). Nutritional methodology in metabolic research with rats. *Annu. Rev. Biochem.* **40**, 549–572.

87. Griffith, W. H., and Wade, N. J. (1940). The interrelationship of choline, cystine, and methionine in the occurrence and prevention of hemorrhagic degeneration in young rats. *J. Biol. Chem.* **132**, 627–637.

88. Harrington, D. D. (1969). Vitamin A deficiency induced hydrocephalus in rabbits. Doctoral Thesis, Massachusetts Institute of Technology, Cambridge, Massachusetts.

89. Harrington, D. D. (1971). University of Kentucky, Lexington (unpublished observations).

90. Harrington, D. D., and Newberne, P. M. (1970). Correlation of maternal blood levels of vitamin A at conception and the incidence of hydrocephalus in newborn rabbits: An experimental animal model. *Lab. Anim. Care* **20**, 675–680.

91. Harris, L. J., Constable, B. J., Howard, A. N., and Leader, A. (1956). Vitamin C economy of rabbits. *Brit. J. Nutr.* **10**, 373–382.

92. Harrison, H. E., and Harrison, H. C. (1961). Intestinal transport of phosphate: Action of vitamin D, calcium and potassium. *Amer. J. Physiol.* **201**, 1007–1012.

93. Hartroft, W. S. (1950). Accumulation of fat in liver cells and in lipodiastaemata preceding experimental dietary cirrhosis. *Anat. Rec.* **106**, 61–87.

94. Hass, G. M., Trueheart, R. E., Taylor, C. B., and Stumpe, M. (1958). An experimental histologic study of hypervitaminosis D. *Amer. J. Pathol.* **34**, Part 1, 395–431.

95. Hayes, K. C., and McCombs, H. L. (1969). Fine structure of the arachnoid granulations in vitamin A deficiency. *Fed. Proc., Fed. Amer. Soc. Exp. Biol.* **28**, 489.

96. Hayes, K. C., McCombs, H. L., and Faherty, T. P. (1970). The fine structure of vitamin A deficiency. I. Parotid duct metaplasia. *Lab. Invest.* **22**, 81–89.

97. Hegsted, D. M., Vitale, J. J., and McGrath, H. (1956). The effect of low temperature and dietary calcium upon magnesium requirement. *J. Nutr.* **58**, 175–188.

98. Herndon, J. F., and Hove, E. L. (1955). Surgical removal of the cecum and its effect on digestion and growth in rabbits. *J. Nutr.* **57**, 261–270.

99. Hill, C. H., Starcher, B., and Kim, C. (1967). Role of copper in the formation of elastin. *Fed. Proc., Fed. Amer. Soc. Exp. Biol.* **26**, Part 1, 129–133.

100. Hogan, A. G., and Hamilton, J. W. (1942). Adequacy of simplified diets for guinea pigs and rabbits. *J. Nutr.* **23**, 533–543.

101. Hogan, A. G., and Ritchie, W. S. (1933). Nutritional requirements of the rabbit. *Proc. Soc. Exp. Biol. Med.* **30**, 1193–1194.

102. Hogan, A. G., and Ritchie, W. S. (1934). Nutritional requirements of rabbits and guinea pigs. *Mo., Agr. Exp. Sta., Res. Bull.* **219**, 1–28.

103. Holley, H. L., and Carlson, W. W. (1955). "Potassium Metabolism in Health and Disease." Grune & Stratton, New York.

104. Horwitt, M. K., Harvey, C. C., Century, B., and Witting, L. A. (1961). Polyunsaturated lipids and tocopherol requirements. *J. Amer. Diet. Ass.* **38**, 231–235.

105. Houpt, T. R. (1963). Urea utilization by rabbits fed a low-protein ration. *Amer. J. Physiol.* **205**, 1144–1150.

106. Hove, E. L., and Copeland, D. H. (1954). Progressive muscular dystrophy in rabbits as a result of chronic choline deficiency. *J. Nutr.* **53**, 391–405.

107. Hove, E. L., Copeland, D. H., and Salmon, W. D. (1954). Choline deficiency in the rabbit. *J. Nutr.* **53**, 377–389.

108. Hove, E. L., and Harris, P. L. (1947). Relative activity of the tocopherols in curing muscular dystrophy in rabbits. *J. Nutr.* **33**, 95–106.

109. Hove, E. L., and Herndon, J. F. (1955). Potassium deficiency in the rabbit as a cause of muscular dystrophy. *J. Nutr.* **55**, 363–374.

110. Hove, E. L., and Herndon, J. F. (1957). Vitamin B₆ deficiency in rabbits. *J. Nutr.* **61**, 127–136.

111. Hove, E. L., and Herndon, J. R. (1957). Growth of rabbits on purified diets. *J. Nutr.* **63**, 193–199.

112. Huang, T. C., Ulrich, H. E., and McCay, C. M. (1954). Antibiotics, growth, food utilization and the use of chromic oxide in studies with rabbits. *J. Nutr.* **54**, 621–630.

113. Hunt, C. E. (1971). University of Alabama in Birmingham, Birmingham (unpublished observations).

114. Hunt, C. E., and Carlton, W. W. (1965). Cardiovascular lesions associated with experimental copper deficiency in the rabbit. *J. Nutr.* **87**, 385–393.

115. Hunt, C. E., Carlton, W. W., and Newberne, P. M. (1970). Interrelationships between copper deficiency and dietary ascorbic acid in the rabbit. *Brit. J. Nutr.* **24**, 61–69.

116. Hunt, C. E., Landesman, J., and Newberne, P. M. (1970). Copper deficiency in chicks: Effects of ascorbic acid on iron, copper, cytochrone oxidase activity, and aortic mucopolysaccharides. *Brit. J. Nutr.* **24**, 607–614.

117. Hurley, L. S. (1967). Studies on nutritional factors in mammalian development. *J. Nutr.* **91**, Part 2, Suppl. 1, 27–38.

118. Hurley, L. S., and Swenerton, H. (1966). Congenital malformations resulting from zinc deficiency in rats. *Proc. Soc. Exp. Biol. Med.* **123**, 692–696.

119. Hurley, L. S., and Swenerton, H. (1971). Lack of mobilization of bone and liver zinc under teratogenic conditions of zinc deficiency in rats. *J. Nutr.* **101**, 597–603.

120. Innes, J. R. M., and Yevich, P. P. (1954). So-called nutritional muscular dystrophy as a cause of "paralysis" in rabbits. *Amer. J. Pathol.* **30**, Part 1, 555–565.

121. Irving, J. T. (1964). Dynamics and function of phosphorus. *In* "Mineral Metabolism: An Advanced Treatise" (C. L. Comar and F. Bronner, eds.) Vol. 2, Part 2A, pp. 249–313. Academic Press, New York.

122. Jackson, S. F., and Fell, H. B. (1963). Epidermal fine structure in embryonic chicken skin during atypical differentiation induced by vitamin A in culture. *Develop. Biol.* **7**, 394–419.

123. Johnson, B. C., and Wolf, G. (1960). The function of vitamin A in carbohydrate metabolism; its role in adrenocorticoid production. *Vitam. Horm.* (New York) **18**, 457–483.

124. Jones, D., Gresham, G. A., Lloyd, H. G., and Howard, A. N. (1965). "Yellow fat" in the wild rabbit. *Nature* (London) **207**, 205–206.

125. Jones, D., Howard, A. N., and Gresham, G. A. (1969). Aetiology of "yellow fat" disease (pansteatitis) in the wild rabbit. *J. Comp. Pathol.* **79**, 329–334.

126. Juneja, H. S., Moudgal, N. R., and Ganguly, J. (1969). Studies on

metabolism of vitamin A. The effect of hormones on gestation in retinonate-fed female rats. *Biochem. J.* **111**, 97–105.

127. Kato, J. (1966). Effects of the administration of vitamin D_2, D_3, parathyroid hormone and calcium on hypocalcification of rabbit dentine and on changes in blood constituents caused by experimental rickets. *Gumma J. Med. Sci.* **15**, 174–193.

128. Kornberg, A., and Endicott, K. M. (1946). Potassium deficiency in the rat. *Amer. J. Physiol.* **145**, 291–298.

129. Kulwich, R. Struglia, L., and Pearson, P. B. (1953). The effect of coprophagy on the excretion of B vitamins by the rabbit. *J. Nutr.* **49**, 639–645.

130. Kunkel, H. O., and Pearson, P. B. (1948). Magnesium in the nutrition of the rabbit. *J. Nutr.* **36**, 657–666.

131. Kunkel, H. O., and Pearson, P. B. (1948). The quantitative requirements of the rat for magnesium. *Arch. Biochem.* **18**, 461–465.

132. Lambert, G. E., Miller, J. P., Olsen, R. T., and Frost, D. V. (1958). Hypercholesteremia and atherosclerosis induced in rabbits by purified high fat rations devoid of cholesterol. *Proc. Soc. Exp. Biol. Med.* **97**, 544–549.

133. Lamming, G. E., Salisbury, G. W., Hays, R. L., and Kendall, K. A. (1954). The effect of incipient vitamin A deficiency on reproduction in the rabbit. I. Decidua, ova and fertilization. *J. Nutr.* **52**, 217–225.

134. Lamming, G. E., Salisbury, G. W., Hays, R. L., and Kendall, K. A. (1954). The effect of incipient vitamin A deficiency on reproduction in the rabbit. II. Embryonic and fetal development. *J. Nutr.* **52**, 227–239.

135. Lamming, G. E., Woollam, D. H. M., and Millen, J. W. (1954). Hydrocephalus in young rabbits associated with maternal vitamin A deficiency. *Brit. J. Nutr.* **8**, 363–369.

136. Lardy, H. A. (1951). The influence of inorganic ions on phosphorylation reactions. *Phosphorus Metab. Symp., 1st, 1951* Vol. 1 pp. 477–499.

137. Leach, R. M. (1967). Role of manganese in the synthesis of mucopolysaccharides. *Fed. Proc., Fed. Amer. Soc. Exp. Biol.* **26**, Part 1, 118–120.

138. Leach, R. M., and Muenster, A. (1962). Studies on the role of manganese in bone formation. I. Effect upon the mucopolysaccharide content of chick bone. *J. Nutr.* **78**, 51–56.

139. Lease, J. G., Parsons, H. T., and Kelly, E. (1937). A comparison in five types of animals of the effects of dietary egg white and of a specific factor given orally or parenterally. *Biochem. J.* **31**, 433–437.

140. Leroy, J. (1926). Nécessité du magnésium pour la croissance de la souris. *C.R. Soc. Biol.* **94**, 431–435.

140a. Lohr, J. E., and McLaren, R. D. (1971). "Yellow fat disease" (pansteatitis) in wild hares in New Zealand. *N. Z. Vet. J.* **19**, 266–269.

141. Lombardi, B. (1971). Effects on choline deficiency on rat hepatocytes. *Fed. Proc., Fed. Amer. Soc. Exp. Biol.* **30**, 139–142.

142. McCarter, A., Riddell, P. E., and Robinson, G. A. (1962). Molybdenosis induced in laboratory rabbits. *Can. J. Biochem. Physiol.* **40**, 1415–1425.

143. McGilvery, R. W. (1970). "Biochemistry; A Functional Approach." Saunders, Philadelphia, Pennsylvania.

144. Mackenzie, C. G., and McCollum, E. V. (1940). The cure of nutritional muscular dystrophy in the rabbit by α-tocopherol and its effect on creatine metabolism. *J. Nutr.* **19**, 345–362.

145. McWard, G. W., Nicholson, L. B., and Poulton, B. R. (1967). Arginine requirement of young rabbit. *J. Nutr.* **92**, 118–120.

146. Madsen, H., and Taylor, E. L. (1939). Does the rabbit chew its cud? *Nature (London)* **143**, 981–982.

147. Mann, I., Pirie, A., Tănsley, K., and Wood, C. (1946). Some effects of vitamin A deficiency on the eye of the rabbit. *Amer. J. Ophthalmol.* **29**, 801–815.

148. Mathieu, L. G., and Smith, S. E. (1961). Phosphorus requirements of growing rabbits. *J. Anim. Sci.* **20**, 510–513.

149. Mattill, H. A., and Golumbic, C. (1942). Vitamin E, cod liver oil and muscular dystrophy. *J. Nutr.* **23**, 625–631.

150. Maynard, L. A., and Loosli, J. K. (1962). "Animal Nutrition," 5th ed., pp. 193–207. McGraw-Hill, New York.

151. Melancon, M. J., and De Luca, H. F. (1970). Vitamin D stimulation of calcium-dependent adenosine triphosphatase in chick intestinal brush borders. *Biochemistry* **9**, 1658–1664.

152. Melancon, M. J., Morii, H., and De Luca, H. F. (1970). Physiologic effects of vitamin D, parathyroid hormone, and calcitonin, *In* "The Fat-Soluble Vitamins" (H. F. De Luca and J. W. Suttie, eds.), pp. 111–123. Univ. of Wisconsin Press, Madison.

153. Mellanby, E. (1935). Lesions of the central and peripheral nervous systems produced in young rabbits by vitamin A deficiency and high cereal intake. *Brain* **58**, 141–173.

154. Mellanby, E. (1950). "A Story of Nutritional Research." Williams & Wilkins, Baltimore, Maryland.

155. Mellanby, M., and Killick, E. M. (1926). A preliminary study of factors influencing calcification processes in the rabbit. *Biochem. J.* **20**, 902–926.

156. Merkusin, V. V. (1966). Rost krol'cat pri raznon urovne proteinovogo pitanija. (Growth of young rabbits on different amounts of protein in the diet.) *Krolikovod. Zverovod.* **2**, 17–19; cited in *Nutr. Abstr. Rev.* **37**, 288 (1967).

157. Millen, J. W., and Dickson, A. D. (1957). The effect of vitamin A upon the cerebrospinal-fluid pressures of young rabbits suffering from hydrocephalus due to maternal hypovitaminosis A. *Brit. J. Nutr.* **11**, 440–446.

158. Millen, J. W., and Woollam, D. H. M. (1956). The effects of the duration of vitamin A deficiency in female rabbits upon the incidence of hydrocephalus in their young. *J. Neurol., Neurosurg. Psychiat.* **19**, 17–20.

159. Millen, J. W., Woollam, D. H. M., and Lamming, G. E. (1954). Congenital hydrocephalus due to experimental hypovitaminosis A. *Lancet* **2**, 679–683.

160. Mitchell, G. V., Seward, C. R., and Fox, M. R. S. (1969). Effect of vitamin A deficiency on mitochondrial lipids in rat livers. *J. Nutr.* **97**, 8–12.

161. Moore, R. A., Bittenger, I., Miller, M. L., and Hellman, L. M. (1942). Abortion in rabbits fed a vitamin K-deficient diet. *Amer. J. Obstet. Gynecol.* **43**, 1007–1012.

162. Moore, T. (1957). "Vitamin A," pp. 192–207. Elsevier, Amsterdam.

163. Moritz, A. R., and Krenz, C. (1930). The relation of the fat-soluble vitamins (A and D) to the development of experimental rickets in rabbits. *J. Nutr.* **2**, 257–264.

164. Morot. C. (1882). Des pelotes stomacal des leporides. *Mem. Soc. Cent. Med. Vet.* **12**, Ser. 1.

165. Morris, E. R., and O'Dell, B. L. (1963). Relationship of excess calcium and phosphorus to magnesium requirement and toxicity in guinea pigs. *J. Nutr.* **81**, 175–181.

166. National Academy of Sciences—National Research Council. (1966). "Nutrient Requirements of Rabbits," Publ. No. 1194. Nat. Acad. Sci., Washington, D.C.

167. Navia, J. M., and Lopez, H. (1970). University of Alabama in Birmingham, Birmingham.

168. Okamoto, M., Bitman, J. Cecil, H. C., Connolly, M. R., Miller, R. W., and Wrenn, T. R. (1962). Replacement and absorption of cerebrospinal fluid in normal and vitamin A-deficient calves. *J. Dairy Sci.* **45**, 882–885.

169. Olcese, O., and Pearson, P. B. (1948). The value of urea in the diet of rabbits. *Proc. Soc. Exp. Biol. Med.* **69**, 377–379.

170. Olcese, O., Pearson, P. B., and Schweigert, B. S. (1948). The synthesis of certain B vitamins by the rabbit. *J. Nutr.* **35**, 577–590.

171. Olcese, O., Pearson, P. B., and Sparks, P. (1949). Intestinal synthesis of niacin and the metabolic interrelationship of tryptophan and niacin in the rabbit. *J. Nutr.* **39**, 93–105.

172. Olson, J. A. (1968). Some aspects of vitamin A metabolism. *Vitam. Horm. (New York)* **26**, 1–63.

173. Olson, J. A. (1969). Metabolism and function of vitamin A. *Fed. Proc., Fed. Amer. Soc. Exp. Biol.* **28**, 1670–1677.

174. Olson, J. A., and Hayaishi, O. (1965). The enzymatic cleavage of β-carotene into vitamin A by soluble enzymes of rat liver and intestine. *Proc. Nat. Acad. Sci. U.S.* **54**, 1364–1370.

175. Olson, J. A., and Lakshmanan, M. R. (1970). Enzymatic transformations of vitamin A, with particular emphasis on carotenoid cleavage. *In* "The Fat-Soluble Vitamins" (H. F. De Luca and J. W. Suttie, eds.), pp. 213–226. Univ. of Wisconsin Press, Madison.

176. Olson, R. E. (1970). Studies of the *in vitro* biosynthesis of vitamin K-dependent clotting proteins. *In* "The Fat-Soluble Vitamins" (H. F. De Luca and J. W. Suttie, eds.), pp. 463–489. Univ. of Wisconsin Press, Madison.

177. Olson, R. E. (1971). Scientific contributions of Wendell H. Griffith to our understanding of the function of choline. *Fed. Proc., Fed. Amer. Soc. Exp. Biol.* **30**, 131–138.

178. Partridge, S. M. (1966). Biosynthesis and nature of elastin structures. *Fed. Proc., Fed. Amer. Soc. Exp. Biol.* **25**, Part 2, 1023–1029.

179. Passmore, R. (1935). A note on a synthetic diet for rabbits. *Biochem. J.* **29**, 2469–2470.

180. Peifer, J. J., and Holman, R. T. (1956). Relation of dietary cholesterol to essential fatty acid deficiency. *Fed. Proc., Fed. Amer. Soc. Exp. Biol.* **15**, 326.

181. Pelc, S. R., and Fell, H. B. (1960). The effect of excess vitamin A on the uptake of labelled compounds by embryonic skin in organ culture. *Exp. Cell Res.* **19**, 99–113.

182. Phillips, P. H., and Bohstedt, G. (1938). Studies on the effects of a bovine blindness-producing ration upon rabbits. *J. Nutr.* **15**, 309–319.

183. Prasad, A. S. (1967). Nutritional metabolic role of zinc. *Fed. Proc., Fed. Amer. Soc. Exp. Biol.* **26**, Part 1, 172–185.

184. Raisz, L. G., and Trummel, C. L. (1970). Role of vitamin D in bone metabolism. *In* "The Fat-Soluble Vitamins" (H. F. De Luca and J. W. Suttie, eds.), pp. 93–99. Univ. of Wisconsin Press, Madison.

185. Reid, J. M., Hove, E. L., Braucher, P. F., and Mickelsen, O. (1963). Thiamine deficiency in rabbits. *J. Nutr.* **80**, 381–385.

186. Reid, M. E. (1962). Nutrient requirements of the guinea pig. *Nat. Acad. Sci.—Nat. Res. Counc., Publ.* **990**, 11–24.

187. Rich, A. R., and Hamilton, J. D. (1940). The experimental production of cirrhosis of the liver by means of a deficient diet. *Bull. Johns Hopkins Hosp.* **66**, 185–198.

188. Ringler, D. H., and Abrams, G. D. (1970). Nutritional muscular dystrophy and neonatal mortality in a rabbit breeding colony. *J. Amer. Vet. Med. Ass.* **157**, 1928–1934.

189. Ringler, D. H., and Abrams, G. D. (1971). Laboratory diagnosis of vitamin E deficiency in rabbits fed a faulty commercial ration. *Lab. Anim. Sci.* **21**, 383–388.

190. Robinson, H. C., Telser, A., and Dorfman, A. (1966). Studies on biosynthesis of the linkage region of chondroitin sulfate-protein complex. *Proc. Nat. Acad. Sci. U.S.* **56**, 1859–1866.

191. Rook, J. A. F., and Storry, J. E. (1962). Magnesium in the nutrition of farm animals. *Nutr. Abstr. Rev.* **32**, 1055–1077.

192. Rosenthal, H. L., and Cravitz, L. (1958). Organ, urine and feces vitamin B_{12} content of normal and starved rabbits. *J. Nutr.* **64**, 281–290.

193. Salmon, W. D., and Newberne, P. M. (1962). Cardiovascular disease in choline deficient rats. *Arch. Pathol.* **73**, 190–209.

194. Schenk, E. A., Gaman, E., and Feigenbaum, A. S. (1966). Spontaneous aortic lesions in rabbits. I. Morphologic characteristics. *Circ. Res.* **19**, 80–88.

195. Schrader, G. A., Prickett, C. O., and Salmon, W. D. (1937). Symptomatology and pathology of potassium and magnesium deficiencies in the rat. *J. Nutr.* **14**, 85–109.

196. Scott, M. L. (1966). Vitamin K in animal nutrition. *Vitam. Horm. (New York)* **24**, 633–647.

197. Scrutton, M. C., Utter, M. F., and Mildvan, A. S. (1966). Pyruvate carboxylase. VI. The presence of tightly bound manganese. *J. Biol. Chem.* **241**, 3480–3487.

198. Shaw, R. K., and Phillips, P. H. (1953). The potassium and sodium requirements of certain mammals. *J.-Lancet* **73**, 176–180.

199. Shrader, R. E.; and Everson, G. J. (1968). Pancreatic pathology in manganese-deficient guinea pigs. *J. Nutr.* **94**, 269–281.

200. Shull, R., Alfin-Slater, R. B., Deuel, H. J., Jr., and Ershoff, B. H. (1957). Comparative effects of α-tocopherol, DPPD and other antioxidants on muscular dystrophy in guinea pig. *Proc. Soc. Exp. Biol. Med.* **95**, 263–265.

201. Simnett, K. I., and Spray, G. H. (1961). The influence of diet on the vitamin B_{12} activity in the serum, urine and faeces of rabbits. *Brit. J. Nutr.* **15**, 555–566.

202. Slade, L. M., and Robinson, D. W. (1970). Nitrogen metabolism in rabbits and guinea pigs. *Proc. Amer. Soc. Anim. Sci.* **21**, 195–200.

203. Smith, S. E., and Ellis, G. H. (1947). Copper deficiency in rabbits: Achromotrichia, alopecia and dermatosis. *Arch. Biochem.* **15**, 81–88.

204. Smith, S. E., and Ellis, G. H. (1947). Studies of the manganese requirement of rabbits. *J. Nutr.* **34**, 33–41.

205. Smith, S. E., Medlicott, M., and Ellis, G. H. (1944). Manganese deficiency in the rabbit. *Arch. Biochem.* **4**, 281–289.

206. Smith, S. E., Medlicott, M., and Ellis, G. H. (1944). The blood picture of iron and copper deficiency anemias in the rabbit. *Amer. J. Physiol.* **142**, 179–181.

207. Sorsby, A., Reading, H. W., and Bunyan, J. (1966). Effect of vitamin A deficiency on the retina of the experimental rabbit. *Nature (London)* **210**, 1011–1015.

208. Southern, H. N. (1940). Coprophagy in the wild rabbit. *Nature (London)* **145**, 262.

209. Spellberg, M. A., and Keeton, R. W. (1940). The production of fatty and fibrotic livers in guinea pigs and rabbits by seemingly adequate diets. *Amer. J. Med. Sci.* **200**, 688–697.

210. Spellberg, M. A., Keeton, R. W., and Ginsberg, R. (1942). Dietary production of hepatic cirrhosis in rabbits. *Arch. Pathol.* **33**, 204–220.

211. Srivastava, U., Devi, A., and Sarker, N. K. (1965). Biochemical changes in progressive muscular dystrophy. III. Nucleic acid, phosphorus and creatine metabilism in the muscle, liver and brain of rabbits maintained with a choline-deficient diet. *J. Nutr.* **86**, 298–302.

212. Suttie, J. W. (1969). Control of clotting factor biosynthesis by vitamin K. *Fed. Proc., Fed. Amer. Soc. Exp. Biol.* **28**, 1696–1701.

213. Swan, K. C., and Salit, P. W. (1941). Lens opacities associated with experimental calcium deficiency. *Amer. J. Ophthalmol.* **24**, Part 1, 611–614.

214. Takada, A., Porta, E. A., and Hartroft, W. S. (1967). The recovery of experimental dietary cirrhosis. I. Functional and structural features. *Amer. J. Pathol.* **51**, 929–957.

215. Tappel, A. L. (1962). Vitamin E as the biological lipid antioxidant. *Vitam. Horm. (New York)* **20**, 493–510.

216. Tappel, A. L., and Zalkin, H. (1959). Inhibition of lipid peroxidation im mitochondria by vitamin E. *Arch. Biochem. Biophys.* **80**, 333–336.

217. Tennyson, V. M., and Pappas, G. D. (1961). Exlectronmicroscope studies of the developing telencephalic chorioid plexus in normal and hydrocephalic rabbits. *In* "Disorders of the Developing Nervous System" (W. S. Fields and M. M. Desmond, eds.), pp. 267–318. Thomas, Springfield, Illinois.

218. Thacker, E. J. (1956). The dietary fat level in the nutrition of the rabbit. *J. Nutr.* **58**, 243–249.

219. Thacker, E. J., and Brandt, C. S. (1955). Coprophagy in the rabbit. *J. Nutr.* **55**, 375–385.

220. Thacker, E. J., and Ellis, G. H. (1948). Liver damage and growth in the rabbit. *J. Nutr.* **36**, 579–593.

221. Thomas, L., McCluskey, R. T., Potter, J. L., and Weissman, G. (1960). Comparison of the effects of papain and vitamin A on cartilage. I. The effect in rabbits. *J. Exp. Med.* **111**, 705–718.

222. Thompson, H. V., and Worden, A. N. (1956). "The Rabbit." Collins, London.
223. Thompson, J. F., and Ellis, G. H. (1947). Is cobalt a dietary essential for the rabbit? *J. Nutr.* **34**, 121–127.
224. Tucker, H. F., and Eckstein, H. C. (1937). The effect of supplementary methionine and cystine on the production of fatty livers by diet. *J. Biol. Chem.* **121**, 479–484.
225. Underwood, E. J. (1971). "Trace Elements in Human and Animal Nutrition," 3rd ed. Academic Press, New York.
226. Vallee, B. L. (1959). Biochemistry, physiology and pathology of zinc. *Physiol. Rev.* **39**, 443–490.
227. Valli, V. E. O., McCarter, A., McSherry, B. J., and Robinson, G. A. (1969). Hematopoeisis and epiphyseal growth zones in rabbits with molybenosis. *Amer. J. Vet. Res.* **30**, 435–445.
228. Van Vleet, J. F., Hall, B. V., and Simon, J. (1967). Vitamin E deficiency: A sequential study by means of light and electron microscopy of the alterations occurring in regeneration of skeletal muscle of affected weanling rabbits. *Amer. J. Pathol.* **51**, 815–830.
229. Van Vleet, J. F., Hall, B. V., and Simon, J. (1968). A sequential light and electron microscopic study of skeletal muscle degeneration in weanling rabbits. *Amer. J. Pathol.* **52**, 1067–1079.
230. Vasington, F. D., Reichard, S. M., and Nason, A. (1960). Biochemistry of vitamin E. *Vitam. Horm. (New York)* **18**, 43–87.
231. Wacker, W. E. C. (1969). The biochemistry of magnesium. *Ann. N. Y. Acad. Sci.* **162**, 717–726.
232. Wacker, W. E. C., and Vallee, B. L. (1958). Magnesium metabolism. *N. Eng. J. Med.* **259**, 431–438.
233. Wald, G. (1935). Carotenoids and the visual cycle. *J. Gen. Physiol.* **19**, 351–371.
234. Wald, G. (1960). The visual function of vitamin A. *Vitam. Horm. (New York)* **18**, 417–430.
235. Walser, M. (1967). Magnesium metabolism. *Ergeb. Physiol., Biol. Chem. Exp. Pharmakol.* **59**, 185–296.
236. Warner, R. G. (1962). Nutrient requirements of the laboratory rat. *Nat. Acad. Sci.—Nat. Res. Counc. Publ.* **990**, 51–95.
237. Wasserman, R. H., Corradino, R. A., and Taylor, A. N. (1969). Binding proteins from animals with possible transport function. *J. Gen. Physiol.* **54**, 114s–137s.
238. Wasserman, R. H., and Taylor, A. N. (1966). Vitamin D₃-induced calcium-binding protein in chick intestinal mucosa. *Science* **152**, 791–793.
239. Wasserman, R. H., and Taylor, A. N. (1969). Some aspects of the intestinal absorption of calcium, with special reference to vitamin D. *In* "Mineral Metabolism: An Advanced Treatise" (C. L. Comar and F. Bronner, eds.), Vol. 3, pp. 321–403. Academic Press, New York.
240. Weatherell, J. A., and Weidmann, S. M. (1959). The skeletal changes of chronic experimental fluorosis. *J. Pathol. Bacteriol.* **78**, 233–255.
241. Webster, B., and Chesney, A. M. (1928). Endemic goitre in rabbits. III. Effects of administration of iodine. *Bull. Johns Hopkins Hosp.* **43**, 291–308.
242. Webster, B., Clawson, T. A., and Chesney, A. M. (1928). Endemic goitre. II. Heat production in goitrous and non-goitrous animals. *Bull. Johns Hopkins Hosp.* **43**, 278–290.
243. White, A., Handler, P., and Smith, E. L. (1968). "Principles of Biochemistry," 4th ed. McGraw-Hill, New York.
244. Wigand, G. (1959). Production of hypercholesterolemia and atherosclerosis in rabbits by feeding different fats without supplementary cholesterol. *Acta Med. Scand.* **166**, Suppl. 351, 1–91.
245. Wiss, O., and Gloor, U. (1960). Vitamin A and lipid metabolism. *Vitam. Horm. (New York)* **18**, 485–498.
246. Witzel, E. W., and Hunt, G. M. (1962). The ultrastructure of the choroid plexus in hydrocephalic offspring from vitamin A-deficient rabbits. *J. Neuropathol. Exp. Neurol.* **21**, 250–262.
247. Wolbach, S. B., and Bessey, O. A. (1940). Relative overgrowth of the central nervous system in vitamin A deficiency in young rats. *Science* **91**, 599–600.
248. Wolf, G., and De Luca, L. (1970). Recent studies on some metabolic functions of vitamin A. *In* "The Fat-Soluble Vitamins" (H. F. De Luca and J. W. Suttie, eds.), pp. 257–265. Univ. of Wisconsin Press, Madison.
249. Wolf, G., and Johnson, B. C. (1960). Vitamin A and mucopolysaccharide biosynthesis. *Vitam. Horm. (New York)* **18**, 439–455.
250. Woodward, D. L., and Reed, D. J. (1969). Effect of magnesium deficiency on electrolyte distribution in the rabbit. *Amer. J. Physiol.* **217**, 1477–1482.
251. Woodward, D. L., and Reed, D. J. (1969). Uptake of ^{28}Mg and ^{45}Ca by tissues of magnesium-deficient rabbits. *Amer. J. Physiol.* **217**, 1483–1486.
252. Wooley, J. G. (1947). Niacin deficiency in rabbits and response to tryptophane and to niacin. *Proc. Soc. Exp. Biol. Med.* **65**, 315–317.
253. Wooley, J. G. (1954). Growth of three- to four-week-old rabbits fed purified and stock rations. *J. Nutr.* **52**, 39–50.
254. Wooley, J. G., and Mickelsen, O. (1954). Effect of potassium, sodium or calcium on the growth of young rabbits fed purified diets containing different levels of fat and protein. *J. Nutr.* **52**, 591–600.
255. Wooley, J. G. and Sebrell, W. H. (1945). Niacin (nicotinic acid), an essential growth factor for rabbits fed a purified diet. *J. Nutr.* **29**, 191–199.
256. Yoshida, T., Pleasants, J. R., Reddy, B. S., and Wostmann, B. S. (1968). Efficiency of digestion in germ free and conventional rabbits. *Brit. J. Nutr.* **22**, 723–737.
257. Zachman, R. D. (1967). The stimulation of RNA synthesis *in vivo* and *in vitro* by retinol (vitamin A) in the intestine of vitamin A-deficient rats. *Life Sci.* **6**, 2207–2213.
258. Zalkin, H., and Tappel, A. L. (1960). Studies of the mechanism of vitamin E action. IV. Lipide peroxidation in the vitamin E-deficient rabbit. *Arch. Biochem. Biophys.* **88**, 113–117.
259. Zalkin, H., Tappel, A. L., Caldwell, K. A., Shibko, S., Desai, I. D., and Holliday, T. A. (1962). Increased lysosomal enzymes in muscular dystrophy of vitamin E-deficient rabbits. *J. Biol. Chem.* **237**, 2678–2682.

CHAPTER 17

Metabolic, Traumatic, Mycotic, and Miscellaneous Diseases of Rabbits

Ronald E. Flatt, Steven H. Weisbroth, and Alan L. Kraus

I. INTRODUCTION

This chapter is composed of disease entities that did not conveniently fit into other chapters and which in our opinion did not justify separate chapters. The placment of an entity in this chapter does not imply a lack of importance of the disease. Quite the opposite is true for some of the diseases included here. Many entities were considered for inclusion in this chapter and some, of necessity, had to be omitted.

II. PREGNANCY TOXEMIA

Pregnancy toxemia is a disease that is seldom recognized in domestic rabbits. Pregnant as well as postparturient and pseudopregnant females may be affected (44–46, 59). In a large outbreak of pregnancy toxemia reported by Greene involving 72 cases, 43 cases (60%) were in pregnant does, 15 cases (21%) were in postparturient does, and 14 cases (19%) were in resting or pseudopregnant females (44). The incidence in multiparous does was approximately 4 times that observed in primiparous does. Affected females were most often obese. Greene (44, 45) also found that Dutch, Polish, and English breeds had the highest incidence of pregnancy toxemia. There was also an increased incidence of pregnancy toxemia in carriers of dwarfism in the Polish breed and cretinism in the Dutch breed.

The clinical appearance of pregnancy toxemia varied greatly with a mild, nearly asymptomatic condition at one end of the spectrum and a severe, rapidly fatal disease at the other end. In the pregnant females, the clinical disease was first observed while nest building was occurring. The clinical signs most commonly observed were depression, dyspnea with acetone odor on the breath in severe cases, and decreased urine production. Abortion, incoordination, convulsions, and coma sometimes preceded death (44, 59). In the mild to moderate cases recovery occurred.

Examination of the blood in severe cases revealed a marked elevation of nonprotein nitrogen. Calcium was decreased, phosphate increased, and tests for acetone were positive. Postmortem examination of animals that died or were killed usually revealed obesity, actively secreting mammary glands (even in the pseudopregnant doe), large corpora lutea in the ovaries, and areas of necrosis in the mesenteric fat. The liver, kidneys, and heart were pale and the adrenal glands were small, pale, and often contained cortical adenomas. The thyroid glands were small and pale and the pituitary gland enlarged. Microscopic examination of the liver revealed a severe fatty change with focal necrosis in some cases. There was fatty change in the kidney tubules and in the heart. There were many fatty vacuoles in the adrenal cortex especially within the thickened zona fasciculata. Cortical adenomas were also present in the adrenals. The follicles of the thyroid were lined by low cuboidal epithelium and filled with pale colloid. The pars distalis contained many acidophils and multiple adenomas. The pars intermedia was thickened.

Like pregnancy toxemia in other domestic animals, the cause of this disease in rabbits is complex. It is apparent that many factors may have an influence in the development of this disease including breed, age (only as it relates to having young), sex, obesity, the number of previous pregnancies, etc. Green (45) felt that no definitive conclu-

sion could be reached concerning etiology; however, he felt that there was considerable evidence that this disease was of hypophyseal origin and related to abnormal function of this gland during terminal pregnancy. Greene also reported that the outbreak of pregnancy toxemia in his rabbits was associated with wide-spread reproductive disturbances (abortions, stillbirths, desertion of young, cannibalism, fetal anomalies) and an increased incidence of uterine tumors. He felt that environmental changes might have led to an endocrine imbalance (primarily through the pituitary) and thereby caused the reproductive disturbances, uterine tumors, and pregnancy toxemia. Attempts to treat pregnancy toxemia in rabbits have not been reported.

III. SHOCK DISEASE

A disease characterized by severe hypoglycemia and sudden death has been described in snowshoe hares (38–43). The disease was first observed when showshoe hares were being trapped for the purpose of being banded and released. Many of the trapped rabbits, especially those that were trapped several times, were found dead in the trap with no evidence of infectious disease or traumatic injury. Similar deaths were observed in showshoe hares brought into the laboratory. The rabbits died during shipment and shortly after arrival at the laboratory (43). Approximately 70 to 100% of the rabbits died within two weeks of arrival at the laboratory. The majority died in the first three days. Further observations led to the conclusion that the disease occurred naturally in uncaptured showshoe hares and that trapping and holding these rabbits was precipitating the onset of clinical disease.

Shock disease was observed in both winter and summer, in both young and adults, and in both males and females (40, 41). It was postulated that shock disease was responsible for the cyclical decimation of snowshoe hares that was known to occur approximately every 10 years (39, 40).

The clinical disease was characterized by convulsion, opisthotonos, dyspnea, grinding of teeth, fixation of the eyes, and death in 10 to 15 minutes. During acute clinical episodes the blood sugar of affected rabbits averaged 15 to 17 mg% compared to 115 mg% in clinically normal animals. Additionally, total white blood counts averaged 3000–4000 with 68–87% polymorphonuclear leukocytes and 13–32% lymphocytes. This was interpreted as a leukopenia with an absolute lymphopenia (43). Normal liver glycogen was measured at 5.5% while affected rabbits had 0.02–0.15% liver glycogen (39). Postmortem examination revealed no specific changes; however, the liver was friable and consistently had extensive fatty change (43). The diagnosis of shock disease was based on demonstration of characteristic clinical signs, low blood sugar levels, and typical postmortem findings as well as the absence of other causes of death. It was proposed that liver damage was the fundamental problem in shock disease and as a result carbohydrate metabolism was deranged leading to inadequate liver glycogen storage. During periods of excitement, adrenaline stimulation quickly exhausted the meager glycogen reserve with hypoglycemia and death following shortly (93).

IV. PYLOROSPASM

Pylorospasm with muscular hypertrophy has been reported several times in the domestic rabbit. It has been seen in Holland (55), England (7), and the United States (117). The cases reported in Holland and England involved substantial numbers of postweaning or young adult animals. Although a satisfactory cause had not been established, these cases were presumed to be due to some noxious substance ingested with the food.

Anorexia, gastric and cecal tympany, constipation, and grinding of the teeth as an indication of intestinal pain were the most consistent signs. The course was short, 1–5 days, with death as a frequent but not invariable termination. Radiographic studies following a barium sulfate meal showed barium was unable to pass through the pylorus. This separation was due to spasmodic contraction of the pylorus (55). In establishing the diagnosis it is necessary, at postmortem, to distinguish between congenital pyloric stenosis and pylorospasm with muscular hypertrophy. The former has not been known to occur in rabbits but presumably would be seen in very young rabbits on an individual basis and without the intense, thickened muscular hypertrophy typical of pylorospasm.

A variety of narcotic and sympathomimetic relaxants were used in unsuccessful attempts to treat the condition (55).

V. MUCOID ENTEROPATHY

A. Introduction

Mucoid enteropathy (ME) has many synonyms including bloat, enteritis, enteritis complex, mucoid enteritis, mucoid diarrhea, scours, and hypoamylasemia. It may be defined as a subacute enteric disease with high mortality. A number of enteric disease entities have, in the past, been clinically grouped into an imprecise symptom complex termed "mucoid enteritis." The authors have taken this occasion to change the name of this condition to "mucoid enteropathy." When examined it is clear that inflammation of the intestines (enteritis) is not part of this disease and the name enteritis serves only to mislead.

As research has progressed in the delineation of the components of this complex, it has become increasingly evident that a number of distinct infectious entities may be associated with, or superimposed on, a now more clearly defined condition (ME). Certain factors have tended in combination to blur distinctions between the component entities. Such factors include the common denominators of diarrhea and dehydration as the chief clinical effects and a high rate of mortality as a common terminus. They share additionally the commonality of the 7- to 10-week age group as the most susceptible age for expression of the disease(s). The association of the weaning age group with enteric diseases has been noted by many (see ahead) and the shift from neonatal to postweaning (adult) nutritional substrates has been proposed as having a role in their pathogenesis. Also tending to introduce confusion is the apparent tendency of these diseases to occur either singly or in variable combination which frequently presents to the clinician a symptom composite with a varying range of expression. Hence the common descriptive term, "mucoid enteritis."

Enteric diseases known to occur either separately or in association with ME to form a complex include Tyzzer's disease, salmonellosis, clostridial enterotoxemia, colibacillosis, and intestinal coccidiosis. In the review that follows the approach will be to characterize uncomplicated ME in the light of current understanding and to indicate the confusing impingement of these other diseases where appropriate. Finally, it must be recognized that all of the entities in the ME complex probably have not been brought to light and identified.

B. Etiology

The etiology of ME is unknown but the major pathological features of the disease enable its classification as an enterotoxin-induced secretory diarrhea (113). It shares many clinical and pathological features with (human) cholera and on this basis has been similarly classified (113). It should be mentioned that experimental *Vibrio cholerae* infections in rabbits are histologically typical of cholera, not rabbit ME (78). Etiological understanding of ME has not yielded, however, to a simple interpretation, e.g., the presence of a causative bacterium like *Vibrio cholerae* in human cholera, nor has the disease been experimentally reproduced with bacterial isolates so as to meet Koch's postulates. Indeed, after 15 years of a research program substantially committed to ME at the Fontana (California) U.S.D.A. Station, Hagen wrote (in 1956) that understanding of the cause of ME has not appreciably advanced since its original observation in 1929 (48).

Many theories have been proposed for the etiology of ME, often without supporting data or corroborative experimental results. McCuistion advanced the idea that weanling domestic rabbits were deprived of natural sources of amylase. This lack, he proposed, caused a condition of hypoamylasemia to develop, which was accompanied by inability to digest polysaccharides. This, in turn, was manifested as ME. No experimental evidence accompanied this claim (68, 69). Beneficial effects were claimed for a nutritional amylase supplement (takadiastase) with which to correct the proposed deficiency. This theory was tested, in part, by Arrington and Wallace, who were unable to demonstrate beneficial effects from the feeding of takadiastase, although they had no ME cases in their experimental groups (3). The hypoamlasemia theory of ME has been generally discarded. Its focusing of attention on the nutritional substrates of this critical age group may have some merit, however.

Escherichia coli has figured prominently in serious interpretations of the cause of ME; prinicipally from two lines of evidence. The first of these relates to the frequent isolation of both enteropathogenic and nonenteropathogenic *E. coli* from field cases of ME in *Oryctolagus* rabbits (36, 47, 61, 73, 114) and also in a similar disease of *Sylvilagus* and *Lepus* rabbits (81, 114, 120). The second line of evidence concerns a more clearly defined enterotoxin-induced diarrheal disease (colibacillosis) known to occur naturally in *Oryctolagus* rabbits (35, 116) (see also Chapter 9). Evidence against *E. coli* as the sole etiological agent is substantial however. The disease (ME) has not been reproduced with *E. coli* cultures or endotoxins. *Escherichia coli* serotypes isolated from rabbits with ME are not consistent or the same as those isolated in rabbit colibacillosis (36). The histopathology of colibacillosis in rabbits, and other species (77, 103), is not consistent with what is observed in ME (113). Colibacillosis (and endotoxemia) is classically febrile (36, 77), whereas ME is typified by subnormal temperatures (113). The role, therefore (if any) of *E. coli* in ME is uncertain at the present. It is possible that its presence in ME is explainable on the basis of opportunistic multiplication in a milleu transformed to the favor of gram-negative bacterial species and without symptomological significance.

Enterotoxemia has long been thought of possible etiological significance in ME (60). In an English review of experience with ME complex (having pathological elements of both Tyzzers disease and ME), it was observed that of 22 affected rabbits, *Clostridium welchii* Type A was isolated from 19 (17). A vaccination program utilizing a *C. welchii* toxoid of Type A,B,C, and D was instituted in breeding does and their progeny. Vaccination failed to influence the mortality due to ME and it was concluded that the clostridial isolates exerted no pathogenic effects. Recent work suggests that *C. perfringens* is capable (experimentally) of inducing diarrheal disease in young rabbits (21, 22). In this regard it appears that the significance of clostridia may

parallel that of *E. coli* as a secondary effect of rather than as a primary cause of ME.

In the same colony described above, Cowie-Whitney related the waxing and waning of mortality due to a universal *Eimeria perforans* infection to periodical changes in the type of coccidiostat used for routine purposes (17). This data was plotted as a function of time over a three-year period. When the curve for ME complex mortality was plotted on the same graph, a perfect correlation for mortality was observed between ME complex and coccidiosis. Cowie-Whitney concluded that intestinal coccidiosis was intimately associated with ME and in some way potentiated or triggered the ME complex. Other authors also have observed simultaneous (and minimal) *Eimeria* infections in rabbits with ME (113).

Recent evidence suggests that ME is infectious (113). The authors induced ME in young rabbits by an oral inoculum prepared from freshly ground ME intestines and intestinal contents. Clinically typical ME followed an incubation period of 9–14 days. The diagnosis was verified by clinical laboratory and histopathological criteria. This appears to be the only report of successful induction by inocula from ME-infected rabbits and should stimulate additional advances utilizing this approach.

It is uncertain if the acute enteritis described by Richter and Hendren (88) and also by Yuill and Hanson (120) in cottontail (*Sylvilagus*) rabbits is a comparative analog of ME in *Oryctolagus* rabbits. The pathological descriptions accord more with either Tyzzer's disease or colibacillosis than they do with ME, although copious quantities of colonic mucus were described by Richter and Hendren (88). An intensive search at the ultramicroscopic level failed to reveal viral agents in the affected rabbits (10).

C. Incidence and Epizootiology

Measured by any standard, ME ranks as one of the most important causes of mortality in rabbit production colonies. Reliable statistics on its incidence are scanty. Many affected rabbits may have clinically mild, transient, or clinically silent episodes (113). Records from the Fontana (California) laboratory involving 15,329 live births over a 15-year period indicate an overall mortality of 18% before weaning in which the greatest cause of death was ME (91). Greenham reported 60–70% mortality due to ME in young rabbits in one outbreak over a 2-year period (47). An incidence of 12.4% of rabbit autopsy cases at the Veterinary College was reported in Hungary by Vetesi (114), who also mentioned field outbreaks varying in mortality from 20 to 30% in large colonies to 100% in some small colonies. J. Cowie-Whitney reported an incidence of 11.1% over a 3-year period (17). Little is known about the epidemiological aspects of ME despite the importance of the disease as a major economic factor in commercial production colonies. There is some suggestion that ME occurs as a consequence of intensive husbandry efforts. The disease was not described in detail until 1949 (52). Cases were known as early as 1929 in pioneering production colonies in California (106, 109) but did not become common until the second World War and the advent of large-scale commercial production units fed on scientifically compounded fortified diets (62–64). The disease appears to be unknown in feral or wild *Oryctolagus* rabbits. There is some evidence that the disease was recognized as early as 1943 in England (37, 76).

It is widely distributed in the United States; indeed, informal estimates indicate that there are no conventional rabbitries entirely free of ME. It has been reported from Canada (104), England (70, 80, 83), Hungary (114), Italy (71), and Germany (61, 73). Although some investigators have indicated pronounced seasonal variation in the incidence (17, 91, 113), there is no general agreement on the pattern and it seems safest to conclude that factors unknown at this time appear to cause inexplicable waxing in the case rate. Males and females appear to be equally susceptible. There is general agreement that the incidence follows a normal distribution curve as a function of age with the great majority of cases occurring in rabbits 7–10 weeks of age although occasional cases may occur as early as 2 weeks, and as late as (or later than) 20 weeks of age (17, 48, 52, 62–64). There is some evidence that the incidence is higher in the progeny of a doe's first litter than in her subsequent litters, and also that the progeny of certain does may be more at risk than those of other does (91). There is at present no explanation for maternal factors that appear to influence the incidence.

D. Clinical Findings

The clinical features of ME were first comprehensively described by Hurt (52) in 1949. A more recent description has been published as well (113). Although generally regarded as an acute disease with a course of 2–3 days, careful observations indicate a course of 7–8 days (subacute) as more accurate (17, 48).

Clinical signs include anorexia, polydipsia, and subnormal temperature (99–102°F in ME compared to the normal 102–104°F). The animal appears depressed, the posture is crouched, and the hair coat is roughened. Weight loss (from diarrhea) is rapid and the rabbit appears thin. The abdomen is bloated, however, due to gas and fluid-filled intestines. The distended portions of the alimentary tract include stomach, duodenum, jejunum, ileum, cecum, and colon. The perineum is usually stained with mucus or light yellow to brown, liquid feces. Paradoxically, hard dry feces may be excreted simultaneously with copious quantities of gelatinous mucus and both may be found in the dropping pan. Squinting of the eyes, cold extremities, and grinding of the teeth are occasionally observed.

Clinical laboratory findings are in the main attributable to dehydration: increased packed cell volume (hematocrit), erythrocyte count, and hemoglobin concentration (113). The sedimentation rate is increased. It is known through a series of intensive investigations by Evans and his colleagues (26–29) that the erythrocytes of ME-infected rabbits are both agglutinated (polyagglutinable) and hemolyzed by sera from normal rabbits. The serum from ME rabbits with abnormal erythrocytes lacked this activity for their own (or other) erythrocytes. Evans proposed that bacterial products (enzymes) entered the splanchnic circulation and in some way so modified the erythrocytic stroma as to make them agglutinable by normal sera. It was proposed that autologous sera of ME-affected rabbits (*in vivo*) rapidly became depleted (or absorbed) of agglutinating and hemolyzing activity. Transfused, ^{51}Cr-labeled erythrocytes were rapidly (50% in 24 hours) cleared from the circulation of normal rabbits. Transfusions of 8–10 ml of ME-affected erythrocytes were sufficient to induce fatal hemolytic reactions. The significance of this finding, as it relates to the pathogenesis of ME, is uncertain. Actual systemic invasion by bacteria could not be demonstrated. The more recent interpretation of the etiology of ME as an enterotoxin-induced diarrhea (113) is strengthened, however, by the observation of systemic effects related to the absorbtion from the gut of bacterial products.

Other laboratory findings include elevated total leukocyte counts with increased numbers of neutrophils at the expense of lymphocytes. Values for serum lipase, glucose, albumin and globulins, phosphorus, and blood urea nitrogen were elevated in ME-affected rabbits (113). Serum sodium, chloride, calcium, and potassium were decreased. The concentration of sodium in jejunal fluid was equivalent to that of serum (113).

E. Gross Pathology

Grossly discernible lesions in ME are limited to the alimentary tract, mesenteric lymph nodes, and gallbladder. In most cases the stomach is disintended with fluid and gas and the mucosa lined with tenacious mucus. The duodenum is generally filled with gas and watery, bile-stained fluids. The jejunal loops are characteristically distended with translucent watery fluid which is recognizable through the thin, unopened intestinal wall. The ileum is occasionally distended like the jejunum but more often is not and contains pasty to partially inspisated contents. In some cases the cecum is impacted with dried contents and accumulations of gas. In rabbits with cecal impactions the proximal sacculated colon may also contain dried contents for the first 1–2 cm. The sacculated colon is distended with a clear, gelatinous mucus as a characteristic finding in ME. The consistency of the mucus varies from that of a loose gel to that of a transparent cast. The nonsacculated (lower or distal) colon is usually filled with gelatinous contents and the rectum may be similarly filled with mucus or empty. Distention of the gallbladder is also a common finding.

F. Microscopic Pathology

The esophagus and stomach appear to be devoid of microscopic lesions (113). Hyperplasia of goblet cells in the doudenum, jejunum, and ileum is a characteristic finding in ME, however, goblet cell hyperplasia is most apparent in sections of ileum where only a moderate number of goblet cells are normally present and an increase more readily recognized. Apart from these findings the small intestine is histologically normal and without inflammatory changes.

Goblet cell hyperplasia may be present in the cecum, appendix, sacculus rotundus, and sacculated and nonsacculated colon but its significance (or occurrence) is difficult to assess because of the normal rich abundance of goblet cells in these tissues. Goblet cell hyperplasia may occasionally be observed in the gallbladder, bile ducts, and pancreatic ducts and also in the epithelium of the trachae (113).

Of interest to comparative pathologists is the role that rabbit ME may play as a model with characteristics of human cystic fibrosis (57a, 113). Cystic fibrosis is now generally recognized to be a disease of defective electrolyte transport and hypersecretion of mucus. Excessive electrolyte loss occurs in sweat and salivary and lacrimal hypersecretions and excessive mucus is produced by intestines, pancreatic and bile ducts, and respiratory epithelium. The disease is also called "mucoviscidosis" in recognition of this primary clinical effect.

G. Diagnosis

Clinical diagnosis of ME is established on the basis of the characteristic signs of dehydration, gelatinous diarrhea, bloating of the abdomen, succussion splash, and associated clinical laboratory findings (113). The diagnosis is further strengthened by the typical postmortem findings of gas- and fluid-filled stomach and small intestines, cecal impaction, and gelatinous accumulations in the colon. Diagnosis is confirmed by microscopic examination of the ileum, cecum, and colon. Demonstration of goblet cell hyperplasia and depletion of acidic mucus in the colonic gland with plugging and dilatation of gland lumens with mucus are definitive findings (113). The observation of mucosal necrosis

and sloughing, submucosal edema, and/or inflammatory cell infiltration indicates complication with other enteric entities.

H. Treatment and Prevention

Recommended treatment for ME appears at the present to be generally palliative and aimed at reducing secondary bacterial complications, rather than specific. Treatments such as withholding feed and water (112), feeding of tea or tannins (65), feeding gypsum plasterboard, changing the diet, and feeling of amylase supplements (68, 69) are concluded to be without foundation, not recommended, and, as in the case of withholding water, may be deleterious (12).

A number of antibiotic, amino acid, and vitamin supplements have been subjected to field trials for analysis of their effect of ME mortality at the Fontana station (11, 13, 14, 48) and elsewhere (47). Results have been inconsistent and often disappointing. The general experience has been that antibiotic supplements have been effective in reducing ME mortality but ineffective in preventing the disease or influencing the morbidity. Diets containing 10 gm aureomycin and 9 mg vitamin B_{12} per ton have been effective in reducing the expected mortality in treated groups to 25–75% of that experienced in untreated controls. The Fontana group has stressed the importance of feeding antibiotic-treated diets for the full 6 weeks between the second and eighth weeks of life for the most consistent results. The clinical efficacy of individual fluid therapy designed to replace fluid and electrolytes lost by diarrhea are unknown at present but suggested here as a rational approach to treatment. The general uncertainty surrounding the etiology and epizootiology of this disease preclude control recommendations at this time.

VI. SUPERFICIAL MYCOSES

Dermatophytoses (ringworm, favus) are uncommonly encountered in the domestic rabbit. They are more frequent in pet or backyard rabbitries where the poor husbandry conditions favoring dermatophytosis are more prevalent and are comparatively rare under laboratory conditions or in commercial rabbitries with satisfactory hygienic standards. They are most often encountered as individual infections of a sporadic nature although epizootics have been reported in rabbit colonies (4, 16, 49). Most reports indicate that younger rabbits are more susceptible than adults.

Ringworm is caused by infection of the epidermis and such adnexal structures as hair follicles and hair shafts with pathogenic dermatophytic fungi. *Trichophyton mentagrophytes* is the dermatophyte most commonly encountered in *Oryctolagus* ringworm infections both in the United States and abroad (2, 4, 33, 49, 50, 54, 75, 115). In fact, infection with the other 4 reported species, *Microsporum gypseum* (23, 50, 118), *M. canis* (23), *M. audouini* (86), and *T. schoenleini* (20, 23, 56, 97) are so infrequent as to be regarded as rare. Inasmuch as *T. schoenleini* and *M. audouini* are now generally regarded as almost exclusively anthropophilic dermatophytes (33, 87), and the reports of these infections (in rabbits) appeared in the older literature when synonymy was rife, their actual status as rabbit pathogens is doubtful. Georg, for example, has concluded (33) that the only diagnosis of *M. audouini* was made in error and actually should have been classified as a *T. mentagrophytes* infection. The differential features of these 5 reported dermatophytic species are summarized in Table I.

The natural occurrence of dermatomycosis in wild lagomorph species has been the subject of several recent surveys (33, 74) and case reports. In these species, also, dermatomycosis is infrequent and usually sporadic. Although several unusual dermatophytes, e.g., *Dermatophilus congolensis* (98) and *M. cookie* (33) have been reported, the most frequent isolate, as in domestic rabbits, appears to be *T. mentagrophytes* (1).

Dermatophytic infections usually arise on or about the head in rabbits. The lesions are pruritic and spread to the paws and other areas of the body occurs as a secondary

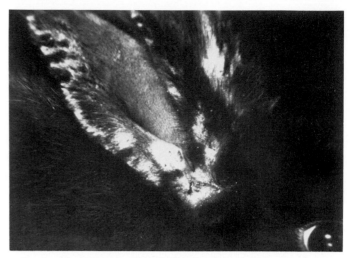

Fig. 1. Black-furred rabbit with patchy alopecia and crusty dermatophytic lesions about the head and ears. (Photograph courtesy of *J. Amer. Vet. Med. Ass.*)

phenomenon. The lesions are generally, but not uniformly, circular and characterized by an underlying inflammatory reaction. The surface of the lesion has a crusty appearance with patchy alopecia (Fig. 1). Histologic sections reveal hyperkeratosis, acanthosis, and diffuse infiltration of the dermis with polymorphonuclear leukocytes. Lymphocytes

TABLE I

DIFFERENTIAL FEATURES OF 5 DERMATOPHYTIC SPECIES REPORTED PATHOGENIC FOR RABBITS[a]

Dermatophyte	Wood's light fluorescence	Microscopic examination of KOH mounts		Colony characteristics	Microscopic cultural features
		Skin scapings	Infected hairs		
Trichophyton mentagrophytes (Syn.: *T. gypseum, T. quinckeanum*)	Nonfluorescent	Mycelia; chains of arthrospores	Large spore (3 to 5 μm) ectothrix; mycelia within hair	Colony usually flat, granular; may be yellow-orange on underside of colony	Macroconidia rare; microconidia abundant as tightly wound spirals, nodular bodies
Trichophyton shoenleini (Syn.: *Achorion Schoenleini*)	Nonfluorescent (occasionally whitish)	Masses of irregular mycelia and athrospores	Hair follicle invasion not reported in animals	Growth slow; surface irregularly heaped, glabrous	Macroconidia absent; chlamydospores numerous: hyphae clubbed at ends (chandeliers)
Microsporum canis (Syn.: *M. lanosum, M. felineum, M. equinum*)	Bright apple-green; fluorescence of infected hairs	Mycelia; chains of arthrospores	Small spore (2 to 3 μm) ectothrix; sheath of spores in mosaic; mycelia within hair parallel to length	Growth rapid, surface cottony; undersurface with yellow-orange pigment	Macroconidia numerous; 6- to 9-celled, often with knob at end
Microsporum gypseum (Syn.: *Achorion gypseum*	Nonfluorescent	Mycelia; chains and masses of arthrospores	Large spore (5 to 8 μm) ectothrix; mycelia in hair parallel to length	Growth rapid, surface powdery; pleomorphic fluffy border; no pigmentation	Macroconidia abundant, 4- to 6-celled, and shorter than those of *M. canis*
Microsporum audouini	Bright apple-green; fluorescence of infected hairs	Mycelia: chains of arthrospores	Small spore (2 to 3 μm) ectothrix; sheath of arthrospores in mosaic pattern; mycelia in hair parallel to length	Growth slow; surface velvety; undersurface nonpigmented or salmon-colored	Mycelia usually sterile: micro- and macroconidia usually absent; no distinctive features

[a]Table courtesy of *J. Amer. Vet. Med. Ass.*

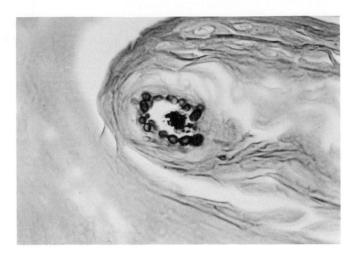

Fig. 2. Photomicrograph of *M. gypseum*-infected rabbit hair in cross section. Note circumferential (ectothrix) arrangement of arthrospores. Gridley fungus stain. (Photograph courtesy of *J. Amer. Vet. Med. Ass.*)

and plasma cells are also located in the dermis and about the hair follicles. Abscessation of hair follicles by secondarily invading bacteria is common. Mycotic elements are not observed with certainty in hematoxylin and eosin stained preparations. Sections prepared with Gridley fungus stain or by the periodic acid-Schiff (PAS) reaction reveal abundant forms predominately associated with hair shafts and hair follicles (Figs. 2 and 3).

Diagnosis of dermatophytosis must involve differentiation from other clinical entities giving a general clinical appearance of crusty alopecia of the head and ears. The clinical entities and differential qualities involved in preliminary clinical diagnosis of dermatophytosis in the rabbit are summarized in Table II. Diagnosis of skin lesions in rab-

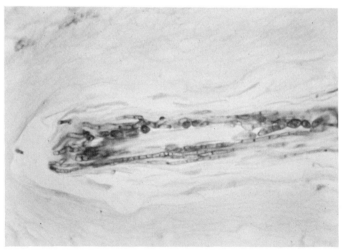

Fig. 3. Photomicrograph of *M. gypseum*-infected rabbit hair in longitudinal section. Note mycelium growth parallel to long axis of hair shaft. Periodic acid-Schiff stain. (Photograph courtesy of *J. Amer. Vet. Med. Ass.*)

bits begins with evaluation of a carefully taken clinical history because some lesions can be given a provisional clinical diagnosis on that basis. Many incidental lesions, e.g., genetic hairlessness, trauma, hair clipping or depilatory hair loss, and fur pulling, can in this way be identified. Observation of type and distribution of lesions is important because lesions associated with pasteurellosis, conjunctivitis, abscesses, ear mites, moist dermatitis, and molt generally form typical clinical entities.

After evaluation of the clinical history and observation of the lesions, a skin scraping is performed and a diagnosis is made. The scraping should come from the periphery of the lesion and be deep enough to draw minimal capillary bleeding. Scrapings are mounted on a glass slide under a coverslip in 10% KOH and observed microscopically. Microscopic evaluation should be performed under reduced illumination. The slide should be examined immediately and after 2 sequential 30-minute intervals. The entire area under the coverslip should be examined. Species identification of parasitic mites can be determined at this level of diagnostic investigation. Dermatophytosis can be diagnosed provisionally by the finding of fungal forms (mycelia or arthrospores, or both) in the macerated cornified epithelium and hairs, but etiological diagnosis must be deferred until more information has been gained. The finding of fungal forms in the skin scraping should be followed up with an evaluation of the lesion under Wood's light (ultraviolet) illumination and culture of scrapings on a suitable medium for isolation of dermatophytes. Reliance cannot be placed on culture alone to substantiate a diagnosis of dermatophytosis because many rodents (and cats) are known to act as asymptomatic and lesionless carriers of dermatophytic fungi (31, 32, 34, 92). Substantiation of the diagnosis must include evidence of epithelial invasion by the fungus such as fungal forms in skin scrapings or histological sections.

Treatment of dermatophytosis involves consideration of its zoonotic hazard (see Chapter 18). Active cases are potentially infectious not only for man, but for other rabbits and other animal species as well. Diagnostically confirmed cases should be promptly isolated from other animals. Untreated cases should be euthanized and the cage and its immediate environment disinfected. Rabbit ringworm caused under experimental conditions by *T. mentagrophytes*, *M. gypseum*, and *M. canis* has been shown to respond well to orally administered griseofulvin (49). In at least one instance an outbreak in a rabbit colony was successfully treated with griseofulvin also (49). Individual rabbits should receive griseofulvin daily at the rate of 25 mg/kg administered in aqueous suspension by gastric intubation. Treatment should be continued for 14 days. Colonies may be mass treated by providing a ration containing 0.375 gm of powdered griseofulvin per pound of feed. The feed should be available for 14 days (49).

TABLE II

CLINICAL ENTITIES AND DIFFERENTIAL QUANLITIES INVOLVED IN PRELIMINARY DIAGNOSIS OF
DERMATOSIS OR ALOPECIA OR BOTH IN DOMESTIC (*Oryctolagus*) RABBITS

Clinical entity	Cause	Differential clinical findings
Snuffles conjunctivitis	Bacterial, usually *Pasteurella multocida*	Scalding of skin with loss of hair from spilling of exudates from medial canthus of the eye. Usually bilateral, clinically typical. Skin scrapings $(-)^a$ for parasitic or fungal forms
Ear mites	*Psoroptes cuniculi*	Ear mite infections, though generally confined to the external auditory canal and skin of the external ear, frequently extend to the skin of the head circumferentially related to the ears. Skin scrapings reveal psoroptic mites
Sarcoptic mange	*Sarcoptes scabiei*	Frequently arises on the head and spreads to the trunk. Deep skin scrapings reveal sarcoptic mites.
Slobbers	Bacterial, various opportunistic forms	Inflammation of skin with alopecia, generally confined to chin and ventral part of neck. Believed to result from chronic wetness of the area with secondary bacterial dermatosis. Skin scrapings (−) for parasitic or fungal forms
Notoedric mange	*Notoedres minor*	Usually encountered as body mange. Deep skin scrapings reveal notoedric mites
Molt	Idiopathic	Occasional source of partial or general alopecia, noninflammatory. Skin appears normal, may begin on the head. Skin scrapings (−) for parasitic or fungal forms
Fur pulling	Breeding females	Nest-building females pull hair from self, after several litters appear to have patchy hair loss. Hairless areas noninflammatory, skin scrapings (−) for parasitic or fungal forms.
Genetic hairlessness	At least several genes involved with various patterns	Noninfrequent source of hairlessness, may be partial or complete, noninflammatory, skin scrapings (−) for parasitic or fungal forms.
Dermatophytosis	*Trichophyton* and *Microsporum* spp.	Generally begins on skin of head or ears; lesions inflammatory, crusty, hairless. Healing and rehairing begins in the lesion center, skin scrapings (+) for fungal forms in epithelial cells and infected hairs.

$^a(-)$ = indication for negative; $(+)$ = indication for positive results. (Table courtesy of *J. Amer. Vet. Med. Ass.*)

VII. DEEP MYCOSES

Rabbit species of all genera appear to be remarkably resistant to deep or systemic mycotic infections (96). A review of the literature in fact reveals only two naturally occurring infections of significance; aspergillosis of the domestic rabbit and adiaspiromycosis of several wild lagomorph species. Several isolated case reports of other infections are in the reference list (5, 99, 108).

Aspergillosis is of only historical significance as a disease of *Oryctolagus* rabbits, particularly as standards of husbandry have advanced in recent times. It is well documented in the older European literature (7, 16 20 59, 97) but has apparently not been observed (or reported) in the Western Hemisphere. The disease is principally associated with *Aspergillus fumigatus*, but *A. niger* occasionally has been isolated as well. Tissue distribution of lesions appears to be confined to the lungs. The course of the disease is chronic with gradual onset of cachexia and dyspnea as the chief clinical findings. Postmortem examination reveals a focal caseating necrosis that must be differentiated from tuberculosis and necrobacillosis (Schmorl's disease). Diagnosis is established by histopathological demonstration of the organism and confirmed by cultural isolation of *Asper-*

gillus. The disease, in rabbits, is similar in most pertinent details to the avain infection, to which the reader is referred for a more thorough presentation (15).

Adiaspiromycosis (haplomycosis) is caused by fungi of the genus *Emmonsia (Haplosporangium)*. Two species, *E. parva* and *E. crescens*, are recognized as important in wild animal infections and have a worldwide distribution. It has been described once as a natural infection in a domestic rabbit (24). A third, and newly described species, *E. brasiliensis* has been reported as a human pathogen in Brazil. *Emmonsia* sp. have been encountered principally in the western United States as pathogens of wild rodents, carnivores, and lagomorphs of the genera *Sylvilagus* and *Ochotona*.

Cultures of *Emmonsia* grown on artificial media grow as typical fungi with profuse spore production. Spores of two sizes, 10–40 μm and 250–500 μm, are produced depending on the species of the isolate, *E. parva* and *E. crescens*, respectively. The large spores (adiaspores) correspond to the form (or stage) found in infected tissues.

Naturally occurring infections appear to be generally benign and limited in tissue distribution to the lungs (53). The degree of infection is dose related and overwhelming experimental infections eventually kill the host. It is assumed that the natural route of infection is by inhalation. There is no evidence of mycelial growth or reproduction in infected tissues but radial expansion of the adiaspores or spherules is commonly observed. The inflammatory reaction evoked by the presence of spherules is mild and granulomatous. It is assumed that infected hosts remain persistently infected for the remainder of the life span. Diagnosis is established on histopathological criteria (demonstration of typical spherules) but should be confirmed by cultural isolation. There are no known methods of prevention or treatment. The reader is referred to an excellent comprehensive review of adiaspiromycosis for a more detailed presentation (53).

III. MASTITIS

Inflammation of the mammary glands of rabbits, also known as blue-breast, occurs occasionally in lactating does and rarely in does in pseudocyesis. Does that are heavy milk producers are said to be predisposed to mastitis. Other predisposing factors include poor sanitation and injury to the breasts. Affected does have fever (104–105°F), decreased appetite, increased thirst, and they become depressed (59, 66, 67, 107, 112). Septicemia and death may follow. Mammary glands involved in the inflammatory process are swollen, hot, and painful. The skin over affected glands is initially pink to red but becomes bluish-purple; therefore, the name "blue-breast." The microscopic changes in the affected mammary glands are similar to acute bacterial mastitis in other species and include congestion, hemorrhage, and dense infiltration of the mammary tissue with polymorphonuclear leukocytes (Fig. 4). The inflammation quickly spreads from breast to breast until all breasts are affected. Mastitis in rabbits usually occurs sporadically but it also may be quite contagious, spreading from doe to doe (66, 100, 107).

Staphylococcus sp. is most often incriminated as the cause of mastitis in rabbits, but *Streptococcus* sp is also thought to be involved with some regularity, (59, 66, 100, 107, 112). Treatment consists of isolating the affected doe and treating her with 50,000–100,000 units of penicillin intramuscularly twice daily for 3 to 5 days. Foster nursing the young of an affected doe on another doe is not recommended because of the probability of spreading the disease to the healthy doe.

IX. ULCERATIVE PODODERMATITIS
(SORE HOCKS)

Traumatic ulcerative dermatitis of the planter surface of the metatarsal region, and less commonly the volar

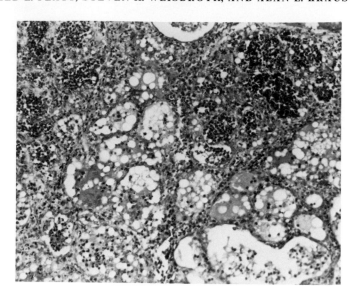

Fig. 4. Photomicrograph of rabbit mammary gland densely infiltrated with polymorphonuclear leukocytes. (Tissue sections provided courtesy of Veterinary Pathology Division, Armed Forces Institute of Pathology.)

surface of the metacarpal-phalangeal region, is commonly referred to as sore hocks. The name "sore hocks" is inaccurate in that the affected parts are not restricted to the hindlegs and the area of the hindlegs commonly affected is not the hock but the metatarsal region.

The cause of this disease is pressure necrosis of the skin, usually resulting from the bearing of heavy body weight on wire floors of cages (80). Immature rabbits and rabbits of smaller breeds are seldom affected (107, 112). The presence of wet cage floors, especially from urine-soaked feces, has been suggested as a cause or predisposing cause of ulcerative pododermatitis (80, 95, 107, 110, 112), but we have observed the disease in rabbits maintained in wire-floored cages which were washed regularly and where fecal material was not allowed to accumulate. Excessive nervousness in rabbits causing them to stamp their feet frequently also may predispose to ulcerative pododermatitis. The form of wire crimp in the floor is also known to be important in influencing development of pododermatitis.

The lesions consist of circumscribed ulcerated areas in the skin covered by a dry crusty scab (Fig. 5). The lesions vary in size but are quite consistently located on the plantar surface of the metatarsal region and occasionally on the volar metacarpal-phalangeal region. Secondary bacterial infection of the dermis adjacent to the ulcerated epithelium may occur and abscesses sometimes form under the crusty debris covering the ulcerated area. *Staphylococcus aureus* is a common cause of these abscesses. Affected rabbits may appear healthy in every regard except for the lesions described; however, anorexia, weight loss, humping of the back, stilted movements, and death have been described.

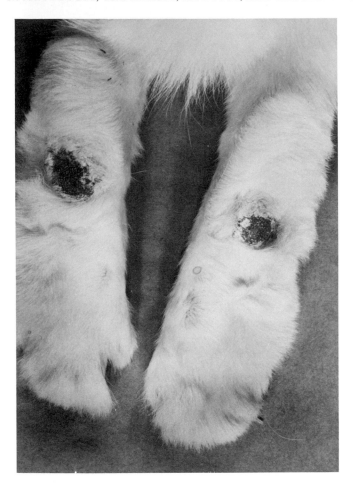

Fig. 5. Plantar surface of the metatarsal area of a rabbit with ulcerative pododermatitis.

X. TRAUMATIC VERTEBRAL FRACTURE (TRAUMATIC VERTEBRAL DISLOCATION, PARALYSIS OF THE HINDQUARTERS, BROKEN BACK)

Posterior paralysis as a consequence of mechanical damage to the spinal cord following vertebral fracture or dislocation occurs quite commonly in the domestic rabbit (90, 105). The injury may be referred to as broken back, paralysis of the hindquarters, traumatic vertebral dislocation, etc. The onset is ordinarily sudden and usually coincides with struggling or inadequate support of the hindquarters when handled. Frequently the incident is not observed (or recognized) and the rabbit is found paralyzed in the cage. The heavy hindquarters twist, when unsupported, about the lumbosacral junction which acts as a fulcrum in applying leverage to the vertebral column. The L_7 vertebral body or its caudal articular processes are the most frequent sites of fractures. Fractures of these areas are more common than dislocation.

Diagnosis may be established on clinical criteria and confirmed by radiography. (Fig. 6) Clinical criteria include complete or partial motor paralysis with loss of skin sensation. If the cord is completely severed (paraplegia), motor control of the anal sphincter and urinary bladder is lost as well. If recognition of the condition is delayed by several days, overfilling of the bladder, elevated blood urea nitrogen, signs of uremia, decubitus ulcers, and staining of the perineum with liquid feces are common findings. Such cases should be recognized as irremediable and humanely euthanized.

Frequently however, the fractured bone ends remain fixed in position and the signs of spinal cord injury are related to transient swelling (edema) at the site of trauma. Over the course of 1–2 weeks resolution of spinal shock restores function to some degree (paresis) which may be compatible with life. Euthanasia need not be recommended in cases where control over the urinary bladder and anal sphincter remains unimpaired and where partial to substantially complete motor recovery follows an observation period of 2 weeks. Frequently an episode of spinal fracture with substantial recovery may be recognized months or years later as overflexion of the sacrum.

XI. MOIST DERMATITIS

Moist dermatitis is a chronic progressive disease of the skin that occurs with sporadic distribution (82, 93, 107, 112). The disease is also known as slobbers, ptyalism, and wet dewlap. It is thought that the dermatosis resulting from chronic wetness of the skin of the chin, intermandibular

Also, the frequent shifting of the body weight from one leg to another or from hind feet to front feet has been observed (107, 112). The lesions on the front feet are thought to be a result of the shifting of weight from sore hindfeet to the front feet. Death of 80 to 90% of the untreated rabbits was observed by one author (110).

The use of solid-bottom cages with frequently changed soft bedding has been used to treat ulcerative pododermatitis. The clean, dry bedding minimizes trauma to the affected parts and allows healing to take place. The use of a resting board placed in the wire-bottom cage has also been recommended to reduce the trauma to the affected parts. Resting boards tend to prevent feces falling through the cage floor and thereby increase the problems of providing adequate sanitation. Topical use of zinc and iodine ointments and 0.2% solutions of aluminum acetate has been described, and antibiotics are occasionally used. If abscesses are present they should be drained and systemic antibiotics used.

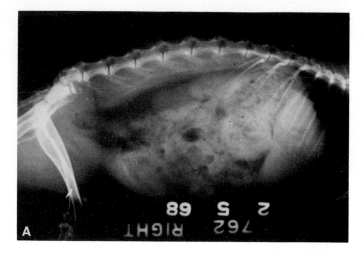

Fig. 6. Radiograph (A) of rabbit with fracture of L_7 vertebral body. Note override of fractured ends indicating probably section of spinal cord. The consequences of cord section, (B), include urinary and fecal incontinence and flaccid paralysis.

space, and cervical ventrum predisposes to secondary bacterial invasion of these tissues. Three general conditions are believed to play a role in chronic wetting of the skin in this area: (a) drooling (ptyalism) from chronic dental diseases, especially malocclusion, (b) continual wetting of the skin during drinking from crocks or water pans, and (c) poor husbandry conditions, especially cold, damp contact bedding. Malocclusion may involve not only the incisors but also the premolars and molars leading to ulceration of the tongue (82).

Any of a variety of bacteria may invade the skin which is made susceptible from chronic wetness. Some have suggested that Schmorl's disease (necrobacillosis) represents a special variant of moist dermatitis with more extensive invasion and tissue destruction (see Chapter 9). As the disease progresses, the gross pathological features con-

sist of inflammation, partial hair loss (alopecia), and ulceration and necrosis of the affected skin. Microscopic examination of these tissues demonstrates irregular patches of ulceration, coagulative necrosis, abscessation, and colonization of the dermal layers with bacteria. The process is surrounded peripherally and in the underlying subcutaneous tissues with cellular elements of both acute and chronic inflammation. Lymphogenous and hematogenous extension may occur, especially to the ventral cervical and pulmonary lymph nodes.

Unless the bacterial invasion has become systemic, the prognosis for individual cases is generally good, if treated. Treatment consists of rectifying the cause of chronic wetting, i.e., clean dry bedding (or suspended wire floor), bottle watering, clipping maloccluded incisors, etc. The hair should be clipped from the affected area and the skin washed well with surgical soap and water. A topical broad spectrum antibiotic (e.g., chloramphenicol ointment) should be applied daily for a 10- to 14-day course. If the infection appears extensive, a course of systemic antibiotics should be administered as well.

XII. HEAT PROSTRATION

The unusual sensitivity of rabbits to heat is probably related to the large ratio of surface area to body weight and the insulating quality of the rabbit's fur. In the summer months, losses due to heat prostration may exceed losses due to other disease in nonenvironmentally controlled quarters.

Clinical signs may include rapid respiration, cyanosis, prostration, and blood-tinged fluid from the nose and mouth. Does near kindling are reported to be most susceptible to this condition (51).

During bouts of especially hot weather, nonenvironmentally controlled quarters may be cooled by such means as water sprays, foggers, and fans. Affected animals may be cooled by placing wet burlap over them (51) or by immersion in a cold water bath until rectal temperature approaches the normal range (72).

Experimentally, Rathore has shown that New Zealand White bucks exposed to environmental temperature of 36.1°C (96.8°F) at 45% humidity for 1 or 2 days were not as fertile as control (unheated) bucks (84) and that abnormal spermatozoa were present in the ejaculate of the heated animals (85).

XIII. HAIR BALLS

Rabbits will occasionally ingest hair by licking or pulling their own fur or that of a pen mate. The hair may accumul-

Fig. 7. Hair ball from the stomach of a New Zealand White rabbit found incidentally at postmortem examination.

ate in the stomach and form round to oval masses called trichobezoars or hairballs (Fig. 7). Generally these hair balls do not cause an obstruction, and they are usually found incidentially during postmortem examination. They may, however, occasionally cause partial or complete obstruction of the digestive tract. In such cases the affected rabbit stops eating and loses weight (7, 51). The mass in the stomach may be palpable in the live rabbit. It may be treated by giving mineral oil via stomach tube.

The reason for the ingestion of the hair is not known, although the lack of sufficient roughage in the diet and boredom have been suggested.

XIV. PLANT TOXICOSIS

The poisoning of rabbits by toxic plants has been reported infrequently in comparison to large farm livestock. This may be because few poisonings occur or because the economic loss is far less in rabbits and professional help is infrequently sought when losses occur. Most reported poisonings occurred before commercial feed was widely used.

The wooly pod or broad-leafed milkweed, *Asclepias eriocarpa* (Fig. 8), has been reported to cause illness in domestic and cottontail rabbits in the Pacific Southwest (107, 111). The leaves and stems, whether fresh or dried, were capable of causing disease. The "milk" or latex was also toxic. Affected rabbits clinically exhibited varying degrees of weakness or paralysis of both front legs and hindlegs and the neck muscles. The affected rabbit may rest its

Fig. 8. *Asclepias eriocarpa*, wooly-pod milkweed. (Photo courtesy of Dr. K. W. Hagen, Jr.)

Fig. 9. Rabbit with wooly-pod milkweed poisoning resting its head on the floor. (Photo courtesy of Dr. K. W. Hagen, Jr.)

head on the floor (Fig. 9) or be unable to lift its head from the floor. Because of the peculiar position of the head the disease has been refused to "head down disease." In addition, there may be drooling, rough hair coat, subnormal temperature, and tarlike feces. At postmortem examination focal hemorrhages were observed on many organs (107, 111). If the milkweed is removed and the rabbit is assisted in eating and drinking, recovery may occur in a few days (107).

Impaired reproductive performance in female rabbits fed ladino clover has been reported (119). The report indicated that complete infertility occurred when this clover was fed both prior to and after breeding. The cause of infertility was not determined but failure of ovulation and implantation was observed. Although the high estrogen content of the clover was considered as a possible cause, the author concluded that the infertility might be due to some other factor (119).

Blount (7) briefly mentions fatal poisonings of rabbits by feeding comfrey (*Symphytum officinale*) and fatal poisoning of rabbits as a result of feeding a mixture of dock, buttercup, and wild parsnip.

XV. BAND KERATOPATHY

Band keratopathy, a degenerative disease of the cornea of man, was first described from a colony of rabbits used to produce experimental uveitis (25). In addition to the uveitis which was produced intentionally, there was an accidental hypervitaminosis D as a result of an error in feed formulation. Band keratopathy was observed only in those rabbits which had both uveitis and hypervitaminosis D. Rabbits with uveitis alone did not develop band keratopathy and likewise it was found that rabbits with hypervitaminosis D but no uveitis did not develop band keratopathy. The development of the corneal disease, therefore, appeared to depend on the concomitant hypercalcemia and uveitis. Widespread deposits of calcium were also present in affected rabbits and this was interpreted as metastatic calcification related to the hypervitaminosis D. The kidneys, arteries, trachea, and smooth muscle in the small intestine contained mineralized deposits. Mineralization in the superficial portion of the cornea was also produced experimentally by laser irradiation of the cornea (30). The cornea became opaque after 12 days and calcareous granules (hydroxyapatite) were observed beneath the corneal epithelium. Doughman *et al.* (19) described 2 models for the production of band keratopathy. The first is like that described above utilizing a combination of immunological uveitis following vitamin D intoxication. Band keratopathy resulted 100% of the time under these conditions. The authors thought this model was best explained by the theory of calciphylaxis. The second model involved the development of band keratopathy in 35% of the eyes injected with polyethylene sulfonate (19).

XVI. PULMONARY EMPHYSEMA

Naturally occurring pulmonary emphysema was observed in rabbits during the course of a study utilizing rabbits as an experimental model (101). As a result of this observation, the naturally occurring disease was studied in 155 rabbits of various ages and breeds. A generalized type of the disease was described in which young rabbits were spared but more than 50% of rabbits over 1½ years old were affected (101). In addition to this generalized type, a localized type was described occurring along the margins of lung lobes. The localized type was considered to be the result of an inflammatory process. Chronic interstitial pneumonia was sometimes associated with both types of the emphysema but the causative organism was not identified by the author (101). It seems very likely that the interstitial pneumonia was the result of an earlier bout of enzootic pneumonia cause by *Pasteurella multocida* (see Chapter 9). The lesions observed in the naturally occurring pulmonary emphysema consisted of fenestrations in the alveolar walls. The fenestrations were considered to be progressive in nature and eventual destruction of the alveolar wall resulted. The lesions were thought to be an atrophy of the alveolar wall. Chronic pulmonary emphysema was reproduced experimentally in rabbits by repeatedly giving inert particulate material intravenously. The lesions produced were thought to be identical with those occurring naturally and these finding supported the concept that ischemic atrophy of alveolar walls is the cause of chronic pulmonary emphysema (102). An electron microscopic study of the alveolar walls in naturally affected rabbits revealed the loss of capillary endothelium, partial or complete filling of capillaries with collagen, and recanalization of some damaged capillaries (8). These findings also suggested that the disease is initiated by occlusion of capillaries and subsequent ischema, leading to breakdown of capillary walls.

XVII. EXTRAUTERINE PREGNANCY (ABDOMINAL PREGNANCY, ECTOPIC PREGNANCY)

Records of extrauterine pregnancy are not uncommon in domestic (*Oryctolagus*) rabbits (6, 9, 18, 24a, 57, 58, 79, 94), and have been reported as occurring in *Sylvilagus* rabbits as well (79). They are believed to occur most commonly, as in other species, as secondary (or false) extrauterine pregnancies in which fetal implantation originates in the uterus. Through some traumatic episode or weakness of the uterine wall, the fetus and its membranes then escape by way of a performation of the wall to lie in the abdominal cavity (89).

Primary extrauterine pregnancy is thought to occur more rarely. In this condition the fertilized ovum escapes the grasp of the fallopian infundibulum and is left free in the abdominal cavity. Implantation usually occurs on the parietal peritoneum with induction of a decidual reaction and vascularization. Primary extrauterine pregnancy has been reported in the rabbit (9, 24a, 57, 79), often with near-term degrees of fetal development.

Extrauterine pregnancy should be suspected when(a) an analysis of the breeding record shows several normal pregnancies with an abrupt change and (b) a firm nonfluctuant mass is palpable in the abdominal cavity. Fetal mummification appears to be an almost invariable consequence of extrauterine pregnancy of either type. The condition is not generally of great clinical significance to the affected rabbit.

REFERENCES

1. Adams, L., Salvin, S. B., and Hadlow, W. J. (1959). Ringworm in a population of snowshoe hares. *J. Mammal.* 37, 94–99.
2. Alteras, I. (1966). Ringworm in rabbit due to *Trichophyton quinckeanu. Mycopathol. Mycol. Appl.* 28, 361–367.
3. Arrington, L. R., and Wallace, L. J. (1966). The effect of diastase upon mucoid enteritis and growth of rabbits. *Vet. Med. & Small Anim. Clin.* 61, 1210–1212.
4. Banks, K. L., and Clarkson, T. B. (1967). Naturally occurring dermatomycosis in the rabbit. *J. Amer. Vet. Med. Ass.* 151, 926–929.
5. Barrington, F. J. F. (1921). A rabbit's kidney affected with a parasitic mould. *J. Pathol. Bacteriol.* 23, 347–348.
6. Bell, W. B. (1910). Primary abdominal pregnancy in a rabbit. *Proc. Roy. Soc. Med.* 4, 228–233.
7. Blount, W. P. (1957). Rabbits' ailments. "Fur and Feathers," pp. 118–120. Idle, Bradford, England.
8. Boatman, E. W., and Martin, H. B. (1965). Electron microscopy in pulmonary emphysema of rabbits. *Amer. Rev. Resp. Dis.* 91, 197–205.
9. Boycott, A. E. (1910). Peritoneal fetuses in a rabbit. *J. Pathol. Bacteriol.* 14; 156–139.
10. Brown, R. C., Richter, C. B., and Bloomer, M. D. (1969). Ultrastructural pathology of an acute fatal enteritis of captive cottontail rabbit. Search for a viral etiologic agent. *Amer. J. Pathol.* 57, 93–126.
11. Casady, R. B., Daman, R. A., and Suitor, A. E. (1961). Effect of supplementary lysine and methionine on enteritis mortality, growth and feed efficiency in young rabbits. *J. Nutr.* 74. 120–124.
12. Casady, R. B., Everson, D. O., Suitor, A. E., and Mize, K. E. (1963). The effect of restricted availability of water on the incidence of enteritis and on growth in young rabbits. *Lab. Anim. Care* 13, 685–688.
13. Casady, R. B., Hagen, K. W., Jr., Bertrand, J. E., and Thomas, H. G. (1964). Effect of zinc bacitracin on the incidence of enteritis and growth in young rabbits. *Clin. Med.* 71, 871–875.
14. Casady, R. B., Hagen, K. W., and Sittman, K. (1969). Effect of high level antibiotic supplementation in the ration on growth and enteritis in young domestic rabbits. *J. Anim. Sci.* 23, 477–480.
15. Chute, H. L. (1972). Fungal infections. *In* "Diseases of Poultry" (M. S. Hofstad *et al.*, eds.), pp. 448–455. Iowa State Univ. Press, Ames.
16. Cohrs, P., Jaffe, R., and Meessen, H., eds. (1958). "Pathologie der laboratoriumstiere," Vol. II, pp. 65–71. Springer-Verlag, Berlin and New York.
17. Cowie-Whitney, J. (1970). Some aspects of the enteritis complex of rabbits. *In* "Nutrition and Disease in Laboratory Animals" (W. D. Tavernor, ed.), pp. 122–131. Baillière, London.
18. Crary, D. D., and Sawin, P. B. (1950). An ectopic pregnancy in the rabbit. *Anat. Rec.* 108, 3.
19. Doughman, D. J., Olson, G. A., Nolan, S. and Hajny, R. (1969). Experimental band keratopathy. *Arch. Ophthalmol.* 81, 264–271.
20. Dumas, J. (1953). "Les Animaux de Laboratoire," pp. 287–288. Editions Medicales, Flammarion, Paris.
21. Duncan, C. L., and Strong, D. H. (1969). Experimental production of diarrhea in rabbits with *Clostridium perfringens. Can. J. Microbiol.* 15, 765–816.
22. Duncan, C. L., and Strong, D. H. (1969). Ileal loop fluid accumulation and production of diarrhea in rabbits by cell-free productions of *Clostridium perfringens. J. Bacteriol.* 100, 86–99.
23. Dvorak, J., Otcenasek, M; (1964). Geophilic, zoophilic and anthropophilic dermatophytes. *Mycopathol. Mycol. Appl.* 23, 295–296.
24. Dvorak, J., Otcenasek, M., and Rasin, K. (1966). Adiaspiromycosis in mice and a laboratory rabbit. *J. Amer. Vet. Med. Ass.* 149; 932.
24a. Eales, N. B. (1932). Abdominal pregnancy in animals with an account of a case of multiple ectopic gestation in a rabbit. *J. Anat.* 67, 108–117.
25. Economon, J. W., Silverstein, A. M., and Zimmerman, L. E. (1963). Band keratopathy in a rabbit colony. *Invest. Ophthalmol.* 2; 361–368.
26. Evans, R. S., Bingham, M., Hickey, M., and Hassett, C. (1959). A hemolytic system associated with mucoid enteritis in rabbits. *J. Ass. Amer. Physicians* 72, 188–199.
27. Evans, R. S., Bingham, M., and Weiser, R. S. (1963). A hemolytic system associated with enteritis in rabbits. I Nature of the cell change and the serum factors concerned. *J. Exp. Med.* 117, 647–661.
28. Evans, R. S., Bingham, M., and Weiser, R. S. (1963). A hemolytic system associated with enteritis in rabbits. II. Studies on the survival of transfused red cells. *J. Lab. Clin. Med.* 62, 559–570.
29. Evans, R. S., Bingham, J., and Weiser, R. S. (1968). A hemolytic system associated with enteritis in rabbits. III. Observations on the epidemiology and pathogenesis. *J. Lab. Clin. Med.* 72, 495–504.
30. Fine, B. S., Berkow, J. W., and Fine, S. (1968). Corneal calcification. *Science* 162, 129–130.
31. Fuentes, C. A., and Aboulafia, R. (1955). *Trichophyton mentagrophytes* from apparently healthy guinea pigs. *Arch. Dermatol. Syph.* 71, 478–480.
32. Fuentes, C. A., Bosch, Z. E., and Boudet, C. C. (1956). Occurrence of *Trichophyton mentagrophytes* and *Microsporum gypseum* on hairs of healthy cats. *J. Invest. Dermatol.* 23, 311–313.
33. Georg, L. K. (1960). Animal ringworm in public health. *v.s., Pub. Health Serv., Publ.* 727, 9–17.
34. Gip, L., and Martin, B. (1964). Occurrence of *Trichophyton mentagrophytes* Asteroid on hairs of guinea pigs without ringworm lesions. *Acta Dermat. -Venereol.* 44, 208–210.
35. Glantz, P. J. (1970). Unclassified *Escherichia coli* Serogroup 0 × 1 isolated from fatal diarrhea of rabbits. *Can. J. Comp. Med.* 34, 47–49.
36. Glantz, P. J. (1971). Serotypes of *Escherichia coli* associated with colibacillosis in neonatal animals. *Ann. N. Y. Acad. Sci.* 175, Art 1; 67–69.
37. Gordon, R. F. (1943). The problems of backyard poultry and rabbits. *Vet. Rec.* 55, 83.
38. Green, R. G., and Larson, C. L. (1938). A description of shock disease in the snowshoe hare. *Amer. J. Hyg.* 28, 190–21.

39. Green, R. G., and Larson, C. L. (1938). Shock disease and the snowshoe hare cycle. *Science* **87**, 298–299.

40. Green, R. G., Larson, C. L., and Bell, J. R. (1939). Shock disease as the cause of the periodic decimation of the snowshoe hare. *Amer. J. Hyg.* **30**, 83–102.

41. Green, R. G., Mather, D. W., and Larson, C. L. (1938). Occurrence of shock disease among young snowshoe hares. *Proc. Soc. Exp. Biol. Med.* **38**, 816–817.

42. Green, R. G., and Shillinger, J. E. (1936). Shock death in hares. *Minn. Wildl. Dis. Invest.* **2**, 16–18.

43. Green, R. G., Shillinger, J. E., and Larson, C. L. (1936). Spontaneous hypoglycemia of hares. A new disease of hares characterized by shock death from low blood sugar. *Minn. Wildl. Dis. Invest.* **2**, 126–198.

44. Greene, H. S. N. (1937). Toxemia of pregnancy in the rabbit. I. Clinical manifestations and pathology. *J. Exp. Med.* **65**, 809–832.

45. Green, H. S. N. (1938). Toxemia of pregnancy in the rabbit. II. Etiological consideration with especial reference to hereditary factors. *J. Exp. Med.* **67**, 369–388.

46. Greene, H. S. N. (1965). Diseases of the rabbit. *In* "The Pathology of Laboratory Animals" (W. E. Ribelin and J. R. McCoy, eds.), pp. 330–350. Thomas, Springfield, Illinois.

47. Greenham, L. W. (1962). Some preliminary observations on rabbit mucoid enteritis. *Vet. Rec.* **74**, 79–85.

48. Hagen, K. W., Jr. (1956). Infectious diseases of the rabbit. *In* "Animal Diseases, Yearbook of Agriculture," pp. 562–563. U. S. Department of Agriculture, Washington, D. C.

49. Hagen, K. W., Jr. (1969). Ringworm in domestic rabbits: Oral treatment with griseofulvin. *Lab. Anim. Care* **19**, 635–638.

50. Hagen, K. W., Jr., and Gorham, J. R. (1972). Dermatomycoses in fur animals: chinchilla, ferret, mink, and rabbit. *Vet. Med. & Small Anim. Clin.* **67**, 43–48.

51. Hagen, K. W., Jr., and Lund, E. E. (1962). Common diseases of domestic rabbits. *U.S. Dep. Agr., Agr. Res. Serv., Publ.* **ARS-45-3**.

52. Hurt, L. M. (1949). *In* "Annual Report," p. 97. Los Angeles County Livestock Department, Los Angels, California.

53. Jellison, W. L. (1970). Adiaspiromycosis. *In* "Infectious Disease of Wild Mammals" (J. W. Davis, L. H. Karstad, and D. O. Trainer, eds.), pp. 321–323. Iowa State Univ. Press, Ames.

54. Kaffka, A., and Reith, H. (1960). *Trichophyton mentagrophytes*-varianten bei laboratoriumstieren. *Zentrabl. Bakteriol., Parasitenk., Infektimskr. Hyg., Abt. I: Orig.* 96–106.

55. Klarenbeek, A. (1946). Spontaneous pylorospasm in rabbits. *J. Amer. Vet. Med. Ass.* **111**, 54.

56. Kral, F. 1955. Classification, symptomatology and recent treatment of animal dermatomycosis (ringworm). *J. Amer. Vet. Med. Ass.* **127**, 395–402.

57. Kuntz, A. A. (1922). A case of abdominal pregnancy with retention of dead fetuses in the rabbit. *Anat. Rec.* **23**, 237–239.

57a. Leader, R. W. (1970). Search for animal models. *In* "Cystic Fibrosis and Related Human and Animal Diseases" (S. Jakowska, ed.), p. 19. Gordon & Breach, New York.

58. Leibold, A. A. 1917. Ectopic pregnancy with mummification of the foetus in a rabbit. *J. Amer. Vet. Med. Ass.* **50**, 614–617.

59. Lesbouyries, G. (1963). "Pathologie du Lapin," pp. 48–49 and 140–205. Librairie Maloine, Société Anonyme d'Editions Médicales et Scientifiques, Paris, France.

60. Lesbouyries, and Berthelon, (1936). Enterotoxemie du lapin. *Bull. Acad. Vet. Fr.* [N.S.] **9**, 74–82.

61. Loliger, H. C., Matthes, S., Schubert, H.-J., and Heckmann, F. (1969). Die akuten dysenterien der jungkaninchen. *Deut. Tieraertzl. Wochenschr.* **76**, 16–20. and 38–41.

62. Los Angeles County Livestock Department. (1940). Annual Report," p. 47. Los Angeles, California.

63. Los Angeles County Livestock Department. (1941). "Annual Report," p. 57. Los Angeles, California.

64. Los Angeles County Livestock Department. (1943). "Annual Report," p. 40. Los Angeles, California.

65. Lund, E. E. (1952). Experimental use of tannins to control enteritis in rabbits. *Amer. Rabbit. J.* **22**, 122–123.

66. Lund, E. E., and Hagen, K. W. (1962). Common diseases of domestic rabbits. *Wash., Ext. Sev., Inst. Agr. Sci., Ext. Bull.* **397**.

67. Lupanov, M. T. (1954). Clinical treatment and prevention of infectious mastitis in rabbits (in Russian). *Karakulevod. Zverovod.* **6**, 52–53.

68. McCuistion, W. R. (1964). Rabbit mucoid enteritis (neonatal hypoamylasemia). *Vet. Med. & Small. Anim. Clin.* **59**, 815–818.

69. McCuistion, W. R. (1965). The mucoid enteritis story. *Small Stock Mag.* **49**, 5 and 23.

70. Mack, R. (1964). Disorders of the digestive tract of domesticated rabbits. *Vet. Bull.* **32**, 191–199.

71. Marcato, P. S., and Sjaban, M. (1967). Sull enteropatia mucosa (enterite mucoide) del coniglio. *Nuova Vet.* **43**, 546–556.

72. March, F. (1938). Experiments in heatstrokes in Iran. *Trans. Roy. Soc. Trop. Med. Hyg.* **32**, 371–394.

73. Matthes, S. (1969). Die darmflora gesunder und dysenteriekranker jungkaninchen. *Zentrabl. Veterinaer med., Reihe B* **16**, 563–570.

74. Menges, R. W., Love, G. J., Smith, W. W., and Georg, L. K. (1957). Ringworm in wild animals in southwestern Georgia. *Amer. J. Vet. Res.* **18**, 672–677.

75. Mohapatra, L. M., Gugnani, H. C., and Shivrajan, K. (1964). Natural infection in laboratory animals due to *Trichophyton mentagrophytes* in India. *Mycopathol. Mycol. Appl.* **24**, 275–280.

76. Muir, R. (1943). The problems of backyard poultry and rabbits. *Vet. Rec.* **55**, 87.

77. Nielsen, N. O., and Clugston, R. E. (1971). Comparison of *E. coli* endotoxin shock and acute experimental edema disease in young pigs. *Ann. N. Y. Acad. Sci.* **176**, Art. 1, 178–189.

78. Norris, H. R., Finkelstein, R. A., Dutta, N. K., and Sprinz, H. (1965). Intestinal manifestation of cholera in infant rabbits. A morphologic study. *Lab. Invest.* **4**, 1428–1436.

79. Nutting, F. (1942). A case of extrauterine pregnancy in the rabbit. *Anat. Rec.* **84**, 215–219.

80. Ostler, D. C. (1961). The diseases of broiler rabbits. *Vet. Rec.* **73**, 1237–1255.

81. Pederson, V. C. (1964). Ulcerative enteritis of the cottontail rabbit. *Diss. Abstr.* **2418**, 3474.

82. Pollock, S. (1951). Slobbers in rabbits. *J. Amer. Vet. Med. Ass.* **119**, 443–444.

83. Pout, D. (1971). Mucoid enteritis in rabbits. *Vet. Rec.* **89**, 214–216.

84. Rathore, A. K. (1970). High temperature exposure of male rabbits: Fertility of does mated to bucks subjected to 1 and 2 days of heat treatment. *Brit. Vet. J.* **126**, 168–172.

85. Rathore, A. K. (1970). High temperature exposure of male rabbits: Sperm morphology of 1 and 2 days heated rabbits. *Indian Vet. J.* **47**, 837–839.

86. Ravaiolo, L., and Tonolo, A. (1956). Infezione de *Microsporum audouini* Gruby nel coniglio. *Rend. Ist. Super. Sanit. (Engl. Ed.)* **19**, 1201–1206.

87. Rebell, G., and Taplin, D. (1970). "Dermatophytes. Their Recognition and Identification," 2nd ed. pp. 10–19. Univ. of Miami Press, Coral Gables, Florida.

88. Richter, C. B., and Hendren, R. L. (1969). Pathology and epidemiology of acute enteritis in captive cottontail rabbits (*Sylvilagus floridanus*). *Pathol. Vet.* **6**, 159–175.

89. Roberts, S. J. (1956). "Veterinary Obstetrics and Genital Diseases," pp. 95–96. Edwards, Ann Arbor, Michigan.

90. Roe, F. J. C., and Stiff, A. L. (1962). Fracture dislocation of lumbar spine occurring spontaneously in rabbits. *J. Anim. Tech. Ass.* **12**, 92–94.

91. Rollins, W. C., and Casady, R. B. (1967). An analysis of preweaning

deaths in rabbits with special emphasis on enteritis and pneumonia. *Anim. Prod.* **9**, 87–92.

92. Rosenthal, S. A., and Wapnick, H. (1963). The value of MacKenzie's hair brush technic in the isolation of *Trichophyton mentagrophytes* from clinically normal guinea pigs. *J. Invest. Dermatol.* **41**, 5–6.

93. Casady, R. B., Sawin, P. B., Van Dam, J. Commercial Rabbit Raising. Agriculture Handbook No. 309, Agricultural Res. Serv., U.S. Dep. Agric., Oct., 1971.

94. Sawin, P. B., and Latimer, H. B. (1950). Rabbit fetuses in the abdominal cavity. *Anat. Rec.* **108**, 3.

95. Schwartz, B., and Shook, W. B. (1928). Rabbit parasites and diseases. *U.S., Dep. Agr., Farmers' Bull.* **1568**.

96. Schwarz. J. (1954). The deep mycoses in laboratory animals. *Proc. Anim. Care Panel* **5**, 37–70.

97. Seifried, O. (1937). "Die Krankheiten des Kaninchens," pp. 116–120. Springer-Verlag, Berlin and New York.

98. Shotts, E. B., Jr., and Kistner, T. P. (1970). Naturally occurring cutaneous streptothricosis in a cottontail rabbit. *J. Amer. Vet. Med. Ass.* **157**, 667–670.

99. Soerensen, B., and Saliba, A. M. (1961). Actinomicose espontanea en coelhos. *Biologico* **27**, 131–134.

100. Soituz, V. (1930). Sur un staphylocoque adapte la glande mammaire de la lapine. *Rev. Pathol. Comp. Hyg. Gen.* **30**, 381–388.

101. Strawbridge, H. T. G. (1960). Chronic pulmonary emphysema (an experimental study). II. Spontaneous pulmonary emphysema in rabbits. *Amer. J. Pathol.* **37**, 309–331.

102. Strawbridge, H. T. G. (1960). Chronic pulmonary emphysema. III. Experimental pulmonary emphysema. *Amer. J. Pathol.* **37**, 391–412.

103. Taylor, J., Wilkins, M. P., and Payne, J. (1961). Relation of rabbit gut reaction to enteropathogenic *Escherichia coli*. *Brit. J. Exp. Pathol.* **42**, 43–52.

104. Taylor, P. A. (1970). Mucoid enteritis in rabbits. *Ont., Dep. Agr. Food, Vet. Serv. Br., Quart. Bull* **5**, 3–4.

105. Templeton, G. S. (1946). Treatment for paralyzed hindquarters. *Amer. Rabbit J.* **16**, 155.

106. Templeton, G. S. (1949). Mucoid enteritis cause of mortality. *Amer. Rabbit. J.* **19**, 115.

107. Templeton, G. S. (1962). "Domestic Rabbit Production," pp. 166–174. Interstate Publishers, Danville, Illinois.

108. Turn, J., and Eveleth, D. F. (1954). Coccidioidomycosis in a North Dakota cottontail rabbit (*Sylvilagus floridanus*). *Proc. N. Dak. Acad. Sci.* **8**, 42–43.

109. Turner, L. M. (1931). This problem of the bloat. *Amer. Rabbit. J.* **1**, 42.

110. Vaida, M. (1959). Die squamose Sohlenhautentzundung der Kaninchen (Pododermatitis Squamosa plantae cuniculorum). *Wien. Tieraertzl. Monatsschr.* **46**, 380–388.

111. Vail, L. (1942). Wooly-pod or broad-leafed milkweed (*Asclepias eriocarpa*) poisoning of rabbits. *N. Amer. Vet.* **23**, 539–542.

112. Vail, E. L., and McKenny, F. D. (1943). Diseases of domestic rabbits. *U.S., Fish Wildl. Serv. Conserv. Bull.* **31**.

113. Van Kruiningen, J. H., and Williams, C. B. (1972). Mucoid enteritis of rabbits. Comparison to cholera and cystic fibrosis. *Vet. Pathol.* **9**, 53–77.

114. Vetesi, F. (1970). A nyul un mucoid enteritise (coli-enterotoxgemiaga). *Magy. Allatorv. Lapja* **25**, 464–471.

115. Vilanova, Y., and Casanovas, M. (1951). Observations cliniques et mycologiques sur une épidémie de trichophyte transmise du lapin à l'homme. *Presse Med.* **59**, 1760–1762.

116. Weber, A., and Manz, J. (1971). Serologische untersuchungen der O-antigene von *E. coli*-stammen isoliert von kaninchen. *Tieraertzl. Wochenschr.* **84**, 441.

117. Weisbroth, S. H. (1973). Unpublished postmortem findings (personal communication).

118. Weisbroth, S. H., and Scher, S. (1971). *Microsporum gypseum* dermatophytosis in a rabbit. *J. Amer. Vet. Med. Ass.* **159**, 629–634.

119. Wright, P. A. (1960). Infertility in rabbits by feeding ladino clover. *Proc. Soc. Exp. Biol. Med.* **105**, 428–430.

120. Yuill, T. M., and Hanson, R. P. (1965). Coliform enteritis of cottontail rabbits. *J. Bacteriol.* **89**, 1–8.

CHAPTER 18

Diseases of Public Health Significance

Fritz P. Gluckstein

I. INTRODUCTION

There are a number of diseases man can acquire from laboratory and domestic rabbits but these animals do not present a major health hazard. Wild rabbits and hares are somewhat more likely to endanger man, primarily because wild leporids are the major source of human tularemia. Complete descriptions of those agents pathogenic for rabbits are found elsewhere in this volume (Chapter 9, 10, 11, 12, and 13).

II. BACTERIAL DISEASES

A. Tularemia

Tularemia, while entirely unknown in laboratory and domestic rabbits, is endemic in wild rabbits and hares. It occurs throughout North America, continental Europe, and in several Asiatic, North African, and South American countries (113).

In North America about 90% of human tularemia cases are the result of contact with wild leporids. Cottontails (*Sylvilagus* spp.) are by far the most important reservoir, while jackrabbits (*Lepus* spp.) play a minor role. Few human cases have been attributed to snowshoe hares (*Lepus americanus*) (71, 72).

Man contracts tularemia by handling infected tissues, blood, or excreta or by eating the undercooked meat of infected wild leporids. It is important to note that *Francisella (Pasteurella) tularensis* will penetrate the conjunctiva and the apparently unbroken skin. Tularemia can also be transmitted through the bites and fluids of ticks and flies infected by rabbits. The wood tick (*Dermacentor andersoni*), the dog tick (*Dermacentor variabilis*), the Lone Star tick (*Amblyomma americanum*), and the deer fly (*Chrysops discalis*) are the arthropods which most commonly transmit the disease to man. In Sweden and Russia mosquitoes of the genus *Aedes* are suspected of spreading the disease to man (36, 123, 152).

Wild rabbits and hares can be a definite hazard to human health and should be handled with great caution in areas where tularemia is endemic. Gloves and masks covering the entire face should be worn while skinning or dressing rabbits, and protective clothing and insect repellant should be used to prevent the bites of arthropods.

B. Plague

While plague occurs primarily in wild rodents (the major reservoir for human infection), wild leporids have also been shown to play a role. Plague in wild rabbits and hares has been reported from the United States, South America, and England (149). In the United States, human cases of plague have been traced to cottontail rabbits and jackrabbits which had become infected through the transfer of fleas from diseased rodents.

Man acquires rabbit-borne plague through the handling of diseased leporids and possibly through the bites of fleas from infected rabbits; modes of infection very similar to those in tularemia (76, 77, 85).

No cases of human plague have been attributed to laboratory or domestic rabbits, but instances of secondary plague in domestic rabbits due to infected rats have been observed (149).

C. Listeriosis

Although *Listeria monocytogenes* was first isolated from laboratory rabbits, naturally occurring listeriosis is rare in laboratory rabbit colonies. The disease is not uncommon in domestic rabbits, particularly in Europe, where it affects primarily young animals and pregnant does (60, 104, 122, 142).

Listeriosis has been reported in hares and wild rabbits in Europe (37, 38, 122), and jackrabbits and snowshoe hares in North America (38).

Very little is known about the epidemiology of listeriosis. No human cases of the disease contracted from rabbits or hares have been documented. However, since *L. monocytogenes* is present in the discharges of affected animals and is quite resistant to environmental factors, rabbits and hares must be considered at least a potential source of human infection. The true role of the domestic and wild rabbit as asymptomatic carriers of *L. monocytogenes* remains to be evaluated.

D. Brucellosis

Laboratory and domestic rabbits are an unlikely source of human brucellosis. Spontaneous brucellosis in laboratory rabbits has been reported only once in the literature (102). The disease seems to be equally rare in domestic rabbits where again only one outbreak has been recorded (109).

Brucellosis occurs in hares (*Lepus europaeus*) and to a lesser degree in wild rabbits (*Oryctolagus cuniculus*) on the continent of Europe. Most cases are caused by *Brucella suis* but infections due to *B. bovis* and *B. melitensis* have also been reported (65, 87, 136). In the United States *B. suis* and possibly *B. melitensis* have been isolated from the black-tailed jackrabbit (*L. californicus*) (137). Hunters and others likely to touch wild hares and rabbits with unprotected hands could become infected.

E. Salmonellosis

A review of the literature suggests that salmonellosis in laboratory and domestic rabbits is more common than is generally believed. There are reports from different parts of the world of *Salmonella* infections involving various serotypes in these animals (25, 30a, 35, 49, 53, 89, 97, 106, 111, 135, 141). One report described a case of human infection contracted from laboratory rabbits (73).

Laboratory and domestic rabbits should not be overlooked as a source of human salmonellosis especially since their role as asymptomatic carriers has not yet been adequately assessed. Small children who come in close contact with the feces of pet rabbits are particularly prone to become infected.

Although salmonellae have been isolated from wild rabbits and hares in Europe and from a cottontail in the United States, salmonellosis is uncommon in wild leporids (42, 134).

F. Pasteurellosis

Pasteurella multocida infections are common in laboratory and domestic rabbits and may assume epizootic proportions in wild leporids in Europe (29, 78, 86, 147). Only one rabbit-borne human case of pasteurellosis, a localized would infection due to a bite, has been recorded (19). This suggests human pasteurellosis, unless bite transmitted is unlikely to be acquired from rabbits (57).

G. Leptospirosis

Leptospirosis has been found in laboratory rat and mouse colonies, but little has been done to search for the disease in laboratory and domestic rabbits. Only one report of leptospirosis in laboratory rabbits could be found in the literature (150). As a result it is difficult to assess the role of laboratory and domestic rabbits as reservoirs of human infection. It should be kept in mind that infected animals may not show any clinical signs while endangering their handlers by shedding leptospirae in their urine. Human infection could result from penetration of mucous membranes, abraded skin, or possibly through ingestion or inhalation.

Leptospirosis of various serotypes is not uncommon in hares and wild rabbits in Britain and central Europe and has also been reported in cottontails in the United States and Canada (7, 30, 47, 51, 62, 138–140). No cases of leptospirosis in man have been traced directly to wild leporids.

H. Yersiniosis (Pseudotuberculosis)

Pseudotuberculosis or yersiniosis caused by *Yersinia (Pasteurella) pseudotuberculosis* is a common disease of hares in Europe where it also occurs in wild and domestic rabbits (21, 130, 144). In North America the disease has been reported in cottontails, jackrabbits, and snowshoe hares as well as domestic rabbits (148).

In man, *Y. pseudotuberculosis* may cause a fatal septicemia but more commonly produces a rather benign mesenteric lymphadenitis that frequently simulates acute appendicitis (20, 79).

Very little is known about the transmission of pseudotuberculosis from animals to man. It is generally presumed that most human infections occur via the oral route during close contact with infected animals or through the ingestion of food contaminated with feces of diseased animals (79, 93, 143).

Although one case of human pseudotuberculosis derived from domestic rabbits has been documented, (130), leporids are unlikely to be a significant source of the disease in man provided ordinary sanitary precautions are observed.

I. Tuberculosis

Despite a high susceptibility to experimental infection, naturally occurring tuberculosis has only rarely been observed in rabbits. Most of the few cases were caused by *Mycobacterium bovis* but infections due to *M. avium* and *M. tuberculosis* have been reported (23, 31, 41, 56, 83, 131). Tuberculosis in wild leporids appears to have been confined to occasional instances in central European hares (147).

The possibility that man may contract tuberculosis from rabbits appears extremely remote, particularly since infection with *M. tuberculosis*, the organism to which man is most susceptible, is exceedingly rare in leporids.

J. Staphylococcal Infections

Staphylococcal infections are not frequent in laboratory and domestic rabbits (16a, 54, 114, 129, 131). Most of these infections in laboratory and domestic rabbits appear to be due to α-hemolytic strains or phage types of human origin, suggesting that human carriers play a significant role in introducing staphylococci into laboratory and domestic rabbit populations which then may infect workers in laboratories or rabbitries.

Staphylococcal infections are considerably more common in hares and wild rabbits (34, 78, 103, 147). This is not a significant human hazard since staphylococci of animal origin only rarely affect man (52).

K. Streptococcal Infections

Streptococcal infections have been reported only in domestic rabbits (83, 121, 131). Such infections when due to streptococci of animal origin (Lancefield Group C) are of no appreciable human hazard since these organisms are usually nonpathogenic or only slightly pathogenic to man. Streptococcal infection of human origin (Lancefield Group A) are more easily acquired by man.

L. Melioidosis

Spontaneous melioidosis (*Pseudomonas pseudomallei*) has been found in laboratory rabbits in Indochina and other parts of southeast Asia (2, 83, 132). No laboratory infections have been reported and it appears likely that man acquires the disease from laboratory rabbits.

The principal sources and modes of infection are not yet fully known. Man, who is not highly susceptible, apparently acquires melioidosis through contact with contaminated water or mud or through the ingestion of contaminated food and water (2, 39).

M. Tyzzer's Disease

Tyzzer's disease (*Bacillus piliformis*) is primarily a disease of rodents but it also occurs in laboratory and domestic rabbits (3, 32, 140a). In 1968 the disease was observed in a rhesus monkey (*Macaca mulatta*) suggesting that other primates including man might be susceptible (100).

N. Pseudomoniasis

Outbreaks of fatal infections due to *Pseudomonas aeruginosa* have been observed in commercial rabbitries and in laboratory rabbit colonies (4, 83, 88).

With the increased use of antibiotics human pseudomoniasis has become more prevalent. *Pseudomonas aeruginosa* is resistant to most antimicrobial agents and tends to become dominant after susceptible organisms have been suppressed (69, 98). The rabbit is a possible source of *Pseudomonas* infections.

O. Other Bacterial Infections

Other bacterial diseases occurring in leporids that are possibly transmissible to man include colibacillosis (*Escherichia coli*) and necrobacillosis (*Fusobacterium necrophorum*) in domestic and wild rabbits, and *Clostridium perfringens* and *Bordetella bronchiseptica* infection in domestic rabbits (50, 80a, 89, 95a, 131, 142a, 147, 154).

III. MYCOTIC DISEASES

Dermatomycosis (ringworm) is the only significant mycotic disease of leporids transmissible to man. The disease can be acquired from laboratory and domestic rabbits through contact with infected animals or contaminated cages and hutches. Children are particularly prone to infection since they are likely to have close contact with pet rabbits (5, 48, 67, 75, 95, 110).

Prevention of rabbit-borne ringworm can be difficult because lesions in affected animals may be inconspicious or even absent (16, 48).

Ringworm has been reported in hares and wild rabbits but no human infections have been attributed to these animals (8).

IV. VIRAL DISEASES

A. Introduction

No thorough assessment of the risk of man acquiring viral infections from rabbits can be made since very little is known about the viral flora of leporids. Those few viral diseases of rabbits that are of public health significance are confined to wild species.

B. California Encephalitis Complex

In recent years isolation and serological evidence suggests inapparent infections with viruses of the California encephalitis complex to be widespread in wild leporids in North America and Europe (17, 33, 74, 128). In North America the snowshoe hare is most frequently infected (26, 59, 155). The viruses are transmitted by mosquitoes primarily of the genus *Aedes*. However, isolation of viruses from wood ticks (*D. andersoni*) and rabbit ticks (*Haemaphysalis leporis-palustris*) suggest that ticks may also serve as vectors (58, 91, 117). Human California encephalitis infections have become a public health problem of increasing importance pointing to the necessity of further study to clarify the role of wild leporids as reservoirs of the infections.

C. Western Equine, Eastern Equine, and St. Louis Encephalitis

Subclinical infections with Western equine and Eastern equine encephalitis and St. Louis encephalitis viruses have been found in several species of wild leporids by isolation and serological tests. The role of the wild leporid as reservoir of these encephalitides has not been established (59, 112, 145 153).

D. Central European Tick-Borne Encephalitis

Antibodies to viruses of central European tick-borne encephalitis have been demonstrated in hares in central Europe. It is not known whether these animals act as a reservoir of the disease (46).

E. Powassan Encephalitis

Neutralizing antibodies to the tick-borne Powassan encephalitis virus were detected in snowshoe hares in Ontario, Canada (90, 92). It is not known whether these animals are a reservoir of the virus.

F. Myxomatosis

There are two reports in the literature suggesting that man may be susceptible to infections with myxoma viruses. One report describes two human subjects who developed a transient conjunctivitis after subcutaneous injection with serum from infected rabbits (119). The other report deals with two children who developed purulent conjunctivitis, palpebral edema, and fever. Their mother was a laboratory attendant handling myxomatous rabbits (116). In neither case was confirmation through isolation or serological test attempted. Attempts to infect rabbits with conjunctival exudate from the two children were unsuccessful. More recent work failed to show either clinical or serological evidence of myxomatosis in humans after intradermal virus injection or prolonged exposure to affected rabbits (40, 64).

G. Reoviruses

Reovirus antibodies have been found in rabbits as well as other animal species. The full role of reoviruses in human and animal diseases is still uncertain, although they are believed to be associated with respiratory and enteric illness (1, 24, 82, 118).

H. Coxsackie Viruses

Neutralizing antibodies to Coxsackie viruses have been demonstrated in cottontail rabbits (99). Coxsackie virus has also been found in the blood of a cottontail (101). Present knowledge does not permit any conclusion as to the role of leporids in the dissemination of Coxsackie viruses.

V. RICKETTSIAL DISEASES

A. Rocky Mountain Spotted Fever

Rocky Mountain spotted fever is endemic throughout most of the United States, in parts of Canada, Mexico, and Central and South America (13, 115). Man acquires the infection through the bites of infected ticks or through contamination of the skin with crushed tissues or feces of infected, ticks. The ticks most commonly serving as vectors are the wood tick (*Dermacentor andersoni*), the dog tick (*D. variabalis*), and the Lone Star tick (*Amblyomma americanum*) all of which feed on wild leporids (105, 115). The rabbit tick (*Haemaphysalis leporis-palustris*) does not bite man but appears to play an important role in the dissemination of *Rickettsia rickettsii* in nature (11, 105, 108).

Antibodies to *R. rickettsii* have been found in cottontails and jackrabbits (70, 105). In a few instances the organism has been isolated from the blood of wild leporids (11, 126). It has not yet been proven that wild rabbits and hares maintain a rickettsemia for any appreciable period of time. However significant correlation between the geographical distribution of cottontails and the occurrence of human cases of Rocky Mountain spotted fever suggests that wild leporids may be a reservoir of the disease (70, 145).

B. Q Fever

Detection of antibodies and isolation of the causative organism have shown that infection with *Coxiella burneti* exists in wild leporids in the United States, Spain, and Morocco (9, 12, 22, 63). Various species of ticks including *D. andersoni* feeding on man as well as rabbits are suspected of being vectors of human Q fever. Thus wild leporids are possible reservoirs of human infection (12, 115). Further study to clarify the role of wild rabbits and hares in the dissemination of Q fever is needed.

VI. PROTOZOAN DISEASES

A. Toxoplasmosis

Toxoplasmosis occurs among domestic, laboratory, and wild leporids throughout the world (44, 55, 120, 127, 146, 147). The disease is frequently asymptomatic and infection can only be detected through microscopic examination of tissues or through serological tests (44, 120). Rabbits are suspected of being a source of human toxoplasmosis. There is a high prevalence of serological reactions among rabbit handlers and rabbit trappers (15). One study showed that a number of people with clinical toxoplasmosis had had contact with domestic rabbits (81).

Aside from the eating of undercooked rabbit meat it is not known how the infection could spread from rabbit to man. Arthropod vectors have been ruled out (43, 68). *Toxoplasma* cysts have been isolated from the urine, saliva, and nasal and conjunctival exudates of rabbits (14). But the rarity with which such isolations have been made coupled with the fact that the cysts are quite labile, surviving only briefly outside the animal's body, makes transmission through excreted cysts a remote possibility.

Recently oocystic forms of *Toxoplasma gondii* closely resembling *Isospora bigemina* have been found in cat feces. The oocysts were resistant to environmental factors and capable of infecting mice and presumably man (45, 124).

Although preliminary attempts to detect *Toxoplasma* oocysts in the feces of rabbits, rats, dogs, sheep, and other species have been unsuccessful (15a, 45), further studies are needed to determine whether other animals aside from the cat are capable of disseminating toxoplasmosis. A full assessment of the role of the rabbit as reservoir of human toxoplasmosis must await the results of such studies.

B. Nosematosis and Pneumocystosis

Two protozoan diseases of possible public health significance affecting rabbits are nosematosis and pneumocystosis.

Encephalitozoonosis (nosematosis) is found in laboratory and domestic rabbits. So far only a single case of apparent human nosematosis has been reported. The transmission and host range of *Encephalitozoon cuniculi* is as yet unknown (40a, 44, 80, 94, 96, 107a, 151).

Pneumocystosis appears in rabbits, but this disease is unlikely to affect man unless he is in a weakened condition or suffering from an impairment of the immune mechanism. Although nothing is known about the transmission of the disease it is presumed that infection occurs through inhalation (6, 13, 18, 44, 125).

VII. HELMINTH DISEASES

The only helminths common to rabbits and man are members of the genus *Trichostrongylus*. Man may acquire the infection from rabbits by eating unwashed green vegetables or chewing grass contaminated with rabbit feces. Trichostrongyls are not known to cause any ill effects in man who appears to be merely an incidental host (10, 28, 84).

VIII. ARTHROPOD INFESTATIONS

Rabbits are not an important source of human arthropod infestations. The primary danger of arthropods acquired from rabbits are the serious diseases these arthropods may carry.

Ixodid ticks, while rarely found in laboratory and domestic rabbits, are common in wild leporids and are potential carriers of tularemia and Rocky Mountain spotted fever. It should be noted that the widely distributed rabbit tick, *Haemaphysalis leporis-palustris*, an efficient vector of tularemia and Rocky Mountain spotted fever among wild animals, rarely if ever bites man (66, 156).

Fleas are not particularly host specific and man may acquire them from domestic and wild leporids. The main danger of human infestation is that the flea is a potential vector of plague (10, 61).

Cheletid and various types of mange mites occurring in domestic, laboratory, and wild rabbits may affect man but the infestations are self limiting, causing a transient dermatitis (108, 133, 156).

Sucking as well as biting lice have a high host specificity making it unlikely that man would acquire pediculosis from leporids (10, 27).

REFERENCES

1. Abinati, F. R. (1964). Respiratory viruses of animals. *In* "Occupational Diseases Acquired from Animals," Continued Educ. Ser. No. 124, pp. 53–71. School of Public Health, University of Michigan, Ann Abor.
2. Alexander, A. D. (1964). Melioidosis. *In* "Foreign Animal Diseases," pp. 129–136. Committee on Foreign Animal Diseases, U.S. Livestock Sanit. Ass.
3. Allen, A. M., Ganaway, F. R., Moore, T. O., and Kinard, R. F. (1965). Tyzzer's disease syndrome in laboratory rabbits. *Amer. J. Pathol.* **46**, 859–882.
4. Alpen, G. R., and Maerz, K. (1969). The incidence of a pathogenic strain of *Pseudomonas* in a rabbit colony. *J. Inst. Anim. Tech.* **28**, 72–74.
5. Alteras, I., and Cojocaru, I. (1969). Human infection by *Trichophyton mentagrophytes* from rabbits. *Mykosen* **12**, 543–544.
6. Areán, V. M. (1971). Pulmonary pneumocystosis. *In* "Pathology of Protozoal and Helminthic Diseases" (R. A. Marcial-Rojas, ed.), pp. 291–317. Williams & Wilkins, Baltimore, Maryland.
7. Asmera, J. (1960). *L. grippotyphosa* u zajice polniho. *Cesk. Epidemiol., Mikrobiol., Immunol.* **9**, 501–504.
8. Austwick, P. K. C. (1969). Mycotic infections. *Symp. Zool. Soc. London* **24**, 249–217.
9. Babudieri, B. (1959). Q fever: A zoonosis. *Advan. Vet. Sci.* **5**, 81–182.
10. Belding, D. L. (1965). "Textbook of Parasitology." Appleton, New York.
11. Bell, J. F. (1970). Rocky Mountain spotted fever. *In* "Infectious Diseases of Wild Mammals" (J. W. Davis, L. H. Karstad, and D. O. Trainer, eds.), pp. 324–333. Iowa State Univ. Press, Ames.

12. Bell, J. F. (1970). Q (query) fever. *In* "Infectious Diseases of Wild Mammals" (J. W. Davis, L. H. Karstad, and D. O. Trainer, eds.), pp. 343–357. Iowa State Univ. Press. Ames.

13. Benenson, A. S., ed. (1970). "Control of Communicable Diseases in Man." Amer. Pub. Health Ass., New York.

14. Berndt, M. (1969). Untersuchungen über die Ausscheidung von *Toxoplasma gondii* bei frisch und latent infizierten Kaninchen und Meerschweinchen. Dissertation, Veterinary Medical Faculty, Free University, Berlin.

15. Beverly, J. K. A., and Beattie, C. P. (1954). Human toxoplasma infection. *J. Hyg.* **52**, 37–46.

15a. Bicknell, S. R. (1972). An unsuccessful attempt to transmit toxoplasmosis to mice from the faeces of artificially infected sheep. *Brit. Vet. J.* **128**, xi–xviii.

16. Bisping, W. (1963). Die Dermatomykosen in ihrer Bedeutung als Zooanthroponosen. *Deut. Med. Wochenschr.* **88**, 584–592.

16a. Blackmore, D. K., and Francis, R. A. (1970). The apparent transmission of staphylococci of human origin to laboratory animals. *J. Comp. Pathol.* **80**, 645–651.

17. Blaskovic, D. (1969). Biology of arboviruses of the California complex and the Bunyamwera group. *In* "Arboviruses of the California Complex and the Bunyamwera Group" (V. Bardos *et al.*, eds.), pp. 21–39. Publ. House Slovak Acad. Sci., Bratislava.

18. Blazek, K. (1960). Die Pneumocystis-Pneumonia beim Feldhasen (*Lepus europaeus*, *Pallas*). *Zentralbl. Allg. Pathol. Pathol. Anat.* **101**, 484–489.

19. Boisvert, P. L., and Fouser, M. D. (1941). Human infection with *Pasteurella lepiseptica* following a rabbit bite. *J. Amer. Vet. Med. Ass.* **116**, 1910–1902.

20. Borg, K., and Thal, E. (1961). Pseudotuberkulosen *Pasteurella* (*pseudotuberculosis*) som zoonos. *Läkartidningen* **58**, 1923–1935.

21. Bouvier, G. (1969). Observations sur les maladies du gibier et des animaux sauvages faites en 1967 et 1968. *Schweiz. Arch. Tierheilk.* **111**, 688–694.

22. Bowen, R. E., McMahon, K. J., and Mitchell, R. W. (1960). Infectious diseases in a black-tailed jack rabbit (*Lepus californicus melanotes*, Mearns). *Trans. Kans. Acad. Sci.* **63**, 276–284.

23. Brack, M. (1966). Spontane durch *Mycobacterium avium* verursachte Tuberkulose beim Hauskaninchen. *Deut. Tieraerztl. Wochenschr.* **73**, 317–321.

24. Bricout, F. (1970). Respiratory enteroviruses and reoviruses. *In* "Clinical Virology" (R. Debre and J. Celers, eds.), pp. 626–636. Saunders, Philadelphia, Pennsylvania.

25. Bruner, D. W., and Moran, A. B. (1949). Salmonella infections of domestic animals. *Cornell Vet.* **39**, 53–63.

26. Burgdorfer, W., Newhouse, V. F., and Thomas, L. A. (1961). Isolation of California encephalitis virus from the blood of a snowshoe hare (*Lepus americanus*) in western Montana. *Amer. J. Hyg.* **73**, 344–349.

27. Busvine, J. R. (1966). "Insects and Hygiene." Methuen, London.

28. Cameron, T. W. M. (1962). Helminths of animals transmissible to man. *Amer. J. Med. Sci.* **243**, 354–381.

29. Carter, G. R. (1967). Pasteurellosis: *Pasteurella multocida* and *Pasteurella hemolytica*. *Advan Vet. Sci.* **11**, 321–379.

30. CDC Zoonoses Surveillance. (1966). "Leptospiral Serotype Distribution Lists." U.S. Dept. of Health, Education and Welfare, Public Health Service, Atlanta, Georgia.

30a. Corazzola, S., Zanin, E., and Bersani, G. (1971). Food poisoning in man following an outbreak of salmonellosis in rabbits. *Vet. Ital.* **22**, 370–373.

31. Coulaud, E. (1924). La tuberculose par contamination naturelle chez le lapin. *Ann. Inst. Pasteur, Paris* **38**, 581–597.

32. Cutlip, R. C., Amtower, W. C., Beall, C. W., and Matthews, P. J.

(1971). An epizootic of Tyzzer's disease in rabbits. *Lab. Anim. Sci.* **21**, 356–361.

33. Danielova, V., Kolman, J. M., Malkova, D., Marhoul, Z., and Smetana, A. (1969). Natural focus of Tahyna virus in South Moravia. Result of virological investigation. *In* "Arboviruses of the California Complex and the Bunyamwera Group" (V. Bardos *et al.*, eds.), pp. 147–150. Publ. House Slovak Acad. Sci., Bratislava.

34. Davis, J. W. (1970). Staphylococcosis in rabbits and hares. *In* "Infectious Diseases of Wild Mammals" (J. W. Davis, L. H. Karstad, and D. O. Trainer, eds.), pp. 283–285. Iowa State Univ. Press, Ames.

35. Duthie, R. C., and Mitchell, C. A. (1931). *Salmonella enteriditis* infection in guinea pigs and rabbits. *J. Amer. Vet. Med. Ass.* **78**, 24–41.

36. Eigelsbach, H., and Hornick, R. B. (1964). Occupational tularemia. *In* "Occupational Diseases Acquired from Animals," Continued Educ. Ser. No. 124, pp. 295–302. School of Public Health, University of Michigan, Ann Arbor.

37. Englert, H. K. (1967). Wildkrankheiten und Humanmedizin. *Landarzt* **43**, 1770–1776.

38. Eveland, W. C. (1970). Listeriosis. *In* "Infectious Diseases of Wild Mammals" (J. W. Davis, L. H. Karstad, and D. O. Trainer, eds.), pp. 273–282. Iowa State Univ. Press, Ames.

39. Felsenfeld, O. (1966). "The Epidemiology of Tropical Diseases." Thomas, Springfield, Illinois.

40. Fenner, F., and Ratcliffe, F. N. (1965). Myxomatosis." Cambridge Univ. Press, London and New York.

40a. Flatt, R. E., and Jackson, S. J. (1970). Renal nosematosis in young rabbits. *Pathol. Vet.* **7**, 492–497.

41. Francis, J. (1958). "Tuberculosis in Animals and Man." Cassel, London.

42. Franklin, J., Simmons, M. L., and Cosgrove, G. E. (1966). A pathogen survey in the Kansas cottontail. *Bull. Wildl. Disease Ass.* **7**, 52–53.

43. Frenkel, J. K. (1970). Pursuing toxoplasma. *J. Infec. Dis.* **122**, 553–559.

44. Frenkel, J. K. (1971). Protozoal diseases of laboratory animals. *In* "Pathology of Protozoal and Helminthic Diseases" (R. A. Marcial-Rojas, ed.), pp. 318–369. Williams & Wilkins, Baltimore, Maryland.

45. Frenkel, J. K., Dubey, J. P., and Miller, N. L. (1970). *Toxoplasma gondii* in cats: Fecal stages identified as coccidian oocysts. *Science* **167**, 893–896.

46. Freymann, R. (1957). "Die Virusencephalitiden in der Sovjetunion und in Mitteleuropa," No. 28. Berichte des Osteuropa-Instituts an der Freien Univ., Berlin.

47. Galton, M. M., Menges, R. W., Shotts, E. B., Nahmias, A. J., and Heath, C. W. (1962). "Leptospirosis, Epidemiology, Clinical Manifestations in Man and Animals and Methods in Laboratory Diagnosis." U. S. Dept. of Health, Education and Welfare, Public Health Service, Communicable Disease Center, Atlanta, Georgia.

48. Georg, L. K. (1960). Animal ringworm in public health. *U.S., Pub. Health Serv., Publ.* **727**.

49. Giorgi, W. (1968). Docencas observados em coelhos durante o quinquenio 1963–1967, no Estado de Sao Paulo. *Biolologico* **34**, 71–82.

50. Glantz, P. J. (1970). Unclassified *Escherichia coli* serogroup OXI isolated from fatal diarrhea of rabbits. *Can. J. Comp. Med.* **34**, 47–49.

51. Gorman, G. W., McKeever, S., and Grimes, R. D. (1962). Leptospirosis in wild mammals from southwestern Georgia. *Amer. J. Trop. Med. Hyg.* **11**, 518–524.

52. Grün, L. (1964). "Staphylokokken in Klinik und Praxis." Wiss. Verlagsges., Stuttgart.

53. Habermann, R. T., and Williams, F. P. (1958). Salmonellosis in laboratory animals. *J. Nat. Cancer Inst.* **20**, 933–941.

54. Hagen, K. W. (1963). Disseminated staphylocococcic infection in young domestic rabbits. *J. Amer. Vet. Med. Ass.* **142**, 1421–1422.

55. Harcourt, R. A. (1967). Toxoplasmosis in rabbits. *Vet. Rec.* **81**, 191–192.

56. Harkins, M. J., and Saleeby, E. (1929). Spontaneous tuberculosis of rabbits. *J. Infec. Dis.* **43**, 554–556.

57. Henderson, A. (1963). *Pasteurella multocida* infection in man; a review of the literature. *Antonie van Leeuwenhoek*; *J. Microbiol. Serol.* **29**, 359–367.

58. Hoff, G. L., Anslow, R. O., Spalatin, J., and Hanson, R. P. (1971). Isolation of Montana snowshoe hare serotype of California encephalitis virus group from a snowshoe hare and *Aedes* mosquitoes. *J. Wildl. Dis.* **71**, 28–33.

59. Hoff, G. L., Yuill, T. M., Iverson, J. O., and Hanson, R. P. (1970). Selected microbial agents in snowshoe hares and other vertebrates of Alberta. *J. Wildl. Dis.* **6**, 472–478.

60. Holmes, R. G. (1961). Listeriosis in rabbits. *Vet. Rec.* **73**, 791.

61. Hopkins, G. H. E. (1957). Host-association of siphonaptera. *In* "First Symposium on Host Specificity Among Parasites of Vertebrates," pp. 64–87. Neuchatel, Switzerland.

62. Horsch, F., Klockmann, J., Janetzky, B., Drechsler, H., and Löbnitz, P. (1970). Untersuchungen von Wildtieren auf Leptospirose. *Monatsh. Veterinaer med.* **25**, 634–639.

63. Horsfall, W. R., and Ferris, D. H. (1962). *Coxiella burnetii*, causative agent of Q fever. Abstract of the literature to 1961. *Wildlife Dis.* No. 28 (microcard).

64. Jackson, E. W., Dorn, C. R., Saito, J. K., and McKercher, D. G. (1966). Absence of serological evidence of myxoma virus infections in humans exposed during an outbreak of myxomatosis. *Nature* (*London*) **211**, 313–314.

65. Jacotot, H., Vallee, A., and Virat, B. (1954). Nouveaux documents sur la brucellose du leivre en France. *Bull. Acad. Vet. Fr.* [N.S.] **27**, 249–253.

66. James, M. T., and Harwood, R. F. (1969). "Herm's Medical Entomology." Macmillan, New York.

67. Jänisch, W., and Koch, H. A. (1965). Einige Beobachtungen über Dermatomykosen bei Versuchtieren. *Z. Versuchstierk.* **6**, 12–17.

68. Janitschke, K. (1971). Die Bedeutung vom Tieren als Infektionsquelle des Menschen mit Toxoplasmen. *Deut. Med. Wochenschr.* **96**, 78–83.

69. Jawetz, E., Melnick, J. L., and Adelberg, E. A. (1970). "Review of Medical Microbiology." Lange Med. Publ., Los Altos, California.

70. Jellison, W. L. (1945). The geographical distribution of Rocky Mountain spotted fever and Nuttall's cottontail in the western United States. *Pub. Health Rep.* **60**, 958–961.

71. Jellison, W. L., Owen, C. R., Frederick, B. J., and Kohls, G. M. (1961). Tularemia and animal populations: Ecology and etiology. *Wildl. Dis.* No. 17 (microcard).

72. Jellison, W. L., and Parker, R. R. (1945). Rodents, rabbits and tularemia in North America: Some zoological and epidemiological considerations. *Amer. J. Trop. Med. Hyg.* **25**, 349–362.

73. Joester, P., and Günther, O. (1954). *Salmonella enteriditis var. Jena* als Erreger einer Kaninchenseuche mit Erkrankung des Tierpflegers. *Monatsh. Tierheilk.* **6**, 71–75.

74. Joubert, L., and Oudar, J. (1969). Les zoonoses arborvirales, leur presence en France. *Rev. Med. Vet.* **120**, 737–764.

75. Kaffka, A., and Rieth, H. (1958). Laboratoriumstiere als Ursache einer Berufsdermatomykose und Massnahmen zur Verhütung weiterer Pilzinfektionen. *Zentralbl. Bakteriol., Parasitenk., Infektionskr. Hyg., Abt. 1: Orig.* **171**, 319–321.

76. Kartman, L. (1960). The role of rabbits in sylvatic plague epidemiology, with special attention to human cases in New Mexico, and use of the fluorescent antibody technique for detection of *Pasteurelle pestis* in field specimens. *Zoonoses Res.* **1**, 1–27.

77. Kartman, L. (1970). Historical and oecological observations on plague. *Trop. Geogr. Med.* **22**, 257–275.

78. Kerschagl, W. (1965). "Wildkrankheiten." Oesterreichischer Jagd- und Fischerei-Verlag, Vienna.

79. Knapp, W. (1968). Die Pseudotuberkulose des Menschen. *Ther. Umsch.* **25**, 195–200.

80. Koller, L. D. (1969). Spontaneous *Nosema cuniculi* infection in laboratory rabbits. *J. Amer. Vet. Med. Ass.* **155**, 1108–1114.

80a. Kristensen, K. H., and Lautrop, J. (1962). En familieepidemi forarsaget af kighostebakterien *Bordetella bronchiseptica*. *Ugeskr. Laeger* **124**, 303–308.

81. Kunstyr, I., Jira, J., Princova, D., and Mika, J. (1970). Survey of toxoplasma antibodies in domestic rabbits. *Folia Parasitol.* (*Praha*) **17**, 277–280.

82. Leers, W. D. (1968). Reoviruses. *In* "Textbook of Virology" (A. J. Rhodes and C. E. Van Rooyen, eds.), pp. 625–629. Williams & Wilkins, Baltimore, Maryland.

83. Lesbouyries, G. (1963). "Pathologie du Lapin." Librairie Maloine, Paris.

84. Levine, N. D. (1968). "Nematode Parasites of Domestic Animals and of Man." Burgess, Minneapolis, Minnesota.

85. Link, V. B. (1950). Plague epizootics in cottontail rabbits. *Pub. Health Rep.* **65**, 696.

86. Loosli, R. (1967). Zoonoses in common laboratory animals. *In* "Husbandry of Laboratory Animals" (M. L. Conalty, ed.), pp. 307–325. Academic Press, New York.

87. McCaughey, W. J. (1969). Brucellosis in wildlife. *Symp. Zool. Soc. London* **24**, 99–105.

88. McDonald, R. A., and Pinhero, A. F. (1967). Water chlorination controls *Pseudomonas aeruginosa* in a rabbitry. *J. Amer. Vet. Med. Ass.* **115**, 863–864.

89. Mack, R. (1962). Disorders of the digestive tract of domesticated rabbits. *Vet. Bull* (*London*) **32**, 191–199.

90. McLean, D. M. (1968). Tick-brone arboviruses of North America. *In* "Textbook of Virology" (A. J. Rhodes and C. E. Van Rooyen, eds.), pp. 734–735. Williams & Wilins, Baltimore, Maryland.

91. McLean, D. M. (1968). California encephalitis group of viruses. *In* "Textbook of Virology" (A. J. Rhodes and C. E. Van Rooyen, eds.), pp. 754–758. Williams & Wilkins, Baltimore, Maryland.

92. McLean, D. M., Ronald, K., Scholten, T. H., and MacPherson, L. W. (1961). Arthropod-borne encephalitis in Ontario. *Wildl. Dis.* No. 18 (microcard).

93. Mair, N. S. (1969). Pseudotuberculosis in free living wild animals. *Symp. Zool. Soc. London* **24**, 107–117.

94. Matsubayashi, H., Koike, T., Mikata, I., Takei, H., and Hagiwara, S. (1959). A case of encephalitozoon-like body infection in man. *AMA Arch. Pathol.* **67**, 181–187.

95. Mayer, H. (1970). Bildbericht, Dermatomykosen bei Kindern durch infizierte Tiere. *Berlin. Müenchen. Tieräerztl. Wochenschr.* **83**, 97.

95a. Mayer, H. (1971). Bordetellainfektionen, ein Problem der Massentierhaltung bei Kaninchen. *Berlin. Muenchen. Tieraertzl. Wochenschr.* **84**, 273–274.

96. Møller, T. (1968). A survey on toxoplasmosis and encephalitozoonsis in laboratory animals. *Z. Versuchstierk.* **10**, 27–38.

97. Morel, P. (1958). Infection du lapin par *Salmonella pullorum*. *Rec. Med. Vet.* **134**, 281–283.

98. Morgan, H. R. (1965). The enteric bacteria. *In* "Bacterial and Mycotic Infections of Man" (R. J. Dubos and J. G. Hirsch, eds.), 3rd ed., pp. 610–648. Lippincott, Philadelphia, Pennsylvania.

99. Morris, J. A., and O'Connor, J. R. (1952). Neutralization of viruses of the coxsackie group by sera of wild rabbits. *Cornell Vet.* **42**, 56–61.

100. Niven, J. S. F. (1968). Tyzzer's disease in laboratory animals. *Z. Versuchstierk.* **10**, 168–175.

101. O'Connor, J. R., and Morris, J. A. (1955). Recovery of Texas-1

type coxsackie virus from blood of wild rabbit and from sewage contaminating rabbit's feeding ground. *Amer. J. Hyg.* **61**, 314–320.

102. Onetto, E., Canessa, E., and Leyton, G. (1938). Infection espontanca en cures y conejos por *Brucella suis. Rev. Inst. Bacteriol. Chile* **6**, 41–47; cited by Pallaske and Krahnert (106).

103. Osebold, J. W., and Gray, D. M. (1960). Disseminated staphylococcal infections in wild jack rabbits (*Lepus californicus*). *J. Infec. Dis.* **106**, 91–94.

104. Ostler, D. C. (1961). The diseases of broiler rabbits. *Vet. Rec.* **73**, 1237–1255.

105. Pagan, E. F., McMahon, K. J., and Bowen, R. E. (1961). Complement fixing antibodies for *R. rickettsii* in serums of black tailed jack rabbits. *Pub. Health Rep.* **76**, 1120–1122.

106. Pallaske, G., and Krahnert, R. (1958). Durch Bakterien und pflanzliche Parasiten hervorgerufenen Infektions-krankheiten. *In* "Pathologie der Laboratoriumstiere" (P. Cohrs, R. Jaffe, and H. Meesen, eds.), Vol. II, pp. 1–83. Springer-Verlag, Berlin and New York.

107. Parker, R. R., Pickens, E. G., Lackman, D. B., Bell, E. J., and Thraikill, F. B. (1951). Isolation and characterization of Rocky Mountain spotted fever rickettsiae from the rabbit tick *Haemaphysalis leporis-palustris*, Packard. *Pub. Health Rep.* **66**, 455–463.

107a. Pattison, M., Clegg, F. G., and Duncan, AL.L (1971). An outbreak of encephalomyelitis in broiler rabbits caused by Nosema cumculi. *Vet. Rec.* **80**, 404–405.

108. Pegg, E. J. (1970). Three ectoparasites of veterinary interest communicable to man. *Med. Biol. Illus.* **20**, 106–110.

109. Pérès, DD, and Granon-Fabré, DD (1935). La melitococcie chez le lapin. *Rev. Gen. Med. Vet.* **64**, 201; cited by Pallaske and Krahnert (106).

110. Pier, A. C. (1964). Dermatophytosis in animals transmissible to man. *In* "Ocupational Diseases Acquired from Animals," Continued Educ. Ser. No. 124, pp. 179–187. School of Public Health, University of Michigan, Ann Arbor.

111. Ray, J. P., and Mallick, B. B. (1970). Public health significance of salmonella infections in laboratory animals. *Indian Vet. J.* **47**, 1033–1037.

112. Reeves, W. C., and Hammon, W. McD. (1962). Epidemiology of the arthropod-borne viral encephalitides in Kern County, California 1943–1952. *Univ. Calif., Berkeley, Publ. Pub. Health* **4**.

113. Reilly, J. R. (1970). Tularemia. *In* "Infectious Diseases of Wild Mammals" (J. W. Davis, L. H. Karstad, and D. O. Trainer, eds.), pp. 175–199. Iowa State Univ. Press, Ames.

114. Renquist, D., and Soave, O. (1969). Staphylococcal pneumonia in a laboratory rabbit: An epidemiologic follow-up study. *J. Amer. Vet. Med. Ass.* **155**, 1221–1223.

115. Rhodes, A. J., and Van Rooyen, C. E. (1968). "Textbook of Virology." Williams & Wilkins, Baltimore, Maryland.

116. Roemmele, O. (1958). Erfolgreiche Schutzimpfung mit Shope'scher Vaccine gegen Myxomatose der Hauskaninchen und Infektionsversuche. *Berlin. Müenchen. Tieräerztl. Wochenschr.* **71**, 128–130.

117. Rosicky, B. (1969). On the ecology of arboviruses of the California complex and the Bunyamwera group. *In* "Arboviruses of the California Complex and the Bunyamwera Group" (V. Bardos *et al.*, eds.), pp. 99–106. Publ. House Slovak Acad. Sci. Bratislava.

118. Sabin, A. B. (1959). Reoviruses. *Science* **130**, 1387–1389.

119. Sanarelli, G. (1898). Das myxomatogene Virus. *Zentralbl. Bakteriol., Parasitenk., Infektionskr. Hyg., Abt. 1: Orig.* **23**, 865–873.

120. Sanger, V. L. (1971). Toxoplasmosis. *In* "Parasitic Diseases of Wild Mammals" (J. W. Davis and R. C. Anderson, eds.), pp. 326–330. Iowa State Univ. Press, Ames.

121. Seelemann, M. (1954). "Biologie der Streptokokken." Hans Carl, Nürnberg.

122. Seeliger, H. P. R. (1961). "Listeriosis." Hafner, New York.

123. Shaughnessy, H. J. (1963). Tularemia. *In* "Diseases Transmitted from Animals to Man" (T. G. Hull, ed.), pp. 588–604. Thomas, Springfield, Illinois.

124. Sheffield, H. G., and Melton, M. L. (1970). *Toxoplasma gondii*, the oocyst, sporozoite, and infection of cultured cells. *Science* **167**, 892–893.

125. Sheldon, W. H. (1959). Subclinical pneumocystis pneumonitis. *AMA J. Dis. Child.* **97**, 287–297.

126. Shirai, A., Bozeman, F. M., Perri, S., Humphries, J. W., and Fuller, S. (1961). Ecology of Rocky Mountain spotted fever. I. *Rickettsia rickettsii* recovered from cottontail rabbit from Virginia. *Proc. Soc. Exp. Biol. Med.* **107**, 211–214.

127. Siim, J. C., Biering-Sørensen, V., and Møller, T. (1963). Toxoplasmosis in domestic animals. *Advan. Vet. Sci.* **8**, 335–429.

128. Simpson, D. I. H. (1969). Arboviruses and free-living wild animals. *Symp. Zool. Soc. London* **24**, 13–28.

129. Soave, O. A. (1963). Diagnosis and control of common diseases of hamsters, rabbits, and monkeys. *J. Amer. Vet. Med. Ass.* **142**, 285–290.

130. Splino, M., Peychl, L., Kyntera, F., and Kotrlik, J. (1969). Isolierung von *Pasteurella pseudotuberculosis* aus Leistenlymphknoten. *Zentralbl. Bakteriol., Parasitenk., Infektionskr. Hyg., Abt. 1: Orig.* **211**, 360–364.

131. Sprehn, C. (1965). "Kaninchenkrankheiten." Oertel & Sporer, Reutlingen.

132. Stanton, A. T., and Fletcher, W. (1925). Melioidosis and its relation to glanders. *J. Hyg.* **23**, 347–363.

133. Tas, J., and van der Hoeden, J. (1964). Scabies. *In* "Zoonoses" (J. van der Hoeden, ed.), pp. 740–748. Elsevier, Amsterdam.

134. Taylor, J. (1969). *Salmonella* in wild animals. *Symp. Zool. Soc. London* **24**, 51–73.

135. Taylor, J., and Atkinson, J. D. (1965). *Salmonella* in laboratory animals. *Med. Res. Counc. (Gt. Brit.), Lab. Anim. Bur., Collect. Pap.* **4**, 57–66.

136. Thomsen, A. (1959). Occurrence of *Brucella* infection in swine and hares, with special regard to the European countries. *Nord. Veterinaer med.* **11**, 709–718.

137. Thorpe, B. D., Sidwell, R. W., Bushman, J. B., Smart, K. L., and Moyes, R. (1965). Brucellosis in wildlife and livestock of west central Utah. *J Amer. Vet. Med. Ass.* **146**, 225–232.

138. Twigg, G. I., Cuerden, M. C., and Hughes, D. M. (1969). Leptospirosis in British wild mammals. *Symp. Zool. Soc. London* **24**, 75–98.

139. Twigg, G. I., Cuerden, C. M., Hughes, D. M., and Medhurst, P. (1969). The leptospirosis reservoir in British wild mammals. *Vet. Rec.* **84**, 424–426.

140. van der Hoeden, J. (1964). Leptospirosis. *In* "Zoonoses" (J. van der Hoeden, ed.), pp. 240–273. Elsevier, Amsterdam.

140a. Van Kruiningen, H. J., and Blodgett, S. B. (1971). Tyzzer's disease in a Connecticut rabbitry. *J. Amer. Vet. Med. Ass.* **158**, 1205–1212.

141. Verma, N. S., and Sharma, S. P. (1969). Salmonellosis in laboratory animals. *Indian Vet. J.* **46**, 1101–1102.

142. Vetesi, F., and Kemenes, F. (1967). Studies on listeriosis in pregnant rabbits. *Acta Vet. Acad. Sci. Hung.* **17**, 27–38.

142a. Weber, A., and Manz, J. (1971). Serologische Untersuchungen der O-Antigene von E. Coli-Stammen isoliert von Kaninchen. *Berlin. Muenchen. Tierartzl. Wocherschr.* **84**, 441–443.

143. Weidenmüller, H. (1959). Zur Rodentiose bei Tier und Mensch. *Tieraerztl. Umsch.* **14**, 256–259.

144. Weidenmüller, H. (1966). Pseudotuberkulose bei Wildtieren. *Tieraerztl. Umsch.* **21**, 447–448.

145. Weinburgh, H. B. (1964). "Field Rodents, Rabbits and Hares: Public Health Importance, Biology, Survey and Control." U. S. Dept. of Health, Education and Welfare, Public Health Service, Communicable Disease Center, Atlanta, Georgia.

146. Werner, H. (1966). Spontane Toxoplasma Infektion beim

Hauskaninchen (*Oryctolagus cuniculus*). *Zentralb. Bakteriol., Parasitenk., Infektionskr. Hyg., Abt. 1: Orig.* **199**, 259–263.

147. Wetzel, R., and Rieck, W. (1962). "Krankheiten des Wildes." Parey, Berlin.

148. Wetzler, T. F., and Hubbard, W. T. (1967). *Pasteurella pseudotuberculosis* in North America. *Symp. Ser. Immunobiol. Stand.* **9**, 33–44.

149. WHO Expert Committee on Plague. (1959). Third Report. *World Health Organ., Tech. Rep. Ser.* **165**.

150. Wojtek, H. L. (1966). Zur Leptospirose der Haus-und Versuchstiere. *Tieraerztl. Umsch.* **21**, 449–455.

151. Yost, D. H. (1958). Encephalitozoon infection in laboratory animals. *J. Nat. Cancer Inst.* **20**, 957–960.

152. Young, L. S., and Sherman, I. L. (1969). Tularemia in the United States: Recent trends and a major epidemic in 1968. *J. Infec. Dis.* **119**, 109–110.

153. Yuill, T. M., and Hanson, R. P. (1964). Serologic evidence of California encephalitis virus and western equine encephalitis virus in snowshoe hares. *Zoonoses Res.* **3**, 153–164.

154. Yuill, T. M., and Hanson, R. P. (1965). Coliform enteritis of cottontail rabbits. *J. Bacteriol* **89**, 1–8.

155. Yuill, T. M., Iversen, J. O., and Hanson, R. P. (1969). Evidence for arbovirus infections in a population of snowshoe hares: A possible mortality factor. *Bull. Wildl. Dis. Ass.* **5**, 248–253.

156. Yunker, C. E. (1964). Infections of laboratory animals potentially dangerous to man: Ectoparasites and other arthropods, with emphasis on mites. *Lab. Anim. Care* **14**, 455–465.

Author Index

F

Faber, J.J., 94(195), *148*

Fabro, S., 98(123), 107(122), 108(122), 115(122), 122(122), 134(124), 137(124), 138(124), 139(122,125), 141(124), 142(122,124), 143(122), *147*

Faherty, T.P., 410(96), *430*

Faigle, J.W., 131(231), *149*

Fainstate, T.D., 102(136), *147*

Fairweather, F.A., 92(81,82), 131(81), 132(81), *146*

Fancher, O.E., 108(235), 115(235), 116(235), 122(235), 132(235), 133(235), 141(235), 142(235), 143(235), *149*

Fardeau, F., 332(84), *370*

Farlow, J.E., 307(62), *311*

Farooq, A., 29(113), *47*

Farrell, G., 93(127), *147*

Faulkner, G.H., 362(85), *370*

Faust, E.C., 296(64), *311*

Faust, P.F., 324(16), *326*

Fay, L.D., 355, 356, 361(206), 365, *373*

Feaster, J.P., *429*

Febvre, H., 245(22), *258*, 361(86), 362(86), 363(86), *370*

Fedoroff, S., 157(14), 158, *164*

Fee, A.R., 86(20), *89*

Feeley, J.C., 206(256), 207(256), *236*

Feigelson, P., 389(95), *399*

Feigenbaum, A.S. 156(60,61), *165*, 394(31,45, 169, *398, 400*, 404(74), 405(74), 406(74, 74a, 411(194), 419(74), 425(194), *429, 432*

Feingold, B.F., *311*

Feldman, W.H., 336, *370*

Felisati, D., 116(128), 118(128), 121(128), 123(128), 142(128), *147*

Fell, H.B., 410, *429, 430, 432*

Felsenfeld, O., 456(39), *459*

Fenner, F., 238(43), 239(35,37,38,40,41,42,43,46,170), 240(43), 241(30,41,43), 242(41,43), 244(35,38,43,170), 245(43), 247(38,43,169,170), 248(36,44,169), *258, 261*, 310(65), *311*, 356, 361(90), 363(90,332), *370, 375*, 395(32,98), *398, 399*, 457(40), *459*

Ferguson, J.H., 57(73), *71*

Ferguson, W.S., 427(66), *429*

Ferm, V.H., 101(348), *151*

Fernandez, V., 94(129), *147*

Fernando, M.A., 318(33), *327*

Fernando, S.T., 325(34), *327*

Ferris, D.H., 457(63), *460*

Ferris, G.F., 299(67), 301(66,67), *311, 312*

Ferry, N.S., 195, 198(58), *232*

Festisov, V.I., 320(35), *327*

Fiala, G., 416(10), *428*

Field, E.J., 77(21), 79(21), *89*

Field, G., *312*

Field, J., 336, 348(274), *374*

Fielder, F.G., 385(189), *401*

Fields, R., 9(111), 14(111), *19*

Fiennes, R.N., *312*

Fillios, L.C., 158(25), *164*

Finch, C.A., 67(17), *70*

Fine, B.S., 448(30), *449*

Fine, S., 448(30), *449*

Finkelstein, R.A., 437(78), *450*

Finzi, C., 94(327), *151*

Firor, W.M., 86(22), *89*

Fischer, K., 170(31,32), *177*

Fischer, R.G., 360(91), *370*

Fischer, W., 169, *177*

Fishbein, M., 103(130), *147*

Fisher, B., 360(92), *371*

Fisher, E.R., 245(94), 246, 247, *258, 259*, 360(92), 362(184), 363(184), *371, 372*

Fisher, H., 32(28), *45*, 404(74), 405, 406(74), 419(74), *428, 429*

Fisher, K.D., 406(67), *429*

Fisk, R.T., 241(86), 242(86), *259*

Fjellstrom, D., 97(237), *149*

Flatt, R.E., 63(37a), *70*, 198(61), 199(61), 202(60), 227(62), 230, *232*, 277(50), *283, 335, 336, 371*, 458(40a), *459*

Fleischer, L., 168, *177*

Fletcher, W., 229(219), *235*, 456(132), *461*

Flir, K., 335, *371*

Florence, L., 274(166), *285*

Floretin, P., 335, *375*

Florey, H.W., 84(18,71), *89, 90*

Florio, R., 222(64), 223, *232*

Flux, J.E.C., 356(96), *371*

Fly, M.N., 86, *90*

Flynn, R.J., 181(37); *191*

Follett, E.A.C., 250(124), 255(23), *258, 260*

Follis, R.H., Jr., 417(69), 419(69), 424(68,69), *429*

Foote, R.H., 39(11), *45*, 84(47,64,65), *89, 90*, 97(53,131), *145, 147*

Forbes, W., 157(48), *165*

Forcum, D.L., 306(70), 307(70), *312*

Ford, B., 298(191), 300(191), 307(191), 308, 309(196), *314*

Ford, C.E., 16(70), *18*

Ford, C.M., 293(71), *312*

Ford, R., 137(91), *146*

Forgeot, P., 228(241), *235*

Forkner, C.E., 347(272), *374*

Forsberg, U., 92(132), *147*

Forstner, M.J., *312*

Foster, H.L., 180, 181(22), 182(12), 190, *191*, 276(75a), *283*

Fouser, M.D., 455(19), *459*

Fouts, J.R., 131(133,196), *147, 148*

Fox, I., 294(225), 306(61), *311, 312, 315*

Fox, M.R.S., 406(160), *431*

Fox, R.R., 9(205), 11(76,77,81,115,116,204), 12(219), 13, 14, 15(53,141), 16(114,141,176), *18, 19, 20, 21*, 25(27), 30, 33(26), *45*, 57, 58(38), 61, 66(38,69), 67(38,69), 68(69), 69, *70, 71*, 75(34), *89*, 103(88), 104(88), 108(88), 109(88), 110(88), 112(88), 113(88), 114(88,360), 121(134), 122(88,134), 123(88), 124(88,439), 126(439), 128(439), 135(134), 143(438,439), *146, 147, 151, 153*, 184, *191*, 200(65), 202, 232, 289(75), *312*, 333(97,98), 335(218), 336, 351(217), 352(98,217), 365(97), *371, 373*, 379(30,122,123,124), 382(38), 383, 384(34), 385(37a), 386(44a), 387(24), 388(183a), 389(42,43,44,102), 390, 391, 392(37,37a), 395(44b), 396(44c), *397, 398, 399, 400, 401*

Foxx, T.S. 292(76), 293(76), 302(76), *312*

Franceschetti, A., 122(135), *147*

Francis, E., 206, 207(66), *232, 234*

Francis, J., 455(41), *459*

Francis, R.A., 455(16a), *459*

Francis, T., 256(48), *258*

Frank, I., *429*

Franklin, J., 455(42), *459*

Franklin, K.J., 93(34), 94(34), *145*

Fraser, C.E.O., 230(218), *235*

Fraser, F.C., 92(137,138), 102(136), 133(137), *147*

Fratta, I.D., 116(139), 117(139), 118(139), 126(139), 142(139), *147*

Freda, V.J., 173(70), *177*

Frederick, B.J., 454(71), *460*

Fredrickson, D.S., 158(26), *164*

Frei, Y.F., 67(85), *71*

Frenkel, J.K., 269(52), 270(52,53,53a), 271(52), 272(51,52,53,53a), 274(52), 275(77), 276(77), 277(52,77), 281(52), 282, *283, 284*, 457(43,44,45), *459*

Fresh, J.W., 57(73), *71*

Freymann, R., 457(46), *459*

Fridhandler, L., 98(313), 134(313), 140(293), *149, 150*

Fried, S.M., 225(68), 226(68), *232*

Friedewald, W.F., 333(100), 359(99, 266, 267), 360(103), 362(102), 368(100), *371, 374*

Friedman, F., 123(63), 136(63), *145*

Friedman, L., 92(303), *150*

Friedman, M., 159(27), 164, *164*

Friedman, R.M., 362(5), *369*

Friedman, W.F., 124(140), 138(140), *147*, 412, 413, *429*

Frier, H.I., 409(72), *429*

Fritz, H., 104(142), 105(142), 109(141), 117(142), 120(142), 122(142), 131(231), 140(142), 143(141), *147, 149*

Fritz, T.E., 181(37), *191*

Frommer, G.P., 29(82), *46*

Frost, D.V., 419(132), *431*

Fry, L.R., 107(375), 135(375), *152*, 385(177), *401*

Frye, C.M., 421, *429*

Fuentes, C.A., 442(31,32), *449*

Fujiwara, K., 218(69,70,71), *232*

Fukutome, S., 325(97), *328*

Fuller, H.S., 295(248), *315*

Fuller, S., 457(126), *461*

Furth, J., 332(316), *375*

G

Gadbois, D.S., 124(362), 136(362), *151*

Gaedeke, R., 282(55), *283*

Gallina, A.M., 409(72), *429*

Galton, M.M., 230(150), *234*, 455(47), *459*

Galvin, T.J., 325(36), *327*

Gaman, E.M., 32(28), *45*, 156(60,61), *165*, 394(31,45,169), *309, 400*, 404(74), 405, 406,

Subject Index

A 4
B 5
C 6
D 7
E 8
F 9
G 0
H 1
I 2
J 3